Current Topics in Behavioral Neurosciences

For further volumes:
http://www.springer.com/series/7854

About this Series

Current Topics in Behavioral Neurosciences provides critical and comprehensive discussions of the most significant areas of behavioral neuroscience research, written by leading international authorities. Each volume offers an informative and contemporary account of its subject, making it an unrivalled reference source. Titles in this series are available in both print and electronic formats.

With the development of new methodologies for brain imaging, genetic and genomic analyses, molecular engineering of mutant animals, novel routes for drug delivery, and sophisticated cross-species behavioral assessments, it is now possible to study behavior relevant to psychiatric and neurological diseases and disorders on the physiological level. The *Behavioral Neurosciences* series focuses on "translational medicine" and cutting-edge technologies. Preclinical and clinical trials for the development of new diagostics and therapeutics as well as prevention efforts are covered whenever possible.

John F. Cryan · Andreas Reif
Editors

Behavioral Neurogenetics

Springer

Editors
John F. Cryan
Department of Anatomy and Neuroscience
University College Cork
Western Gateway Building
Western Rd., Cork
Ireland

Prof. Dr. Andreas Reif
Department of Psychiatry
and Psychotherapy
University of Würzburg
Füchsleinstrasse 15
97080 Würzburg
Germany

ISSN 1866-3370 ISSN 1866-3389 (electronic)
ISBN 978-3-642-27858-7 ISBN 978-3-642-27859-4 (eBook)
DOI 10.1007/978-3-642-27859-4
Springer Heidelberg New York Dordrecht London

Library of Congress Control Number: 2012936979

Printed on acid-free paper

Springer is part of Springer Science+Business Media (www.springer.com)

Preface

The field of behavioural neurogenetics has developed significantly over the past two decades. This has been largely driven by technical advances in the field of molecular genetics both in model systems and in clinical analysis. Indeed, it is hoped that the elucidation and ongoing functionalisation of the human genome may provide new insights into the aetiology, course and, ultimately, treatment of psychiatric illnesses.

This book covers a wide array of topics relevant to behavioural genetics from both a preclinical and clinical standpoint. Indeed in juxtaposing both areas of research the reader will appreciate the true translational nature of the field. Topics covered range from technical advances in genetic analysis in humans and animals to specific descriptions of advances in schizophrenia, attention disorders, depression and anxiety disorders, autism, aggression, neurodegeneration and neurodevelopmental disorders. The importance of gene–environment interactions is emphasised and the role of neuroimaging in unravelling the functional consequences of genetic variability described.

Part I of this book focuses on advances in the basic sciences of behavioural neurogenetics with a strong emphasis on animal models of psychiatric illness. It opens with a chapter by Robert Gerlai highlighting the use of model organisms and specifically zebrafish (*Danio rerio*) in modelling complex human psychiatric disease and its applicability for behavioural genetic studies. Lisa Tarantino follows by giving the state of the art on forward genetic approaches to elucidate the contribution of genetic variation to complex behavioural phenotypes. Carola and Gross illuminate the relative contribution of environment and genetics to psychopathology and how this is informing translational studies in animals and humans. Miczek and colleagues discuss the neurobiological mechanisms underlying aggression and how it can be modified in genetically modified mice. Wegener, Mathe and Neumann describe the utility of various selectively bred rodent strains to ask key questions regarding the underlying pathophysiology of depression and anxiety. Next up is a chapter on how genetic manipulations in rodent models have allowed for analysis of the impact of specific roles of glutamate receptors and transporters in cognitive and emotional behaviours shown to be altered by stress. The final two chapters are relevant to schizophrenia. First, O'Tuathaigh and colleagues describe how phenotypic characterisation of genetic models of candidate risk genes and/or putative pathophysiological processes implicated in schizophrenia, as well as examination of epidemiologically relevant gene × environment interactions in these models, can illuminate molecular and pathobiological mechanisms involved in schizophrenia. Powell, on the other hand, reviews the literature on genetic models of sensorimotor gating as they apply to schizophrenia and other neuropsychiatric disorders and discusses the utility of prepulse inhibition as a tool in phenotyping mutant mouse models.

Part II of this volume focuses on advances in clinical genetic analysis as applied to various neuropsychiatric disorders. Bayes and colleagues describe how second generation sequencing technologies are generating unprecedented amounts of sequence data very rapidly and at relatively limited costs. They also describe the challenges associated with such data generation in terms of data interpretation, analysis and management in addition to highlighting where such technologies are moving to in the future. Hakonarson and colleagues follow on with a description of the role of copy number variations in a number of neurodevelopmental disorders including autism, attention-deficit/hyperactivity disorder and schizophrenia. They also discuss relevant methodological considerations for such analysis. The application of neuroimaging has transformed modern neuroscience research, thus Hariri and colleagues describe how such approaches can be used to understand the interplay between genes and behaviour in shaping individual variability in brain function. Christine Freitag, Philip Asherson and Joahnnes Hebebrand discuss in detail the behavioural neurogenetics of childhood disorders including autism spectrum disorders, attention deficit/hyperactivity disorder, nocturnal enuresis and obesity. A chapter follows this on the use of new technologies to identify genes relevant to Schizophrenia by Dan Rujescu. As highlighted in the preclinical section there is growing appreciation of gene–environment as an emerging area in psychiatry research. Katja Karg and Srijan Sen give a comprehensive introduction to the field from a clinical context emphasising theoretical and practical problems that are worth considering. The behavioural neurogenetics of affective and anxiety disorders is expertly reviewed by Katharina Domschke and Andreas Reif in the subsequent chapter. This is followed up by a discussion by Quinn and colleagues of the contribution of variable number tandem repeat polymorphisms to a range of psychiatric disorders. Individual variability in response to stimulant drugs has long been known and Amy Hart, Harriet De Wit and Abraham Palmer examine the evidence for the contribution of genetic polymorphisms to this response. The penultimate chapter in the book focuses on the cognitive genetics of psychiatric disorders and reviews evidence for the heritability of the main cognitive phenotypes and early progress in the field using cytogenetic, linkage and candidate gene-based research methodologies. The book closes with a chapter from Daniela Galimberti and Elio Scarpini on the behavioural genetics of neurodegenerative disorders with a special focus on susceptibility genes for Alzheimer's Disease and Frontotemporal Lobar Degeneration.

These chapters, either individually or as a whole, provide a broad overview of the current status of a rapidly evolving and exciting field. Both the basic scientist and clinician alike will value this volume. It will also be of use to the novice to the field, to whom it will serve as an in-depth introduction to this area of research. Finally, it is worth noting that the advances made in behavioural neurogenetics to date have been very promising and we are particularly optimistic that the parallel advances in both basic and clinical neurogenetics fields will lead to a better understanding and eventual treatment strategies for complex neuropsychiatric disorders where there remains a large unmet medical need.

Cork, Ireland John F. Cryan
Würzburg, Germany Andreas Reif

Contents

Part I Preclinical Behavioral Genetics

Using Zebrafish to Unravel the Genetics of Complex Brain Disorders 3
Robert Gerlai

Forward Genetic Approaches to Understanding Complex Behaviors 25
Lisa M. Tarantino and Amy F. Eisener-Dorman

Mouse Models of the 5-HTTLPR × Stress Risk Factor for Depression 59
Valeria Carola and Cornelius Gross

Behavioral and Pharmacogenetics of Aggressive Behavior 73
Aki Takahashi, Isabel M. Quadros, Rosa M. M. de Almeida and Klaus A. Miczek

Selectively Bred Rodents as Models of Depression and Anxiety 139
Gregers Wegener, Aleksander A. Mathe and Inga D. Neumann

Stress-Induced Deficits in Cognition and Emotionality: A Role for Glutamate 189
Carolyn Graybeal, Carly Kiselycznyk and Andrew Holmes

Susceptibility Genes for Schizophrenia: Mutant Models, Endophenotypes and Psychobiology 209
Colm M. P. O'Tuathaigh, Lieve Desbonnet, Paula M. Moran and John L. Waddington

Genetic Models of Sensorimotor Gating: Relevance to Neuropsychiatric Disorders 251
Susan B. Powell, Martin Weber and Mark A. Geyer

Part II Clinical Behavioral Genetics

Applications of Second Generation Sequencing Technologies in Complex Disorders 321
Mònica Bayés, Simon Heath and Ivo Glynne Gut

Rare Genomic Deletions and Duplications and their Role in Neurodevelopmental Disorders 345
Joseph T. Glessner, John J. M. Connolly and Hakon Hakonarson

Toward a Mechanistic Understanding of How Variability in Neurobiology Shapes Individual Differences in Behavior 361
Ryan Bogdan, Justin M. Carré and Ahmad R. Hariri

Behavioural Genetics of Childhood Disorders 395
Christine M. Freitag, Philip Asherson and Johannes Hebebrand

Schizophrenia Genes: On the Matter of Their Convergence 429
Dan Rujescu

Gene × Environment Interaction Models in Psychiatric Genetics 441
Katja Karg and Srijan Sen

Behavioral Genetics of Affective and Anxiety Disorders 463
Katharina Domschke and Andreas Reif

Behavioural Genetics of the Serotonin Transporter 503
K. Haddley, V. J. Bubb, G. Breen, U. M. Parades-Esquivel and J. P. Quinn

Genetic Factors Modulating the Response to Stimulant Drugs in Humans 537
Amy B. Hart, Harriet de Wit and Abraham A. Palmer

The Cognitive Genetics of Neuropsychiatric Disorders 579
A. Corvin, G. Donohoe, A. Hargreaves, L. Gallagher and M. Gill

Behavioral Genetics of Neurodegenerative Disorders 615
Daniela Galimberti and Elio Scarpini

Index 633

Part I
Preclinical Behavioral Genetics

Using Zebrafish to Unravel the Genetics of Complex Brain Disorders

Robert Gerlai

Abstract The zebrafish has been prominently utilized in developmental biology for the past three decades and numerous genetic tools have been developed for it. Due to the accumulated genetic knowledge the zebrafish has now been considered an excellent research tool in other disciplines of biology too, including behavioral neuroscience and behavior genetics. Given the complexity of the vertebrate brain in general and the large number of human brain disorders whose mechanisms remain mainly unmapped in particular, there is a substantial need for appropriate laboratory research organisms that may be utilized to model such diseases and facilitate the analysis of their mechanisms. The zebrafish may have a bright future in this research field. It offers a compromise between system complexity (it is a vertebrate similar in many ways to our own species) and practical simplicity (it is small, easy to keep, and it is prolific). These features have made zebrafish an excellent choice, for example, for large scale mutation and drug screening. Such approaches may have a chance to tackle the potentially large number of molecular targets and mechanisms involved in complex brain disorders. However, although promising, the zebrafish is admittedly a novel research tool and only few empirical examples exist to support this claim. In this chapter, first I briefly review some of the rapidly evolving genetic methods available for zebrafish. Second, I discuss some promising examples for how zebrafish have been used to model and analyze molecular mechanisms of complex brain disorders. Last, I present some recently developed zebrafish behavioral paradigms that may have relevance for a spectrum of complex human brain disorders including those associated with abnormalities of learning and memory, fear and anxiety, and social behavior. Although at this point co-application

---

R. Gerlai (✉)
Department of Psychology, University of Toronto Mississauga,
3359 Mississauga Road North, Rm 3035,
Mississauga, ON L5L 1C6, Canada
e-mail: robert_gerlai@yahoo.com

Curr Topics Behav Neurosci (2012) 12: 3–24
DOI: 10.1007/7854_2011_180

Published Online: 17 January 2012

of the genetics and behavioral approaches is rare with zebrafish, I argue that the rapid accumulation of knowledge in both of these disciplines will make zebrafish a prominent research tool for the genetic analysis of complex brain disorders.

Keywords Zebrafish · High throughput behavioral screening · Fetal alcohol syndrome · Alcoholism · Learning and memory · Fear and anxiety

Contents

1 Zebrafish, a Novel Research Tool with Some Advantages 4
2 Zebrafish, the Favorite of Geneticists 5
3 Reverse Genetic Tools 6
4 Forward Genetic Tools 8
5 Parkinson's Disease 10
6 Tauopathy and Alzheimer's Disease 10
7 The Anxiety "Cluster" 11
8 Autism Spectrum Disorders 12
9 Schizophrenia 12
10 Expanding Our Horizon: The Need for Sophisticated Behavioral Test Paradigms in Zebrafish Research 13
11 Learning and Memory 14
12 Fear and Anxiety 17
13 Social Behavior 19
14 Concluding Remarks 21
References 21

1 Zebrafish, a Novel Research Tool with Some Advantages

Numerous laboratory organisms are available for one to employ in the analysis of how genes influence behavior or how certain biochemical processes affect the functioning of the brain. For example, rats have been traditionally utilized to screen libraries of compounds to identify drugs that may be beneficial in a range of human brain and behavioral disorders both in academic and biopharmaceutical preclinical research. The house mouse has been employed in a range of behavior genetic applications including those that probe the effects of specific genes (reverse genetics) and their roles in behavior and brain function [e.g. Gerlai et al. (1995), Pekhletski et al. (1996)]. But other, simpler, laboratory organisms including the flat worm (Giles and Rankin 2009), the sea slug (Bailey and Kandel 2008), or the fruit fly (Sokolowski 2001) have also been successfully utilized to study the biological bases of brain function and behavior. Compared to these laboratory model organisms the zebrafish is quite novel. I use the term "model" here in a very loose sense and only mean that the reason for the use of animal species in the laboratory is to isolate and mimic some aspects of complex biological phenomena, a reductionist approach that may yield results faster than in human due to the simpler features of the studied organism and to the precisely controlled laboratory conditions. In this sense, zebrafish may be an ideal model organism.

The zebrafish strikes an optimal compromise between system complexity and practical simplicity. It is a vertebrate species with a physiology (Alsop and Vijayan 2008), brain anatomy (Tropepe and Sive 2003), and neurochemistry (Chatterjee and Gerlai 2009) characteristic of the prototypical vertebrate, and thus translationally relevant to our own species. Most importantly, the nucleotide sequence of zebrafish genes is often found highly similar (70–80% homology) to the mammalian (and human) counterparts, and the amino acid sequence of functionally relevant domains of its proteins has been found to be even more evolutionarily conserved, i.e. similar to mammalian sequences (Reimers et al. 2004; Renier et al. 2007). It is thus quite probable that a gene identified in zebrafish as involved in particular functions/dysfunctions of its brain, will have a human homolog serving similar functions and vica versa. Briefly, the translational relevance of zebrafish research is expected to be high.

The number of animals one can screen is a crucial factor in forward genetics where one does not know which and how many genes may influence the phenotypical function in question. In case of brain disorders or behavioral function, the number of such genes may be quite large and thus one may have to analyze thousands of mutants to tackle this complexity and identify appropriate mutations and thus the genes involved. One can, of course, generate the same number of zebrafish and mice for screening purposes but there are several reasons why zebrafish may be preferred. First, a single female zebrafish can produce 200 offspring at every spawning and can spawn multiple times a week. Second, zebrafish is small (4 cm long) and is highly social and thus a large number of subjects may be housed cheaply in a small animal holding room. For example, a standard zebrafish stand-alone high density rack system (e.g. Aquatic Ecosystems Inc, FL, or Aquaneering Inc. CA) with six shelves and about 23 liter tanks per shelf, can house about 2,000 zebrafish, and a 40 m^2 standard vivarium room may be fitted up with up to 10 such racks. Briefly, the same room that may house a couple of hundred mice can have about 20,000 zebrafish in it. Therefore, when it comes to large scale screening, the zebrafish has a definite cost advantage. Given the relative simplicity of this vertebrate species and the fact that it is a phylogenetically older "design" compared to mammals, one may also argue that it may allow the analysis of fundamental core mechanisms of the chosen brain function. Last, adding zebrafish to the list of already well studied vertebrates (e.g. the mouse and the rat) should facilitate cross species comparison and finding common characteristics and mechanisms, which should also enhance our ability to translate the findings to human.

2 Zebrafish, the Favorite of Geneticists

Genetics is one of the strengths of zebrafish and excellent reviews have been published on numerous genetic techniques developed for this species (Amsterdam and Hopkins 2006; Chen and Ekker 2004; Lekven et al. 2000; Patton and Zon 2001).

Here I will discuss these techniques only briefly. In addition to sophisticated gene expression analyses including quantitative reverse transcriptase polymerase chain reaction (q-RT-PCR) and DNA microarrays (gene chip), both reverse genetic and forward genetic methods are available with zebrafish, although the former, the forward genetic approaches, have been more prevalent in zebrafish research. Reverse genetic analysis allows one to study the phenotypical effects of targeted manipulation of known genes. The main goal of forward genetic studies, on the other hand, is to discover novel genes by the introduction of random mutations.

3 Reverse Genetic Tools

Among the reverse genetic tools, TILLING has been employed successfully in zebrafish (Moens et al. 2008). Targeting induced local lesions in genomes (TILLING) allows the identification of mutations in specific genes of interest in chemically mutagenized zebrafish populations. The method was first described for mutation detection in Arabidopsis about a decade ago but since then it has successfully been adopted for zebrafish too (Moens et al. 2008). The essence of the TILLING method is the screening of chemically mutagenized populations of zebrafish using the polymerase chain reaction (PCR) for mutations in the gene of interest. Unlike in gene targeting with the use of homologous recombination in embryonic stem cells employed in the mouse (Capecchi 1989), and most recently in the rat (Tong et al. 2010), the actual mutagenesis conducted in TILLING is random, i.e. not targeted. The "trick" of TILLING is then the identification of the mutation(s) in the target gene. The identification of the mutation may be achieved using two different approaches. One is the resequencing of every single mutagenized genome (the gene of interest and its sequence is known and thus alterations in the sequence can be detected), a brute force approach that is becoming increasingly feasible with the ever improving speed of sequencing methods. The other method is based upon the use of cell, a plant-specific extracellular glycoprotein that cleaves heteroduplex DNA at single nucleotide mismatches (resulting from the introduced point mutation). Using fluorescently labeled primers to detect cell cleavage products on a LiCor acrylamide slab gel, cell can identify a heterozygous single nucleotide mismatch [for further details of the TILLING methodology and its use, see Moens et al. (2008)].

Another approach that is principally a reverse genetic method, i.e. it is also aimed at the characterization of the function of known genes, is a knock down method using morpholinos (Bill et al. 2009). Morpholinos are antisense oligonucleotides (usually 25 bases long) composed of a phosphorodiamidate backbone with a morpholine ring and the same bases as DNA (Bill et al. 2009). Morpholinos are injected into the target cells and act by steric hindrance to block ribosome entry and hence prevent protein production. The antisense technology has been long employed in mammalian species but the efficiency and specificity of the oligonucleotide approach has been controversial. In zebrafish, however, the morpholino approach has been successfully employed. For example, due to the altered backbone of the morpholinos they are not affected by nucleases and are, therefore, highly resistant to breakdown in vivo. Furthermore, because of their small size and unusual chemistry, morpholinos remain undetected for the immune system. Morpholinos are traditionally introduced into the yolk of 1–8 cell-stage embryos in which the cytoplasmic bridges connecting the embryonic cells allow rapid diffusion of the hydrophilic morpholinos leading to ubiquitous delivery. The analysis of the result of the gene expression knock down is usually limited to the embryonic stage of zebrafish. However, examples already exist suggesting that modified morpholino chemistry (the VIVO-morpholino, which allows penetration of the oligonucleotide into adult zebrafish cells) may be successfully employed in adult zebrafish as well [e.g. Kim et al. (2010)]. This is a crucial novel development considering that most complex brain disorders and higher behavioral functions can be best modeled, observed, and analyzed in the adult zebrafish.

RNA-interference, or RNAi, is yet another intriguing possibility for targeted modification of gene expression in zebrafish. RNAi is a transcriptional gene silencing mechanism-induced by short (21–23 bases long) double stranded RNA whose main mechanisms are believed to be gene expression regulation via miRNA's (endogeneous microRNA's) and defense against viral genetic material mediated by dsRNAs (double stranded RNAs), terms that represent structurally indistinguishable mRNA species. Irrespective of the physiological function, a few years ago it was realized that the cell's RNAi mechanism could be utilized for the induction of targeted knock down of gene expression by delivering double stranded short RNA sequences specific to the chosen target gene. Although zebrafish cells have been shown to possess the RNAi machinery, the functional consequences, especially the specificity of the RNAi-based manipulation, have been questioned [for review see Skromne and Prince (2008)] and thus whether this technology will lead to success in zebrafish remains to be seen. Another technology, transgenic methods, however, may offer a currently existing true and tried alternative.

In the mammalian neurobehavioral genetics field, transgenic methods have been perhaps the most fruitful reverse genetic approaches. Transgenic technologies have also been successfully employed with zebrafish [for a recent review see Skromne and Prince (2008)]. These techniques make use of a variety of methods (e.g. enzymatic approaches, transposons, and retroviruses) to enable the delivery and increase the efficiency of incorporation of foreign DNA into the genome of zebrafish thereby generating stable transgenic fish lines in which the foreign DNA is expressed. Irrespective of the mode of delivery, transgenic zebrafish may be divided into two main classes: transgenic overexpressors and dominant negative transgenics. In the former, overexpression of the transgene is achieved, for example, by delivering multiple copies of the transgene or using a strong promoter, and is used to test the functional consequences of the excess amount of translated gene product. In the latter, the expressed transgene product interferes with or blocks the function of the endogenous gene and thus allows one to test the effect of loss-of-function at the phenotypical level.

In addition to constitutive transgene expression, inducible expression systems are also available in zebrafish. Given that zebrafish tolerate a broad range of temperatures (in its natural geographical range temperatures may vary between 10 and 35°C), heatshock promoters have been successfully employed to induce transgene expression in a temporally controlled manner, and the use of focal heating has also allowed the induction of transgene expression in a spatially restricted manner, at least in superficial structures. Furthermore, the Gal4-UAS system (Scott et al. 2007), well developed for the fruit fly, the tetracycline transactivator system (Huang et al. 2005), as well as the Cre/loxP system (Langenau et al. 2005) used, for example, in the temporal and spatial control of knock out of genes in the mouse, have all been utilized in the control of transgene expression in zebrafish.

Although the classical knock out technology based upon homologous recombination in embryonic stem cells as employed in the mouse (Capecchi 1989) and most recently in the rat (Tong et al. 2010) is not yet available in the zebrafish, research in this direction is also progressing. For example, germline transmitting embryonic stem cells have been isolated from zebrafish (Fan and Collodi 2006) and methods alternative to classical gene targeting are also being explored. For example, nuclear transfer of genetically modified cultured embryonic fibroblast cells has been achieved in zebrafish, which implies that targeted genetic modification for in vivo analysis of gene function may be possible via this route (Lee et al. 2002). Furthermore, zinc finger nuclease-based knockout technology is also being developed for zebrafish (Ekker 2008). Zinc finger nucleases are genetically engineered restriction enzymes that cut the DNA sequence of interest according to their specific design. Zebrafish embryos injected with the specific custom designed zinc finger nuclease-encoding mRNA are reared to adulthood and crossed with wild type fish. As much as 25% of the resulting offspring has been shown to transmit the induced mutation, usually a frameshift allele in the germline (Ekker 2008). Last, a gene-breaking transposon-based method to generate mutations is also being developed for zebrafish (Sivasubbu et al. 2006) to mention but a few reverse genetic technologies.

The range of reverse genetic approaches discussed above is somewhat misleading, however. Although they do demonstrate how fast zebrafish genetics is evolving, many of these methodologies are not mature enough for reliable use even for embryonic or developmental biology phenotypes, the focus of most of these investigations. For complex neurobehavioral traits, virtually none of these methods have been employed.

4 Forward Genetic Tools

Unlike reverse genetic approaches, forward genetics has provided decades of consistent success with zebrafish. The first two large scale comprehensive forward genetic screens the Tubingen (Haffter et al. 1996) and Boston (Driever et al. 1996) screens were conducted about 15 years ago and set the stage for subsequent screening studies. Since then, most forward genetic studies have utilized a chemical mutagen, ethyl-nitroso-urea (ENU), which is expected to induce single nucleotide point mutations, and when dosed appropriately, one mutation per genome on average. The advantage of ENU mutagenesis is that ENU is efficient and with it one can achieve a good coverage of the entire genome, i.e. can expect to hit a large proportion of genes as long as a large enough number of animals (thousands) are mutagenized and analyzed. The Achilles heel of ENU mutagenesis, however, is the subsequent linkage analysis, which requires several generations of crosses and cumbersome mapping. Nevertheless, due to the availability of high resolution markers developed for zebrafish, the genes carrying the induced mutations can be successfully identified using linkage analysis-based positional cloning (Knapik 2000; Patton and Zon 2001). Another concern with ENU-based forward genetic analysis is that the point mutation induced by ENU is often recessive and thus may not be observable unless bred into a homozygous form, which requires three generations of breeding to create an F3 or a backcross segregating population (Patton and Zon 2001). Although not without technical complications, an alternative that can speed up the generation of recessive homozygous mutants does exist, it is gynogenesis. For gynogenesis one may use heat shock or pressure to manipulate the cell division cycle at the earliest embryonic stage leading to haploid to diploid genome conversion and allowing the generation of homozygous fish in essentially one step without the cumbersome breeding [for review of experimental examples using gynogenesis, see Patton and Zon (2001)].

ENU has been the most frequently employed mutagen for zebrafish, but other mutagenesis methods have also been successfully utilized. Perhaps the most promising among them is retrovirus mediated insertional mutagenesis (Amsterdam and Hopkins 2006). This method has the advantage over ENU because insertion of the viral genetic material into the zebrafish genome not only induces a mutation but by leaving the unique viral sequences in the genome allows fast and efficient localization and cloning of the mutated gene (Amsterdam and Hopkins 2006).

A potential concern with forward genetic approaches in zebrafish is that this species has undergone a partial genome duplication event in its evolutionary past. The concern is that the genes whose locus is on the duplicated region of the genome (approximately 20% of the genome) may be able to compensate for the induced mutation and could mask its effects if the sister gene remained active and if its function has not changed much since the gene duplication event. However, some argue that the partial "tetraploidy" may be viewed as an advantage in zebrafish for forward genetics especially when it comes to the analysis of the genetics of complex traits: the induced mutations are not expected to be lethal and may only have small quantitative effects on the phenotype, thus making functional phenotypical identification and characterization feasible. Whether these arguments turn out to be correct will have to be seen. But the few already existing examples suggest that we have some reason for optimism. Below I review such examples organized according to their disease relevance.

5 Parkinson's Disease

Parkinson's disease is the second most prevalent human neurodegenerative disease, which primarily affects the nigro-striatal dopaminergic system. Numerous genes have been identified associated with Parkinson's disease in humans three of which are believed to underlie early onset of Parkinson's Disease: PARK2 encoding Parkin, PINK1 encoding PTEN-induced putative kinase 1, and PARK7 encoding DJ-1 [for review see Bandmann and Burton (2010)]. The zebrafish orthologs of the human genes have been identified and, for example, the PARK2 gene product is known to be a 458 amino acid long protein which is 75% similar to the human protein with functionally important regions having 93% similarity. The similarities between the other two human genes with their zebrafish orthologs are also substantial. Morpholino-induced knock down of expression of these genes have led to significant dopaminergic neuron loss and/or to increased sensitivity of these neurons to toxins [although the selectivity of the effects have been questioned in the case of PARK6, reviewed by Bandmann and Burton (2010)]. Behavioral impairments resulting from the knock down have not been shown in zebrafish with one exception: Morpholino antisense knock down induced reduction of PINK1 expression led to loss of tyrosine hydroxylase staining in distinct groups of dopaminergic neurons in the zebrafish brain and when challenged with normally sub-effective concentrations of 1-methyl-4-phenyl-1,2,3,6-tetrapyradine (MPTP) the affected larval zebrafish showed reduced locomotion. Detailed swim path analysis of these larvae was not conducted, however.

6 Tauopathy and Alzheimer's Disease

Another class of neurodegenerative diseases is the tauopathies. These diseases are all defined by their underlying molecular pathology: the presence of hyperphosphorylated insoluble forms of the microtubule associated protein Tau, which is deposited in neurons in the form of observable neurofibrillary tangles. Most of these diseases are sporadic [e.g. progressive supranuclear palsy (PSP), corticobasal degeneration (CBD), and Pick's disease (PiD)], but neurofibrillary tangles have also been associated with familial early onset of Alzheimer's disease. Both transient as well as stable transgenic zebrafish lines overexpressing Tau have been generated and in both structures resembling neurofibrillary tangles have been reported [for review see Bandmann and Burton (2010)]. Behavioral analysis of the effects of these changes has not been conducted except for a rudimentary, observation-based, judgment as to whether tactile stimulation induced a normal escape response in the 48 h post-fertilization embryo expressing a mutant form of Tau (Paquet et al. 2009).

Alzheimer's disease is often mentioned under tauopathies but there are other prominent known genetic factors associated with this disease too. Alzheimer's disease is the most common form of dementia affecting over 25 million people worldwide (Querfurth and LaFerla 2010). The histological hallmarks of the disease are neurofibrillary tangles (mentioned above) and, perhaps even more characteristically, amyloid plaques. Although the familial (heritable) form of the disease only makes up 4–5% of all cases, clearly demonstrating the importance of environmental factors, the genetic analysis of Alzheimer's disease has led to major discoveries and clarified some of the biological mechanisms core to the disease. The genes encoding amyloid-β precursor protein (APP), and presenilins PS1 and PS2 have been found to underlie familial Alzheimer's disease, and the $\varepsilon 4$ allele of the gene encoding apolipoprotein E (ApoE) has been found to be a risk factor in sporadic (non-familial) Alzheimer's cases. The APP, PS1, PS2, and ApoE genes have all been identified in zebrafish and have been found to be highly homologous to their mammalian, and human counterparts, with certain functionally relevant regions approaching 100% identity with human (e.g. the transmembrane region of APP). Furthermore, some of the components of the γ-secretase complex mediating the processing and cleavage of APP, which may lead to the generation of the toxic Aβ (40, 42) peptide have started to be examined in zebrafish (reviewed in Xia 2010). To characterize the involvement of APP in zebrafish embryonic development, the expression of the APP gene was reduced using morpholinos, which led to significant shortening of the body length of the embryos (Joshi et al. 2009). The involvement of overexpression of APP (as in Down's patients, for example, who also develop Alzheimer's disease) or of the expression of mutant forms of APP (identified in familial Alzheimer's cases), or the role of presinilins have not been investigated in zebrafish, nor has been any studies conducted for the potential behavioral consequences of transgenic manipulations of these genes. The function of ApoE and its potential role in Alzheimer's disease related abnormalities also has not been investigated with zebrafish.

7 The Anxiety "Cluster"

Perhaps the largest cluster of human brain disorders in terms of prevalence are neuropsychiatric conditions including anxiety and stress disorders, depression, obsessive compulsive disorders, and several forms of phobias. Although both academic and pharmaceutical and biotechnology companies have been studying the potential mechanisms of these disorders and several treatment options have been developed, these diseases still represent an enormous unmet medical need. This is because the causative factors (both genetic and environmental) behind these diseases are difficult to trace and the mechanisms of the diseases are also complex. A recent review paints an optimistic picture as to the potential use of zebrafish in the analysis and modeling of neuropsychiatric conditions (Mathur and Guo 2010) and I share this optimism. For example, zebrafish possess a glucocorticoid and a mineralocorticoid receptor that have been cloned and sequenced, and has a corticoid signaling pathway highly similar to that of mammals (Amsterdam and Hopkins 2006; Denver 2009). Also, the major components necessary for success, including the genetic tools discussed above and the novel behavioral paradigms, some of which are presented below, already exist for zebrafish. Nevertheless, one must also acknowledge that these studies with zebrafish are only starting now and thus have not produced major breakthroughs.

8 Autism Spectrum Disorders

The zebrafish has also been suggested for the analysis of the mechanisms of autism spectrum disorders (Tropepe and Sive 2003). The number of genetic factors underlying autism spectrum disorders is believed to be much less than in such neuropsychiatric conditions as anxiety or schizophrenia, and thus animal genetic models have been generated with much hope [for a review, see e.g. Gerlai and Gerlai (2003)]. Importantly, several of the genes implicated in the human disease have been shown to have homologs in zebrafish [for a most recent review, see Mathur and Guo (2010)]. It may therefore be possible to recapitulate some aspects of autism spectrum disorders by selectively targeting these genes and testing the effect of the genetic or pharmacological manipulations on developmental as well as behavioral characteristics in zebrafish. It is also notable that the zebrafish is a highly social species, and the novel behavioral paradigms that are being developed to induce and quantify social behavioral responses in zebrafish may also contribute to this research, for example, by allowing large scale mutaganesis screening-based identification of molecular mechanisms involved in vertebrate social behavior.

9 Schizophrenia

Schizophrenia is a neurodevelopmental disorder that often manifests first during adolescence. Unlike in the case of autism spectrum disorders the number of genes involved in schizophrenia may be extremely large (hundreds) and many of these genes may only have a minor "predisposing" effect, which have hindered the unraveling of the mechanisms of the disease. Nevertheless, some of the genes implicated in schizophrenia have been identified in zebrafish. For example, DISC1, a schizophrenia susceptibility gene, has been shown to play roles in cell migration and differentiation in the zebrafish neural crest (Drerup et al. 2009) as well as in the development of oligodendrocytes and neuronal lineages developing from olig2 expressing precursor cells. In addition to delineating the cellular and molecular roles of some schizophrenia associated genes, there is already one example for a genetic manipulation to affect zebrafish behavior. SHANK3 is a synaptic scaffolding protein whose gene was recently identified to carry mutations in some patients suffering from schizophrenia and was also found in autistic patients. Morpholino-induced knock down of the expression of the corresponding gene resulted in robust morphological abnormalities as well as impaired swimming in response to tactile stimulation in the zebrafish larva (Gauthier et al. 2010). It may be noted, however, that the specificity of the morpholino-induced changes may be questionable given that the attempt to rescue the phenotype by injection of wild type or mutant SHANK3 mRNA sequences led only to partial success at best. Last, psychopharmacological experiments have already started to be utilized with zebrafish in the analysis and modeling of schizophrenia. One behavioral endophenotype often argued to be an important aspect of schizophrenia is reduced prepulse inhibition, or PPI. Prepulse inhibition is believed to be a measure of sensory gating. It is induced by employing a weak stimulus (the prepulse), which is expected to inhibit the reaction to a subsequent stronger startling stimulus (the pulse). Larval zebrafish exhibit PPI of the acoustic startle response similarly to what has been demonstrated in rodents [reviewed in Mathur and Guo (2010)]. PPI can be disrupted by dopamine agonists in the zebrafish larvae, an alteration that is reversed by antipsychotic drugs similarly to the mammalian situation (Braff et al. 2001). In addition to these promising psychopharmacology results, a forward genetic screen has already isolated a mutant "Ophelia", which exhibited reduced PPI (Burgess and Granato 2007). In summary, the first examples showing how zebrafish may be utilized in the investigation of the genetic mechanisms of complex brain disorders already exist. However, in most of these studies the behavioral consequences of the employed experimental manipulations were not analyzed or were studied in a rudimentary manner. This is, in general, the current weakness of the zebrafish as an experimental tool: its genetics and neurobiology have been traditionally powerful but its behavioral characteristics are largely unmapped because only a few behavioral test paradigms are available (Sison et al. 2006).

10 Expanding Our Horizon: The Need for Sophisticated Behavioral Test Paradigms in Zebrafish Research

Sophisticated behavioral paradigms may be crucial for two main reasons: first, the construct and face validity of the genetic (reverse genetics) or pharmacological models may only be fully established using behavioral tests; and two, behavioral paradigms may represent unbiased screening tools for forward genetic applications (Gerlai 2002). Arguably, behavioral analysis can efficiently probe a broad spectrum of brain functions in a large number of subjects. It is not limited to particular brain regions or neurobiological mechanisms, and it is simple and cheap to conduct (Gerlai and Clayton 1999). Arguably, high throughput behavioral screens may be able to systematically reveal mutation or drug-induced functional changes in the brain. Fortunately, for the past few years a clear upsurge of zebrafish behavioral studies is evident, indicating that behavioral neuroscience and behavior genetics has started to acknowledge the utility of this species. Below I discuss some of these recent studies focusing on the question of how behavioral analysis may be utilized for the discovery of novel genes and compounds affecting brain dysfunction associated with complex human brain disorders. Three main behavioral focus areas will be represented below: learning and memory, which is relevant for a number of neurodegenerative diseases including Alzheimer's disease; fear and anxiety, which are important behavioral responses and behavioral states relevant for a spectrum of neuropsychiatric conditions; and social behavior, whose abnormalities may be important in the analysis of autism spectrum disorders and schizophrenia, to mention but the two most important diseases in this domain.

11 Learning and Memory

Learning and memory has been extensively studied by scholars of several scientific fields. Numerous mechanistic questions related to how learning occurs, and what memory is, have been successfully tackled. For example, by now a large number of genes and biochemical mechanisms underlying learning and memory have been identified (Sweatt 2010). Can zebrafish add anything to this wealth of knowledge? Although hundreds of genes involved in learning and memory have been identified, the mechanisms of these complex processes are far from being understood. It is likely that the number of undiscovered genes that play roles in learning and memory is large. For example, according to conservative estimates, most vertebrate genomes contain about 30,000 genes. Recent microarray studies suggest that at least 50% of all the genes of the genome are expressed in the vertebrate brain (e.g. in zebrafish), i.e. about 15,000 genes (Pan et al. 2010). Given that plasticity is perhaps the most complex aspect of brain function, it is likely that a large proportion of these genes, i.e. potentially thousands of them, are involved in some mechanisms subserving plasticity, i.e. learning and memory.

A number of laboratories have realized that the cheap and easy to breed zebrafish may offer a solution for high throughput screening which would be costly with traditional laboratory rodents. Investigators have started to characterize the cognitive capabilities of zebrafish and have already developed several test methods that can measure learning and memory efficiently and fast [for examples see Sison et al. (2006), Sison and Gerlai (2010)]. The key in these paradigms concerns automatability. Even if one needs to employ several repeated training trials, as is the case in most learning paradigms, if these trials are administered in an automated manner, and if the behavioral responses that reflect learning and memory performance are easy to measure and do not require the constant presence and attention of an experimenter, the paradigm may be run in multiple test apparati in parallel and thus become high throughput.

A successful high throughput learning task design utilizes moving (animated) images of conspecifics, which are shown on a computer screen placed by each side of the experimental tank (Pather and Gerlai 2009). Previously, access to view a shoal (group) of zebrafish has been shown to represent a reward for experimental zebrafish and that this visual stimulus (the sight of a group of zebrafish) can support good learning performance (Al-Imari and Gerlai 2008). Subsequently, it has been demonstrated that computer animated images of zebrafish can serve as a positive reinforcement (Gerlai et al. 2009a). The simple manner with which the reward could be administered allowed the development of an automated learning task (Pather and Gerlai 2009). The task is also simple. For a short period of time [20 s in Pather and Gerlai (2009)] the image of the moving shoal is shown and then it is turned off for 90 s. After this 90 s no-image period the image of the moving shoal is shown on the opposite side of the test tank for again 20 s, and the sequence repeats itself. As a result of the alternating image presentation sides, zebrafish have to make a choice during the no-image period as to whether they stay close to the side where the image was just shown, or move to the opposite side, where the image will appear. The natural tendency of zebrafish is to stay close to its conspecifics and thus initially zebrafish spend the highest amount of time near the side where the image was shown last. However, as the training proceeds, zebrafish spend increasingly longer amount of time near the side that *will* show the image. There are several important points to make about these results. First, the motivation to stay close to conspecifics does not habituate over time and thus the experimental subjects remain motivated to perform in this task, a major advantage compared to the use of food, which satiates zebrafish quickly. Second, the stimulus is a visual cue administered precisely using consumer grade (i.e. cheap) video-equipment. Third, the behavioral response (distance from stimulus screen) is easy to quantify using video-tracking systems and/or motion detectors (e.g. photocell detector arrays). Fourth, multiple trials [in Pather and Gerlai (2009), thirty trials] can be administered without the intervention by the experimenter. The fish stays in the test tank and is given the stimuli and their responses are measured repeatedly across the continuous sequence of trials. As a result of all these above features, the paradigm is fully automated and thus multiple set-ups can be run in parallel. Although the 30 trials required 3,300 s (55 min) per experimental fish (Giles and Rankin 2009), one can easily set up several such test apparati. Briefly, the throughput of the task can be dramatically increased by scaling up. In our facility a 20 m^2 test room could be easily fitted with 50 such test apparati, i.e. in an eight work hour day, one can test 400 zebrafish in a single room using this learning task, a sufficiently high throughput even for large scale mutagenesis screens (Haffter and Nüsslein-Volhard 1996).

The above paradigm is new and thus there are numerous unexplored questions one may need to address. For example, we do not know whether zebrafish can forecast the future, i.e. whether their performance improvement was due to better timing of their responses (knowing when and where the shoaling image will appear in the near future). It is possible that the performance improvement of the fish was simply due to acquisition of CS-US association: disappearance of the stimulus on one side serving as the conditioned stimulus predicted the reappearance of the unconditioned stimulus, the shoal image, on the opposite side. There are many questions that concern possible optimization of the task as well: is the 20/90 s stimulus/no-stimulus interval ratio the best? Could longer tanks (the original experiment was conducted in a 50 cm long tank, a distance that can be easily traversed by the fast zebrafish) be more appropriate allowing more sensitive detection of performance improvements/deficits? Also, how would other stimulus/no-stimulus schedules (random versus fixed ratio, increasing stringency versus constant) affect the behavior of the fish? Last, mechanistic questions as to what neuroanatomical structures subserve the task, what drugs may influence performance in it, and how sensitive it may be to detect mutation-induced changes all will have to be explored. Clearly, there are many questions when one introduces a new paradigm. Nevertheless, the above example demonstrates how one can utilize species-specific perceptual, motor and motivational characteristics to design relatively simple and high throughput behavioral test methods that may allow addressing many of the above questions in the future.

The above learning task have an important temporal component, the delay between the stimulus presented on one versus the other side of the tank and as such may allow the analysis of a complex forms of learning in zebrafish known to be associated with the hippocampus in mammals, trace conditioning (McEchron and Disterhoft 1999) and/or acquisition of relational memory (Cohen et al. 1997). Although fish do not have a structure whose circuitry resembles that of the mammalian hippocampus, they do possess a brain region, the lateral pallium that is believed to be a structure homologous to the mammalian hippocampus (Vargas et al. 2009). Furthermore, fish without the classical mammalian hippocampal circuitry have also been found to be able to learn spatial learning paradigms, a class of tasks that is associated with hippocampal function in mammals (Salas et al. 1996). Spatial learning has also been demonstrated in zebrafish (Sison and Gerlai 2010), however, the spatial task employed (learning to find a particular location in a plus maze) was extremely time consuming. It required many repeated trials which could be administered only manually. Could one design a high throughput spatial task for zebrafish?

This question was answered in a recent study (Gómez-Laplaza and Gerlai 2010) that demonstrated good learning performance of zebrafish in a latent-learning paradigm. The paradigm consisted of two phases, a long training phase and a brief probe trial. During the training phase zebrafish were allowed to explore a complex maze which consisted of a starting chamber that was connected to a goal chamber by a left and right tunnel. Zebrafish were allowed to explore the maze in groups of ten (a shoal) for 16 consecutive days, each day once for 50 min. Allowing zebrafish to swim around the maze in ten-member shoals facilitated active exploration and reduced passive fear responses. During the exploration of the maze certain shoals were allowed to go through only one of the tunnels, i.e. there was a set of fish for which only the left tunnel was open and the right tunnel was blocked and another for which the left tunnel was blocked and only the right tunnel was open, and yet another group for which both tunnels were open, a spatial exploration task. The second part of the paradigm was a short (10 min long) probe trial, during which both tunnels of the maze were open, a shoal of stimulus fish was placed inside a transparent container and into the goal chamber of the maze, and the experimental fish were tested singly in the maze. Given the social nature of zebrafish, the experimental subject was highly motivated to get as close to the stimulus fish in the goal chamber as possible. Which route, the right versus the left tunnel, the experimental fish took was video-recorded and analyzed. The results showed that those fish that experienced the right tunnel open during the maze exploration phase of the paradigm also used the right tunnel during the probe trial, those fish that experienced the left tunnel open used the left tunnel during the probe trial and those fish that experienced both the left and right tunnel open chose randomly. Why is this paradigm high throughput? Although the exploration phase of the paradigm took 16 days, because the fish were not monitored and their behavior was not analyzed, one could set up a large number of mazes and train a large number of fish every day. The probe trial was conducted for every fish separately, but it lasted only for 10 min per fish and the swim path of the fish could be quantified using automated video-tracking techniques (Blaser and Gerlai 2006). Thus this phase of the paradigm could also be made high throughput. Furthermore, given the spatial nature of the task, this paradigm is likely to be capable of tapping into complex forms of learning and memory.

There are again many questions about this novel paradigm. What motivates the fish to learn the maze? In other words, why fish remember the tunnel they explored before? This form of learning is termed latent learning because apparently there is no external experimenter controlled motivator (positive or negative reinforcement) presented. However, it has been argued (Gómez-Laplaza and Gerlai 2010), based on prior supporting evidence, that exploration of novelty itself is rewarding in this task and the novel aspect of the maze is what kept the fish motivated to explore and learn. The results of this study also suggested that learning in this paradigm was likely based upon acquiring and remembering external visual cues, i.e. spatial learning, a hypothesis that will need to be proven in the future. But again, despite the novel aspect of the task and the fact that there may be numerous questions one could explore with it, the paradigm does appear to be appropriate for high throughput screening of learning and mnemonic characteristics of zebrafish and mutation-induced changes in these characteristics.

There are numerous human disorders associated with memory loss and/or impairment of cognitive function, perhaps the most devastating and prevalent is Alzheimer's disease discussed above. But milder forms of memory problems, mild cognitive impairment (MCI) and age-related memory decline also affect a large percentage of the aging human population in the twenty-first century. Given the large unmet medical need associated with these diseases and the potential complexity of the genetic mechanisms underlying them (Haffter et al. 1996), the importance of appropriate screening tools with which mutation-induced changes in learning and memory processes may be identified is unquestionable.

12 Fear and Anxiety

Fear (induced by particular negative stimuli) and anxiety (a more diffuse and prolonged behavioral state not associated with particular induction stimuli) affects a large percentage of the human population (Weisberg 2009) and despite concerted efforts by pharmaceutical research companies and academic laboratories and despite the existence of several drugs, proper treatment is still not available for a proportion of patients. It has been argued by several researchers that zebrafish may be successfully utilized to study and model some of the mechanisms of vertebrate fear and/or anxiety (Gerlai 2010). For example, fear responses have been reliably induced in zebrafish using a chemical cue, the alarm substance (Speedie and Gerlai 2008). Alarm substances have been shown to elicit fear and panic reactions in a broad range of fish species (Speedie and Gerlai 2008). These substances, which are produced by epidermal club cells in the skin of many fish species, are released when the skin is cut or damaged. In nature, the alarm substance is believed to signal danger, perhaps the presence of an actively hunting predator (piscivore fish species or a bird of prey). In the laboratory, the alarm substance has been successfully utilized to experimentally induce fear responses. Zebrafish have also been shown to reliably respond to this chemical cue with alarm reactions that include erratic movements (zig-zagging), jumping (or leaping) and freezing (complete immobility) (Speedie and Gerlai 2008). From the perspective of mutagenesis screening, however reliable these responses may seem, the alarm substance approach suffers from a major disadvantage. This substance has to be extracted from the skin of conspecifics which entails cutting or homogenizing the skin of freshly sacrificed fish and washing, diluting the extract. Because of the variability inherent in this extraction process, the exact dose and potency of the substance can not be ascertained across multiple experiments (multiple extractions). Recently, however, zebrafish has been shown to respond to a synthetic alarm substance that shares a key chemical structural element with that of natural alarm substances from several fish species (Parra et al. 2009). Hypoxantin-3-N-Oxide, H3NO, has been found to induce alarm reactions in zebrafish similar to those elicited by the natural alarm substance (Parra et al. 2009). Thus, it is now possible to precisely control the dose of the alarm substance and reduce unwanted experimental error variation, a crucial requirement for high throughput mutagenesis screens.

Although the synthetic alarm substance, H3NO, now allows precise and replicable fear induction, this method suffers from a drawback. Olfactory cues are notoriously difficult to work with. The onset (delivery) of the cue and now also its dose can be precisely controlled, however, its offset (washout) is difficult to achieve. In most behavioral paradigms, experimenters want to introduce the subject to its test chamber (tank) and let the subject habituate, establish a stable baseline behavior before administering the cue (the alarm substance in this case). This allows pre- and post-cue delivery periods to be compared and thus is a more powerful experimental design. Ideally, after the delivery of the cue and recording the effects of this delivery, one would like to turn it off and again compare periods during and after cue delivery. But this is quite cumbersome with olfactory cues. Furthermore, even if the experimental paradigm does not require turning off the cue during the behavioral session, the cue may be difficult to remove from the tank for the next subject. Residual amounts of the alarm substance may remain in the test tank even after emptying and refilling the test tank. As even trace amounts of the alarm substance may influence the behavior of the fish, this olfactory cue is difficult to work with especially when one wants to run a large number of fish as required for mutagenesis screening.

To circumvent the above issues, cues of other modalities have been tried for the induction of fear responses. Zebrafish, being a diurnal vertebrate, has excellent vision and respond well to visual cues. Zebrafish have been demonstrated to respond differentially to the sight of live fish according to whether the fish species shown were predatory or harmless and whether they were sympatric (coinhabiting the geographical region) or allopatric with zebrafish (Bass and Gerlai 2008). The latter study also demonstrated that zebrafish uniquely responded to a sympatric predator, the Indian leaf fish (*Nandus nandus*) and that the sight (solely visual stimuli) of the predator was sufficient to induce a maximal fear response (erratic movements and jumps). Utilizing this finding, subsequently zebrafish have been found to exhibit significant antipredatory responses not only to the sight of live Indian leaf fish but also to animated (moving) computer images of this species (Gerlai et al. 2009b). In this latter paradigm, both the presentation of stimuli and the recording and analysis of the fear responses were conducted in an automated computerized manner, i.e. the test paradigm was scalable and thus potentially appropriate for high throughput screening. Although numerous parameters of this automated fear paradigm will have to be optimized (e.g. size of the test tank, size and speed of movement of the predator image, presence or absence of hiding places, level of illumination, etc.), the results demonstrate the feasibility of high throughput screening for agents (mutations or pharmaceutical compounds) that may have fear altering properties.

13 Social Behavior

The last behavioral focus area I discuss in this paper is social behavior. Social behavior is a common term for a range of complex behavioral phenomena from agonistic (aggressive) encounters to reproductive (courtship) behaviors. Here I focus on a behavior within this broad range termed affiliative behavior, social cohesion or group forming. Affiliative or group forming behaviors are characteristic of our own species. Humans tend to form groups, which in modern history led to the development of the complex society where a set of rules govern. We are particularly sensitive to social signals and tend to spontaneously follow a large number of complex social rules. Briefly, being social is an inherent human trait. There are numerous human disorders that are associated with abnormalities in social behavior, one prominent example is the autism spectrum disorders (ASD). Treatment for ASD and other forms of abnormal social behaviors is lacking for two main reasons. One, the mechanisms underlying these diseases are unclear. Two, the mechanisms underlying social behaviors in general are not understood. Laboratory model organisms have been proposed to speed up the discovery of such mechanisms [for review see Gerlai and Gerlai (2003)]. The question as to whether autism may be modeled using animals is not trivial, however. Some may be skeptical and say that in order to model autism in animals one would need to understand its mechanism first in humans, so what is the use of animal research? Nevertheless, it is becoming clearer that even such complex phenomena as social behavior has not only face but also construct validity in animal models, i.e. not only looks similar in animals but may also be mechanistically similar to that of our own species. Briefly, it may make sense to study social behavior in vertebrates other than humans, discover the underlying mechanisms in the laboratory organism, and look for translational aspects of the work, i.e. human homologs. Zebrafish is perhaps the most social vertebrate model organism currently under study in the laboratory. Zebrafish are found swimming in groups in nature, a behavior that they maintain under the artificial confines of the laboratory (Engeszer et al. 2007). It is this swimming together response, or shoaling, that may be an excellent behavioral phenomenon to study from a translational perspective. Answering such questions as to what neurobiological mechanisms (circuits, synaptic processes, biochemical interactions) underlie group forming or social cohesion in zebrafish may help us understand human social behavior and ultimately perhaps the mechanisms of the abnormalities of human social behavior. The first step in this research could be the characterization of social behavior in zebrafish followed by the development of behavioral test paradigms that could detect mutation or drug-induced changes in brain function at the level of social behavior. Below I present some examples of recent discoveries with zebrafish that may be useful to make the first steps in this direction.

Zebrafish forms groups and swims in group formation but due to unavailability of appropriate behavioral quantification methods, the complexities of this behavior were not properly described in the past. By now, however, methods have been designed that allow the quantification of numerous parameters of shoaling behavior, including moment to moment changes of the distances among every possible pairs of fish within the shoal (Miller and Gerlai 2007, 2008). A periodic (cyclical) fluctuation of shoal cohesion has been discovered in zebrafish (Miller and Gerlai 2008). Analysis of shoaling is now further developed to allow high throughput automated tracking of multiple fish and thus the precise description of how the entire shoal behaves. This method may enable one to screen for mutations but would require the use of a group of fish that carry the same mutation, which would necessitate breeding an extra generation (i.e. testing not the individual mutant fish but its offspring). Perhaps a faster behavioral screening method may be to test single fish and its response to social stimuli. The disadvantage of the latter approach is that complex group dynamics may not be detected but the advantage is that the test would save the extra generation of breeding.

Testing responses of individual fish to social stimuli has been achieved with an experimental set up similar to the predator visual stimulus paradigm (Gerlai et al. 2009a). Here the computer monitor placed on the side of the experimental tank shows animated (moving) images of zebrafish (five fish in this case). Each fish on the monitor moves independently and in different randomized directions and with a speed that changes from second to second while remaining within the range of the speed of normally swimming zebrafish. This artificial "shoal" elicits a robust behavioral response. The single experimental fish placed in the test tank usually does not exhibit a preference for any sides of the tank, explores the entire tank, and thus its position when averaged over a period of time (e.g. for one minute intervals) ends up to be in the middle of the tank, which is 25 cm away from the computer screen in case of a 50 cm long tank. However, as soon as the computer screen shows the artificial shoal, the experimental zebrafish moves closer to the computer screen and on average stays about 10 cm away from it, a distance that is similar to what has been obtained with freely moving zebrafish in a real shoal (Miller and Gerlai 2008). Given that the visual stimulus that elicits the response is computer controlled and the subject's distance from the stimulus screen is recorded using computerized video-tracking, the entire test paradigm is automated, i.e. does not require the presence of the experimenter during the behavioral recording session. The paradigm therefore is high throughput and has utility in screening for mutation or drug-induced changes in social behavior. Indeed, this paradigm has been already used to detect strain (genetic) differences between populations of zebrafish, and alcohol and dopamine receptor antagonist-induced changes in social behavior [Gerlai et al. (2009a) and unpublished results].

14 Concluding Remarks

The excellent genetic tools developed for zebrafish have already provided promising results in the analysis of complex brain disorders. Importantly, several genes implicated in a number of human brain disorders have been shown to have zebrafish homologs. The function of these genes in embryonic development, and in a few cases in behavioral responses has started to be investigated. The genetic tools are constantly refined. Increasing number of genetic markers is becoming available for linkage analysis-based gene localization in random mutagenesis. Reverse genetic tools are also rapidly developing. In addition, but also very importantly, numerous novel behavioral paradigms are being developed and our understanding of the behavioral responses and capabilities of zebrafish has been exponentially increasing over the past few years suggesting that efficient and high throughput phenotypical screening applications (Gerlai 2002) are becoming reality for zebrafish. Given the fact that behavior is the output of the brain and that vertebrates share numerous biological features, including nucleotide sequence homologies, it is likely that zebrafish neurobehavioral genetics will facilitate the identification of numerous genes and compounds leading to the understanding and better treatment of human brain disorders.

Acknowledgments Supported by NIH/NIAAA (USA) and NSERC (Canada).

References

Al-Imari L, Gerlai R (2008) Conspecifics as reward in associative learning tasks for zebrafish (*Danio rerio*). Behav Brain Res 189:216–219

Alsop D, Vijayan MM (2008) Development of the corticosteroid stress axis and receptor expression in zebrafish. Am J Physiol Regul Integr Comp Physiol 294:R711–R719

Amsterdam A, Hopkins N (2006) Mutagenesis strategies in zebrafish for identifying genes involved in development and disease. Trends Genet 22:473–478
Bailey CH, Kandel ER (2008) Synaptic remodeling, synaptic growth and the storage of long-term memory in Aplysia. Prog Brain Res 169:179–198
Bandmann O, Burton EA (2010) Genetic zebrafish models of neurodegenerative diseases. Neurobiol Dis 40:58–65
Bass SLS, Gerlai R (2008) Zebrafish (*Danio rerio*) responds differentially to stimulus fish: the effects of sympatric and allopatric predators and harmless fish. Behav Brain Res 186:107–117
Bill BR, Petzold AM, Clark KJ, Schimmenti LA, Ekker SC (2009) A primer for morpholino use in zebrafish. Zebrafish 6:69–77
Blaser R, Gerlai R (2006) Behavioral phenotyping in zebrafish: comparison of three behavioral quantification methods. Behav Res Meth 38:456–469
Braff DL, Geyer MA, Light GA, Sprock J, Perry W, Cadenhead KS et al (2001) Impact of prepulse characteristics on the detection of sensorimotor gating deficits in schizophrenia. Schizophr Res 49:171–178
Burgess HA, Granato M (2007) Sensorimotor gating in larval zebrafish. J Neurosci 27:4984–4994
Capecchi MR (1989) Altering the genome by homologous recombination. Science 244:1288–1292
Chatterjee D, Gerlai R (2009) High precision liquid chromatography analysis of dopaminergic and serotoninergic responses to acute alcohol exposure in zebrafish. Behav Brain Res 200:208–213
Chen E, Ekker SC (2004) Zebrafish as a genomics research model. Curr Pharm Biotechnol 5: 409–413
Cohen NJ, Poldrack RA, Eichenbaum H (1997) Memory for items and memory for relations in the procedural/declarative memory framework. Memory 5:131–178
Denver RJ (2009) Structural and functional evolution of vertebrate neuroendocrine stress systems. Ann N Y Acad Sci 1163:1–16
Drerup CM, Wiora HM, Topczewski J, Morris JA (2009) Disc1 regulates foxd3 and sox10 expression, affecting neural crest migration and differentiation. Development 136:2623–2632
Driever W, Solnica-Krezel L, Schier AF, Neuhauss, SCF, Malicki J, Stemple DL, Stainier DYR, Zwartkruis F, Abdelilah S, Rangini Z, Belak J, Boggs C (1996) A genetic screen for mutations affecting embryogenesis in zebrafish. Development 123:37–46
Ekker SC (2008) Zinc finger-based knockout punches for zebrafish genes. Zebrafish 5:121–123
Engeszer RE, Patterson LB, Rao AA, Parichy DM (2007) Zebrafish in the wild: a review of natural history and new notes from the field. Zebrafish 4:21–40
Fan L, Collodi P (2006) Zebrafish embryonic stem cells. Methods Enzymol 418:64–77
Gauthier J, Champagne N, Lafreniere RG, Xiong L, Spiegelman D, Brustein E et al (2010) De novo mutations in the gene encoding the synaptic scaffolding protein SHANK3 in patients ascertained for schizophrenia. Proc Natl Acad Sci USA 107:7863–7868
Gerlai R (2010) Zebrafish antipredatory responses: a future for translational research? Behav Brain Res (in press)
Gerlai R (2002) Phenomics: fiction or the future? Trends Neurosci 25:506–509
Gerlai J, Gerlai R (2003) Autism: a large unmet medical need and a complex research problem. Physiol Behav 79:461–470
Gerlai R, Clayton NS (1999) Analysing hippocampal function in transgenic mice: an ethological perspective. Trends Neurosci 22:47–51
Gerlai R, Chatterjee D, Pereira T, Sawashima T, Krishnannair R (2009a) Acute and chronic alcohol dose: population differences in behavior and neurochemistry of zebrafish. Genes Brain Behav 8:586–599
Gerlai R, Fernandes Y, Pereira T (2009b) Zebrafish (*Danio rerio*) responds to the animated image of a predator: towards the development of an automated aversive task. Behav Brain Res 201:318–324
Gerlai R, Wojtowicz JM, Marks A, Roder J (1995) Over-expression of a calcium binding protein, S100ß, in astrocytes alters synaptic plasticity and impairs spatial learning in transgenic mice. Learn Mem 2:26–39

Giles AC, Rankin CH (2009) Behavioral and genetic characterization of habituation using *Caenorhabditis elegans*. Neurobiol Learn Mem 92:139–146
Gómez-Laplaza LM, Gerlai R (2010) Latent Learning in Zebrafish (*Danio rerio*). Behav Brain Res 208:509–515
Haffter P, Nüsslein-Volhard C (1996) Large scale genetics in a small vertebrate, the zebrafish. Int J Dev Biol 40:221–227
Haffter P, Granato M, Brand M, Mullins MC, Hammerschmidt M, Kane DA, Odenthal J, Van Eeden FJM, Jiang YJ, Heisenberg CP, Kelsh RN, Furutaniseiki M, Vogelsang E, Beuchle D, Schach U, Fabian C, Nüsslein-Volhard C (1996) The identification of genes with unique and essential function in the development of the zebrafish, *Danio rerio*. Development 123:1–36
Huang CJ, Jou TS, Ho YL, Lee WH, Jeng YT, Hsieh FJ, Tsai HJ (2005) Conditional expression of a myocardium-specific transgene in zebrafish transgenic lines. Dev Dyn 233:1294–1303
Joshi P, Liang JO, Dimonte K, Sullivan J, Pimplikar SW (2009) Amyloid precursor protein is required for convergent-extension movements during zebrafish development. Dev Biol 335:1–11
Kim S, Radhakrishnan UP, Rajpurohit SK, Kulkarni V, Jagadeeswaran P (2010) Vivo-Morpholino knockdown of alpha IIb: a novel approach to inhibit thrombocyte function in adult zebrafish. Blood Cells Mol Dis 44:169–174
Knapik EW (2000) ENU mutagenesis in zebrafish—from genes to complex diseases. Mamm Genome 11:511–519
Langenau DM, Feng H, Berghmans S, Kanki JP, Kutok JL, Look AT (2005) Cre/lox-regulated transgenic zebrafish model with conditional myc-induced T cell acute lymphoblastic leukemia. Proc Natl Acad Sci USA 102:6068–6073
Lee KY, Huang H, Ju B, Yang Z, Lin S (2002) Cloned zebrafish by nuclear transfer from long-term-cultured cells. Nat Biotech 20:795–799
Lekven AC, Helde KA, Thorpe CJ, Rooke R, Moon RT (2000) Reverse genetics in zebrafish. Physiol Genomics 2:37–48
Mathur P, Guo S (2010) Use of zebrafish as a model to understand mechanisms of addiction and complex neurobehavioral phenotypes. Neurobiol Dis 40:66–72
McEchron MD, Disterhoft JF (1999) Hippocampal encoding of non-spatial trace conditioning. Hippocampus 9:385–396
Miller N, Gerlai R (2008) Oscillations in shoal cohesion in zebrafish (*Danio rerio*). Behav. Brain Res 193:148–151
Miller N, Gerlai R (2007) Quantification of shoaling behaviour in zebrafish (*Danio rerio*). Behav Brain Res 184:157–166
Moens CB, Donn TM, Wolf-Saxon ER, Ma TP (2008) Reverse genetics in zebrafish by TILLING. Brief Funct Genomic Proteomic 7:454–459
Pan Y, Razak Z, Mo K, Westwood JT, Gerlai R (2010) Chronic alcohol exposure induced gene expression changes in the zebrafish brain. Genes Brain Behav (in press)
Parra KV, Adrian JC Jr, Gerlai R (2009) The synthetic substance hypoxanthine 3-N-oxide elicits alarm reactions in zebrafish (*Danio rerio*). Behav. Brain Res 205:336–341
Paquet D et al (2009) A zebrafish model of tauopathy allows in vivo imaging of neuronal cell death and drug evaluation. J Clin Invest 119:1382–1395
Pather S, Gerlai R (2009) Shuttle box learning in zebrafish. Behav Brain Res 196:323–327
Patton EE, Zon LI (2001) The art and design of genetic screens: zebrafish. Nat Rev Genet 2:956–966
Pekhletski R, Gerlai R, Overstreet L, Huang X-P, Agopyan N, Slater NT, Roder J, Hampson DR (1996) Impaired motor learning and short-term synaptic plasticity in mice lacking mGluR4 metabotropic glutamate receptors. J Neurosci 16:6364–6373
Querfurth HW, LaFerla FM (2010) Alzheimer's disease. N Engl J Med 362:329–344
Reimers MJ, Hahn ME, Tanguay RL (2004) Two zebrafi sh alcohol dehydrogenases share common ancestry with mammalian class I, II, IV, and V alcohol dehydrogenase genes but have distinct functional characteristics. J Biol Chem; 279:38303–38312
Renier C, Faraco JH, Bourgin P, Motley T, Bonaventure P, Rosa F, Mignot E (2007) Genomic and functional conservation of sedative-hypnotic targets in the zebrafish. Pharmacogen Genomics 17:237–253

Salas C, Rodríguez F, Vargas JP, Durán E, Torres B (1996) Spatial learning and memory deficits after telencephalic ablation in goldfish trained in place and turn maze procedures. Behav Neurosci 110:965–980
Scott EK, Mason L, Arrenberg AB, Ziv L, Gosse NJ, Xiao T, Chi NC, Asakawa K, Kawakami K, Baier H (2007) Targeting neural circuitry in zebrafish using GAL4 enhancer trapping. Nat Methods 4:323–326
Sison M, Gerlai R (2010) Associative learning in zebrafish (*Danio rerio*) in the plus maze. Behav Brain Res 207:99–104
Sison M, Cawker J, Buske C, Gerlai R (2006) Fishing for genes of vertebrate behavior: zebra fish as an upcoming model system. Lab Animal 35:33–39
Sivasubbu S, Balciunas D, Davidson AE, Pickart MA, Hermanson SB, Wangensteen KJ, Wolbrink DC, Ekker SC (2006) Gene-breaking transposon mutagenesis reveals an essential role for histone H2afza in zebrafish larval development. Mech Dev 123:513–529
Skromne I, Prince VE (2008) Current perspectives in zebrafish reverse genetics: moving forward. Dev Dyn 237:861–882
Sokolowski MB (2001) Drosophila: genetics meets behaviour. Nat Rev Genet 2:879–890
Speedie N, Gerlai R (2008) Alarm substance induced behavioral responses in zebrafish (*Danio rerio*) Behav. Brain Res 188:168–177
Sweatt JD (2010) Mechanisms of memory. 2nd edn, Elsevier, Amsterdam, p 343
Tong C, Li P, Wu NL, Yan Y, Ying QL (2010) Production of p53 gene knockout rats by homologous recombination in embryonic stem cells. Nature 467:211–213
Tropepe V, Sive HL (2003) Can zebrafish be used as a model to study the neurodevelopmental causes of autism? Genes Brain Behav 2:268–281
Vargas JP, López JC, Portavella M (2009) What are the functions of fish brain pallium? Brain Res Bull 79:436–440
Weisberg RB (2009) Overview of generalized anxiety disorder: epidemiology, presentation, and course. J Clin Psychiatry 70(Suppl 2):4–9
Xia W (2010) Exploring Alzheimer's disease in zebrafish. J Alzheimer's Dis 20:981–990

Forward Genetic Approaches to Understanding Complex Behaviors

Lisa M. Tarantino and Amy F. Eisener-Dorman

Abstract Assigning function to genes has long been a focus of biomedical research. Even with complete knowledge of the genomic sequences of humans, mice and other experimental organisms, there is still much to be learned about gene function and control. Ablation or overexpression of single genes using knockout or transgenic technologies has provided functional annotation for many genes, but these technologies do not capture the extensive genetic variation present in existing experimental mouse populations. Researchers have only recently begun to truly appreciate naturally occurring genetic variation resulting from single nucleotide substitutions, insertions, deletions, copy number variation, epigenetic changes (DNA methylation, histone modifications, etc.) and gene expression differences and how this variation contributes to complex phenotypes. In this chapter, we will discuss the benefits and limitations of different forward genetic approaches that capture the genetic variation present in inbred mouse strains and present the utility of these approaches for mapping QTL that influence complex behavioral phenotypes.

Keywords Mouse · Genetics · Behavior · Mapping · QTL · Mutagenesis

Contents

1 The Importance of Phenotype 27
2 Forward Genetics and Inbred Strains: The QTL Approach 27
3 QTL Mapping: The Basics 28
4 Initial QTL Identification: Standard Mapping Populations 29

L. M. Tarantino (✉) · A. F. Eisener-Dorman
Department of Psychiatry, University of North Carolina,
Chapel Hill, NC 27516, USA
e-mail: lisat@med.unc.edu

Curr Topics Behav Neurosci (2012) 12: 25–58
DOI: 10.1007/7854_2011_189

Published Online: 2 February 2012

4.1 Recombinant Inbred Lines 29
4.2 Intercross and Backcross Mapping Populations 32
4.3 Consomic Lines or Chromosome Substitution Strains 34
5 QTL Fine Mapping 35
5.1 Congenic Strains 35
5.2 Advanced Intercross Lines 36
5.3 Heterogeneous Stocks 37
5.4 Outbred Mice 39
5.5 The Collaborative Cross 40
6 Haplotype Association Mapping 41
7 Mutagenesis 43
8 Beyond QTL Analysis: Finding the Quantitative Trait Gene 47
9 The Future of Forward Genetics 47
References 48

Animal models, and rodents in particular, have been used extensively to study behaviors that model human psychiatric diseases. While rats have been particularly well-suited to this research, the mouse has recently emerged as the primary model for studying genetic and genomic aspects of human disease (Cryan and Holmes 2005). This is due, in large part, to the availability of a vast array of genetic and genomic resources for such endeavors in the mouse.

The ability to introduce genes into the mouse (transgenics) and retain tissue-specific expression patterns as well as the ability to disrupt gene function (knockouts) represented a significant advance in a field that has now become known as "functional genomics." Reverse genetic approaches such as these have been useful for identification and analysis of genes believed to increase risk for schizophrenia (Desbonnet et al. 2009), anxiety (Holmes et al. 2003; Lesch et al. 2003), depression (Savitz et al. 2009; Cryan and Slattery 2010) and autism (Sudhof 2008) but are not without limitations. The types of genetic lesions introduced are often severe, completely knocking out gene function over the course of the life-span, and may not accurately reflect the genetic basis of many complex human diseases. The ability to make conditional knockouts that limit loss-of-function to specific tissues or developmental timepoints has overcome some of these limitations. However, the reverse genetic approach is generally hypothesis-driven, depending on some prior knowledge regarding a gene's role in a disease or pathway. Understandably, the genes interrogated are often those in pathways that are already targeted by current drug treatments, making the approach inherently biased. More comprehensive phenotyping might result in identification of gene function not yet attributed to a specific gene. Recently, large-scale reverse genetic projects have been launched in the United States (knockout mouse project (KOMP), Texas A&M Institute for Genomic Medicine (TIGM)), Europe (European Conditional Mouse Mutagenesis (EUCOMM)) and Canada (North American Conditional Mouse Mutagenesis (NorCOMM)) to systematically interrogate the role of every gene by producing targeted knockouts and analyzing the resulting mutants in a variety of phenotypic assays (Collins et al. 2007; Gailus-Durner et al. 2009). These efforts to catalog gene function in a non-hypothesis-driven, unbiased manner will surely provide new insights into gene function. However, studying induced mutants exclusively fails to account for naturally occurring variation, genetic modifiers and gene-gene interactions that may be of primary importance in complex human diseases. In addition, recent research in human populations indicates that loss-of-function variants can occur far more frequently than expected with no discernable effect on health (Macarthur and Tyler-Smith 2010), suggesting that null mutations may not be the best approach for modeling many human diseases.

1 The Importance of Phenotype

In the earliest days of mouse genetics, the phenotype was the primary starting point for genetic analysis. Observations of spontaneous, chemical- or radiation-induced mutations progressed to identification of linkage groups and, eventually, to chromosomal assignment (for excellent reviews, see (Lyon 2002; Paigen 2003)). As molecular tools advanced, starting with the identification of genetic markers and progressing to complete sequencing of the mouse genome, the means for exact gene localization were realized. Knowledge regarding the organization of the genome increased along with an appreciation for the complexities therein, resulting in a renewed interest in the study of complex traits from a systems biology perspective. Recent years have witnessed a resurgence of forward genetic approaches that consider naturally occurring genetic and phenotypic variation and place more emphasis on the genetic complexities that are likely to explain a large portion of human phenotypic variation.

As opposed to reverse genetics, forward genetic studies start with the measurement or observation of a phenotype followed by mapping of the causative loci or genes. A major advantage of this approach is that the role of the gene or genes in a specific biological process is/are proven a priori. In addition, forward genetic approaches are, by definition, unbiased in terms of gene identification since the phenotype is the primary level of measurement. The term "forward genetics" applies to any phenotype-driven mapping approach, including the use of standard inbred strain crosses and haplotype association analysis—both of which result in the identification of quantitative trait loci (QTL).

In this chapter, we will describe forward genetic approaches for gene identification and functional analysis, detailing the advantages, disadvantages, successes and failures of several commonly used forward genetic models.

2 Forward Genetics and Inbred Strains: The QTL Approach

Early mouse geneticists, including pioneers of the field William Castle and Clarence Little, utilized visible characteristics such as coat color and tumor susceptibility to study Mendelian inheritance. These studies eventually led to the development of inbred mouse strains in the early 1900s (reviewed in (Crow 2002)). These strains are the starting point for most forward genetic approaches. Inbred strains are produced by brother-sister mating for 20 or more consecutive generations to fix all loci in a homozygous state. There are now hundreds of inbred strains available, and while mice within a single inbred strain are genetically identical, substantial genetic and phenotypic variation exists across strains. This phenotypic and genetic variation has been used for decades to further our understanding of the genetic basis of human disease. Inbred lines and the crosses derived from them have been instrumental in the search for genes and genomic loci that contribute to the phenotypic variability for complex behavioral traits. These loci, termed quantitative trait loci, or QTL, are regions of the genome that are associated with the phenotypic expression of complex traits, and the genomic locations of these regions can be identified using a process known as QTL mapping.

3 QTL Mapping: The Basics

In its simplest form, QTL mapping can be described as a statistical association analysis between phenotype and genotype at specific locations across the genome. In order to achieve this, two things are necessary—phenotypic and genetic variation. Phenotypic variation is easily observed among inbred strains (Hamilton and Frankel 2001), and experimental crosses between inbred strains provide both phenotypic and genetic variation. Genetic variation across the genome is measured using a set of markers that can distinguish between the two or more strains used in an experimental cross. The source of these markers has changed over the years with advances in technology and knowledge of genomic structure. Genetic markers initially took the form of linked visible characteristics such as coat color and progressed to restriction fragment length polymorphisms (RFLPs). With the advent of the polymerase chain reaction (PCR), scientists had access to a trove of genetic markers called simple sequence length polymorphisms (SSLPs) that were more numerous and provided dense genetic maps. Single nucleotide polymorphisms (SNPs) are single nucleotide changes that are also detectable by PCR, and millions of them exist between inbred strains, providing even greater resolution for genetic mapping. SNPs are now the primary source of genetic variation measured in mapping studies. Genetic markers need not be causative alleles and simply serve as a means of ascertaining allelic state at a genomic location.

Once genotypes and phenotypes are collected, association of genotype with phenotype can be performed. This aim can be accomplished by single-marker analysis using basic statistical techniques (ANOVA, marker regression) and comparing marker status at each genomic location with phenotype. In the late 1980s a technique for analyzing the interval between two genotyped markers was developed by Lander and Botstein (1989), called interval mapping (IM). This technique involves calculating the likelihood of a particular genotype based on the recombination frequency and genetic distance between two markers. A LOD score is calculated to indicate the strength of the association between the phenotype and genotype. This technique has been refined over the years—Jansen (1993) and Zeng (1993) developed an algorithm for composite interval mapping (CIM) that allows for interval mapping in combination with multiple-regression analysis to account for background QTL. The specifics of QTL mapping algorithms are beyond the scope of this chapter, but several books are available that provide a thorough discussion (Broman and Sen 2009; Wu et al. 2010). Linkage and mapping programs are freely available, including mapmaker QTL (Lander et al. 1987), QTL Cartographer (Basten et al. 1994, 2002), map manager QTX (Manly et al. 2001) and R/qtl (Broman et al. 2003) or its graphical user interface, J/qtl (Smith et al. 2009). Most mapping programs were designed to analyze the basic mapping populations described below, but special attention must be given to populations frequently used for fine mapping QTL.

4 Initial QTL Identification: Standard Mapping Populations

4.1 Recombinant Inbred Lines

Donald Bailey and Benjamin Taylor developed RI lines at the Jackson Laboratory with the goal of creating a recombinant population for linkage mapping. RI lines are created by intercrossing the F1 hybrids of two inbred progenitor strains to generate F2 progeny, which are then sib-mated for at least 20 generations, resulting in homozygosity at each locus and a mixture of the parental alleles providing genetic variation necessary for QTL mapping. Each RI line consists of animals that are genetically identical but distinct from other RI lines in the set (Fig. 1c) (Bailey 1971).

The first murine RI panel, the CXB line, was constructed by Bailey from a BALB/cBy × C57BL/6By cross. Bailey used the seven lines resulting from the initial crosses to identify loci associated with histocompatibility factors and coat color (Taylor 1978). Twelve CXB lines now exist and have been used to map QTL that influence exploratory activity (Blizard and Bailey 1979; Crabbe et al. 1982; Neiderhiser et al. 1992), circadian rhythms of locomotion (Schwartz and Zimmerman 1990), avoidance (Neiderhiser et al. 1992) and sleep behaviors (Tafti et al. 1997). Taylor developed more RI panels, such as the BXH (C57BL/6J × C3H/HeJ) and AKXL (AKR/J × C57L/J) RI sets, which were used in the study of murine leukemia virus (Taylor et al. 1971), drug-induced seizures (Taylor 1976) and lipopolysaccharide response (Watson et al. 1977). There are now RI panels for crosses between A/J and C57BL/6J (AXB and BXA), B6(Cg)-Tyrc-2 J/J and C3H/HeJ (BXH) and C57BL/6J and DBA/2J (BXD), as well as RI lines derived from mice selectively bred for righting response to ethanol—the Long Sleep and Short Sleep mice, ILS/IbgTejJ and ISS/IbgTejJ (ILSXISS) (http://jaxmice.jax.org).

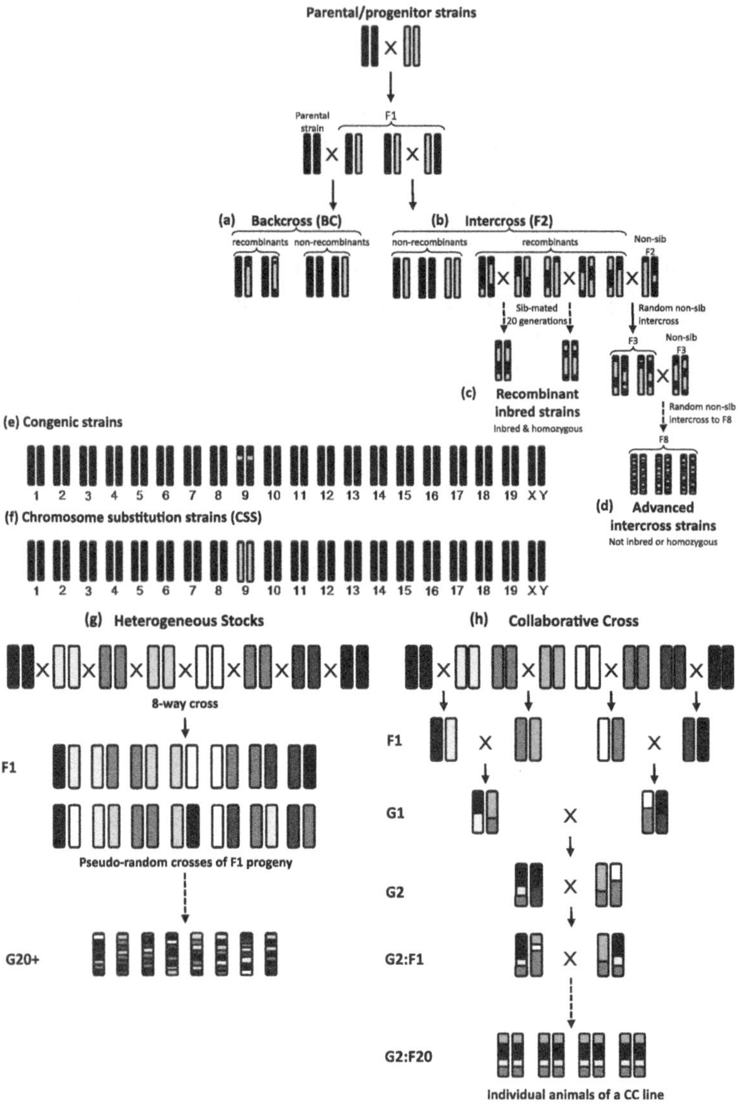

The BXD RI panel is by far the most frequently used, beginning with an initial set of 25 lines (Taylor et al. 1977) but now consisting of 79 strains (Peirce et al. 2004) (http://www.genenetwork.org). QTL for alcohol (Cunningham 1995; Phillips et al. 1994, 1995; Rodriguez et al. 1994, 1995) and drug-related (Alexander et al. 1996; Grisel et al. 1997; Jones et al. 1999; Mogil et al. 1997; Phillips et al. 1998) behaviors were among the first to be dissected using BXD lines. More recently, concerted efforts have been made to characterize the full BXD panel for neurobehavioral phenotypes, particularly those relating to pain,

◀**Fig. 1** **a–h** Breeding schemes of various genetic mapping populations. Backcross mice (**a**) are generated by crossing two inbred strains to recover F1 mice and then backcrossing the F1 mice to one of the parental strains, resulting in one copy of a chromosome that has undergone meiotic recombination and another copy from the inbred strain to which the F1 was backcrossed. Intercross mice (**b**) result from crossing F1 mice, resulting in two recombined chromosomes. Recombinant inbred strains (**c**) are produced by continuous brother-sister mating among F2s, resulting in homozygosity at each locus but with a reassortment of alleles from each parental strain. Advanced intercross lines (**d**) are also produced by crossing F2s, but inbreeding is avoided by restricting breeding partners to exclude common parents or grandparents and a large breeding population is maintained to avoid allelic drift. Congenic strains (**e**) have a small portion of one chromosome from a donor strain introgressed onto a recipient background strain, while chromosome substitution strains (**f**) have an entire donor chromosome introgressed onto a recipient background strain (examples of a congenic interval on Chr 9 (**e**) and a Chr 9 CSS (**f**) are shown). Heterogeneous stocks (**g**) are similar to advanced intercross lines but are derived from eight different inbred strains to increase genetic diversity. The collaborative cross (**h**) also has eight parental strains, but the breeding scheme is similar to recombinant inbred lines to maximize recombination while breeding each CC line to homozygosity

drug addiction, anxiety, stress and locomotor activity (Philip et al. 2010). One innovative approach uses brain-specific gene expression data from a microarray analysis of the BXD panel to map neurobehavioral QTL (Bao et al. 2007).

RI lines offer several advantages for QTL mapping. RI lines are genetic reference populations—genetically stable over time and infinitely reproducible. Therefore, each line must be genotyped only once, and that data becomes available to anyone using the line in the future (Taylor 1978). Comparisons of phenotypic data can be made across laboratories enabling genetic correlations and investigation of similarities among phenotypes. Phenotypic data can be examined retrospectively as marker density increases and more phenotypes are collected. This process has been automated by Williams and colleagues, who established the GeneNetwork database that provides tools for RI-QTL mapping and correlations among more than 1,200 phenotype datasets, including gene expression (Wang et al. 2003). RI lines are also particularly well-suited for mapping QTL for phenotypes for which genetic variance is small compared to environmental variance. This feature of RI lines is derived from the ability to conduct repeated sampling from individuals with the same genetic background, resulting in strain means that reduce environmental noise and more accurately reflect the strain phenotype.

The ability to map QTL for complex traits is dependent upon the availability of a sufficiently large RI panel that provides enough power to detect QTL of small effect size (Gora-Maslak et al. 1991). However, the RI strains were developed to investigate Mendelian traits, and the small number of strains per line has limited their usefulness for identifying QTL for complex phenotypes. Prior to the development of dense sets of genetic markers, the utility of RIs was also limited by the availability of loci for mapping. The expansion of the BXD line, along with the development of a more dense set of polymorphic markers, now allows for mapping of QTL responsible for 10% of the phenotypic variance to approximately 1–2 megabases (Mb) (Peirce et al. 2004).

However, even the largest RI sets have several characteristics that mitigate their usefulness in complex trait analysis. Existing RI panels were produced by crossing only two inbred strains and from only a limited number of inbred strains, resulting in limited genetic variability both within and across panels. Both the advantages and disadvantages of RI lines were considered when planning the collaborative cross (CC) (Churchill et al. 2004), an experimental population akin to RI strains but with added genetic variability. The CC is discussed in more detail below.

4.2 Intercross and Backcross Mapping Populations

The limited power for mapping provided by existing RI lines led to the use of other types of mapping populations—either for replication of RI-identified QTL or as alternative crosses for initial QTL identification. The most commonly used QTL mapping populations are intercross (F2) or backcross (BC) lines produced by breeding two inbred strains known to differ for the complex trait(s) of interest. The resulting F1 hybrid progeny are either intercrossed to produce an F2 generation (Fig. 1b) or backcrossed to one of the parental strains to produce a BC generation (Fig. 1a). At each locus, an F2 animal will have one of three genotypes: homozygous for either of the parental alleles or heterozygous, possessing one allele from each of the parental strains. A BC animal has only two genotype possibilities: heterozygous or homozygous for the allele of the parental strain to which the F1 progeny were backcrossed.

Unlike RI animals, intercross and backcross progeny are genetically heterogeneous; each F2 or BC individual within the population is characterized by a unique pattern of genome-wide recombination. Therefore, each set of animals must be genotyped and the informativeness of these genotypes is limited to the set of phenotypes collected for any particular set of animals. When genotyping costs were prohibitive, selective genotyping of animals from the top and bottom 15% of the phenotypic distribution was frequently performed using DNA pools (Darvasi and Soller 1994; Wang and Paterson 1994). The genetic (DNA) contribution of each individual to the pool was often weighted toward individuals at the phenotypic extremes (Taylor et al. 1999, 2001). Individuals in each pool were genotyped only at loci for which a significant or suggestive QTL were identified, and the entire mapping population was only genotyped for those loci that proved significant. As genotyping costs have decreased, the use of pooling has waned.

The number of markers required to genotype either an F2 or BC is relatively low based on the limited number of recombinations in these populations. A polymorphic marker every 40 Mb provides adequate coverage, as each marker will sweep 20 Mb in each direction. The choice of which type of cross to use is dependent on the genetic architecture of the phenotype of interest. Using an F2 mapping population is recommended to obtain an overview of the number, location and estimated effect sizes of both additive and dominant QTL segregating in the population, as well as QTL interactions (Darvasi 1998). Approximately 30% fewer F2 animals are required to detect additive QTL in comparison with a BC population due to the increased number of recombinations in the F2. However, BC mapping populations are generally considered more efficient for QTL identification, specifically for QTL with dominant effects. The decreased genetic variance in a BC—a population with a 1:1 genotypic ratio and only two prospective genotypes at any given locus—results in fewer genetic interactions than in an F2. The LOD threshold for significance is therefore lower for a BC, and gene effects are more prominent because the decreased genetic variance results in fewer genetic interactions in the BC (Darvasi 1998; Silver 1995).

Regardless of the cross type, QTL mapping has been successful in identifying thousands of loci. Conventional F2 populations derived from various inbred parental strains have been used to identify QTL for behaviors such as contextual fear conditioning (Wehner et al. 1997), spatial learning (Steinberger et al. 2003), alcohol preference (Fernandez et al. 2000) and consumption (Boyle and Gill 2008), anxiety (Bailey et al. 2008; Eisener-Dorman et al. 2010; Henderson et al. 2004; Turri et al. 2001a; Turri et al. 2004), home cage activity (Turri et al. 2001b; Umemori et al. 2009), ethanol-induced (Hitzemann et al. 1998) and cocaine-induced (Boyle and Gill 2009) locomotor activation and locomotor activity in the open field (Bailey et al. 2008; Eisener-Dorman et al. 2010; Gershenfeld et al. 1997; Kelly et al. 2003; Koyner et al. 2000). QTL for locomotor activity in the open field (Eisener-Dorman et al. 2010), alcohol preference (Melo et al. 1996) and body weight regulation (Zhang and Gershenfeld 2003) have been identified using BC populations.

The deluge of QTL for various phenotypes caused concerns regarding the possibility of numerous false positives (Type I Error). A paper published in 1995 proposed thresholds for reporting significant or suggestive QTL loci (Lander and Kruglyak 1995). Some researchers worried that the thresholds were too restrictive and that it would be cost prohibitive to phenotype and genotype populations of the size necessary to result in the levels of significance proposed (Yoon 1996). Others recommended the use of multiple strategies, including provisional mapping and replication of suggestive and significant QTL in independent populations and pooling results (Atkins 2001). At the time, however, the significance levels proposed by Lander and Kruglyak were almost universally adopted when reporting QTL. Advances in computational power now make it more common to generate significance thresholds based on individual experimental populations by running permutation tests (Churchill and Doerge 1994), and this function is present in most QTL mapping software packages.

Large F2 or BC populations have substantial power for detecting QTL, but they also have limitations. Neither F2 nor BC mapping panels are capable of providing high-resolution QTL localization due to the relatively low number of recombinations present in these populations. This results in coarsely mapped QTL with broad peaks that can encompass half of a chromosome and contain hundreds of candidate genes. Because the QTL regions identified are so large, the individual effects of closely-linked QTL are difficult to dissect. QTL comprised of multiple closely-linked, small-effect QTL may fail to be detected if the loci have opposing phenotypic effects. Consequently, the fine mapping necessary to narrow QTL intervals and, ultimately, identify quantitative trait genes (QTG) must be carried out in specialized populations.

4.3 Consomic Lines or Chromosome Substitution Strains

Consomic lines, or chromosome substitution strains (CSS), are produced by introgressing a single chromosome from a donor strain onto a recipient strain background by repeated backcrossing for at least 10 generations (Fig. 1f). A full CSS panel consists of 21 lines—one for each of the 19 autosomes and 2 sex chromosomes. Established CSS lines are genetic reference populations that allow for relatively fast localization of a QTL or group of QTL to a single chromosome and have several advantages over standard mapping populations. Because statistical comparisons are made across 21 CSS lines and the background strain rather than across hundreds of markers in hundreds of F2 or BC animals, the P-value required for significance (corrected for multiple comparisons) in a CSS experiment will be lower. The total number of mice necessary for mapping QTL to a single chromosome in a CSS panel is only slightly lower than in an F2 cross based on sample size estimates required to detect a QTL that accounts for a similar amount of the variance in either population (Belknap 2003). However, no genotyping is required in the initial scan, making the CSS more economical. In addition, the effect size of a QTL will be amplified in the CSS due to the presence of only homozygotes.

These advantages make the CSS an attractive option for initial identification of QTL. However, the end result is an entire chromosome, or multiple chromosomes, on which reside QTL that must be localized and fine-mapped. CSS provide a starting population that requires fewer generations of breeding to produce interval-specific congenic strains or mice for recombinant progeny testing. Since QTL were identified and present on a homogeneous genetic background in the CSS, their effect size should be sufficient for recovery in congenic lines. However, if a significant result on a single chromosome represents a cluster of two or more QTL, the effect might be lost upon construction of congenic or subcongenic lines, as described below.

Constructing a CSS panel is costly and takes years of breeding. Realistically, therefore, most researchers are limited to existing CSS panels and the genetic and phenotypic variability that exists therein. Three CSS panels have been reported in the literature, beginning with the B6. A panel (A/J introgressed on to C57BL/6J) (Singer et al. 2004). This panel has been the most extensively used for mapping QTL for behavioral traits such as prepulse inhibition (Leussis et al. 2009; Petryshen et al. 2005), anxiety (Singer et al. 2005) and sleep-related epilepsy (Strohl et al. 2007) and other complex phenotypes, such as onset of puberty (Nathan et al. 2006) and resistance to diet-induced obesity (Buchner et al. 2008).

Takada et al. produced a CSS panel with MSM/Ms introgressed onto C57BL/6J (Takada et al. 2008) that has been used to identify QTL for emotionality (Takahashi et al. 2008), social interactions (Takahashi et al. 2010) and home cage activity (Nishi et al. 2010). Finally, Gregorova et al. (Gregorova et al. 2008) have introgressed PWD/Ph onto C57BL/6J and have characterized the lines for blood chemistry-related phenotypes.

5 QTL Fine Mapping

The identification of QTL is fairly straightforward, as evidenced by the thousands that now exist in online databases and in the literature (Flint et al. 2005). However, identifying the gene, or genes, underlying a QTL peak has been a persistent limitation of commonly used QTL methods. The limited resolution of standard QTL crosses (RIs, F2s, BCs) results in identification of large genomic regions, usually > 20 Mb, containing hundreds of genes. To identify the quantitative trait gene (QTG), the QTL region must be narrowed considerably before a manageable list of candidate genes can be interrogated. Fine mapping of QTL has been the focus of a great deal of research in recent years (Flint et al. 2005). Initially, congenic and subcongenic strains were pursued and have resulted in some success. However, partitioning QTL in this way is not always possible, and efforts toward the use of primary crosses have resulted in the development of advanced intercross lines (AILs), heterogeneous stocks (HS), outbred mice and the CC.

5.1 Congenic Strains

Congenic strains are derived from repeated backcrossing of a donor strain to a recipient strain until only a small region of the QTL-containing donor strain chromosome is introgressed onto the recipient strain background (Fig. 1e). This process, aided by genotyping the QTL region of interest, takes 13 generations (12 backcrosses followed by an intercross to obtain founders), but the process has been accelerated by marker-assisted selection of both the introgressed region and the remaining recipient background in a strategy called "speed congenics" (Markel et al. 1997; Wakeland et al. 1997). Consequently, it is now possible to produce congenics in less than half the number of generations. QTL intervals in congenic strains can be further narrowed by backcrossing the congenic founders to break up the introgressed region, thereby producing subcongenics.

There have been some successes in using congenic and subcongenic mice to narrow QTL regions. Berrettini and colleagues identified a morphine preference QTL (*Mop2*) in a C57BL/6J (B6) × DBA/2J (D2) F2 (Berrettini et al. 1994). The 20-Mb QTL was confirmed using a marker-assisted breeding strategy to generate a pair of reciprocal congenic mouse strains—the B6 QTL interval was introgressed onto the D2 recipient background (D2.B6-*Mop2*), and the D2 QTL interval was introgressed onto the B6 recipient background (B6.D2-*Mop2*) (Ferraro et al. 2005). Heterozygous D2.B6-*Mop2* congenic mice were backcrossed to D2 mice, thereby fragmenting the 28.8-Mb QTL interval of the congenic into smaller, discrete intervals of 9.5 and 17.2 Mb in the resulting subcongenic strains (Doyle et al. 2008). Mice from one of these two subcongenic strains exhibited morphine preferences similar to both the D2.B6-*Mop2* congenic and the B6 parental strain mice, thereby narrowing the Chr 10 QTL interval to a 9.5-Mb region containing 39 genes (Doyle et al. 2008). In another study, the obesity-related QTL *Fob3* on Chr 15 was identified in an F2 cross originating from two outbred mouse lines selected for either high or low body fat (fat and lean lines) (Horvat et al. 2000). One mouse, selected because it was recombinant within the *Fob3* interval, was backcrossed for 10 generations to the Fat line to create two subcongenic strains with the Lean line *Fob3* QTL interval introgressed onto the Fat line recipient background (Stylianou et al. 2004). Further backcrossing of the *Fob3* subcongenic strains to the Fat line yielded additional recombinants, allowing the isolation of two smaller QTL (*Fob3a* and *Fob3b*) within the original *Fob3* QTL interval (Stylianou et al. 2004). Continued backcrossing of select recombinants within the 22.39-Mb *Fob3b* interval generated six additional subcongenic strains with overlapping donor intervals. Four subcongenic strains were intercrossed with the Fat line to generate four F2 mapping populations, which identified two closely-linked QTL, *Fob3b1* (4.98 Mb) and *Fob3b2* (7.68 Mb), within the *Fob3b* QTL interval (Prevorsek et al. 2010).

However, QTL effects often disappear during the construction of congenics or subcongenics. Reasons for this loss of QTL effect could be the removal of the QTL from a heterogeneous background, thereby eliminating epistatic interactions that increase QTL effect size. It is also likely that QTL that disappear during congenic production might actually be groups of two or more smaller effect size QTL that appear as a single peak in an F2 or BC but do not have sufficient power alone to result in a statistically significant change in phenotype (Legare et al. 2000; Legare and Frankel 2000).

5.2 Advanced Intercross Lines

Although the ability to restrict genetic heterogeneity to only two strains has advantages (Cheng et al. 2010), the limited number of recombinations present in F2 and BC mice also limits the resolution of these mapping resources. AILs are produced by intercrossing two parental strains to obtain an F2 and then intercrossing each successive generation (Fig. 1d). AILs have the advantage of expanding the genetic map by increasing the number of breakpoints, thereby increasing the recombination fraction between markers. The increased number of recombinations in an F10 can reduce the size of a QTL interval by five-fold in comparison with a similarly-sized population of F2 animals (Darvasi and Soller 1995). However, the increase in resolution comes at the cost of power to detect

QTL. This loss in power is due to the increased number of markers necessary to account for the reduction in linkage disequilibrium between markers and genetic drift that results in altered allele frequencies. The latter can be controlled somewhat by choosing an appropriate breeding strategy (Rockman and Kruglyak 2008). In general, maintaining large population sizes and controlling for inbreeding can mitigate concerns about power in an AIL.

One of the largest AILs is the LG/J × SM/J (LG, SM) AIL produced by Cheverud and colleagues at Washington University in St. Louis, Missouri (Norgard et al. 2008). The F9-F10 LG, SM AILs have been used successfully to identify dozens of QTL for bone-length (Kenney-Hunt et al. 2008; Norgard et al. 2008, 2009). More recently, F34 LG, SM AILs were used to replicate and refine QTL from previous studies. Norgard et al. were able to replicate almost 80% of QTL identified in both F2-F3 and F9-F10 populations. However, QTL interval reduction in the F34 was less than expected compared to previous studies. QTL intervals ranged from 0.6 to 14 Mb with half of the QTL having confidence intervals from 2 to 5 Mb. The authors postulated that family structure bias inflated the QTL peaks and decreased the confidence interval sizes reported in earlier studies. An alternative explanation is the presence of multiple linked loci at the QTL peaks—a problem that is encountered with any fine mapping population.

Cheng et al. (2010) took advantage of the power of an F2 and the mapping resolution of the LG, SM AIL to identify QTL for methamphetamine sensitivity. QTL were identified for both methamphetamine- and saline-induced locomotor behavior in a large LG × SM F2 and the F34 LG, SM AIL. Both populations were also combined and analyzed as one large intercross. Cheng et al. found that the population structure present in AILs must be considered when mapping to reduce Type I errors. By using a mixed model that considered relatedness of individuals in the AIL for both mapping and significance threshold determination, the authors were able to identify several QTL intervals at sub-centimorgan resolution—one of which contains only a single gene that is now being assessed for its role in methamphetamine sensitivity. Samocha et al. (2010) used a similar approach to map QTL for acoustic startle response, habituation and prepulse inhibition of the startle response. A large set of F2 identified QTL for multiple behaviors. The population size of the AIL used in this study did not provide enough power to detect significant QTL on its own, but combined analysis of both the F2 and AIL populations narrowed QTL intervals identified initially in the F2. Taken together, these studies indicate that the approach of using both an F2 and AILs, along with taking into account population structure, is an effective method for identification and fine-mapping of QTL.

5.3 Heterogeneous Stocks

Heterogeneous Stocks (HS) lines are produced by pseudo-random breeding over multiple generations with the aim of increasing the number of meiotic crossovers and, thus, the genetic resolution (Fig. 1g). The term "heterogeneous stock" is commonly used to describe two independently established eight-way inbred mouse crosses. The Boulder HS was established in 1970 from an eight-way cross between C57BL/6, BALB/c, RIII, AKR, DBA/2, I, A/J and C3H strains and has been breeding for more than 60 generations (McClearn et al. 1970). Initially, the Boulder HS mice were conceived as a normative population of mice with increased genetic variation in comparison to individual inbred strains (McClearn et al. 1970). The Northport HS, derived from A/J, AKR/J, BALB/cJ, C3H/HeJ, C57BL/6J, CBA/J, DBA/2J and LP/J strains, was established in 1994 as a stock for selective breeding (Hitzemann et al. 1991) and has been breeding for more than 50 generations (Demarest et al. 2001; Hitzemann et al. 1994). More recently, two additional HS lines have been reported. A collaborative cross HS (CC-HS), derived from the same eight founder strains that were used to establish the CC, was recently reported by the Hitzemann laboratory and is currently at the 12th generation of outbreeding (Iancu et al. 2010). In addition, a cross between four inbred strains, C57BL/6J, DBA/2J, BALB/cJ and LP/J, called the HS4 has also been produced and maintained in the Hitzemann laboratory for 19 generations (Malmanger et al. 2006).

It is estimated that HS mice at 60 generations of random breeding could increase mapping resolution by 30-fold over an F2 or BC. Talbot et al. (1999) used the Boulder HS to fine map a QTL for open field behavior identified on Chr 1 in previous studies (Caldarone et al. 1997; Flint et al. 1995; Gershenfeld et al. 1997; Wehner et al. 1997) and were able to refine the chromosomal location to a 1.6-Mb region. Using single-marker association analysis, they also replicated a QTL on Chr 12 but failed to replicate three additional QTL on Chrs 1, 10 and 15. The inability to replicate QTL in the HS was attributed to the inability to distinguish between identical alleles contributed by different progenitor strains. This shortcoming was overcome by the development of a multipoint mapping model that takes into account information from flanking markers and progenitor haplotypes. This mapping algorithm, called HAPPY, is available for download or use as a web-based program (http://www.well.ox.ac.uk/~rmott/happy.html). Using the multipoint mapping method, all three previously undetected loci were mapped.

The Northport HS has been used to fine map a QTL on Chr 2 for ethanol-induced locomotor activity. The QTL was initially identified in the BXD RIs and replicated in a B6XD2 F2. Using a G32-35 HS, Demarest et al. (2001) were able to replicate the Chr 2 QTL and resolve the region into three separate peaks.

Mott and Flint (2002) also developed a technique, called the inbred-outbred cross, for using HS mice to both detect and fine map QTL. The technique involves generating an F2 cross using HS and a genetically-distinct line (inbred strain, knockout, transgenic), detecting QTL using standard methods (low marker resolution) and then dense genotyping in candidate QTL regions to take advantage of the increased resolution of the HS-contributed genetic material. Through simulations, Mott and Flint showed that a 5% effect size QTL could be detected and fine mapped with 50% probability to within 6 Mb by genotyping 1,500 animals.

HS mice have been used primarily to fine map previously identified QTL (Demarest et al. 2001; Malmanger et al. 2006; Talbot et al. 1999; Turri et al. 1999), although their increased genetic variability and expanded genetic map also make them suitable for genome-wide association studies. However, several procedural and analytical hurdles had to be addressed before genome-wide QTL identification in the HS could be realized. First, because of the expansion of the genetic map, 100 times more markers and 10 times more animals are required to have the power to detect QTL with 5% effect or less (Flint et al. 2005; Valdar et al. 2006b), making genome-wide mapping in the HS an expensive proposition. Also, unknown selective pressures and random fluctuations in allele frequencies during production and/or maintenance of the stock may affect the resolving power of the HS. These issues are detailed in the first publication of genome-wide association in the Northport HS (Valdar et al. 2006b). Costs for line production and genotyping were reduced by collecting over 100 phenotypes in parallel for each mouse (Solberg et al. 2006), and model-averaging techniques were developed to analyze multiple QTL models and overcome genotype correlations (linkage disequilibrium) and family structure. Hundreds of QTL for dozens of phenotypes were detected with confidence intervals averaging 2.8 Mb—a substantial improvement over standard F2 crosses.

5.4 Outbred Mice

Outbred stocks are closed populations of genetically variable animals that are bred to maintain maximum heterozygosity (Chia et al. 2005). Dozens of outbred stocks exist, and like HS mice, they offer the advantage of increased mapping resolution but suffer some of the same drawbacks—increased sample size and marker density are required for QTL detection. In addition, outbred stocks (with the exception of HS lines) do not have the advantage of progenitor allele information to derive the origin of alleles in the offspring. This limitation introduces difficulties in the use of these stocks for QTL mapping. Only one example of a behavioral QTL has been published thus far using outbred stocks. The *Rgs2* gene that influences anxiety in mice was identified using outbred MF1 mice (Yalcin et al. 2004). Yalcin et al. determined that the haplotype patterns in MF1 mice were very similar to those found in standard inbred strains. Thus, they were able to use similar techniques for mapping as those used with HS mice.

Aldinger et al. (2009) recently examined genetic variation and population structure in CD-1 outbred stocks and determined that CD-1 mice are reasonably outbred, polymorphic at a significant number of loci and resemble human populations in terms of complex genetic history. Although CD-1 mice have population substructure that may affect mapping results, the authors believe this structure could be exploited and that the CD-1 mice represent a valuable genetic mapping resource. Williams et al. (2009) used CD-1 mice to confirm the effect of a duplication of the *Glo1* gene on anxiety-related behavior in mice. However, outbred mice have not yet been used for genome-wide association mapping of behavioral traits. Perhaps as genomic and analytical resources are developed, these untapped resources of genetic variability will become utilized more frequently in the identification and fine mapping of QTL.

5.5 The Collaborative Cross

In the late 1990s and early 2000s, the discussion in the mouse genetics community started to focus on the problem of QTG identification; it was determined that current mouse resources were not suitable for tackling such a difficult problem. Various standard mapping populations had the features necessary for identifying complex trait loci, but not one mapping resource had all of the features that were necessary if such an endeavor was to be successful down to the point of gene identification. Mouse geneticists began to envision a reference population of mice, the CC, that would capitalize on the advantages of existing resources by offering genetic diversity, mapping power and high-resolution. The CC would also provide a platform for systems genetic studies and modeling of complex networks, including gene by gene and gene by environment interactions. The CC was to be derived from eight divergent inbred strains (A/J, C57BL/6J, 129S1/SvImJ, NOD/LtJ, NZO/HlLt, CAST/EiJ, PWK/PhJ and WSB/EiJ) to maximize genetic diversity and ensure that the resulting population would provide the phenotypic diversity necessary to study any trait of interest. Like the AIL and HS lines, the CC would be subjected to multiple generations of breeding to increase meiotic recombination and provide the necessary mapping resolution for fine mapping QTL (Fig. 1h). However, the CC would be inbred and, upon completion, would provide a fully-genotyped reference population that could be provided to individual investigators. Finally, the CC would be sufficiently large ($\sim$1,000 lines) to provide the power necessary to map QTL of relatively small-effect size.

The initial and largest crosses to produce the CC were begun in 2005 at the Oak Ridge National Laboratory (ORNL). Separate, smaller breeding populations were located in Tel Aviv, Israel and Western Australia. In late 2008, the Department of Energy announced that the ORNL mouse genetics program was being phased out, and the ORNL CC population was transferred to the University of North Carolina (UNC). The Tel Aviv population has recently joined the ORNL lines at UNC. Several CC lines have already reached the inbred state due to an acceleration of the process using marker-assisted inbreeding. At least 100 lines will be completed by the end of 2012, and additional lines will follow over the next several years (personal communication, Darla Miller).

It is likely that the final population of the CC will be closer to 500 rather than 1,000 lines, as first proposed. Simulation studies have shown that 500 lines provide both adequate power and fine resolution for QTL mapping (Broman 2005; Valdar et al. 2006a). A multi-stage strategy has also been proposed using an initial mapping panel of 100 strains followed by a second set of 100 strains that have informative allele combinations and recombination events based on QTL localization from the first stage (Churchill et al. 2004). This strategy will make the CC more accessible for smaller laboratories. The completion of fully-genotyped CC-RI lines would also enable the production of many more CC recombinant inbred intercrosses (CC-RIX) produced by crossing different CC lines. Each CC-RIX mouse will inherit a chromosome from each CC parent and, thus, genotype information can be extrapolated. Heterozygosity will be recaptured in the CC-RIX lines, making them a more realistic model for human populations and allowing for more accurate modeling of complex human diseases.

6 Haplotype Association Mapping

Inbred strain surveys are an important initial step in understanding the genetic and phenotypic architecture of complex traits. Until recently, however, strain surveys had become less frequently used, meaning that researchers had a limited set of data on which to base experiments. In 2000, the Mouse Phenome Project was started as an effort to collect phenotypic data from a core set of inbred strains. These data would be stored in a central location, be publicly available on the Mouse Phenome Database (MPD; http://phenome.jax.org/ and Bogue et al. (2007)) and would offer investigators a wide range of phenotypes with which to inform research decisions for complex trait analysis (Bogue 2003; Paigen and Eppig 2000).

The effort to collect phenotype data for inbred strains coincided with efforts to increase the number of single nucleotide polymorphisms (SNPs) identified in inbred mouse lines and construct a SNP haplotype map of the mouse similar to the HapMap being constructed in humans (2003) (http://hapmap.ncbi.nlm.nih.gov/). This convergence of phenotype data and SNP genotype data presented an obvious opportunity to take advantage of both the phenotypic and genotypic diversity among inbred strains. In the context of genetic mapping, inbred strains offer several advantages over existing mapping populations. Inbred strains are, of course, a reference population of mice enabling collection and storage of phenotypic and genetic data that can be utilized repeatedly. In addition, inbred strains, like RIs, allow for repeated sampling within a strain to reduce environmental variance. Inbred strains also have increased genetic and phenotypic diversity in comparison to RI or standard F2 or BC populations, as well as a dense genetic map due to increased recombination.

In 2001, Grupe et al. used existing inbred strain phenotype data from the MPD and over 3,000 SNPs identified in their laboratory and at the Whitehead Institute (Lindblad-Toh et al. 2000) to conduct "in silico QTL mapping" in mouse. The basic approach is straightforward—as described previously, inbred strains are genetically identical within a strain but genetically diverse across strains. Therefore, associating SNP genotypes across the genome with strain phenotypes, similar to an RI-QTL study, can be expected to yield chromosomal regions that are associated with the trait of interest. The in silico mapping technique was successful in identifying QTL that overlapped with previously published reports using standard crosses (Grupe et al. 2001). Nevertheless, subsequent commentary on the method exposed several weaknesses, including lack of power, insufficient genetic variability and failure to control for both Type I and II errors (Chesler et al. 2001; Darvasi 2001). Darvasi (2001) estimated that between 40 and 150 strains would be necessary to identify a QTL affecting a quantitative trait with a heritability of 50%, and the number of strains used in the Grupe study ranged from 4 to 8. However, Grupe et al. argued that this estimate was based on power calculations using Lander and Schork significance levels based on an infinite density of genetic markers and was not applicable to their study.

Regardless, it was generally agreed that in silico mapping, more recently termed haplotype association mapping (HAM), might be a useful tool for identification of QTL. Three factors were necessary for its successful implementation: dense SNP coverage across the genome for more than just a few strains, phenotype data for multiple strains and appropriate analysis tools for genotype/phenotype associations. Pletcher et al. (2004) addressed two of these problems by identifying over 10,000 SNPs in 48 strains and developing a HAM algorithm called SNPster (http://snpster.gnf.org) to perform genotype/phenotype associations across the genome using a three-SNP sliding haplotype window. Using this method, the authors were able to identify QTL peaks for several Mendelian traits, including coat color and retinal degeneration. They were also able to replicate previously identified QTL for more complex traits such as saccharin preference, high-density lipoprotein (HDL) levels and gallstone formation. In addition, most map locations spanned intervals of 1 Mb or less—a significant improvement over standard F2 or BC mapping.

However, several hurdles still remained for HAM. As more dense SNP panels were genotyped and more was known about the genetic structure of the inbred strains, it became clear that an extensive family structure existed that could lead to spurious associations. In addition, only a limited number of strains existed for which there was both phenotype and genotype data, resulting in limited power to detect QTL for complex phenotypes. McClurg et al. (2007) attempted to address family structure by calculating a genetic similarity matrix along with a weighted bootstrap method to decrease the significance of nonspecific associations. In addition, gene expression data were used to optimize the power of the SNPster algorithm to identify cis-acting expression QTL that are considered, by some, to be a highly-enriched set of true positives with a low false-positive rate (Chesler et al. 2006). Additional analysis programs have been developed to perform association mapping in inbred strains, including Efficient Mixed-Model Association (EMMA; (Kang et al. 2008)) and hmmSNP, which uses a hidden Markov model (HMM)-based algorithm (Tsaih and Korstanje 2009). Each program uses a different algorithm and has a slightly different way of dealing with population structure. However, most programs give qualitatively similar results (unpublished data, Tim Wiltshire).

The full power of HAM has not yet been realized, as the resources for performing such analyzes are still being compiled. The quantity of SNPs in mice now number in the millions across more than 100 inbred strains. The number of phenotypes in the mouse phenome database continue to grow, with over 2,000 phenotypes stored and an average of 19 mouse strains tested per phenotype. However, the HAM technique has been used successfully to narrow down previously identified QTL intervals (Burgess-Herbert et al. 2009; Cervino et al. 2005; Harrill et al. 2009; Park et al. 2003; Wang et al. 2005) and to identify new QTL (Bopp et al. 2010; Liao et al. 2004; Liu et al. 2006; Pletcher et al. 2004).

Few HAM studies have been published for behavior. Webb et al. conducted a whole genome association study for prepulse inhibition and identified QTL on Chrs 1 and 13 (Webb et al. 2009) that overlap with genes that alter PPI when knocked out or regions of the human genome that contain genes that have been implicated in schizophrenia. Miller et al. (2010) measured tail suspension, an animal model of behavioral despair or depression, in 33 inbred strains of mice and identified four QTLs, including one that overlapped a human region that contains a locus associated with major depressive disorder and bipolar disorder.

Initial reports of QTL mapping using HAM resulted in concern about the fate of standard QTL studies (Chesler et al. 2001; Darvasi 2001). However, these concerns have not been realized. In general, HAM is viewed as one step in QTL identification that can be used for initial identification of QTL regions that must then be replicated using standard approaches. More commonly, HAM can be used to both replicate and narrow QTL that were identified using standard approaches (Burgess-Herbert et al. 2008, 2009; DiPetrillo et al. 2005). This technique is particularly effective when using strains that are genetically similar. Although genetic similarity may result in phenotypic similarity, QTL can be detected even when using parental strains that are phenotypically similar (Bailey et al. 2008; Eisener-Dorman et al. 2010). This is due to transgressive segregation, the reshuffling of alleles that occurs when two strains are crossed (Rieseberg et al. 1999). QTL identified in these crosses may be fine mapped to relatively small genomic regions using haplotype comparisons since regions of shared haplotype can be eliminated from consideration (Bailey et al. 2008; Eisener-Dorman et al. 2010).

As more phenotype data are collected for larger numbers of inbred strains and more information is gathered on the genomic structure of the laboratory mouse, HAM analysis will become an even more vital tool with which to identify complex disease genes.

7 Mutagenesis

In the late 1990s, with QTL accumulating in the literature and very few identified genes, the scientific community was looking for alternative approaches to identification of complex disease genes. Mutagenesis, the induction of mutations by chemicals or radiation, was particularly attractive. Mutagenesis in the mouse was not new—radiation-induced mutagenesis experiments to determine the genetic effects of radiation on mammals had been ongoing at the Department of Energy's Oak Ridge National Laboratory (ORNL) since just after World War II. The use of *N*-ethyl-*N*-nitrosourea (ENU) as a chemical mutagen in mice began in 1978 when it was shown to have a mutation rate 12 times higher in male spermatogonia than that of X-rays (Justice 2004).

ENU is an alkylating agent that causes single base pair mutations—usually A–T to T–A transversions or A–T to G–C transitions. ENU induces random heritable mutations at a rate of approximately 1.4×10^{-6} per nucleotide site

(Takahasi et al. 2007), although this rate varies depending on the protocol and the species or inbred strain background. Because of the random nature of mutations resulting from ENU, it was proposed as an unbiased way to induce a single mutation that would affect a complex phenotype. Thus, unlike transgenic and knockout mouse technologies, which target a specific gene of interest, a strength of the ENU mutagenesis approach is that it is not hypothesis-driven and is not limited to the study of known genes previously associated with specific biological pathways. Downstream analysis for ENU studies included mapping of the causative mutation followed by identification of the genetic lesion. The presence of only one causative mutation was anticipated to render the gene easier to identify.

In 1997, several large-scale ENU mutagenesis centers were set up in the UK and Germany to produce and screen ENU-generated mutants using batteries of phenotypic tests, including dysmorphologies, behavior, neurological abnormalities, immunology, clinical chemistry and hearing (Hrabe de Angelis et al. 2000; Nolan et al. 2000). In 2000, the NIH awarded grant funding to three large-scale mutagenesis centers in the United States at the University of Tennessee at Memphis, Northwestern University and the Jackson Laboratory with a mandate on production, identification and dissemination of ENU-induced mutants displaying alterations in nervous system function and behavior (collectively called the Neuromice.org Consortium). Additional ENU mutagenesis centers were formed outside the auspices of NIH, both small and large-scale, in academia and industry (for a complete list see (Cordes 2005)).

The mutagenesis centers quickly began producing hundreds of mutants for dozens of disease-related domains. However, it quickly became obvious that identification of the causative mutation would not be as straightforward as hoped. There have been many successes in gene identification for ENU-induced mutants in the areas of immunology (Hoebe and Beutler 2008; Sandberg et al. 2005; Tabeta et al. 2006; Theodoratos et al. 2010), development (Herron et al. 2002; Garcia–Garcia et al. 2005; Stottmann et al. 2009) and metabolism (Lloyd et al. 2005, 2006, 2010; Wilkes et al. 2009). ENU-disrupted genes responsible for neurological traits that might be considered less susceptible to environmental fluctuations have also been successfully identified in the areas of deafness (Grillet et al. 2009; Mackenzie et al. 2009; Parker et al. 2010; Schwander et al. 2009a, b), ataxia (Sharkey et al. 2009; Swanson et al. 2010; Xie et al. 2010) and epilepsy (Frankel et al. 2009; Tokuda et al. 2011). Mapping mutants for more complex behavioral phenotypes, on the other hand, has proven problematic.

The basic steps in ENU mutagenesis (Fig. 2) include induction of mutations by injection of the mutagen in male mice. These G0 mice are then bred with wildtype females, and the offspring of that cross, the G1, can be screened for dominant mutations. Further breeding of the G1 males to wildtype females produces G2 females that can be bred back to the G1 male. The resulting G3 animals will carry recessive mutations at a rate of one per eight G3s. Once an outlier is identified, the standard course of action is to prove heritability by crossing the animal back to a wildtype, produce F1 mice and then intercross to recover the mutation in the F2 progeny—one-quarter of which should exhibit a fully-penetrant, recessive

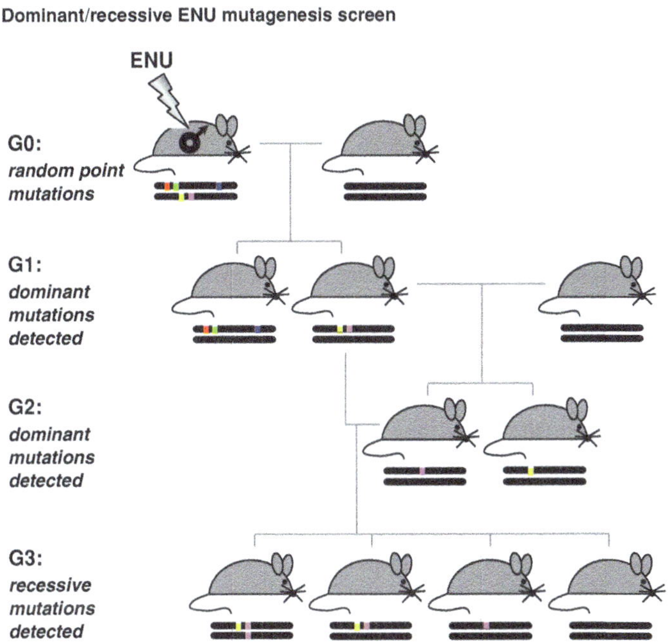

Fig. 2 Standard ENU mutagenesis. The standard ENU mutagenesis scheme includes injecting male mice (G0) with ENU. Doses vary depending upon strain, but generally, three weekly injections of 85–100 mg/kg ENU works well in B6 mice. ENU-treated males become infertile following treatment but many will regain fertility after 10–12 weeks. Loss of fertility is often used as an indicator that the ENU was effective. After regaining fertility, ENU-treated males are bred to wildtype females of the same strain or a different strain. G1 animals produced from this cross can be tested for dominant mutations. G1 mice will inherit a mutagenized genome from the father and the mutations will differ between G1s. To recover recessive mutations, the G1 females are crossed back to the G1 father to generate G2s. Two copies of any mutations that the father and daughter share will be recovered in the G3 at a rate of one mutation per eight G3s. For this reason, G3 pedigrees are often screened for phenotypes in order to recover at least one or two affected animals with which to propagate the new mutant colony

phenotype. If no F2 mice exhibit the expected phenotype, it is assumed that the original outlier had multiple alleles contributing to the phenotype (ENU produces one mutation approximately every 1–2 Mb; Kile and Hilton 2005) that were dissociated by outcrossing OR that the original outlier displayed an abnormal phenotype due to non-genetic reasons (environmental variability, etc.). If the ratio of affected mice in the F2 is lower than expected, reduced penetrance may be an issue that needs to be considered during the mapping process. If, however, the heritability cross proves successful, the mutants enter into a mapping funnel that starts with outcrossing a proven mutant to a different inbred mapping strain and F2 or BC animals are produced and phenotyped. Here we start to enter into familiar “QTL” territory—if the phenotype produced by the mutation is too weak to be observed on a mixed background or interacts with QTL in the mapping cross, the phenotype may disappear and mapping attempts will be unsuccessful.

One way to deal with a non-robust phenotype is to cross the mutant to a number of inbred strains in order to find a heterogeneous background that allows for the recovery of the mutant phenotype in the mapping cross, but this increases mapping costs. Alternatively, sufficient SNP data are now available that allow mapping to strains that are closely related to the background strain and might reduce the number of interacting QTL (Bailey et al. 2008; Eisener-Dorman et al. 2010; Xia et al. 2010). The mutant phenotype might also be preserved by limiting the amount of the mapping strain background by backcrossing the F1 animals to a mutant. This strategy also increases the number of animals in the mapping cross that carry the causative mutation. Finally, a contemporaneous population of mice not carrying the mutation can be produced, phenotyped and genotyped to provide a control for both choosing outliers in the mutant cross and identifying background QTL.

Even with these strategies, however, identification of genes in ENU mutants that exhibit behavioral anomalies has been painfully slow, even when the phenotype appears robust. Many mutants with abnormal behavioral profiles have been reported (Rastan et al. 2004; Reijmers et al. 2006; Hamre et al. 2007; Mathews et al. 2009), yet gene identification for ENU behavioral mutants has been slow to materialize.

Recent publications, however, indicate that ENU-induced behavioral mutants are beginning to yield genes. In 2007, Keays et al. identified an ENU-induced mutation in the guanosine triphosphate (GTP) binding pocket of alpha-1 tubulin (*Tuba1*). *Tuba1* mice exhibited hyperactive behavior but also had a secondary correlated phenotype—body weight. The presence of a secondary phenotype that was more stable and reproducible than the hyperactivity phenotype aided in fine mapping and identifying the gene. Speca et al. (2010) recently reported on the identification of an ENU-induced nonsense mutation in the Unc-79 gene in a mouse mutant (Lightweight) initially identified in a screen for animals with enhanced locomotor activity. The screen was conducted on a sensitized genetic background (mice heterozygous null for dopamine transporter), but the Lightweight mutation acted independently of this background. As the name implies, this mutant also had a correlated phenotype of low body weight. Peaks for both weight and locomotor activity overlapped, and the use of both phenotypes contributed to the identification of the causative ENU-induced mutation. Finally, an ENU-generated missense mutation in the *Grin1* gene was identified in a mouse mutant that displayed increased spontaneous locomotor activity (Furuse et al. 2010). It should be noted that all three of these mutants were identified in screens for dominant ENU mutations, and all were initially generated on a mixed genetic background (i.e. ENU mutagenized males were crossed to females of a different inbred strain). These results suggest that dominant mutations are ultimately easier to identify than recessive mutations. Furthermore, using a mixed genetic background in the primary ENU screen may increase the chance of recovering mutant mice in the mapping cross, thereby hastening identification of the mutated gene.

The goals of the NIH-funded ENU centers included generation and distribution of mutagenized mice with the idea that individual researchers could follow up on phenotypes of interest and map the mutated genes. Although gene identification of ENU behavioral mutants has been slow, improvements in next-generation sequencing technologies and decreased costs may accelerate the process. Deep sequencing of ENU-generated mutants, in combination with or independent of mapping information, could yield the causative mutations. Several large libraries of cryopreserved sperm and/or embryos from G1 mice are also available at RIKEN (Yoshiki et al. 2009), Harwell (Glenister and Thornton 2000), the Australian Phenomics Network and the German ENU mutagenesis center. As sequencing prices decrease, these libraries could be screened to identify mutations genome-wide. Cryo-recovery and phenotyping of mice with mutations in specific genes would complement efforts such as the KOMP.

8 Beyond QTL Analysis: Finding the Quantitative Trait Gene

The ultimate goal of QTL mapping is identification of the underlying polymorphism that can provide insight into the biology of the phenotype/disease. Regardless of how a QTL is identified, members of the complex trait community have determined that several methods are appropriate and necessary for QTL validation (Abiola et al. 2003). These criteria include relating the gene function to the QTL phenotype, identifying allelic polymorphisms or assessing gene homology to determine if the sequence is evolutionarily conserved across species. Manipulation of the gene of interest is also advantageous for QTL validation and includes the study of knockout, knockin, transgenic or otherwise genetically-altered mouse models. Genetic or functional complementation of different inbred strains or a genetically-deficient mouse model serves as an additional confirmation of the QTL (Kono et al. 2003; Yalcin et al. 2004).

Identification of the QTG represents the beginning of a new phase of analysis that explores the mechanisms underlying alterations in gene expression and biological pathways (Flint 2003). For example, *Rgs2* was mapped as an anxiety QTL and was later genetically dissected and confirmed using quantitative complementation (Yalcin et al. 2004). These findings have since been translated to human anxiety research, with the human RGS2 ortholog correlating with introversion and social anxiety, thereby becoming a potential target for the treatment of social anxiety disorders (Smoller et al. 2008).

9 The Future of Forward Genetics

As knowledge of the genomic sequence and organization of both the human and mouse genomes increases, an appreciation for the complexities that will likely explain the genetic components of complex behaviors has become more apparent. Single-gene, reverse genetic approaches like knockouts and transgenics still have much to offer toward functional annotation of genes. However, it is now widely accepted that naturally occurring genetic variation, along with epigenetic effects and environmental variation, may act synergistically to increase risk for complex neuropsychiatric diseases. Although the complex trait community has made consistent advances in the search for QTL that influence behavior, the progress from QTL to QTG has been painstakingly slow. However, genomic tools in the mouse are being developed at a rapid pace and promise to transform complex trait analysis. For example, dense SNP maps now available make it possible to narrow QTL regions using haplotype comparisons—the substantial reduction of a QTL interval that used to take years can now be accomplished in hours depending on the parental strains utilized (Eisener-Dorman et al. 2010).

New sequencing technologies are driving down costs and allowing for more accurate sequencing of entire genomes. As acquiring sequence becomes less cost prohibitive, the ability to sequence individual mice in a specific genomic interval or genome-wide will become a real option even for smaller laboratories. This will be especially useful for gene identification in ENU mutagenized lines.

Using new sequencing technologies, the Sanger Institute is sequencing the entire genome from 17 different inbred mouse strains, including the CC parental strains, as part of its Mouse Genomes Project (http://www.sanger.ac.uk/resources/mouse/genomes/). These data will provide an invaluable resource for mapping and haplotype analysis in the CC as it comes online. These sequence data, along with the increased genetic variation and mapping resolution offered by the CC, will provide access to QTL that have been unrepresented in current mapping populations.

This is a particularly exciting time for mouse genetics as resources and technology advance at a rapid pace. Behavioral scientists are poised to gain new insights into the biology and genetic control of complex behaviors with the ultimate goal of translational studies in humans.

References

Abiola O, Angel JM, Avner P, Bachmanov AA, Belknap JK, Bennett B, Blankenhorn EP, Blizard DA, Bolivar V, Brockmann GA, Buck KJ, Bureau JF, Casley WL, Chesler EJ, Cheverud JM, Churchill GA, Cook M, Crabbe JC, Crusio WE, Darvasi A, de Haan G, Dermant P, Doerge RW, Elliot RW, Farber CR, Flaherty L, Flint J, Gershenfeld H, Gibson JP, Gu J, Gu W, Himmelbauer H, Hitzemann R, Hsu HC, Hunter K, Iraqi FF, Jansen RC, Johnson TE, Jones BC, Kempermann G, Lammert F, Lu L, Manly KF, Matthews DB, Medrano JF, Mehrabian M, Mittlemann G, Mock BA, Mogil JS, Montagutelli X, Morahan G, Mountz JD, Nagase H, Nowakowski RS, O'Hara BF, Osadchuk AV, Paigen B, Palmer AA, Peirce JL, Pomp D, Rosemann M, Rosen GD, Schalkwyk LC, Seltzer Z, Settle S, Shimomura K, Shou S, Sikela JM, Siracusa LD, Spearow JL, Teuscher C, Threadgill DW, Toth LA, Toye AA, Vadasz C, Van Zant G, Wakeland E, Williams RW, Zhang HG, Zou F (2003) The nature and identification of quantitative trait loci: a community's view. Nat Rev Genet 4:911–916

Aldinger KA, Sokoloff G, Rosenberg DM, Palmer AA, Millen KJ (2009) Genetic variation and population substructure in outbred CD-1 mice: implications for genome-wide association studies. PLoS One 4:e4729

Alexander RC, Wright R, Freed W (1996) Quantitative trait loci contributing to phencyclidine-induced and amphetamine-induced locomotor behavior in inbred mice. Neuropsychopharmacology 15:484–490

Atkins JKBAL (2001) The replicability of QTLs for murine alcohol preference drinking behavior across eight independent studies. Mamm Genome 12: 893–899

Bailey DW (1971) Recombinant-inbred strains. An aid to finding identity, linkage, and function of histocompatibility and other genes. Transplantation 11:325–327

Bailey JS, Grabowski-Boase L, Steffy BM, Wiltshire T, Churchill GA, Tarantino LM (2008) Identification of QTL for locomotor activation and anxiety using closely-related inbred strains. Genes Brain Behav 7:761–769

Bao L, Peirce JL, Zhou M, Li H, Goldowitz D, Williams RW, Lu L, Cui Y (2007) An integrative genomics strategy for systematic characterization of genetic loci modulating phenotypes. Hum Mol Genet 16:1381–1390

Basten CJ, Weir BS, Zeng Z-B (1994) Zmap—a QTL cartographer. In: Smith C, Gavora JS, Benkel B, Chesnais J, Fairfull W, Gibson JP, Kennedy BW, Burnside EB (eds) 5th world congress on genetics applied to livestock production: computing strategies and software. Organizing Committee, 5th world congress on genetics applied to livestock production, Guelph, Ontario, Canada, pp 65–66

Basten CJ, Weir BS, Zeng Z-B (2002) QTL Cartographer, version 1.16. Department of Statistics, North Carolina State University, Raleigh, NC

Belknap JK (2003) Chromosome substitution strains: some quantitative considerations for genome scans and fine mapping. Mamm Genome 14:723–732

Berrettini WH, Ferraro TN, Alexander RC, Buchberg AM, Vogel WH (1994) Quantitative trait loci mapping of three loci controlling morphine preference using inbred mouse strains. Nat Genet 7:54–58

Blizard DA, Bailey DW (1979) Genetic correlation between open-field activity and defecation: analysis with the CXB recombinant-inbred strains. Behav Genet 9:349–357

Bogue M (2003) Mouse phenome project: understanding human biology through mouse genetics and genomics. J Appl Physiol 95:1335–1337

Bogue MA, Grubb SC, Maddatu TP, Bult CJ (2007) Mouse phenome database (MPD). Nucleic Acids Res 35:D643–D649

Bopp SE, Ramachandran V, Henson K, Luzader A, Lindstrom M, Spooner M, Steffy BM, Suzuki O, Janse C, Waters AP, Zhou Y, Wiltshire T, Winzeler EA (2010) Genome wide analysis of inbred mouse lines identifies a locus containing Ppar-gamma as contributing to enhanced malaria survival. PLoS One 5: e10903

Boyle AE, Gill KJ (2008) Confirmation of provisional quantitative trait loci for voluntary alcohol consumption: genetic analysis in chromosome substitution strains and F2 crosses derived from A/J and C57BL/6J progenitors. Pharmacogenet Genomics 18:1071–1082

Boyle AE, Gill KJ (2009) A verification of previously identified QTLs for cocaine-induced activation using a panel of B6.A chromosome substitution strains (CSS) and A/J × C57Bl/6J F2 mice. Psychopharmacology (Berl) 207(2):325–334

Broman KW (2005) The genomes of recombinant inbred lines. Genetics 169:1133–1146

Broman KW, Sen S (2009) A guide to QTL mapping with R/qtl. Springer, New York

Broman KW, Wu H, Sen S, Churchill GA (2003) R/qtl: QTL mapping in experimental crosses. Bioinformatics 19:889–890

Buchner DA, Burrage LC, Hill AE, Yazbek SN, O'Brien WE, Croniger CM, Nadeau JH (2008) Resistance to diet-induced obesity in mice with a single substituted chromosome. Physiol Genomics 35:116–122

Burgess-Herbert SL, Cox A, Tsaih SW, Paigen B (2008) Practical applications of the bioinformatics toolbox for narrowing quantitative trait loci. Genetics 180:2227–2235

Burgess-Herbert SL, Tsaih SW, Stylianou IM, Walsh K, Cox AJ, Paigen B (2009) An experimental assessment of in silico haplotype association mapping in laboratory mice. BMC Genet 10:81

Caldarone B, Saavedra C, Tartaglia K, Wehner JM, Dudek BC, Flaherty L (1997) Quantitative trait loci analysis affecting contextual conditioning in mice. Nat Genet 17:335–337

Cervino AC, Li G, Edwards S, Zhu J, Laurie C, Tokiwa G, Lum PY, Wang S, Castellani LW, Lusis AJ, Carlson S, Sachs AB, Schadt EE (2005) Integrating QTL and high-density SNP analyses in mice to identify Insig2 as a susceptibility gene for plasma cholesterol levels. Genomics 86:505–517

Cheng R, Lim JE, Samocha KE, Sokoloff G, Abney M, Skol AD, Palmer AA (2010) Genome-wide association studies and the problem of relatedness among advanced intercross lines and other highly recombinant populations. Genetics 185:1033–1044

Chesler EJ, Bystrykh L, De Haan G, Cooke MP, Su AI, Manly KF, Williams RW (2006) Reply to normalization procedures and detection of linkage signal in genetical-genomics experiments. Nat Genet 38:856–858

Chesler EJ, Rodriguez-Zas SL, Mogil JS (2001) In silico mapping of mouse quantitative trait loci. Science 294:2423

Chia R, Achilli F, Festing MF, Fisher EM (2005) The origins and uses of mouse outbred stocks. Nat Genet 37:1181–1186

Churchill GA, Airey DC, Allayee H, Angel JM, Attie AD, Beatty J, Beavis WD, Belknap JK, Bennett B, Berrettini W, Bleich A, Bogue M, Broman KW, Buck KJ, Buckler E, Burmeister M, Chesler EJ, Cheverud JM, Clapcote S, Cook MN, Cox RD, Crabbe JC, Crusio WE, Darvasi A, Deschepper CF, Doerge RW, Farber CR, Forejt J, Gaile D, Garlow SJ, Geiger H, Gershenfeld H, Gordon T, Gu J, Gu W, de Haan G, Hayes NL, Heller C, Himmelbauer H, Hitzemann R, Hunter K, Hsu HC, Iraqi FA, Ivandic B, Jacob HJ, Jansen RC, Jepsen KJ, Johnson DK, Johnson TE, Kempermann G, Kendziorski C, Kotb M, Kooy RF, Llamas B, Lammert F, Lassalle JM, Lowenstein PR, Lu L, Lusis A, Manly KF, Marcucio R, Matthews D, Medrano JF, Miller DR, Mittleman G, Mock BA, Mogil JS, Montagutelli X, Morahan G, Morris DG, Mott R, Nadeau JH, Nagase H, Nowakowski RS, O'Hara BF, Osadchuk AV, Page GP, Paigen B, Paigen K, Palmer AA, Pan HJ, Peltonen-Palotie L, Peirce J, Pomp D, Pravenec M, Prows DR, Qi Z, Reeves RH, Roder J, Rosen GD, Schadt EE, Schalkwyk LC, Seltzer Z, Shimomura K, Shou S, Sillanpaa MJ, Siracusa LD, Snoeck HW, Spearow JL, Svenson K, Tarantino LM, Threadgill D, Toth LA, Valdar W, de Villena FP, Warden C, Whatley S, Williams RW, Wiltshire T, Yi N, Zhang D, Zhang M, Zou F (2004) The collaborative cross, a community resource for the genetic analysis of complex traits. Nat Genet 36:1133–1137

Churchill GA, Doerge RW (1994) Empirical threshold values for quantitative trait mapping. Genetics 138:963–971

Collins FS, Rossant J, Wurst W (2007) A mouse for all reasons. Cell 128:9–13

Cordes SP (2005) N-ethyl-N-nitrosourea mutagenesis: boarding the mouse mutant express. Microbiol Mol Biol Rev 69:426–439

Crabbe JC, Rigter H, Kerbusch S (1982) Analysis of behavioural responses to an ACTH analog in CXB/By recombinant inbred mice. Behav Brain Res 4:289–314

Crow JF (2002) C. C. Little, cancer and inbred mice. Genetics 161:1357–1361

Cryan JF, Holmes A (2005) The ascent of mouse: advances in modelling human depression and anxiety. Nat Rev Drug Discov 4:775–790

Cryan JF, Slattery DA (2010) GABAB receptors and depression. Current status. Adv Pharmacol 58:427–451

Cunningham CL (1995) Localization of genes influencing ethanol-induced conditioned place preference and locomotor activity in BXD recombinant inbred mice. Psychopharmacology (Berl) 120:28–41

Darvasi A (1998) Experimental strategies for the genetic dissection of complex traits in animal models. Nat Genet 18:19–24

Darvasi A (2001) In silico mapping of mouse quantitative trait loci. Science 294:2423

Darvasi A, Soller M (1994) Selective DNA pooling for determination of linkage between a molecular marker and a quantitative trait locus. Genetics 138:1365–1373

Darvasi A, Soller M (1995) Advanced intercross lines, an experimental population for fine genetic mapping. Genetics 141:1199–1207

Demarest K, Koyner J, McCaughran J Jr, Cipp L, Hitzemann R (2001) Further characterization and high-resolution mapping of quantitative trait loci for ethanol-induced locomotor activity. Behav Genet 31:79–91

Desbonnet L, Waddington JL, O'Tuathaigh CM (2009) Mutant models for genes associated with schizophrenia. Biochem Soc Trans 37:308–312

DiPetrillo K, Wang X, Stylianou IM, Paigen B (2005) Bioinformatics toolbox for narrowing rodent quantitative trait loci. Trends Genet 21:683–692

Doyle GA, Furlong PJ, Schwebel CL, Smith GG, Lohoff FW, Buono RJ, Berrettini WH, Ferraro TN (2008) Fine mapping of a major QTL influencing morphine preference in C57BL/6 and DBA/2 mice using congenic strains. Neuropsychopharmacology 33:2801–2809

Eisener-Dorman AF, Grabowski-Boase L, Steffy BM, Wiltshire T, Tarantino LM (2010) Quantitative trait locus and haplotype mapping in closely related inbred strains identifies a locus for open field behavior. Mamm Genome 21:231–246

Fernandez JR, Tarantino LM, Hofer SM, Vogler GP, McClearn GE (2000) Epistatic quantitative trait loci for alcohol preference in mice. Behav Genet 30:431–437

Ferraro TN, Golden GT, Smith GG, Martin JF, Schwebel CL, Doyle GA, Buono RJ, Berrettini WH (2005) Confirmation of a major QTL influencing oral morphine intake in C57 and DBA mice using reciprocal congenic strains. Neuropsychopharmacology 30:742–746

Flint J (2003) Analysis of quantitative trait loci that influence animal behavior. J Neurobiol 54:46–77

Flint J, Corley R, DeFries JC, Fulker DW, Gray JA, Miller S, Collins AC (1995) A simple genetic basis for a complex psychological trait in laboratory mice. Science 269:1432–1435

Flint J, Valdar W, Shifman S, Mott R (2005) Strategies for mapping and cloning quantitative trait genes in rodents. Nat Rev Genet 6:271–286

Frankel WN, Yang Y, Mahaffey CL, Beyer BJ, O'Brien TP (2009) Szt2, a novel gene for seizure threshold in mice. Genes Brain Behav 8:568–576

Furuse T, Wada Y, Hattori K, Yamada I, Kushida T, Shibukawa Y, Masuya H, Kaneda H, Miura I, Seno N, Kanda T, Hirose R, Toki S, Nakanishi K, Kobayashi K, Sezutsu H, Gondo Y, Noda T, Yuasa S, Wakana S (2010) Phenotypic characterization of a new Grin1 mutant mouse generated by ENU mutagenesis. Eur J Neurosci 31:1281–1291

Gailus-Durner V, Fuchs H, Adler T, Aguilar Pimentel A, Becker L, Bolle I, Calzada-Wack J, Dalke C, Ehrhardt N, Ferwagner B, Hans W, Holter SM, Holzlwimmer G, Horsch M, Javaheri A, Kallnik M, Kling E, Lengger C, Morth C, Mossbrugger I, Naton B, Prehn C, Puk O, Rathkolb B, Rozman J, Schrewe A, Thiele F, Adamski J, Aigner B, Behrendt H, Busch DH, Favor J, Graw J, Heldmaier G, Ivandic B, Katus H, Klingenspor M, Klopstock T, Kremmer E, Ollert M, Quintanilla-Martinez L, Schulz H, Wolf E, Wurst W, de Angelis MH (2009) Systemic first-line phenotyping. Methods Mol Biol 530:463–509

Garcia–Garcia MJ, Eggenschwiler JT, Caspary T, Alcorn HL, Wyler MR, Huangfu D, Rakeman AS, Lee JD, Feinberg EH, Timmer JR, Anderson KV (2005) Analysis of mouse embryonic patterning and morphogenesis by forward genetics. Proc Natl Acad Sci USA 102:5913–5919

Gershenfeld HK, Neumann PE, Mathis C, Crawley JN, Li X, Paul SM (1997) Mapping quantitative trait loci for open-field behavior in mice. Behav Genet 27:201–210

Glenister PH, Thornton CE (2000) Cryoconservation—archiving for the future. Mamm Genome 11:565–571

Gora-Maslak G, McClearn GE, Crabbe JC, Phillips TJ, Belknap JK, Plomin R (1991) Use of recombinant inbred strains to identify quantitative trait loci in psychopharmacology. Psychopharmacology (Berl) 104:413–424

Gregorova S, Divina P, Storchova R, Trachtulec Z, Fotopulosova V, Svenson KL, Donahue LR, Paigen B, Forejt J (2008) Mouse consomic strains: exploiting genetic divergence between Mus m. musculus and Mus m. domesticus subspecies. Genome Res 18:509–515

Grillet N, Schwander M, Hildebrand MS, Sczaniecka A, Kolatkar A, Velasco J, Webster JA, Kahrizi K, Najmabadi H, Kimberling WJ, Stephan D, Bahlo M, Wiltshire T, Tarantino LM, Kuhn P, Smith RJ, Muller U (2009) Mutations in LOXHD1, an evolutionarily conserved stereociliary protein, disrupt hair cell function in mice and cause progressive hearing loss in humans. Am J Hum Genet 85:328–337

Grisel JE, Belknap JK, O'Toole LA, Helms ML, Wenger CD, Crabbe JC (1997) Quantitative trait loci affecting methamphetamine responses in BXD recombinant inbred mouse strains. J Neurosci 17:745–754

Grupe A, Germer S, Usuka J, Aud D, Belknap JK, Klein RF, Ahluwalia MK, Higuchi R, Peltz G (2001) In silico mapping of complex disease-related traits in mice. Science 292:1915–1918

Hamilton BA, Frankel WN (2001) Of mice and genome sequence. Cell 107:13–16

Hamre KM, Goldowitz D, Wilkinson S, Matthews DB (2007) Screening for ENU-induced mutations in mice that result in aberrant ethanol-related phenotypes. Behav Neurosci 121: 665–678

Harrill AH, Watkins PB, Su S, Ross PK, Harbourt DE, Stylianou IM, Boorman GA, Russo MW, Sackler RS, Harris SC, Smith PC, Tennant R, Bogue M, Paigen K, Harris C, Contractor T, Wiltshire T, Rusyn I, Threadgill DW (2009) Mouse population-guided resequencing reveals that variants in CD44 contribute to acetaminophen-induced liver injury in humans. Genome Res 19:1507–1515

Henderson ND, Turri MG, DeFries JC, Flint J (2004) QTL analysis of multiple behavioral measures of anxiety in mice. Behav Genet 34:267–293

Herron BJ, Lu W, Rao C, Liu S, Peters H, Bronson RT, Justice MJ, McDonald JD, Beier DR (2002) Efficient generation and mapping of recessive developmental mutations using ENU mutagenesis. Nat Genet 30:185–189

Hitzemann B, Dains K, Kanes S, Hitzemann R (1994) Further studies on the relationship between dopamine cell density and haloperidol-induced catalepsy. J Pharmacol Exp Ther 271:969–976

Hitzemann R, Cipp L, Demarest K, Mahjubi E, McCaughran J Jr (1998) Genetics of ethanol-induced locomotor activation: detection of QTLs in a C57BL/6J × DBA/2J F2 intercross. Mamm Genome 9:956–962

Hitzemann R, Dains K, Bier-Langing CM, Zahniser NR (1991) On the selection of mice for haloperidol response and non-response. Psychopharmacology (Berl) 103:244–250

Hoebe K, Beutler B (2008) Forward genetic analysis of TLR-signaling pathways: an evaluation. Adv Drug Deliv Rev 60:824–829

Holmes A, Murphy DL, Crawley JN (2003) Abnormal behavioral phenotypes of serotonin transporter knockout mice: parallels with human anxiety and depression. Biol Psychiatry 54:953–959

Horvat S, Bunger L, Falconer VM, Mackay P, Law A, Bulfield G, Keightley PD (2000) Mapping of obesity QTLs in a cross between mouse lines divergently selected on fat content. Mamm Genome 11:2–7

Hrabe de Angelis MH, Flaswinkel H, Fuchs H, Rathkolb B, Soewarto D, Marschall S, Heffner S, Pargent W, Wuensch K, Jung M, Reis A, Richter T, Alessandrini F, Jakob T, Fuchs E, Kolb H, Kremmer E, Schaeble K, Rollinski B, Roscher A, Peters C, Meitinger T, Strom T, Steckler T, Holsboer F, Klopstock T, Gekeler F, Schindewolf C, Jung T, Avraham K, Behrendt H, Ring J, Zimmer A, Schughart K, Pfeffer K, Wolf E, Balling R (2000) Genome-wide, large-scale production of mutant mice by ENU mutagenesis. Nat Genet 25:444–447

Iancu OD, Darakjian P, Walter NA, Malmanger B, Oberbeck D, Belknap J, McWeeney S, Hitzemann R (2010) Genetic diversity and striatal gene networks: focus on the heterogeneous stock-collaborative cross (HS-CC) mouse. BMC Genomics 11:585

International HapMap Consortium (2003) The international HapMap project. Nature 426:789–796

Jansen RC (1993) Interval mapping of multiple quantitative trait loci. Genetics 135:205–211

Jones BC, Tarantino LM, Rodriguez LA, Reed CL, McClearn GE, Plomin R, Erwin VG (1999) Quantitative-trait loci analysis of cocaine-related behaviours and neurochemistry. Pharmacogenetics 9:607–617

Justice MJ (2004) From the atomic age to the genome project. Genetica 122:3–7

Kang HM, Zaitlen NA, Wade CM, Kirby A, Heckerman D, Daly MJ, Eskin E (2008) Efficient control of population structure in model organism association mapping. Genetics 178: 1709–1723

Keays DA, Tian G, Poirier K, Huang GJ, Siebold C, Cleak J, Oliver PL, Fray M, Harvey RJ, Molnar Z, Pinon MC, Dear N, Valdar W, Brown SD, Davies KE, Rawlins JN, Cowan NJ, Nolan P, Chelly J, Flint J (2007) Mutations in alpha-tubulin cause abnormal neuronal migration in mice and lissencephaly in humans. Cell 128:45–57

Kelly MA, Low MJ, Phillips TJ, Wakeland EK, Yanagisawa M (2003) The mapping of quantitative trait loci underlying strain differences in locomotor activity between 129S6 and C57BL/6J mice. Mamm Genome 14:692–702

Kenney-Hunt JP, Wang B, Norgard EA, Fawcett G, Falk D, Pletscher LS, Jarvis JP, Roseman C, Wolf J, Cheverud JM (2008) Pleiotropic patterns of quantitative trait loci for 70 murine skeletal traits. Genetics 178:2275–2288

Kile BT, Hilton DJ (2005) The art and design of genetic screens: mouse. Nat Rev Genet 6:557–567

Kono DH, Park MS, Theofilopoulos AN (2003) Genetic complementation in female (BXSB × NZW) F2 mice. J Immunol 171:6442–6447

Koyner J, Demarest K, McCaughran J Jr, Cipp L, Hitzemann R (2000) Identification and time dependence of quantitative trait loci for basal locomotor activity in the BXD recombinant inbred series and a B6D2 F2 intercross. Behav Genet 30:159–170

Lander E, Kruglyak L (1995) Genetic dissection of complex traits: guidelines for interpreting and reporting linkage results. Nat Genet 11:241–247

Lander ES, Botstein D (1989) Mapping mendelian factors underlying quantitative traits using RFLP linkage maps. Genetics 121:185–199

Lander ES, Green P, Abrahamson J, Barlow A, Daly MJ, Lincoln SE, Newberg LA (1987) MAPMAKER: an interactive computer package for constructing primary genetic linkage maps of experimental and natural populations. Genomics 1:174–181

Legare ME, Bartlett FS 2nd, Frankel WN (2000) A major effect QTL determined by multiple genes in epileptic EL mice. Genome Res 10:42–48

Legare ME, Frankel WN (2000) Multiple seizure susceptibility genes on chromosome 7 in SWXL-4 congenic mouse strains. Genomics 70:62–65

Lesch KP, Zeng Y, Reif A, Gutknecht L (2003) Anxiety-related traits in mice with modified genes of the serotonergic pathway. Eur J Pharmacol 480:185–204

Leussis MP, Frayne ML, Saito M, Berry EM, Aldinger KA, Rockwell GN, Hammer RP Jr, Baskin-Hill AE, Singer JB, Nadeau JH, Sklar P, Petryshen TL (2009) Genomic survey of prepulse inhibition in mouse chromosome substitution strains. Genes Brain Behav 8:806–816

Liao G, Wang J, Guo J, Allard J, Cheng J, Ng A, Shafer S, Puech A, McPherson JD, Foernzler D, Peltz G, Usuka J (2004) In silico genetics: identification of a functional element regulating H2-Ealpha gene expression. Science 306:690–695

Lindblad-Toh K, Winchester E, Daly MJ, Wang DG, Hirschhorn JN, Laviolette JP, Ardlie K, Reich DE, Robinson E, Sklar P, Shah N, Thomas D, Fan JB, Gingeras T, Warrington J, Patil N, Hudson TJ, Lander ES (2000) Large-scale discovery and genotyping of single-nucleotide polymorphisms in the mouse. Nat Genet 24:381–386

Liu P, Wang Y, Vikis H, Maciag A, Wang D, Lu Y, Liu Y, You M (2006) Candidate lung tumor susceptibility genes identified through whole-genome association analyses in inbred mice. Nat Genet 38:888–895

Lloyd DJ, Bohan S, Gekakis N (2006) Obesity, hyperphagia and increased metabolic efficiency in Pc1 mutant mice. Hum Mol Genet 15:1884–1893

Lloyd DJ, Hall FW, Tarantino LM, Gekakis N (2005) Diabetes insipidus in mice with a mutation in Aquaporin-2. PLoS Genet 1:e20

Lloyd DJ, Wheeler MC, Gekakis N (2010) A point mutation in Sec61alpha1 leads to diabetes and hepatosteatosis in mice. Diabetes 59:460–470

Lyon MF (2002) A personal history of the mouse genome. Annu Rev Genomics Hum Genet 3:1–16

Macarthur DG, Tyler-Smith C (2010) Loss-of-function variants in the genomes of healthy humans. Hum Mol Genet 19:R125–R130

Mackenzie FE, Parker A, Parkinson NJ, Oliver PL, Brooker D, Underhill P, Lukashkina VA, Lukashkin AN, Holmes C, Brown SD (2009) Analysis of the mouse mutant cloth-ears shows a role for the voltage-gated sodium channel Scn8a in peripheral neural hearing loss. Genes Brain Behav 8:699–713

Malmanger B, Lawler M, Coulombe S, Murray R, Cooper S, Polyakov Y, Belknap J, Hitzemann R (2006) Further studies on using multiple-cross mapping (MCM) to map quantitative trait loci. Mamm Genome 17:1193–1204

Manly KF, Cudmore RH Jr, Meer JM (2001) Map manager QTX, cross-platform software for genetic mapping. Mamm Genome 12:930–932

Markel P, Shu P, Ebeling C, Carlson GA, Nagle DL, Smutko JS, Moore KJ (1997) Theoretical and empirical issues for marker-assisted breeding of congenic mouse strains. Nat Genet 17:280–284

Mathews TA, Brookshire BR, Budygin EA, Hamre K, Goldowitz D, Jones SR (2009) Ethanol-induced hyperactivity is associated with hypodopaminergia in the 22-TNJ ENU-mutated mouse. Alcohol 43:421–431

McClearn GE, Wilson J, Meredith W (1970) The use of isogenic and heterogenic mouse stock in behavioral research. In: Lindzey G, Thiessen D (eds) Contributions to behavioral-genetic analysis: the mouse as a prototype. Appleton-Century-Crofts, New York, pp 3–22

McClurg P, Janes J, Wu C, Delano DL, Walker JR, Batalov S, Takahashi JS, Shimomura K, Kohsaka A, Bass J, Wiltshire T, Su AI (2007) Genomewide association analysis in diverse inbred mice: power and population structure. Genetics 176:675–683

Melo JA, Shendure J, Pociask K, Silver LM (1996) Identification of sex-specific quantitative trait loci controlling alcohol preference in C57BL/6 mice. Nat Genet 13:147–153

Miller BH, Schultz LE, Gulati A, Su AI, Pletcher MT (2010) Phenotypic characterization of a genetically diverse panel of mice for behavioral despair and anxiety. PLoS One 5:e14458

Mogil JS, Richards SP, O'Toole LA, Helms ML, Mitchell SR, Kest B, Belknap JK (1997) Identification of a sex-specific quantitative trait locus mediating nonopioid stress-induced analgesia in female mice. J Neurosci 17:7995–8002

Mott R, Flint J (2002) Simultaneous detection and fine mapping of quantitative trait loci in mice using heterogeneous stocks. Genetics 160:1609–1618

Nathan BM, Hodges CA, Palmert MR (2006) The use of mouse chromosome substitution strains to investigate the genetic regulation of pubertal timing. Mol Cell Endocrinol 254–255:103–108

Neiderhiser JM, Plomin R, McClearn GE (1992) The use of CXB recombinant inbred mice to detect quantitative trait loci in behavior. Physiol Behav 52:429–439

Nishi A, Ishii A, Takahashi A, Shiroishi T, Koide T (2010) QTL analysis of measures of mouse home-cage activity using B6/MSM consomic strains. Mamm Genome 21:477–485

Nolan PM, Peters J, Strivens M, Rogers D, Hagan J, Spurr N, Gray IC, Vizor L, Brooker D, Whitehill E, Washbourne R, Hough T, Greenaway S, Hewitt M, Liu X, McCormack S, Pickford K, Selley R, Wells C, Tymowska-Lalanne Z, Roby P, Glenister P, Thornton C, Thaung C, Stevenson JA, Arkell R, Mburu P, Hardisty R, Kiernan A, Erven A, Steel KP, Voegeling S, Guenet JL, Nickols C, Sadri R, Nasse M, Isaacs A, Davies K, Browne M, Fisher EM, Martin J, Rastan S, Brown SD, Hunter J (2000) A systematic, genome-wide, phenotype-driven mutagenesis programme for gene function studies in the mouse. Nat Genet 25:440–443

Norgard EA, Jarvis JP, Roseman CC, Maxwell TJ, Kenney-Hunt JP, Samocha KE, Pletscher LS, Wang B, Fawcett GL, Leatherwood CJ, Wolf JB, Cheverud JM (2009) Replication of long-bone length QTL in the F9–F10 LG, SM advanced intercross. Mamm Genome 20: 224–235

Norgard EA, Roseman CC, Fawcett GL, Pavlicev M, Morgan CD, Pletscher LS, Wang B, Cheverud JM (2008) Identification of quantitative trait loci affecting murine long bone length in a two-generation intercross of LG/J and SM/J Mice. J Bone Miner Res 23:887–895

Paigen K (2003) One hundred years of mouse genetics: an intellectual history. I. The classical period (1902–1980). Genetics 163:1–7

Paigen K, Eppig JT (2000) A mouse phenome project. Mamm Genome 11:715–717

Park YG, Clifford R, Buetow KH, Hunter KW (2003) Multiple cross and inbred strain haplotype mapping of complex-trait candidate genes. Genome Res 13:118–121

Parker A, Hardisty-Hughes RE, Wisby L, Joyce S, Brown SD (2010) Melody, an ENU mutation in Caspase 3, alters the catalytic cysteine residue and causes sensorineural hearing loss in mice. Mamm Genome 21:565–576

Peirce JL, Lu L, Gu J, Silver LM, Williams RW (2004) A new set of BXD recombinant inbred lines from advanced intercross populations in mice. BMC Genet 5:7

Petryshen TL, Kirby A, Hammer RP Jr, Purcell S, O'Leary SB, Singer JB, Hill AE, Nadeau JH, Daly MJ, Sklar P (2005) Two quantitative trait Loci for prepulse inhibition of startle identified on mouse chromosome 16 using chromosome substitution strains. Genetics 171:1895–1904

Philip VM, Duvvuru S, Gomero B, Ansah TA, Blaha CD, Cook MN, Hamre KM, Lariviere WR, Matthews DB, Mittleman G, Goldowitz D, Chesler EJ (2010) High-throughput behavioral phenotyping in the expanded panel of BXD recombinant inbred strains. Genes Brain Behav 9:129–159

Phillips TJ, Crabbe JC, Metten P, Belknap JK (1994) Localization of genes affecting alcohol drinking in mice. Alcohol Clin Exp Res 18:931–941

Phillips TJ, Huson M, Gwiazdon C, Burkhart-Kasch S, Shen EH (1995) Effects of acute and repeated ethanol exposures on the locomotor activity of BXD recombinant inbred mice. Alcohol Clin Exp Res 19:269–278

Phillips TJ, Huson MG, McKinnon CS (1998) Localization of genes mediating acute and sensitized locomotor responses to cocaine in BXD/Ty recombinant inbred mice. J Neurosci 18:3023–3034

Pletcher MT, McClurg P, Batalov S, Su AI, Barnes SW, Lagler E, Korstanje R, Wang X, Nusskern D, Bogue MA, Mural RJ, Paigen B, Wiltshire T (2004) Use of a dense single nucleotide polymorphism map for in silico mapping in the mouse. PLoS Biol 2:e393

Prevorsek Z, Gorjanc G, Paigen B, Horvat S (2010) Congenic and bioinformatics analyses resolved a major-effect Fob3b QTL on mouse Chr 15 into two closely linked loci. Mamm Genome 21:172–185

Rastan S, Hough T, Kierman A, Hardisty R, Erven A, Gray IC, Voeling S, Isaacs A, Tsai H, Strivens M, Washbourne R, Thornton C, Greenaway S, Hewitt M, McCormick S, Selley R, Wells C, Tymowska-Lalanne Z, Roby P, Mburu P, Rogers D, Hagan J, Reavill C, Davies K, Glenister P, Fisher EM, Martin J, Vizor L, Bouzyk M, Kelsell D, Guenet JL, Steel KP, Sheardown S, Spurr N, Gray I, Peters J, Nolan PM, Hunter AJ, Brown SD (2004) Towards a mutant map of the mouse–new models of neurological, behavioural, deafness, bone, renal and blood disorders. Genetica 122:47–49

Reijmers LG, Coats JK, Pletcher MT, Wiltshire T, Tarantino LM, Mayford M (2006) A mutant mouse with a highly specific contextual fear-conditioning deficit found in an N-ethyl-N-nitrosourea (ENU) mutagenesis screen. Learn Mem 13:143–149

Rieseberg LH, Archer MA, Wayne RK (1999) Transgressive segregation, adaptation and speciation. Heredity 83(Pt 4):363–372

Rockman MV, Kruglyak L (2008) Breeding designs for recombinant inbred advanced intercross lines. Genetics 179:1069–1078

Rodriguez LA, Plomin R, Blizard DA, Jones BC, McClearn GE (1994) Alcohol acceptance, preference, and sensitivity in mice. I. Quantitative genetic analysis using BXD recombinant inbred strains. Alcohol Clin Exp Res 18:1416–1422

Rodriguez LA, Plomin R, Blizard DA, Jones BC, McClearn GE (1995) Alcohol acceptance, preference, and sensitivity in mice. II. Quantitative trait loci mapping analysis using BXD recombinant inbred strains. Alcohol Clin Exp Res 19:367–373

Samocha KE, Lim JE, Cheng R, Sokoloff G, Palmer AA (2010) Fine mapping of QTL for prepulse inhibition in LG/J and SM/J mice using F(2) and advanced intercross lines. Genes Brain Behav 9:759–767

Sandberg ML, Sutton SE, Pletcher MT, Wiltshire T, Tarantino LM, Hogenesch JB, Cooke MP (2005) c-Myb and p300 regulate hematopoietic stem cell proliferation and differentiation. Dev Cell 8:153–166

Savitz J, Lucki I, Drevets WC (2009) 5-HT(1A) receptor function in major depressive disorder. Prog Neurobiol 88:17–31

Schwander M, Lopes V, Sczaniecka A, Gibbs D, Lillo C, Delano D, Tarantino LM, Wiltshire T, Williams DS, Muller U (2009a) A novel allele of myosin VIIa reveals a critical function for the C-terminal FERM domain for melanosome transport in retinal pigment epithelial cells. J Neurosci 29:15810–15818

Schwander M, Xiong W, Tokita J, Lelli A, Elledge HM, Kazmierczak P, Sczaniecka A, Kolatkar A, Wiltshire T, Kuhn P, Holt JR, Kachar B, Tarantino L, Muller U (2009b) A mouse model for nonsyndromic deafness (DFNB12) links hearing loss to defects in tip links of mechanosensory hair cells. Proc Natl Acad Sci USA 106:5252–5257

Schwartz WJ, Zimmerman P (1990) Circadian timekeeping in BALB/c and C57BL/6 inbred mouse strains. J Neurosci 10:3685–3694

Sharkey LM, Cheng X, Drews V, Buchner DA, Jones JM, Justice MJ, Waxman SG, Dib-Hajj SD, Meisler MH (2009) The ataxia3 mutation in the N-terminal cytoplasmic domain of sodium channel Na(v)1.6 disrupts intracellular trafficking. J Neurosci 29:2733–2741

Silver LM (1995) Mouse genetics concepts and applications. Oxford University Press, Oxford

Singer JB, Hill AE, Burrage LC, Olszens KR, Song J, Justice M, O'Brien WE, Conti DV, Witte JS, Lander ES, Nadeau JH (2004) Genetic dissection of complex traits with chromosome substitution strains of mice. Science 304:445–448

Singer JB, Hill AE, Nadeau JH, Lander ES (2005) Mapping quantitative trait loci for anxiety in chromosome substitution strains of mice. Genetics 169:855–862

Smith R, Sheppard K, DiPetrillo K, Churchill G (2009) Quantitative trait locus analysis using J/qtl. Methods Mol Biol 573:175–188

Smoller JW, Paulus MP, Fagerness JA, Purcell S, Yamaki LH, Hirshfeld-Becker D, Biederman J, Rosenbaum JF, Gelernter J, Stein MB (2008) Influence of RGS2 on anxiety-related temperament, personality, and brain function. Arch Gen Psychiatry 65:298–308

Solberg LC, Valdar W, Gauguier D, Nunez G, Taylor A, Burnett S, Arboledas-Hita C, Hernandez-Pliego P, Davidson S, Burns P, Bhattacharya S, Hough T, Higgs D, Klenerman P, Cookson WO, Zhang Y, Deacon RM, Rawlins JN, Mott R, Flint J (2006) A protocol for high-throughput phenotyping, suitable for quantitative trait analysis in mice. Mamm Genome 17:129–146

Speca DJ, Chihara D, Ashique AM, Bowers MS, Pierce-Shimomura JT, Lee J, Rabbee N, Speed TP, Gularte RJ, Chitwood J, Medrano JF, Liao M, Sonner JM, Eger EI, 2nd, Peterson AS, McIntire SL (2010) Conserved role of unc-79 in ethanol responses in lightweight mutant mice. PLoS Genet 6(8):e1001057

Steinberger D, Reynolds DS, Ferris P, Lincoln R, Datta S, Stanley J, Paterson A, Dawson GR, Flint J (2003) Genetic mapping of variation in spatial learning in the mouse. J Neurosci 23:2426–2433

Stottmann RW, Tran PV, Turbe-Doan A, Beier DR (2009) Ttc21b is required to restrict sonic hedgehog activity in the developing mouse forebrain. Dev Biol 335:166–178

Strohl KP, Gallaugher L, Lynn A, Friedman L, Hill A, Singer JB, Lander ES, Nadeau J (2007) Sleep-related epilepsy in the A/J mouse. Sleep 30:169–176

Stylianou IM, Christians JK, Keightley PD, Bunger L, Clinton M, Bulfield G, Horvat S (2004) Genetic complexity of an obesity QTL (Fob3) revealed by detailed genetic mapping. Mamm Genome 15:472–481

Sudhof TC (2008) Neuroligins and neurexins link synaptic function to cognitive disease. Nature 455:903–911

Swanson DJ, Steshina EY, Wakenight P, Aldinger KA, Goldowitz D, Millen KJ, Chizhikov VV (2010) Phenotypic and genetic analysis of the cerebellar mutant tmgc26, a new ENU-induced ROR-alpha allele. Eur J Neurosci 32:707–716

Tabeta K, Hoebe K, Janssen EM, Du X, Georgel P, Crozat K, Mudd S, Mann N, Sovath S, Goode J, Shamel L, Herskovits AA, Portnoy DA, Cooke M, Tarantino LM, Wiltshire T, Steinberg BE, Grinstein S, Beutler B (2006) The Unc93b1 mutation 3d disrupts exogenous antigen presentation and signaling via Toll-like receptors 3, 7 and 9. Nat Immunol 7:156–164

Tafti M, Franken P, Kitahama K, Malafosse A, Jouvet M, Valatx JL (1997) Localization of candidate genomic regions influencing paradoxical sleep in mice. Neuroreport 8:3755–3758
Takada T, Mita A, Maeno A, Sakai T, Shitara H, Kikkawa Y, Moriwaki K, Yonekawa H, Shiroishi T (2008) Mouse inter-subspecific consomic strains for genetic dissection of quantitative complex traits. Genome Res 18:500–508
Takahashi A, Nishi A, Ishii A, Shiroishi T, Koide T (2008) Systematic analysis of emotionality in consomic mouse strains established from C57BL/6J and wild-derived MSM/Ms. Genes Brain Behav 7:849–858
Takahashi A, Tomihara K, Shiroishi T, Koide T (2010) Genetic mapping of social interaction behavior in B6/MSM consomic mouse strains. Behav Genet 40:366–376
Takahasi KR, Sakuraba Y, Gondo Y (2007) Mutational pattern and frequency of induced nucleotide changes in mouse ENU mutagenesis. BMC Mol Biol 8:52
Talbot CJ, Nicod A, Cherny SS, Fulker DW, Collins AC, Flint J (1999) High-resolution mapping of quantitative trait loci in outbred mice. Nat Genet 21:305–308
Taylor BA (1976) Genetic analysis of susceptibility to isoniazid-induced seizures in mice. Genetics 83:373–377
Taylor BA (1978) Recombinant inbred strains: use in gene mapping. In: Morse HC (ed) Origins of inbred Mice: proceedings of a workshop, Bethesda, Maryland, 14–16 February. Academic Press, New York, pp 423–438
Taylor BA, Bedigian HG, Meier H (1977) Genetic studies of the Fv-1 locus of mice: linkage with Gpd-1 in recombinant inbred lines. J Virol 23:106–109
Taylor BA, Meier H, Myers DD (1971) Host-gene control of C-type RNA tumor virus: inheritance of the group-specific antigen of murine leukemia virus. Proc Natl Acad Sci USA 68:3190–3194
Taylor BA, Tarantino LM, Phillips SJ (1999) Gender-influenced obesity QTLs identified in a cross involving the KK type II diabetes-prone mouse strain. Mamm Genome 10:963–968
Taylor BA, Wnek C, Schroeder D, Phillips SJ (2001) Multiple obesity QTLs identified in an intercross between the NZO (New Zealand obese) and the SM (small) mouse strains. Mamm Genome 12:95–103
Theodoratos A, Whittle B, Enders A, Tscharke DC, Roots CM, Goodnow CC, Fahrer AM (2010) Mouse strains with point mutations in TAP1 and TAP2. Immunol Cell Biol 88:72–78
Tokuda S, Mahaffey CL, Monks B, Faulkner CR, Birnbaum MJ, Danzer SC, Frankel WN (2011) A novel Akt3 mutation associated with enhanced kinase activity and seizure susceptibility in mice. Hum Mol Genet 20(5):988–999
Tsaih SW, Korstanje R (2009) Haplotype association mapping in mice. Methods Mol Biol 573: 213–222
Turri MG, Datta SR, DeFries J, Henderson ND, Flint J (2001a) QTL analysis identifies multiple behavioral dimensions in ethological tests of anxiety in laboratory mice. Curr Biol 11:725–734
Turri MG, DeFries JC, Henderson ND, Flint J (2004) Multivariate analysis of quantitative trait loci influencing variation in anxiety-related behavior in laboratory mice. Mamm Genome 15:69–76
Turri MG, Henderson ND, DeFries JC, Flint J (2001b) Quantitative trait locus mapping in laboratory mice derived from a replicated selection experiment for open-field activity. Genetics 158:1217–1226
Turri MG, Talbot CJ, Radcliffe RA, Wehner JM, Flint J (1999) High-resolution mapping of quantitative trait loci for emotionality in selected strains of mice. Mamm Genome 10:1098–1101
Umemori J, Nishi A, Lionikas A, Sakaguchi T, Kuriki S, Blizard DA, Koide T (2009) QTL analyses of temporal and intensity components of home-cage activity in KJR and C57BL/6J strains. BMC Genet 10:40
Valdar W, Flint J, Mott R (2006a) Simulating the collaborative cross: power of quantitative trait loci detection and mapping resolution in large sets of recombinant inbred strains of mice. Genetics 172:1783–1797
Valdar W, Solberg LC, Gauguier D, Burnett S, Klenerman P, Cookson WO, Taylor MS, Rawlins JN, Mott R, Flint J (2006b) Genome-wide genetic association of complex traits in heterogeneous stock mice. Nat Genet 38:879–887

Wakeland E, Morel L, Achey K, Yui M, Longmate J (1997) Speed congenics: a classic technique in the fast lane (relatively speaking). Immunol Today 18:472–477
Wang GL, Paterson AH (1994) Assessment of DNA pooling strategies for mapping of QTLs. Theor Appl Genet 88:355–361
Wang J, Liao G, Usuka J, Peltz G (2005) Computational genetics: from mouse to human? Trends Genet 21:526–532
Wang J, Williams RW, Manly KF (2003) WebQTL: web-based complex trait analysis. Neuroinformatics 1:299–308
Watson J, Riblet R, Taylor BA (1977) The response of recombinant inbred strains of mice to bacterial lipopolysaccharides. J Immunol 118:2088–2093
Webb BT, McClay JL, Vargas-Irwin C, York TP, van den Oord EJ (2009) In silico whole genome association scan for murine prepulse inhibition. PLoS One 4:e5246
Wehner JM, Radcliffe RA, Rosmann ST, Christensen SC, Rasmussen DL, Fulker DW, Wiles M (1997) Quantitative trait locus analysis of contextual fear conditioning in mice. Nat Genet 17:331–334
Wilkes JJ, Lloyd DJ, Gekakis N (2009) Loss-of-function mutation in myostatin reduces tumor necrosis factor alpha production and protects liver against obesity-induced insulin resistance. Diabetes 58:1133–1143
Williams RT, Lim JE, Harr B, Wing C, Walters R, Distler MG, Teschke M, Wu C, Wiltshire T, Su AI, Sokoloff G, Tarantino LM, Borevitz JO, Palmer AA (2009) A common and unstable copy number variant is associated with differences in Glo1 expression and anxiety-like behavior. PLoS One 4:e4649
Wu R, Ma C, Casella G (2010) Statistical genetics of quantitative traits: linkage, maps and QTL. Springer, New York
Xia Y, Won S, Du X, Lin P, Ross C, La Vine D, Wiltshire S, Leiva G, Vidal SM, Whittle B, Goodnow CC, Koziol J, EM YM, Beutler B (2010) Bulk segregation mapping of mutations in closely related strains of mice. Genetics 186:1139–1146
Xie G, Harrison J, Clapcote SJ, Huang Y, Zhang JY, Wang LY, Roder JC (2010) A new Kv1.2 channelopathy underlying cerebellar ataxia. J Biol Chem 285:32160–32173
Yalcin B, Willis-Owen SA, Fullerton J, Meesaq A, Deacon RM, Rawlins JN, Copley RR, Morris AP, Flint J, Mott R (2004) Genetic dissection of a behavioral quantitative trait locus shows that Rgs2 modulates anxiety in mice. Nat Genet 36:1197–1202
Yoon CK (1996) J. NIH Res. 8:23–24
Yoshiki A, Ike F, Mekada K, Kitaura Y, Nakata H, Hiraiwa N, Mochida K, Ijuin M, Kadota M, Murakami A, Ogura A, Abe K, Moriwaki K, Obata Y (2009) The mouse resources at the RIKEN BioResource center. Exp Anim 58:85–96
Zeng ZB (1993) Theoretical basis for separation of multiple linked gene effects in mapping quantitative trait loci. Proc Natl Acad Sci USA 90:10972–10976
Zhang S, Gershenfeld HK (2003) Genetic contributions to body weight in mice: relationship of exploratory behavior to weight. Obes Res 11:828–838

Mouse Models of the 5-HTTLPR × Stress Risk Factor for Depression

Valeria Carola and Cornelius Gross

Abstract The incidence of mood disorders is known to be influenced by both genetic as well as environmental factors. Increasingly, however it is becoming clear that few genetic and environmental factors act alone, but that instead they regularly act in concert to determine predisposition to psychiatric disorders. Quite a few cases now have been reported in which stratification of subjects by exposure to environmental pathogens has been shown to alter the association between specific genetic variants and mental illness. The best studied of such measured gene-by-environment risk factors for mental illness is the increased risk for major depression reported among persons carrying the short variant (S allele) of a functional polymorphism in the serotonin transporter (5-HTT, SLC6A4) gene promoter and who have been exposed to stressful life events. Recently, a large number of laboratories have tried to model the interaction between 5-HTTLPR genotype and early/adult stress in mouse. Findings from their studies have helped to define the rodent orthologs of the environmental stressors and behavioral traits involved in risk for depression. Furthermore, several of these studies attempted to identify changes in molecular substrates that might underlie the 5-HTT x stress risk factor, pointing to the hippocampus and frontal cortex as critical brain structures involved in the interaction between 5-HTT gene variation and early and adult stress, respectively. These results will serve to help inform clinical research

V. Carola (✉)
Santa Lucia Foundation, Via del Fosso di Fiorano 64/65, 00143 Rome, Italy
e-mail: valeria.carola@uniroma1.it

V. Carola
Department of Psychology, University of Rome, "La Sapienza", Via dei Marsi 78, Rome, Italy

C. Gross
Mouse Biology Unit, European Molecular Biology Laboratory (EMBL), Via Ramarini 32, Monterotondo, Italy

Curr Topics Behav Neurosci (2012) 12: 59–72
DOI: 10.1007/7854_2011_190

Published Online: 14 February 2012

into the origins of major depression and other mental illnesses with interacting genetic and environmental risk factors.

Keywords Gene-by-environment · Serotonin transporter · Mouse models · Early/adult stress risk factor · Anxiety-depression

Contents

1 Challenges of Designing Mouse Models of the 5-HTTLPR × Stress Risk Factor 60
2 Mouse Models of the 5-HTTLPR × Early Stress Risk Factor for Depression 63
3 Mouse Models of the 5-HTTLPR × Adult Stress Risk Factor for Depression 65
4 Physiological and Molecular Substrates of the 5-HTT × Stress Risk Factor in Mice 66
5 Promises of Mouse Models of Gene × Environment Risk Factors for Clinical Research 68
References 70

1 Challenges of Designing Mouse Models of the 5-HTTLPR × Stress Risk Factor

Risk for mood disorders is moderated by genetic as well as environmental factors. Twin studies suggest that genetic factors contribute only moderately to the variance in incidence of these disorders, while environmental factors presumably explain the majority of their incidence. Increasingly, however, it is becoming clear that few genetic and environmental factors act alone, but that instead they function in concert to determine predisposition to mental illness. Several cases now have been documented in which stratification of subjects by exposure to measured environmental pathogens, ranging from childhood maltreatment to cannabis use, have been shown to alter the association between specific genetic variants and mental illness (Caspi and Moffitt 2006).

Arguably the best studied of such measured gene-by-environment risk factors for mental illness is the increased risk for major depression reported among persons carrying the short variant (S allele) of a functional polymorphism in the serotonin transporter (5-HTT, SLC6A4) gene promoter (called serotonin transporter gene-linked polymorphic region, 5-HTTLPR) and who have been exposed to stressful life events. Originally, the short variant was associated with a small, but significant increase in anxiety-related personality traits, such as neuroticism and disagreeableness (Lesch et al. 1996). However, larger effects of the short variant were found following stratification of subjects by exposure to either childhood maltreatment or adult stress. When environmental exposure was taken into account a significant interaction between 5-HTTLPR genotype and stress exposure was uncovered (Caspi et al. 2003). Subjects carrying one or two of the short variants (S/S or S/L individuals) demonstrated increased risk for major depression at age 26 when exposed to either childhood maltreatment or stressful life events in the five preceding years. L/L individuals, on the other hand, showed low incidence of depression regardless of environmental exposure. Intriguingly, they observed significant interactions between the short variant and both early (age 0–11) and adult (age 21–26) stress, suggesting that genetic influences on serotonin homeostasis fundamentally altered the physiological or emotional response of an individual to stress throughout life. The increased risk for depression in stressed S allele carriers was later replicated by several laboratories although there have also been failures and controversy (Risch et al. 2009; Kaufman et al. 2004, 2010; for review, see Uher et al. 2011). Studies in non-ill subjects have also shown evidence of a similar 5-HTTLPR × stress effects on physiological and behavioral traits, including heart rate reactivity and anxiety (Caspi et al. 2010; Williams et al. 2009). Thus, the 5-HTTLPR S allele is likely to act by favoring basic stress coping strategies that subsequently carry an increased risk for depression.

Using electroencephalogram and functional magnetic resonance imaging (fMRI), several groups were able to identify brain regions that showed significantly altered neural activity in response to emotional stimuli in 5-HTTLPR short variant carriers (Hariri et al. 2002; Heinz et al. 2005; Pezawas et al. 2005; Canli et al. 2005, 2006). The most common finding has been an apparent increase in amygdala fMRI signal reactivity to emotional stimuli (Hariri et al. 2002; Canli et al. 2005). However, several studies have suggested that this is a reflection of an increase in baseline neural activity in S allele carriers in a much wider range of forebrain areas, an interpretation supported by perfusion imaging techniques (Canli et al. 2006; Heinz et al. 2007). Several fMRI studies have also shown altered functional connectivity between frontal cortical regions and amygdala in S carriers (Heinz et al. 2005; Pezawas et al. 2005). In one of these studies (Canli et al. 2006) subjects were explicitly stratified by exposure to life stress events. Neural activity in a wide range of forebrain regions showed evidence of a significant 5-HTTLPR × life stress effect following visual exposure to emotional words. In many brain regions, including amygdala and hippocampus, neural responses in S allele carriers increased with life stress events while they decreased in L/L allele carriers so that under high stress load the genotype effect on neural activation was magnified. In other brain nuclei, including regions implicated in imitative behavior such as superior temporal gyrus and superior parietal lobule, L/L subjects experiencing high stress showed increased neural activity responses compared to S allele carriers. Such regions might mediate the stress resilience seen in L/L persons. Moreover, a similar 5-HTTLPR × life stress effect was seen on absolute baseline brain activity as measured by perfusion imaging (Canli et al. 2006), suggesting that 5-HTTLPR genotype moderates the effect of life stress on stable levels of neural activity.

These studies have transformed the psychiatric genetics field and have compelled researchers to incorporate information about environmental exposure into the study of genetic risk factors. The new approach to human association studies has also had a significant impact on the pre-clinical research (Caspi and Moffitt 2006). For the very first time, researchers applying animal models to study mood disorders have a genetic variant in their hands that under the right environmental conditions is expected to show significant and reproducible associations with depression-relevant traits. This is a significant advance because until then no reproducible measured genetic risk factor for mood disorders or associated traits existed.

The mouse has become the premier mammalian model organism for genetic studies due to the amenability of its genome to targeted genetic manipulation. However, because the upstream promoter region harboring the 5-HTTLPR variant is not present in rodents, designing an appropriate mouse model of the 5-HTTLPR × life stress risk factor for depression is not straightforward. This deficit means that studying the polymorphism in mice requires "humanization" of the mouse gene. However, although present gene targeting technology makes it relatively straightforward to produce a constitutive knockout of a gene or to replace a gene with a variant, the need for a selectable marker (typically a neomycin resistance gene) during embryonic stem (ES) cell targeting means it is difficult to seamlessly replace a specific sequence in the mouse genome without leaving behind a mark of the engineering process. Consequently, precise "humanization" of the mouse promoter region is tricky and has so far not been reported. As a compromise researchers have used constitutive 5-HTT knockout mice (Bengel et al. 1998) as an approximative model either in the gene's homozygous, complete knockout, or its heterozygous, partially reduced, form. Studies on homozygous 5-HTT knockout mice have been extremely helpful in better understanding the physiological function of the transporter, while heterozygous mice are arguably a better model of the human promoter polymorphism because the 50% reduction of 5-HTT mRNA in heterozygous knockouts mimics that are found in 5-HTTLPR S allele carriers when compared to L allele carriers (Lesch et al. 1996; Bengel et al. 1998).

The use of heterozygous mice as a model of 5-HTTLPR, however, remains problematic because it remains unclear how and in what way 5-HTTLPR affects gene expression. Data documenting an effect of 5-HTTLPR on gene expression in humans derive primarily from studies in cultured non-neuronal cell lines (Lesch et al. 1996) and only a few studies have examined 5-HTT gene expression in L and S allele carriers in neuronal cell lines or postmortem brain tissue (Greenberg et al. 1999; Mann et al. 2000; Eley et al. 2004; Sugden et al. 2009). Moreover, autoradiography and positron emission tomography studies using 5-HTT selective ligands have failed to find consistent differences in 5-HTT protein binding between the variants, suggesting a potential differential effect on mRNA and protein expression. One possibility is that 5-HTTLPR controls transcription in a cell-type specific or temporally controlled manner and that these features have masked differences in gene expression. It may be, for example, that 5-HTTLPR controls transcription primarily during brain development, when serotonin is known to have a critical role in brain wiring (Trowbridge et al. 2010; Ferreira et al. 2010) but only weakly in adulthood. More controversially, it is possible that 5-HTTLPR affects 5-HTT expression in only a subset of cells expressing the gene. Such issues could be addressed using humanized mice carrying the L and S alleles. A more faithful mouse model of 5-HTTLPR could also serve as a test platform to understand the transcription factor binding and chromatin remodeling mechanisms that are presumed to mediate the transcriptional effects of this variant and about which little or nothing is known.

2 Mouse Models of the 5-HTTLPR × Early Stress Risk Factor for Depression

The first study to incorporate measured environmental stress in the study of behavioral phenotypes in 5-HTT knockout mice involved daily exposure to foot shock (Carroll et al. 2007). From postnatal day 7–13, mouse pups were exposed daily to three mild footshocks over a period of 150 s and then tested in adulthood in a battery of anxiety and depression-related behavioral tests (elevated plus-maze, light/dark test, open field, forced swim test). Although 5-HTT knockout mice showed increased avoidance in anxiety tests and increased behavioral despair in the forced swim test, no interaction between genotype and environmental stressor was detected.

Using a similar battery of behavioral tests a more recent study (Carola et al. 2008) showed that levels of maternal care as measured by maternal licking and grooming of pups during the first 2 weeks of life could significantly modulate the effect of heterozygous 5-HTT knockout mutations on anxiety and depression behavior. Mice were raised by mothers that were genetically identical but nevertheless provided high and low maternal care (C57BL/6 × BALB/c vs. BALB/c × C57BL/6, respectively). Heterozygous 5-HTT knockouts showed equivalent level of avoidance and behavioral despair as wild-type littermates when exposed to high levels of maternal care, but when exposed to low maternal care showed significantly more avoidance and behavioral despair than controls. Furthermore, heterozygous knockout mice exposed to low maternal care also showed enhanced fear in response to partially conditioned cues, suggesting enhanced cognitive processing of ambiguous threatening cues. This study suggested that novelty avoidance, behavioral despair, and ambiguous fear conditioning in mice may be useful surrogates of risk for depression in humans (Fig. 1). It also suggested that environmental stressors that perturb critical social cues such as maternal care might be better models of childhood maltreatment than physical stressors such as foot shock. It is important to keep in mind, however, that low maternal care in mice and childhood maltreatment in humans act on only partially overlapping stages of brain development (Fig. 1). Because mice are born relatively prematurely compared to primates the first week of life in rodents when maternal care is most pronounced corresponds to the late gestational period in primates (Clancy et al. 2007). Thus, rodent studies using maternal care as a surrogate for childhood experiences must be interpreted in light of the caveat that the neural substrates of these manipulations are not necessarily equivalent across species.

Another study evaluated anxiety behavior in heterozygous and homozygous 5-HTT knockout mice following exposure to adverse olfactory stimuli across both pre- and post-natal periods (Heiming et al. 2009). During pregnancy and lactation mothers were exposed to soiled bedding from the cage of an unfamiliar adult male. Because adult males tend to attack and kill pups that are not their own, the olfactory cues were assumed to be aversive to both the mother and her offspring. Consistent with previous studies, homozygous 5-HTT knockout mice showed

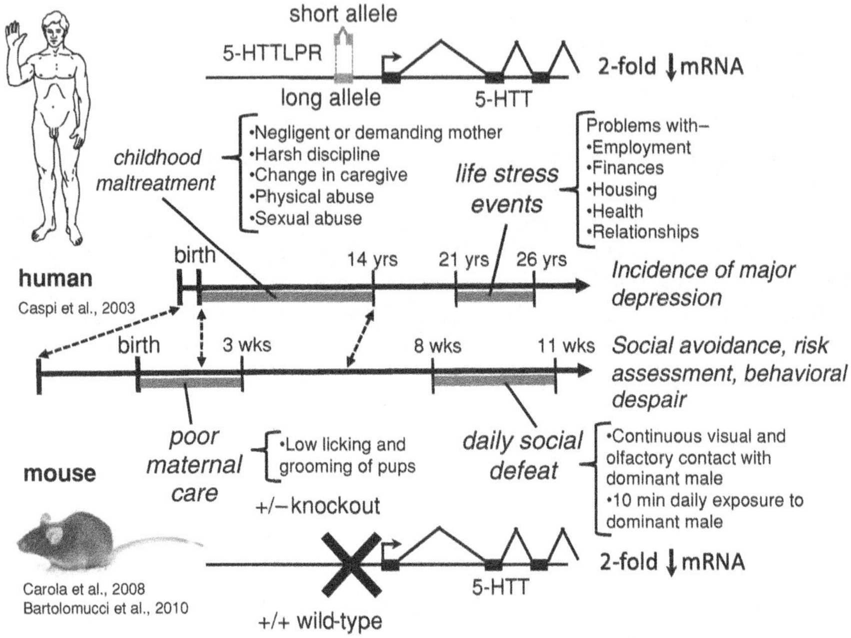

Fig. 1 Mouse models of 5-HTTLPR × early and adult stress risk factors for depression. In humans, the combination of either childhood maltreatment or adult stress and the 5-HTLPR S allele results in increased risk for major depression (Caspi et al. 2003). In mice, the 5-HTTLPR S allele was modeled by a heterozygous knockout (–/+) mutation in 5-HTT that shows a similar twofold decrease in mRNA expression. Childhood maltreatment was modeled by low maternal care (Carola et al. 2008) and adult stressors were modeled by daily social defeat (Bartolomucci et al. 2010). In both cases the low-expressing heterozygous knockout mice were more susceptible to the effects of stress on anxiety and depression-related behavior, mimicking the human findings. These studies suggest that reduced 5-HTT function is associated with increased responsivity to environmental stressors in mice and humans (dotted arrows indicate orthologous developmental time periods)

increased anxiety in the open field, elevated plus maze, and dark-light tests. Exposure to unfamiliar male odor was also associated with increased anxiety in all mice. However, the environmental effect was significant in homozygous knockout mice and not in wild-type mice, indicating greater susceptibility of the knockout to threatening olfactory cues than the wild-type. This study confirmed a role for serotonin homeostasis in regulating the long-term effects of early exposure to environmental stressors. Furthermore, these findings are intriguing because they highlight a susceptibility that spans embryonic and early postnatal development and thus points to a relatively non-specific modulation by serotonin of environmental responsivity during development. It will be important to determine whether environmental pathogens in humans are similarly effective across development in modulating the effect of 5-HTTLPR genotype.

3 Mouse Models of the 5-HTTLPR × Adult Stress Risk Factor for Depression

A much larger number of studies have tried to model the interaction between 5-HTTLPR genotype and adult stress. The first study used a chronic mild stress procedure involving daily exposure to several stressors (forced swimming, wet bedding, light cycle reversal, unfamiliar cage mate, tilted cage, restraint stress, no bedding material, etc.) each day for 4 weeks (Joeyen-Waldorf et al. 2009). Chronic mild stress was associated with increased anxiety in the novelty suppressed feeding test and this effect was potentiated in both heterozygous and homozygous 5-HTT knockout female mice, while males showed no genotype-by-environment interaction effect. A similar genotype by stress effect was observed following exposure of heterozygous and homozygous 5-HTT knockout mice to a winner or loser experience in a resident-intruder paradigm over a period of 3 days (Jansen et al. 2010). Both winners and losers exhibited increased anxiety behavior in the elevated plus maze, dark-light test, and open field demonstrating an anxiogenic effect of adverse social exposure regardless of outcome. Homozygous, but not heterozygous 5-HTT knockout mice showed exaggerated anxiety when exposed to either repeated winning or losing. These data suggest that 5-HT homeostasis modulates responses to a wider range of social experiences that are previously thought and warrants closer investigation in future human association studies.

Although heterozygous 5-HTT knockout mice did not show interactions with repeated winning or losing in the above study, they did show enhanced response to a continuous psychosocial stress paradigm where mice were exposed to brief daily defeat by an aggressive male mouse with whom they were housed in continuous sensory contact (Bartolomucci et al. 2010), possibly due to the greater intensity of the later stressor. Following 3 weeks of psychosocial stress mice were tested for social interaction toward an unfamiliar male mouse in a novel environment. Stressed mice showed strong avoidance of the unfamiliar mouse, but this effect was significantly greater in heterozygous knockout mice than wild-type littermates. An important feature of this study was the monitoring of behavioral and physiological responses to stress in the home cage on a daily basis. Heterozygous 5-HTT knockout mice showed similar responses to stress as wild-type littermates, including decreased home cage locomotion between encounters, increased body weight, and increased temperature responses to daily social defeat. These data suggest that 5-HTT genotype modulates the generalization of avoidance behavior rather than the direct physiological and behavioral consequences of stress. Such an interpretation is in line with a role of 5-HTT in determining not the physical responses to stress, but in modifying stress coping strategies. Moreover, these findings confirm that a 50% decrease in 5-HTT function is sufficient to modulate the response to environmental stressors if those stressors are sufficiently prolonged or salient. Continuous exposure to psychosocial stressors in particular was most effective in interacting with serotonin homeostasis to increase susceptibility to anxiety and depression behavior.

4 Physiological and Molecular Substrates of the 5-HTT × Stress Risk Factor in Mice

As discussed above, fMRI studies have identified numerous brain regions whose activity is modulated by 5-HTTLPR genotype and life stress events and which are thus implicated in mediating the effect of this risk factor on behavioral traits. In particular, neural activity in both amygdala and hippocampus was shown to be significantly modulated by 5-HTTLPR × stress, with S allele subjects showing a positive correlation between life stress events and L/L subjects showing a negative correlation (Canli et al. 2006). Neural activity in the superior temporal gyrus and superior parietal lobule, on the other hand, showed a positive correlation with life stress events in L/L subjects, suggesting a possible role of this structure in the stress resilience of this genotype. In the same study, the authors also presented evidence that resting state brain activity in hippocampus and amygdala as measured while the subjects fixated on a spot was also significantly enhanced in stressed S allele carriers when compared to all other groups. The modulation of resting state brain activity is important as it suggests that serotonin signaling moderates the capacity of stress to reset resting brain activity, and raises the possibility that changes in neural activity homeostasis in selected brain regions may underlie the 5-HTTLPR × stress risk factor. Unfortunately, this hypothesis has not yet been examined in animal models.

Nevertheless, two mouse model studies did attempt to identify changes in molecular substrates that might underlie the 5-HTT × stress risk factor. In the first study, where heterozygous 5-HTT knockout mice were shown to be more susceptible to the anxiogenic and depressogenic effects of low maternal care (Carola et al. 2008), molecular substrates were identified in the adult that correlated with the effect of genotype (G), early environment (eE), and genotype × environment (G×eE; Table 1). First, levels of GABA-A receptors (as measured by binding to the benzodiazepine flunetrazepam) were reduced in the central nucleus of the amygdala of mice exposed to low maternal care regardless of the genotype. This finding replicated earlier data in rats exposed to low maternal care (Caldji et al. 2003) and suggested that decreased inhibition in this nucleus might underlie the increased risk assessment behavior associated with low maternal care. Second, serotonin turnover as measured by the ratio of the serotonin metabolite 5-hydroxyindole-acetic acid (5-HIAA) to serotonin (5-HT) in the hippocampus was decreased in heterozygous 5-HTT knockout mice regardless of environmental exposure. This finding was consistent with earlier studies showing decreased turnover and serotonin neuron firing in this genotype (Mathews et al. 2004). The absence of an effect of environment on serotonin turnover suggested that stress acts downstream of serotonin homeostasis to affect neural circuit function. Finally, levels of brain-derived neurotrophic factor (BDNF) mRNA in the CA1 region of the hippocampus (but not in other parts of the hippocampus) showed an interacting effect of genotype and environment, with heterozygous 5-HTT knockouts exposed to low maternal care showing the highest BDNF levels. Intriguingly, elevated

Table 1 Physiological and Molecular Correlates of the 5-HTT × Stress Risk Factor in Mice. Summary of the effects of gene (G), environment (E), and gene-by-environment (G×E) on molecular substrates as reported by studies measuring anxiety and depression-related behavior in 5-HTT +/+ and ± mice exposed to low maternal care (Carola et al. 2008, 2011) and chronic psychosocial stress (Bartolomucci et al. 2010). Note that molecular substrates following exposure to low maternal care were studied at two time points, postnatal day 10 and adulthood (– no effect; n.d. not determined)

Molecular substrate		Low maternal care		Chronic psychosocial stress
		Adult	Postnatal day 10	
FCtx	5-HT turnover	G	GxE	GxE
AMY	GABA-A receptor	E	-	n.d.
HIP	5-HT turnover	G	G	G
	BDNF mRNA	GxE	E	-
	AMPA receptor	-	GxE	n.d.

BDNF levels in forebrain are associated with elevated anxiety behavior (Yee et al. 2007) and are a candidate molecular substrate for the elevated neural activity seen in this structure in human 5-HTTLPR S allele carriers exposed to life stress events (Canli et al. 2006). Moreover, genetic variation in BDNF itself moderated the long-term effects of maternal care on anxiety behavior (Carola and Gross 2010). In humans too a non-synonymous variant in BDNF (Val66Met) with reduced functionality is able to reverse the neural activity correlates of associated with the 5-HTTLPR S allele (Pezawas et al. 2008). Consistent with a causal role for these neural substrates in behavior, levels of GABA-A receptor binding, serotonin turnover, and BDNF expression were significantly correlated with anxiety behavior in the mice and estimated to each explain about 15–30% of the variance of relevant behavioral measures (Carola et al. 2008).

A follow-up study by the same group (Carola et al. 2011) investigated the origins of these changes by examining molecular substrates in young mice during the period of maternal care (postnatal day 10). Unlike in adult mice, BDNF mRNA in the hippocampus was elevated in all low maternal care offspring regardless of genotype (Table 1). This finding suggests that the low functioning 5-HTT variant may act to maintain elevated BDNF expression induced by low maternal care. Thus, 5-HTT genotype may not modulate the immediate impact of environment on neurophysiology, but rather to control its perseverance. As a result, 5-HTTLPR S allele carriers might be more vulnerable to repeated stressors and, consequently, recurrent illness, a feature supported by recent epidemiological evidence (Uher et al. 2011). At the same time, levels of excitatory neurotransmitter receptors (of the AMPA class) in the hippocampus at postnatal day 10 showed a G×eE effect and were elevated in heterozygous 5-HTT knockout mice exposed to low maternal care. These differences did not persist to adulthood (V. Carola, unpublished observations) and thus enhanced excitatory neurotransmission in hippocampus may serve as an immediate substrate for early stress and 5-HTT variation that subsequently triggers long-term G×eE effects in this or other brain circuits.

Neural substrates of the 5-HTT × adult stress risk factor were investigated in one study (Bartolomucci et al. 2010). These authors observed that although genotype alone determined serotonin turnover in most brain regions, in the frontal cortex serotonin homeostasis showed a significant GxaE effect, with the lowest serotonin turnover in heterozygous 5-HTT knockout mice experiencing chronic social defeat (Table 1). Intriguingly, serotonin turnover in the frontal cortex was best correlated with levels of 5-HTT protein in the amygdala, suggesting cross-talk between these two connected structures. This finding contrasts with serotonin turnover data from experiments with 5-HTT × early stress (Carola et al. 2008), where no such interaction was observed. Why frontal cortical serotonin turnover is more plastic in adulthood than in childhood following stress is not clear, but could reflect differences in the predominant circuits engaged by stress at these ages or by the longer time since childhood stress. This difference suggests that the 5-HTTLPR × early stress and adult stress risk factors are associated with different patterns of neural activity and that these forms of depression may show differential therapeutic responses. It also suggests that, at least following adult stress, altered serotonin homeostasis may directly explain the altered activity and/or connectivity of frontal cortical circuits seen in 5-HTTLPR S carriers by fMRI.

5 Promises of Mouse Models of Gene × Environment Risk Factors for Clinical Research

Animal models of gene by environment risk factors for mental illness hold several promises for the pre-clinical research aimed at developing new diagnostic markers and therapeutic approaches. One important outcome of a successful animal model is its ability to point out homologous animal and human behavior. Until now, the primary avenue to match animal and human behavioral traits associated with mental illness was pharmacological validation, in which drugs that are used to treat behavioral disorders in humans induce homologous behavioral changes in animals. However, pharmacological validation has several problems. The benzodiazepine class of anxiolytics, for example, is frequently used to validate behavioral tests of anxiety in rodents, and these drugs cause increased exploration of novel environments and reductions in contextual fear (Garner et al. 2009). However, because these drugs act non-selectively to boost GABA-A receptor function throughout the brain, they act on a wide range of behaviors beyond anxiety and just because a particular rodent behavior is modulated by benzodiazepines does not mean it reflects anxiety. In the field of schizophrenia the situation is even less clear since it is not obvious how to measure psychotic symptoms in animals. Some behaviors such as head twitches and scratches selectively induced in mice by hallucinogenics have been reported (González-Maeso et al. 2007), but the relevance of these behaviors to the positive symptoms of schizophrenia remain speculative.

For these reasons, a genetic validation approach using a mouse model incorporating environmental exposure could offer a parallel avenue for the identification of human–animal behavioral homologs. An excellent example for the use of a genetic validation approach is a mouse engineered to carry a deficiency of a 1.3 Mb region on mouse chromosome 16 whose orthologous region in humans when deleted results in a 30% incidence of schizophrenia (Stark et al. 2008). These mice have deficits in working memory; a negative symptom found in many schizophrenics and electrophysiological recordings in these mice during a working memory task revealed a deficit in connectivity between hippocampus and medial prefrontal cortex, two regions known to contribute to working memory (Sigurdsson et al. 2010). Poor connectivity between these structures is now a candidate for an endophenotype in the wider population of schizophrenics. As this example shows, genetic models offer a powerful approach to validate both behavior and their associated physiological and molecular substrates. Moreover, because genetic models (whether by modeling the gene alone or a gene × environment risk factor) model the etiology of the disorder rather than its treatment, they are likely to be more specific and informative.

Mouse genetic models have the advantage that current gene targeting technology allows for essentially unrestricted temporal and spatial regulation of transgene expression. Thus, once a genetic model is constructed, follow-up studies can use conditional gene targeting to map the circuitry and time periods involved with a resolution not achievable with pharmacological manipulations. Such approaches then allow for the mapping of G×E risk factors to specific brain circuitry as a prelude to understanding the cellular and network mechanisms involved. Moreover, genetic manipulation can be used to pinpoint the brain circuitry critical for the processing of stressful environmental stimuli with a specificity not possible with the direct manipulation of the stress itself. In this way, mouse models of G×E risk factors can help unravel the molecular mechanisms of susceptibility to stress, invariably a major risk factor across all mental illnesses.

In summary, mouse models of the 5-HTTLPR × stress risk factor for depression have helped to define the rodent orthologs of the environmental stressors and behavioral traits involved in risk for depression. Mouse models have also been able to pinpoint several specific physiological and molecular deficits involved that can now serve to inform clinical research into the origins of major depression. Nevertheless, much still needs to be done to expand on these findings. Additional early stressors (e.g. social defeat, negligence) should be tested for their ability to precipitate depression-related behaviors and determine whether differences between the neural substrates of early and adult stressors are due to their timing or quality. More careful time course analyses are also needed to pinpoint the critical periods for different stressors. These studies will need to be accompanied by physiological, molecular, and anatomical studies aimed at aligning mouse and human development in the postnatal period so as to better understand the human developmental transitions implicated. Work on aligning rodent and primate development during the prenatal period should be used as a model for such research (Clancy et al. 2007).

A more faithful genetic model of the human 5-HTTLPR is needed to help answer outstanding questions about where and when this polymorphism affects

5-HTT expression and the mechanism by which it acts to alter transcription. Conditional genetic and/or pharmacological (Ansorge et al. 2004) manipulations of 5-HTT will be needed to identify the cell types and developmental periods involved. Pharmacological blockade of 5-HTT has been used to restrict the critical window of 5-HTT activity on depression-related behavior in adulthood to the first 2 weeks of mouse development (Ansorge et al. 2004). However, it remains to be determined whether the 5-HTT × stress effects will map to a similar period, particularly as the first weeks of mouse development corresponds to mid-gestation in humans, rather than the postnatal period implicated in human association studies.

Finally, further work on identifying physiological and molecular substrates of the 5-HTT × stress effects is needed. How is brain activity altered during exposure to stressful or anxiogenic stimuli? Is there evidence for similar widespread effects of the 5-HTT × stress risk factor on brain activity in mice and humans? What are the serotonin receptors and circuit mechanisms involved in such effects? Are there developmental consequences of the 5-HTT × stress risk factor on brain wiring that might explain treatment resistance and/or relapse in susceptible individuals? Can novel therapeutic targets be identified in the circuits specifically affected by the 5-HTT × stress risk factor? And finally, it will be imperative to show that mechanisms identified in rodent studies can be extrapolated back to the clinic so as to develop improved depression medications.

References

Ansorge MS, Zhou M, Lira A, Hen R, Gingrich JA (2004) Early-life blockade of the 5-HT transporter alters emotional behavior in adult mice. Science 306:879–881

Bartolomucci A, Carola V, Pascucci T, Puglisi-Allegra S, Cabib S, Lesch KP, Parmigiani S, Palanza P, Gross C (2010) Increased vulnerability to psychosocial stress in heterozygous serotonin transporter knockout mice. Dis Model Mech 3:459–470

Bengel D, Murphy DL, Andrews AM, Wichems CH, Feltner D, Heils A, Mossner R, Westphal H, Lesch KP (1998) Altered brain serotonin homeostasis and locomotor insensitivity to 3, 4-methylenedioxymethamphetamine ("Ecstasy") in serotonin transporter-deficient mice. Mol Pharmacol 53:649–655

Caldji C, Diorio J, Meaney MJ (2003) Variations in maternal care alter GABA(A) receptor subunit expression in brain regions associated with fear. Neuropsychopharmacology 28:1950–1959

Canli T, Omura K, Haas BW, Fallgatter A, Constable RT, Lesch KP (2005) Beyond affect: a role for genetic variation of the serotonin transporter in neural activation during a cognitive attention task. Proc Natl Acad Sci U S A 102:12224–12229

Canli T, Qiu M, Omura K, Congdon E, Haas BW, Amin Z, Herrmann MJ, Constable RT, Lesch KP (2006) Neural correlates of epigenesis. Proc Natl Acad Sci U S A 103:16033–16038

Carola V, Gross C (2010) BDNF moderates early environmental risk factors for anxiety in mouse. Genes Brain Behav 9:379–389

Carola V, Frazzetto G, Pascucci T, Audero E, Puglisi-Allegra S, Cabib S, Lesch KP, Gross C (2008) Identifying molecular substrates in a mouse model of the serotonin transporter × environment risk factor for anxiety and depression. Biol Psychiatry 63:840–846

Carola V, Pascucci T, Puglisi-Allegra S, Cabib S, Gross C (2011) Effect of the interaction between the serotonin transporter gene and maternal environment on developing mouse brain. Behav Brain Res 217:188–194

Carroll JC, Boyce-Rustay JM, Millstein R, Yang R, Wiedholz LM, Murphy DL, Holmes A (2007) Effects of mild early life stress on abnormal emotion-related behaviors in 5-HTT knockout mice. Behav Genet 37:214–222

Caspi A, Moffitt T (2006) Gene–environment interactions in psychiatry: joining forces with neuroscience. Nat Rev Neurosci 7:583–590

Caspi A, Sugden K, Moffitt TE, Taylor A, Craig IW, Harrington H, McClay J, Mill J, Martin J, Braithwaite A, Poulton R (2003) Influence of life stress on depression: moderation by a polymorphism in the 5-HTT gene. Science 301:386–389

Caspi A, Hariri A, Holmes A, Uher R, Moffitt T (2010) Genetic sensitivity to the environment: the case of the serotonin transporter gene and its implications for studying complex diseases and traits. Am J Psychiatry 167:509–527

Clancy B, Finlay BL, Darlington RB, Anand KJ (2007) Extrapolating brain development from experimental species to humans. Neurotoxicology 28:931–937

Eley TC, Sugden K, Corsico A, Gregory AM, Sham P, McGuffin P, Plomin R, Craig IW (2004) Gene-environment interaction analysis of serotonin system markers with adolescent depression. Mol Psychiatry 9:908–915

Ferreira TA, Iacono LL, Gross CT (2010) Serotonin receptor 1A modulates actin dynamics and restricts dendritic growth in hippocampal neurons. Eur J Neurosci 32:18–26

Garner M, Möhler H, Stein DJ, Mueggler T, Baldwin DS (2009) Research in anxiety disorders: from the bench to the bedside. Eur Neuropsychopharmacol 19:381–390

González-Maeso J, Weisstaub NV, Zhou M, Chan P, Ivic L, Ang R, Lira A, Bradley-Moore M, Ge Y, Zhou Q, Sealfon SC, Gingrich JA (2007) Hallucinogens recruit specific cortical 5-HT(2A) receptor-mediated signaling pathways to affect behavior. Neuron 53:439–452

Greenberg BD, Tolliver TJ, Huang SJ, Li Q, Bengel D, Murphy DL (1999) Genetic variation in the serotonin transporter promoter region affects serotonin uptake in human blood platelets. Am J Med Genet 88:83–87

Hariri AR, Mattay VS, Tessitore A, Kolachana B, Fera F, Goldman D, Egan MF, Weinberger DR (2002) Serotonin transporter genetic variation and the response of the human amygdala. Science 297:400–403

Heiming RS, Jansen F, Lewejohann L, Kaiser S, Schmitt A, Lesch KP, Sachser N (2009) Living in a dangerous world: the shaping of behavioral profile by early environment and 5-HTT genotype. Front Behav Neurosci 3:26

Heinz A, Braus DF, Smolka MN, Wrase J, Puls I, Hermann D, Klein S, Grüsser SM, Flor H, Schumann G, Mann K, Büchel C (2005) Amygdala-prefrontal coupling depends on a genetic variation of the serotonin transporter. Nat Neurosci 8:20–21

Heinz A, Smolka MN, Braus DF, Wrase J, Beck A, Flor H, Mann K, Schumann G, Büchel C, Hariri AR, Weinberger DR (2007) Serotonin transporter genotype (5-HTTLPR): effects of neutral and undefined conditions on amygdala activation. Biol Psychiatry 61:1011–1014

Jansen F, Heiming RS, Lewejohann L, Touma C, Palme R, Schmitt A, Lesch KP, Sachser N (2010) Modulation of behavioural profile and stress response by 5-HTT genotype and social experience in adulthood. Behav Brain Res 207:21–29

Joeyen-Waldorf J, Edgar N, Sibille E (2009) The roles of sex and serotonin transporter levels in age- and stress-related emotionality in mice. Brain Res 1286:84–93

Kaufman J, Yang BZ, Douglas-Palumberi H, Houshyar S, Lipschitz D, Krystal JH, Gelernter J (2004) Social supports and serotonin transporter gene moderate depression in maltreated children. Proc Natl Acad Sci U S A 101:17316–17321

Kaufman J, Gelernter J, Kaffman A, Caspi A, Moffitt T (2010) Arguable assumptions, debatable conclusions. Biol Psychiatry 67:e19–e20

Lesch KP, Bengel D, Heils A, Sabol SZ, Greenberg BD, Petri S, Benjamin J, Muller CR, Hamer DH, Murphy DL (1996) Association of anxiety-related traits with a polymorphism in the serotonin transporter gene regulatory region. Science 274:1527–1531

Mann JJ, Huang YY, Underwood MD, Kassir SA, Oppenheim S, Kelly TM, Dwork AJ, Arango V (2000) A serotonin transporter gene promoter polymorphism (5-HTTLPR) and prefrontal cortical binding in major depression and suicide. Arch Gen Psychiatry 57:729–738

Mathews TA, Fedele DE, Coppelli FM, Avila AM, Murphy DL, Andrews AM (2004) Gene dose-dependent alterations in extraneuronal serotonin but not dopamine in mice with reduced serotonin transporter expression. J Neurosci Methods 140:169–181

Pezawas L, Meyer-Lindenberg A, Drabant EM, Verchinski BA, Munoz KE, Kolachana BS, Egan MF, Mattay VS, Hariri AR, Weinberger DR (2005) 5-HTTLPR polymorphism impacts human cingulate-amygdala interactions: a genetic susceptibility mechanism for depression. Nat Neurosci 8:828–834

Pezawas L, Meyer-Lindenberg A, Goldman AL, Verchinski BA, Chen G, Kolachana BS, Egan MF, Mattay VS, Hariri AR, Weinberger DR (2008) Evidence of biologic epistasis between BDNF and SLC6A4 and implications for depression. Mol Psychiatry 13:709–716

Risch N, Herrell R, Lehner T, Liang KY, Eaves L, Hoh J, Griem A, Kovacs M, Ott J, Merikangas KR (2009) Interaction between the serotonin transporter gene (5-HTTLPR), stressful life events, and risk of depression: a meta-analysis. JAMA 301:2462–2471

Sigurdsson T, Stark KL, Karayiorgou M, Gogos JA, Gordon JA (2010) Impaired hippocampal-prefrontal synchrony in a genetic mouse model of schizophrenia. Nature 464:763–767

Stark KL, Xu B, Bagchi A, Lai WS, Liu H, Hsu R, Wan X, Pavlidis P, Mills AA, Karayiorgou M, Gogos JA (2008) Altered brain microRNA biogenesis contributes to phenotypic deficits in a 22q11-deletion mouse model. Nat Genet 40:751–760

Sugden K, Tichopad A, Khan N, Craig IW, D'Souza UM (2009) Genes within the serotonergic system are differentially expressed in human brain. BMC Neurosci 10:50

Trowbridge S, Narboux-Nême N, Gaspar P (2010) Genetic models of serotonin (5-HT) depletion: what do they tell us about the developmental role of 5-HT. Anat Rec (Hoboken) 294: 1615–1623

Uher R, Caspi A, Houts R, Sugden K, Williams B, Poulton R, Moffitt TE (2011) Serotonin transporter gene moderates childhood maltreatment's effects on persistent but not single-episode depression: replications and implications for resolving inconsistent results. J Affect Disord (in press)

Williams LM, Gatt JM, Schofield PR, Olivieri G, Peduto A, Gordon E (2009) 'Negativity bias' in risk for depression and anxiety: brain-body fear circuitry correlates, 5-HTT-LPR and early life stress. Neuroimage 47:804–814

Yee BK, Zhu SW, Mohammed AH, Feldon J (2007) Levels of neurotrophic factors in the hippocampus and amygdala correlate with anxiety- and fear-related behaviour in C57BL6 mice. J Neural Transm 114:431–444

Behavioral and Pharmacogenetics of Aggressive Behavior

Aki Takahashi, Isabel M. Quadros, Rosa M. M. de Almeida and Klaus A. Miczek

Abstract Serotonin (5-HT) has long been considered as a key transmitter in the neurocircuitry controlling aggression. Impaired regulation of each subtype of 5-HT receptor, 5-HT transporter, synthetic and metabolic enzymes has been linked particularly to impulsive aggression. The current summary focuses mostly on recent findings from pharmacological and genetic studies. The pharmacological treatments and genetic manipulations or polymorphisms of a specific target (e.g., 5-HT$_{1A}$ receptor) can often result in inconsistent results on aggression, due to "phasic" effects of pharmacological agents versus "trait"-like effects of genetic manipulations. Also, the local administration of a drug using the intracranial microinjection technique has shown that activation of specific subtypes of 5-HT receptors (5-HT$_{1A}$ and 5-HT$_{1B}$) in mesocorticolimbic

A. Takahashi
Mouse Genomics Resource Laboratory,
National Institute of Genetics (NIG),
Mishima, Shizuoka, Japan

I. M. Quadros
Departamento de Psicobiologia, Universidade Federal de São Paulo (UNIFESP),
São Paulo-SP, Brazil

R. M. M. de Almeida
Instituto de Psicologia, Universidade Federal do Rio Grande do Sul (UFRGS),
Porto Alegre-RS, Brazil

K. A. Miczek (✉)
Departments of Psychology, Psychiatry, Pharmacology and Neuroscience,
Tufts University, Bacon Hall, 530 Boston Ave,
Medford and Boston, MA 02155, USA
e-mail: klaus.miczek@tufts.edu

Curr Topics Behav Neurosci (2012) 12: 73–138
DOI: 10.1007/7854_2011_191

Published Online: 2 February 2012

areas can reduce species-typical and other aggressive behaviors, but the same receptors in the medial prefrontal cortex or septal area promote escalated forms of aggression. Thus, there are receptor populations in specific brain regions that preferentially modulate specific types of aggression. Genetic studies have shown important gene-environment interactions; it is likely that the polymorphisms in the genes of 5-HT transporters or rate-limiting synthetic and metabolic enzymes of 5-HT (e.g., MAOA) determine the vulnerability to adverse environmental factors that escalate aggression. We also discuss the interaction between the 5-HT system and other systems. Modulation of 5-HT neurons in the dorsal raphe nucleus by GABA, glutamate and CRF profoundly regulate aggressive behaviors. Also, interactions of the 5-HT system with other neuropeptides (arginine vasopressin, oxytocin, neuropeptide Y, opioid) have emerged as important neurobiological determinants of aggression. Studies of aggression in genetically modified mice identified several molecules that affect the 5-HT system directly (e.g., Tph2, 5-HT$_{1B}$, 5-HT transporter, Pet1, MAOA) or indirectly [e.g., BDNF, neuronal nitric oxide (nNOS), αCaMKII, Neuropeptide Y]. The future agenda delineates specific receptor subpopulations for GABA, glutamate and neuropeptides as they modulate the canonical aminergic neurotransmitters in brainstem, limbic and cortical regions with the ultimate outcome of attenuating or escalating aggressive behavior.

Keywords Aggression · Serotonin · Trait · State · Neuropeptides · GABA

Contents

1 Introduction 75
2 Definition and Measurement of Aggression 76
3 Aggressive "Trait" Versus "States" 81
4 5-HT Receptors 84
4.1 5-HT1 Family 84
4.2 5-HT2 Family 100
5 Serotonin Transporter (5-HTT) 102
5.1 Pharmacological Approach 102
5.2 Genetic Approach 103
6 Monoamine Oxidase A 104
6.1 Pharmacological Approach 104
6.2 Genetic Approach 104
7 Modulation of Serotonergic Activity by Other Systems 106
7.1 Excitatory and Inhibitory Amino acids 107
7.2 Neuropeptides 109
7.3 Other Molecules that Directly or Indirectly Affect 5-HT Pathways and Aggression 112
8 Concluding Remarks 114
References 116

1 Introduction

One of the oldest roots of the neurogenetic analysis of aggressive behavior can be traced to the domestication of feral animals, as illustrated experimentally in the elevated brain levels of serotonin in tame silver foxes (e.g., Popova et al. 1991). No other transmitter system has been more consistently implicated in the neurobiological mechanisms mediating impulsive aggressive behavior than serotonin (Miczek et al. 2002, 2007; Takahashi et al. 2011). Serotonin emerged as the focus in the initial phase of "top-down" genetics of aggressive behavior, when the brains of individuals that were selected for high aggressive traits revealed lower levels of serotonin, its major acidic metabolite and some of the receptor or transporter molecules upon which this amine acts (Brown et al. 1979; Garattini et al. 1967; Linnoila et al. 1983). It continues to be the key transmitter system of interest in the more recent phase of "bottom-up" genetics, when the gene for a specific receptor or transporter was deleted and the behavioral phenotype indicated a higher than normal level of aggressive behavior (Bouwknecht et al. 2001a; Saudou et al. 1994).

The current understanding of serotonergic mechanisms in aggressive behavior has to accommodate two different but interacting sources of influence. Classically, serotonin has been studied for its role in the predisposition to engage in impulsive violent and antisocial behavior. The question of interest continues to be which features of serotonergic transmission—its rate of synthesis, uptake, metabolism, receptor and transporter expression—are characteristic of an aggressive heritable trait that runs in families (Brunner et al. 1993a). It has become clear that the seductively simple serotonin deficiency hypothesis that links low serotonin activity to the propensity for aggressive behavior is being replaced by more sophisticated and detailed schemes. Recent clinical and preclinical data point to important allelic differences in the genes for specific serotonin receptor subtypes, transporter molecules and metabolic enzymes in impulsively aggressive individuals (de Boer and Koolhaas 2005; Lesch and Merschdorf 2000; Miczek et al. 2002, 2007; Nelson and Chiavegatto 2001; Takahashi et al. 2011).

A second role for serotonin is to modulate aggressive behavior by its phasic activity, particularly during the anticipation of aggressive and defensive acts and at the termination of an aggressive bout. Superimposed on the serotonergic tone are transient changes in release, impulse flow and receptor activation that are synchronized with the initiation and termination of bouts of aggressive and defensive acts (Ferrari et al. 2003). Phasic serotonergic activity appears to be linked to bouts of several species-normative consummatory behaviors ranging from feeding, drinking and sexual acts to fighting and to pathological excesses of these activities. Neuroplastic changes in serotonergic pathways from the dorsal raphé neurons to the neo- and paleocortical terminals result from repeated aggressive experiences as reflected in serotonin receptor regulation, firing rate and release. To further our understanding of the interaction between tonic and phasic serotonergic activities ideally requires real-time measurements over the course of a circadian cycle and arranging conditions that provide a challenge behaviorally and neurochemically.

During the last decade, epigenetic mechanisms for the expression of certain genes that encode for serotonergic metabolic enzymes, transporter proteins and receptors have been scrutinized for their role in impulsively violent individuals. Stressful experiences such as abusive maltreatment during a critical developmental period can lead young adults to engage in more impulsive and violent behavior, particularly in those individuals who are characterized by the expression of a certain allelic variety of Monoamine Oxidase A, an important metabolic enzyme for serotonin (Caspi et al. 2002).

The last decade has seen also a concerted effort to translate more readily between preclinical data and clinical observations, and vice versa. This is evident from both an increasing focus on escalated, atypical forms of aggressive behavior in animal models, and by efforts to define aggressive behaviors using operational and functional definitions and assessments in clinical settings (DSM-V).

2 Definition and Measurement of Aggression

Most research on the neurogenetics of serotonin and aggression in both human and veterinary medicine seeks to understand and control pathological aggression (Volavka et al. 2005), while laboratory experiments using animal models typically deal with adaptive, species-typical aggressive behavior. When studying aggression in animals, it is important to consider the ethological significance of the behavior, including its phylogenetic and ontogenetic development and the functions it serves for individuals and the species. Aggressive signals, postures and acts are used to help an animal obtain specific goals, or to defend against threats actual attacks (Miczek et al. 2002). These behaviors occur when individuals compete for food, water and other resources necessary for survival and reproduction (resident-intruder aggression), when they defend their territory or offspring (territorial and maternal aggression), or in response to frustration or fear (Miczek et al. 2001). Engaging in aggressive behavior is often beneficial to the individual and the species. For example, dominance hierarchies are established and maintained through confrontations between rival males (Fig. 1), and so-called isolation-induced aggression mimics many elements of the behavior used by resident males to exclude other breeding males from their home territory and their mates (Brain and Benton 1979; Miczek et al. 2001, Table 1). However, it should be noted that isolated housing results in social avoidance in a large proportion of a sample of mice, while aggressive behavior is seen in varying numbers of isolated animals depending on species and strain.

Aggressive behavior can be classified as offensive or defensive based upon the distal and proximal conditions that precipitated it, the topography of the behavior, and its consequences (Blanchard and Blanchard 1977; Brain 1979). Defensive behaviors occur in response to threatening or fear-inducing stimuli and often result in escapes (Brain 1979). In male rodents, specific defensive behaviors include escape, freezing, defensive postures and threats, which occur in response to attacks by either predators or conspecifics (Blanchard et al. 2003; Pellis and Pellis 1988; Rasia-Filho

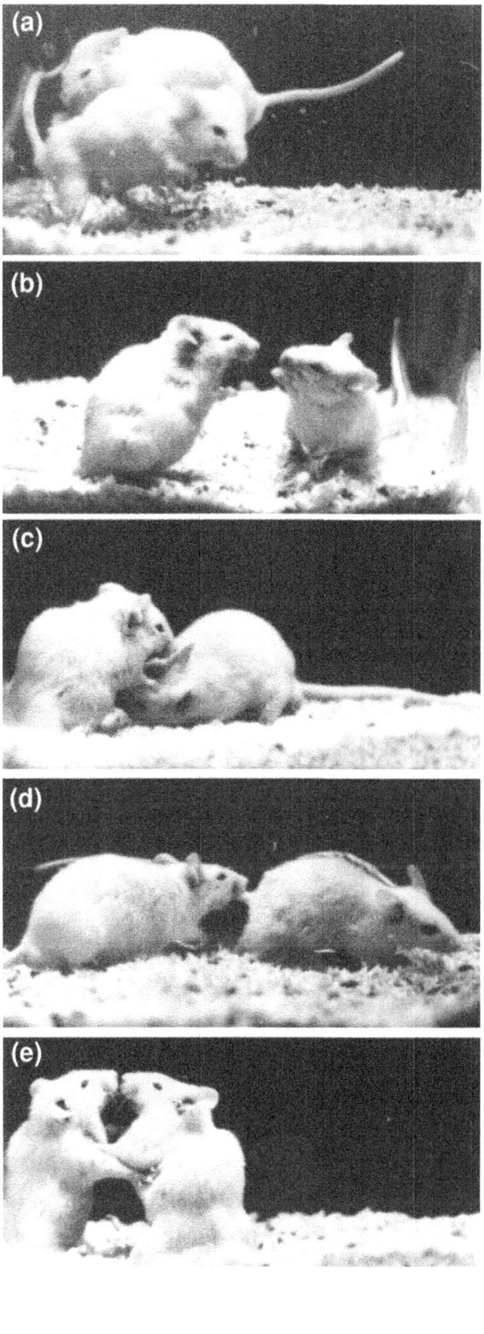

Fig. 1 Mouse agonistic behavior. Behaviors of resident and intruder mice engaged in an aggressive confrontation: **a** the resident leaps and bites the intruder as the intruder attempts to escape; **b** the resident (*right*) threatens as the intruder (*left*) holds a defensive upright posture; **c** the resident investigates the intruder's anogenital region; **d** the resident pursues the fleeing intruder; **e** both resident and intruder engage in a mutual upright defensive posture. *Reprinted with permission from Miczek and O'Donnell* (1978)

et al. 2008). Postpartum female rodents engage in maternal aggression, which includes both defensive and offensive elements, to protect their newborn offspring from male intruders (Lucion and de Almeida 1996; Parmigiani et al. 1998).

Table 1 Types of aggressive behavior in preclinical models

A. Species-typical aggressive behavior

	Situational or experimental variable	Agonistic behavioral measurements	References
Dominant resident, mainly in primates and rats	In a stable colony where dominance hierarchy is to be established and maintained	Frequency and duration of agonistic acts, postures and displays including displacement, threat, pursuit and fight	Bennett et al. (2002) Bernstein et al. (1974) Fairbanks et al. (1999) Higley et al. (1996) Mehlman et al. (1994) Steiniger (1950) Vandenbergh (1967)
Territorial resident (*resident-intruder* test), mainly in mice and hamsters	Requires an established territory. In the laboratory, home-cage of experimental male (*resident*) where it is pair-housed with a female. A male stimulus animal (*intruder*) that is group housed with other males is introduced into resident's cage	Frequency of attack bite, sideways threat, tail-rattle, pursuit, upright posture. Latency to the first bite	Crawley et al. (1975) Eibl-Eibesfeldt (1950) Miczek and O'Donnell (1978) Van Oortmerssen and Bakker (1981)
Maternal aggression, mainly in rats, mice and hamsters	Home-cage of lactating females from postpartum day 1–7. Either male or female of intruder is introduced into dam's cage	Frequency of attack bite (especially directed at the snout and the face), sideways threat, tail-rattle, pursuit, upright posture. Latency to the first bite	Haney et al. (1989) Hurst (1987) Lonstein and Gammie (2002) Noirot et al. (1975) Sgoifo et al. (1992)
Female aggression, mainly in primates and rodents	Dominance hierarchy among female monkeys. In the laboratory settings, female rodent pair-housed with a breeding male. Sexually mature female is introduced as an intruder	Harrassing attacks by dominant female. Frequency of attack bite, sideways threat, tail-rattle, pursuit, upright posture. Latency to the first bite	DeBold and Miczek (1981) Palanza et al. (2005) Smuts (1986) Zitzman et al. (2005)
Isolation-induced aggression (similar to territorial aggression)	Male isolated for 24 h to 8 weeks prior to *resident-intruder* encounter	Frequency of attack bite, sideways threat, tail-rattle, pursuit, upright posture. Latency to the first bite	Cairns and Nakelski (1971) Malick (1979) Valzelli and Bernasconi (1979) Yen et al. (1959)

(continued)

Table 1 (continued)

B. Escalated aggressive behavior

	Situational or experimental variable	Agonistic behavioral measurements	References
Alcohol-heightened aggression, mainly in rats and mice	Ethanol (1.0 g/kg) is injected intraperitoneally or orally or self-administered before the resident-intruder encounter	Frequency of attack bite, sideways threat, tail-rattle, pursuit, upright posture. Latency to the first bite Targets of attack bites (head, dorsal areas, ventral areas, appendages)	Blanchard et al. (1987) Miczek and de Almeida (2001) Miczek et al. (1992, 1998a) Peeke and Figler (1981)
Social provocation (instigation), mainly in hamsters, mice, and rats	A resident male is exposed to another male behind a protective screen in his home-cage prior to the resident-intruder encounter		Fish et al. (1999) Heiligenberg (1974) Potegal (1991) Potegal and Tenbrink (1984)
Frustration-heightened aggression	A resident male is positively reinforced. The reinforcement is abruptly omitted before the resident-intruder encounter		Berkowitz (1993) De Almeida and Miczek (2002)
Aggression induced by low glucocorticoids	Resident rats are adrenalectomized and implanted with a low-dose corticosterone pellet		Haller et al. (2001)
Affective defense ("rage"), mainly in cats	Electrical stimulation (0.2–0.8 mA, 63 Hz, 1 ms per half cycle duration) delivered in medial hypothalamus or midbrain periaqueductal gray	Hissing, arching of the back, retraction of the ears, piloerection, unsheathing of the claws, papillary dilatation and paw striking	Hess (1954) Leyhausen (1979) Siegel et al. (1999)

"Violence" is a controversial concept among ethologists. The term has been used to describe escalated, pathological and abnormal forms of aggression characterized by prolonged and frequent attack bites and aggressive behavior with brief latencies (Miczek et al. 2002, 2003). The differences are quantitative rather than qualitative, with violence marked by shorter attack latencies and higher frequencies and longer durations of fighting than adaptive aggression. Violence has been described to be qualitatively different from adaptive aggression. For instance, attack bites aimed at vulnerable parts of the opponent's body are considered abnormally aggressive, when adaptive aggression consists of less-injurious bites directed at the intruder's back and flanks (Haller et al. 2005). Additional qualitative distinctions that have been proposed include context-independent attacks, which are directed at an opponent regardless of its sex, responsiveness (free-living/anaesthetized/dead) or the test environment (home/neutral cage) (Koolhaas 1978), and lack of ritualistic behaviors, quantified as attack/threat (A/T) ratios (Haller et al. 2005). Therefore in principle "violence" could refer either to quantitatively escalated or hyper-aggression, or to qualitatively abnormal forms of aggression, or even rarely to aggression that is both escalated and abnormal [for review see Natarajan and Caramaschi (2010)].

Human and non-human aggressive behaviors have some common features, but most animal aggression is less complex. Social norms establish boundaries for what is accepted as appropriate aggressive behavior, but the array of inappropriate interpersonal behaviors classified as violence is a serious social and mental health issue (Ferris et al. 2008). Numerous psychiatric disorders defined in the DSM-IV R and the new DSM-V (scheduled for publication in May 2013), including schizophrenia, brief reactive psychosis, anxiety disorder, adjustment disorder, impulse control disorder, antisocial personality disorder, attention deficit disorder, mania/depression, PTSD, autism and substance abuse, specify aggressive behavior among their symptoms (Boles and Miotto 2003; Raine 2002; Rydén et al. 2009; Volavka et al. 2005).

It can be useful to classify human aggressive behavior as either defensive, premeditated or impulsive-hostile in nature (Stoff and Vitiello 1996; Vitiello and Stoff 1997). The premeditated (e.g., predatory and instrumental) and impulsive forms of aggression are especially likely to be diagnosed as pathological and in need of treatment. Impulsive, but not premeditated, aggression is linked to biological and environmental causes, as well as pharmacological or psychological treatment response factors, by a growing body of empirical data (Coccaro et al. 2010).

Aggression as a behavioral state in humans has usually been measured by provoking subjects in competitive situations with fictitious opponents and then giving them opportunities to engage in quantifiable responses that are defined as aggressive [see Table 2, for review see Miczek et al. (2002)]. Human aggression as a trait is assessed using psychometric measures including inventories, questionnaires and scales. The role of 5-HT in human aggression has been successfully investigated using such laboratory techniques (Table 2), but relating laboratory indices of aggression to actual violence and aggression outside the laboratory remains a critical challenge for such approaches; identifying subtypes of human aggression with psychometric and laboratory methods is also difficult. The psychometric instruments used to differentiate between individuals with contrasting aggressive traits such as the impulsive–reactive–hostile–affective subtype versus the controlled–proactive–instrumental–predatory subtype (Stoff and Vitiello 1996) are summarized in Table 2.

3 Aggressive "Trait" Versus "States"

A frequently reiterated theory, based on early clinical and preclinical studies, links impulsive, hostile and violent behavior to a serotonin deficiency (Brown and Goodwin 1986; Goldman et al. 1992; Lesch and Merschdorf 2000; Linnoila and Virkkunen 1992; Mann 1999; Valzelli 1977). Individuals exhibiting such behavior may benefit from pharmacological treatments aimed at inhibiting 5-HT transporters (e.g., SSRIs such as fluoxetine or citalopram), activating 5-HT_{1A} receptors (e.g., buspirone) or blocking 5-HT_{2A} receptors (e.g., risperidone). Acute treatment with these drugs induces phasic changes in 5-HT function that are associated with their transient anti-aggressive effects. In vivo microdialysis allows transient changes in extracellular 5-HT levels to be monitored in anticipation of, during, and after aggressive encounters in rats. One study reported reduced 5-HT levels in the prefrontal cortex during and after the aggressive confrontation, while no changes were detected in the nucleus accumbens, another terminal region (Van Erp and Miczek 2000, Fig. 2). By contrast, chronic treatment with these anti-aggressive compounds may promote as yet undefined neuroadaptive changes in 5-HT function such as autoreceptor desensitization, which may in turn be associated with the emergence of therapeutic effects.

On the other hand, aggression as a "trait" is the focus of genetic studies. While these aggressive traits are clearly polygenic, it is remarkable that several studies have found an interaction between genotypes such as TPH2, MAO-A and 5-HTT polymorphisms and environmental triggers such as social stress underlying an increased likelihood of violent outbursts (see below). For example, a SNP in TPH2 gene (A2051C) has been linked to aggressive behavior in rhesus monkeys. Monkeys with the AA/AC genotype that were reared without their mother (peer-reared) showed increased aggressive acts compared to those with a CC genotype, but the difference disappeared when infants of both genotypes were reared by their mothers (Chen et al. 2010). This review will focus on the interaction between salient environmental events and genes and the subsequent effects on aggressive behaviors.

Gene–gene interactions are also of interest and need to be explored. For example, Passamonti et al. (2008) showed interactions between 5-HTT and MAOA polymorphisms that affected the activity of the anterior cingulate cortex, a brain area that has been implicated in impulsivity, including impulsive aggression. Many other genes may have subtle effects on aggressive phenotypes, and stronger effects may emerge as a result of complex epistatic interactions between those

Table 2 Experimental protocols for assessing 5-HT effects on human aggressive behavior

A. Experimental manipulations

Experimental manipulation	Measurement	Trait/state	References
Aggressive responses toward a competitor are measured in the form of electric shock settings	Activate buttons at 5–10 settings, each corresponding to a different intensity or duration of electric shock	State	Buss (1961) Godlaski and Giancola (2009)
A fictitious instigator or competitor is the target of aggressive responses that are measured in the form of electric shock deliveries	Setting of electric shock level on a scale from 1–10	State	Chermack and Giancola (1997) Taylor (1967)
The subjects are provoked by having points subtracted in a competitive task. The point losses are attributed to a fictitious opponent, but are actually random. Subjects responded by retaliation of point subtractions (=aggressive responses)	Number of point subtractions from a fictitious competitor	State	Cherek and Heistad (1971) Cherek and Lane (1999) Gowin et al. (2010)
Aggression was defined as delivery of electric shocks to a fictitious opponent	Uses a modified version of the Buss aggression machine. Setting of shock level on a scale from 1–5	State	Giancola et al. (2009) Zeichner and Pihl (1979)

B. Psychometric inventories

Psychometric assessment	Instrument	Trait/state	References
Aggression, impulsivity and hostility are measured by Minnesota multiphasic personality inventory (MMPI)	Inventory	Trait	McKinley et al. (1948) Nagtegaal and Rassin (2004)

(continued)

Table 2 (continued)

B. Psychometric inventories

Psychometric assessment	Instrument	Trait/state	References
Buss–Durkee hostility inventory (BDHI), a self-rating scale of anger and hostility. Sixty six items with false/true answers; also contains seven scales: assault, indirect aggression, irritability, negativism, resentment, suspicion and verbal aggression	Inventory	Trait	Buss and Durkee (1957)
Anger and anxiety are measured by state–trait anger expression inventory (STAXI)	Inventory	State/trait	Kim et al. (2009b) Spielberg et al. (1973)
Aggression is measured by beck anxiety inventory and beck depression inventory	Inventory	Trait	Beck et al. (1961) Lamar et al. (2009)

genes (Miczek et al. 2001). In the past 15 years, most rodent studies on genetics and aggressive behavior have used conventional knockout techniques, in which the expression of a gene is completely deleted, affecting the whole body through all developmental stages and inducing compensatory changes in other genes (trait-like change; see Table 3). Newer, more subtle methods, including conditional knockout, viral vector microinfusion, and drug-inducible knockout techniques, can produce transient and local changes in gene expression, allowing the study of more "phasic" changes in gene expression and how they influence aggression. Such studies may explain some of the discrepancies in the results from previous genetic and pharmacological studies of 5-HT function and aggression.

4 5-HT Receptors

Considerable evidence suggests that serotonin receptors regulate aggressive behaviors in various animal species. Among the multiple 5-HT receptor subtypes, it is mainly the first two families of serotonin receptors (5-HT_1 and 5-HT_2) which have been studied in regard to their role on aggression (Miczek et al. 2002; Olivier 2004). There is some, but less, information on the involvement of 5-HT_3 receptors on aggressive behaviors (McKenzie-Quirk et al. 2005; Ricci et al. 2004; Rudissaar et al. 1999). Further development of adequate pharmacological tools that can activate or inhibit a certain subtype of 5-HT receptor specifically is required to study the role of other receptor subtypes (e.g., 5-HT5, 6, 7, 1e, 1f) on aggression.

4.1 5-HT1 Family

4.1.1 Pharmacological Approach

The 5-HT_{1A} receptor partial agonist buspirone has been used clinically to reduce general anxiety. It was shown that buspirone also reduces aggressive behavior in mentally retarded patients (Kavoussi et al. 1997; Ratey et al. 1991), and this compound has been used for the management of aggressive outbursts associated with neuropsychiatric disorders in adults and children (Connor and Steingard 1996; Pabis and Stanislav 1996). Similar findings have been reported in preclinical studies; systemic administration of 5-HT_{1A} receptor agonists (e.g., 8-OH-DPAT, repinotan, alnespirone) dose-dependently decrease aggressive behaviors in a wide range of animal species, including fish, amphibians, birds, rodents, guinea pigs and non-human primates (Bell and Hobson 1994; Blanchard et al. 1988; Clotfelter et al. 2007; de Boer et al. 1999; de Boer and Koolhaas 2005; Dompert et al. 1985; Haug et al. 1990; Joppa et al. 1997; Lindgren and Kantak 1987; McMillen et al. 1988; Miczek et al. 1998b; Muehlenkamp et al. 1995; Nikulina et al. 1992; Olivier et al. 1992; Sanchez et al. 1993; Sperry et al. 2003; Ten Eyck 2008; Tompkins

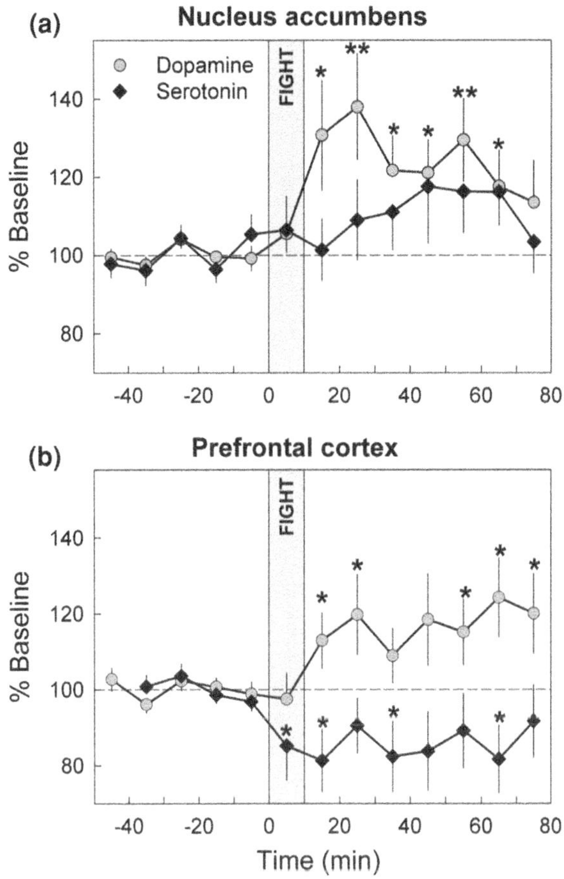

Fig. 2 Dopamine and serotonin during aggression. Measurements of extracellular dopamine and serotonin via in vivo microdialysis in resident male rats before, during and after a confrontation with an intruder. **a** In the nucleus accumbens (*top panel*), dopamine levels (*gray circles*) rise and remain elevated after the confrontation, while serotonin levels (*black diamonds*) do not significantly change. **b** In the prefrontal cortex (*bottom panel*), dopamine levels rise after the confrontation, while serotonin decline and remain lower after the confrontation. Samples were collected every 10 min and levels are expressed as mean percent of baseline ±SEM. Baseline was measured for 50 min before the fight. The vertical *light gray bar* indicates the occurrence of the 10-min fight. *Asterisks* and *double asterisks* represent significant differences from baseline (*dashed line*) at the $p < 0.05$ and 0.01 levels, respectively. *Reprinted with permission from Van Erp and Miczek* (2000)

et al. 1980). One exception are fruit flies (*Drosophila melanogaster*), as they showed increased aggressive behavior after 8-OH-DPAT treatment (Johnson et al. 2009). Selective antagonists of 5-HT_{1A} receptors such as WAY-100635 successfully reversed the effect of 5-HT_{1A} agonists, while the administration of the antagonist by itself did not show any effect on aggression (de Boer and Koolhaas 2005; Mendoza et al. 1999; Miczek et al. 1998b). Notably, in laboratory studies the anti-aggressive effects of most of 5-HT_{1A} agonists are accompanied by nonspecific

Table 3 Genes and aggressive behavior in mice

Gene	Abb	Chr	Background strain (generations of backcross)	Type of aggression	Effects	Serotonin function	References
Plasmacytoma expressed transcript 1	Pet1 (Fev)	1	Mix C57BL/6 and 129 Sv	Isolation-induced resident aggression	Increased	90% reduction of brain 5-HT	Hendricks et al. (2003)
Arginine vasopressin receptor 1B	V1bR	1	Mix C57BL/6 and 129/SvJ	Isloation-induced aggression	Suppressed		Wersinger et al. (2002)
				Neutral cage aggression	Suppressed		
Aadenosine A1 receptor	A1AR	1	Mix C57BL and 129/OlaHsd	Isolation-induced resident aggression	Increased		Gimenez-Llort et al. (2002)
Regulators of G protein signaling 2	Rgs2	1	C57BL/6J (N5)	Social dominance test	Suppressed		Oliveira-dos-Santos et al. (2000)
v-abl Abelson murine leukemia viral oncogene homolog 2 (Abelson-related gene)	Arg	1	Mix C57BL/6 and 129/SvJ	Resident aggression	Suppressed		Koleske et al. (1998)
Glutamic acid decarboxylase 2	GAD_{65}	2	Mix C57BL/6 and CBA2	Isolation-induced resident aggression	Suppressed		Stork et al. (2000)
Calcium channel, voltage-dependent, N type, α1B subunit	Cav2.2	2	Mix C57BL/6J and 129S4/SvJae	Isolation-induced resident aggression	Increased	Increased hypothalamus 5-HT	Kim et al. (2009a)

(continued)

Table 3 (continued)

Gene	Abb	Chr	Background strain (generations of backcross)	Type of aggression	Effects	Serotonin function	References
Dopamine β-hydroxylase	Dbh	2	Mix C57BL/6J and 129/SvEv	Isolation-induced resident aggression	Suppressed		Marino et al. (2005)
Brain-derived neurotrophic factor [±]	BDNF	2	C57BL/6J (>N10)	Isolation-induced resident aggression	Increased	Decreased brain 5-HT	Lyons et al. (1999)
β2-microglobulin	β2 m	2	129S2/SvPas (Mix C57BL/6J?)	Resident aggression	Suppressed		Loconto et al. (2003)
Oxytocin	Oxt	2	Mix C57BL/6J and 129SvEv	Isolation-induced resident aggression	Increased		Winslow et al. (2000)
				Resident aggression	Increased		
					Suppressed		DeVries et al. (1997)
Membrane metallo endopeptidase (enkephalinase)	NEP	3	Mix C57BL/6J and 129 Sv	Resident aggression	Increased		Fischer et al. (2000)
Cannabinoid receptor 1	CB1	4	CD1 (N15)	Isolation-induced resident aggression	Increased[a]		Martin et al. (2002)
Preproenkephalin	Enk	4	Mix CD1 and 129	Isolation-induced resident aggression	Increased[a]		Konig et al. (1996)

(continued)

Table 3 (continued)

Gene	Abb	Chr	Background strain (generations of backcross)	Type of aggression	Effects	Serotonin function	References
Endothelial nitric oxide synthase	eNOS	5	Mix C57BL/6 and 129 Sv	Resident aggression	Suppressed	Increased 5-HT turnover	Demas et al. (1999)
				Neutral cage aggression	Suppressed		
Interleukin-6	IL-6	5	Mix C57BL/6, 129/SvEv	Isolation-induced resident aggression	Increased[b]	No difference in brain 5-HT concentration	Alleva et al. (1998)
Adrenergic receptor, α2c	Naα_{2C}	5	C57BL/6J (N7)	Isolation-induced resident aggression	Increased		Sallinen et al. (1998)
Neuronal nitric oxide synthase	nNOS	5	Mix C57BL/6J, 129 Sv, DBA2	Resident aggression	Increased	Reduced brain 5-HT turnover	Nelson et al. (1995)
				Maternal aggression	Suppressed		Gammie and Nelson (1999)
Acetylcholinesterase	AChE	5	129S6/SvEvTac	Home-cage hierarchy	Suppressed		Duysen et al. (2002)
Adenylate cyclase activating polypeptide 1 receptor 1	PAC1	6	Mix C57BL/6J and 129 Sv	Resident aggression	Suppressed		Nicot et al. (2004)
Tachykinin receptor 1	NK-1r	6	Mix C57BL/6 and 129 Sv	Isolation-induced resident aggression	Suppressed		De Felipe et al. (1998)
A cluster of vomeronasal receptor genes	V1R$_{a/b}$	6	129/SvEv	Maternal aggression	Suppressed		Del Punta et al. (2002)

(continued)

Table 3 (continued)

Gene	Abb	Chr	Background strain (generations of backcross)	Type of aggression	Effects	Serotonin function	References
Oxytocin receptor	Oxtr	6	Mix C57BL/6 and 129 Sv	Isloation-induced aggression	Increased		Takayanagi et al. (2005)
				Cage-mate injury	Increased		
Histamine receptor H1	H1	6	Mix C57BL/6J and 129	Isolation-induced resident aggression	Suppressed[b]	Increased 5-HT turnover	Yanai et al. (1998)
Amyloid β(A4) precursor protein-binding, family A, member 2	X11L	7	C57BL/6	Competitive feeding	Subordinate	Increased Hypothalamus 5-HT	Sano et al. (2009)
Transient receptor potential cation channel, subfamily C, member 2	TRP2	7	Mix C57BL/6J and 129Sv	Resident aggression	Suppressed		Stowers et al. (2002)
Neuropeptide Y receptor Y1	Y1	8	Mix C57BL/6 and 129SvJ	Resident aggression	Increased	Reduced TPH mRNA expression in the raphe nuclei	Karl et al. (2004)
Prostaglandin E receptor 1 (subtype EP1)	EP1	8	C57BL/6 (N5)	Neutral cage aggression	Increased	No difference in 5-HT turnover	Matsuoka et al. (2005)
Norepinephrine transporter	NET	8	Mix C57BL/6J and 129SvJ	Isolation-induced resident aggression	Increased[a]		Haller et al. (2002)
Neural cell adhesion molecule 1	NCAM1	9	C57BL/6J (>N5)	Isolation-induced resident aggression	Increased		Stork et al. (1997)

(continued)

Table 3 (continued)

Gene	Abb	Chr	Background strain (generations of backcross)	Type of aggression	Effects	Serotonin function	References
Aromatase (Cyp19a1)	Ar	9	Mix C57BL/6 and 129S/SvEv	Resident aggression	Suppressed		Matsumoto et al. (2003)
				Aggression toward female	Increased		
5-hydroxytryptamine (serotonin) receptor 1B	5-HT_{1B}	9	129S2/SvPas	Resident aggression	Increased	Deleted 5-HT_{1B} receptor expression	Saudou et al. (1994)
				Maternal aggression	Increased		Brunner and Hen (1997)
Estrogen receptor-α	ERα	10	Mix C57BL/6J and 129	Resident aggression	Suppressed		Ogawa et al. (1998b)
				Female aggression	Increased		Ogawa et al. (1998a)
Fyn tryrosine kinase	Fyn	10	Mix C57BL/6 and CBA	Isolation-induced resident aggression	Suppressed		Miyakawa et al. (2001)

(continued)

Table 3 (continued)

Gene	Abb	Chr	Background strain (generations of backcross)	Type of aggression	Effects	Serotonin function	References
Nuclear receptor subfamily 2, group E, member 1	Nr2e1	10	C57BL/6J (> N6)	Resident aggression	Increased		Young et al. (2002)
Adenosine A2a receptor	A2AR	10	CD1 (N4)	Isolation-induced resident aggression	Increased		Ledent et al. (1997)
Tryptophan hydroxylase 2	Tph2	10	FVB/N (N7)	Home-cage injury	Increased	Reduced brain 5-HT	Alenina et al. (2009)
Glutamate receptor, ionotropic, AMPA1 (α1)	GluR-A	11	C57BL/6J (N5)	Isolation-induced aggression	Suppressed	No difference in brain 5-HT level	Vekovischeva et al. (2004)
				Neutral cage aggression	Suppressed		
5-hydroxytryptamine (serotonin) transporter	SERT	11	C57BL/6J (N8)	Isolation-induced resident aggression	Suppressed	Increased extracelluler 5-HT	Holmes et al. (2002)
Estrogen Receptor-β	ERβ	12	Mix C57BL/6J and 129	Resident aggression	Increased[a]		Ogawa et al. (1999)
Dopamine transporter	DAT	13	Mix C57BL/6J and 129SvJ	Dyadic encounter aggression	Increased		Rodriguiz et al. (2004)
5-hydroxytryptamine (serotonin) receptor 1A	5-HT$_{1A}$	13	129S1/Sv	Resident aggression	Suppressed	Deleted 5-HT$_{1A}$ receptor expression	Zhuang et al. (1999)
Catechol-O-methyltransferase 1 [±]	COMT	16	Mix C57BL/6J and 129SvJ	Neutral cage aggression	Increased	No difference in brain 5-HT level	Gogos et al. (1998)

(continued)

Table 3 (continued)

Gene	Abb	Chr	Background strain (generations of backcross)	Type of aggression	Effects	Serotonin function	References
Calcium/calmodulin-dependent protein kinase II alpha [±]	αCaMKII	18	Mix C57BL/6, 129/OU, BALB/c	Defensive aggression	Increased	Reduced 5-HT release in the dorsal raphe nucleus	Chen et al. (1994)
Melanocortin-5 receptor	MC5R	18	C57BL/6 (>N7)	Neutral cage aggression	Suppressed		Morgan et al. (2004)
Monoamine oxidase A	MAOA	X	C3H/HeJ	Resident aggression	Increased	Increased brain 5-HT up to ninefold	Cases et al. (1995)
				Cage-mate injury	Increased		
Cyclic nucleotide–gated channel a2	Cnga2	X	Mix C57BL/6J and 129SvEv	Resident aggression	Suppressed		Mandiyan et al. (2005)
Guanosine diphosphate (GDP) dissociation inhibitor 1	Gdi1	X	C57BL/6J (N5)	Isolation-induced resident aggression	Suppressed		D'Adamo et al. (2002)
Androgen receptor[c]	AR	X	C57BL/6 and 129SvEv	Isolation-induced resident aggression	Suppressed[b]		Raskin et al. (2009)

[±] Data obtained from heterozygote of knockout mouse, [a] behavioral change only at the first encounter, [b] behavior change after repeated exposure, [c] Nestine-Cre-LoxP conditional knockout mice that selectively lack AR expression in the nervous system

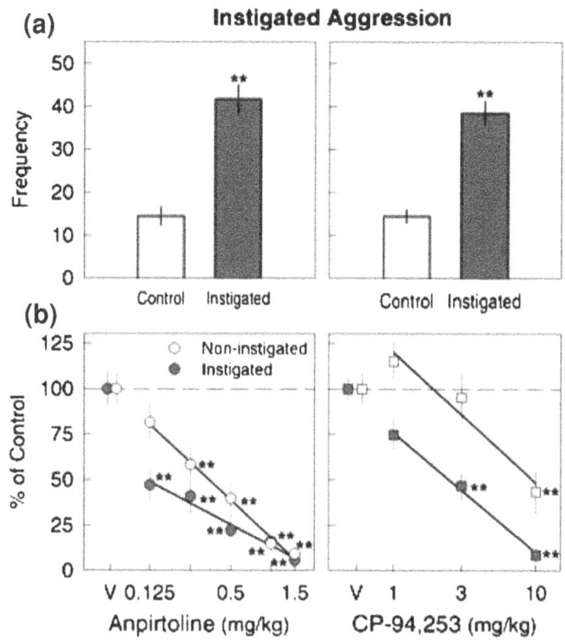

Fig. 3 **a** Effects of social instigation on aggressive behavior by a resident mouse toward a male intruder. *Bars* represent the mean frequency ±SEM (*vertical lines*) of attack bites under control (*light gray*) and instigated (*dark gray*) conditions. *Asterisks* denote statistical significance from control (*double asterisks* $P < 0.01$). **b** Preferential reduction of instigated aggressive behavior by the 5-HT$_{1B}$ agonist anpirtoline (*left panel, filled circles*) and CP-94,253 (*right panel, filled squares*). Symbols represent the mean frequency of attack bites, expressed as a percentage of vehicle (*V*) baseline, ±SEM. *Light gray* symbols represent non-instigated fighting and *dark gray* symbols represent instigated levels of fighting. *Asterisks* denote significance from vehicle baseline ($P < 0.05$). *Adapted from Fish et al.* (1999) *and de Almeida and Miczek* (2002)

behavioral side effects including sedation, motor inactivity, stereotypic behavior or reduced social interest (de Boer and Koolhaas 2005; Miczek et al. 1998b; Olivier et al. 1995). However, some 5-HT$_{1A}$ agonists (i.e., alnespirone and S-15535) reduced aggressive behavior specifically without changing the other non-aggressive behaviors, at least in a wild-derived rat strain (de Boer et al. 1999, 2000; de Boer and Koolhaas 2005). These compounds are suggested to have different effects on 5-HT receptors on presynaptic and postsynaptic terminals (de Boer and Koolhaas 2005), and thus it is possible that a subpopulation of 5-HT$_{1A}$ receptors on which those compounds acts is involved in anti-aggressive effect specifically.

The 5-HT$_{1B}$ receptor agonists seem more specific in their anti-aggressive effect than 5-HT$_{1A}$ agonists. In mice and rats, the systemic administration of 5-HT$_{1B}$ agonists (e.g., CP-94253, CP-93129, CGS-12066B) reduces aggressive behavior without sedation, or motor or sensory impairment (de Almeida et al. 2001a; de Almeida and Miczek 2002; de Boer and Koolhaas 2005; Fish et al. 1999; Miczek et al. 2002, 2004; Mos et al. 1993; Olivier et al. 1990; Olivier 2004, Fig. 3).

These effects were antagonized by $5\text{-HT}_{1B/1D}$ antagonist GR-127935, further confirming that the anti-aggressive effects of these compounds were mediated by 5-HT_{1B} receptors (de Boer and Koolhaas 2005). However, there are no clinically approved drugs that target 5-HT_{1B} receptors. There is an amino acid difference in the binding domain of 5-HT_{1B} receptors of humans and rodents, and therefore the pharmacological sensitivity and specificity of 5-HT_{1B} agonists in rodents may not compare with those in humans (Olivier 2004).

The precise functions and sites of action for the anti-aggressive effects of 5-HT_{1A} and 5-HT_{1B} receptors still remain to be resolved. One site of inhibitory action for 5-HT is the 5-HT_1 autoreceptors that are localized at either the somata or the synaptic terminal of serotonin neurons. Stimulation of 5-HT_1 autoreceptors inhibits serotonergic neural activities and therefore reduces 5-HT impulse flow and release. On the other hand, 5-HT_1 receptors are also localized at the postsynaptic terminal as heteroreceptors at the projection sites of serotonin neurons. By contrast, the effect of agonists that act on 5-HT_1 heteroreceptors can be interpreted as increasing 5-HT neurotransmission.

Microdialysis studies have shown that both systemic and intra-raphé administration of 5-HT_{1A} and 5-HT_{1B} agonists decreased extracellular levels of 5-HT in the forebrain areas including striatum, hippocampus and frontal cortex (Adell et al. 2001; Bonvento et al. 1992; De Groote et al. 2003; Dekeyne et al. 2000; Gobert et al. 1998; Hjorth and Sharp 1991; Johnson et al. 2001; Knobelman et al. 2000; Kreiss and Lucki 1994; Sprouse and Aghajanian 1987). These data suggest that the inhibition of 5-HT release by 5-HT_{1A} and 5-HT_{1B} agonists is mediated by the 5-HT_1 autoreceptors. Note that these findings suggest that compounds which decrease 5-HT neurotransmission concurrently reduce aggressive behavior, which challenges the serotonin deficiency hypothesis of aggression.

On the other hand, the importance of the 5-HT_{1A} and 5-HT_{1B} heretoreceptors on projection sites has been emphasized in studies using lesion or depletion of raphé 5-HT neurons either by systemic administration of the tryptophan hydroxylase inhibitor PCPA or by intracerebral injection of the 5-HT neurotoxin, 5,7 -dihydroxytryptamine (5,7-DHT). Since these neurotoxic treatments destroy the function of serotonergic neurons, one can observe the effect of 5-HT_1 agonists on postsynaptic receptors exclusively. If only somatodendritic and presynaptic autoreceptors are the site of action of 5-HT_{1A} and 5-HT_{1B} agonists, PCPA or 5,7-DHT treatment should eliminate the anti-aggressive effect of 5-HT_{1A} and 5-HT_{1B} agonists. However, lesions or depletion of 5-HT neurons did not affect the anti-aggressive effects of 5-HT_{1A} and 5-HT_{1B} agonists (de Almeida et al. 2001b; Miczek et al. 1998b) or even had pro-aggressive effects (Sanchez and Hyttel 1994; Sijbesma et al. 1991). While these data may suggest postsynaptic 5-HT_1 receptors as critical sites of action, it is important to note that these pharmacological depletions spared a subpopulation of receptors.

The intracranial microinjection technique has been used to determine the critical brain regions underlying the anti-aggressive effects of 5-HT_{1A} and 5-HT_{1B} receptor agonists (Table 4). 5-HT in the mammalian central nervous system derives mainly from the dorsal and median raphé nuclei (DRN and MRN,

Table 4 Modulation of aggressive behaviors after local infusion of drugs targeting 5-HT receptors in selected brain regions

Brain region	Type of aggression, species	Target; drugs and doses	Pharmacological effects	References
DRN	Resident aggression, male rats	5-HT_{1A}: 8-OH-DPAT, 1–10 μg $5\text{-HT}_{1A}/5\text{-HT}_{1B}$: Eltopronazine, 1–30 μg (agonists)	↓ aggressive behavior, with inactivity and decreased social interaction	Mos et al. (1993)
	Resident aggression, male rats	5-HT_{1A}: Alnespirone, 25 μg (agonist)	↓ aggression; no side effects	Van der Vegt et al. (2003)
	Alcohol-escalated aggression, male mice	5-HT_{1A}: 8-OH-DPAT, 1.0 μg 5-HT_{1B}: CP-94253, 1.0 μg (agonists)	8-OH-DPAT and CP-94253: ↓ baseline aggression, with reduced motor activity; no effects on alcohol-related aggression	Faccidomo et al. (2008)
	Schedule-heightened aggression, male mice	5-HT_{1B}: CP-93129, 0.1–1.0 μg (agonist)	↓ escalated aggression; reduced walking behavior	Bannai et al. (2007)
	Alcohol-escalated aggression, male mice	5-HT_{1B}: CP-93129, 0.1–1.0 μg (agonist)	↓ baseline and alcohol-related aggression (0.5–1.0 μg); concomitant reduction in motor activity	Faccidomo et al. 2012
	Maternal aggression, rats	5-HT_{1A}: 8-OH-DPAT, 0.56 μg (agonist)	↑ maternal aggression (0.56 μg); no motor effects DPAT-escalated aggression prevented by infusion of CP-93129 (1.0 μg) into the orbitofrontal cortex	Veiga et al. (2011)
MRN	Maternal aggression, rats	5-HT_{1A}: 8-OH-DPAT, 0.2–2.0 μg (agonist)	↓ maternal aggression; no side effects	De Almeida and Lucion (1997)
PAG	Maternal aggression, rats	*Dorsal PAG* 5-HT_{1A}: 8-OH-DPAT, 0.2–2.0 μg (agonist)	↓ maternal aggression (0.2–2.0 μg); no side effects	de Almeida and Lucion (1997)

(continued)

Table 4 (continued)

Brain region	Type of aggression, species	Target; drugs and doses	Pharmacological effects	References
	Maternal aggression, rats	*Dorsal PAG* $5\text{-HT}_{2A/2C}$: α-methyl-5-HT maleate, 0.2–1.0 μg (agonist) $5\text{-HT}_{2A/2C}$: ketanserin, 1.0 μg (antagonist)	α-methyl-5-HT maleate: ↓ maternal aggression; no motor effects Ketanserin: no effects on aggression, decreased motor activity	de Almeida et al. (2005)
	Hypothalamic-stimulated defensive aggression, cats	*PAG* 5-HT_{1A}: 8-OH-DPAT[a], 0.016 ng–1.0 μg 5-HT_{2C}: DOI[a], 3.57 ng–0.54 μg (agonists)	8-OH-DPAT: ↓ defensive hissing (0.66–1.0 μg), effect prevented by antagonist p-MPPI. No motor effect DOI: facilitation of defensive hissing (0.54 μg)	Shaikh et al. (1997)
Septal nuclei	Maternal aggression, rats	*Medial septal nucleus* 5-HT_{1A}: 8-OH-DPAT, 0.2–2.0 μg (agonist)	↑ maternal aggression (0.2–0.5 μg); reduced activity only with highest dose (2.0 μg)	de Almeida and Lucion (1997)
	Maternal aggression, rats	*Medial septal nucleus* $5\text{-HT}_{2A/2C}$: alpha-methyl-5-HT maleate, 0.2–1.0 μg (agonist) $5\text{-HT}_{2A/2C}$: ketanserin, 1.0 μg (antagonist)	No effects on aggressive or non-aggressive behaviors (agonist) Ketanserin: no effects on aggression, but decreased motor activity	de Almeida et al. (2005)
	Resident aggression, male mice (castrated males with hormonal replacement)	*Lateral septal nucleus* 5-HT_{1A}: 8-OH-DPAT[a], 0.33 ng 5-HT_{1B}: CGS12066[a], 18.0 ng (agonists)	8-OH-DPAT: no behavioral effects CGS: ↓ aggression with CGS, only in androgen-replacement condition; no side effects	Cologer-Clifford et al. (1997)
	Alcohol-escalated aggression, male mice	*Lateral septal nucleus* 5-HT_{1B}: CP-94253, 1.0 μg (agonist)	No effects on alcohol-related or baseline aggression	Faccidomo et al. (2008)

(continued)

Table 4 (continued)

Brain region	Type of aggression, species	Target; drugs and doses	Pharmacological effects	References
Hypothalamus	Resident aggression, male mice (castrated males with hormonal replacement)	*Medial pre-optic area* 5-HT_{1A}: 8-OH-DPAT[a], 0.16 ng 5-HT_{1B}: CGS12066[a], 9.0 ng (agonists)	8-OH-DPAT: ↓ aggression in androgen- and estrogen-replacement conditions; no motor effects CGS: ↓ aggression in androgen- and estrogen-replacement conditions; no motor effects	Cologer-Clifford et al. (1997)
	Vasopressin-escalated aggression, male hamsters	*Anterior hypothalamus* 5-HT_{1A}: 8-OH-DPAT[a], 0.03–3.28 ng 5-HT_{1B}: CGS12066[a], 4.5–45.0 ng (agonists)	8-OH-DPAT: ↓ escalated aggression (0.33, 3.28 ng); no apparent motor effects 5-HT_{1B}: no effects on aggression (trends toward inhibition)	Ferris et al. (1999)
	PAG-stimulated defensive aggression, cats	*Medial hypothalamus* 5-HT_{1A}: 8-OH-DPAT[a], 0.03–1.0 μg 5-HT_{2C}: DOI[a], 0.036–0.36 μg (agonists)	8-OH-DPAT: ↓ defensive hissing (0.33–1.0 μg); effect prevented by antagonist p-MPPI DOI: facilitates defensive hissing (0.36 μg); effect prevented by antagonist LY-53,857	Hassanain et al. (2003)
Amygdala	Maternal aggression, rats	*Corticomedial amygdaloid nucleus* 5-HT_{1A}: 8-OH-DPAT, 0.2–2.0 μg (agonist)	↓ maternal aggression (0.5–2.0 μg); no side effects	De Almeida and Lucion (1997)
Prefrontal cortex	Resident aggression, male mice	5-HT_{1B}: CP-94253, 1.0 μg (agonist)	No effects on aggressive or non-aggressive behaviors	de Almeida et al. (2006)
	Alcohol-escalated aggression, male mice	5-HT_{1B}: CP-94253, 0.5–1.0 μg (agonist)	↑ alcohol-related aggression (1.0 μg); mild increases in activity; no effects on baseline aggression	Faccidomo et al. (2008)

(continued)

Table 4 (continued)

Brain region	Type of aggression, species	Target; drugs and doses	Pharmacological effects	References
	Alcohol-escalated aggression, male mice	$5\text{-}HT_{1B}$: CP-93129, 0.1–1.0 μg (agonist)	↓ baseline and alcohol-related aggression (0.1–1.0 μg); no motor effects	Faccidomo et al. 2012
Orbitofrontal cortex	Resident aggression, male mice	$5\text{-}HT_{1B}$: CP-94253, 0.1–1.0 μg (agonist)	↓ aggression (0.56–1.0 μg); no side effects. CP-94253 effects prevented by 1B antagonist, GR-127935.	De Almeida et al. (2006)
	Maternal aggression, rats	$5\text{-}HT_{1B}$: CP-93129, 1.0 μg $5\text{-}HT_{1B}$: CP-94253, 0.56–1.0 μg (agonists)	CP-93129: ↓ maternal aggression (1.0 ug); no side effects CP-94253: no behavioral effect	Veiga et al. (2007)
	Alcohol-escalated aggression, male mice	$5\text{-}HT_{1B}$: CP-94253, 0.5–1.0 μg (agonist)	No effects on alcohol-related or baseline levels of aggression	Faccidomo et al. (2008)
	Social instigation-escalated aggression, male mice	$5\text{-}HT_{1A}$: 8-OH-DPAT, 0.1–1.0 μg $5\text{-}HT_{1B}$: CP-93129, 0.1–1.0 μg (agonists)	8-OH-DPAT: ↓ instigated aggression (0.56–1.0 μg); no side effects CP-93129: ↓ instigated aggression (only 0.1 μg); no side effects	Centenaro et al. (2008)

DRN dorsal raphé nucleus, *MRN* median raphé nucleus, *PAG* periaqueductal gray area

[a] doses calculated based on the following molecular weights (MW) of the compounds (as available at Sigma–Aldrich): 8-OH-DPAT hydrobromide, MW = 328.29; CGS-12066 maleate salt, MW = 450.41; DOI hydrochloride, MW = 357.62

respectively). Activation of 5-HT$_{1A}$ and 5-HT$_{1B}$ receptors in the DRN with microinfusion of selective receptor agonists consistently reduced aggressive behavior in rats and mice, but with concomitant reduction of motor activity and social interactions (Bannai et al. 2007; Faccidomo et al. 2008; Mos et al. 1993; Van Der Vegt et al. 2003). Infusion of a 5-HT$_{1A}$ agonist into the MRN also reduced aggressive behavior of lactating female rats (de Almeida and Lucion 1997). In projection sites of 5-HT neurons, 5-HT$_{1B}$ receptors likely modulate 5-HT release from synaptic terminals as autoreceptors, whereas both 5-HT$_{1A}$ and 5-HT$_{1B}$ receptors modulate postsynaptic neurons (Olivier et al. 1992). In most studies, local activation of 5-HT$_{1A}$ and 5-HT$_{1B}$ in projection regions (e.g., medial preoptic area, lateral septum, orbitofrontal cortex, anterior hypothalamus, medial hypothalamus, periacqueductal gray) reduced aggressive behavior in various procedures and species (see Table 4). Interestingly, under conditions that may escalate aggression, such as consumption of moderate doses of alcohol or maternal aggression, a 5-HT$_{1A}$ or 5-HT$_{1B}$ agonist further increased levels of aggressive behavior when infused into the medial prefrontal cortex (Faccidomo et al. 2008) or the medial septal area (de Almeida and Lucion 1997), respectively. Further studies are required to delineate the mechanisms for such pro-aggressive effects.

4.1.2 Genetic Approach

The 5-HT$_{1B}$ receptor was the first molecule to be linked to aggression by using the gene knockout technique. Male mice with disrupted 5-HT$_{1B}$ receptor expression (Htr1b$^{-/-}$) increased isolation-induced aggressive behavior compared to the wild-type mice (Bouwknecht et al. 2001b; Saudou et al. 1994). However, the aggressive behavior of the background strain of mice (129/Sv-ter) was very low, close to zero, and thus the frequency of attack bites in Htr1b$^{-/-}$ was low and the latency to initiate fighting was very long compared to other strains of mice. These mice displayed behavioral disinhibition in other behavioral tests including hyperlocomotor activity (Brunner et al. 1999; Ramboz et al. 1995), drug intake (Crabbe et al. 1996; Rocha et al. 1998), measures of anxiety-like behavior (Brunner et al. 1999; Malleret et al. 1999) and autonomic hyperreactivity to novelty (Bouwknecht et al. 2001a). Females of Htr1b$^{-/-}$ increased their aggressive behavior during the postpartum period (Brunner and Hen 1997). These results suggest a role for 5-HT$_{1B}$ receptors in the inhibition of some aggressive and impulsive behaviors. The level of tissue 5-HT concentration in Htr1b$^{-/-}$ mice was lower than wild-type in nucleus accumbens and spinal cord (Ase et al. 2000).

There are several polymorphisms in the human 5-HT$_{1B}$ receptor (HTR1B) gene. The most frequently studied HTR1B polymorphism in relation to aggressive phenotype is a single nucleotide polymorphism (SNP) G861C. This SNP had significant linkage with aggressive behavior in antisocial alcoholism (Lappalainen et al. 1998). Specifically, 861C, the SNP with lower 5-HT$_{1B}$ receptor expression (Huang et al. 1999), was related to antisocial behavior. This SNP is actually a synonymous polymorphism, in which no amino acid change occurred, and thus this SNP seems not to have any function by itself. Thus, G861C polymorphism may have a linkage with other functional polymorphisms that causes the expression change of 5-HT$_{1B}$ receptors. Jensen et al. (2009) showed that a SNP (A1997G) in a non-coding regulatory region of the HTR1B gene is actually related to the 5-HT$_{1B}$ expression pattern. This SNP is in the binding site for the microRNA miR-96 that inhibits the translation or degrades the HTR1B mRNA. College students homozygous for the A-allele, who have reduced 5-HT$_{1B}$ receptor expression, reported a greater history of aggressive behaviors than G-allele individuals. Also, another SNP in the 5′UTR region of the HTR1B gene, A161T, was found to correlate significantly with a history of aggression in subjects who completed violent suicides (Zouk et al. 2007). Individuals with the T161 locus had more lifetime aggressive behaviors, and again this polymorphism was linked to reduced transcriptional activity of 5-HT$_{1B}$ receptors (Sun et al. 2002). However, these associations between the HTR1B polymorphisms (G861C, G261T, or C129T) and aggression and antisocial behavior are not seen in other studies (Huang et al. 1999; Kranzler et al. 2002; New et al. 2001; Sinha et al. 2003; Van den Berg et al. 2008). Based on the findings from pharmacological studies and gene knockout and polymorphism studies, it is likely that the 5-HT$_{1B}$ receptor has an important role in the inhibition of certain types of aggression.

In contrast to the strong pharmacological evidence implicating the 5-HT$_{1A}$ receptor in aggression, no linkage has yet been reported between the 5-HT$_{1A}$ gene polymorphism and aggression. Also, deletion of 5-HT$_{1A}$ gene (Htr1a$^{-/-}$) reduced aggressive behavior in mice (Zhuang et al. 1999), which is the complete opposite of the findings with 5-HT$_{1A}$ agonists. However, there is evidence for a correlation between 5-HT$_{1A}$ receptor expression and aggression. A human PET study found a higher 5-HT$_{1A}$ receptor distribution in prefrontal cortex of subjects with higher aggression scores based on a self-report questionnaire (Witte et al. 2009). Also, mice selected for short latency to attack showed higher 5-HT$_{1A}$ receptor expression and receptor sensitivity (Korte et al. 1996; Van Der Vegt et al. 2001; van Riel et al. 2002; Veenema et al. 2005), while rats selected for higher defensive reactions showed reduced 5-HT$_{1A}$ receptor expression in several brain areas (Popova et al. 1998). It is possible that the polymorphisms which directly or indirectly affect 5-HT$_{1A}$ receptor transcription may be associated with either aggressive or defensive responses.

4.2 5-HT2 Family

4.2.1 Pharmacological Approach

Atypical antipsychotic agents (e.g., risperidone) with significant antagonist action at 5-HT$_{2A}$ receptors have been successfully used to reduce aggressive outbursts in patients diagnosed with various neuropsychiatric disorders (Buckley et al. 1997;

Buitelaar et al. 2001; Czobor et al. 1995; De Deyn et al. 1999; Fava 1997; Keck et al. 2000; Zarcone et al. 2001). However, some reports cast doubt on the routine use of antipychotics (Swanson et al. 2008; Tyrer et al. 2008). In one study, the placebo control group showed the greatest reduction in aggressively challenging behavior compared to antipsychotic drug treatments in people with intellectual disability (Tyrer et al. 2008). In animal models, selective 5-HT_{2A} antagonists (e.g., ketanserin, ritanserin and MDL 100907) reduce aggressive behaviors in a behaviorally non-specific manner (Rodriguez-Arias et al. 1998; Sakaue et al. 2002; Shih et al. 1999; White et al. 1991).

Activation of 5-HT_{2A} and 5-HT_{2C} receptors by DOI and other substituted phenylisopropylamines also reduces aggressive behavior in several species including flies, amphibians, mice and rats (Bonson et al. 1994; de Almeida and Lucion 1994; Johnson et al. 2009; Muehlenkamp et al. 1995; Olivier et al. 1995; Sanchez et al. 1993; Ten Eyck 2008). However, the anti-aggressive effects of 5-HT_2 ligands are accompanied by sedative effects in the same dose range. Local infusion of a 5-$HT_{2A/2C}$ agonist into the PAG reduces maternal aggression in rats (de Almeida et al. 2005), whereas microinjections into the medial hypothalamus and into the PAG increased defensive aggression in cats (Hassanain et al. 2003; Shaikh et al. 1997, see Table 4). This latter effect is likely linked to the role of 5-$HT_{2A/2C}$ receptors in anxiety-like behavior (Lucki and Wieland 1990; Nogueira and Graeff 1995). The development of a more selectively acting pharmacological tools will allow a more adequate differentiation of 5-HT_2 receptor subtypes, and promises to dissociate the anti-aggressive and sedative effects.

4.2.2 Genetic Approach

Platelet 5-HT_{2A} receptor binding is increased in patients with personality disorders and in a psychiatric population with greater lifetime aggression scores (Coccaro et al. 1997; McBride et al. 1994). Positive correlation between impulsive physical aggression and 5-HT_{2A} receptor expressions in orbitofrontal cortex has been reported using PET (Rosell et al. 2010) and a similar finding was reported in a postmortem study of suicide victims (Mann et al. 1986; Oquendo et al. 2006). However, another PET study reported opposite changes in 5-HT_{2A} receptor expression (Meyer et al. 2008). It is possible that polymorphisms that affect the level of expression of 5-HT_{2A} receptors can be associated with self-directed aggression. In some samples, a significant linkage was found between polymorphisms in the 5-HT_{2A} receptor (HTR2A) gene, T102C, A1438G and His452Tyr, and aggressive-impulsive trait or adolescent-onset antisocial behavior in humans (Assal et al. 2004; Bjork et al. 2002; Burt and Mikolaiewski 2008; Nomura et al. 2006), but others have reported no such link between aggression and HTR2A polymorphisms (Khait et al. 2005; Van den Berg tet al. 2008). Again, the successful pharmacotherapeutic management of aggressive patients using compounds with affinity for 5-HT_{2A} receptors would suggest that violence-prone individuals may be characterized by distinctive HTR2A polymorphisms.

Linkage of a minor polymorphism in the 5-HT_{2B} receptor (HTR2B) gene with antisocial behavior was reported recently. The HTR2B Q20* allele, which is found exclusively in the Finnish population, contains a stop codon in HTR2B and blocked the expression of the 5-HT_{2B} receptor. Males with the Q20* allele are prone to show more impulsive violence or impulsive suicidal behavior interacting strongly with alcohol than control (Bevilacqua et al. 2010). Male mice with disrupted 5-HT_{2B} receptors ($Ht2b^{-/-}$) also showed increased impulsive behavior. Interestingly, these mice had three times higher testosterone level in the cerebrospinal fluid, which is also consisted with the Q20* individuals among humans.

5 Serotonin Transporter (5-HTT)

5.1 Pharmacological Approach

Aggressive behavior in humans and animals can be reduced and prevented by blocking serotonin transporter molecules, which in turn presumably increases 5-HT levels in the brain. Clinically, chronic treatment (i.e., >3 weeks) with selective serotonin reuptake inhibitors (SSRIs) has been shown to reduce aggressive outbursts and violent behavior in psychiatric patients (Barkan et al. 2006; Blader 2006; Bond 2005; Coccaro and Kavoussi 1997; New et al. 2004; Reist et al. 2003; Walsh and Dinan 2001). However, SSRIs have occasionally been reported to increase the incidence of aggressive and suicidal behavior, and the causes of these paradoxical effects remain unknown (Spigset 1999; Troisi et al. 1995).

Both acute and chronic SSRI treatment can produce a dose-dependent reduction in aggressive behavior in animal models (Carrillo et al. 2009; Delville et al. 1996; Olivier et al. 1989; Pinna et al. 2003). When given acutely to rodents and non-human primates, several SSRIs including fluoxetine, fluvoxamine and sertraline reduced aggression in various contexts (Abbadie et al. 1994; Carrillo et al. 2009; Cutler et al. 1997; Delville et al. 1996; Fairbanks et al. 2001; Ferris et al. 1997; Fuller 1996; Ho et al. 2001; Sanchez and Meier 1997). Mice treated with the SSRI citalopram daily for three weeks no longer showed increased levels of aggression after consuming a moderate dose of alcohol, and baseline levels of aggression also showed modest reductions (Caldwell and Miczek 2008). On the other hand, the very low levels of agonistic behavior typically seen in laboratory rats may be restored to species-typical levels by chronic SSRI treatment (Mitchell et al. 1991; Mitchell 2005; Mitchell and Redfern 1992, 1997). Thus, SSRIs are more likely to exhibit anti-aggressive effects in conditions of escalated agonistic behavior, such as fighting enhanced by alcohol (Caldwell and Miczek 2008).

Mechanistically, extracellular levels of 5-HT in the prefrontal cortex of rats are elevated by both acute and chronic administration of citalopram or the more potent isomer escitalopram, implying that SSRIs' effects on aggression and other mood disorders is a result of increased cortical 5-HT (Ceglia et al. 2004). However, the anti-aggressive effects of fluoxetine, another SSRI, may be mediated primarily by acting on neurosteroids and GABA transmission, and only secondarily by elevating 5-HT (Pinna et al. 2003, 2006). In addition, long-term SSRI treatment probably recruits both pre- and post-synaptic mechanisms and neuroplastic events that may also have therapeutic actions (Benmansour et al. 1999; Blier and de Montigny 1998; Ceglia et al. 2004; Pineyro et al. 1994).

5.2 Genetic Approach

The serotonin-transporter-gene-linked polymorphic region (5-HTTLPR) is a variation in the length of the 5′-flanking transcriptional control region (promoter) of the 5-HTT gene that affects the gene's transcriptional activity. The variation has been found in humans (Heils et al. 1996) and also in great apes and rhesus monkeys (Lesch et al. 1997). The short-length (*s*) allele reduces 5-HTT expression in vitro compared to the long length homozygote (*l/l*) and lowers the prolactin response to clomipramine in humans, indicating reduced 5-HT function (Heils et al. 1995; Lesch et al. 1996; Whale et al. 2000). Studies in humans have found that males and females with one or two copies of the *s* allele (*s/s*, *s/l*) exhibit more hostility, aggression, anxiety and depression and lower agreeableness than *l/l* homozygotes (Lesch and Merschdorf 2000). "Type 2" alcoholics who displayed high impulsivity and antisocial behaviors showed a higher frequency of the *s* allele than either "Type 1" alcoholics without antisocial behavior or healthy controls (Hallikainen et al. 1999). As in humans, rhesus monkeys with the *s* allele were more aggressive than *l/l* individuals (Jarrell et al. 2008; Lesch and Merschdorf 2000).

5-HTT polymorphism also exhibits a genotype-environment interaction. Rhesus monkeys with the *s* allele that were peer-reared without their mothers had lower CSF levels of 5-hydroxyindoleacetic acid (5HIAA) than *l/l* individuals, but this difference was not present in maternally reared monkeys (Bennett et al. 2002). Peer-reared male monkeys also showed both altered CSF 5HIAA levels and an increase in aggression-related behavior (Higley et al. 1991; Kraemer et al. 1989). Humans carrying the *s* allele exhibited more suicidal ideations or attempts in response to stressful life events than *l/l* homozygotes, but did not differ in less stressful situations (Caspi et al. 2003). Therefore, it is possible that animals with the *s* allele are more vulnerable to stressful challenges, and subsequently escalate their aggressive behaviors toward others and themselves. However, the presence of the *s* allele did not affect human aggression consistently across sexes (Cadoret et al. 2003) or cultures (Baca-Garcia et al. 2004).

Deletion of the 5-HTT gene in mice (Slc6a4) produced seemingly contrary results. Wild-type C57BL/6J mice attacked more and showed shorter latencies to start fighting in the resident-intruder test than homozygote and heterozygote 5-HTT knockouts (Holmes et al. 2002, Table 3). 5-HTT knockout mice have higher extracellular 5-HT concentrations and lower 5-HT uptake in the forebrain compared to wild-type (Mathews 2004). Knockout mice with reduced or absent 5-HTT function also exhibited more than 50 phenotypic changes, including alterations of behavioral, physiological, morphological and sensory functions (Murphy and Lesch 2008), and those pleiotropic changes in various other phenotypes may contribute to the reduction of aggression in these mice. Comparable findings on 5-HTT and aggression have also been reported in rats; 5-HTT knockout rats on a Wistar/Crl background showed reduced offensive behaviors and longer attack latencies compared to wild-types (Homberg 2007). Thus genetic ablation of 5-HTT has consistently been shown to reduce aggressive behaviors in rodents.

6 Monoamine Oxidase A

6.1 Pharmacological Approach

Inhibition of MAO_A leads to a reduction in the oxidative metabolism of monoamines, presumably making 5-HT and other monoamines more available in the brain. Although the importance of MAO inhibitors as antidepressants was soon recognized, only a few studies have evaluated the effects of MAO inhibitors on aggression in preclinical models (Miczek 1987). Non-selective inhibitors of both MAO_A and MAO_B (e.g., phenelzine, isocarboxazid, tranylcypromine) produce acute anti-aggressive effects only at doses that also produce sedation and alter other non-aggressive behaviors (DaVanzo et al. 1966; Sofia 1969; Valzelli et al. 1967; Welch and Welch 1968). Non-selective MAO inhibitors or selective MAO_B inhibitors can be clinically useful for treating patients with personality disorders who exhibit suicidal tendencies and impulsive aggression, but the drugs also have a profile of undesirable side effects (Hollander 1999; Raj 2004).

6.2 Genetic Approach

The gene for monoamine oxidase A (MAOA) was the first to be identified as a possible determinant for pathological aggression in humans, and it has remained the focus of most genetic and epigenetic studies. A Dutch family with a syndrome of borderline mental retardation and dysregulated impulsive aggression was identified by Brunner et al. (1993b). Aggressive outbursts were seen in all affected males in the kindred, and some exhibited aberrant sexual behavior, attempted murder and arson. Linkage and sequence analyses identified one missense mutation in the MAOA gene on the X chromosome, so that MAOA function was completely disturbed; the affected males had more serotonin and lower levels of norepinephrine, dopamine and 5-HT metabolites in their urine (Brunner et al. 1993a). MAOA is also an important factor in animal aggression. Aggressive behaviors (skin wounds among cage mates and briefer attack latency in the resident-intruder test) escalated in male mice with a disrupted MAOA gene on either

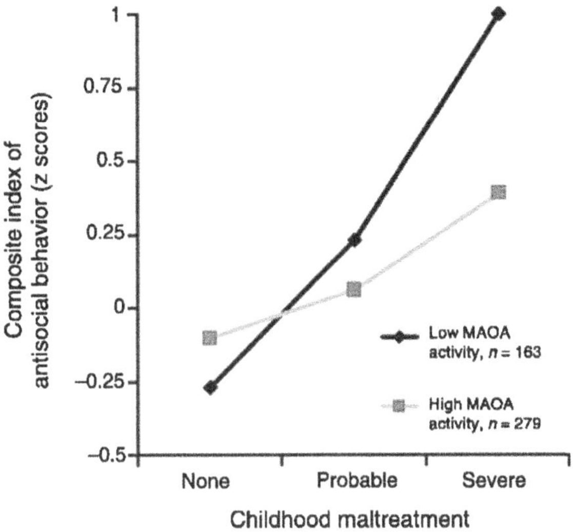

Fig. 4 Means on the composite index of antisocial behavior as a function of MAOA activity and a childhood history of maltreatment. MAOA activity is the gene expression level associated with allelic variants of the functional promoter polymorphism, grouped into low and high activity; childhood maltreatment is grouped into three categories of increasing severity. The antisocial behavior composite is standardized (z score) to a $M = 0$ and $SD = 1$; group differences are interpretable in SD unit differences (d). *Reprinted with permission from Caspi et al.* (2002)

C3H/He or 129 Sv background compared to wild-type mice (Cases et al. 1995; Scott et al. 2008). Mice with an MAOA deficiency also showed a large increase in 5-HT and norepinephrine and a slight elevation in dopamine in the brain and liver (Cases et al. 1995; Kim et al. 1997, Table 3); the behavioral changes in the MAOA-deficient mice are probably caused by this change in 5-HT function. Escalated aggression in MAOA mutant mice was blocked by 5-HT_{2A} receptor antagonists such as ketanserin and MDL100907 (Shih et al. 1999). Some of the behavioral and brain structural abnormalities in the MAOA-deficient mice were ameliorated when 5-HT was depleted by PCPA early in the developmental process (Cases et al. 1995, 1996).

MAOA expression can also be affected by variable-number tandem repeat (VNTR) polymorphism on the upstream region of the MAOA gene. The number of tandem repeats determines MAOA levels: alleles with 3.5 or 4 repeats have 2–10 times higher transcription than 3–5 repeat alleles in vitro (Denney et al. 1999; Sabol et al. 1998). A powerful interaction between MAOA genotype and environment on aggressive behavior has been reported (Caspi et al. 2002, Fig. 4). Individuals with low MAOA expression (MAOA-L) polymorphisms who suffered from abuse, neglect or traumatic life events in the first 15 years of their lives were more likely to have a history of adolescent conduct disorder, violence and criminal arrests, and also scored higher on aggressive disposition in a self-report questionnaire compared to MAOA-L individuals without abuse or trauma, or individuals with higher MAOA expression (MAOA-H) (Caspi et al. 2002; Foley et al. 2004; Frazzetto et al. 2007; Kim-Cohen et al. 2006; Weder et al. 2009; Widom and Brzustowicz 2006). The effect of MAOA genotype disappeared if all subjects were lumped together regardless of rearing environment (Fresan et al. 2007), and MAOA-H individuals sometimes reported higher aggression in interviews and on questionnaires (Manuck et al. 2000, 2002). The finding that individuals with the MAOA-L allele are vulnerable to environmental factors, showing a high propensity to engage in aggressive behaviors when in a stressful environment, is consistent in males but not females (Sjoberg et al. 2007). Rhesus monkeys have a similar repeat length variation polymorphism in the MAOA gene (rhMAOA-LPR) which is also linked to aggression. Monkeys with a low-activity allele who were reared by their mother showed more aggressive behavior and attained higher dominance rank than monkeys with the low-activity allele who were peer-reared separately from their parents (Newman et al. 2005). Higley and Suomi (1986) attributed this inhibition of aggression to increased fear and anxiety in peer-reared monkeys.

Substantial differences in both volume and activity of limbic system and neocortical areas between MAOA-L and MAOA-H individuals have been found in neuroimaging studies [see Buckholtz and Meyer-Lindenberg (2008) for a review]. In healthy male human volunteers with the MAOA-L variant, fMRI showed smaller limbic and orbitofrontal volumes and higher activity in amygdala and hippocampus during aversive recall (Meyer-Lindenberg et al. 2006), which may be related to violent behavior. Lower MAOA activity in cortical and subcortical brain areas is associated with elevated aggression as measured by a self-report questionnaire, with no effect of MAOA polymorphism (Alia-Klein et al. 2008). These data show that the MAOA activity is one determinant of the propensity to aggression. The interaction between MAOA polymorphism and stressful social experiences is especially critical, since it can escalate aggression and can also change relevant brain structures.

7 Modulation of Serotonergic Activity by Other Systems

The 5-HT neurons in the raphé nuclei are modulated by other amines, acids, peptides and steroids (Adell et al. 2002). Several efforts have been undertaken to uncover the nature of the neural systems that modulate 5-HT neurons to promote escalated aggressive behaviors. Here we will focus briefly on inhibitory and excitatory neurotransmitters and some neuropeptides in terms of their interaction with 5-HT system. Especially, dorsal raphé nucleus (DRN), the largest 5-HT nucleus in the brain, will be highlighted because of its important role for aggression (Bannai et al. 2007; Faccidomo et al. 2008; Jacobs and Cohen 1976; Koprowska and Romaniuk 1997; Mos et al. 1993; Sijbesma et al. 1991; Van Der Vegt et al. 2003; Vergnes et al. 1986).

7.1 Excitatory and Inhibitory Amino acids

7.1.1 γ-Aminobutyric Acid

γ-Aminobutyric Acid (GABA) is the major inhibitory neurotransmitter in the brain. Large numbers of GABA interneurons and distal GABAergic afferents are found in the DRN (Belin et al. 1983; Gervasoni et al. 2000; Nanopoulos et al. 1982; Wang et al. 1992), and both $GABA_A$ and $GABA_B$ receptors are expressed in the DRN (Bowery et al. 1987). In vivo electrophysiology studies have shown that the activation of either $GABA_A$ or $GABA_B$ receptors can inhibit the cell firing of serotonergic neurons (Colmers and Williams 1988; Gallager and Aghajanian 1976; Innis and Aghajanian 1987; Judge et al. 2004). On the other hand, in vivo microdialysis studies have shown that the 5-HT release may be differentially modulated by $GABA_A$ receptors and $GABA_B$ receptors depending on the projection sites (Tao et al. 1996). In addition, microinjection of a $GABA_B$ receptor agonist into the DRN can induce either increases or decreases of 5-HT neuronal activity (Abellan et al. 2000; Takahashi et al. 2010b; Tao et al. 1996).

Systemic administrations of positive modulators of $GABA_A$ receptors such as benzodiazepines, barbiturates and neurosteroids exert dose-dependent biphasic effects on aggressive behaviors, from escalation to inhibition (Miczek et al. 2003). Low to moderate doses of $GABA_A$ positive modulators escalate aggression in mice, rats, pigs and monkeys (Arnone and Dantzer 1980; Cole and Wolf 1970; Ferrari et al. 1997; Fish et al. 2001; Gourley et al. 2005; Miczek 1974; Miczek and O'Donnell 1980; Mos and Olivier 1989; Rodgers and Waters 1985; Weerts and Miczek 1996). On the other hand, pharmacological activation of $GABA_A$ receptors in the DRN inhibits aggressive behaviors in rats (Van Der Vegt et al. 2003) but has no effect in mice (Takahashi et al. 2010a). However, we found an interaction between alcohol and $GABA_A$ receptor in the DRN on aggressive behavior. Only under the influence of alcohol, local administration of $GABA_A$ receptor agonist muscimol in the DRN also heightened aggressive behaviors (Takahashi et al. 2010a). Therefore, $GABA_A$ modulation of serotonergic neurons may connect to drug-induced escalated aggression (e.g., alcohol, benzodiazepines) but not to species-typical aggression.

$GABA_B$ receptors in the DRN are also involved in the escalation of aggression in mice. Pharmacological activation of $GABA_B$ receptors in the DRN escalates intermale aggression (Takahashi et al. 2010b). The pro-aggressive effect of $GABA_B$ agonist baclofen was blocked by selective agonists of $GABA_B$ receptors (phaclofen, CGP54626). Additionally, the systemic administration of baclofen also escalates aggression in male mice. Therefore, both $GABA_A$ and $GABA_B$ receptors are involved in escalated aggression via different mechanism to modulate different types of aggression. In vivo microdialysis showed that $GABA_B$ activation in the DRN increased extracellular 5-HT level in the medial prefrontal cortex (Takahashi et al. 2010b, Fig. 5). This result suggests that the phasic activation of 5-HT system may be able to promote certain types of escalated aggressive behaviors in mice.

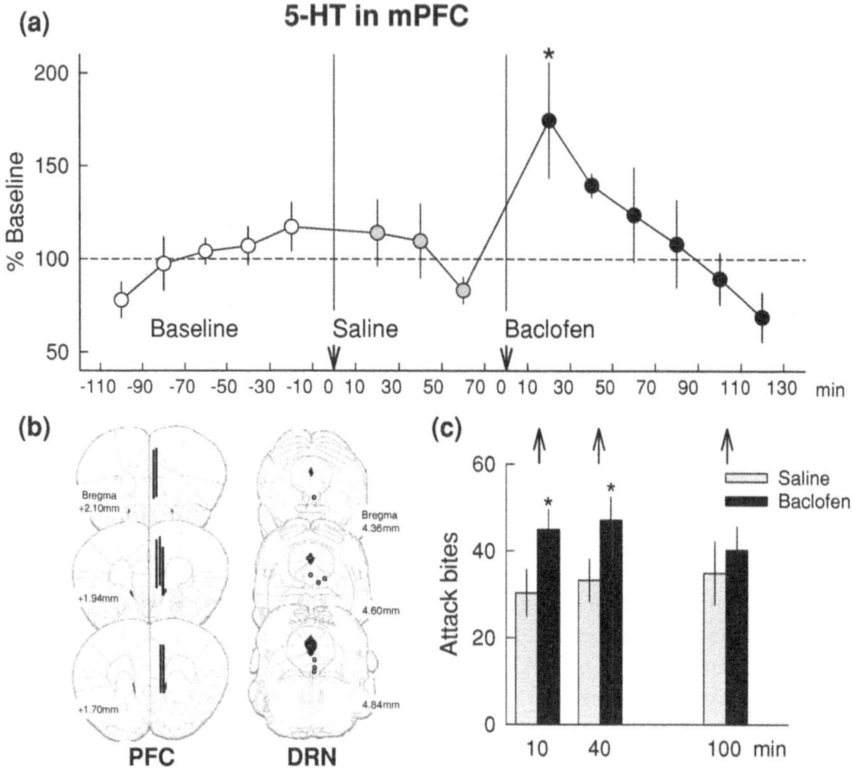

Fig. 5 Extracellular 5-HT concentration in the medial prefrontal cortex (mPFC) of mice after $GABA_B$ receptor activation in the dorsal raphe nucleus (DRN). **a** Baclofen microinjected into the DRN increased the 5-HT level in the mPFC whereas saline injection did not change the 5-HT level. Twenty-minute samples were collected: five samples for baseline, three samples after saline injection and six samples after baclofen (0.06 nmol) injection. Data are means ± SEM expressed as percentage of baseline ($n = 7$); *aisterisks* = $p < 0.05$ compared to baseline. **b** Histological representation of probe placement in the mPFC for the microdialysis (*vertical bars*: 2 mm probe membrane) and drug injection site in the DRN (*circles*). **c** The effect of 0.06 nmol baclofen (*black bars*) or saline (*gray bars*) on attack bites at different post-injection intervals (10, 40 and 100 min, corresponding to fractions 9, 11 and 14 in the microdialysis, respectively). Escalated attack bites were observed both 10 and 40 min after the intra-DRN baclofen injection. Values are means ±SEM; *asterisks* = $p < 0.05$compared to corresponding vehicle control. *From Takahashi et al.* (2010b)

7.1.2 Glutamate

The DRN receives glutamate input by the descending projections from the lateral habenula, periaqueductal gray, lateral hypothalamus, interpeduncular nucleus and medial prefrontal cortex (Aghajanian and Wang 1977; Behzadi et al. 1990; Kalen et al. 1986; Maciewicz et al. 1981). Both the N-metyl-D-aspartate (NMDA) and α-amino-3-hydroxy-5-methyl-4-isoxazolaproprionate/kainate (AMPA/kainate), ionotropic receptors for glutamate, are localized on serotonergic neurons and increase the 5-HT release in the DRN and its projection areas (Celada et al. 2001; Pallotta et al. 1998; Tao et al. 1996; Tao and Auerbach 2000; Vandermaelen et al. 1986). Metabotropic receptors of glutamate (mGluRs) are also found in the DRN, but their function on serotonergic neurotransmission and aggressive behavior needs to be characterized by further investigations. Systemic administrations of classic antagonists of NMDA receptors, including phencyclidine (PCP) and dizocilpine (MK-801), can increase aggressive behavior (Burkhalter and Balster 1979; Krsiak 1974; McAllister 1990; Musty and Consroe 1982; Rewerski et al. 1971; Wilmot et al. 1987), while other studies find that these compounds are suppressive and sedative due to their marked side effects (Belozertseva and Bespalov 1999; Lang et al. 1995; Miczek and Haney 1994; Tyler and Miczek 1982). Memantine and neramaxane block the channel of NMDA receptors with different characteristics and promote alcohol-heightened aggression in resident mice confronting an intruder. Anatomically discrete analysis is required to identify the sites of action for NMDA receptors that produce escalated aggressive behavior. The 5-HT system is one of the candidates, especially the descending glutamatergic projection from the medial prefrontal cortex (mPFC) to the DRN will be interesting to investigate. The prefrontal cortex (PFC) is implicated in the emotion regulation including aggression (Davidson et al. 2000; Miczek et al. 2007). This mPFC-DRN glutamatergic projection is involved in the controllability or emotion regulation (Amat et al. 2005) and it is possible that this pathway is also involved in the regulation of aggression.

7.2 Neuropeptides

7.2.1 Corticotropin-Releasing Factor

The DRN is innervated by corticotropin-releasing factor (CRF) immunoreactive fibers, and is the site for both subtypes of CRF receptors, CRF1 and CRF2 (Chalmers et al. 1995; Potter et al. 1994; Swanson et al. 1983). CRF, CRF receptors and other peptides of the CRF family (i.e., urocortins), play key modulatory roles on DRN 5-HT neurons (Valentino and Commons 2005). Electrophysiological and microdialysis studies consistently report that i.c.v. or intra-DRN microinjections of CRF, or drugs targeting CRF receptors, exert potent modulatory control over 5-HT neural firing (Kirby et al. 2000; Lowry et al. 2000), and 5-HT output to limbic, striatal and prefrontal cortical regions (Amat et al. 2004, 2005; Forster et al. 2008; Lukkes et al. 2008; Meloni et al. 2008; Price et al. 1998; Price and Lucki 2001).

CRF, CRF1 and CRF2 receptors are implicated in maternal and inter-male aggression in mice and hamsters (D'Anna et al. 2005; Farrokhi et al. 2004; Gammie et al. 2004, 2005, 2007; Gammie and Stevenson 2006). In rats, i.c.v. or intra-amygdala infusions of low doses of CRF itself facilitated or induced pro-aggressive effects (Elkabir et al. 1990; Tazi et al. 1987). Under conditions of escalated aggression promoted by moderate doses of alcohol in male mice, CRF1 receptors are a promising target for pharmacological intervention. Systemic administration of the antagonists of CRF1 receptors reduce alcohol-heightened aggression, but also reduce baseline levels of aggressive behavior (Quadros et al. 2009b). When locally administered into the DRN, CRF1 antagonists (e.g., CP-154526 or MTIP) prevent the escalated levels of aggression observed after consumption of alcohol, with no detectable side effects on other behaviors. Remarkably, such anti-aggressive effects of CRF1 antagonists can be abolished with the infusion of 8-OH-DPAT into the DRN, which transiently slows 5-HT impulse flow. On the other hand, microinfusion of a CRF2 antagonist (e.g., Astressin-2B) into the DRN escalates aggressive behavior (Quadros et al. 2009a).

The modulation of aggressive behaviors by CRF systems depends on the species such as mice or rats, and type of aggression (species-typical, maternal or escalated aggression). Initial evidence suggests the 5-HT cells in the DRN as one of the critical sites for such modulation in escalated aggression that is promoted by alcohol, with presumably opposing roles for CRF1 and CRF2 receptors.

7.2.2 Neuropeptide Y

Neuropeptide Y (NPY) controls primarily food intake, energy balance and metabolic regulation (Herzog 2003). In addition to the critical role for energy homeostasis, NPY is also implicated in aggressive behavior (Emeson and Morabito 2005). Male mice with deleted expression of Y1 receptor ($Y1^{-/-}$) showed obesity and reduced energy homeostasis (Kushi et al. 1998). These animals also exhibited increased aggressive behaviors in the resident-intruder test (Karl et al. 2004). However, escalated aggression was observed only in the home-cage but not in the novel environment in $Y1^{-/-}$ mice. Both NPY and its receptors can be found in the DRN (Martel et al. 1990; O'Donohue et al. 1985; Saria et al. 1984), and NPY inhibits both hyperpolarizing and depolarizing slow synaptic potentials in the DRN (Kombian and Colmers 1992). $Y1^{-/-}$ mice showed reduced expression of tryptophan hydroxylase mRNA in the raphé nuclei, and also 5-HT_{1A} agonists could suppress the heightened aggression of $Y1^{-/-}$ mice (Karl et al. 2004). These results suggest that the disruption of Y1 receptor expression specifically increases territorial aggression by altering 5-HT function.

7.2.3 Arginine Vasopressin

Arginine vasopressin (AVP) is a neuropeptide that modulates a variety of social behaviors including pair-bonding, social recognition, maternal behavior, and aggression (Albers and Bamshad 1998; Coccaro et al. 1998; Ferris et al. 1992;

Goodson 2008; Koolhaas et al. 1990; Neumann et al. 2010; Winslow and Insel 1993). Selective antagonists of vasopressin V1a receptors (SRX251, [d(CH2) 5Tyr(Me)AVP]) inhibited inter-male aggression (Ferris et al. 2006; Ferris and Potegal 1988), suggesting the involvement of V1a receptors in aggressive behaviors. Evidence points to a critical role of the interaction between AVP and 5-HT in certain types of aggressive behaviors. In humans, a positive correlation has been observed between AVP concentrations in the CSF and the life history of aggression, a composite measure of trait aggression. Also, there was a positive correlation between AVP concentrations and prolactin responses to a challenge with d-fenfluramine (Coccaro et al. 1998). This result indicates that individuals that have higher aggression ratings tend to have a high AVP concentration in the CSF and a hyporesponsive 5-HT system. Neuronal interactions between AVP containing neurons and 5-HT neurons are found in the anterior hypothalamus (Ferris et al. 1997, 1999), and this AVP-5-HT link is implicated in aggression. Microinjection of AVP into the anterior or lateral hypothalamus increased aggressive behavior in hamsters, and systemic fluoxetine, a 5-HTT inhibitor, blocked the pro-aggressive effect of AVP (Delville et al. 1996; Ferris et al. 1997) Therefore, 5-HT may have an inhibitory effect on the AVP-heightened aggression. By contrast, mice with disrupted Ca^{2+} channel expression ($Cav2^{-/-}$) showed escalated level of aggressive behavior and also higher AVP concentration in the CSF. However, these animals showed over-activation of the dorsal 5-HT neurons and increased 5-HT concentration in the hypothalamus (Kim et al. 2009a). Further investigation will uncover how AVP and 5-HT interact and whether there are specific types of aggression that require the activity of either 5-HT or AVP systems independently.

7.2.4 Oxytocin

Oxytocin, which has a structure similar to vasopressin, differing in only two amino acid positions, also plays an important role in the control of social behaviors like vasopressin. Administration of the agonists of oxytocin receptor increased aggressive behavior in lactating females (Bosch et al. 2005; Ferris et al. 1992) but reduced territorial aggression in males (Harmon et al. 2002). Mice in which the gene for oxytocin and its receptor were knocked out, showed enhanced aggressive behaviors in both males and lactating females (Ragnauth et al. 2005; Takayanagi et al. 2005; Winslow et al. 2000) but see (DeVries et al. 1997). In humans, the level of oxytocin in the CSF is inversely correlated with the life history of aggression (Lee et al. 2009). Oxytocin and its receptor are expressed in the DRN, and the effect of local infusion of oxytocin can facilitate passive avoidance in rats (Kovács et al. 1979). However, the exact interactions between the oxytocin and serotonergic systems in their modulatory effects on aggressive behavior remain to be discovered.

7.3 Other Molecules that Directly or Indirectly Affect 5-HT Pathways and Aggression

7.3.1 Tryptophan Hydroxylase

Variations in Tryptophan hydroxylase (TPH) activity can directly affect 5-HT neurotransmission. Polymorphisms in the TPH1 or TPH2 gene change the enzyme activity and 5-HT level (Jonsson et al. 1997) for TPH1 gene; (Cichon et al. 2008; Lin et al. 2007; Zhang et al. 2004, 2005) for TPH2 gene. Two SNPs in tight linkage disequilibrium, A218C and A779C, located in the intron of TPH1 gene [which are often referred to as U (upper) and L (lower) allele], have been shown to be associated with state and trait anger as assessed by interview and self-report questionnaires (Manuck et al. 1999; Rujescu et al. 2002). Lower CSF 5-HIAA levels have been observed in healthy men, but not women, with the U allele (Jonsson et al. 1997), whereas the lowest CSF 5-HIAA was observed in LL carriers diagnosed with antisocial alcoholism (Nielsen et al. 1994). Apparently, both U and L alleles may relate to different types of aggression. Hennig et al. (2005) performed a factor analysis on the Buss-Durkee Hostility Inventory (BDHI) and found two factors, named as "neurotic hostility" and "aggressive hostility". Neurotic hostility is characterized by irritability, resentment, verbal hostility and guilt, whereas aggressive hostility represents a more "cold" aggression including assault and negativism but low guilt. Individuals with the UU allele showed higher aggressive hostility but slightly lower neurotic hostility (Hennig et al. 2005), whereas LL homozygote showed higher level of impulsivity in the measurements which focused more on neurotic hostility (New et al. 1998). This may explain controversial results about the association between the polymorphisms and aggression; depending on which dimensions of aggressive behavior are measured, the association pattern may change.

In mouse models, the focus has been on the Tph2 gene, because Tph2 is preferentially expressed in the brain, whereas Tph1 is more widely distributed throughout the body (Walther et al. 2003). Mice with an R439H point mutation in the Tph2 gene have strongly reduced 5-HTP, 5-HT and its acidic metabolite levels in (Beaulieu et al. 2008). This mutant mouse showed increased aggressive behaviors in the social interaction test in a neutral arena where typically very low levels of aggression are seen (Beaulieu et al. 2008). In contrast, another polymorphism in Tph2, C1473G, which inhibits Tph2 activity in the midbrain, reduced the intensity of aggression in mice (Osipova et al. 2009). The mouse data on the R439H mutation in the Tph2 gene may be relevant to a large association study in children with attention deficit/hyperactivity disorder (ADHD), which identified a TPH2 gene polymorphism associated with impulsivity (Oades et al. 2008).

7.3.2 Pet-1

Pet-1 (also known as Fev), one of the transcription factors, is specifically expressed in the serotonergic raphé neurons, and has a critical role in 5-HT neural development. Deletion of Pet-1 expression reduced 5-HT levels in the forebrain, and also depleted expression of TPH, 5-HTT, and the vesicular monoamine transporter 2 (Vmat2). In the resident-intruder test, Pet-1 knockout mice (Pet-1-/-) engaged in a higher frequency and intensity of attacks toward a conspecific male (Hendricks et al. 2003). These increases in aggressive behavior in Pet-1-/- mice are embedded in broad behavioral disruptions that also extended to maternal and anxiety-like behaviors (Hendricks et al. 2003; Lerch-Haner et al. 2008), and it is possible that those other behavioral changes promote indirectly aggressive behaviors. Since Pet-1 expresses specifically in the 5-HT neurons, the promoter or enhancer regions of this gene are useful for gene manipulation of 5-HT neurons (Scott et al. 2005).

7.3.3 Brain-Derived Neurotrophic Factor

Brain-derived neurotrophic factor (BDNF) has several important roles in neuronal survival, development, differentiation and plasticity. Mice with decreased BDNF expression including knockout ($BDNF^{+/-}$) and conditional knockout ($BDNF^{2L/1LNes-Cre}$ and $BDNF^{2L/2LCk-Cre}$) all showed increased inter-male aggression (Chan et al. 2006; Lyons et al. 1999). Higher extracellular levels of 5-HT in the hippocampus was observed in $BDNF^{+/-}$ mice compared to wild-type (Deltheil et al. 2008). However, fluoxetine could reduce heightened aggressive behavior in $BDNF^{+/-}$ mice (Lyons et al. 1999). All mutants changed $5\text{-}HT_{2A}$ receptor expression, however $BDNF^{+/-}$ showed increased $5\text{-}HT_{2A}$ expression in the lateral frontal cortex and hypothalamus (Lyons et al. 1999) whereas $BDNF^{2L/1LNes-Cre}$ and $BDNF^{2L/2LCk-Cre}$ exhibit reduced $5\text{-}HT_{2A}$ receptor expression in the prefrontal cortex (Chan et al. 2006; Rios et al. 2006). A SNP in the BDNF gene, Val66Met, has attracted considerable interest because of its association with mood disorders and hippocampal function in humans (Egan et al. 2003; Neves-Pereira et al. 2002). Knockin mice with this human Met allele also showed increased aggressive behavior and changed the response to SSRI treatment (Chen et al. 2006). Conditional BDNF knockout mice using kainate receptor promoter lack BDNA expression especially in hippocampal CA3 area. These mice also showed heightened aggression compared to wild-type, suggesting an important role of hippocampal BDNF in aggression (Ito et al. 2011). In contrast to the consistent results on aggressive behavior among BDNF mutant mice, studies on polymorphisms of the BDNF gene in aggressive behavior in humans remain to be resolved. No association was observed between Val66Met polymorphism and proneness to violence in a Chinese male sample (Tsai et al. 2005). Other SNPs in the BDNF gene may be associated with high impulsivity in children with ADHD (Oades et al. 2008). A further role of BDNF is evident in neuroadaptions in the

VTA-accumbens pathway of mice and rats that are repeatedly attacked and show evidence for defeat, submission and disrupted social interactions (Berton et al. 2006; Miczek et al. 2011).

7.3.4 Neuronal Nitric Oxide

Nitric oxide, a free radical gas which diffuses across membranes, is involved in several cellular functions [for review, see Calabrese et al. (2007)]. Mice lacking neuronal nitric oxide synthase (nNOS$^{-/-}$) show various deficits in their physiological development and also behavior (Huang et al. 1993). nNOS$^{-/-}$ males, but not females, showed higher duration of aggressive behavior and also displayed much fewer submissive postures compared to wild-types (Nelson et al. 1995). Serotonergic dysfunctions were observed in the nNOS$^{-/-}$ mice, specifically reduced 5-HT turnover in the brain and deficient 5-HT$_{1A}$ and 5-HT$_{1B}$ receptor function (Chiavegatto et al. 2001). Escalated aggression in the nNOS$^{-/-}$ was rescued by 5-HTP treatment which increased 5-HT level and turnover. These findings point to an important role of nitric oxide for the normal 5-HT function, and thus increased aggression nNOS$^{-/-}$ may be induced by changing 5-HT activity.

7.3.5 α-Calcium-Calmodulin Kinase II

α-Calcium-calmodulin kinase II (α-CaMKII) is a neural specific enzyme and has been shown to be involved in long-term potentiation (LTP) (Silva et al. 1992). Heterozygotes of α-CaMKII knockout mice showed escalated defensive aggression but not offensive aggression (Chen et al. 1994). In the resident-intruder test, resident α-CaMKII heterozygotes showed similar aggressive behavior as wild-type mice. In contrast, when the mutant was tested as an intruder, they exhibited highly defensive reactions toward the resident. Reduced 5-HT release was observed in the dorsal raphé of α-CaMKII mutant in vitro, and thus the changed 5-HT function may be associated with defensive aggression in this mouse.

8 Concluding Remarks

- Activation of 5-HT$_{1A}$, 5-HT$_{1B}$ and 5-HT$_{2A/2C}$ receptors in mesocorticolimbic areas reduce species-typical and other aggressive behaviors. In contrast, agonists at 5-HT$_{1A}$ and 5-HT$_{1B}$ receptors in the medial prefrontal cortex or septal area can increase aggressive behavior under specific conditions. 5-HT exerts a complex role in aggression that depends on (1) the subtypes of receptors and the brain in which they are expressed (2) the types of aggressive behaviors, and

(3) the trait characteristics of the human subject or the mouse (i.e., *Mus musculus* is a pugnacious species and many inbred strains are quite placid).

- Based on genetic studies, it is likely that the 5-HT_{1B} receptor has an important role in the inhibition of aggression. For other receptor subtypes such as 5-HT_{1A} or 5-HT_{2A} receptors, the association between aggressive behavior and polymorphisms in receptor genes all failed to have a reliable relationship compared to relatively consistent pharmacological data. It may be important to determine where in the brain of aggressive individuals the gene expression is changed by the polymorphisms.
- An important role of MAOA in aggressive behaviors is supported by genetic studies of deleterious mutation of MAOA gene in humans and knockout mice. In contrast to the serotonin deficiency hypothesis, these results suggest that chronically increased 5-HT levels due to reduced MAOA function—trait-like change—may promote or intensify escalated aggressive displays.
- Consistent anti-aggressive effects of SSRI treatment in several species strongly point to the serotonin transporter as a promising therapeutic target for management of impulsive, escalated aggression. More studies are required to determine the precise neurobiological mechanisms recruited for the anti-aggressive effects of SSRIs to emerge, especially as a result of chronic treatment that induces pre- and post-synaptic neuroadaptive processes.
- The genetic studies in human and nonhuman primates also suggest 5-HTT polymorphisms such as the short allele as a risk factor for violent traits, which seems to be particularly relevant in combination with environmental stress. Results from transgenic mice lacking 5-HTT are more difficult to interpret due to the wide variety of behavioral functions that are affected by the genetic manipulation.
- Gene polymorphism studies of MAOA, 5-HTT and Tph2 revealed critical gene-environment interactions as a risk factor for violent traits. Individuals with a certain allele are particularly prone to engage in violent behavior when they have a history of early life maltreatment, but the effect disappeared when they are reared in an environment with low stress.
- Modulatory mechanisms of serotonergic neurons that induce escalated aggression have gradually been identified. So far, evidence points to the modulations of activity of dorsal raphé nucleus by GABA, glutamate, CRF and its autoreceptors as being particularly relevant to the display of escalated aggression. Genetic analyses of aggressive individuals have identified several molecules that affect the 5-HT system directly (e.g., Tph2, 5-HT1B, 5-HT transporter, Pet1, MAOA) or indirectly (e.g., Neuropeptide Y, αCaMKII, NOS, BDNF).

In the current phase of advanced molecular genetic studies, it is necessary to remind ourselves to match them with more sophisticated behavioral methodologies of aggression. It is evident that the neurogenetic mechanisms differentiate types and patterns of aggressive behavior requiring higher resolution. The current focus on brain serotonin takes advantage of the most intensively studied neural system that has been implicated in the neurogenetics of aggression. It remains astounding how such trace amounts of this indole amine can engender such profound changes in aggressive traits and states.

References

Abbadie C, Honore P, Fournie-Zaluski MC, Roques BP, Besson JM (1994) Effects of opioids and non-opioids on c-fos-like immunoreactivity induced in rat lumbar spinal cord neurons by noxious heat stimulation. Eur J Pharmacol 258:215–227

Abellan MT, Jolas T, Aghajanian GK, Artigas F (2000) Dual control of dorsal raphe serotonergic neurons by $GABA_B$ receptors. Electrophysiological and microdialysis studies. Synapse 36:21–34

Adell A, Celada P, Abellan MT, Artigas F (2002) Origin and functional role of the extracellular serotonin in the midbrain raphe nuclei. Brain Res Rev 39:154–180

Adell A, Celada P, Artigas F (2001) The role of 5-HT_{1B} receptors in the regulation of serotonin cell firing and release in the rat brain. J Neurochem 79:172–182

Aghajanian GK, Wang RY (1977) Habenular and other midbrain raphe afferents demonstrated by a modified retrograde tracing technique. Brain Res 122:229–242

Albers HE, Bamshad M (1998) Role of vasopressin and oxytocin in the control of social behavior in Syrian hamsters (*Mesocricetus auratus*). Prog Brain Res 119:395–408

Alenina N, Kikic D, Todiras M, Mosienko V, Qadri F, Plehm R, Boye P, Vilianovitch L, Sohr R, Tenner K, Hortnagl H, Bader M (2009) Growth retardation and altered autonomic control in mice lacking brain serotonin. Proc Natl Acad Sci USA 106:10332–10337

Alia-Klein N, Goldstein RZ, Kriplani A, Logan J, Tomasi D, Williams B, Telang F, Shumay E, Biegon A, Craig IW, Henn F, Wang GJ, Volkow ND, Fowler JS (2008) Brain monoamine oxidase a activity predicts trait aggression. J Neurosci 28:5099–5104

Alleva E, Cirulli F, Bianchi M, Bondiolotti GP, Chiarotti F, De Acetis L, Panerai AE (1998) Behavioural characterization of interleukin-6 overexpressing or deficient mice during agonistic encounters. Eur J Neurosci 10:3664–3672

Amat J, Baratta MV, Paul E, Bland ST, Watkins LR, Maier SF (2005) Medial prefrontal cortex determines how stressor controllability affects behavior and dorsal raphe nucleus. Nat Neurosci 8:365–371

Amat J, Tamblyn JP, Paul ED, Bland ST, Amat P, Foster AC, Watkins LR, Maier SF (2004) Microinjection of urocortin 2 into the dorsal raphe nucleus activates serotonergic neurons and increases extracellular serotonin in the basolateral amygdala. Neuroscience 129:509–519

Arnone M, Dantzer R (1980) Effects of diazepam on extinction induced aggression in pigs. Pharmacol Biochem Behav 13:27–30

Ase AR, Reader TA, Hen R, Riad M, Descarries L (2000) Altered serotonin and dopamine metabolism in the CNS of serotonin 5-HT_{1A} or 5-HT_{1B} receptor knockout mice. J Neurochem 75:2415–2426

Assal F, Alarcon M, Solomon EC, Masterman D, Geschwind DH, Cummings JL (2004) Association of the serotonin transporter and receptor gene polymorphisms in neuropsychiatric symptoms in Alzheimer disease. Arch Neurol 61:1249–1253

Baca-Garcia E, Vaquero C, Diaz-Sastre C, García-Resa E, Saiz-Ruiz J, Fernández-Piqueras J, de Leon J (2004) Lack of association between the serotonin transporter promoter gene polymorphism and impulsivity or aggressive behavior among suicide attempters and healthy volunteers. Psychiatry Res 126:99–106

Bannai M, Fish EW, Faccidomo S, Miczek KA (2007) Anti-aggressive effects of agonists at $5\text{-}HT_{1B}$ receptors in the dorsal raphe nucleus of mice. Psychopharmacology 193:295–304
Barkan T, Peled A, Modai I, Weizman A, Rehavi M (2006) Characterization of the serotonin transporter in lymphocytes and platelets of schizophrenia patients treated with atypical or typical antipsychotics compared to healthy individuals. Eur Neuropsychopharmacol 16: 429–436
Beaulieu JM, Zhang X, Rodriguiz RM, Sotnikova TD, Cools MJ, Wetsel WC, Gainetdinov RR, Caron MG (2008) Role of GSK3β in behavioral abnormalities induced by serotonin deficiency. Proc Nat Acad Sci USA 105:1333–1338
Beck AT, Ward CH, Mendelsohn M, Mock J, Erbaugh J (1961) An inventory for measuring depression. Arch Gen Psychiatry 4:561–571
Behzadi G, Kalen P, Parvopassu F, Wiklund L (1990) Afferents to the median raphe nucleus of the rat: retrograde cholera toxin and wheat germ conjugated horseradish peroxidase tracing, and selective D-[^{3}H] aspartate labelling of possible excitatory amino acid inputs. Neuroscience 37:77–100
Belin MF, Nanopoulos D, Didier M, Aguera M, Steinbusch H, Verhofstad A, Maitre M, Pujol JF (1983) Immunohistochemical evidence for the presence of gamma-aminobutyric acid and serotonin in one nerve cell. A study on the raphe nuclei of the rat using antibodies to glutamate decarboxylase and serotonin. Brain Res 275:329–339
Bell R, Hobson H (1994) $5\text{-}HT_{1A}$ receptor influences on rodent social and agonistic behavior: a review and empirical study. Neurosci Biobehav Rev 18:325–338
Belozertseva IV, Bespalov AY (1999) Effects of NMDA receptor channel blockade on aggression in isolated male mice. Aggress Behav 25:381–396
Benmansour S, Cecchi M, Morilak DA, Gerhardt GA, Javors MA, Gould GG, Frazer A (1999) Effects of chronic antidepressant treatments on serotonin transporter function, density, and mRNA level. J Neurosci 19:10494–10501
Bennett AJ, Lesch KP, Heils A, Long JC, Lorenz JG, Shoaf SE, Champoux M, Suomi SJ, Linnoila MV, Higley JD (2002) Early experience and serotonin transporter gene variation interact to influence primate CNS function. Mol Psychiatry 7:118–122
Berkowitz L (1993) Aggression. Its causes, consequences and control. Mc Graw Hill, New York
Bernstein IS, Rose RM, Gordon TP (1974) Behavioral and environmental events influencing primate testosterone levels. J Hum Evol 3:517–525
Berton O, McClung CA, DiLeone RJ, Krishnan V, Renthal W, Russo SJ, Graham D, Tsankova NM, Bolanos CA, Rios M, Monteggia LM, Self DW, Nestler EJ (2006) Essential role of BDNF in the mesolimbic dopamine pathway in social defeat stress. Science 311:864–868
Bevilacqua L, Doly S, Kaprio J, Yuan Q, Tikkanen R, Paunio T, Zhou Z, Wedenoja J, Maroteaux L, Diaz S, Belmer A, Hodgkinson CA, Dell'osso L, Suvisaari J, Coccaro E, Rose RJ, Peltonen L, Virkkunen M, Goldman D (2010) A population-specific HTR2B stop codon predisposes to severe impulsivity. Nature 468:1061–1066
Bjork JM, Moeller FG, Dougherty DM, Swann AC, Machado MA, Hanis CL (2002) Serotonin 2a receptor T102C polymorphism and impaired impulse control. Am J Med Genet 114:336–339
Blader JC (2006) Pharmacotherapy and postdischarge outcomes of child inpatients admitted for aggressive behavior. J Clin Psychopharmacol 26:419–425
Blanchard DC, Rodgers RJ, Hendrie CA, Hori K (1988) 'Taming' of wild rats (*Rattus rattus*) by $5HT_{1A}$ agonists buspirone and gepirone. Pharmacol Biochem Behav 31:269–278
Blanchard RJ, Blanchard CD (1977) Aggressive behavior in the rat. Behav Biol 21:197–224
Blanchard RJ, Hori K, Blanchard DC, Hall J (1987) Ethanol effects on aggression of rats selected for different levels of aggressiveness. Pharmacol Biochem Behav 27:641–644
Blanchard RJ, Wall PM, Blanchard DC (2003) Problems in the study of rodent aggression. Horm Behav 44:161–170
Blier P, de Montigny C (1998) Possible serotonergic mechanisms underlying the antidepressant and anti-obsessive-compulsive disorder responses. Biol Psychiatry 44:313–323
Boles SM, Miotto K (2003) Substance abuse and violence—a review of the literature. Aggress Violent Behav 8:155–174
Bond AJ (2005) Antidepressant treatments and human aggression. Eur J Pharmacol 526:218–225

Bonson KR, Johnson RG, Fiorella D, Rabin RA, Winter JC (1994) Serotonergic control of androgen-induced dominance. Pharmacol Biochem Behav 49:313–322

Bonvento G, Scatton B, Claustre Y, Rouquier L (1992) Effect of local injection of 8-OH-DPAT into the dorsal or median raphe nuclei on extracellular levels of serotonin in serotonergic projection areas in the rat brain. Neurosci Lett 137:101–104

Bosch OJ, Meddle SL, Beiderbeck DI, Douglas AJ, Neumann ID (2005) Brain oxytocin correlates with maternal aggression: link to anxiety. J Neurosci 25:6807–6815

Bouwknecht JA, Hijzen TH, van der GJ, Maes RA, Hen R, Olivier B (2001a) Absence of 5-HT_{1B} receptors is associated with impaired impulse control in male 5-HT_{1B} knockout mice. Biol Psychiatry 49:557–568

Bouwknecht JA, van der GJ, Hijzen TH, Maes RA, Hen R, Olivier B (2001b) Corticosterone responses in 5-HT_{1B} receptor knockout mice to stress or 5-HT_{1A} receptor activation are normal. Psychopharmacology 153:484–490

Bowery NG, Hudson AL, Price GW (1987) $GABA_A$ and $GABA_B$ receptor site distribution in the rat central nervous system. Neuroscience 20:365–383

Brain P, Benton D (1979) The interpretation of physiological correlates of differential housing in laboratory rats. Life Sci 24:99–115

Brain PF (1979) Hormones, drugs and aggression. Eden Press, Montreal

Brown GL, Goodwin FK (1986) Cerebrospinal fluid correlates of suicide attempts and aggression. In: Mann JJ (ed) Psychobiology of suicidal behavior (Annals of the New York Academy of Sciences, Vol 487). New York Academy of Sciences, New York, pp 175–220

Brown GL, Goodwin FK, Ballenger JC, Goyer PF, Major LF (1979) Aggression in humans correlates with cerebrospinal fluid amine metabolites. Psychiatry Res 1:131–139

Brunner D, Buhot MC, Hen R, Hofer M (1999) Anxiety, motor activation, and maternal-infant interactions in $5HT_{1B}$ knockout mice. Behav Neurosci 113:587–601

Brunner D, Hen R (1997) Insights into the neurology of impulsive behavior from serotonin receptor knockout mice. Ann N Y Acad Sci 836:81–105

Brunner HG, Nelen M, Breakefield XO, Ropers HH, van Oost BA (1993a) Abnormal behavior associated with a point mutation in the structural gene for monoamine oxidase A. Science 262:578–580

Brunner HG, Nelen MR, Vanzandvoort P, Abeling NGGM, Vangennip AH, Wolters EC, Kuiper MA, Ropers HH, Vanoost BA (1993b) X-linked borderline mental-retardation with prominent behavioral disturbance—phenotype, genetic localization, and evidence for disturbed monoamine metabolism. Am J Hum Genet 52:1032–1039

Buckholtz JW, Meyer-Lindenberg A (2008) MAOA and the neurogenetic architecture of human aggression. Trends Neurosci 31:120–129

Buckley PF, Ibrahim ZY, Singer B, Orr B, Donenwirth K, Brar PS (1997) Aggression and schizophrenia: efficacy of risperidone. J Am Acad Psychiatry Law 25:173–181

Buitelaar JK, van der Gaag RJ, Cohen-Kettenis P, Melman TM (2001) A randomized controlled trial of risperidone in the treatment of aggression in hospitalized adolescents with subaverage cognitive abilities. J Clin Psychiatry 62:239–248

Burkhalter JE, Balster RL (1979) The effects of phencyclidine on isolation-induced aggression in mice. Psychol Rep 45:571–576

Burt SA, Mikolaiewski AJ (2008) Preliminary evidence that specific candidate genes are associated with adolescent-onset antisocial behavior. Aggress Behav 34:437–445

Buss AH (1961) The psychology of aggression. Wiley and Sons, New York

Buss AH, Durkee A (1957) An inventory for assessing different kinds of hostility. J Consult Psychol 21:343–349

Cadoret RJ, Langbehn D, Caspers K, Troughton EP, Yucuis R, Sandhu HK, Philibert R (2003) Associations of the serotonin transporter promoter polymorphism with aggressivity, attention deficit, and conduct disorder in an adoptee population. Compr Psychiat 44:88–101

Cairns RB, Nakelski JS (1971) On fighting in mice: ontogenetic and experiential determinants. J Comp Physiol Psychol 74:354–364

Calabrese V, Mancuso C, Calvani M, Rizzarelli E, Butterfield DA, Stella AM (2007) Nitric oxide in the central nervous system: neuroprotection versus neurotoxicity. Nat Rev Neurosci 8: 766–775
Caldwell EE, Miczek KA (2008) Long-term citalopram maintenance in mice: selective reduction of alcohol-heightened aggression. Psychopharmacology 196:407–416
Carrillo M, Ricci LA, Coppersmith GA, Melloni RH Jr (2009) The effect of increased serotonergic neurotransmission on aggression: a critical meta-analytical review of preclinical studies. Psychopharmacology 205:349–368
Cases O, Seif I, Grimsby J, Gaspar P, Chen K, Pournin S, Muller U, Aguet M, Babinet C, Shih JC, De Maeyer E (1995) Aggressive behavior and altered amounts of brain serotonin and norepinephrine in mice lacking MAOA. Science 268:1763–1766
Cases O, Vitalis T, Seif I, DeMaeyer E, Sotelo C, Gaspar P (1996) Lack of barrels in the somatosensory cortex of monoamine oxidase A-deficient mice: role of a serotonin excess during the critical period. Neuron 16:297–307
Caspi A, McClay J, Moffitt TE, Mill J, Martin J, Craig IW, Taylor A, Poulton R (2002) Role of genotype in the cycle of violence in maltreated children. Science 297:851–854
Caspi A, Sugden K, Moffitt TE, Taylor A, Craig IW, Harrington H, McClay J, Mill J, Martin J, Braithwaite A, Poulton R (2003) Influence of life stress on depression: moderation by a polymorphism in the 5-HTT gene. Science 301:386–389
Ceglia I, Acconcia S, Fracasso C, Colovic M, Caccia S, Invernizzi RW (2004) Effects of chronic treatment with escitalopram or citalopram on extracellular 5-HT in the prefrontal cortex of rats: role of 5-HT_{1A} receptors. Br J Pharmacol 142:469–478
Celada P, Puig MV, Casanovas JM, Guillazo G, Artigas F (2001) Control of dorsal raphe serotonergic neurons by the medial prefrontal cortex: involvement of serotonin-1A, $GABA_A$, and glutamate receptors. J Neurosci 21:9917–9929
Centenaro LA, Vieira K, Zimmermann N, Miczek KA, Lucion AB, de Almeida RM (2008) Social instigation and aggressive behavior in mice: role of 5-HT_{1A} and 5-HT_{1B} receptors in the prefrontal cortex. Psychopharmacology 201:237–248
Chalmers DT, Lovenberg TW, De Souza EB (1995) Localization of novel corticotropin-releasing factor receptor (CRF2) mRNA expression to specific subcortical nuclei in rat brain: comparison with CRF1 receptor mRNA expression. J Neurosci 15:6340–6350
Chan JP, Unger TJ, Byrnes J, Rios M (2006) Examination of behavioral deficits triggered by targeting *Bdnf* in fetal or postnatal brains of mice. Neuroscience 142:49–58
Chen C, Rainnie DG, Greene RW, Tonegawa S (1994) Abnormal fear response and aggressive behavior in mutant mice deficient for α-calcium-calmodulin kinase II. Science 265:291–294
Chen GL, Novak MA, Meyer JS, Kelly BJ, Vallender EJ, Miller GM (2010) The effect of rearing experience and TPH2 genotype on HPA axis function and aggression in Rhesus monkeys: a retrospective analysis. Horm Behav 57:184–191
Chen ZY, Jing DQ, Bath KG, Ieraci A, Khan T, Siao CJ, Herrera DG, Toth M, Yang C, McEwen BS, Hempstead BL, Lee FS (2006) Genetic variant BDNF (Val66Met) polymorphism alters anxiety-related behavior. Science 314:140–143
Cherek DR, Heistad GT (1971) Fixed-interval induced aggression. Psychon Sci 25:7–8
Cherek DR, Lane SD (1999) Laboratory and psychometric measurements of impulsivity among violent and nonviolent female parolees. Biol Psychiatry 46:273–280
Chermack ST, Giancola PR (1997) The relation between alcohol and aggression: an integrated biopsychosocial conceptualization. Clin Psychol Rev 17:621–649
Chiavegatto S, Dawson VL, Mamounas LA, Koliatsos VE, Dawson TM, Nelson RJ (2001) Brain serotonin dysfunction accounts for aggression in male mice lacking neuronal nitric oxide synthase. Proc Natl Acad Sci USA 98:1277–1281
Cichon S, Winge I, Mattheisen M, Georgi A, Karpushova A, Freudenberg J, Freudenberg-Hua Y, Babadjanova G, Van Den Bogaert A, Abramova LI, Kapiletti S, Knappskog PM, McKinney J, Maier W, Abou Jamra R, Schulze TG, Schumacher J, Propping P, Rietschel M, Haavik J, Nothen MM (2008) Brain-specific tryptophan hydroxylase 2 (TPH2): a functional Pro206Ser

substitution and variation in the 5 '-region are associated with bipolar affective disorder. Hum Mol Genet 17:87–97

Clotfelter ED, O'Hare EP, McNitt MM, Carpenter RE, Summers CH (2007) Serotonin decreases aggression via 5-HT_{1A} receptors in the fighting fish *Betta splendens*. Pharmacol Biochem Behav 87:222–231

Coccaro EF, Kavoussi RJ (1997) Fluoxetine and impulsive aggressive behavior in personality-disordered subjects. Arch Gen Psychiatry 54:1081–1088

Coccaro EF, Kavoussi RJ, Hauger RL, Cooper TB, Ferris CF (1998) Cerebrospinal fluid vasopressin levels: correlates with aggression and serotonin function in personality-disordered subjects. Arch Gen Psychiatry 55:708–714

Coccaro EF, Kavoussi RJ, Sheline YI, Berman ME, Csernansky JG (1997) Impulsive aggression in personality disorder correlates with platelet 5- HT_{2A} receptor binding. Neuropsychopharmacology 16:211–216

Coccaro EF, Lee R, Kavoussi RJ (2010) Aggression, suicidality, and intermittent explosive disorder: serotonergic correlates in personality disorder and healthy control subjects. Neuropsychopharmacology 35:435–444

Cole HF, Wolf HH (1970) Laboratory evaluation of aggressive behavior of the grasshopper mouse (Onychomys). J Pharm Sci 59:969–971

Colmers WF, Williams JT (1988) Pertussis toxin pretreatment discriminates between pre- and postsynaptic actions of baclofen in rat dorsal raphe nucleus in vitro. Neurosci Lett 93:300–306

Cologer-Clifford A, Simon NG, Lu SF, Smoluk SA (1997) Serotonin agonist-induced decreases in intermale aggression are dependent on brain region and receptor subtype. Pharmacol Biochem Behav 58:425–430

Connor DF, Steingard RJ (1996) A clinical approach to the pharmacotherapy of aggression in children and adolescents. In: Understanding aggressive behavior in children. Ann N Y Acad Sci 794:290–307

Crabbe JC, Phillips TJ, Feller DJ, Hen R, Wenger CD, Lessov CN, Schafer GL (1996) Elevated alcohol consumption in null mutant mice lacking 5-HT_{1B} serotonin receptors. Nat Genet 14:98–101

Crawley JN, Schleidt WM, Contrera JF (1975) Does social environment decrease propensity to fight in male mice? Behav Biol 15:73–83

Cutler MG, Rodgers RJ, Jackson JE (1997) Behavioural effects in mice of subchronic buspirone, ondansetron, and tianeptine. I. Social interactions. Pharmacol Biochem Behav 56:287–293

Czobor P, Volavka J, Meibach RC (1995) Effect of risperidone on hostility in schizophrenia. J Clin Psychopharmacol 15:243–249

D'Adamo P, Welzl H, Papadimitriou S, Raffaele dB, Tiveron C, Tatangelo L, Pozzi L, Chapman PF, Knevett SG, Ramsay MF, Valtorta F, Leoni C, Menegon A, Wolfer DP, Lipp HP, Toniolo D (2002) Deletion of the mental retardation gene Gdi1 impairs associative memory and alters social behavior in mice. Hum Mol Genet 11:2567–2580

D'Anna KL, Stevenson SA, Gammie SC (2005) Urocortin 1 and 3 impair maternal defense behavior in mice. Behav Neurosci 119:1061–1071

DaVanzo JP, Daugherty M, Ruckart R, Kang L (1966) Pharmacological and biochemical studies in isolation-induced fighting mice. Psychopharmacologia 9:210–219

Davidson RJ, Putnam KM, Larson CL (2000) Dysfunction in the neural circuitry of emotion regulation–a possible prelude to violence. Science 289:591–594

de Almeida RM, Rosa MM, Santos DM, Saft DM, Benini Q, Miczek KA (2006) 5-HT_{1B} receptors, ventral orbitofrontal cortex, and aggressive behavior in mice. Psychopharmacology 185:441–450

de Almeida RMM, Faccidomo S, Fish E, Miczek KA (2001a) Inhibition of alcohol-heightened aggression by action at post-synaptic 5-HT1b receptors in male mice. Aggress Behav 3:234–235

de Almeida RMM, Giovenardi M, da Silva SP, de Oliveira VP, Stein DJ (2005) Maternal aggression in Wistar rats: effect of 5-$HT_{2A/2C}$ receptor agonist and antagonist microinjected

into the dorsal periaqueductal gray matter and medial septum. Braz J Med Biol Res 38:597–602
de Almeida RMM, Lucion A (1994) Effects of intracerebroventricular administration of 5-HT receptor agonists on the maternal aggression of rats. Eur J Neurosci 264:445–448
de Almeida RMM, Lucion AB (1997) 8-OH-DPAT in the median raphe, dorsal periaqueductal gray and corticomedial amygdala nucleus decreases, but the medial septal area it can increase maternal aggressive behavior in rats. Psychopharmacology 134:392–400
de Almeida RMM, Miczek KA (2002) Aggression escalated by social instigation or by discontinuation of reinforcement ("frustration") in mice: inhibition by anpirtoline—a 5-HT_{1B} receptor agonist. Neuropsychopharmacology 27:171–181
de Almeida RMM, Nikulina EM, Faccidomo S, Fish EW, Miczek KA (2001b) Zolmitriptan–a 5-$HT_{1B/D}$ agonist, alcohol, and aggression in mice. Psychopharmacology 157:131–141
de Boer SF, Koolhaas JM (2005) 5-HT_{1A} and 5-HT_{1B} receptor agonists and aggression: a pharmacological challenge of the serotonin deficiency hypothesis. Eur J Pharmacol 526:125–139
de Boer SF, Lesourd M, Mocaer E, Koolhaas JM (1999) Selective antiaggressive effects of alnespirone in resident-intruder test are mediated via 5-$hydroxytryptamine_{1A}$ receptors: a comparative pharmacological study with 8-hydroxy-2-dipropylaminotetralin, ipsapirone, buspirone, eltoprazine, and WAY-100635. J Pharmacol Exp Ther 288:1125–1133
de Boer SF, Lesourd M, Mocaer E, Koolhaas JM (2000) Somatodendritic 5-HT_{1A} autoreceptors mediate the anti-aggressive actions of 5-HT_{1A} receptor agonists in rats: An ethopharmacological study with S-15535, alnespirone, and WAY-100635. Neuropsychopharmacology 23:20–33
De Deyn PP, Rabheru K, Rasmussen A, Bocksberger JP, Dautzenberg PL, Eriksson S, Lawlor BA (1999) A randomized trial of risperidone, placebo, and haloperidol for behavioral symptoms of dementia. Neurology 53:946–955
De Felipe C, Herrero JF, O'Brien JA, Palmer JA, Doyle CA, Smith AJH, Laird JMA, Belmonte C, Cervero F, Hunt SP (1998) Altered nocieption, analgesia and aggression in mice lacking the receptor for substance P. Nature 392:394–397
De Groote L, Olivier B, Westenberg HG (2003) Role of 5-HT_{1B} receptors in the regulation of extracellular serotonin and dopamine in the dorsal striatum of mice. Eur J Pharmacol 476:71–77
DeBold JF, Miczek KA (1981) Sexual dimorphism in the hormonal control of aggressive behavior of rats. Pharmacol Biochem Behav 14:89–93
Dekeyne A, Brocco M, Adhumeau A, Gobert A, Millan MJ (2000) The selective serotonin 5-HT1A receptor ligand, S15535, displays anxiolytic-like effects in the social interaction and Vogel models and suppresses dialysate levels of 5-HT in the dorsal hippocampus of freely-moving rats. A comparison with other anxiolytic agents. Psychopharmacology 152:55–66
Del Punta K, Leinders-Zufall T, Rodriguez I, Jukam D, Wysocki CJ, Ogawa S, Zufall F, Mombaerts P (2002) Deficient pheromone responses in mice lacking a cluster of vomeronasal receptor genes. Nature 419:70–74
Deltheil T, Guiard BP, Guilloux JP, Nicolas L, Delomenie C, Reperant C, Le Maitre E, Leroux-Nicollet I, Benmansour S, Coudore F, David DJ, Gardier AM (2008) Consequences of changes in BDNF levels on serotonin neurotransmission, 5-HT transporter expression and function: studies in adult mice hippocampus. Pharmacol Biochem Behav 90:174–183
Delville Y, Mansour KM, Ferris CF (1996) Serotonin blocks vasopressin-facilitated offensive aggression: interactions within the ventrolateral hypothalamus of golden hamsters. Physiol Behav 59:813–816
Demas GE, Kriegsfeld LJ, Blackshaw S, Huang P, Gammie SC, Nelson RJ, Snyder SH (1999) Elimination of aggressive behavior in male mice lacking endothelial nitric oxide synthase. J Neurosci 19:1–5
Denney RM, Koch H, Craig IW (1999) Association between monoamine oxidase A activity in human male shin fibroblasts and genotype of the MAOA promoter-associated variable number tandem repeat. Hum Genet 105:542–551

DeVries AC, Young WS, Nelson RJ (1997) Reduced aggressive behaviour in mice with targeted disruption of the oxytocin gene. J Neuroendocrinol 9:363–368

Dompert WU, Glaser T, Traber J (1985) 3H-TVX Q 7821: identification of 5-HT_1 binding sites as target for a novel putative anxiolytic. Naunyn Schmiedebergs Arch Pharmacol 328:467–470

Duysen EG, Stribley JA, Fry DL, Hinrichs SH, Lockridge O (2002) Rescue of the acetylcholinesterase knockout mouse by feeding a liquid diet; phenotype of the adult acetylcholinesterase deficient mouse. Brain Res Dev Brain Res 137:43–54

Egan MF, Kojima M, Callicott JH, Goldberg TE, Kolachana BS, Bertolino A, Zaitsev E, Gold B, Goldman D, Dean M, Lu B, Weinberger DR (2003) The BDNF val66met polymorphism affects activity-dependent secretion of BDNF and human memory and hippocampal function. Cell 112:257–269

Eibl-Eibesfeldt I (1950) Beiträge zur Biologie der Haus- und der Ährenmaus nebst einigen Beobachtungen an anderen Nagern. Z Tierpsychol 7:558–587

Elkabir DR, Wyatt ME, Vellucci SV, Herbert J (1990) The effects of separate or combined infusions of corticotrophin-releasing factor and vasopressin either intraventricularly or into the amygdala on aggressive and investigative behaviour in the rat. Regul Pept 28:199–214

Emeson RB, Morabito MV (2005) Food fight: the NPY-serotonin link between aggression and feeding behavior. Sci STKE 2005:pe12

Faccidomo S, Bannai M, Miczek KA (2008) Escalated aggression after alcohol drinking in male mice: dorsal raphe and prefrontal cortex serotonin and 5-HT_{1B} receptors. Neuropsychopharmacology 33:2888–2899

Faccidomo S, Quadros IM, Takahashi A, Fish EW, Miczek KA (2012) Infralimbic and dorsal raphé microinjection of the 5-HT_{1B} receptor agonist CPG–93,129: attenuation of aggressive behavior in CFW male mice. Psychopharmacology. doi: 10.1007/s00213-011-2629-1 (published online 12 January 2012)

Fairbanks LA, Fontenot MB, Phillips-Conroy JE, Jolly CJ, Kaplan JR, Mann JJ (1999) CSF monoamines, age and impulsivity in wild grivet monkeys (*Cercopithecus aethiops aethiops*). Brain Behav Evol 53:305–312

Fairbanks LA, Melega WP, Jorgensen MJ, Kaplan JR, McGuire MT (2001) Social impulsivity inversely associated with CSF 5-HIAA and fluoxetine exposure in vervet monkeys. Neuropsychopharmacology 24:370–378

Farrokhi C, Blanchard DC, Griebel G, Yang M, Gonzales C, Markham C, Blanchard RJ (2004) Effects of the CRF1 antagonist SSR125543A on aggressive behaviors in hamsters. Pharmacol Biochem Behav 77:465–469

Fava M (1997) Psychopharmacologic treatment of pathologic aggression. Psychiatr Clin North Am 20:427–451

Ferrari PF, Parmigiani S, Rodgers RJ, Palanza P (1997) Differential effects of chlordiazepoxide on aggressive behavior in male mice: the influence of social factors. Psychopharmacology 134:258–265

Ferrari PF, Van Erp AMM, Tornatzky W, Miczek KA (2003) Accumbal dopamine and serotonin in anticipation of the next aggressive episode in rats. Eur J Neurosci 17:371–378

Ferris CF, Lu SF, Messenger T, Guillon CD, Heindel N, Miller M, Koppel G, Robert BF, Simon NG (2006) Orally active vasopressin V1a receptor antagonist, SRX251, selectively blocks aggressive behavior. Pharmacol Biochem Behav 83:169–174

Ferris CF, Melloni RH, Koppel G, Perry KW, Fuller RW, Delville Y (1997) Vasopressin/serotonin interactions in the anterior hypothalamus control aggressive behavior in golden hamsters. J Neurosci 17:4331–4340

Ferris CF, Pilapil CG, Hayden-Hixson D, Wiley RG, Koh ET (1992) Functionally and anatomically distinct populations of vasopressinergic magnocellular neurons in the female golden hamster. J Neuroendocrinol 4:193–205

Ferris CF, Potegal M (1988) Vasopressin receptor blockade in the anterior hypothalamus suppresses aggression in hamsters. Physiol Behav 44:235–239

Ferris CF, Stolberg T, Delville Y (1999) Serotonin regulation of aggressive behavior in male golden hamsters (*Mesocricetus auratus*). Behav Neurosci 113:804–815

Ferris CF, Stolberg T, Kulkarni P, Murugavel M, Blanchard R, Blanchard DC, Febo M, Brevard M, Simon NG (2008) Imaging the neural circuitry and chemical control of aggressive motivation. BMC Neurosci 9:111

Fischer HS, Zernig G, Schuligoi R, Miczek KA, Hauser KF, Gerard C, Saria A (2000) Alterations within the endogenous opioid system in mice with targeted deletion of the neutral endopeptidase ("enkephalinase") gene. Regul Pept 96:53–58

Fish EW, Faccidomo S, DeBold JF, Miczek KA (2001) Alcohol, allopregnanolone and aggression in mice. Psychopharmacology 153:473–483

Fish EW, Faccidomo S, Miczek KA (1999) Aggression heightened by alcohol or social instigation in mice: reduction by the 5-HT_{1B} receptor agonist CP-94,253. Psychopharmacology 146:391–399

Foley DL, Eaves LJ, Wormley B, Silberg JL, Maes HH, Kuhn J, Riley B (2004) Childhood adversity, monoamine oxidase A genotype, and risk for conduct disorder. Arch Gen Psychiat 61:738–744

Forster GL, Pringle RB, Mouw NJ, Vuong SM, Watt MJ, Burke AR, Lowry CA, Summers CH, Renner KJ (2008) Corticotropin-releasing factor in the dorsal raphe nucleus increases medial prefrontal cortical serotonin via type 2 receptors and median raphe nucleus activity. Eur J Neurosci 28:299–310

Frazzetto G, Di Lorenzo G, Carola V, Proietti L, Sokolowska E, Siracusano A, Gross C, Troisi A (2007) Early trauma and increased risk for physical aggression during adulthood: the moderating role of MAOA genotype. PLoS ONE 2:e486

Fresan A, Camarena B, Apiquian R, Aguilar A, Urraca N, Nicolini H (2007) Association study of MAO-A and DRD4 genes in schizophrenic patients with aggressive behavior. Neuropsychobiology 55:171–175

Fuller RW (1996) The influence of fluoxetine on aggressive behavior. Neuropsychopharmacology 14:77–81

Gallager DW, Aghajanian GK (1976) Effect of antipsychotic drugs on the firing of dorsal raphe cells. II. Reversal by picrotoxin. Eur J Pharmacol 39:357–364

Gammie SC, Bethea ED, Stevenson SA (2007) Altered maternal profiles in corticotropin-releasing factor receptor 1 deficient mice. BMC Neurosci 8:17

Gammie SC, Hasen NS, Stevenson SA, Bale TL, D'Anna KL (2005) Elevated stress sensitivity in corticotropin-releasing factor receptor 2 deficient mice decreases maternal, but not intermale aggression. Behav Brain Res 160:169–177

Gammie SC, Negron A, Newman SM, Rhodes JS (2004) Corticotropin-releasing factor inhibits maternal aggression in mice. Behav Neurosci 118:805–814

Gammie SC, Nelson RJ (1999) Maternal aggression is reduced in neuronal nitric oxide synthase-deficient mice. J Neurosci 19:8027–8035

Gammie SC, Stevenson SA (2006) Intermale aggression in corticotropin-releasing factor receptor 1 deficient mice. Behav Brain Res 171:63–69

Garattini S, Giacalone E, Valzelli L (1967) Isolation, aggressiveness and brain 5-hydroxytryptamine turnover. J Pharm Pharmacol 19:338–339

Gervasoni D, Peyron C, Rampon C, Barbagli B, Chouvet G, Urbain N, Fort P, Luppi PH (2000) Role and origin of the GABAergic innervation of dorsal raphe serotonergic neurons. J Neurosci 20:4217–4225

Giancola PR, Levinson CA, Corman MD, Godlaski AJ, Morris DH, Phillips JP, Holt JC (2009) Men and women, alcohol and aggression. Exp Clin Psychopharmacol 17:154–164

Giménez-Llort L, Fernández-Teruel A, Escorihuela RM, Fredholm BB, Tobeña A, Pekny M, Johansson B (2002) Mice lacking the adenosine A_1 receptor are anxious and aggressive, but are normal learners with reduced muscle strength and survival rate. Eur J Neurosci 16: 547–550

Gobert A, Rivet JM, Audinot V, Newman-Tancredi A, Cistarelli L, Millan MJ (1998) Simultaneous quantification of serotonin, dopamine and noradrenaline levels in single frontal cortex dialysates of freely-moving rats reveals a complex pattern of reciprocal auto- and heteroreceptor-mediated control of release. Neuroscience 84:413–429

Godlaski AJ, Giancola PR (2009) Executive functioning, irritability, and alcohol-related aggression. Psychol Addict Behav 23:391–403

Gogos JA, Morgan M, Luine V, Santha M, Ogawa S, Pfaff D, Karayiorgou M (1998) Catechol-O-methyltransferase-deficient mice exhibit sexually dimorphic changes in catecholamine levels and behavior. Proc Natl Acad Sci USA 95:9991–9996

Goldman D, Dean M, Brown GL, Bolos AM, Tokola R, Virkkunen M, Linnoila M (1992) D2 dopamine receptor genotype and cerebrospinal fluid homovanillic acid, 5-hydroxyindoleacetic acid and 3-methoxy-4-hydroxyphenylglycol in alcoholics in Finland and the United States. Acta Psychiatr Scand 86:351–357

Goodson JL (2008) Nonapeptides and the evolutionary patterning of sociality. Prog Brain Res 170:3–15

Gourley SL, DeBold JF, Yin W, Cook J, Miczek KA (2005) Benzodiazepines and heightened aggressive behavior in rats: reduction by $GABA_A/\alpha_1$ receptor antagonists. Psychopharmacology 178:232–240

Gowin JL, Swann AC, Moeller FG, Lane SD (2010) Zolmitriptan and human aggression: interaction with alcohol. Psychopharmacology 210:521–531

Haller J, Bakos N, Rodriguiz RM, Caron MG, Wetsel WC, Liposits Z (2002) Behavioral responses to social stress in noradrenaline transporter knockout mice: effects on social behavior and depression. Brain Res Bull 58:279–284

Haller J, Mikics E, Halasz J, Toth M (2005) Mechanisms differentiating normal from abnormal aggression: glucocorticoids and serotonin. Eur J Pharmacol 526:89–100

Haller J, van de Schraaf J, Kruk MR (2001) Deviant forms of aggression in glucocorticoid hyporeactive rats: a model for 'pathological' aggression? J Neuroendocrinol 13:102–107

Hallikainen T, Saito T, Lachman HM, Volavka J, Pohjalainen T, Ryynanen OP, Kauhanen J, Syvalahti E, Hietala J, Tiihonen J (1999) Association between low activity serotonin transporter promoter genotype and early onset alcoholism with habitual impulsive violent behavior. Mol Psychiatry 4:385–388

Haney M, DeBold JF, Miczek KA (1989) Maternal aggression in mice and rats towards male and female conspecifics. Aggress Behav 15:443–453

Harmon AC, Huhman KL, Moore TO, Albers HE (2002) Oxytocin inhibits aggression in female Syrian hamsters. J Neuroendocrinol 14:963–969

Hassanain M, Bhatt S, Siegel A (2003) Differential modulation of feline defensive rage behavior in the medial hypothalamus by $5\text{-}HT_{1A}$ and $5\text{-}HT_2$ receptors. Brain Res 981:201–209

Haug M, Wallian L, Brain PF (1990) Effects of 8-OH-DPAT and fluoxetine on activity and attack by female mice towards lactating intruders. Gen Pharmacol 21:845–849

Heiligenberg W (1974) Processes governing behavioral states of readiness. Adv Study Behav 5:173–200

Heils A, Teufel A, Petri S, Seemann M, Bengel D, Balling U, Riederer P, Lesch KP (1995) Functional promoter and polyadenylation site mapping of the human serotonin (5-HT) transporter gene. J Neural Transm Gen Sect 102:247–254

Heils A, Teufel A, Petri S, Stober G, Riederer P, Bengel D, Lesch KP (1996) Allelic variation of human serotonin transporter gene expression. J Neurochem 66:2621–2624

Hendricks TJ, Fyodorov DV, Wegman LJ, Lelutiu NB, Pehek EA, Yamamoto B, Silver J, Weeber EJ, Sweatt JD, Deneris ES (2003) Pet-1 ETS gene plays a critical role in 5-HT neuron development and is required for normal anxiety-like and aggressive behavior. Neuron 37:233–247

Hennig J, Reuter M, Netter P, Burk C, Landt O (2005) Two types of aggression are differentially related to serotonergic activity and the A779C TPH polymorphism. Behav Neurosci 119:16–25

Herzog H (2003) Neuropeptide Y and energy homeostasis: insights from Y receptor knockout models. Eur J Pharmacol 480:21–29

Hess WR (1954) Das Zwischenhirn. Benno Schwabe & Co., Basel

Higley JD, Mehlman PT, Poland RE, Taub DM, Vickers J, Suomi SJ, Linnoila M (1996) CSF testosterone and 5-HIAA correlate with different types of aggressive behaviors. Biol Psychiatry 40:1067–1082

Higley JD, Suomi SJ (1986) Parental behavior in primates. In: Sluckin W (ed) Parental behavior in animals and humans. Blackwell Press, Oxford, pp 152–207

Higley JD, Suomi SJ, Linnoila M (1991) CSF monoamine metabolite concentrations vary according to age, rearing, and sex, and are influenced by the stressor of social separation in rhesus monkeys. Psychopharmacology 103:551–556

Hjorth S, Sharp T (1991) Effect of the 5-HT_{1A} receptor agonist 8-OH-DPAT on the release of 5-HT in dorsal and median raphe-innervated rat brain regions as measured by in vivo microdialysis. Life Sci 48:1779–1786

Ho HP, Olsson M, Westberg L, Melke J, Eriksson E (2001) The serotonin reuptake inhibitor fluoxetine reduces sex steroid-related aggression in female rats: an animal model of premenstrual irritability? Neuropsychopharmacology 24:502–510

Hollander E (1999) Managing aggressive behavior in patients with obsessive-compulsive disorder and borderline personality disorder. J Clin Psychiatry 60(Suppl 15):38–44

Holmes A, Murphy DL, Crawley JN (2002) Reduced aggression in mice lacking the serotonin transporter. Psychopharmacology 161:160–167

Homberg JR (2007) Serotonin transporter deficiency in rats improves inhibitory control but not behavioural flexibility. Eur J Neurosci 26:2066–2073

Huang PL, Dawson TM, Bredt DS, Snyder SH, Fishman MC (1993) Targeted disruption of the neuronal nitric oxide synthase gene. Cell 75:1273–1286

Huang YY, Grailhe R, Arango V, Hen R, Mann JJ (1999) Relationship of psychopathology to the human $serotonin_{1B}$ genotype and receptor binding kinetics in postmortem brain tissue. Neuropsychopharmacology 21:238–246

Hurst JL (1987) Behavioral variation in wild house mice *Mus domesticus* Rutty: a quantitative assessment of female social organization. Anim Behav 35:1846–1857

Innis RB, Aghajanian GK (1987) Pertussis toxin blocks 5-HT_{1A} and $GABA_B$ receptor-mediated inhibition of serotonergic neurons. Eur J Pharmacol 143:195–204

Ito W, Chehab M, Thakur S, Li J, Morozov A (2011) BDNF-restricted knockout mice as an animal model for aggression. Genes Brain Behav 10:365–374

Jacobs BL, Cohen A (1976) Differential behavioral effects of lesions of the median or dorsal raphe nuclei in rats: open field and pain-elicited aggression. J Comp Physiol Psychol 90: 102–108

Jarrell H, Hoffman JB, Kaplan JR, Berga S, Kinkead B, Wilson ME (2008) Polymorphisms in the serotonin reuptake transporter gene modify the consequences of social status on metabolic health in female rhesus monkeys. Physiol Behav 93:807–819

Jensen KP, Covault J, Conner TS, Tennen H, Kranzler HR, Furneaux HM (2009) A common polymorphism in serotonin receptor 1B mRNA moderates regulation by miR-96 and associates with aggressive human behaviors. Mol Psychiatry 14:381–389

Johnson DE, Rollema H, Schmidt AW, McHarg AD (2001) Serotonergic effects and extracellular brain levels of eletriptan, zolmitriptan and sumatriptan in rat brain. Eur J Pharmacol 425:203–210

Johnson O, Becnel J, Nichols CD (2009) Serotonin 5-HT_2 and 5-HT_{1A}-like receptors differentially modulate aggressive behaviors in Drosophila melanogaster. Neuroscience 158: 1292–1300

Jonsson EG, Goldman D, Spurlock G, Gustavsson JP, Nielsen DA, Linnoila M, Owen MJ, Sedvall GC (1997) Tryptophan hydroxylase and catechol-O-methyltransferase gene polymorphisms: relationships to monoamine metabolite concentrations in CSF of healthy volunteers. Eur Arch Psychiatry Clin Neurosci 247:297–302

Joppa MA, Rowe RK, Meisel RL (1997) Effects of serotonin 1A or 1B receptor agonists on social aggression in male and female Syrian hamsters. Pharmacol Biochem Behav 58:349–353

Judge SJ, Ingram CD, Gartside SE (2004) GABA receptor modulation of 5-HT neuronal firing: characterization and effect of moderate in vivo variations in glucocorticoid levels. Neurochem Int 45:1057–1065

Kalen P, Pritzel M, Nieoullon A, Wiklund L (1986) Further evidence for excitatory amino acid transmission in the lateral habenular projection to the rostral raphe nuclei: lesion-induced decrease of high affinity glutamate uptake. Neurosci Lett 68:35–40

Karl T, Lin S, Schwarzer C, Sainsbury A, Couzens M, Wittmann W, Boey D, von Horsten S, Herzog H (2004) Y1 receptors regulate aggressive behavior by modulating serotonin pathways. Proc Natl Acad Sci USA 101:12742–12747

Kavoussi R, Armstead P, Coccaro E (1997) The neurobiology of impulsive aggression. Psychiatr Clin North Am 20:395–403

Keck PE Jr, Strakowski SM, McElroy SL (2000) The efficacy of atypical antipsychotics in the treatment of depressive symptoms, hostility, and suicidality in patients with schizophrenia. J Clin Psychiatry 61(Suppl 3):4–9

Khait VD, Huang YY, Zalsman G, Oquendo MA, Brent DA, Harkavy-Friedman JM, Mann JJ (2005) Association of serotonin 5-HT_{2A} receptor binding and the T102C polymorphism in depressed and healthy Caucasian subjects. Neuropsychopharmacology 30:166–172

Kim C, Jeon D, Kim YH, Lee CJ, Kim H, Shin HS (2009a) Deletion of N-type Ca^{2+} channel Cav2.2 results in hyperaggressive behaviors in mice. J Biol Chem 284:2738–2745

Kim JJ, Shih JC, Chen K, Chen L, Bao SW, Maren S, Anagnostaras SG, Fanselow MS, DeMaeyer E, Seif I, Thompson RF (1997) Selective enhancement of emotional, but not motor, learning in monoamine oxidase A-deficient mice. Proc Nat Acad Sci USA 94: 5929–5933

Kim YR, Jahng JW, Min SK (2009b) Association between the serotonin transporter gene (5-HTTLPR) and anger-related traits in Korean schizophrenic patients. Neuropsychobiology 59:165–171

Kim-Cohen J, Caspi A, Taylor A, Williams B, Newcombe R, Craig IW, Moffitt TE (2006) MAOA, maltreatment, and gene-environment interaction predicting children's mental health: new evidence and a meta-analysis. Mol Psychiatry 11:903–913

Kirby LG, Rice KC, Valentino RJ (2000) Effects of corticotropin-releasing factor on neuronal activity in the serotonergic dorsal raphe nucleus. Neuropsychopharmacology 22:148–162

Knobelman DA, Kung HF, Lucki I (2000) Regulation of extracellular concentrations of 5-hydroxytryptamine (5-HT) in mouse striatum by 5-HT_{1A} and 5-HT_{1B} receptors. J Pharmacol Exp Ther 292:1111–1117

Koleske AJ, Gifford AM, Scott ML, Nee M, Bronson RT, Miczek KA, Baltimore D (1998) Essential roles for the Abl and Arg tyrosine kinases in neurulation. Neuron 21:1259–1272

Kombian SB, Colmers WF (1992) Neuropeptide Y selectively inhibits slow synaptic potentials in rat dorsal raphe nucleus in vitro by a presynaptic action. J Neurosci 12:1086–1093

König M, Zimmer AM, Steiner H, Holmes PV, Crawley JN, Brownstein MJ, Zimmer A (1996) Pain responses, anxiety and aggression in mice deficient in pre-proenkephalin. Nature 383:535–538

Koolhaas JM (1978) Hypothalamically induced intraspecific aggressive behaviour in the rat. Exp Brain Res 32:365–375

Koolhaas JM, Van der Brink THC, Roozendaal B, Boorsma F (1990) Medial amygdala and aggressive behavior: interaction between testosterone and vasopressin. Aggress Behav 16:223–229

Koprowska M, Romaniuk A (1997) Behavioral and biochemical alterations in median and dorsal raphe nuclei lesioned cats. Pharmacol Biochem Behav 56:529–540

Korte SM, Meijer OC, De Kloet ER, Buwalda B, Keijser J, Sluyter F, van Oortmerssen G, Bohus B (1996) Enhanced 5-HT_{1A} receptor expression in forebrain regions of aggressive house mice. Brain Res 736:338–343

Kovács GL, Bohus B, Versteeg DH, De Kloet ER, de Wied D (1979) Effect of oxytocin and vasopressin on memory consolidation: sites of action and catecholaminergic correlates after local microinjection into limbic-midbrain structures. Brain Res 175:303–314

Kraemer GW, Ebert MH, Schmidt DE, McKinney WT (1989) A longitudinal study of the effect of different social rearing conditions on cerebrospinal fluid norepinephrine and biogenic amine metabolites in rhesus monkeys. Neuropsychopharmacology 2:175–189

Kranzler HR, Hernandez-Avila CA, Gelernter J (2002) Polymorphism of the 5-HT_{1B} receptor gene (HTR1B): Strong within-locus linkage disequilibrium without association to antisocial substance dependence. Neuropsychopharmacology 26:115–122

Kreiss DS, Lucki I (1994) Differential regulation of serotonin (5-HT) release in the striatum and hippocampus by 5-HT_{1A} autoreceptors of the dorsal and median raphe nuclei. J Pharmacol Exp Ther 269:1268–1279

Krsiak M (1974) Isolation-induced timidity in mice as a measure of anxiolytic activity of drugs. Act Nerv Super (Praha) 16:241–242

Kushi A, Sasai H, Koizumi H, Takeda N, Yokoyama M, Nakamura M (1998) Obesity and mild hyperinsulinemia found in neuropeptide Y-Y1 receptor-deficient mice. Proc Natl Acad Sci USA 95:15659–15664

Lamar M, Cutter WJ, Rubia K, Brammer M, Daly EM, Craig MC, Cleare AJ, Murphy DG (2009) 5-HT, prefrontal function and aging: fMRI of inhibition and acute tryptophan depletion. Neurobiol Aging 30:1135–1146

Lang A, Harro J, Soosaar A, Koks S, Volke V, Oreland L, Bourin M, Vasar E, Bradwejn J, Mannisto PT (1995) Role of N-methyl-D-aspartic acid and cholecystokinin receptors in apomorphine-induced aggressive behaviour in rats. Naunyn Schmiedebergs Arch Pharmacol 351:363–370

Lappalainen J, Long JC, Eggert M, Ozaki N, Robin RW, Brown GL, Naukkarinen H, Virkkunen M, Linnoila M, Goldman D (1998) Linkage of antisocial alcoholism to the serotonin 5-HT_{1B} receptor gene in 2 populations. Arch Gen Psychiatry 55:989–994

Ledent C, Vaugeois JM, Schiffmann SN, Pedrazzini T, El Yacoubi M, Vanderhaeghen JJ, Costentin J, Heath JK, Vassart G, Parmentier M (1997) Aggressiveness, hypoalgesia and high blood pressure in mice lacking the adenosine A_{2a} receptor. Nature 388:674–678

Lee R, Ferris C, Van de Kar LD, Coccaro EF (2009) Cerebrospinal fluid oxytocin, life history of aggression, and personality disorder. Psychoneuroendocrinology 34:1567–1573

Lerch-Haner JK, Frierson D, Crawford LK, Beck SG, Deneris ES (2008) Serotonergic transcriptional programming determines maternal behavior and offspring survival. Nat Neurosci 11:1001–1003

Lesch KP, Bengel D, Heils A, Sabol SZ, Greenberg BD, Petri S, Benjamin J, Mueller CR, Hamer DH, Murphy DL (1996) Association of anxiety-related traits with a polymorphism in the serotonin transporter gene regulatory region. Science 264:1537–1551

Lesch KP, Merschdorf U (2000) Impulsivity, aggression, and serotonin: a molecular psychobiological perspective. Behav Sci Law 18:581–604

Lesch KP, Meyer J, Glatz K, Flugge G, Hinney A, Hebebrand J, Klauck SM, Poustka A, Poustka F, Bengel D, Mossner R, Riederer P, Heils A (1997) The 5-HT transporter gene-linked polymorphic region (5-HTTLPR) in evolutionary perspective: alternative biallelic variation in rhesus monkeys. J Neural Transm 104:1259–1266

Leyhausen P (1979) Cat behavior: predatory and social behavior of domestic and wild cats. Garland STPM Press, New York

Lin YMJ, Chao SC, Chen TM, Lai TJ, Chen JS, Sun HS (2007) Association of functional polymorphisms of the human tryptophan hydroxylase 2 gene with risk for bipolar disorder in Han Chinese. Arch Gen Psychiatry 64:1015–1024

Lindgren T, Kantak KM (1987) Effects of serotonin receptor agonists and antagonists on offensive aggression in mice. Aggress Behav 13:87–96

Linnoila M, Virkkunen M, Scheinin M, Nuutila A, Rimon R, Goodwin FK (1983) Low cerebrospinal fluid 5-hydroxyindoleacetic acid concentration differentiates impulsive from nonimpulsive violent behavior. Life Sci 33:2609–2614

Linnoila VMI, Virkkunen M (1992) Aggression, suicidality, and serotonin. J Clin Psychiatry 53:46–51

Loconto J, Papes F, Chang E, Stowers L, Jones EP, Takada T, Kumanovics A, Fischer LK, Dulac C (2003) Functional expression of murine V2R pheromone receptors involves selective association with the M10 and M1 families of MHC class Ib molecules. Cell 112:607–618

Lonstein JS, Gammie SC (2002) Sensory, hormonal, and neural control of maternal aggression in laboratory rodents. Neurosci Biobehav Rev 26:869–888

Lowry CA, Rodda JE, Lightman SL, Ingram CD (2000) Corticotropin-releasing factor increases in vitro firing rates of serotonergic neurons in the rat dorsal raphe nucleus: evidence for activation of a topographically organized mesolimbocortical serotonergic system. J Neurosci 20:7728–7736

Lucion AB, de Almeida RMM (1996) On the dual nature of maternal aggression in rats. Aggress Behav 22:365–373

Lucki I, Wieland S (1990) 5-Hydroxytryptamine-1A receptors and behavioral responses. Neuropsychopharmacology 3:481–493

Lukkes JL, Forster GL, Renner KJ, Summers CH (2008) Corticotropin-releasing factor 1 and 2 receptors in the dorsal raphe differentially affect serotonin release in the nucleus accumbens. Eur J Pharmacol 578:185–193

Lyons WE, Mamounas LA, Ricaurte GA, Coppola V, Reid SW, Bora SH, Wihler C, Koliatsos VE, Tessarollo L (1999) Brain-derived neurotrophic factor-deficient mice develop aggressiveness and hyperphagia in conjunction with brain serotonergic abnormalities. Proc Natl Acad Sci USA 96:15239–15244

Maciewicz R, Foote WE, Bry J (1981) Excitatory projection from the interpeduncular nucleus to central superior raphe neurons. Brain Res 225:179–183

Malick JB (1979) The pharmacology of isolation-induced aggressive behavior in mice. In: Essman WB (ed) Current Developments in Psychopharmacology. SP Medical and Scientific Books, New York, pp 1–27

Malleret G, Hen R, Guillou JL, Segu L, Buhot MC (1999) 5-HT_{1B} receptor knock-out mice exhibit increased exploratory activity and enhanced spatial memory performance in the Morris water maze. J Neurosci 19:6157–6168

Mandiyan VS, Coats JK, Shah NM (2005) Deficits in sexual and aggressive behaviors in Cnga2 mutant mice. Nat Neurosci 8:1660–1662

Mann JJ (1999) Role of the serotonergic system in the pathogenesis of major depression and suicidal behavior. Neuropsychopharmacology 21:99S–105S

Mann JJ, Stanley M, McBride PA, McEwen BS (1986) Increased serotonin$_2$ and β-adrenergic receptor binding in the frontal cortices of suicide victims. Arch Gen Psychiatry 43:954–959

Manuck SB, Flory JD, Ferrell RE, Dent KM, Mann JJ, Muldoon MF (1999) Aggression and anger-related traits associated with a polymorphism of the tryptophan hydroxylase gene. Biol Psychiatry 45:603–614

Manuck SB, Flory JD, Ferrell RE, Mann JJ, Muldoon MF (2000) A regulatory polymorphism of the monoamine oxidase-A gene may be associated with variability in aggression, impulsivity, and central nervous system serotonergic responsivity. Psychiatry Res 95:9–23

Manuck SB, Flory JD, Muldoon MF, Ferrell RE (2002) Central nervous system serotonergic responsivity and aggressive disposition in men. Physiol Behav 77:705–709

Marino MD, Bourdelat-Parks BN, Liles LC, Weinshenker D (2005) Genetic reduction of noradrenergic function alters social memory and reduces aggression in mice. Behav Brain Res 161:197–203

Martel JC, Fournier A, St PS, Quirion R (1990) Quantitative autoradiographic distribution of [125I] Bolton-Hunter neuropeptide Y receptor binding sites in rat brain. Comparison with [125I] peptide YY receptor sites. Neuroscience 36:255–283

Martin M, Ledent C, Parmentier M, Maldonado R, Valverde O (2002) Involvement of CB1 cannabinoid receptors in emotional behaviour. Psychopharmacology 159:379–387

Mathews TA (2004) Gene dose-dependent alterations in extraneuronal serotonin but not dopamine in mice with reduced serotonin transporter expression. J Neurosci Methods 140:169–181

Matsumoto T, Honda S, Harada N (2003) Alteration in sex-specific behaviors in male mice lacking the aromatase gene. Neuroendocrinology 77:416–424

Matsuoka Y, Furuyashiki T, Yamada K, Nagai T, Bito H, Tanaka Y, Kitaoka S, Ushikubi F, Nabeshima T, Narumiya S (2005) Prostaglandin E receptor EP1 controls impulsive behavior under stress. Proc Natl Acad Sci USA 102:16066–16071

McAllister KH (1990) Ethological analysis of the effects of MK-801 upon aggressive male mice: similarity to chlordiazepoxide. Pharmacol Biochem Behav 37:101–106

McBride PA, Brown RP, DeMeo M, Keilp J, Mieczkowski T, Mann JJ (1994) The relationship of platelet 5-HT_2 receptor indices to major depressive disorder, personality traits, and suicidal behavior. Biol Psychiatry 35:295–308

McKenzie-Quirk SD, Girasa KA, Allan AM, Miczek KA (2005) 5-HT_3 receptors, alcohol and aggressive behavior in mice. Behav Pharmacol 16:163–170

McKinley JC, Hathaway SR, Meehl PE (1948) The Minnesota multiphasic personality inventory: the K-scale. J Consult Psychol 12:20–31

McMillen BA, DaVanzo EA, Scott SM, Song AH (1988) N-alkyl-substituted aryl-piperazine drugs: relationship between affinity for serotonin receptors and inhibition of aggression. Drug Dev Res 12:53–62

Mehlman PT, Higley JD, Faucher I, Lilly AA, Taub DM, Vickers J, Suomi SJ, Linnoila M (1994) Low CSF 5-HIAA concentrations and severe aggression and impaired impulse control in nonhuman primates. Am J Psychiatry 151:1485–1491

Meloni EG, Reedy CL, Cohen BM, Carlezon WA Jr (2008) Activation of raphe efferents to the medial prefrontal cortex by corticotropin-releasing factor: correlation with anxiety-like behavior. Biol Psychiatry 63:832–839

Mendoza DL, Bravo HA, Swanson HH (1999) Antiaggresive and anxiolytic effects of gepirone in mice, and their attenuation by WAY 100635. Pharmacol Biochem Behav 62:499–509

Meyer JH, Wilson AA, Rusjan P, Clark M, Houle S, Woodside S, Arrowood J, Martin K, Colleton M (2008) Serotonin$_{2A}$ receptor binding potential in people with aggressive and violent behaviour. J Psychiatry Neurosci 33:499–508

Meyer-Lindenberg A, Buckholtz JW, Kolachana B, Hariri AR, Pezawas L, Blasi G, Wabnitz A, Honea R, Verchinski B, Callicott JH, Egan M, Mattay V, Weinberger DR (2006) Neural mechanisms of genetic risk for impulsivity and violence in humans. Proc Nat Acad Sci USA 103:6269–6274

Miczek KA (1974) Intraspecies aggression in rats: effects of *d*-amphetamine and chlordiazepoxide. Psychopharmacologia 39:275–301

Miczek KA (1987) The psychopharmacology of aggression. In: Iversen LL, Iversen SD, Snyder SH (eds) Handbook of psychopharmacology, vol 19. New directions in behavioral pharmacology. Plenum, New York, pp 183–328

Miczek KA, Barros HM, Sakoda L, Weerts EM (1998a) Alcohol and heightened aggression in individual mice. Alcohol Clin Exp Res 22:1698–1705

Miczek KA, de Almeida RMM (2001) Oral drug self-administration in the home cage of mice: alcohol-heightened aggression and inhibition by the 5-HT_{1B} agonist anpirtoline. Psychopharmacology 157:421–429

Miczek KA, de Almeida RMM, Kravitz EA, Rissman EF, de Boer SF, Raine A (2007) Neurobiology of escalated aggression and violence. J Neurosci 27:11803–11806

Miczek KA, Faccidomo S, de Almeida RMM, Bannai M, Fish EW, DeBold JF (2004) Escalated aggressive behavior: new pharmacotherapeutic approaches and opportunities. Ann N Y Acad Sci 1036:336–355

Miczek KA, Fish EW, De Bold JF (2003) Neurosteroids, $GABA_A$ receptors, and escalated aggressive behavior. Horm Behav 44:242–257

Miczek KA, Fish EW, DeBold JF, de Almeida RMM (2002) Social and neural determinants of aggressive behavior: pharmacotherapeutic targets at serotonin, dopamine and γ-aminobutyric acid systems. Psychopharmacology 163:434–458

Miczek KA, Haney M (1994) Psychomotor stimulant effects of *d*-amphetamine, MDMA and PCP: aggressive and schedule-controlled behavior in mice. Psychopharmacology 115:358–365

Miczek KA, Hussain S, Faccidomo S (1998b) Alcohol-heightened aggression in mice: attenuation by 5-HT_{1A} receptor agonists. Psychopharmacology 139:160–168

Miczek KA, Maxson SC, Fish EW, Faccidomo S (2001) Aggressive behavioral phenotypes in mice. Behav Brain Res 125:167–181

Miczek KA, Nikulina EM, Takahashi A, Covington HE, III, Yap JJ, Boyson CO, Shimamoto A, de Almeida RM (2011) Gene expression in aminergic and peptidergic cells during aggression and defeat: relevance to violence, depression and drug abuse. Behav Genet, Online First

Miczek KA, O'Donnell JM (1978) Intruder-evoked aggression in isolated and nonisolated mice: effects of psychomotor stimulants and L-dopa. Psychopharmacology 57:47–55
Miczek KA, O'Donnell JM (1980) Alcohol and chlordiazepoxide increase suppressed aggression in mice. Psychopharmacology 69:39–44
Miczek KA, Weerts EM, Tornatzky W, DeBold JF, Vatne TM (1992) Alcohol and "bursts" of aggressive behavior: ethological analysis of individual differences in rats. Psychopharmacology 107:551–563
Mitchell PJ (2005) Antidepressant treatment and rodent aggressive behaviour. Eur J Pharmacol 526:147–162
Mitchell PJ, Fletcher A, Redfern PH (1991) Is antidepressant efficacy revealed by drug-induced changes in rat behaviour exhibited during social interaction? Neurosci Biobehav Rev 15:539–544
Mitchell PJ, Redfern PH (1992) Acute and chronic antidepressant drug treatments induce opposite effects in the social behavior of rats. J Psychopharmacol 6:241–257
Mitchell PJ, Redfern PH (1997) Potentiation of the time-dependent, antidepressant-induced changes in the agonistic behaviour of resident rats by the 5-HT_{1A} receptor antagonist, WAY-100635. Behav Pharmacol 8:585–606
Miyakawa T, Yagi T, Takao K, Niki H (2001) Differential effect of Fyn tyrosine kinase deletion on offensive and defensive aggression. Behav Brain Res 122:51–56
Morgan C, Thomas RE, Ma W, Novotny MV, Cone RD (2004) Melanocortin-5 receptor deficiency reduces a pheromonal signal for aggression in male mice. Chem Senses 29: 111–115
Mos J, Olivier B (1989) Quantitative and comparative analyses of pro-aggressive actions of benzodiazepines in maternal aggression of rats. Psychopharmacology 97:152–153
Mos J, Olivier B, Poth M, Van Oorschot R, Van Aken H (1993) The effects of dorsal raphé administration of eltoprazine, TFMPP and 8-OH-DPAT on resident intruder aggression in the rat. Eur J Pharmacol 238:411–415
Muehlenkamp F, Lucion A, Vogel WH (1995) Effects of selective serotenergic agonists on aggressive behavior in rats. Pharmacol Biochem Behav 50:671–674
Murphy DL, Lesch KP (2008) Targeting the murine serotonin transporter: insights into human neurobiology. Nat Rev Neurosci 9:85–96
Musty RE, Consroe PF (1982) Phencyclidine produces aggressive behavior in rapid eye movement sleep- deprived rats. Life Sci 30:1733–1738
Nagtegaal MH, Rassin E (2004) The usefulness of the thought suppression paradigm in explaining impulsivity and aggression. Pers Individ Dif 37:1233–1244
Nanopoulos D, Belin MF, Maitre M, Vincendon G, Pujol JF (1982) Immunocytochemical evidence for the existence of GABAergic neurons in the nucleus raphe dorsalis. Possible existence of neurons containing serotonin and GABA. Brain Res 232:375–389
Natarajan D, Caramaschi D (2010) Animal violence demystified. Front. Behav Neurosci 4:9
Nelson RJ, Chiavegatto S (2001) Molecular basis of aggression. Trends Neurosci 24:713–719
Nelson RJ, Demas GE, Huang PL, Fishman MC, Dawson VL, Dawson TM, Snyder SH (1995) Behavioural abnormalities in male mice lacking neuronal nitric oxide synthase. Nature 378:383–386
Neumann ID, Veenema AH, Beiderbeck DI (2010) Aggression and anxiety: social context and neurobiological links. Front Behav Neurosci 4:12
Neves-Pereira M, Mundo E, Muglia P, King N, Macciardi F, Kennedy JL (2002) The brain-derived neurotrophic factor gene confers susceptibility to bipolar disorder: evidence from a family-based association study. Am J Hum Gene 71:651–655
New AS, Buchsbaum MS, Hazlett EA, Goodman M, Koenigsberg HW, Lo J, Iskander L, Newmark R, Brand J, O'Flynn K, Siever LJ (2004) Fluoxetine increases relative metabolic rate in prefrontal cortex in impulsive aggression. Psychopharmacology 176:451–458
New AS, Gelernter J, Goodman M, Mitropoulou V, Koenigsberg H, Silverman J, Siever LJ (2001) Suicide, impulsive aggression, and HTR1B genotype. Biol Psychiatry 50:62–65

New AS, Gelernter J, Yovell Y, Trestman RL, Nielsen DA, Silverman J, Mitropoulou V, Siever LJ (1998) Tryptophan hydroxylase genotype is associated with impulsive-aggression measures: a preliminary study. Am J Med Genet 81:13–17

Newman TK, Syagailo YV, Barr CS, Wendland JR, Champoux M, Graessle M, Suomi SJ, Higley JD, Lesch KP (2005) Monoamine oxidase A gene promoter variation and rearing experience influences aggressive behavior in rhesus monkeys. Biol Psychiatry 57:167–172

Nicot A, Otto T, Brabet P, Dicicco-Bloom EM (2004) Altered social behavior in pituitary adenylate cyclase-activating polypeptide type I receptor-deficient mice. J Neurosci 24:8786–8795

Nielsen DA, Goldman D, Virkkunen M, Tokola R, Rawlings R, Linnoila M (1994) Suicidality and 5-hydroxyindoleacetic acid concentration associated with a tryptophan-hydroxylase polymorphism. Arch Gen Psychiatry 51:34–38

Nikulina EM, Avgustinovich DF, Popova NK (1992) Role of $5HT_{1A}$ receptors in a variety of kinds of aggressive behavior in wild rats and counterparts selected for low defensiveness towards man. Aggress Behav 18:357–364

Nogueira RL, Graeff FG (1995) Role of 5HT receptor subtypes in the modulation of dorsal periaqueductal gray generated aversion. Pharmacol Biochem Behav 52:1–6

Noirot E, Goyens J, Buhot MC (1975) Aggressive behavior of pregnant mice towards males. Horm Behav 6:9–17

Nomura M, Kusumi I, Kaneko M, Masui T, Daiguji M, Ueno T, Koyama T, Nomura Y (2006) Involvement of a polymorphism in the $5\text{-}HT_{2A}$ receptor gene in impulsive behavior. Psychopharmacology 187:30–35

O'Donohue TL, Chronwall BM, Pruss RM, Mezey E, Kiss JZ, Eiden LE, Massari VJ, Tessel RE, Pickel VM, DiMaggio DA (1985) Neuropeptide Y and peptide YY neuronal and endocrine systems. Peptides 6:755–768

Oades RD, Lasky-Su J, Christiansen H, Faraone SV, Sonuga-Barke EJS, Banaschewski T, Chen W, Anney RJL, Buitelaar JK, Ebstein RP, Franke B, Gill M, Miranda A, Roeyers H, Rothenberger A, Sergeant JA, Steinhausen HC, Taylor EA, Thompson M, Asherson P (2008) The influence of serotonin- and other genes on impulsive behavioral aggression and cognitive impulsivity in children with attention-deficit/hyperactivity disorder (ADHD): findings from a family-based association test (FBAT) analysis. Behav Brain Funct 4:48

Ogawa S, Chan J, Chester AE, Gustafsson JA, Korach KS, Pfaff DW (1999) Survival of reproductive behaviors in estrogen receptor β gene-deficient (βERKO) male and female mice. Proc Nat Acad Sci USA 96:12887–12892

Ogawa S, Eng V, Taylor J, Lubahn DB, Korach KS, Pfaff DW (1998a) Roles of estrogen receptor-α gene expression in reproduction-related behaviors in female mice. Endocrinology 139:5070–5081

Ogawa S, Washburn TF, Taylor J, Lubahn DB, Korach KS, Pfaff DW (1998b) Modifications of testosterone-dependent behaviors by estrogen receptor-α gene disruption in male mice. Endocrinology 139:5058–5069

Oliveira-dos-Santos AJ, Matsumoto G, Snow BE, Bai D, Houston FP, Whishaw IQ, Mariathasan S, Sasaki T, Wakeham A, Ohashi PS, Roder JC, Barnes CA, Siderovski DP, Penninger JM (2000) Regulation of T cell activation, anxiety, and male aggression by RGS2. Proc Nat Acad Sci USA 97:12272–12277

Olivier B (2004) Serotonin and aggression. Ann N Y Acad Sci 1036:382–392

Olivier B, Mos J, Rasmussen D (1990) Behavioural pharmacology of the serenic, eltoprazine. Rev Drug Metab Drug Interact 8:31–83

Olivier B, Mos J, Tulp MTM, van der Poel AM (1992) Animal models of anxiety and aggression in the study of serotonergic agents. In: Langer SZ (ed) Serotonin receptor subtypes: pharmacological significance and clinical implications. Karger, Basel, pp 67–79

Olivier B, Mos J, Van der Heyden J, Hartog J (1989) Serotonergic modulation of social interactions in isolated male mice. Psychopharmacology 97:154–156

Olivier B, Mos J, Van Oorschot R, Hen R (1995) Serotonin receptors and animal models of aggressive behavior. Pharmacopsychiatry 28:80–90

Oquendo MA, Russo SA, Underwood MD, Kassir SA, Ellis SP, Mann JJ, Arango V (2006) Higher postmortem prefrontal 5-HT_{2A} receptor binding correlates with lifetime aggression in suicide. Biol Psychiatry 59:235–243

Osipova DV, Kulikov AV, Popova NK (2009) C1473G polymorphism in mouse tph2 gene is linked to tryptophan hydroxylase-2 activity in the brain, intermale aggression, and depressive-like behavior in the forced swim test. J Neurosci Res 87:1168–1174

Pabis DJ, Stanislav SW (1996) Pharmacotherapy of aggressive behavior. Ann Pharmacother 30:278–287

Palanza P, Della Seta D, Ferrari PF, Parmigiani S (2005) Female competition in wild house mice depends upon timing of female/male settlement and kinship between females. Anim Behav 69:1259–1271

Pallotta M, Segieth J, Whitton PS (1998) N-methyl-d-aspartate receptors regulate 5-HT release in the raphe nuclei and frontal cortex of freely moving rats: differential role of 5-HT_{1A} autoreceptors. Brain Res 783:173–178

Parmigiani S, Ferrari PF, Palanza P (1998) An evolutionary approach to behavioral pharmacology: using drugs to understand proximate and ultimate mechanisms of different forms of aggression in mice. Neurosci Biobehav Rev 23:143–153

Passamonti L, Cerasa A, Gioia MC, Magariello A, Muglia M, Quattrone A, Fera F (2008) Genetically dependent modulation of serotonergic inactivation in the human prefrontal cortex. Neuroimage 40:1264–1273

Peeke HVS, Figler MH (1981) Modulation of aggressive behavior in fish by alcohol and congeners. Pharmacol Biochem Behav 14(Suppl 1):79–84

Pellis SM, Pellis VC (1988) Play-fighting in the Syrian golden hamster *Mesocricetus auratus* Waterhouse, and its relationship to serious fighting during postweaning development. Dev Psychobiol 21:323–337

Pineyro G, Blier P, Dennis T, de Montigny C (1994) Desensitization of the neuronal 5-HT carrier following its long-term blockade. J Neurosci 14:3036–3047

Pinna G, Costa E, Guidotti A (2006) Fluoxetine and norfluoxetine stereospecifically and selectively increase brain neurosteroid content at doses that are inactive on 5-HT reuptake. Psychopharmacology 186:362–372

Pinna G, Dong E, Matsumoto K, Costa E, Guidotti A (2003) In socially isolated mice, the reversal of brain allopregnanolone down-regulation mediates the anti-aggressive action of fluoxetine. Proc Natl Acad Sci USA 100:2035–2040

Popova NK, Avgustinovich DF, Kolpakov VG, Plyusnina IZ (1998) Specific [H-3]8-OH-DPAT binding in brain regions of rats genetically predisposed to various defense behavior strategies. Pharmacol Biochem Behav 59:793–797

Popova NK, Voitenko NN, Kulikov AV, Avgustinovich DF (1991) Evidence for the involvement of central serotonin in mechanism of domestication of silver foxes. Pharmacol Biochem Behav 40:751–756

Potegal M (1991) Attack priming and satiation in female golden hamsters: tests of some alternatives to the aggression arousal interpretation. Aggress Behav 17:327–335

Potegal M, Tenbrink L (1984) Behavior of attack-primed and attack-satiated female golden hamsters (*Mesocricetus auratus*). J Comp Psychol 98:66–75

Potter E, Sutton S, Donaldson C, Chen R, Perrin M, Lewis K, Sawchenko PE, Vale W (1994) Distribution of corticotropin-releasing factor receptor mRNA expression in the rat brain and pituitary. Proc Natl Acad Sci USA 91:8777–8781

Price ML, Curtis AL, Kirby LJ, Valentino RJ, Lucki I (1998) Effects of corticotropin-releasing factor on brain serotonergic activity. Neuropsychopharmacology 18:492–502

Price ML, Lucki I (2001) Regulation of serotonin release in the lateral septum and striatum by corticotropin-releasing factor. J Neurosci 21:2833–2841

Quadros IM, Miguel TT, DeBold JF, Miczek KA (2009a) Opposing action of CRF1 vs. CRF2 receptors in the dorsal raphé: modulation of alcohol-heightened aggression. 2009 Neuroscience meeting planner. Society for neuroscience, Chicago, Program No.445.5/T8

Quadros IM, Takahashi A, Miczek KA (2009b) Serotonin and aggression. In: Müller CP, Jacobs BL (eds) Handbook of the behavioral neurobiology of serotonin. Academic Press, pp 687–714

Ragnauth AK, Devidze N, Moy V, Finley K, Goodwillie A, Kow LM, Muglia LJ, Pfaff DW (2005) Female oxytocin gene-knockout mice, in a semi-natural environment, display exaggerated aggressive behavior. Genes Brain Behav 4:229–239

Raine A (2002) Biosocial studies of antisocial and violent behavior in children and adults: a review. J Abnorm Child Psychol 30:311–326

Raj YP (2004) Psychopharmacology of borderline personality disorder. Curr Psychiatry Rep 6:225–231

Ramboz S, Saudou F, Amara DA, Belzung C, Segu L, Misslin R, Buhot MC, Hen R (1995) 5-HT_{1B} receptor knock out: behavioral consequences. Behav Brain Res 73:305–312

Rasia-Filho AA, Giovenardi M, de Almeida RM (2008) Drugs and aggression. Recent Pat CNS Drug Discov 3:40–49

Raskin K, de Gendt K, Duittoz A, Liere P, Verhoeven G, Tronche F, Mhaouty-Kodja S (2009) Conditional inactivation of androgen receptor gene in the nervous system: effects on male behavioral and neuroendocrine responses. J Neurosci 29:4461–4470

Ratey J, Sovner R, Parks A, Rogentine K (1991) Buspirone treatment of aggression and anxiety in mentally retarded patients: a multiple-baseline, placebo lead-in study. J Clin Psychiatry 52:159–162

Reist C, Nakamura K, Sagart E, Sokolski KN, Fujimoto KA (2003) Impulsive aggressive behavior: open-label treatment with citalopram. J Clin Psychiatry 64:81–85

Rewerski W, Kostowski W, Piechocki T, Rylski M (1971) The effects of some hallucinogens on aggressiveness of mice and rats. I. Pharmacology 5:314–320

Ricci LA, Grimes JM, Melloni RH Jr (2004) Serotonin type 3 receptors modulate the aggression-stimulating effects of adolescent cocaine exposure in Syrian hamsters (*Mesocricetus auratus*). Behav Neurosci 118:1097–1110

Rios M, Lambe EK, Liu RJ, Teillon S, Liu JH, Akbarian S, Roffler-Tarlov S, Jaenisch R, Aghajanian GK (2006) Severe deficits in 5-HT_{2A}-mediated neurotransmission in BDNF conditional mutant mice. J Neurobiol 66:408–420

Rocha BA, Scearce-Levie K, Lucas JJ, Hiroi N, Castanon N, Crabbe JC, Nestler EJ, Hen R (1998) Increased vulnerability to cocaine in mice lacking the serotonin-1B receptor. Nature 393:175–178

Rodgers RJ, Waters AJ (1985) Benzodiazepines and their antagonists: a pharmacoethological analysis with particular reference to effects on "aggression". Neurosci Biobehav Rev 9:21–35

Rodriguez-Arias M, Minarro J, Aguilar MA, Pinazo J, Simon VM (1998) Effects of risperidone and SCH 23390 on isolation-induced aggression in male mice. Eur Neuropsychopharmacol 8:95–103

Rodriguiz RM, Chu R, Caron MG, Wetsel WC (2004) Aberrant responses in social interaction of dopamine transporter knockout mice. Behav Brain Res 148:185–198

Rosell DR, Thompson JL, Slifstein M, Xu X, Frankle WG, New AS, Goodman M, Weinstein SR, Laruelle M, Abi-Dargham A, Siever LJ (2010) Increased serotonin 2A receptor availability in the orbitofrontal cortex of physically aggressive personality disordered patients. Biol Psychiatry 67:1154–1162

Rudissaar R, Pruus K, Skrebuhhova T, Allikmets L, Matto V (1999) Modulatory role of 5- HT_3 receptors in mediation of apomorphine-induced aggressive behaviour in male rats. Behav Brain Res 106:91–96

Rujescu D, Giegling I, Bondy B, Gietl A, Zill P, Moller HJ (2002) Association of anger-related traits with SNPs in the TPH gene. Mol Psychiatry 7:1023–1029

Rydén E, Thase ME, Straht D, Aberg-Wistedt A, Bejerot S, Landen M (2009) A history of childhood attention-deficit hyperactivity disorder (ADHD) impacts clinical outcome in adult bipolar patients regardless of current ADHD. Acta Psychiatr Scand 120:239–246

Sabol SZ, Hu S, Hamer D (1998) A functional polymorphism in the monoamine oxidase A gene promoter. Hum Genet 103:273–279

Sakaue M, Ago Y, Sowa C, Sakamoto Y, Nishihara B, Koyama Y, Baba A, Matsuda T (2002) Modulation by 5-HT_{2A} receptors of aggressive behavior in isolated mice. Jpn J Pharmacol 89:89–92

Sallinen J, Haapalinna A, Viitamaa T, Kobilka BK, Scheinin M (1998) Adrenergic α_{2c} receptors modulate the acoustic startle reflex, prepulse inhibition, and aggression in mice. J Neurosci 18:3035–3042

Sanchez C, Arnt J, Hyttel J, Moltzen EK (1993) The role of serotonergic mechanisms in inhibition of isolation-induced aggression in male mice. Psychopharmacology 110:53–59

Sanchez C, Hyttel J (1994) Isolation-induced aggression in mice: effects of 5- hydroxytryptamine uptake inhibitors and involvement of postsynaptic 5-HT_{1A} receptors. Eur J Pharmacol 264:241–247

Sanchez C, Meier E (1997) Behavioral profiles of SSRIs in animal models of depression, anxiety and aggression. Psychopharmacology 129:197–205

Sano Y, Ornthanalai VG, Yamada K, Homma C, Suzuki H, Suzuki T, Murphy NP, Itohara S (2009) X11-like protein deficiency is associated with impaired conflict resolution in mice. J Neurosci 29:5884–5896

Saria A, Theodorsson-Norheim E, Lundberg JM (1984) Evidence for specific neuropeptide Y-binding sites in rat brain synaptosomes. Eur J Pharmacol 107:105–107

Saudou F, Amara DA, Dierich A, Lemeur M, Ramboz S, Segu L, Buhot MC, Hen R (1994) Enhanced aggressive behavior in mice lacking 5-HT_{1B} receptor. Science 265:1875–1878

Scott AL, Bortolato M, Chen K, Shih JC (2008) Novel monoamine oxidase A knock out mice with human-like spontaneous mutation. NeuroReport 19:739–743

Scott MM, Wylie CJ, Lerch JK, Murphy R, Lobur K, Herlitze S, Jiang W, Conlon RA, Strowbridge BW, Deneris ES (2005) A genetic approach to access serotonin neurons for in vivo and in vitro studies. Proc Natl Acad Sci USA 102:16472–16477

Sgoifo A, Stilli D, Musso E, Mainardi D, Parmigiani S (1992) Offensive and defensive bite-target topographies in attacks by lactating rats. Aggress Behav 17:47–52

Shaikh MB, De Lanerolle NC, Siegel A (1997) Serotonin 5-HT_{1A} and 5-$HT_{2/1C}$ receptors in the midbrain periaqueductal gray differentially modulate defensive rage behavior elicited from the medial hypothalamus of the cat. Brain Res 765:198–207

Shih JC, Ridd MJ, Chen K, Meehan WP, Kung MP, Seif I, De Maeyer E (1999) Ketanserin and tetrabenazine aggression in mice lacking monoamine oxidase A. Brain Res 835:104–112

Siegel A, Roeling TAP, Gregg TR, Kruk MR (1999) Neuropharmacology of brain-stimulation-evoked aggression. Neurosci Biobehav Rev 23:359–389

Sijbesma H, Schipper J, De Kloet ER, Mos J, Van Aken H, Olivier B (1991) Postsynaptic 5-HT_1 receptors and offensive aggression in rats: a combined behavioural and autoradiographic study with eltoprazine. Pharmacol Biochem Behav 38:447–458

Silva AJ, Paylor R, Wehner JM, Tonegawa S (1992) Impaired spatial learning in α-calcium-calmodulin kinase-II mutant mice. Science 257:206–211

Sinha R, Cloninger CR, Parsian A (2003) Linkage disequilibrium and haplotype analysis between serotonin receptor 1B gene variations and subtypes of alcoholism. Am J Med Genet B Neuropsychiatr Genet 121B:83–88

Sjoberg RL, Nilsson KW, Wargelius HL, Leppert J, Lindstrom L, Oreland L (2007) Adolescent girls and criminal activity: role of MAOA-LPR genotype and psychosicial factors. Am J Med Genet B Neuropsychiatr Genet 144B:159–164

Smuts BB (1986) Sexual competition and mate choice. In: Smuts BB, Cheney DL, Seyfarth RM, Wrangham RW, Struhsaker TT (eds) Primate societies. University of Chicago Press, Chicago, pp 385–399

Sofia RD (1969) Structural relationship and potency of agents which selectively block mouse killing (muricide) behavior in rats. Life Sci 8:1201–1210

Sperry TS, Thompson CK, Wingfield JC (2003) Effects of acute treatment with 8-OH-DPAT and fluoxetine on aggressive behaviour in male song sparrows (Melospiza melodia morphna). J Neuroendocrinol 15:150–160

Spielberg CD, Sharma S, Singh M (1973) Development of Hindi edition of state-trait anxiety inventory. Indian J Psychol 48:11–20
Spigset O (1999) Adverse reactions of selective serotonin reuptake inhibitors—reports from a spontaneous reporting system. Drug Saf 20:277–287
Sprouse JS, Aghajanian GK (1987) Electrophysiological responses of serotoninergic dorsal raphe neurons to 5-HT_{1A} and 5-HT_{1B} agonists. Synapse 1:3–9
Steiniger F (1950) Beiträg zur Soziologie und sonstigen Biologie der Wanderratte. Z Tierpsychol 7:356–379
Stoff DM, Vitiello B (1996) Role of serotonin in aggression of children and adolesecents: biochemical and phamacological studies. In: Stoff DM (ed) Aggression and violence: genetic. Neurobiological and Biosocial Perspectives, Lawrence Erlbaum Associates, Mahwah, pp 101–123
Stork O, Ji FY, Kaneko K, Stork S, Yoshinobu Y, Moriya T, Shibata S, Obata K (2000) Postnatal development of a GABA deficit and disturbance of neural functions in mice lacking GAD65. Brain Res 865:45–58
Stork O, Welzl H, Cremer H, Schachner M (1997) Increased intermale aggression and neuroendocrine response in mice deficient for the neural cell adhesion molecule (NCAM). Eur J Neurosci 9:1117–1125
Stowers L, Holy TE, Meister M, Dulac C, Koentges G (2002) Loss of sex discrimination and male–male aggression in mice deficient for TRP2. Science 295:1493–1500
Sun HF, Chang YT, Fann CS, Chang CJ, Chen YH, Hsu YP, Yu WY, Cheng AT (2002) Association study of novel human serotonin 5-HT_{1B} polymorphisms with alcohol dependence in Taiwanese Han. Biol Psychiatry 51:896–901
Swanson JW, Swartz MS, Van Dorn RA, Volavka J, Monahan J, Stroup TS, Mcevoy JP, Wagner HR, Elbogen EB, Lieberman JA (2008) Comparison of antipsychotic medication effects on reducing violence in people with schizophrenia. Br J Psychiatry 193:37–43
Swanson LW, Sawchenko PE, Rivier J, Vale WW (1983) Organization of ovine corticotropin-releasing factor immunoreactive cells and fibers in the rat brain: an immunohistochemical study. Neuroendocrinology 36:165–186
Takahashi A, Kwa C, DeBold JF, Miczek KA (2010a) $GABA_A$ receptors in the dorsal raphé nucleus of mice: escalation of aggression after alcohol consumption. Psychopharmacology 211:467–477
Takahashi A, Quadros IM, de Almeida RM, Miczek KA (2011) Brain serotonin receptors and transporters: initiation vs. termination of escalated aggression. Psychopharmacology 213:183–212
Takahashi A, Shimamoto A, Boyson CO, DeBold JF, Miczek KA (2010b) $GABA_B$ receptor modulation of serotonin neurons in the dorsal raphe nucleus and escalation of aggression in mice. J Neurosci 30:11771–11780
Takayanagi Y, Yoshida M, Bielsky IF, Ross HE, Kawamata M, Onaka T, Yanagisawa T, Kimura T, Matzuk MM, Young LJ, Nishimori K (2005) Pervasive social deficits, but normal parturition, in oxytocin receptor-deficient mice. Proc Natl Acad Sci USA 102:16096–16101
Tao R, Auerbach SB (2000) Regulation of serotonin release by GABA and excitatory amino acids. J Psychopharmacol 14:100–113
Tao R, Ma Z, Auerbach SB (1996) Differential regulation of 5-hydroxytryptamine release by $GABA_A$ and $GABA_B$ receptors in midbrain raphe nuclei and forebrain of rats. Br J Pharmacol 119:1375–1384
Taylor SP (1967) Aggressive behavior and physiological arousal as a function of provocation and the tendency to inhibit aggression. J Personality 35:297–310
Tazi A, Dantzer R, Le Moal M, Rivier J, Vale W, Koob GF (1987) Corticotropin-releasing factor antagonist blocks stress-induced fighting in rats. Regul Pep 18:37–42
Ten Eyck GR (2008) Serotonin modulates vocalizations and territorial behavior in an amphibian. Behav Brain Res 193:144–147
Tompkins EC, Clemento AJ, Taylor DP, Perhach JL Jr (1980) Inhibition of aggressive behavior in rhesus monkeys by buspirone. Res Commun Psychol Psychiatr Behav 5:337–352

Troisi A, Vicario E, Nuccetelli F, Ciani N, Pasini A (1995) Effects of fluoxetine on aggressive behavior of adult inpatients with mental retardation and epilepsy. Pharmacopsychiatry 28:1–4

Tsai SJ, Liao DL, Yu YW, Chen TJ, Wu HC, Lin CH, Cheng CY, Hong CJ (2005) A study of the association of (Val66Met) polymorphism in the brain-derived neurotrophic factor gene with alcohol dependence and extreme violence in Chinese males. Neurosci Lett 381:340–343

Tyler CB, Miczek KA (1982) Effects of phencyclidine on aggressive behavior in mice. Pharmacol Biochem Behav 17:503–510

Tyrer P, Oliver-Africano PC, Ahmed Z, Bouras N, Cooray S, Deb S, Murphy D, Hare M, Meade M, Reece B, Kramo K, Bhaumik S, Harley D, Regan A, Thomas D, Rao B, North B, Eliahoo J, Karatela S, Soni A, Crawford M (2008) Risperidone, haloperidol, and placebo in the treatment of aggressive challenging behaviour in patients with intellectual disability: a randomised controlled trial. Lancet 371:57–63

Valentino RJ, Commons KG (2005) Peptides that fine-tune the serotonin system. Neuropeptides 39:1–8

Valzelli L (1977) About a specific neurochemistry of aggressive behavior. In: Delgado JMR (ed) Behavioral neurochemistry. Spectrum Publications, Inc., New York, pp 113–132

Valzelli L, Bernasconi S (1979) Aggressiveness by isolation and brain serotonin turnover changes in different strains of mice. Neuropsychobiology 5:129–135

Valzelli L, Giacalone E, Garattini S (1967) Pharmacological control of aggressive behavior in mice. Eur J Pharmacol 2:144–146

Van den Berg L, Vos-Loohuis M, Schilder MBH, van Oost BA, Hazewinkel HAW, Wade CM, Karlsson EK, Lindblad-Toh K, Liinamo AE, Leegwater PAJ (2008) Evaluation of the serotonergic genes htr1A, htr1B, htr2A, and slc6A4 in aggressive behavior of golden retriever dogs. Behav Genet 38:55–66

Van Der Vegt BJ, de Boer SF, Buwalda B, de Ruiter AJ, de Jong JG, Koolhaas JM (2001) Enhanced sensitivity of postsynaptic serotonin-1A receptors in rats and mice with high trait aggression. Physiol Behav 74:205–211

Van Der Vegt BJ, Lieuwes N, van de Wall EH, Kato K, Moya-Albiol L, Martinez-Sanchis S, de Boer SF, Koolhaas JM (2003) Activation of serotonergic neurotransmission during the performance of aggressive behavior in rats. Behav Neurosci 117:667–674

Van Erp AMM, Miczek KA (2000) Aggressive behavior, increased accumbal dopamine, and decreased cortical serotonin in rats. J Neurosci 20:9320–9325

van Oortmerssen GA, Bakker TCM (1981) Artificial selection for short and long attack latencies in wild *Mus musculus domesticus*. Behav Genet 11:115–126

van Riel E, Meijer OC, Veenema AH, Joels M (2002) Hippocampal serotonin responses in short and long attack latency mice. J Neuroendocrinol 14:234–239

Vandenbergh JG (1967) The development of social structure in free-ranging rhesus monkeys. Behaviour 29:179–194

Vandermaelen CP, Matheson GK, Wilderman RC, Patterson LA (1986) Inhibition of serotonergic dorsal raphe neurons by systemic and iontophoretic administration of buspirone, a non-benzodiazepine anxiolytic drug. Eur J Pharmacol 129:123–130

Veenema AH, Sijtsma B, Koolhaas JM, De Kloet ER (2005) The stress response to sensory contact in mice: genotype effect of the stimulus animal. Psychoneuroendocrinology 30:550–557

Veiga CP, Miczek KA, Lucion AB, de Almeida RMM (2007) Effect of 5-HT_{1B} receptor agonists injected into the prefrontal cortex on maternal aggression in rats. Braz J Med Biol Res 40:825–830

Veiga CP, Miczek KA, Lucion AB, de Almeida RMM (2011) Social instigation and aggression in postpartum female rats: role of 5-HT_{1A} and 5-HT_{1B} receptors in the dorsal raphé nucleus and prefrontal cortex. Psychopharmacology 213:475–487

Vekovischeva OY, Aitta-aho T, Echenko O, Kankaanpaa A, Seppala T, Honkanen A, Sprengel R, Korpi ER (2004) Reduced aggression in AMPA-type glutamate receptor GluR-A subunit-deficient mice. Genes Brain Behav 3:253–265

Vergnes M, Depaulis A, Boehrer A (1986) Parachlorophenylalanine-induced serotonin depletion increases offensive but not defensive aggression in male rats. Physiol Behav 36:653–658

Vitiello B, Stoff DM (1997) Subtypes of aggression and their relevance to child psychiatry. J Am Acad Child Adolesc Psychiatry 36:307–315

Volavka J, Czobor P, Citrome L, McQuade RD, Carson WH, Kostic D, Hardy S, Marcus R (2005) Efficacy of aripiprazole against hostility in schizophrenia and schizoaffective disorder: data from 5 double-blind studies. J Clin Psychiatry 66:1362–1366

Walsh MT, Dinan TG (2001) Selective serotonin reuptake inhibitors and violence: a review of the available evidence. Acta Psychiatr Scand 104:84–91

Walther DJ, Peter JU, Bashammakh S, Hortnagl H, Voits M, Fink H, Bader M (2003) Synthesis of serotonin by a second tryptophan hydroxylase isoform. Science 299:76

Wang QP, Ochiai H, Nakai Y (1992) GABAergic innervation of serotonergic neurons in the dorsal raphe nucleus of the rat studied by electron microscopy double immunostaining. Brain Res Bull 29:943–948

Weder N, Yang BZ, Douglas-Palumberi H, Massey J, Krystal JH, Gelernter J, Kaufman J (2009) MAOA genotype, maltreatment, and aggressive behavior: the changing impact of genotype at varying levels of trauma. Biol Psychiatry 65:417–424

Weerts EM, Miczek KA (1996) Primate vocalizations during social separation and aggression: effects of alcohol and benzodiazepines. Psychopharmacology 127:255–264

Welch BL, Welch AS (1968) Rapid modification of isolation-induced aggressive behavior and elevation of brain catecholamines and serotonin by the quick-acting monoamine-oxidase inhibitor pargyline. Commun Behav Biol 1:347–351

Wersinger SR, Ginns EI, O'Carroll AM, Lolait SJ, Young WS III (2002) Vasopressin V1b receptor knockout reduces aggressive behavior in male mice. Mol Psychiatry 7:975–984

Whale R, Quested DJ, Laver D, Harrison PJ, Cowen PJ (2000) Serotonin transporter (5-HTT) promoter genotype may influence the prolactin response to clomipramine. Psychopharmacology 150:120–122

White SM, Kucharik RF, Moyer JA (1991) Effects of serotonergic agents on isolation-induced aggression. Pharmacol Biochem Behav 39:729–736

Widom CS, Brzustowicz LM (2006) MAOA and the "cycle of violence": Childhood abuse and neglect, MAOA genotype, and risk for violent and antisocial behavior. Biol Psychiatry 60:684–689

Wilmot CA, Vander Wende C, Spoerlein MT (1987) The effects of phencyclidine on fighting in differentially housed mice. Pharmacol Biochem Behav 28:341–346

Winslow JT, Hearn EF, Ferguson J, Young LJ, Matzuk MM, Insel TR (2000) Infant vocalization, adult aggression, and fear behavior of an oxytocin null mutant mouse. Horm Behav 37:145–155

Winslow JT, Insel TR (1993) Effects of central vasopressin administration to infant rats. Eur J Pharmacol 233:101–107

Witte AV, Floel A, Stein P, Savli M, Mien LK, Wadsak W, Spindelegger C, Moser U, Fink M, Hahn A, Mitterhauser M, Kletter K, Kasper S, Lanzenberger R (2009) Aggression is related to frontal serotonin-1A receptor distribution as revealed by PET in healthy subjects. Hum Brain Mapp 30:2558–2570

Yanai K, Son LZ, Endou M, Sakurai E, Nakagawasai O, Tadano T, Kisara K, Inoue I, Watanabe T, Watanabe T (1998) Behavioral characterization and amounts of brain monoamines and their metabolites in mice lacking histamine H_1 receptors. Neuroscience 87:479–487

Yen CY, Stanger RL, Millman N (1959) Ataractic suppression of isolation-induced aggressive behavior. Arch Int Pharmacodyn Ther 123:179–185

Young KA, Berry ML, Mahaffey CL, Saionz JR, Hawes NL, Chang B, Zheng QY, Smith RS, Bronson RT, Nelson RJ, Simpson EM (2002) Fierce: a new mouse deletion of Nr2e1; violent behaviour and ocular abnormalities are background-dependent. Behav Brain Res 132:145–158

Zarcone JR, Hellings JA, Crandall K, Reese RM, Marquis J, Fleming K, Shores R, Williams D, Schroeder SR (2001) Effects of risperidone on aberrant behavior of persons with developmental disabilities: I. A double-blind crossover study using multiple measures. Am J Ment Retard 106:525–538

Zeichner A, Pihl RO (1979) Effects of alcohol and behavior contingencies on human-aggression. J Abnorm Psychol 88:153–160

Zhang XD, Beaulieu JM, Sotnikova TD, Gainetdinov RR, Caron MG (2004) Tryptophan hydroxylase-2 controls brain serotonin synthesis. Science 305:217

Zhang XD, Gainetdinov RR, Beaulieu JM, Sotnikova TD, Burch LH, Williams RB, Schwartz DA, Krishnan KRR, Caron MG (2005) Loss-of-function mutation in tryptophan hydroxylase-2 identified in unipolar major depression. Neuron 45:11–16

Zhuang X, Gross C, Santarelli L, Compan V, Trillat AC, Hen R (1999) Altered emotional states in knockout mice lacking 5-HT_{1A} or 5-HT_{1B} receptors. Neuropsychopharmacology 21:S52–S60

Zitzman D, DeBold JF, Miczek KA (2005) Positive modulation of the $GABA_A$ receptor heightens aggression in ovariectomized, nulliparous female mice. 2005 Neuroscience meeting planner, Program No. 76.15, Washington DC: Society for Neuroscience.

Zouk H, McGirr A, Lebel V, Benkelfat C, Rouleau G, Turecki G (2007) The effect of genetic variation of the serotonin 1B receptor gene on impulsive aggressive behavior and suicide. Am J Med Genet B Neuropsychiatr Genet 144B:996–1002

Selectively Bred Rodents as Models of Depression and Anxiety

Gregers Wegener, Aleksander A. Mathe and Inga D. Neumann

Abstract Stress related diseases such as depression and anxiety have a high degree of co morbidity, and represent one of the greatest therapeutic challenges for the twenty-first century. The present chapter will summarize existing rodent models for research in psychiatry, mimicking depression- and anxiety-related diseases. In particular we will highlight the use of selective breeding of rodents for extremes in stress-related behavior. We will summarize major behavioral, neuroendocrine and neuronal parameters, and pharmacological interventions, assessed in great detail in two rat model systems: The Flinders Sensitive and Flinders Resistant Line rats (FSL/FRL model), and rats selectively bred for high (HAB) or low (LAB) anxiety related behavior (HAB/LAB model). Selectively bred rodents also provide an excellent tool in order to study gene and environment interactions. Although it is generally accepted that genes and environmental factors determine the etiology of mental disorders, precise information is limited: How rigid is the genetic disposition? How do genetic, prenatal and postnatal influences interact to shape adult disease? Does the genetic predisposition determine the vulnerability to prenatal and

G. Wegener (✉)
Centre for Psychiatric Research, Aarhus University Hospital,
8240 Risskov, Denmark
e-mail: wegener@dadlnet.dk

G. Wegener
Unit for Drug Research and Development, School of Pharmacy (Pharmacology),
North-West University, Potchefstroom, South Africa

A. A. Mathe
Department of Clinical Neuroscience, Karolinska Institutet,
Stockholm, Sweden

I. D. Neumann
Department of Behavioural and Molecular Neuroscience,
University of Regensburg, Regensburg, Germany

Curr Topics Behav Neurosci (2012) 12: 139–187

DOI: 10.1007/7854_2011_192

Published Online: 18 February 2012

postnatal or adult stressors? In combination with modern neurobiological methods, these models are important to elucidate the etiology and pathophysiology of anxiety and affective disorders, and to assist in the development of new treatment paradigms.

Keywords Animal models · Selective breeding · Depression · Anxiety

Contents

1 Introduction 140
2 Selectively Bred Models on Depression 143
2.1 The Flinders Sensitive and Resistant Line Rat 143
2.2 Learned Helplessness Rats (cLH/cNLH) 155
2.3 Fawn Hooded Rats 155
2.4 Wistar-Kyoto Rats 156
2.5 Swim Low-Active/Swim High-Active Rats 156
2.6 High/Low Stress Reactivity Mice 157
3 Selectively Bred Models on Anxiety 157
3.1 Rats Selectively HAB and LAB Anxiety-Related Behavior 157
3.2 HAB LAB Mice 169
3.3 Floripa H and L Rats 170
3.4 Maudsley Reactive and Nonreactive Rats 170
3.5 High/Low Avoidance Rats: RHA/RLA; SHA/Bru and SLA/Bru 171
4 Conclusions 171
References 172

1 Introduction

It has been estimated that 127 million Europeans out of a population of 466 million currently live with a brain disorder, with total annual costs (of brain disorders in Europe) of €386 billion in 2004 (Andlin-Sobocki et al. 2005). Of these, mental disorders constitute about 60% of the total costs reflecting the large socioeconomic burden of these diseases. These numbers greatly emphasize the importance of developing new strategies in treating mental disorders.

Stress-related diseases such as depression and anxiety, having a high degree of co-morbidity, represent one of the greatest therapeutic challenges for the twenty-first century. Although it is generally accepted that genes and environmental factors determine depression, precise information is limited: How rigid is the genetic disposition? How do genetic, pre- and post-natal influences interact to shape adult depression? Does the genetic predisposition determine the vulnerability to pre- and post-natal or adult stressors?

When depression, and in some degree anxiety, precipitates, the dominating etiological hypotheses have focused at a dysregulation in the serotonergic and noradrenergic system. This is emphasized by the mechanism of action of the currently marketed antidepressants, which almost exclusively act by direct modulation of these two systems. Unfortunately, the efficacy of the presently clinically used antidepressant drugs are low, only approximately 30–35% after subtracting the placebo effects. Thus, there exists a major unmet medical need, the resolution of which is contingent on elucidating the disease etiology and pathogenesis.

Therefore, good model systems are needed to answer fundamental neurobiological questions and to predict responses to novel therapeutic agents. Animal models can be assessed on the basis of five major criteria (Willner 1984; Geyer and Markou 1995): face validity (how well the model resembles the disease/condition), construct validity (how well the model is consistent with theoretical rationale), etiological validity (how identical are the etiologies of the disease (phenomenon) in the animal model and in humans), convergent/discriminant validity (the degree to which a test correlates with other tests that attempt to measure the same construct/ the degree to which a test measures aspects of a phenomenon that are different from other aspects of the phenomenon that other tests assess (Campbell and Fiske 1959), and predictive validity (how well the model responds favorably to clinically established drugs). An optimal model fulfills all the criteria. However, a model can even be useful, even if not all conditions are met (Geyer and Markou 1995).

Genetic selection for behavioral and other phenotypic characteristics is a core feature in evolution, and crucial for survival of any species. Selective breeding is the process of breeding plants or animals for particular genetic traits, which is desired by the researcher. Selective breeding has proven to be a valuable tool in the advancement of science. In fact, one of the most commonly used animals in laboratory research, the Wistar rat, may be considered—in the strict terms of selective breeding—an example of selective breeding. This strain was developed at the Wistar Institute in 1906 for use in biological and medical research, and it was the first rat strain developed to serve as a model organism at a time when biological laboratories primarily used mice (Clause 1998; Lindsay and Baker 2006). Several of the laboratory rat strains used today originate from the original Wistar colony established by Donaldson, Greenman, and King (Clause 1998; Lindsay and Baker 2006).

With this development, it is therefore not surprising that several selectively bred animal models also have been established in neuroscience and psychiatric research.

Selectively bred models are essential for studying underlying mechanisms of the disease and have been established for various psychopathological entities/ phenotypes, as it appears from Table 1.

The present chapter will highlight the use of selective breeding in order to establish animal models for research in psychiatry, with special focus on depression and anxiety disorders. We will highlight two rat models: the Flinders Sensitive and Flinders Resistant Line (FSL/FRL) and rats selectively bred for high (HAB) or low (LAB) anxiety-related behavior (HAB/LAB model). A very brief description of other existing selectively bred rodents, modeling depression and anxiety is also given. For the remaining models on e.g., schizophrenia and epilepsy, a detailed description can be found in the references listed in Table 1.

Table 1 Overview of some selectively bred animal models

Disease/phenotype	Model	Species	Reference
ADHD/impulsivity	Naples high/low excitability (NHE/NLE)	Rats	Sadile et al. (1988); Cerbone et al. (1993); Viggiano et al. (2003)
Aggression	Novosibirsk	Rats	Naumenko et al. (1989)
	Turku	Mice	Sandnabba (1996)
	SAL/LAL	Mice	van Oortmerssen and Bakker (1981)
Alcoholism	HAD/LAD	Rats	Li et al. (1993); Li and Lumeng (1977); Murphy et al. (2002)
	Fawn hooded (FH/Wjd)	Rats	Rezvani et al. (1990), (1991), (2007)
Anxiety	Floripa H and L	Rats	Ramos et al. (1998), (2002), (2003)
	HAB/LAB	Rats	*See Main Text* Liebsch et al. (1998b); Landgraf and Wigger (2002); Landgraf et al. (2007)
	HAB/LAB	Mice	Kessler et al. (2011); Kromer et al. (2005)
	Maudsley MRS/MNS	Rats	Broadhurst (1960); Blizard and Adams (2002)
	High/low avoidance: syracuse SHA/BRU and SLA/Bru Roman HA/LA	Rats	Brush et al. (1985), (1989); Brush (2003)
Depression	WKY	Rats	Pare and Redei (1993); Lahmame et al. (1997); Will et al. (2003)
	FSL/FRL	Rats	*See Main Text* Overstreet et al. (2005); Overstreet (1986)
	SwLo/SwHi	Rats	West and Weiss (1998b)
	LR/HR	Mice	Touma et al. (2008)
	cLH/cNLH	Rats	Vollmayr and Henn (2001); Vollmayr et al. (2001)
	Fawn hooded (FH/Wjd)	Rats	Rezvani et al.(2002)
Epilepsy	WAG/Rij	Rats	Van Luijtelaar and Coenen (1986), (1989); Coenen and Van Luijtelaar (2003)
	Gaers	Rats	Vergnes et al. (1982); Marescaux et al. (1992); Marescaux et al. (1984)
Schizophrenia	APO-SUS/APO-UNSUS	Rats	Ellenbroek and Cools (2000); Costall and Naylor (1973); Ellenbroek and Cools (2002)

Table 2 Symptoms in depressed individuals which can be modeled in FSL rats

Symptom/activity	Patients	FSL rats
Suicidal ideas	Frequent	Cannot be modeled
Activity	Psychomotor retardation	Reduced bar pressing for rewards
Anhedonia	Yes	Yes (following stress)
Appetite	Reduced	Reduced
Weight	Weight loss	Lower body weight
Cognitive performance	Reduced	Reduced/normal (dependent on test)
REM Sleep	Elevated	Elevated
Anxiety	Not a core feature	No anxiety
HPA axis dysregulation	Yes	Yes
Treatment response (see text)	Yes	Yes
Killer T-cell activity	Reduced	Reduced
Cardiovascular morbidity	Increased	Increased

2 Selectively Bred Models on Depression

2.1 The Flinders Sensitive and Resistant Line Rat

The Flinders Line rats were established by selective breeding for differential responses to the anticholinesterase agent, diisopropyl fluorophosphate (DFP), at Flinders University in Adelaide, Australia. The original rationale was to breed a rat strain that would be genetically resistant to irreversible anticholinesterase agents, DFP. However, the selective breeding of Sprague–Dawley (SD) rats, resulted in a line more sensitive to DFP, the Flinders Sensitive Line (FSL), whereas the Flinders Resistant Line (FRL) rats were not more resistant than an outbred control (Overstreet et al. 1979; Russell et al. 1982). Being less tolerant to DFP, the FSL rat were also found to be more sensitive to drugs targeting the cholinergic system, in particular effects of directly acting muscarinic receptor agonists (Russell and Overstreet 1987; Overstreet and Russell 1982; Overstreet 1986) and to have more muscarinic receptors in several brain regions (Overstreet and Russell 1984).

As it also was reported that depressed individuals were more sensitive to cholinergic agonists than normal controls, defined by behavior, neuroendocrine measures and sleep (Janowsky et al. 1980, 1994; Risch et al. 1981), it was suggested that the FSL rat might be a model for depression.

Today there are now breeding colonies of the FSL rats in Australia, Canada, Denmark, Greece, Israel, Mexico, South Africa, Sweden and United States.

2.1.1 Key Features of the FSL Rat Depression Model

As mentioned, the FSL line phenotypically resembles a number of depression symptoms and has been a useful tool to elucidate the endophenotype of depression. Indeed, extensive work has demonstrated that many of the core symptoms of depression can be reproduced in the FSL strain; the more salient characteristics are shown in Table 1.

Depression-Related Behavior

Several observations of the unmotivated and motivated behavior of the FSL rat suggest that it exhibits psychomotor retardation, a key behavioral characteristic of depressed individuals (Lecrubier 2006). In particular, the FSL rat is less active in a novel open field (Overstreet and Russell 1982; Overstreet et al. 1986), bar-presses at a low rate for water or food reward (Overstreet and Russell 1982; Bushnell et al. 1995), and does not complete food-motivated nonmatching- to-sample learning trials in a timely manner (Bushnell et al. 1995).

Importantly, the FSL rats show increased immobility in the forced swim test (FST), which is the prototypic screening tool for depression-like behavior in rodents (Overstreet and Russell 1982; Schiller et al. 1992; El Khoury et al. 2006). Especially of interest for the predictive validity of the model, these behaviors are reversible by chronic but not acute treatment with antidepressants.

Anhedonia, the inability to experience pleasure, is often regarded as a core symptom of depression. Interestingly, under basal conditions the FSL compared to FRL rats did not show signs of anhedonia, and signs of anhedonia were only found when FSL rats were exposed to chronic mild stress (Pucilowski et al. 1993; Matthews et al. 1996), supporting this model as being a candidate for Gene × Environment studies. We have replicated these findings, and found that group housed FSL rats display a higher level of anhedonia following chronic mild stress exposure, when compared with the FRL rats (Mathé et al. unpublished results).

A reduction in appetite is a classical symptom seen in most depressed individuals, while an increase in appetite and weight gain is observed in fewer. Appetite and food intake in the FSL and FRL rats has not been studied in detail, but the FSL rat weighs less than the FRL rat and in a recent study the FSL were found to consume less food than FRL (Abildgaard et al. 2010). It, therefore, appears that the FSL rats have a decreased appetite thus resembling the reduced appetite in depressed individuals.

Anxiety-Related Behavior

Anxiety is not considered a core feature of depression, but there exist a high degree of co-morbidity of anxiety with depression. Therefore, the behavior of FSL was examined in the classical test of anxiety-like behavior, the elevated plus maze (EPM), which is an unconditioned test for anxiety in rodents, and works by creating a conflict between an animal's exploratory drive and its fear of open and brightly-lit areas. Under baseline conditions no differences were discovered between the FSL and FRL lines (Overstreet et al. 1995). Treatment with a benzodiazepine exerted a comparable anxiolytic effect in both FSL and FRL rats and did not differentiate between the two strains (Schiller et al. 1991; Mathé et al. unpublished data).

However, our own recent results demonstrated that FSL rats had a reduced level of unconditioned anxiety on the EPM compared to the FRL rats (Abildgaard et al. 2010). Specifically, the FSL spent more time on the open arms and had a higher level of full entries onto open arms. Similar findings have been described in young FSL rats compared to SD rats (Braw et al. 2006). It is, however, of interest to note that FSL rats did exhibit some anxiogenic behavior in the social interaction task (Overstreet et al. 2004b), which may reflect enhanced social anxiety, or alternatively, reduced social motivation and social withdrawal.

Taken together, as anxiety does not seem to be a prominent feature of the FSL strain, the FSL rats seem to be a model for depression without comorbidity of anxiety. However, as anxiety can be judged from multiple paradigms, further studies are warranted.

Cognition

Cognitive disturbances in depressed individuals can involve both learning difficulties and memory loss. Most learning and memory studies on FSL rats used foot shock as the motivating stimulus, where the FSL rat show greater difficulty in acquiring a shock-motivated, active avoidance task (Overstreet et al. 1990). On the other hand, the FSL rat exhibited normal memory of a shock-motivated passive avoidance task (Overstreet et al. 1992; Russell et al. 1982). Also in a food-motivated task similar completion rates between the FSL and the FRL were obtained, achieved by reducing the size of the food pellet in the FSL rats (Bushnell et al. 1995). Although the FSL rats did not perform this task as rapidly as the FRL rats, they chose the correct bar just as efficiently (Bushnell et al. 1995). Thus, there is no definitive evidence for cognitive disturbances in the FSL rats under basal conditions, which is further underlined by a recent study from Aarhus, where the FSL show similar spatial memory abilities as the FRL in the Morris Water Maze (Wegener et al. unpublished).

Pain

Affective disorders have been repeatedly linked with alterations in thermal and visceral pain perception (Haug et al. 2004; Vedolin et al. 2009; Robinson et al. 2009). However, only a limited number of studies have been carried out in the FSL model. In a model with partial denervation of the sciatic nerve (PSL model), which produces a chronic decrease in touch and heat withdrawal thresholds (allodynia) and an increased response to noxious mechanical and heat stimuli (hyperalgesia, Fujioka et al. 2001; Fumagalli et al. 2007), the FSL rats expressed significantly lower levels of tactile allodynia and less heat hyperalgesia following PSL injury compared to SD rats

(Shir et al. 2001). These studies have been carried out using a denervation model, and no results for basal pain parameters are available.

Sleep Patterns

Sleeping disorders are very closely associated with depressive disorders, with both insomnia and hypersomnia being observed. Two pronounced changes associated with depression are increases in rapid eye movement (REM) sleep and decreases in slow wave sleep (Jindal et al. 2002; Thase et al. 1995; Benca 1996; Benca et al. 1992; Adrien 2002). Basal sleep recordings in the FSL have demonstrated that the FSL rat exhibited a reduced latency to—and greater amount of REM sleep than the FRL rats (Benca et al. 1996; Shiromani et al. 1991), with no differences in slow wave sleep patterns. Thus, the FSL rat resembled depressed individuals with regard to the elevated REM sleep, but not with regard to the reduced slow wave sleep (Jindal et al. 2002; Benca et al. 1992).

Hypothalamic-Pituitary Adrenal Axis

Distinct changes in the Hypothalamic-Pituitary Adrenal (HPA) axis reactivity and cortisol levels in severe depression (melancholia) are well documented (Keck and Holsboer 2001). However, studies in the FSL/FRL model are conflicting: Whereas no differences in corticosterone levels under basal conditions or in response to a chronic mild stressor have been found (Ayensu et al. 1995), in another study, the FSL rats had significantly lower plasma ACTH concentrations compared with the FRL, but still with no differences in plasma corticosterone concentrations between the two groups (Owens et al. 1991). In the brain, it was found that the density of anterior pituitary CRF receptor binding sites was elevated in the FSL rats compared with the FRL (Owens et al. 1991).

This finding is further substantiated by later studies, where FSL and FRL, were subjected to 1 h acute restraint and the effects of the stress exposure, including possible strain specific changes were studied (Zambello et al. 2008). Under basal conditions, no significant differences between FSL and FRL rats in the CRH mRNA expression were found. However, an upregulation of the CRH mRNA hybridization signal was detected in the central amygdala of the stressed FRL, compared to the non-stressed FRL rats (Zambello et al. 2008). Following these findings, it was hypothesized that, since a hypoactive mechanism of response to stressful stimuli in the FSL rats was present, lack of amygdala CRH activation following stress could suggest a subtype of allostatic load, which may alter the interpretation of environmental stimuli by FSL rats and consequently influence their behavioral response to stressful situations (Zambello et al. 2008). However, these suggestions require further examinations.

Monoamine Metabolism

The monoaminergic hypothesis of depression (Schildkraut 1965) suggests that there are distinct abnormalities of the serotonergic system in depressed individuals, although several inconsistencies exist. For example, both serotonergic $5HT_{1A}$ receptor overactivity (Arango et al. 1995) as well as $5HT_2$ receptor underactivity have been reported (Mikuni et al. 1991).

In the FSL, several differences in the serotonin synthesis, $5HT_{1A}$ receptor sensitivity and the density of the serotonin transporter were found compared to the FRL strains (Overstreet et al. 1994; Kanemaru et al. 2009; Nishi et al. 2009; Kovacevic et al. 2010). However, whether these abnormalities are comparable to those seen in depressed individuals are not known, and there are some data suggesting inverse pharmacological responses in FSL rats and depressed individuals. For example, studies have shown that the FSL rats are more sensitive to the hypothermic effects of 5-HT1A receptor agonists (Wallis et al. 1988; Overstreet et al. 1994), but depressed individuals are usually less sensitive to these effects of similar agents (Lesch 1991). Moreover, in depressed individuals, both increases (Reddy et al. 1992), decreases (Asberg et al. 1984) or no changes (Roy et al. 1985) in the serotonin metabolite 5-hydroxy-indoleacetic acid (5-HIAA) in cerebrospinal fluid have been reported. How this may relate to the FSL model remains to be established.

The psychomotor retardation and anhedonia-like features following chronic mild stress (CMS) in the FSL rats may suggest the dopaminergic system to be involved. This has been supported by a few studies on dopamine metabolism and release from selected brain regions of FSL rats (Zangen et al. 2001; Yadid et al. 2001), and in behavioral responses to dopaminergic agents (Crocker and Overstreet 1991). However, it is not clear how these findings can be translated to human pathology.

Nitric Oxide Signaling

The atypical neurotransmitter nitric oxide (NO) possesses both neuroprotective and neurodestructive properties (Dawson and Dawson 1996; McCaslin and Oh 1995). Nitric oxide has been implicated in the psychopathology of depression, as postmortem studies on brains from the Stanley Consortium (Bethesda, MD, USA) have demonstrated that patients suffering from depression have an increase in NO synthase-immunoreactivity in the CA1 hippocampal area (Oliveira et al. 2008). By virtue of its unpaired electron, NO promotes the formation of free radicals and has been linked to various neurodegenerative processes (Ischiropoulos and Beckman 2003). Drugs that affect the major NO pathways have also been shown to possess antidepressant-like properties (Wegener and Volke 2010), and antidepressants have been shown to affect the NO signaling (Wegener et al. 2003). Investigation of NO signaling in the FSL rats did not reveal any baseline FSL–FRL differences in hippocampal constitutive NO synthase (cNOS) activity and neuronal nitric oxide

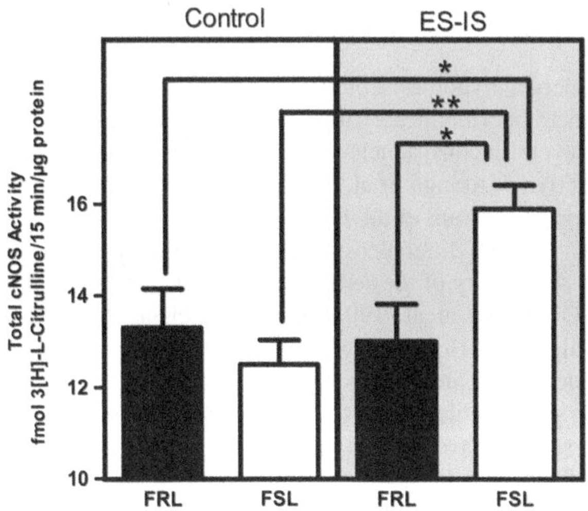

Fig. 1 Hippocampal constitutive nitric oxide synthase (cNOS) activity data under basal conditions [FSL ($n = 8$), FRL ($n = 7$)], and following escapable stress/inescapable stress (ES–IS) [FSL ($n = 8$), FRL ($n = 7$)]. Following ES–IS, cNOS activity is significantly elevated in FSL rats compared to unstressed FSL controls (** $p < 0.005$) and versus pre- and post-stress FRL animals (* $p < 0.05$). Pre- and post-stress activity levels for FRL rats did not differ from one another. Values shown are means + S.E.M. Reprinted from (Wegener et al. 2010) with permission. © Cambridge University Press

synthase (nNOS) protein levels. However, following exposure to stress in the escapable stress/inescapable stress paradigm, the FSL strain showed a larger activation of the cNOS system (Fig. 1), confirming the NMDA–NO cascade as an important vulnerability factor in the depression-like phenotype of the FSL rat (Wegener et al. 2010). Furthermore, several distinct agents affecting NO synthesis have also been shown to be effective antidepressants in the FSL (Wegener et al. unpublished observations, see Table 3).

Neuropeptides

Neuropeptide Y (NPY) is one of the most abundant peptides in the mammalian brain, interacting with the noradrenaline, serotonin and dopamine systems with effects on multiple brain functions. Several findings suggest that NPY plays an important role in the pathophysiology of depression and anxiety (Mathe et al. 2007; Heilig 2004).

Studying NPY in FSL has revealed marked similarities between vicissitudes of NPY in the rat depression models and human subjects. Thus, we have shown decreased NPY levels in the hippocampus of the FSL rats (Jimenez-Vasquez et al. 2000), and NPY protein and mRNA were found reduced in the CA1-2 regions and the dentate gyrus of FSL rats compared with FRL (Jimenez-Vasquez et al. 2000).

Table 3 Effect of some antidepressants (common and experimental), in the forced swim test (FST) in FSL (See also Overstreet et al. (2005); and Overstreet (1993))

Class	Drug	Response in FST	References
TCA	Desipramine	+	Overstreet et al. (2004b); Overstreet and Griebel (2004); Overstreet et al. (2010a); Overstreet et al. (2004a); Overstreet et al. (2008)
	Imipramine	+	(Schiller et al. (1992); Liebenberg et al. (2010)
	Nortryptyline	+	Petersen et al. (2009)
NaSSA	Nefazodone	+	Dremencov et al. (2004)
SSRI	Citalopram	+	Overstreet et al. (2004b)
	Escitalopram	+	El Khoury et al. (2006); Wegener et al. (unpublished)
	Paroxetine	+	Zangen et al. (2001); Zangen et al. (2002); Zangen et al. (1999), (1997)
	Sertraline	+	Pucilowski and Overstreet (1993)
	Fluoxetine	+	Overstreet and Griebel (2004); Overstreet et al. (2004a); Overstreet et al. (2008)
NOS inhibitors	Methylene blue	+	(Wegener et al. (unpublished)
	L-NAME	+	(Wegener et al. (unpublished)
	7-Nitroindazole	+	(Wegener et al. (unpublished)
NPY R5 antagonist	Lu AA33810	+	Walker et al. (2009)
CRF antagonist	CP-154,526	–	Overstreet et al. (2004b)
	SSR125543	+	Overstreet and Griebel (2004)
NK2 antagonists	Saredutant	+	Overstreet et al. (2010b)
Melatonin antagonists	S 20304	+	Overstreet et al. (1998)
PDE inhibitors	Rolipram	+	Overstreet et al. (1989)
	Sildenafil	+ (together with atropine)	Liebenberg et al. (2010)
Miscellaneous	ECS	+	Jimenez-Vasquez et al. (2007); Wegener et al. (unpublished)
	Tianeptine	+	Wegener et al. (unpublished)
	Ketamine	+	Wegener et al. (unpublished)
	Exercise	+	Bjornebekk et al. (2005)
	Nerve growth factor	+	Overstreet et al. 2010a)
	Nemifitide	+	Overstreet et al. (2004a)
	Amibegron	+	Overstreet et al. (2008)
	Lu AA21004	+	Mørk et al. (2012)
	Inositol	+	Einat et al. (2002)

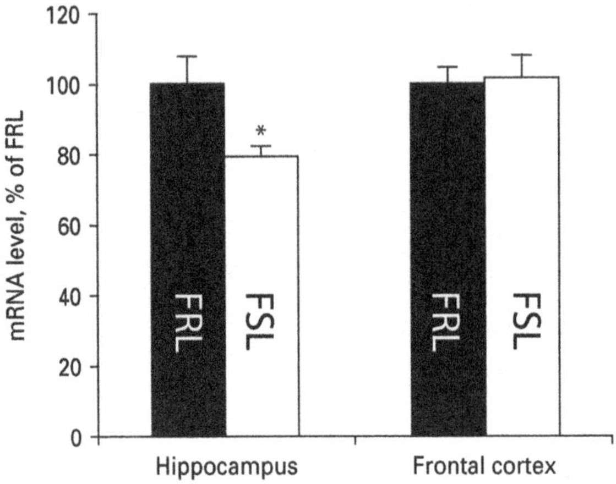

Fig. 2 Messenger RNA samples from hippocampus and frontal cortex of FSL (%; $n = 9$) and FRL (&; $n = 9$) rats were used for quantification of the expression levels of BDNF using real-time qPCR. Values for each individual were normalized with the geometric mean of the reference genes Ywhaz and Hmbs in the hippocampus and Ywhaz and Actb in the frontal cortex. Plotted data show mean group values + S.E.M. of mRNA expression as % of FRL rats. * Indicates significant between-group differences ($p < 0.05$). Reprinted from (Elfving et al. 2010a) with permission. © Cambridge University Press

In contrast, local NPY-Y1 receptor binding was increased, indicating functional significance of the changes in NPY availability (Jimenez-Vasquez et al. 2000a, 2000b; Husum et al. 2001; Caberlotto et al. 1999; Mathe et al. 2007).

Consistent with these data are the findings that antidepressants, lithium and ECS, all increase NPY expression in selected brain regions, and that the increases are larger in the FSL compared with the FRL strain (Husum et al. 2003; Husum et al. 2001; Jimenez Vasquez et al. 2000; Jimenez-Vasquez et al. 2007). For instance, reduced NPY levels in the CSF of depressed patients and altered NPY and NPY receptors mRNA expression in post-mortem brains have been shown (Widdowson et al. 1992; Olsson et al. 2004; Hou et al. 2006; Heilig 2004; Caberlotto et al. 1999; Caberlotto and Hurd 2001). Conversely, increased NPY concentrations in CSF following successful treatment of depressed in-patients with citalopram or ECT have been reported (Nikisch and Mathe 2008; Nikisch et al. 2005).

Neurotrophic Factors

Several studies have found decreased serum or plasma brain-derived neurotrophic factor (BDNF) levels in depressed patients, and a positive correlation between BDNF reduction and the severity of the disease has also been observed (Shimizu et al. 2003; Karege et al. 2002, 2005; Aydemir et al. 2006). Moreover, in post-mortem hippocampal tissue, increased levels of BDNF immunoreactivity have been reported in subjects treated with antidepressants compared to untreated subjects (Chen et al. 2001). These findings constitute the rationale for studying BDNF also in the FSL model of depression. In a recent study from Aarhus, BDNF expression in the hippocampus was significantly decreased in the FSL compared with FRL rats (Fig. 2), while no differences were found in the frontal cortex or CSF (Elfving et al. 2010a). Contraintuitively, BDNF levels in serum and whole blood of the FSL rats were significantly increased compared with FRL rats (Elfving et al. 2010a). Whether this finding is relevant, or a peculiarity of the FSL, remains to be fully established. However, recent studies underline that multiple factors must be taken into consideration when correlating serum BDNF with clinical state (Elzinga et al. 2011; Bus et al. 2011; Gass and Hellweg 2010; Sartorius et al. 2009). Nevertheless, the regulation of the BDNF levels in hippocampus, serum, and whole blood in FSL and FRL rats adds to the hypothesis that neurotrophic factors may be related to the pathophysiology of depression.

Similar to findings with BDNF, we have recently characterized vascular endothelial growth factor (VEGF) in the FSL rats. VEGF protein, but not mRNA, expression in the hippocampus and frontal cortex were found to be significantly decreased in the FSL compared with FRL rats, while no differences were found in the striatum, hypothalamus or serum (Elfving et al. 2010b).

Neurogenesis and Cell Proliferation

Hippocampal neurogenesis has been implicated in the etiology of depression and has been suggested to constitute the final common mechanism underlying antidepressant treatments (Santarelli et al. 2003). In order to further explore the hypothesis that reduction in hippocampal neurogenesis contributes to the etiology of depression, which was essentially based on studies on healthy rats exposed to repeated or chronic stressors, the FSL model was tested under a variety of circumstances (Petersen et al. 2008, 2009; Husum et al. 2006; Bjornebekk et al. 2007).

We found that adult FSL rats have significantly more BrdU-immunoreactive (IR) cells in the dentate gyrus compared with FRL, and aging caused an exacerbated loss of these cell types in the FSL. FSL animals treated chronically with nortriptyline, there was no apparent effect on the number of BrdU-IR cells, although it significantly decreased the immobility time in the FST (Petersen et al. 2009). Taken together, these results clearly demonstrate a dissociation of the effects of antidepressants on behavior in the FST and cell proliferation. Thus SSRIs and tricyclics can decrease immobility in the FST without affecting the cytogenesis and, conversely, increased cytogenesis is not necessarily reflected in decreased depression-like behavior in FST. These data are of importance since they indicate that changes in cell proliferation may be sufficient, but are not necessary for antidepressant effects of currently used antidepressants in the FSL/FRL model.

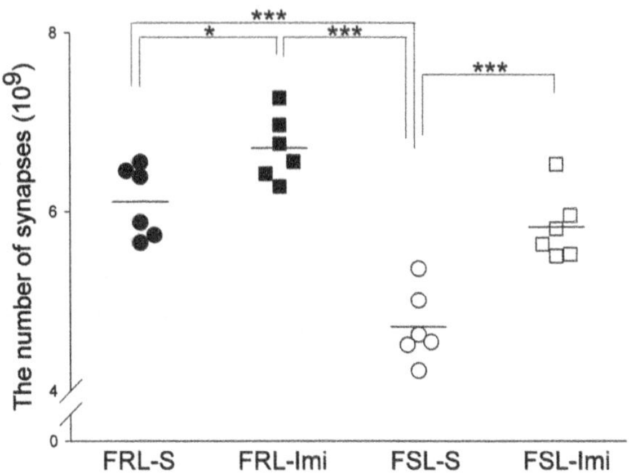

Fig. 3 Effect of Imipramine on the non-perforated spine synapses in FSL and FRL rats. The total number of spine synapses in the CA1 stratum radiatum was significantly smaller in the FSL rats compared to the FRL rats. Following 3 weeks of Imipramine (Imi, 15 mg/kg/day) there were a significant increase in the spine synapses in both FSL and FRL, thereby normalizing the FSL spine synapse numbers. (* $p < 0.05$; *** $p < 0.001$). Modified from (Chen et al. 2010) with permission. © John Wiley & Sons

Also other aspects of brain remodeling have been proposed to be essential for development of disease. Thus, both the hippocampal volume (Videbech and Ravnkilde 2004) and synaptic morphology may play a role (Nestler et al. 2002). Therefore, we have investigated changes in hippocampal volume, neuron and synapse numbers in the FSL and FRL following chronic imipramine therapy, using design-based stereological methods (Chen et al. 2010). We found that the volume and the number of neurons and synapses were significantly smaller in the FSL saline group compared with the FRL saline group, a feature which was reversed following imipramine treatment (Fig. 3). Our experiments illustrate the importance of using a disease model to study cell proliferation and effects of treatments that could potentially be translated to human condition.

Cardiovascular and Metabolic Function

Depression is a well-known risk factor for the development of ischemic heart disease and is associated with increased cardiovascular morbidity and mortality (Barefoot and Schroll 1996; Egede et al. 2005; Hemingway and Marmot 1999; Rugulies 2002). Major depression doubles the risk of adverse cardiovascular events within 12 months in patients with newly diagnosed coronary heart disease (Carney et al. 1988) and increases the risk of mortality after acute myocardial infarction (Frasure-Smith et al. 1993). The presence of diabetes has been found to double the risk of co-morbid depression (Anderson et al. 2001), and a meta-analysis has shown that depression increases the risk of developing type 2 diabetes in adults by 37% (Knol et al. 2006). In a study comparing the myocardial responsiveness to ischemia/reperfusion injury and the effects of ischemic pre-conditioning in hearts from FSL rats using SD rats as controls, it was observed that the myocardial infarct size was significantly larger in the FSL rats than in the SD rats following ischemia/reperfusion injury, but have maintained cardioprotective mechanism following ischemic preconditioning (Solskov et al. 2010). In the same study, it was also demonstrated that FSL were hyperinsulinemic, with a strong tendency in different levels of fasting glucose levels compared with SD rats (Solskov et al. 2010).

However, in a recent study performed in Aarhus, we have not been able to detect any difference in fasting glucose levels in FSL compared with FRL (Abildgaard et al. 2010). Metabolic stress induced by a high fat diet increased insulin levels during an oral glucose tolerance test in both FSL and FRL, with fasting blood glucose levels significantly increased by high fat diet in the FSL rat (Abildgaard et al. 2010). Interestingly, the metabolic changes were associated with increased depression-like behavior in the FST and cognitive impairments in the object recognition test in the FSL only. These findings confirm the FSL as a model with a greater metabolic susceptibility, and further highlight the usefulness of the model in translational interdisciplinary depression research.

2.1.2 Gene × Environment Interactions

Interactions with the environment, which can have negative or positive consequences, have also been found to be a major determinant of disease. For example, psychosocial stress in adulthood impairs the health condition of the individual (Lupien et al. 2009; Bale et al. 2010; Reber et al. 2007), whereas social support or exercise exert beneficial effects on somatic and mental health (Brene et al. 2007; Dishman et al. 2006; Dunn and Dishman 1991; Young 1979; Neumann 2009).

The consequences of acute, subchronic or chronic stress are largely dependent on the individual (and genetically determined) stress susceptibility, and there is good evidence that FSL and FRL as well as HAB and LAB rats provide good models to study gene × environment interactions. For example, as mentioned above, a subchronic adult stress paradigm significantly upregulates the NO signaling pathway in the FSL only (Wegener et al. 2010), and metabolic stress more severely impacts the FSL compared with the FRL (Abildgaard et al. 2010). In another series of experiments, we compared adult female FSL and SD rats in a paradigm of 7 weeks of social isolation at the age of 29 weeks, and observed increased number of BrdU-IR cells in the FSL, whereas it had no impact in the SD strain (Bjornebekk et al. 2007). Other environmental stimuli may be experienced positive. Thus, we have examined the effect of physical activity using running wheels. We observed that voluntary wheel running had antidepressant effects and selectively altered NPY and NPY Y1 receptor and opiate expression in the FSL, but not FRL, rats further supporting a role of NPY in their phenotype

(Bjornebekk et al. 2006, 2010). These findings parallel and support the results from human studies (Russo-Neustadt et al. 1999; Ransford 1982).

In addition to adult stress exposure, gene × environment interactions have been described with respect to early-life stress, either prenatally or postnatally. Thus, a large number of human and animal studies show a strong association between an adverse fetal or immediate postnatal environment and behavioral and emotional development later in life (Abe et al. 2007; Maccari et al. 2003; Nagano et al. 2008; O'Connor et al. 2002; Tazumi et al. 2005; Van Den Bergh et al. 2005). Stressful experiences during early life have been hypothesized to enhance susceptibility (eventually triggered by adult stress) for mental illness (Cottrell and Seckl 2009; Fumagalli et al. 2007; Maynard et al. 2001).

Prenatal stress studies have not yet been carried out in FSL/FRL. However, the classical post-natal stress paradigm, maternal separation, in FSL and FRL have been demonstrated to exacerbate the depression-like behavior of the FSL, but not the FRL (El Khoury et al. 2006). Treatment with escitalopram selectively decreased depression-like behavior in the FST in both maternally non-separated and separated FSL, but not FRL rats (El Khoury et al. 2006). Maternal separation in FSL has been also been found to reduce NPY in dorsal hippocampus of both female and male FSL rats compared with FRL rats (Jimenez-Vasquez et al. 2001; Wortwein et al. 2006).

In another study, we analyzed hippocampal synaptic transmission and plasticity in vivo and ionotropic receptors for glutamate in FSL and FRL rats subjected to maternal separation. A strong inhibition of long-term potentiation (LTP) and lower synaptic expression of NR1 subunit of the NMDA receptor were found in FSL rats (Ryan et al. 2009), and unexpectedly maternal separation induced a remodeling of synaptic plasticity only in FSL rats, reducing inhibition of LTP accompanied by marked increase of synaptic NR1 subunit and GluR2/3 subunits of AMPA receptors (Ryan et al. 2009). This finding is in line with the demonstration that maternal separation increased the hippocampal cell number, while consistently with this increase, chronic escitalopram treatment reduced the cell number (Petersen et al. 2008; Husum et al. 2008).

In a study of basal differences in synaptic signaling between FSL and FRL rats, as well as on consequences of maternal separation in adulthood, it was found that the FSL rats showed basal differences in the interaction/activation of distinct synaptic mediators purified hippocampal synaptosomes (Musazzi et al. 2010). In addition, following maternal separation, the FSL rats displayed a blunted response of the mediators, suggesting a synaptic dysfunction in the FSL animals (Musazzi et al. 2010). Escitalopram treatment restored some but not all alterations observed in FSL rats after early-life stress, suggesting that early gene-environment interaction may cause life-long synaptic changes affecting the course of depression-like behavior and response to drugs (Musazzi et al. 2010).

Finally, using an open-ended approach based on a proteomic analysis of serum, maternal separation was found to induce changes in inflammation and transport proteins in FSL rats (Carboni et al. 2010), changes that were partly reversed following treatment with escitalopram or nortriptyline (Carboni et al. 2010). No comparison between early-life stress in FSL and FRL was carried out.

These experiments underline that the consequences of environmental factors are strongly determined by the genetic background, suggesting that a genetically shaped phenotype can be further modulated by environmental factors.

2.1.3 Response to Treatment

A detailed review of the different studies, where the FSL rat has been used to test for the antidepressant-like effects of drugs lies beyond the scope of this text, and only a brief overview is given in the Table 3. A detailed review of the classical antidepressants and selective serotonin reuptake inhibitors (SSRIs) as well as a variety of novel agents that presumably have different actions from the well-characterized antidepressants, can be found elsewhere (Overstreet 2002, 2005).

2.2 Learned Helplessness Rats (cLH/cNLH)

The learned helplessness (LH) paradigm is a well characterized rat model of depression, in which the animals are exposed to uncontrollable and unpredictable aversive events, i.e., foot shock (Overmier and Seligman 1967). The model has good face and predictive validity, including alterations in HPA axis activity and REM sleep characteristic of depression (Maier 1991; Breier et al. 1987; Henn and Vollmayr 2005). However, in outbred rats not the entire proportion of animals become helpless. Therefore, breeding of helpless lines from Harlan SD outbred rats was initiated in 1990 to achieve a higher yield of helpless animals following inescapable shock-training (Vollmayr and Henn 2001; Henn and Vollmayr 2005). This resulted in congenitally learned helpless (cLH) rats exhibiting a helpless phenotype without exposure to uncontrollable shock, and a congenitally not learned helpless (cNLH) strain being resistant to the effects of inescapable shock (Vollmayr and Henn 2001).

2.3 Fawn Hooded Rats

A high degree of comorbidity between alcoholism and depression have been reported (Merikangas and Gelernter 1990; Cloninger et al. 1979). The Fawn-Hooded (FH/Wjd) rat is an inbred strain of rat, originally selected from a background of platelet serotonin storage abnormality (Tschopp and Zucker 1972). The rat has been reported to exhibit both high immobility in the FST, elevated serum corticosterone and high voluntary ethanol intake, measures that have been linked with depression and alcoholism in humans (Rezvani et al. 2002, 2007). For example, the FH/Wjd rat drinks up to 6 g/kg 10% ethanol per day, and responds to drugs that are effective in humans with a reduction in alcohol intake (Rezvani et al. 1999). Interestingly, the exaggerated immobility in the FST and the hypercorticosterone levels can be also attenuated following chronic antidepressant treatments (Aulakh et al. 1988, 1993; Rezvani et al. 1999). The FH/Wjd also exhibits abnormalities in the central serotonergic function (Aulakh et al. 1994; Bendotti and Samanin 1987; Arora et al. 1983; Dumbrille-Ross and Tang 1981), but whether these serotonergic abnormalities contribute to both behaviors remains to be determined. In addition, the first results showing decreased NPY in hippocampus in a model of depression were obtained in the FH, changes that were reversed following ECS (Mathe et al. 1998). These findings were of great heuristic value as they led to subsequent identification of reduced NPY expression in other, both genetic and environmental models, including the FSL/FRL as described above.

2.4 Wistar-Kyoto Rats

The Wistar-Kyoto (WKY) rat strain was developed as the normotensive control strain for the spontaneously hypertensive rat, and bred from the Wistar strain starting in 1963 (Okamoto and Aoki 1963). The WKY presents with hormonal, behavioral, and physiological measures that mimic those found in depressed patients, such as increased immobility in the FST (Lahmame et al. 1997; Rittenhouse et al. 2002; Paré 1992, 1994) and dysregulation of the HPA and hypothalamic–pituitary–thyroid axes (Solberg et al. 2001; Redei et al. 1994; Gómez et al. 1996). However, the WKY responds with variable degree to antidepressants (López-Rubalcava and Lucki 2000; Lahmanie and Armario 1996; Lahmame et al. 1997), and has therefore been proposed as a model of treatment-resistant depression (Lahmame et al. 1997). Therefore, the model has been further developed into 'WKY most immobile' (WMI) and 'WKY least immobile' (WLI) rats (Will et al. 2003).

2.5 Swim Low-Active/Swim High-Active Rats

Since low motor activity and a condition of passive stress coping in a swim test have been proposed to represent depression-like behavior in the rat, SD rats were bred in accordance with the motor-activities starting in 1987 (Weiss et al. 1998). Two rat lines have been obtained, Swim Low-Active (SwLo) and Swim High-Active (SwHi) rats, which differ dramatically in FST behavior. The SwLo rats show little struggling and much floating, while SwHi rats show the reverse (Weiss et al. 1998). Importantly, when SwLo rats were given antidepressant, chronic but not acute administration increased swim-test activity of SwLo rats (West and Weiss 1998a). Information on neurotransmitter involvement was limited, but studies suggest involvement of both glutamatergic (Tabb et al. 2007) and dopaminergic (West et al. 1999a, 1999b) mechanisms as well as alterations of the stress axis (Gutman et al. 2008).

2.6 High/Low Stress Reactivity Mice

As mentioned before, dysfunctions (hyper- or hypo-activity) of the HPA axis may play a prominent role in the development of major depressive disorders (De Kloet et al. 1998; Holsboer 2000; Bale 2006). Therefore, attempts of generating animal models mimicking these neuroendocrine core symptoms have been made in order to unravel parameters underlying increased or decreased stress reactivity (Touma et al. 2008). Mice expressing a hyper- or a hyporeactivity of the HPA axis were selected for the 'high reactivity' (HR) and the 'low reactivity' (LR) breeding line. Compared with LR animals, the HR males and females were 'hyperactive' in some behavioral paradigms (Touma et al. 2008), resembling symptoms of restlessness and agitation often seen in melancholic depression. On the neuroendocrine level, the circadian rhythm of glucocorticoid secretion revealed a flattened diurnal rhythm (Touma et al. 2008), mimicking findings from patients suffering from melancholic depression (Deuschle et al. 1997; Keller et al. 2006).

3 Selectively Bred Models on Anxiety

3.1 Rats Selectively HAB and LAB Anxiety-Related Behavior

An adequate level of innate anxiety and fear is essential for survival of individuals and species. Naturally, there exists a wide individual range in trait anxiety: from extremely low to extremely high. Similarly, in humans, anxiety-related pathologies including generalized anxiety, panic disorders or social phobia, reflect extremes in trait anxiety with significant contributions of adverse life events shaping the individual anxiety phenotype. In order to reveal neuroendocrine, neurochemical and neurogenetic mechanisms of a complex behavioral phenotypes such as anxiety, and in order to identify potential targets for psychotherapy, we have established and extensively studied selectively bred HAB and LAB rats (Landgraf and Wigger 2002; Landgraf et al. 2007; Neumann et al. 2010). This approach is particularly promising to further our understanding of genetic mechanisms underlying anxiety-related disorders. Other relevant rodent models for anxiety-related behavior include exposure to early-life stress (Wigger and

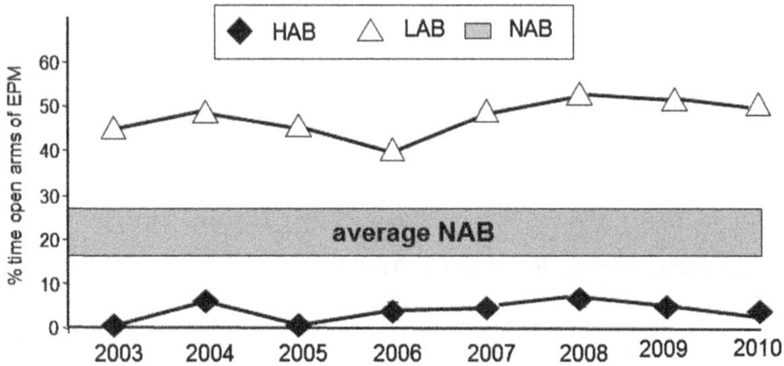

Fig. 4 Anxiety-related behavior of male and female HAB, LAB and non-selected NAB rats on the elevated plus-maze (EPM) which is consistent over the years between 2003 and 2010

Neumann 1999; Huot et al. 2001), chronic stress in adulthood (Barrot et al. 2005; Reber et al. 2007; 2008), or transgenic modifications (Bale 2006; Mantella et al. 2003). Recently, the Landgraf group succeeded in establishing also mice lines selectively bred for high (M-HAB) and low (M-LAB) anxiety-related behavior (see Table 1; Landgraf et al. 2007).

Since 1993, we have selectively and bi-directionally bred outbred Wistar rats for high (HAB) versus low (LAB) anxiety-related behavior based on their behavioral performance on the EPM at the Max Planck Institute of Psychiatry in Munich and, since 2002, at the University of Regensburg (Landgraf and Wigger 2002; Landgraf et al. 2007; Neumann et al. 2010; see Fig. 4). Male and female HAB and LAB rats are selected at the age of 9 weeks for further breeding only, if the percentage of time spent on the open arms of the elevated plus maze (EPM) is below 5% and above 40–45%, respectively. For experimental purposes, HAB rats with an anxiety level of less than 10% and LAB rats with more than 35% time on the open arms during testing at the age of 9 weeks are used.

3.1.1 Key Features of the HAB/LAB Model

Anxiety-Related Behavior

The behavioral profile of HAB and LAB rats with respect to the selection criteria has been reliable and robust over the last 10–15 years (Liebsch et al. 1998b; Neumann et al. 2010; see Fig. 4), is present over all seasons, independent of sex or age and could be confirmed in different European laboratories. The extremes in anxiety could be confirmed in a battery of relevant behavioral tests, including the EPM, the open field, the light dark box and the holeboard (Slattery and Neumann

2010; Ohl et al. 2001; Henniger et al. 2000). In addition, HAB rats are unable to properly extinguish inappropriate fear in tests for conditioned anxiety despite similar acquisition of fear (Muigg et al. 2008). Thus, although bred for high innate non-conditioned anxiety on the EPM, this finding reveals that HAB rats also display exaggerated responses to conditioned fear, which makes them a potentially useful model for posttraumatic stress disorder. Also, LAB rats were able to extinguish this fear faster than non-selected Wistar rats (NAB), again showing robust trait differences between these lines.

Effects of Anxiolytic Treatment

The predictive validity of the HAB/LAB model is substantial, and anxiolytic agents reverse or attenuate the high anxiety phenotype seen in HAB rats.

Acute treatment with the reference anxiolytic drug, diazepam (1 mg/kg, i.p), reduced anxiety in HAB rats on the EPM or in the light–dark box, whereas it was without effect in LAB rats (Liebsch et al. 1998a; Jochum et al. 2007). In addition to its effects on anxiety-related behavior in HAB rats, diazepam also corrected the abnormal pain sensitivity seen in HAB rats (Jochum et al. 2007). Similar to diazepam, we have repeatedly found a reliable and profound anxiolytic effect of chlordiazepoxide (20 mg/kg, i.p.) with relatively low individual differences in responsiveness (Slattery; Beiderbeck and Neumann, unpublished).

Also, manipulation of various relevant anxiogenic and anxiolytic neuropeptide systems of the brain, such as arginine vasopressin, CRH, oxytocin and neuropeptide S, respectively, was found to be effective in reducing the high level of anxiety seen in HAB rats.

Depression-Like Behavior

Importantly, mimicking the high degree of comorbidity of anxiety and depression mentioned above, HAB rats are also characterized by depression-related behavior in the FST, independent of sex (Keck et al. 2003b; Slattery and Neumann 2010; Frank and Landgraf 2008).

Effects of Antidepressant Treatment

Chronic treatment with antidepressant drugs reduces the depression-like status in HAB rats. Treatment of male HAB rats with the SSRI paroxetine during a period of 8 weeks markedly increased active stress coping in the FST to a level similar to that seen in LAB rats. The reversal of the depression-like phenotype was accompanied by a reduction of the high level of hypothalamic vasopressin expression and normalization of the Dexamethasone-suppression/CRH-challenge test (DEX/CRH test) described below (Keck et al. 2003a). However, paroxetine did not alter vasopressin expression or any anxiety-related behavior assessed in LABs, underlining the impact of the genetic predisposition to trait anxiety and comorbid depression-like behavior on drug effects.

Recently, landmark studies have shown that inactivation of Brodmann Area 25 (BA25) using deep brain stimulation alleviated depressive symptoms in severely depressed patients (Mayberg et al. 2005). Interestingly, transient pharmacological inactivation using muscimol of the infralimbic cortex, the rodent correlate of BA25, decreased the high inborn depression-like behavior of the HAB rats supporting their face and predictive validity for affective disorders (Slattery et al. 2010). It remains to be determined whether this region may also be involved in their innate anxiety phenotype.

Several lines of evidence resulting from both preclinical and clinical studies support the view that repetitive transcranial magnetic stimulation (rTMS) of left frontal brain regions exerts antidepressant effects (Post and Keck 2001). Thus, acute, subchronic or long-term rTMS sessions reduced the duration of immobility in rodents in the Porsolt swim test, exert neuroprotective effects both in vitro and in vivo and change the expression of BDNF and cholecystokinin similar to those reported after antidepressant drug treatment (Post and Keck 2001). To test for rTMS efficacy in a psychopathological rat model, HAB and LAB rats received stimuli over two 3-day series (Keck et al. 2001a). The stimulation point was set at the left frontal cortex in order to mimic clinical conditions. Repetitive transcranial magnetic stimulation increased active stress coping in HAB rats, rendering these animals indistinguishable from LAB rats. The rTMS-induced shift in HAB animals towards active stress coping was markedly higher than has previously been reported in "normal" Wistar rats (Zyss et al. 1997). Furthermore, rTMS treatment resulted in a significant attenuation of the neuroendocrine hyper-response of the HPA axis to and acute ethologically relevant stressors characteristic for HAB rats, whereas their high anxiety level remained unchanged.

Cognition

Emotionality and cognition are closely inter-related, as the assessment of environmental stimuli, in particular of potentially dangerous situations, is dependent on the acquisition and storage of information. To further test this hypothesis, the cognitive performance of HAB and LAB rats has been estimated on the modified holeboard, where HABs showed a clearly improved declarative memory performance indicating a better learning strategy despite (or because of) a more passive performance during the test (Ohl et al. 2002). No differences in working memory in a visual-spatial task were found (Ohl et al. 2002).

In confirmation, line-dependent differences were also seen with respect to social memory and cognition abilities, as HAB, but not LAB, males could distinguish between a known and an unknown juvenile after a 30-min inter-exposure interval (Landgraf and Wigger 2002). The lack of social preference seen in LABs (Lukas and Neumann, unpublished) may underlie their impaired social memory.

To which extent the high central vasopressinergic drive (Wigger et al. 2004; Bosch et al. 2006; Bosch and Neumann 2010) contributes to the improved learning and memory performance in HAB rats, remains to be shown. Brain vasopressin is an important neuromodulator promoting cognitive functions (Engelmann et al. 1996) and may thus facilitate the storage of adverse emotional situations and shape future stress coping.

Social Behaviors

Other behavioral differences, which have been established over the years of selective breeding for high versus low trait anxiety, include a variety of social behaviors (Neumann et al. 2010), which is of interest as several psychopathologies often are accompanied by various abnormalities in social interactions including aggression, social phobia, or impaired social bonding.

Inter-Male Aggression

Male HAB and LAB rats differ in inter-male aggression with LAB males showing an extreme high level of offensive behavior in the resident-intruder test (Veenema et al. 2007), a finding which is also reflected by an abnormal aggression of LAB males displayed towards females and anesthetized males compared with NAB rats (Neumann et al. 2010). In contrast, HAB males rather display an intermediate level of intermale aggression (Beiderbeck et al. 2007; Neumann et al. 2010; Veenema et al. 2007). Thus, the selective breeding for low trait anxiety resulted in a socio-behavioral phenotype characterized not only by low anxiety and fear responses, reduced risk assessment and an active stress coping style, but also by abnormal social behaviors in different social settings, thus providing an excellent model for studying mechanisms of pathological aggression.

Social Phobia

In a relevant behavioral setup for social phobia/social preference, NAB and HAB rats show social preference. In this test, rats are allowed to explore a small wired cage placed into the cage of the experimental rat. Social preference is seen, when the animal explores the small cage longer and more often, when it contains a conspecific animal. Here, LABs, in general, do not search for social contact, display reduced contact to cage mates (Ohl et al. 2001) and do not show social reference, which can also be interpreted as social phobia (Lukas and Neumann, unpublished).

Maternal Care

Line-differences in social behavior are also found in females with respect to maternal care and maternal defense behavior (maternal aggression). Here, HAB dams are more protective towards their pups, leave the nest less often, show more arched back nursing and more maternal aggression (Bosch et al. 2005; Bosch and Neumann 2010; Neumann et al. 2005a). A significant contribution of high activity of the brain vasopressin system seen in HAB dams to their maternal behavior profile could recently be identified (Bosch and Neumann 2008, 2010). Interestingly, a correlation between high maternal trait anxiety and the intensity of maternal care has also been found in mouse dams bred for high and low anxiety and, again, line-dependent differences in brain vasopressin appear to underlie the behavioral differences (Kessler et al. 2011).

Pain

Similar to depressed patients, HAB rats have been shown to exhibit decreased sensitivity to thermal pain (Jochum et al. 2007). Acute administration of diazepam partly reversed the abnormal pain response. Also, treatment of male HAB rats with the SSRI citalopram, daily for 8 weeks, reduced anxiety-related behavior as assessed on the EPM and reversed the abnormal thermal pain response in HAB rats (Jochum et al. 2007). As, in contrast to the thermal pain sensitivity, the sensitivity to visceral pain has been found to be elevated in patients suffering from affective disorders (Haug et al. 2004), it would be of interest to test visceral pain threshold in HAB versus LAB rats.

Sleep

In depressed patients, sleep disturbances are a common symptom, as mentioned above. Analysis of sleep patterns in HAB and LAB rats revealed similar circadian fluctuation in sleep-wake behavior, but differences in their spontaneous sleep–wake behavior. HAB rats spend less time awake and more time in non-REM sleep (Lancel et al. 2002). This difference is particularly pronounced during darkness and this is in accord with the observed decreased locomotor activity of HAB rats during the nighttime (Liebsch et al. 1998b). The larger amount of non-REM sleep in the HAB group was not associated with an increased length of non-REM episodes, but with a greater number of sleep episodes, suggesting a higher non-REM sleep fragmentation. HAB rats also displayed less pre-REM sleep and REM sleep than LAB rats during the light period. Thus, the sleep pattern of HAB rats seems to be opposite to that found in depression, and further studies are needed.

Hypothalamic-Pituitary Adrenal Axis

The behavioral differences between HAB and LAB rats are accompanied by distinct neuroendocrine underpinnings. Although basal levels of plasma ACTH or corticosterone reflecting basal activity of the HPA axis do not differ, the responsiveness of the HPA axis to a mild emotional (and non-social) stressor is more pronounced in HAB compared with LAB males (and with non-selected Wistar rats; Landgraf et al. 1999; Neumann et al. 2010), thus resembling psychiatric patients (Holsboer 2000). However, the neuroendocrine response to social stimuli is aggravated in LABs, paralleling their abnormal social behavioral responses described above (Neumann et al. 2010; Veenema et al. 2007). HAB rats also show an aberrant hormonal secretion pattern during the DEX/CRH test (Keck et al. 2002; Neumann et al. 2010), a clinical test used for the neuroendocrine characterization of depressed patients (Ising et al. 2005). Intravenous administration of DEX revealed DEX-non-suppression and, thus, impairment of negative feedback regulation in HAB rats. Subsequent administration of CRH resulted in pronounced ACTH and corticosterone responses, which are absent in LAB and NAB rats. This indicates a contribution of endogenous vasopressin, which—together with exogenous CRH—triggered pituitary secretion of ACTH despite (impaired) DEX-suppression. Consequently, and in support of this hypothesis, an acute intravenous administration of a vasopressin V1a receptor antagonist abolished the ACTH and corticosterone hyper-response and normalized the DEX/CRH test outcome in HAB rats, pointing towards a significant contribution of the endogenous brain vasopressin system in these neuroendocrine abnormalities (Keck et al. 2002). Furthermore, chronic administration of paroxetine over 8 weeks prevented the abnormal DEX/CRH response in HAB rats with a concomitant attenuation of the vasopressin hyperdrive in the hypothalamic paraventricular nucleus (PVN, Keck et al. 2003b).

Neuronal Activity

In addition to neuroendocrine responsiveness, neuronal responses within brain regions belonging to the anxiety/fear circuitry to anxiogenic, social or pharmacological stimuli also differ between HAB and LAB rats (Muigg et al. 2007; Salchner et al. 2006; Salome et al. 2004; Frank et al. 2006). For example, in response to airjet stimulation, an escape-provoking stimulus, HAB rats show a higher neuronal activity in various hypothalamic areas including the medial preoptic and anterior hypothalamic areas, and in the nucleus accumbens as estimated by quantification of the expression of the early immediate gene *fos* (Salome et al. 2004). Similarly, in response to forced swimming, neuronal responses within selected cortical, septal, and hypothalamic areas were more pronounced in HAB males. These neuronal responses could be attenuated after chronic paroxetine treatment (Muigg et al. 2007), which could underlie the reduction in depression-like behavior seen after antidepressive treatment in HAB rats.

Neuropeptides

Vasopressin

Given the importance of brain vasopressin in the regulation of anxiety, depression, and neuroendocrine stress coping (Landgraf and Neumann 2004; Frank and Landgraf 2008), the vasopressin gene was considered a candidate gene in high trait anxiety. Indeed, high vasopressin mRNA expression within the parvocellular part of the PVN both under basal conditions and in response to stressor exposure is a reproducible characteristics for male and female HAB rats (Keck et al. 2003b;

Wigger et al. 2004; Bosch et al. 2006; Bosch and Neumann 2010; Frank and Landgraf 2008). Consequently, increased vasopressin immunoreactivity and local neuropeptide release were both found in the HAB hypothalamus. In addition, vasopressin V1a receptor binding is also elevated within the lateral septum of HAB rats, a region relevant for the regulation of anxiety and social behaviors (Keck et al. 2003b).

In line with our hypothesis of a substantial contribution of high endogenous vasopressin activity to the high anxiety and depression-like phenotype of HAB rats, blockade of vasopressin V1a receptors within the hypothalamic PVN reduced their anxiety level and resulted in a more active coping style (Wigger et al. 2004). Further, the V1a antagonist (i.v.) normalized the pathological outcome of the DEX/CRH test in male HABs (Keck et al. 2002). These findings, reflecting both construct and predictive validity, confirm the involvement of endogenous central vasopressin in the behavioral and neuroendocrine phenomena of high trait anxiety and depression.

Interestingly, line-dependent differences in brain vasopressin also appear to contribute to differences in social behavior, i.e. intermale aggression as well as maternal behavior (Beiderbeck et al. 2007; Veenema et al. 2007; Bosch et al. 2010; Bosch and Neumann 2008).

Given the robust increase in brain vasopressin activity in HAB rats in parallel to their anxiogenic and hyperresponsive neuroendocrine phenotypes, underlying genetic mechanisms are likely. Indeed, 10 single nucleotide polymorphisms (SNPs) within the vasopressin promoter were found between the lines. In addition, a single base pair substitution has been identified in the first intron of the vasopressin gene of HAB rats itself (Murgatroyd et al. 2004). One of the SNPs identified was found to be embedded in a potential transcription factor binding site (CArG box), the locus of binding to the transcriptional repressor CBF-A. Functional relevance of the SNP was identified by in vitro DNA binding assay and revealed that CBF-A binding to the CArG box derived from the HAB allele was indeed diminished, resulting in an attenuated transcriptional repression of the vasopressin gene. Thus, this genetic mechanism may underlie vasopressin over-expression in the PVN of HAB rats (Murgatroyd et al. 2004; Landgraf et al. 2007). Interestingly, out of 100 outbred Wistar rats, the HAB allele was found in 3 rats (heterozygeous) indicating a gene frequency of 1.5% in the general Wistar rat population.

In addition to vasopressin, other brain neuropeptides such as CRH, oxytocin, prolactin, NPY, or neuropeptide S are important neuromodulators of emotionality, in particular of anxiety- and depression-related behaviors. Thus, an endophenotype of high or low anxiety accompanied by differences in active or passive stress coping style is likely to be accompanied by differences in the activity of several endogenous neuropeptide and other neurotransmitter systems.

Corticotropin-Releasing Hormone

Corticotropin-Releasing Hormone (CRH) exerts anxiogenic and depression-like effects, and CRH mRNA expression has recently been found to be up-regulated within the PVN of HAB rats compared with LAB rats (Bosch et al. 2006). In contrast, line-dependent differences in CRH receptor binding could not be identified. Thus, differences in endogenous CRH system activity are likely to contribute to the emotional phenotype of HAB rats, as suggested in depressed patients (Nemeroff 2004; Keck and Holsboer 2001). Indeed, acute peripheral administration of the non-peptide CRH 1 receptor antagonist R121919 reduced anxiety levels on the EPM in HAB, but not LAB, rats and reduced stress-induced corticotropin secretion in both rat lines (Keck et al. 2001b).

A high level of anxiety could also be due to an attenuation of endogenous anxiolytic neuropeptides such as oxytocin (Neumann et al. 2000; Waldherr and Neumann 2007), prolactin (Torner et al. 2001; Donner et al. 2007) and/or neuropeptide S (NPS; Xu et al. 2004).

Oxytocin

Besides its capacity to modulate complex social behaviors, oxytocin is an established anxiolytic neuropeptide of the brain (Blume et al. 2008; Neumann 2008) and has antidepressive properties (Slattery and Neumann 2010). However, differences in central oxytocin expression or release were not found between HAB and LAB rats, except in lactation (Bosch et al. 2007).

In order to study potential anxiolytic or antidepressive effects of oxytocin in a psychopathological animal model, HAB and LAB rats were treated i.c.v. with either oxytocin or an oxytocin receptor antagonist (Slattery and Neumann 2010), which was only effective in female but not male rats: chronic oxytocin reduced the high anxiety level of HAB females on the EPM, whereas chronic i.c.v. treatment with the oxytocin antagonist increased anxiety only in female LAB rats without any effect in HABs or males. In contrast, acute manipulation of the oxytocin system did not alter anxiety-related behavior independent of sex and trait anxiety. Also, passive/active stress coping in the FST was not altered by any manipulation of the oxytocin system. Thus, chronic oxytocin seems to be a promising therapeutic strategy in particular for the treatment of anxiety disorders in women.

Prolactin

Prolactin has distinct anxiolytic properties, and the brain prolactin is involved in the regulation of anxiety and stress coping (Torner et al. 2004; Bunck et al. 2009; Ditzen et al. 2010). Plasma prolactin levels were found elevated in HAB rats in response to a mild emotional stressor (Neumann et al. 1998; Landgraf et al. 1999). The behavioral significance of this finding needs to be studied, as plasma prolactin does not reflect intracerebral prolactin release patterns (Torner et al. 2004), but peripheral prolactin can cross the blood brain barrier. However, differences in brain prolactin expression, release or receptor binding have not been studied in HAB and LAB rats until now.

Neuropeptide S

Neuropeptide S (NPS) is another powerful anxiolytic neuropeptide (Xu et al. 2004; Leonard et al. 2008; Vitale et al. 2008), and our preliminary results indicate substantial genetic and activity differences of the endogenous brain NPS system between HAB and LAB rats. For example, the latter express more NPS receptors in the hypothalamus (Slattery, Wegener, Naik, Mathé, Neumann, unpublished). In order to confirm the anxiolytic properties described in non-selected male mice (Xu et al. 2004) and rats (Vitale et al. 2008) in a psychopathological animal model, we acutely treated HAB and LAB rats with i.c.v. NPS 45 min prior to EPM testing can (Slattery et al. 2008). Indeed, preliminary evidence suggest that NPS reversed the high trait anxiety of HABs and exerted modest antidepressant effects (Slattery et al. 2008). Further, i.c.v. NPS improves consolidation of extinguished learned fear (Sartori et al. 2009).

Other Neurochemical Differences

Serotonin Disturbances in serotonergic neurotransmission are likely to contribute to the pathophysiology of anxiety and depression disorders and to underlie the hyperactivity of the HPA axis (Holsboer 2000). In HAB rats, serotonin 1A receptor expression was found to be reduced in the hippocampus, whereas the expression of the serotonin transporter binding sites was increased. Further, the basal availability of extracellular serotonin as estimated in microdialysates did not differ between the lines, but serotonin release in response to emotional stress was abolished in HAB rats. Chronic paroxetine markedly increased the stress-induced rise in hippocampal serotonin release, but did not alter receptor expression (Keck et al. 2005). Thus, the reduced raphe-hippocampal serotonergic transmission of HAB rats, which is evident both at the presynaptic (release) and postsynaptic (receptor) level, are likely to contribute to their high emotionality.

Hippocampal Neurogenesis

Consistent with the generally elevated stress responsiveness including HPA axis hyperreactivity and impaired negative feedback functions, we found reduced hippocampal cell survival and neurogenesis in 43-days old male HAB rats (Lucassen et al. 2009). Specifically, the number of newly generated surviving (BrdU-positive) cells in the subgranular cell layer/subgranular zone of the hippocampal dentate gyrus was found to be lower in HAB versus LAB rats. Further, the number of hippocampal doublecortin-positive cells reflecting neurogenesis is lower in HAB rats (Lucassen et al. 2009).

These results show that the high level of anxiety and activity of the HPA axis may affect cell survival in HAB rats, which may, in turn, also be partly responsible for their behavioral phenotype.

3.1.2 Gene × Environment Interactions

Early life stress, such as prenatal and immediately postnatal stress is a well-characterized risk factor for the development of affective disorders in adulthood,

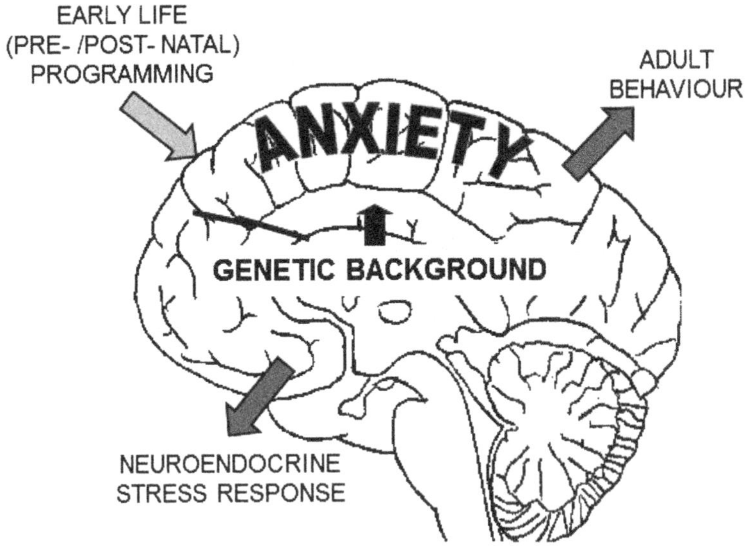

Fig. 5 Gene × early environment interactions shape adult behavioral and neuroendocrine stress responses as seen in rats with genetic determination of trait anxiety

and there are well documented interactions of genetic and environmental factors (Caspi et al. 2003; Fig. 5, see also Sect. 2.1.2). Selectively bred rodents with clear genetic determinants are valuable models for studying gene × environment interactions. Both prenatal stress as well as postnatal maternal separation resulted in line-dependent behavioral effects seen in adult HAB and LAB rats.

Gene × prenatal environment interactions: Surprisingly, after exposure to prenatal stress between pregnancy days 4 and 18 (exposure of the pregnant dam to maternal defeat by an unknown lactating resident daily for 45 min between pregnancy days 4 and 10, and to restraint between pregnancy days 11 and 18 for 60 min daily) adult male HAB rats became less anxious (see Fig. 6). This was confirmed in two independent tests for anxiety-related behavior, i.e. the EPM and the modified holeboard (Bosch et al. 2006). The opposite behavioral consequences of prenatal stress were accompanied by opposing effects on central vasopressin and CRH mRNA expression: whereas the genetically determined high level of hypothalamic vasopressin mRNA expression of HAB rats was not altered by prenatal stress, it was elevated in early life stressed LAB male offspring (Fig. 6). Similarly, the genetic difference in CRH expression within the PVN with high levels in unstressed HAB compared with LAB controls was also at least partly abolished by prenatal stress. Further, after prenatal stress the high HPA axis response to a mild stressor found in HAB rats was reduced and found to be indistinguishable from unstressed and prenatally stressed LABs (Bosch et al. 2006).

Opposing effects of prenatal stress were also found with respect to hippocampal neurogenesis, which has been shown to be stress-sensitive and implicated in depression (Pittenger and Duman 2008). As mentioned above, the survival of

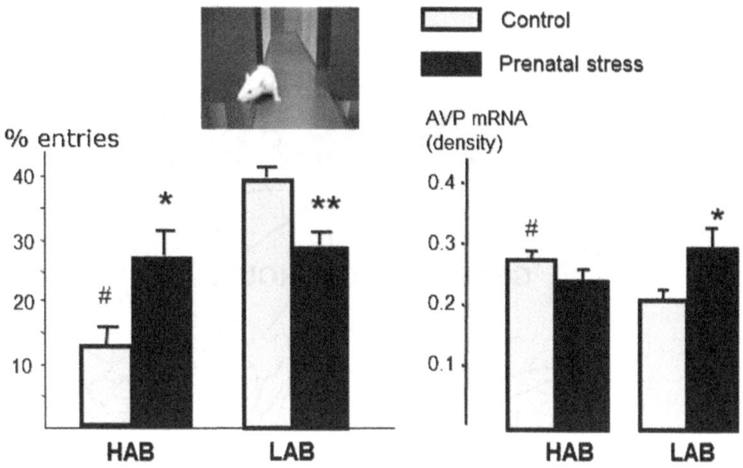

Fig. 6 Opposite effects of prenatal stress on anxiety-related behavior on the EPM (% entries open arms; left) and basal vasopressin mRNA expression in the hypothalamic PVN (right) in male HAB and LAB adult offspring indicating gene × early environment interactions. # versus LAB; * versus respective control. Adapted from (Bosch et al. 2006)

newly generated hippocampal cells is lower in young unstressed HAB compared with LAB males (day 43 of life). Interestingly, while prenatal stress caused a further reduction in the number of BrdU and doublecortin positive cells in the subgranular zone of the hippocampal dentate gyrus in HAB rats, it did not affect this parameter in the LAB rats (Lucassen et al. 2009). Although detailed mechanisms underlying the opposing effects of prenatal stress in HAB and LAB rats are unknown, line-dependent differences in the activity of the placental enzyme 11-beta hydroxysteroid dehydrogenase type 2, which catalyzes maternal corticosterone to inert 11-dehydrocorticosterone are likely to contribute (Lucassen et al. 2009).

Gene × postnatal environment interactions: Opposite effects on emotionality and neuroendocrine responsiveness and, consequently, approximation of the HAB and LAB behavioral and neuroendocrine phenotypes were also found after immediate postnatal stress, i.e. after maternal separation (Fig. 7). Daily 3-h separation of HAB and LAB offspring from the mother between postnatal days 2 and 15 reduced anxiety in adult HABs as seen on the modified holeboard, but rather increased (EPM) or had no effect (holeboard) in LABs (Neumann et al. 2005b). Further, the HPA axis hyper-responses seen in HAB control rats became attenuated after postnatal stress, whereas maternal separation did not significantly alter neuroendocrine responses in LAB rats (Neumann et al. 2005b; Fig. 7).

These experiments underline that the consequences of environmental factors are strongly determined by the genetic background. Vice versa, even a robust genetically determined individual behavioral phenotype can be shaped by environmental factors; likely via epigenetic mechanisms (Murgatroyd et al. 2009).

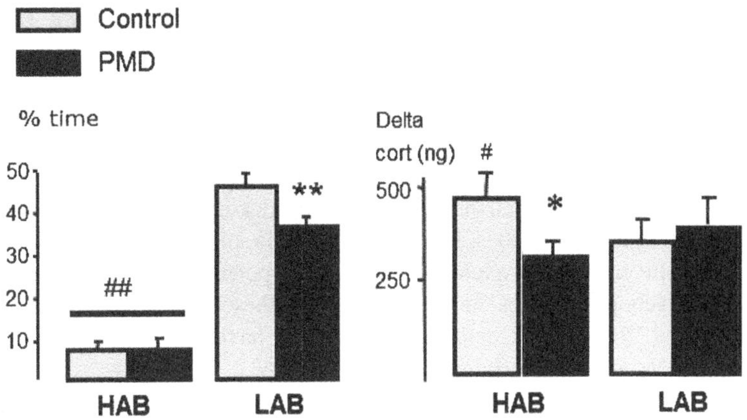

Fig. 7 Opposite effects of postnatal stress (periodic maternal separation; PMD) on anxiety-related behavior on the EPM (% time open arms; left) and plasma corticosterone response to novel environment exposure (elevated platform, right) in HAB and LAB adult offspring indicating gene × early environment interactions. # versus LAB; * versus respective control. Adapted from (Neumann et al. 2005b)

The mitigating effect of early life experiences on behavioral and neuroendocrine parameters in rats representing extremes in trait anxiety may also reflect an evolutionary benefit, as the genetic variability among individuals of a species is sustained, while maintaining adequate responses to potentially dangerous stimuli in adulthood. Our results in HAB and LAB rats after pre- and postnatal stress exposure further indicate that gene × environment interactions can be found at behavioral, neuroendocrine, neuronal and gene levels indicating their complexity.

3.2 HAB LAB Mice

Although various selectively bred rat lines are useful tools for studying behavioral and especially neuroendocrine parameters of depression and anxiety, as well as environmental factors shaping innate stress coping style, genetic studies such as the functional analysis of candidate genes underlying, for example, high trait anxiety or depression-related behavior, are a priori limited in rats. Therefore, Landgraf and co-workers also generated mice selectively bred for high (M-HAB) versus low (M-LAB) anxiety-related behavior (Kromer et al. 2005). After 9 generations of continuous breeding (sibling mating), a robust behavioral divergence had been achieved. High trait anxiety has been confirmed on the EPM and in the light–dark box; further, M-HAB mice pups show more ultrasound vocalization, which was reversed by diazepam (Kromer et al. 2005). In agreement to what has been found in HAB and LAB rats, M-LAB mice are more active in several tests for depression-like behavior (FST, tail suspension). Also, as seen in HAB and LAB rats, differences in the hypothalamic expression of vasopressin are likely to underlie their trait anxiety and also line-dependent social behavior (Kessler et al. 2011), but here M-LABs show signs of central diabetes insipidus, i.e. low vasopressin expression and availability compared with M-HAB and non-selected CD1 mice (Kessler et al. 2007). A SNP in exon 1 of the vasopressin gene of LAB mice causes an amino acid substitution in the signal peptide of the vasopressin precursor, and is likely to impair processing and trafficking of the precursor (Bunck et al. 2009; Kessler et al. 2007). Besides vasopressin, differences in the expression of the cytosolic enzyme glyoxalase-I, which is of potential interest in the context of various psychopathologies, have been found in these mice (Kromer et al. 2005; Hambsch et al. 2010). Thus, selectively bred mice have a high potential to reveal novel candidate genes underlying high trait anxiety.

3.3 Floripa H and L Rats

The Floripa H and L rat lines, have been developed based on selection for high and low locomotion in the central aversive area of an open field, an experimental measure of fearfulness in rodents (Ramos and Mormede 1998). The Floripa H and L lines differ from each other not only for the selected behavior, but also for other experimental indices of anxiety, such as the approach towards the open arms of the elevated plus maze and the white compartment of the black/white box (Ramos et al. 2003). In addition, compared with Floripa L, the Floripa H rats show less depression-like behavior in the FST (Hinojosa et al. 2006), suggesting the Floripa rats to be a combined model of both anxiety and depression. The Floripa L female rats consumed more ethanol than their Floripa H counterparts at concentrations of 6 and 10% in a two-bottle choice protocol (Izídio and Ramos 2007), however the connection to the anxiety-like phenotype remains to be determined.

3.4 Maudsley Reactive and Nonreactive Rats

Based on reactivity in the open field, with defecation and urination being the central variables, two lines of rats, later termed the Maudsley high reactive (high defecation, MHR) and low, nonreactive (low defecation, MLR) rats, were bred from Wistar in the 1950s aiming to model the human personality dimension of emotionality (Broadhurst 1957, 1960, 1962, 1975). The overall conclusion of the several studies published were that Maudsley reactive rats yielded higher scores in several tests of anxiety-like/avoidance behavior than non-reactive animals (Blizard and Adams 2002). Although this approach was useful to study individual differences in anxiety-like/avoidance behavior, it has not been pursued intensively in recent years (Pawlak et al. 2008; Blizard and Adams 2002).

3.5 *High/Low Avoidance Rats: RHA/RLA; SHA/Bru and SLA/Bru*

Several rat breeding lines have been selected for learning to actively avoid foot shocks in a two-way shuttle-box. The Roman high avoidance (RHA) and Roman low avoidance (RLA) Wistar rats (Bignami 1965), the Long–Evans rats (Brush et al. 1979, 1985, 1989; Brush 2003), and Syracuse high and low avoidance rats (SHA/Bru and SLA/Bru) are the most prominent examples. Although these attempts differ in nature, some of the common characteristics include differences in learning and memory, and a number of emotion-related behavioral and neuroendocrine characteristics. A detailed review can be found in (Brush 2003).

4 Conclusions

In conclusion, the existing data on selectively bred rodent models, in particular of the FSL/FRL and HAB/LAB rats reviewed above, reveal the importance of rodent breeding lines for studying neurobiological, neuroendocrine and genetic mechanisms underlying anxiety- and depression-related diseases. Essentially, the development of potentially novel therapeutic strategies targeting brain neuropeptide systems such as NPY, vasopressin, CRH, oxytocin or neuropeptide S will be enabled and promoted using such relevant and complementary animal models. Moreover, selectively bred rat and mouse lines provide an important tool in order to provide further evidence for gene × environment interactions demonstrating differential vulnerabilities, for example to prenatal or immediate postnatal adverse life events. The combination of genetic models with various stress paradigms is likely to mimic the human situation more accurately.

Given their behavioral and neuroendocrine phenotype, the neurobiological mechanisms underlying their anxiety- and depression-related behavior, as well as successful pharmacological attempts to reverse the psychopathological phenotype, the FSL/FRL and HAB/LAB rats fulfill the requirements of face, construct and predictive validity of an animal model. Therefore, they should be further exploited to discover potential novel therapeutic strategies.

Acknowledgments GW was supported by grants from The Danish Medical Research Council (grant 271-08-0768) and the Research Foundation of County Midtjylland. AAM was supported from the Swedish Medical Research Council (grant 10414) and the Karolinska Institutet. IDN was supported by DFG, BMBF and Bayerische Forschungsstiftung.

References

Abe H, Hidaka N, Kawagoe C, Odagiri K, Watanabe Y, Ikeda T, Ishizuka Y, Hashiguchi H, Takeda R, Nishimori T, Ishida Y (2007) Prenatal psychological stress causes higher emotionality, depression-like behavior, and elevated activity in the hypothalamo-pituitary-adrenal axis. Neurosci Res 59(2):145–151

Abildgaard A, Solskov L, Volke V, Harvey BH, Lund S, Wegener G (2010) A high-fat diet exacerbates depressive-like behavior in the flinders sensitive line (FSL) rat, a genetic model of depression. Psychoneuroendocrinology. doi: 10.1016/j.psyneuen.2010.09.004

Adrien J (2002) Neurobiological bases for the relation between sleep and depression. Sleep Med Rev 6(5):341–351

Anderson RJ, Freedland KE, Clouse RE, Lustman PJ (2001) The prevalence of comorbid depression in adults with diabetes: a meta-analysis. Diabetes Care 24(6):1069–1078

Andlin-Sobocki P, Jonsson B, Wittchen HU, Olesen J (2005) Cost of disorders of the brain in Europe. Eur J Neurol 12(Suppl 1):1–27

Arango V, Underwood MD, Gubbi AV, Mann JJ (1995) Localized alterations in pre- and postsynaptic serotonin binding sites in the ventrolateral prefrontal cortex of suicide victims. Brain Res 688(1–2):121–133

Arora RC, Tong C, Jackman HL (1983) Serotonin uptake and imipramine binding in blood platelets and brain of Fawn-hooded and Sprague-Dawley rats. Life Sci 33(5):437–442

Asberg M, Bertilsson L, Martensson B (1984) CSF monoamine metabolites in melancholia. Acta Psychiatrica Scandinavica 69(3):201–219

Aulakh CS, Wozniak KM, Hill JL, Devane CL, Tolliver TJ, Murphy DL (1988) Differential neuroendocrine responses to the 5-HT agonist m-chlorophenylpiperazine in Fawn-hooded rats relative to Wistar and Sprague-Dawley rats. Neuroendocrinology 48(4):401–406

Aulakh CS, Hill JL, Murphy DL (1993) Attenuation of hypercortisolemia in Fawn-hooded rats by antidepressant drugs. Eur J Pharmacol 240(1):85–88

Aulakh CS, Tolliver T, Wozniak KM, Hill JL, Murphy DL (1994) Functional and biochemical evidence for altered serotonergic function in the Fawn-hooded rat strain. Pharmacol Biochem Behav 49(3):615–620

Aydemir C, Yalcin ES, Aksaray S, Kisa C, Yildirim SG, Uzbay T, Goka E (2006) Brain-derived neurotrophic factor (BDNF) changes in the serum of depressed women. Prog Neuropsychopharmacol Biol Psychiatry 30(7):1256–1260

Ayensu WK, Pucilowski O, Mason GA, Overstreet DH, Rezvani AH, Janowsky DS (1995) Effects of chronic mild stress on serum complement activity, saccharin preference, and corticosterone levels in flinders lines of rats. Physiol Behav 57(1):165–169

Bale TL (2006) Stress sensitivity and the development of affective disorders. Horm Behav 50(4):529–533

Bale TL, Baram TZ, Brown AS, Goldstein JM, Insel TR, McCarthy MM, Nemeroff CB, Reyes TM, Simerly RB, Susser ES, Nestler EJ (2010) Early life programming and neurodevelopmental disorders. Biol Psychiatry 68(4):314–319

Barefoot JC, Schroll M (1996) Symptoms of depression, acute myocardial infarction, and total mortality in a community sample. Circulation 93(11):1976–1980

Barrot M, Wallace DL, Bolanos CA, Graham DL, Perrotti LI, Neve RL, Chambliss H, Yin JC, Nestler EJ (2005) Regulation of anxiety and initiation of sexual behavior by CREB in the nucleus accumbens. Proc Natl Acad Sci USA 102(23):8357–8362

Beiderbeck DI, Neumann ID, Veenema AH (2007) Differences in intermale aggression are accompanied by opposite vasopressin release patterns within the septum in rats bred for low and high anxiety. Eur J Neurosci 26(12):3597–3605

Benca RM (1996) Sleep in psychiatric disorders. Neurol Clin 14(4):739–764

Benca RM, Obermeyer WH, Thisted RA, Gillin JC (1992) Sleep and psychiatric disorders. A meta-anal Arch Gen Psychiatry 49(8):651–668 discussion 669–670

Benca RM, Overstreet DE, Gilliland MA, Russell D, Bergmann BM, Obermeyer WH (1996) Increased basal REM sleep but no difference in dark induction or light suppression of REM sleep in flinders rats with cholinergic supersensitivity. Neuropsychopharmacology 15(1): 45–51

Bendotti C, Samanin R (1987) The role of putative 5-HT(1A) and 5-HT(1B) receptors in the control of feeding in rats. Life Sci 41(5):635–642

Bignami G (1965) Selection for high rates and low rates of avoidance conditioning in the rat. Anim Behav 13(2–3):221–227

Bjornebekk A, Mathe AA, Brene S (2005) The antidepressant effect of running is associated with increased hippocampal cell proliferation. Int J Neuropsychopharmacol 8(3):357–368

Bjornebekk A, Mathe AA, Brene S (2006) Running has differential effects on NPY, opiates, and cell proliferation in an animal model of depression and controls. Neuropsychopharmacology 31(2):256–264

Bjornebekk A, Mathe AA, Gruber SH, Brene S (2007) Social isolation increases number of newly proliferated cells in hippocampus in female flinders sensitive line rats. Hippocampus 17(12):1193–1200

Bjornebekk A, Mathe AA, Brene S (2010) The antidepressant effects of running and escitalopram are associated with levels of hippocampal NPY and Y1 receptor but not cell proliferation in a rat model of depression. Hippocampus 20(7):820–828

Blizard DA, Adams N (2002) The maudsley reactive and nonreactive strains: a new perspective. Behav Genet 32(5):277–299

Blume A, Bosch OJ, Miklos S, Torner L, Wales L, Waldherr M, Neumann ID (2008) Oxytocin reduces anxiety via ERK1/2 activation: local effect within the rat hypothalamic paraventricular nucleus. Eur J Neurosci 27(8):1947–1956

Bosch OJ, Neumann ID (2008) Brain vasopressin is an important regulator of maternal behavior independent of dams' trait anxiety. Proc Natl Acad Sci USA 105(44):17139–17144

Bosch OJ, Neumann ID (2010) Vasopressin released within the central amygdala promotes maternal aggression. Eur J Neurosci 31(5):883–891

Bosch OJ, Meddle SL, Beiderbeck DI, Douglas AJ, Neumann ID (2005) Brain oxytocin correlates with maternal aggression: link to anxiety. J Neurosci 25(29):6807–6815

Bosch OJ, Kromer SA, Neumann ID (2006) Prenatal stress: opposite effects on anxiety and hypothalamic expression of vasopressin and corticotropin-releasing hormone in rats selectively bred for high and low anxiety. Eur J Neurosci 23(2):541–551

Bosch OJ, Musch W, Bredewold R, Slattery DA, Neumann ID (2007) Prenatal stress increases HPA axis activity and impairs maternal care in lactating female offspring: implications for postpartum mood disorder. Psychoneuroendocrinology 32(3):267–278

Bosch OJ, Pfortsch J, Beiderbeck DI, Landgraf R, Neumann ID (2010) Maternal behaviour is associated with vasopressin release in the medial preoptic area and bed nucleus of the stria terminalis in the rat. J Neuroendocrinol 22(5):420–429

Braw Y, Malkesman O, Dagan M, Bercovich A, Lavi-Avnon Y, Schroeder M, Overstreet DH, Weller A (2006) Anxiety-like behaviors in pre-pubertal rats of the flinders sensitive line (FSL) and Wistar-Kyoto (WKY) animal models of depression. Behav Brain Res 167(2):261–269

Breier A, Albus M, Pickar D, Zahn TP, Wolkowitz OM, Paul SM (1987) Controllable and uncontrollable stress in humans: alterations in mood and neuroendocrine and psychophysiological function. Am J Psychiatry 144(11):1419–1425

Brene S, Bjornebekk A, Aberg E, Mathe AA, Olson L, Werme M (2007) Running is rewarding and antidepressive. Physiol Behav 92(1–2):136–140

Broadhurst PL (1957) Determinants of emotionality in the rat. I. Situational factors. Br J Psychol (London, England : 1953) 48(1):1–12

Broadhurst PL (1960) Experiments in psychogenetics: applications of biometrical genetics to the inheritance of behaviour. In: Eysenck HJ (ed) Experiments in personality. Psychogenetics and psychopharmacology, vol 1. Routledge & Kegan Paul, London, pp 1–102

Broadhurst PL (1962) A note on further progress in a psychogenetic selection experiment. Psychol Rep 10:65–66

Broadhurst PL (1975) The maudsley reactive and nonreactive strains of rats: a survey. Behav Genet 5(4):299–319

Brush FR (2003) Selection for differences in avoidance learning: the syracuse strains differ in anxiety, not learning ability. Behav Genet 33(6):677–696

Brush FR, Froehlich JC, Sakellaris PC (1979) Genetic selection for avoidance behavior in the rat. Behav Genet 9(4):309–316

Brush FR, Baron S, Froehlich JC, Ison JR, Pellegrino LJ, Phillips DS, Sakellaris PC, Williams VN (1985) Genetic differences in avoidance learning by rattus norvegicus: escape/avoidance responding, sensitivity to electric shock, discrimination learning, and open-field behavior. J comp Psychol (Washington, DC: 1983) 99(1):60–73

Brush FR, Blanchard RJ, Blanchard DC (1989) Social dominance and response to a natural predator by rats selectively bred for differences in shuttle-box avoidance learning. Ethoexperimental Approaches to the study of behavior, pp 411–417

Bunck M, Czibere L, Horvath C, Graf C, Frank E, Kessler MS, Murgatroyd C, Muller-Myhsok B, Gonik M, Weber P, Putz B, Muigg P, Panhuysen M, Singewald N, Bettecken T, Deussing JM, Holsboer F, Spengler D, Landgraf R (2009) A hypomorphic vasopressin allele prevents anxiety-related behavior. PLoS One 4(4):e5129

Bus BA, Molendijk ML, Penninx BJ, Buitelaar JK, Kenis G, Prickaerts J, Elzinga BM, Voshaar RC (2011) Determinants of serum brain-derived neurotrophic factor. Psychoneuroendocrinology 36(2):228–239

Bushnell PJ, Levin ED, Overstreet DH (1995) Spatial working and reference memory in rats bred for autonomic sensitivity to cholinergic stimulation: acquisition, accuracy, speed, and effects of cholinergic drugs. Neurobiol Learn Mem 63(2):116–132

Caberlotto L, Hurd YL (2001) Neuropeptide Y Y(1) and Y(2) receptor mrna expression in the prefrontal cortex of psychiatric subjects. Relationship of Y(2) subtype to suicidal behavior. Neuropsychopharmacology 25(1):91–97

Caberlotto L, Jimenez P, Overstreet DH, Hurd YL, Mathe AA, Fuxe K (1999) Alterations in neuropeptide Y levels and Y1 binding sites in the flinders sensitive line rats, a genetic animal model of depression. Neurosci Lett 265(3):191–194

Campbell DT, Fiske DW (1959) Convergent and discriminant validation by the multitrait-multimethod matrix. Psychol Bull 56(2):81–105

Carboni L, Becchi S, Piubelli C, Mallei A, Giambelli R, Razzoli M, Mathe AA, Popoli M, Domenici E (2010) Early-life stress and antidepressants modulate peripheral biomarkers in a gene-environment rat model of depression. Prog Neuropsychopharmacol Biol Psychiatry 34(6):1037–1048

Carney RM, Rich MW, Freedland KE, Saini J, TeVelde A, Simeone C, Clark K (1988) Major depressive disorder predicts cardiac events in patients with coronary artery disease. Psychosom Med 50(6):627–633

Caspi A, Sugden K, Moffitt TE, Taylor A, Craig IW, Harrington H, McClay J, Mill J, Martin J, Braithwaite A, Poulton R (2003) Influence of life stress on depression: moderation by a polymorphism in the 5-HTT gene. Science 301(5631):386–389

Cerbone A, Pellicano MP, Sadile AG (1993) Evidence for and against the naples high- and low-excitability rats as genetic model to study hippocampal functions. Neurosci Biobehav Rev 17(3):295–303

Chen B, Dowlatshahi D, MacQueen GM, Wang JF, Young LT (2001) Increased hippocampal BDNF immunoreactivity in subjects treated with antidepressant medication. Biol Psychiatry 50(4):260–265

Chen F, Madsen TM, Wegener G, Nyengaard JR (2010) Imipramine treatment increases the number of hippocampal synapses and neurons in a genetic animal model of depression. Hippocampus 20(12):1376–1384

Clause BT (1998) The Wistar Institute Archives: rats (not mice) and history. http://www.amphilsoc.org/mendel/1998.htm#Clause. Accessed 9th Sept 2010

Cloninger CR, Reich T, Wetzel R (1979) Alcoholism and affective disorders: familial associations and genetic models. alcoholism and affective disorders: clinical, genetic and biochemical studies. In: Goodwin DW, Erickson CK (eds) Alcoholism and affective disorders. SP Medical and Scientific Books, New York, pp 57–86

Coenen AML, Van Luijtelaar ELJM (2003) Genetic animal models for absence epilepsy: a review of the WAG/Rij strain of rats. Behav Genet 33(6):635–655

Costall B, Naylor RJ (1973) The role of telencephalic dopaminergic systems in the mediation of apomorphine stereotyped behaviour. Eur J Pharmacol 24(1):8–24

Cottrell EC, Seckl JR (2009) Prenatal stress, glucocorticoids and the programming of adult disease. Front Behav Neurosci 3:19

Crocker AD, Overstreet DH (1991) Dopamine sensitivity in rats selectively bred for increases in cholinergic function. Pharmacol Biochem Behav 38(1):105–108

Dawson VL, Dawson TM (1996) Nitric oxide in neuronal degeneration. Proc Soc Exp Biol Med 211(1):33–40

De Kloet ER, Vreugdenhil E, Oitzl MS, Joëls M (1998) Brain corticosteroid receptor balance in health and disease. Endocr Rev 19(3):269–301

Deuschle M, Schweiger U, Weber B, Gotthardt U, Körner A, Schmider J, Standhardt H, Lammers CH, Heuser I (1997) Diurnal activity and pulsatility of the hypothalamus-pituitary-adrenal system in male depressed patients and healthy controls. J Clin Endocrinol Metab 82(1):234–238

Dishman RK, Berthoud HR, Booth FW, Cotman CW, Edgerton VR, Fleshner MR, Gandevia SC, Gomez-Pinilla F, Greenwood BN, Hillman CH, Kramer AF, Levin BE, Moran TH, Russo-Neustadt AA, Salamone JD, Van Hoomissen JD, Wade CE, York DA, Zigmond MJ (2006) Neurobiology of exercise. Obesity (Silver Spring) 14(3):345–356

Ditzen C, Varadarajulu J, Czibere L, Gonik M, Targosz BS, Hambsch B, Bettecken T, Kessler MS, Frank E, Bunck M, Teplytska L, Erhardt A, Holsboer F, Muller-Myhsok B, Landgraf R, Turck CW (2010) Proteomic-based genotyping in a mouse model of trait anxiety exposes disease-relevant pathways. Mol Psychiatry 15(7):702–711

Donner N, Bredewold R, Maloumby R, Neumann ID (2007) Chronic intracerebral prolactin attenuates neuronal stress circuitries in virgin rats. Eur J Neurosci 25(6):1804–1814

Dremencov E, Gispan-Herman I, Rosenstein M, Mendelman A, Overstreet DH, Zohar J, Yadid G (2004) The serotonin-dopamine interaction is critical for fast-onset action of antidepressant treatment: in vivo studies in an animal model of depression. Prog Neuro-Psychopharmacol Biol Psychiatry 28(1):141–147

Dumbrille-Ross A, Tang SW (1981) Absence of high-affinity [3H]imipramine binding in platelets and cerebral cortex of Fawn-hooded rats. Eur J Pharmacol 72(1):137–138

Dunn AL, Dishman RK (1991) Exercise and the neurobiology of depression. Exerc Sport Sci Rev 19:41–98

Egede LE, Nietert PJ, Zheng D (2005) Depression and all-cause and coronary heart disease mortality among adults with and without diabetes. Diabetes Care 28(6):1339–1345

Einat H, Belmaker RH, Zangen A, Overstreet DH, Yadid G (2002) Chronic inositol treatment reduces depression-like immobility of flinders sensitive line rats in the forced swim test. DepressAnxiety 15(3):148–151

El Khoury A, Gruber SH, Mork A, Mathe AA (2006) Adult life behavioral consequences of early maternal separation are alleviated by escitalopram treatment in a rat model of depression. Prog Neuropsychopharmacol Biol Psychiatry 30(3):535–540

Elfving B, Plougmann PH, Muller HK, Mathe AA, Rosenberg R, Wegener G (2010a) Inverse correlation of brain and blood BDNF levels in a genetic rat model of depression. Int J Neuropsychopharmacol 13(5):563–572

Elfving B, Plougmann PH, Wegener G (2010b) Differential brain, but not serum VEGF levels in a genetic rat model of depression. Neurosci Lett 474(1):13–16

Ellenbroek BA, Cools AR (2000) Animal models for the negative symptoms of schizophrenia. Behav Pharmacol 11(3–4):223–233

Ellenbroek BA, Cools AR (2002) Apomorphine susceptibility and animal models for psychopathology: genes and environment. Behav Genet 32(5):349–361

Elzinga BM, Molendijk ML, Oude Voshaar RC, Bus BA, Prickaerts J, Spinhoven P, Penninx BJ (2011) The impact of childhood abuse and recent stress on serum brain-derived neurotrophic factor and the moderating role of BDNF val(66)met. Psychopharmacology (Berl) 214(1):319–328
Engelmann M, Wotjak CT, Neumann I, Ludwig M, Landgraf R (1996) Behavioral consequences of intracerebral vasopressin and oxytocin: focus on learning and memory. Neurosci Biobehav Rev 20(3):341–358
Frank E, Landgraf R (2008) The vasopressin system—from antidiuresis to psychopathology. Eur J Pharmacol 583(2–3):226–242
Frank E, Salchner P, Aldag JM, Salome N, Singewald N, Landgraf R, Wigger A (2006) Genetic predisposition to anxiety-related behavior determines coping style, neuroendocrine responses, and neuronal activation during social defeat. Behav Neurosci 120(1):60–71
Frasure-Smith N, Lesperance F, Talajic M (1993) Depression following myocardial infarction: impact on 6-month survival. J Am Med Assoc 270(15):1819–1825
Fujioka T, Fujioka A, Tan N, Chowdhury GMI, Mouri H, Sakata Y, Nakamura S (2001) Mild prenatal stress enhances learning performance in the non-adopted rat offspring. Neuroscience 103(2):301–307
Fumagalli F, Molteni R, Racagni G, Riva MA (2007) Stress during development: impact on neuroplasticity and relevance to psychopathology. Prog Neurobiol 81(4):197–217
Gass P, Hellweg R (2010) Peripheral brain-derived neurotrophic factor (BDNF) as a biomarker for affective disorders? Int J Neuropsychopharmacol 13(1):1–4
Geyer MA, Markou A (1995) Animal models of psychiatric disorders. In: Bloom FE, Kupfer DJ (eds) Psychopharmacology: the fourth generation of progress, Raven Press, New York, pp 787–798
Gómez F, Lahmame A, De Kloet ER, Armario A (1996) Hypothalamic-pituitary-adrenal response to chronic stress in five inbred rat strains: differential responses are mainly located at the adrenocortical level. Neuroendocrinology 63(4):327–337
Gutman DA, Coyer MJ, Boss-Williams KA, Owens MJ, Nemeroff CB, Weiss JM (2008) Behavioral effects of the CRF1 receptor antagonist R121919 in rats selectively bred for high and low activity in the swim test. Psychoneuroendocrinology 33(8):1093–1101
Hambsch B, Chen BG, Brenndorfer J, Meyer M, Avrabos C, Maccarrone G, Liu RH, Eder M, Turck CW, Landgraf R (2010) Methylglyoxal-mediated anxiolysis involves increased protein modification and elevated expression of glyoxalase 1 in the brain. J Neurochem 113(5): 1240–1251
Haug TT, Mykletun A, Dahl AA (2004) The association between anxiety, depression, and somatic symptoms in a large population: the HUNT-II study. Psychosom Med 66(6):845–851
Heilig M (2004) The NPY system in stress, anxiety and depression. Neuropeptides 38(4): 213–224
Hemingway H, Marmot M (1999) Psychosocial factors in the aetiology and prognosis of coronary heart disease: systematic review of prospective cohort studies. Br Med J 318(7196):1460–1467
Henn FA, Vollmayr B (2005) Stress models of depression: forming genetically vulnerable strains. Neurosci Biobehav Rev 29(4–5):799–804
Henniger MS, Ohl F, Holter SM, Weissenbacher P, Toschi N, Lorscher P, Wigger A, Spanagel R, Landgraf R (2000) Unconditioned anxiety and social behaviour in two rat lines selectively bred for high and low anxiety-related behaviour. Behav Brain Res 111(1–2):153–163
Hinojosa FR, Spricigo L Jr, Izídio GS, Brüske GR, Lopes DM, Ramos A (2006) Evaluation of two genetic animal models in behavioral tests of anxiety and depression. Behav Brain Res 168(1):127–136
Holsboer F (2000) The corticosteroid receptor hypothesis of depression. Neuropsychopharmacology 23(5):477–501
Hou C, Jia F, Liu Y, Li L (2006) CSF serotonin, 5-hydroxyindolacetic acid and neuropeptide y levels in severe major depressive disorder. Brain Res 1095(1):154–158
Huot RL, Thrivikraman KV, Meaney MJ, Plotsky PM (2001) Development of adult ethanol preference and anxiety as a consequence of neonatal maternal separation in long evans rats and reversal with antidepressant treatment. Psychopharmacology (Berl) 158(4):366–373

Husum H, Vasquez PA, Mathe AA (2001) Changed concentrations of tachykinins and neuropeptide Y in brain of a rat model of depression: lithium treatment normalizes tachykinins. Neuropsychopharmacology 24(2):183–191

Husum H, Van Kammen D, Termeer E, Bolwig G, Mathe A (2003) Topiramate normalizes hippocampal NPY-LI in flinders sensitive line 'depressed' rats and upregulates NPY, galanin, and CRH-LI in the hypothalamus: implications for mood-stabilizing and weight loss-inducing effects. Neuropsychopharmacology 28(7):1292–1299

Husum H, Aznar S, Hoyer-Hansen S, Larsen MH, Mikkelsen JD, Moller A, Mathe AA, Wortwein G (2006) Exacerbated loss of cell survival, neuropeptide Y-immunoreactive (IR) cells, and serotonin-IR fiber lengths in the dorsal hippocampus of the aged flinders sensitive line "depressed" rat: implications for the pathophysiology of depression? J Neurosci Res 84(6):1292–1302

Husum H, Wortwein G, Andersson W, Bolwig TG, Mathe AA (2008) Gene-environment interaction affects substance p and neurokinin a in the entorhinal cortex and periaqueductal grey in a genetic animal model of depression: implications for the pathophysiology of depression. Int J Neuropsychopharmacol 11(1):93–101

Ischiropoulos H, Beckman JS (2003) Oxidative stress and nitration in neurodegeneration: cause, effect, or association? J Clin Invest 111(2):163–169

Ising M, Kunzel HE, Binder EB, Nickel T, Modell S, Holsboer F (2005) The combined dexamethasone/CRH test as a potential surrogate marker in depression. Prog Neuropsychopharmacol Biol Psychiatry 29(6):1085–1093

Izídio GS, Ramos A (2007) Positive association between ethanol consumption and anxiety-related behaviors in two selected rat lines. Alcohol 41(7):517–524

Janowsky DS, Risch C, Parker D, Huey L, Judd L (1980) Increased vulnerability to cholinergic stimulation in affective-disorder patients [proceedings]. Psychopharmacol Bull 16(4):29–31

Janowsky DS, Overstreet DH, Nurnberger JI Jr (1994) Is cholinergic sensitivity a genetic marker for the affective disorders? Am J Med Genet 54(4):335–344

Jimenez Vasquez PA, Salmi P, Ahlenius S, Mathe AA (2000a) Neuropeptide y in brains of the flinders sensitive line rat, a model of depression. Effects of electroconvulsive stimuli and d-amphetamine on peptide concentrations and locomotion. Behav Brain Res 111(1–2):115–123

Jimenez-Vasquez PA, Overstreet DH, Mathe AA (2000b) Neuropeptide Y in male and female brains of flinders sensitive line, a rat model of depression. Effects of electroconvulsive stimuli. J Psychiatr Res 34(6):405–412

Jimenez-Vasquez PA, Mathe AA, Thomas JD, Riley EP, Ehlers CL (2001) Early maternal separation alters neuropeptide Y concentrations in selected brain regions in adult rats. Brain Res Dev Brain Res 131(1–2):149–152

Jimenez-Vasquez PA, Diaz-Cabiale Z, Caberlotto L, Bellido I, Overstreet D, Fuxe K, Mathe AA (2007) Electroconvulsive stimuli selectively affect behavior and neuropeptide Y (NPY) and NPY Y(1) receptor gene expressions in hippocampus and hypothalamus of flinders sensitive line rat model of depression. Eur Neuropsychopharmacol 17(4):298–308

Jindal RD, Thase ME, Fasiczka AL, Friedman ES, Buysse DJ, Frank E, Kupfer DJ (2002) Electroencephalographic sleep profiles in single-episode and recurrent unipolar forms of major depression: II. Comparison during remission. Biol Psychiatry 51(3):230–236

Jochum T, Boettger MK, Wigger A, Beiderbeck D, Neumann ID, Landgraf R, Sauer H, Bar KJ (2007) Decreased sensitivity to thermal pain in rats bred for high anxiety-related behaviour is attenuated by citalopram or diazepam treatment. Behav Brain Res 183(1):18–24

Kanemaru K, Nishi K, Hasegawa S, Diksic M (2009) Chronic citalopram treatment elevates serotonin synthesis in flinders sensitive and flinders resistant lines of rats, with no significant effect on Sprague-Dawley rats. Neurochem Int 54(5–6):363–371

Karege F, Perret G, Bondolfi G, Schwald M, Bertschy G, Aubry JM (2002) Decreased serum brain-derived neurotrophic factor levels in major depressed patients. Psychiatry Res 109(2):143–148

Karege F, Bondolfi G, Gervasoni N, Schwald M, Aubry JM, Bertschy G (2005) Low brain-derived neurotrophic factor (BDNF) levels in serum of depressed patients probably results from lowered platelet bdnf release unrelated to platelet reactivity. Biol Psychiatry 57(9): 1068–1072

Keck ME, Holsboer F (2001) Hyperactivity of crh neuronal circuits as a target for therapeutic interventions in affective disorders. Peptides 22(5):835–844

Keck ME, Welt T, Post A, Muller MB, Toschi N, Wigger A, Landgraf R, Holsboer F, Engelmann M (2001a) Neuroendocrine and behavioral effects of repetitive transcranial magnetic stimulation in a psychopathological animal model are suggestive of antidepressant-like effects. Neuropsychopharmacology 24(4):337–349

Keck ME, Welt T, Wigger A, Renner U, Engelmann M, Holsboer F, Landgraf R (2001b) The anxiolytic effect of the CRH(1) receptor antagonist R121919 depends on innate emotionality in rats. Eur J Neurosci 13(2):373–380

Keck ME, Wigger A, Welt T, Muller MB, Gesing A, Reul JM, Holsboer F, Landgraf R, Neumann ID (2002) Vasopressin mediates the response of the combined dexamethasone/CRH test in hyper-anxious rats: implications for pathogenesis of affective disorders. Neuropsychopharmacology 26(1):94–105

Keck ME, Welt T, Muller MB, Landgraf R, Holsboer F (2003a) The high-affinity non-peptide crh1 receptor antagonist R121919 attenuates stress-induced alterations in plasma oxytocin, prolactin, and testosterone secretion in rats. Pharmacopsychiatry 36(1):27–31

Keck ME, Welt T, Muller MB, Uhr M, Ohl F, Wigger A, Toschi N, Holsboer F, Landgraf R (2003b) Reduction of hypothalamic vasopressinergic hyperdrive contributes to clinically relevant behavioral and neuroendocrine effects of chronic paroxetine treatment in a psychopathological rat model. Neuropsychopharmacology 28(2):235–243

Keck ME, Sartori SB, Welt T, Muller MB, Ohl F, Holsboer F, Landgraf R, Singewald N (2005) Differences in serotonergic neurotransmission between rats displaying high or low anxiety/depression-like behaviour: effects of chronic paroxetine treatment. J Neurochem 92(5):1170–1179

Keller J, Flores B, Gomez RG, Solvason HB, Kenna H, Williams GH, Schatzberg AF (2006) Cortisol circadian rhythm alterations in psychotic major depression. Biol Psychiatry 60(3): 275–281

Kessler MS, Murgatroyd C, Bunck M, Czibere L, Frank E, Jacob W, Horvath C, Muigg P, Holsboer F, Singewald N, Spengler D, Landgraf R (2007) Diabetes insipidus and, partially, low anxiety-related behaviour are linked to a SNP-associated vasopressin deficit in lab mice. Eur J Neurosci 26(10):2857–2864

Kessler MS, Bosch OJ, Bunck M, Landgraf R, Neumann ID (2011) Maternal care differs in mice bred for high vs. low trait anxiety: impact of brain vasopressin and cross-fostering. Soc Neurosci 6(2):156–168

Knol MJ, Twisk JWR, Beekman ATF, Heine RJ, Snoek FJ, Pouwer F (2006) Depression as a risk factor for the onset of type 2 diabetes mellitus. A meta-analysis. Diabetologia 49(5):837–845

Kovacevic T, Skelin I, Diksic M (2010) Chronic fluoxetine treatment has a larger effect on the density of a serotonin transporter in the flinders sensitive line (FSL) rat model of depression than in normal rats. Synapse 64(3):231–240

Kromer SA, Kessler MS, Milfay D, Birg IN, Bunck M, Czibere L, Panhuysen M, Putz B, Deussing JM, Holsboer F, Landgraf R, Turck CW (2005) Identification of glyoxalase-i as a protein marker in a mouse model of extremes in trait anxiety. J Neurosci 25(17):4375–4384

Lahmame A, Del Arco C, Pazos A, Yritia M, Armario A (1997) Are wistar-kyoto rats a genetic animal model of depression resistant to antidepressants? Eur J Pharmacol 337(2–3):115–123

Lahmanie A, Armario A (1996) Differential responsiveness of inbred strains of rats to antidepressants in the forced swimming test: are wistar kyoto rats an animal model of subsensitivity to antidepressants? Psychopharmacology 123(2):191–198

Lancel M, Muller-Preuss P, Wigger A, Landgraf R, Holsboer F (2002) The CRH1 receptor antagonist R121919 attenuates stress-elicited sleep disturbances in rats, particularly in those with high innate anxiety. J Psychiatr Res 36(4):197–208

Landgraf R, Neumann ID (2004) Vasopressin and oxytocin release within the brain: a dynamic concept of multiple and variable modes of neuropeptide communication. Front Neuroendocrinol 25(3–4):150–176

Landgraf R, Wigger A (2002) High vs low anxiety-related behavior rats: an animal model of extremes in trait anxiety. Behav Genet 32(5):301–314

Landgraf R, Wigger A, Holsboer F, Neumann ID (1999) Hyper-reactive hypothalamo-pituitary-adrenocortical axis in rats bred for high anxiety-related behaviour. J Neuroendocrinol 11(6):405–407

Landgraf R, Kessler MS, Bunck M, Murgatroyd C, Spengler D, Zimbelmann M, Nussbaumer M, Czibere L, Turck CW, Singewald N, Rujescu D, Frank E (2007) Candidate genes of anxiety-related behavior in HAB/LAB rats and mice: focus on vasopressin and glyoxalase-I. Neurosci Biobehav Rev 31(1):89–102

Lecrubier Y (2006) Physical components of depression and psychomotor retardation. J Clin Psychiatry 67(Suppl 6):23–26

Leonard SK, Dwyer JM, Sukoff Rizzo SJ, Platt B, Logue SF, Neal SJ, Malberg JE, Beyer CE, Schechter LE, Rosenzweig-Lipson S, Ring RH (2008) Pharmacology of neuropeptide s in mice: therapeutic relevance to anxiety disorders. Psychopharmacology (Berl) 197(4):601–611

Lesch KP (1991) 5-HT_{1A} receptor responsivity in anxiety disorders and depression. Prog Neuropsychopharmacol Biol Psychiatry 15:723–733

Li TK, Lumeng L (1977) Alcohol metabolism of inbred strains of rats with alcohol preference and nonpreference. Alcohol Aldehyde Metab Sys 3:625–633

Li TK, Lumeng L, Doolittle DP (1993) Selective breeding for alcohol preference and associated responses. Behav Genet 23(2):163–170

Liebenberg N, Harvey BH, Brand L, Brink CB (2010) Antidepressant-like properties of phosphodiesterase type 5 inhibitors and cholinergic dependency in a genetic rat model of depression. Behav Pharmacol 21(5–6):540–547

Liebsch G, Linthorst AC, Neumann ID, Reul JM, Holsboer F, Landgraf R (1998a) Behavioral, physiological, and neuroendocrine stress responses and differential sensitivity to diazepam in two Wistar rat lines selectively bred for high- and low-anxiety-related behavior. Neuropsychopharmacology 19(5):381–396

Liebsch G, Montkowski A, Holsboer F, Landgraf R (1998b) Behavioral profiles of two Wistar rat lines selectively bred for high or low anxiety-related behaviour. Behav Brain Res 94(2): 301–310

Lindsay JR, Baker HJ (2006) Historical foundations. In: Suckow MA, Weisbroth SH (eds) The laboratory rat. Elsevier Academic Press, Burlington, pp 1–52

López-Rubalcava C, Lucki I (2000) Strain differences in the behavioral effects of antidepressant drugs in the rat forced swimming test. Neuropsychopharmacology 22(2):191–199

Lucassen PJ, Bosch OJ, Jousma E, Kromer SA, Andrew R, Seckl JR, Neumann ID (2009) Prenatal stress reduces postnatal neurogenesis in rats selectively bred for high, but not low, anxiety: Possible key role of placental 11beta-hydroxysteroid dehydrogenase type 2. Eur J Neurosci 29(1):97–103

Lupien SJ, McEwen BS, Gunnar MR, Heim C (2009) Effects of stress throughout the lifespan on the brain, behaviour and cognition. Nat Rev Neurosci 10(6):434–445

Maccari S, Darnaudery M, Morley-Fletcher S, Zuena AR, Cinque C, Van Reeth O (2003) Prenatal stress and long-term consequences: implications of glucocorticoid hormones. Neurosci Biobehav Rev 27(1–2):119–127

Maier SF (1991) Stressor controllability, cognition and fear. In: Madden J (ed) Neurobiology, learning, emotion and affect. Raven Press, New York, pp 155–193

Mantella RC, Vollmer RR, Li X, Amico JA (2003) Female oxytocin-deficient mice display enhanced anxiety-related behavior. Endocrinology 144(6):2291–2296

Marescaux C, Micheletti G, Vergnes M (1984) A model of chronic spontaneous petit mal-like seizures in the rat: comparison with pentylenetetrazol-induced seizures. Epilepsia 25(3): 326–331

Marescaux C, Vergnes M, Depaulis A (1992) Genetic absence epilepsy in rats from strasbourg—a review. J Neural Transmis Suppl 35:37–69

Mathe AA, Jimenez PA, Theodorsson E, Stenfors C (1998) Neuropeptide y, neurokinin a and neurotensin in brain regions of fawn hooded "depressed", Wistar, and Sprague Dawley rats.

Effects of electroconvulsive stimuli. Prog Neuropsychopharmacol Biol Psychiatry 22(3): 529–546

Mathe AA, Husum H, El Khoury A, Jimenez-Vasquez P, Gruber SH, Wortwein G, Nikisch G, Baumann P, Agren H, Andersson W, Sodergren A, Angelucci F (2007) Search for biological correlates of depression and mechanisms of action of antidepressant treatment modalities. Do neuropeptides play a role? Physiol Behav 92(1–2):226–231

Matthews K, Baldo BA, Markou A, Lown O, Overstreet DH, Koob GF (1996) Rewarding electrical brain stimulation: similar thresholds for flinders sensitive line hypercholinergic and flinders resistant line hypocholinergic rats. Physiol Behav 59(6):1155–1162

Mayberg HS, Lozano AM, Voon V, McNeely HE, Seminowicz D, Hamani C, Schwalb JM, Kennedy SH (2005) Deep brain stimulation for treatment-resistant depression. Neuron 45(5): 651–660

Maynard TM, Sikich L, Lieberman JA, LaMantia AS (2001) Neural development, cell–cell signaling, and the "two-hit" hypothesis of schizophrenia. Schizophr Bull 27(3):457–476

McCaslin PP, Oh S (1995) Nitric oxide and glutamate receptors. CNS neurotransmitters and neuromodulators: glutamate. CRC Press, New York, pp 159–179

Merikangas KR, Gelernter CS (1990) Comorbidity for alcoholism and depression. Psychiatric Clin N Am 13(4):613–632

Mikuni M, Kusumi I, Kagaya A, Kuroda Y, Mori H, Takahashi K (1991) Increased 5-HT-2 receptor function as measured by serotonin-stimulated phosphoinositide hydrolysis in platelets of depressed patients. Prog Neuro-Psychopharmacol Biol Psychiatry 15(1):49–61

Muigg P, Hoelzl U, Palfrader K, Neumann I, Wigger A, Landgraf R, Singewald N (2007) Altered brain activation pattern associated with drug-induced attenuation of enhanced depression-like behavior in rats bred for high anxiety. Biol Psychiatry 61(6):782–796

Muigg P, Hetzenauer A, Hauer G, Hauschild M, Gaburro S, Frank E, Landgraf R, Singewald N (2008) Impaired extinction of learned fear in rats selectively bred for high anxiety–evidence of altered neuronal processing in prefrontal-amygdala pathways. Eur J Neurosci 28(11): 2299–2309

Murgatroyd C, Wigger A, Frank E, Singewald N, Bunck M, Holsboer F, Landgraf R, Spengler D (2004) Impaired repression at a vasopressin promoter polymorphism underlies overexpression of vasopressin in a rat model of trait anxiety. J Neurosci 24(35):7762–7770

Murgatroyd C, Patchev AV, Wu Y, Micale V, Bockmuhl Y, Fischer D, Holsboer F, Wotjak CT, Almeida OF, Spengler D (2009) Dynamic DNA methylation programs persistent adverse effects of early-life stress. Nat Neurosci 12(12):1559–1566

Murphy JM, Stewart RB, Bell RL, Badia-Elder NE, Carr LG, McBride WJ, Lumeng L, Li TK (2002) Phenotypic and genotypic characterization of the Indiana University rat lines selectively bred for high and low alcohol preference. Behav Genet 32(5):363–388

Musazzi L, Mallei A, Tardito D, Gruber SH, El Khoury A, Racagni G, Mathe AA, Popoli M (2010) Early-life stress and antidepressant treatment involve synaptic signaling and erk kinases in a gene-environment model of depression. J Psychiatr Res 44(8):511–520

Mørk A, Pehrson A, Tottrup Brennum L, Moller Nielsen S, Zhong H, Lassen AB, Miller S, Westrich L, Boyle NJ, Sanchez C, Weide Fischer C, Liebenberg N, Wegener G, Bundgaard C, Hogg S, Bang-Andersen B, Bryan Stensbol T (2012) Pharmacological effects of Lu AA21004: a novel multimodal compound for the treatment of major depressive disorder. J Pharmacol Exp Ther. doi:10.1124/jpet.111.189068

Nagano M, Ozawa H, Suzuki H (2008) Prenatal dexamethasone exposure affects anxiety-like behaviour and neuroendocrine systems in an age-dependent manner. Neurosci Res 60(4):364–371

Naumenko EV, Popova NK, Nikulina EM, Dygalo NN, Shishkina GT, Borodin PM, Markel AL (1989) Behavior, adrenocortical activity, and brain monoamines in Norway rats selected for reduced aggressiveness towards man. Pharmacol Biochem Behav 33(1):85–91

Nemeroff CB (2004) Early-life adversity, CRF dysregulation, and vulnerability to mood and anxiety disorders. Psychopharmacol Bull 38(Suppl 1):14–20

Nestler EJ, Barrot M, DiLeone RJ, Eisch AJ, Gold SJ, Monteggia LM (2002) Neurobiology of depression. Neuron 34(1):13–25

Neumann ID (2008) Brain oxytocin: a key regulator of emotional and social behaviours in both females and males. J Neuroendocrinol 20(6):858–865

Neumann ID (2009) The advantage of social living: brain neuropeptides mediate the beneficial consequences of sex and motherhood. Front Neuroendocrinol 30(4):483–496

Neumann ID, Wigger A, Liebsch G, Holsboer F, Landgraf R (1998) Increased basal activity of the hypothalamo-pituitary-adrenal axis during pregnancy in rats bred for high anxiety-related behaviour. Psychoneuroendocrinology 23(5):449–463

Neumann ID, Torner L, Wigger A (2000) Brain oxytocin: differential inhibition of neuroendocrine stress responses and anxiety-related behaviour in virgin, pregnant and lactating rats. Neuroscience 95(2):567–575

Neumann ID, Kromer SA, Bosch OJ (2005a) Effects of psycho-social stress during pregnancy on neuroendocrine and behavioural parameters in lactation depend on the genetically determined stress vulnerability. Psychoneuroendocrinology 30(8):791–806

Neumann ID, Wigger A, Kromer S, Frank E, Landgraf R, Bosch OJ (2005b) Differential effects of periodic maternal separation on adult stress coping in a rat model of extremes in trait anxiety. Neuroscience 132(3):867–877

Neumann ID, Veenema AH, Beiderbeck DI (2010) Aggression and anxiety: social context and neurobiological links. Front Behav Neurosci 4:12

Nikisch G, Mathe AA (2008) CSF monoamine metabolites and neuropeptides in depressed patients before and after electroconvulsive therapy. Eur Psychiatry 23(5):356–359

Nikisch G, Agren H, Eap CB, Czernik A, Baumann P, Mathe AA (2005) Neuropeptide y and corticotropin-releasing hormone in CSF mark response to antidepressive treatment with citalopram. Int J Neuropsychopharmacol 8(3):403–410

Nishi K, Kanemaru K, Diksic M (2009) A genetic rat model of depression, flinders sensitive line, has a lower density of 5-HT(1A) receptors, but a higher density of 5-HT(1B) receptors, compared to control rats. Neurochem Int 54(5–6):299–307

O'Connor TG, Heron J, Golding J, Beveridge M, Glover V (2002) Maternal antenatal anxiety and children's behavioural/emotional problems at 4 years. Report from the Avon longitudinal study of parents and children. Br J Psychiatry 180(JUNE):502–508

Ohl F, Toschi N, Wigger A, Henniger MS, Landgraf R (2001) Dimensions of emotionality in a rat model of innate anxiety. Behav Neurosci 115(2):429–436

Ohl F, Roedel A, Storch C, Holsboer F, Landgraf R (2002) Cognitive performance in rats differing in their inborn anxiety. Behav Neurosci 116(3):464–471

Okamoto K, Aoki K (1963) Development of a strain of spontaneously hypertensive rats. Jpn Circ J 27:282–293

Oliveira RM, Guimaraes FS, Deakin JF (2008) Expression of neuronal nitric oxide synthase in the hippocampal formation in affective disorders. Braz J Med Biol Res 41(4):333–341

Olsson A, Regnell G, Traskman-Bendz L, Ekman R, Westrin A (2004) Cerebrospinal neuropeptide Y and substance P in suicide attempters during long-term antidepressant treatment. Eur Neuropsychopharmacol 14(6):479–485

Overmier JB, Seligman ME (1967) Effects of inescapable shock upon subsequent escape and avoidance responding. J Comp Physiol Psychol 63(1):28–33

Overstreet DH (1986) Selective breeding for increased cholinergic function: development of a new animal model of depression. Biol Psychiatry 21(1):49–58

Overstreet DH (1993) The flinders sensitive line rats: a genetic animal model of depression. Neurosci Biobehav Rev 17(1):51–68

Overstreet DH (2002) Behavioral characteristics of rat lines selected for differential hypothermic responses to cholinergic or serotonergic agonists. Behav Genet 32(5):335–348

Overstreet DH, Griebel G (2004) Antidepressant-like effects of CRF1 receptor antagonist SSR125543 in an animal model of depression. Eur J Pharmacol 497(1):49–53

Overstreet DH, Russell RW (1982) Selective breeding for diisopropyl fluorophosphate-sensitivity: behavioural effects of cholinergic agonists and antagonists. Psychopharmacology (Berl) 78(2):150–155

Overstreet DH, Russell RW (1984) Selective breeding for differences in cholinergic function: sex differences in the genetic regulation of sensitivity to the anticholinesterase, DFP. Behav Neural Biol 40(2):227–238

Overstreet DH, Russell RW, Helps SC, Messenger M (1979) Selective breeding for sensitivity to the anticholinesterase DFP. Psychopharmacology (Berl) 65(1):15–20

Overstreet DH, Janowsky DS, Gillin JC, Shiromani PJ, Sutin EL (1986) Stress-induced immobility in rats with cholinergic supersensitivity. Biol Psychiatry 21(7):657–664

Overstreet DH, Double K, Schiller GD (1989) Antidepressant effects of rolipram in a genetic animal model of depression: Cholinergic supersensitivity and weight gain. Pharmacol Biochem Behav 34(4):691–696

Overstreet DH, Rezvani AH, Janowsky DS (1990) Impaired active avoidance responding in rats selectively bred for increased cholinergic function. Physiol Behav 47(4):787–788

Overstreet DH, Russell RW, Hay DA, Crocker AD (1992) Selective breeding for increased cholinergic function: biometrical genetic analysis of muscarinic responses. Neuropsychopharmacology 7(3):197–204

Overstreet DH, Rezvani AH, Pucilowski O, Gause L, Janowsky DS (1994) Rapid selection for serotonin-1A sensitivity in rats. Psychiatr Genet 4(1):57–62

Overstreet DH, Pucilowski O, Rezvani AH, Janowsky DS (1995) Administration of antidepressants, diazepam and psychomotor stimulants further confirms the utility of flinders sensitive line rats as an animal model of depression. Psychopharmacology (Berl) 121(1):27–37

Overstreet DH, Pucilowski O, Retton MC, Delagrange P, Guardiola-Lemaitre B (1998) Effects of melatonin receptor ligands on swim test immobility. Neuroreport 9(2):249–253

Overstreet DH, Hlavka J, Feighner JP, Nicolau G, Freed JS (2004a) Antidepressant-like effects of a novel pentapeptide, nemifitide, in an animal model of depression. Psychopharmacology (Berl) 175(3):303–309

Overstreet DH, Keeney A, Hogg S (2004b) Antidepressant effects of citalopram and CRF receptor antagonist CP-154,526 in a rat model of depression. Eur J Pharmacol 492(2–3):195–201

Overstreet DH, Friedman E, Mathe AA, Yadid G (2005) The flinders sensitive line rat: a selectively bred putative animal model of depression. Neurosci Biobehav Rev 29(4–5):739–759

Overstreet DH, Stemmelin J, Griebel G (2008) Confirmation of antidepressant potential of the selective beta3 adrenoceptor agonist amibegron in an animal model of depression. Pharmacol Biochem Behav 89(4):623–626

Overstreet DH, Fredericks K, Knapp D, Breese G, McMichael J (2010a) Nerve growth factor (NGF) has novel antidepressant-like properties in rats. Pharmacol Biochem Behav 94(4): 553–560

Overstreet DH, Naimoli VM, Griebel G (2010b) Saredutant, an NK2 receptor antagonist, has both antidepressant-like effects and synergizes with desipramine in an animal model of depression. Pharmacol Biochem Behav 96(2):206–210

Owens MJ, Overstreet DH, Knight DL, Rezvani AH, Ritchie JC, Bissette G, Janowsky DS, Nemeroff CB (1991) Alterations in the hypothalamic-pituitary-adrenal axis in a proposed animal model of depression with genetic muscarinic supersensitivity. Neuropsychopharmacology 4(2):87–93

Paré WP (1992) Learning behavior, escape behavior, and depression in an ulcer susceptible rat strain. Integr Physiol Behav Sci 27(2):130–141

Paré WP (1994) Open field, learned helplessness, conditioned defensive burying, and forced-swim tests in WKY rats. Physiol Behav 55(3):433–439

Pare WP, Redei E (1993) Depressive behavior and stress ulcer in Wistar Kyoto rats. J Physiol Paris 87(4):229–238

Pawlak CR, Ho YJ, Schwarting RKW (2008) Animal models of human psychopathology based on individual differences in novelty-seeking and anxiety. Neurosci Biobehav Rev 32(8): 1544–1568

Petersen A, Wortwein G, Gruber SH, Mathe AA (2008) Escitalopram reduces increased hippocampal cytogenesis in a genetic rat depression model. Neurosci Lett 436(3):305–308

Petersen A, Wortwein G, Gruber SH, El-Khoury A, Mathe AA (2009) Nortriptyline mediates behavioral effects without affecting hippocampal cytogenesis in a genetic rat depression model. Neurosci Lett 451(2):148–151
Pittenger C, Duman RS (2008) Stress, depression, and neuroplasticity: a convergence of mechanisms. Neuropsychopharmacology 33(1):88–109
Post A, Keck ME (2001) Transcranial magnetic stimulation as a therapeutic tool in psychiatry: what do we know about the neurobiological mechanisms? J Psychiatr Res 35(4):193–215
Pucilowski O, Overstreet DH (1993) Effect of chronic antidepressant treatment on responses to apomorphine in selectively bred rat strains. Brain Res Bull 32(5):471–475
Pucilowski O, Overstreet DH, Rezvani AH, Janowsky DS (1993) Chronic mild stress-induced anhedonia: greater effect in a genetic rat model of depression. Physiol Behav 54(6): 1215–1220
Ramos A, Mormede P (1998) Stress and emotionality: a multidimensional and genetic approach. Neurosci Biobehav Rev 22(1):33–57
Ramos A, Mellerin Y, Mormède P, Chaouloff F (1998) A genetic and multifactorial analysis of anxiety-related behaviours in lewis and SHR intercrosses. Behav Brain Res 96(1–2):195–205
Ramos A, Kangerski AL, Basso PF, Da Silva Santos JE, Assreuy J, Vendruscolo LF, Takahashi RN (2002) Evaluation of lewis and SHR rat strains as a genetic model for the study of anxiety and pain. Behav Brain Res 129(1–2):113–123
Ramos A, Correia EC, Izídio GS, Brüske GR (2003) Genetic selection of two new rat lines displaying different levels of anxiety-related behaviors. Behav Genet 33(6):657–668
Ransford CP (1982) A role for amines in the antidepressant effect of exercise: a review. Med Sci Sports Exerc 14(1):1–10
Reber SO, Birkeneder L, Veenema AH, Obermeier F, Falk W, Straub RH, Neumann ID (2007) Adrenal insufficiency and colonic inflammation after a novel chronic psycho-social stress paradigm in mice: implications and mechanisms. Endocrinology 148(2):670–682
Reber SO, Obermeier F, Straub RH, Veenema AH, Neumann ID (2008) Aggravation of DSS-induced colitis after chronic subordinate colony (CSC) housing is partially mediated by adrenal mechanisms. Stress 11(3):225–234
Reddy PL, Khanna S, Subhash MN, Channabasavanna SM, Sridhara Rama Rao BS (1992) CSF amine metabolites in depression. Biol Psychiatry 31(2):112–118
Redei E, Pare WP, Aird F, Kluczynski J (1994) Strain differences in hypothalamic-pituitary-adrenal activity and stress ulcer. Am J Physiol 266(2 Pt 2):R353–R360
Rezvani AH, Overstreet DH, Janowsky DS (1990) Genetic serotonin deficiency and alcohol preference in the fawn hooded rats. Alcohol Alcohol 25(5):573–575
Rezvani AH, Overstreet DH, Janowsky DS (1991) Drug-induced reductions in ethanol intake in alcohol preferring and fawn-hooded rats. Alcohol Alcohol Suppl 1:433–437
Rezvani AH, Overstreet DH, Yang Y, Clark E Jr (1999) Attenuation of alcohol intake by extract of hypericum perforatum (St. John's Wort) in two different strains of alcohol-preferring rats. Alcohol Alcohol 34(5):699–705
Rezvani AH, Parsian A, Overstreet DH (2002) The fawn-hooded (fh/wjd) rat: a genetic animal model of comorbid depression and alcoholism. Psychiatr Genet 12(1):1–16
Rezvani AH, Overstreet DH, Cleves M, Parsian A (2007) Further genetic characterization of the fawn-hooded (Fh/Wjd) rat, an animal model of comorbid depression and alcoholism. Psychiatric Genet 17(2):77–83
Risch SC, Kalin NH, Janowsky DS (1981) Cholinergic challenges in affective illness: behavioral and neuroendocrine correlates. J Clin Psychopharmacol 1(4):186–192
Rittenhouse PA, López-Rubalcava C, Stanwood GD, Lucki I (2002) Amplified behavioral and endocrine responses to forced swim stress in the Wistar-Kyoto rat. Psychoneuroendocrinology 27(3):303–318
Robinson MJ, Edwards SE, Iyengar S, Bymaster F, Clark M, Katon W (2009) Depression and pain. Front Biosci 14:5031–5051
Roy A, Pickar D, Linnoila M (1985) Cerebrospinal fluid monoamine and monoamine metabolite concentrations in melancholia. Psychiatry Res 15(4):281–292

Rugulies R (2002) Depression as a predictor for coronary heart disease: a review and meta-analysis. Am J Prevent Med 23(1):51–61

Russell RW, Overstreet DH (1987) Mechanisms underlying sensitivity to organophosphorus anticholinesterase compounds. Prog Neurobiol 28(2):97–129

Russell RW, Overstreet DH, Messenger M, Helps SC (1982) Selective breeding for sensitivity to DFP: generalization of effects beyond criterion variables. Pharmacol Biochem Behav 17(5):885–891

Russo-Neustadt A, Beard RC, Cotman CW (1999) Exercise, antidepressant medications, and enhanced brain derived neurotrophic factor expression. Neuropsychopharmacology 21(5):679–682

Ryan B, Musazzi L, Mallei A, Tardito D, Gruber SH, El Khoury A, Anwyl R, Racagni G, Mathe AA, Rowan MJ, Popoli M (2009) Remodelling by early-life stress of nmda receptor-dependent synaptic plasticity in a gene-environment rat model of depression. Int J Neuropsychopharmacol 12(4):553–559

Sadile AG, Gironi Carnevale UA, Vitullo E, Cioffi LA, Welzl H, Bättig K (1988) Maze learning of the naples high- and low-excitability rat lines. Adv Biosci 70:177–180

Salchner P, Sartori SB, Sinner C, Wigger A, Frank E, Landgraf R, Singewald N (2006) Airjet and FG-7142-induced fos expression differs in rats selectively bred for high and low anxiety-related behavior. Neuropharmacology 50(8):1048–1058

Salome N, Salchner P, Viltart O, Sequeira H, Wigger A, Landgraf R, Singewald N (2004) Neurobiological correlates of high (HAB) versus low anxiety-related behavior (LAB): Differential FOS expression in HAB and LAB rats. Biol Psychiatry 55(7):715–723

Sandnabba NK (1996) Selective breeding for isolation-induced intermale aggression in mice: associated responses and environmental influences. Behav Genet 26(5):477–488

Santarelli L, Saxe M, Gross C, Surget A, Battaglia F, Dulawa S, Weisstaub N, Lee J, Duman R, Arancio O, Belzung C, Hen R (2003) Requirement of hippocampal neurogenesis for the behavioral effects of antidepressants. Science 301(5634):805–809

Sartori S, Hauschild M, Muigg P, Slattery DA, Biederbeck D, Mathe AA, Wegener G, Neumann ID, Singewald N (2009) Modulation of impaired fear extinction by neuropeptides in a psychopathological animal model of trait anxiety In: Society for Neuroscience, Chicago, IL, 2009. Society for Neuroscience, Neuroscience Meeting Planner

Sartorius A, Hellweg R, Litzke J, Vogt M, Dormann C, Vollmayr B, Danker-Hopfe H, Gass P (2009) Correlations and discrepancies between serum and brain tissue levels of neurotrophins after electroconvulsive treatment in rats. Pharmacopsychiatry 42(6):270–276

Schildkraut JJ (1965) The catecholamine hypothesis of affective disorders: A review of supporting evidence. Am J Psychiatry 122(5):509–522

Schiller GD, Daws LC, Overstreet DH, Orbach J (1991) Lack of anxiety in an animal model of depression with cholinergic supersensitivity. Brain Res Bull 26(3):433–435

Schiller GD, Pucilowski O, Wienicke C, Overstreet DH (1992) Immobility-reducing effects of antidepressants in a genetic animal model of depression. Brain Res Bull 28(5):821–823

Shimizu E, Hashimoto K, Okamura N, Koike K, Komatsu N, Kumakiri C, Nakazato M, Watanabe H, Shinoda N, Okada S, Iyo M (2003) Alterations of serum levels of brain-derived neurotrophic factor (BDNF) in depressed patients with or without antidepressants. Biol Psychiatry 54(1):70–75

Shir Y, Zeltser R, Vatine JJ, Carmi G, Belfer I, Zangen A, Overstreet D, Raber P, Seltzer Z (2001) Correlation of intact sensibility and neuropathic pain-related behaviors in eight inbred and outbred rat strains and selection lines. Pain 90(1–2):75–82

Shiromani PJ, Klemfuss H, Lucero S, Overstreet DH (1991) Diurnal rhythm of core body temperature is phase advanced in a rodent model of depression. Biol Psychiatry 29(9):923–930

Slattery DA, Neumann ID (2010) Chronic icv oxytocin attenuates the pathological high anxiety state of selectively bred Wistar rats. Neuropharmacology 58(1):56–61

Slattery DA, Finger BC, Malumby R, Wegener G, Mathe AA, Neumann ID (2008) Neuropeptide s alters affective behaviour in high (HAB) but not low anxiety-related (LAB) rats In: Society for neuroscience, Washington, D.C., 2008, Society for Neuroscience 25(10):1295-303

Slattery DA, Neumann I, Cryan JF (2010) Transient inactivation of the infralimbic cortex induces antidepressant-like effects in the rat. J Psychopharmacol 25(10):1295–1303

Solberg LC, Olson SL, Turek FW, Redei E (2001) Altered hormone levels and circadian rhythm of activity in the WKY rat, a putative animal model of depression. Am J Physiol—Regul Integr Comp Physiol 281(3):R786–R794

Solskov L, Lofgren B, Pold R, Kristiansen SB, Nielsen TT, Overstreet DH, Schmitz O, Botker HE, Lund S, Wegener G (2010) Evaluation of the relationship between hyperinsulinaemia and myocardial ischaemia/reperfusion injury in a rat model of depression. Clin Sci (Lond) 118(4):259–267

Tabb K, Boss-Williams KA, Weiss JM, Weinshenker D (2007) Rats bred for susceptibility to depression-like phenotypes have higher kainic acid-induced seizure mortality than their depression-resistant counterparts. Epilepsy Res 74(2–3):140–146

Tazumi T, Hori E, Uwano T, Umeno K, Tanebe K, Tabuchi E, Ono T, Nishijo H (2005) Effects of prenatal maternal stress by repeated cold environment on behavioral and emotional development in the rat offspring. Behav Brain Res 162(1):153–160

Thase ME, Kupfer DJ, Buysse DJ, Frank E, Simons AD, McEachran AB, Rashid KF, Grochocinski VJ (1995) Electroencephalographic sleep profiles in single-episode and recurrent unipolar forms of major depression: I. Comparison during acute depressive states. Biol Psychiatry 38(8):506–515

Torner L, Toschi N, Pohlinger A, Landgraf R, Neumann ID (2001) Anxiolytic and anti-stress effects of brain prolactin: improved efficacy of antisense targeting of the prolactin receptor by molecular modeling. J Neurosci 21(9):3207–3214

Torner L, Maloumby R, Nava G, Aranda J, Clapp C, Neumann ID (2004) In vivo release and gene upregulation of brain prolactin in response to physiological stimuli. Eur J Neurosci 19(6):1601–1608

Touma C, Bunck M, Glasl L, Nussbaumer M, Palme R, Stein H, Wolferstatter M, Zeh R, Zimbelmann M, Holsboer F, Landgraf R (2008) Mice selected for high versus low stress reactivity: a new animal model for affective disorders. Psychoneuroendocrinology 33(6): 839–862

Tschopp TB, Zucker MB (1972) Hereditary defect in platelet function in rats. Blood 40(2): 217–226

Van Den Bergh BRH, Mulder EJH, Mennes M, Glover V (2005) Antenatal maternal anxiety and stress and the neurobehavioural development of the fetus and child: links and possible mechanisms. A Rev Neurosci Biobehav Rev 29(2):237–258

Van Luijtelaar ELJM, Coenen AML (1986) Two types of electrocortical paroxysms in an inbred strain of rats. Neurosci Lett 70(3):393–397

Van Luijtelaar ELJM, Coenen AML (1989) The WAG/RIJ model for generalized absence seizures. Adv Epileptol 17:78–83

van Oortmerssen GA, Bakker TC (1981) Artificial selection for short and long attack latencies in wild mus musculus domesticus. Behav Genet 11(2):115–126

Vedolin GM, Lobato VV, Conti PC, Lauris JR (2009) The impact of stress and anxiety on the pressure pain threshold of myofascial pain patients. J Oral Rehabil 36(5):313–321

Veenema AH, Torner L, Blume A, Beiderbeck DI, Neumann ID (2007) Low inborn anxiety correlates with high intermale aggression: link to acth response and neuronal activation of the hypothalamic paraventricular nucleus. Horm Behav 51(1):11–19

Vergnes M, Marescaux C, Micheletti G (1982) Spontaneous paroxysmal electroclinical patterns in rat: a model of generalized non-convulsive epilepsy. Neurosci Lett 33(1):97–101

Videbech P, Ravnkilde B (2004) Hippocampal volume and depression: a meta-analysis of mri studies. Am J Psychiatry 161(11):1957–1966

Viggiano D, Vallone D, Ruocco LA, Sadile AG (2003) Behavioural, pharmacological, morpho-functional molecular studies reveal a hyperfunctioning mesocortical dopamine system in an animal model of attention deficit and hyperactivity disorder. Neurosci Biobehav Rev 27(7): 683–689

Vitale G, Filaferro M, Ruggieri V, Pennella S, Frigeri C, Rizzi A, Guerrini R, Calo G (2008) Anxiolytic-like effect of neuropeptide S in the rat defensive burying. Peptides 29(12): 2286–2291

Vollmayr B, Henn FA (2001) Learned helplessness in the rat: improvements in validity and reliability. Brain Res Brain Res Protoc 8(1):1–7

Vollmayr B, Faust H, Lewicka S, Henn FA (2001) Brain-derived-neurotrophic-factor (BDNF) stress response in rats bred for learned helplessness. Mol Psychiatry 6(4):471–474, 358

Waldherr M, Neumann ID (2007) Centrally released oxytocin mediates mating-induced anxiolysis in male rats. Proc Natl Acad Sci USA 104(42):16681–16684

Walker MW, Wolinsky TD, Jubian V, Chandrasena G, Zhong H, Huang X, Miller S, Hegde LG, Marsteller DA, Marzabadi MR, Papp M, Overstreet DH, Gerald CP, Craig DA (2009) The novel neuropeptide y y5 receptor antagonist Lu AA33810 [n-[[trans-4-[(4,5-dihydro[1]benzothiepino[5,4-d]thiazol-2-yl)amino]cycloh exyl]methyl]-methanesulfonamide] exerts anxiolytic- and antidepressant-like effects in rat models of stress sensitivity. J Pharmacol Exp Ther 328(3):900–911

Wallis E, Overstreet DH, Crocker AD (1988) Selective breeding for increased cholinergic function: increased serotonergic sensitivity. Pharmacol Biochem Behav 31(2):345–350

Wegener G, Volke V (2010) Nitric oxide synthase inhibitors as antidepressants. Pharmaceuticals 3(1):273–299

Wegener G, Volke V, Harvey BH, Rosenberg R (2003) Local, but not systemic, administration of serotonergic antidepressants decreases hippocampal nitric oxide synthase activity. Brain Res 959(1):128–134

Wegener G, Harvey BH, Bonefeld B, Muller HK, Volke V, Overstreet DH, Elfving B (2010) Increased stress-evoked nitric oxide signalling in the flinders sensitive line (FSL) rat: a genetic animal model of depression. Int J Neuropsychopharmacol 13(4):461–473

Weiss JM, Cierpial MA, West CH (1998) Selective breeding of rats for high and low motor activity in a swim test: toward a new animal model of depression. Pharmacol Biochem Behav 61(1):49–66

West CH, Weiss JM (1998a) Effects of antidepressant drugs on rats bred for low activity in the swim test. Pharmacol Biochem Behav 61(1):67–79

West CHK, Weiss JM (1998b) Effects of antidepressant drugs on rats bred for low activity in the swim test. Pharmacol Biochem Behav 61(1):67–79

West CH, Bonsall RW, Emery MS, Weiss JM (1999a) Rats selectively bred for high and low swim-test activity show differential responses to dopaminergic drugs. Psychopharmacology (Berl) 146(3):241–251

West CH, Boss-Williams KA, Weiss JM (1999b) Motor activation by amphetamine infusion into nucleus accumbens core and shell subregions of rats differentially sensitive to dopaminergic drugs. Behav Brain Res 98(1):155–165

Widdowson PS, Ordway GA, Halaris AE (1992) Reduced neuropeptide Y concentrations in suicide brain. J Neurochem 59(1):73–80

Wigger A, Neumann ID (1999) Periodic maternal deprivation induces gender-dependent alterations in behavioral and neuroendocrine responses to emotional stress in adult rats. Physiol Behav 66(2):293–302

Wigger A, Sanchez MM, Mathys KC, Ebner K, Frank E, Liu D, Kresse A, Neumann ID, Holsboer F, Plotsky PM, Landgraf R (2004) Alterations in central neuropeptide expression, release, and receptor binding in rats bred for high anxiety: critical role of vasopressin. Neuropsychopharmacology 29(1):1–14

Will CC, Aird F, Redei EE (2003) Selectively bred wistar-kyoto rats: an animal model of depression and hyper-responsiveness to antidepressants. Mol Psychiatry 8(11):925–932

Willner P (1984) The validity of animal models of depression. Psychopharmacology (Berl) 83(1): 1–16

Wortwein G, Husum H, Andersson W, Bolwig TG, Mathe AA (2006) Effects of maternal separation on neuropeptide y and calcitonin gene-related peptide in "depressed" flinders sensitive line rats: a study of gene-environment interactions. Prog Neuropsychopharmacol Biol Psychiatry 30(4):684–693

Xu YL, Reinscheid RK, Huitron-Resendiz S, Clark SD, Wang Z, Lin SH, Brucher FA, Zeng J, Ly NK, Henriksen SJ, de Lecea L, Civelli O (2004) Neuropeptide s: a neuropeptide promoting arousal and anxiolytic-like effects. Neuron 43(4):487–497

Yadid G, Overstreet DH, Zangen A (2001) Limbic dopaminergic adaptation to a stressful stimulus in a rat model of depression. Brain Res 896(1–2):43–47

Young RJ (1979) The effect of regular exercise on cognitive functioning and personality. Br J Sports Med 13(3):110–117

Zambello E, Jimenez-Vasquez PA, El Khoury A, Mathe AA, Caberlotto L (2008) Acute stress differentially affects corticotropin-releasing hormone mRNA expression in the central amygdala of the "depressed" flinders sensitive line and the control flinders resistant line rats. Prog Neuropsychopharmacol Biol Psychiatry 32(3):651–661

Zangen A, Overstreet DH, Yadid G (1997) High serotonin and 5-hydroxyindoleacetic acid levels in limbic brain regions in a rat model of depression: Normalization by chronic antidepressant treatment. J Neurochem 69(6):2477–2483

Zangen A, Overstreet DH, Yadid G (1999) Increased catecholamine levels in specific brain regions of a rat model of depression: normalization by chronic antidepressant treatment. Brain Res 824(2):243–250

Zangen A, Nakash R, Overstreet DH, Yadid G (2001) Association between depressive behavior and absence of serotonin-dopamine interaction in the nucleus accumbens. Psychopharmacology (Berl) 155(4):434–439

Zangen A, Nakash R, Roth-Deri I, Overstreet DH, Yadid G (2002) Impaired release of beta-endorphin in response to serotonin in a rat model of depression. Neuroscience 110(3):389–393

Zyss T, Gorka Z, Kowalska M, Vetulani J (1997) Preliminary comparison of behavioral and biochemical effects of chronic transcranial magnetic stimulation and electroconvulsive shock in the rat. Biol Psychiatry 42(10):920–924

Stress-Induced Deficits in Cognition and Emotionality: A Role for Glutamate

Carolyn Graybeal, Carly Kiselycznyk and Andrew Holmes

Abstract Stress is associated with a number of neuropsychiatric disorders, many of which are characterized by altered cognition and emotionality. Rodent models of stress have shown parallel behavioral changes such as impaired working memory, cognitive flexibility and fear extinction. This coincides with morphological changes to pyramidal neurons in the prefrontal cortex, hippocampus and amygdala, key cortical regions mediating these behaviors. Increasing evidence suggests that alteration in the function of the glutamatergic system may contribute to the pathology seen in neuropsychiatric disorders. Stress can alter glutamate transmission in the prefrontal cortex, hippocampus and amygdala and altered glutamate transmission has been linked to neuronal morphological changes. More recently, genetic manipulations in rodent models have allowed for subunit-specific analysis of the role of AMPA and NMDA receptors as well as glutamate transporters in behaviors shown to be altered by stress. Together these data point to a role for glutamate in mediating the cognitive and emotional changes observed in neuropsychiatric disorders. Furthering our understanding of how stress affects glutamate receptors and related signaling pathways will ultimately contribute to the development of improved therapeutics for individuals suffering from neuropsychiatric disorders.

Keywords Glutamate · Stress · Emotionality · Cognition

C. Graybeal · C. Kiselycznyk · A. Holmes (✉)
National Institute on Alcohol Abuse and Alcoholism,
National Institutes of Health, 5625 Fishers Lane, Room 2N09,
Rockville, MD 20852-9411, USA
e-mail: holmesan@mail.nih.gov

Curr Topics Behav Neurosci (2012) 12: 189–207
DOI: 10.1007/7854_2011_193

Published Online: 20 January 2012

Contents

1 Introduction 190
2 Stress-Induced Executive Dysfunction in Humans and Rodents 190
3 Structural Changes in Response to Stress 193
4 Molecular Mechanisms of Structural Changes: Focus on Glutamate 195
5 Glutamate in Rodent Behavioral Correlates of Emotion and Cognition 196
6 Conclusions 199
References 200

1 Introduction

Stress is associated with a number of neuropsychiatric disorders either as a trigger, as with post-traumatic stress disorder (PTSD), or a risk factor, as with drug addiction or depression. Thus understanding the neurobiological consequences of stress is critical to treating these diseases. The first part of this chapter focuses on behavioral changes in animal models of stress and structural changes in key cortical regions mediating these behaviors. The second part focuses on the role of the glutamatergic system in mediating these changes and, through genetic approaches, some of the advances made in our understanding of the molecular underpinning of stress and mental disorders.

2 Stress-Induced Executive Dysfunction in Humans and Rodents

Stress is a known risk factor of a number of neuropsychiatric disorders such as depression, PTSD and addiction (Hammen 2005; Lupien et al. 2009; Schneiderman et al. 2005; Sinha 2008). Stress can be defined as the expense or "allostatic load" on the organism's homeostatic regulating system (McEwen 2000a). When faced with a stressor, the hypothalamic-pituitary-adrenal (HPA) axis is activated, resulting in the release of corticosteroids (de Kloet et al. 2005; Joels et al. 2007; McEwen 2000b). Though short bouts of stress can be tolerated or even be beneficial, prolonged exposure to stress can be detrimental (McEwen 2000a). Early life stress, such as an unstable childhood, is a strong predictor for the development of depression or addictive behavior in adulthood (Lupien et al. 2009). A stressful life event, a death in the family or chronic stress such as poverty is strongly linked with the development of PTSD, depression and alcoholism (Hammen 2005; Schneiderman et al. 2005; Sinha 2008).

Characteristic of these stress-related neuropsychiatric disorders are deficits in emotional regulation and cognition (Ferreri et al. 2011). Patients with panic disorder or PTSD are unable to normally regulate their emotions. These individuals express greater aversive feelings toward negative stimuli, find themselves preoccupied with worry and are impaired in their ability to suppress feelings of fear (Blechert et al. 2007; Michael et al. 2007). Individuals with attention deficit hyperactivity disorder (ADHD), depression or schizophrenia have deficits in certain learning and memory functions such as spatial working memory and cognitive flexibility (McLean et al. 2004; Murphy et al. 2003; Murray et al. 2008; Rogers et al. 2004; Waltz and Gold 2007). Using various stress paradigms, animal models have begun to show evidence for a direct link between stress and emotional or cognitive dysfunction.

In rodents, exposure to stress can result in increased anxiety- and depression-related behaviors (Sterner and Kalynchuk 2010). Adult rodents that experienced maternal separation as pups, a model of early life stress, have heightened corticosterone secretion following stress and show greater anxiety in the elevated plus-maze (e.g. Eiland and McEwen 2012; Holmes et al. 2005; Huot et al. 2001). Rodents exposed to chronic unpredictable stress (CUS) or repeated restraint stress, or simply injected with corticosterone, show elevated anxiety- like behaviors in the light/dark exploration and the elevated plus-maze test, and a heightened startle response relative to non-stressed controls (e.g. Mozhui et al. 2010; Pego et al. 2008). Cross-strain comparisons indicate the effect of stress is more pronounced in mouse strains with greater trait anxiety-like behavior, suggesting genetic background influences stress susceptibility (Mizoguchi et al. 2000). In addition to heightened anxiety, stress can evoke a behavioral profile reminiscent of depression. Rodents exposed to a stressor or those that received corticosterone orally or systemically show reduced social interaction and increased behavioral despair (e.g. Berton et al. 2006; Gourley et al. 2008; Shirayama et al. 2002; Wood et al. 2008). History of maternal separation or oral corticosterone delivery can decrease sucrose consumption or responding in an appetitive progressive ratio test, measures of anhedonia (Gourley et al. 2009, 2008; Huot ct al. 2001).

Also sensitive to stress are fear conditioning and extinction, measures of emotional learning and regulation (Rodrigues et al. 2009). Stress prior to conditioning has a potentiating effect on fear learning resulting in greater freezing during fear conditioning and post-conditioning fear recall tests (Rau et al. 2005; Sandi et al. 2001; Wood et al. 2008; Yamamoto et al. 2009). A series of experiments examining the effect of footshock stress prior to fear conditioning found that stressed rats showed enhanced fear recall when tested the following day (Rau et al. 2005). Using variants of this basic design, fear generalization and reinstatement were ruled out as alternate explanations for the increase in fear. Rather it appears that stress strengthens the fear memory, which may explain the impairing effects of stress on fear extinction, the learned suppression and the fear response (Corcoran and Quirk 2007; Quirk and Mueller 2008). Rats exposed to restraint stress or given oral corticosterone fail to extinguish the fear response, retaining a greater level of freezing after extinction training as compared to non-stressed subjects (Baran et al. 2009; Gourley et al. 2009; Miracle et al. 2006). In mice, a mere three days of 10 min forced swim stress was sufficient to impair fear extinction (Izquierdo et al. 2006). Stress does not have to immediately precede fear conditioning to affect learning. Rat pups exposed to five days of footshock stress and then tested on fear conditioning as adults fail to extinguish the fear response showing that stress can have long-lasting repercussions on cognition (Judo et al. 2010).

The effects of stress are not limited to cognitive processes associated with emotionality (Arnsten 2009; Holmes and Wellman 2009; Sterner and Kalynchuk 2010). Working memory, the ability to transiently store, recall and utilize recently learned information is vulnerable to stress. Mice exposed to four weeks of cold water submersion are impaired in a delayed alternation T-maze task (Mizoguchi et al. 2000). In rats, four weeks or merely six days of CUS result in impaired spatial working memory in the Morris water maze (MWM) (Cerqueira et al. 2005). This deficit is replicated by four weeks of daily injections of dexamethasone, an agonist of the glucocorticoid receptor, one of the two main corticosterone receptors, suggesting that the effects of stress are mediated at least in part by glucocorticoid receptor activity (Cerqueira et al. 2005).

Spatial learning itself is disrupted by stress. Rats that experienced maternal separation were impaired in an object placement task which examines the ability to discriminate between objects placed in the same or a novel location (Eiland and McEwen 2012). A single session of restraint stress or a corticosterone injection resulted in mice having longer escape latencies in circular hole board maze (Schwabe et al. 2010). In the hidden platform version of the MWM, tail shock stressed rats had greater escape latencies than non-stressed rats (Kim et al. 2001). Predator stress, where the rodent is exposed to either the odor of or an actual predator, impairs spatial working memory in the radial arm version of the MWM (RAWM) (Diamond et al. 1999; Park et al. 2008). In the RAWM, up to six swim paths are available from a center start point, one of which leads to the escape platform. An effect of stress was not present when just four arms were available. However, increasing the difficulty of the task by presenting six arms reveals an effect of stress, suggesting that observable effects of stress may be dependent on task demands (Diamond et al. 1999). Pretreatment with the antidepressants agomelatine or tianeptine can rescue this effect (Campbell et al. 2008; Conboy et al. 2009).

Stress also interferes with cognitive flexibility, the ability to adjust previously learned behavior in response to changing task demands. Rats subjected to either four weeks of CUS or dexamethasone injections or six just days of CUS were impaired on a spatial reversal variant of the MWM, where the escape platform is switched to the quadrant opposite from where it was located during training (Cerqueira et al. 2007; Cerqueira et al. 2005). Two weeks of CUS impaired reversal and extra-dimension set shifting in a texture-odor attention set-shifting task (Bondi et al. 2008). In this rodent adaptation of human cognitive flexibility tasks, the rodent learns to discriminate between digging textures or odors to locate a hidden reward. This stress-induced set-shifting impairment is robust, having been replicated with a number of different stress protocols (Bondi et al. 2008; Lapiz-Bluhm et al. 2009; Liston et al. 2006) and has been shown to be rescued with the antidepressants desipramine and citalopram (Bondi et al. 2008; Lapiz-Bluhm et al. 2009).

From the subset of work highlighted here, there is strong evidence to support the link between stress exposure and altered emotionality and cognitive function. Similar to what is seen in clinical populations, in rodent models stress increases anxiety- and depression-like behaviors, and facilitates the acquisition of fear while impairing control over fear expression. Stress also impairs certain cognitive functions such as spatial learning, working memory and cognitive flexibility. How stress may be affecting these changes via morphological and molecular alterations will be discussed in the following sections.

3 Structural Changes in Response to Stress

Key substrates mediating emotionality and cognition include the prefrontal cortex, hippocampus and amygdala. Changes in cortical volume have been recorded in these regions in human neuropsychiatric patients suggesting possible loci for the behavioral and cognitive pathologies (Sterner and Kalynchuk 2010; van Harmelen et al. 2010). These regions are anatomically interconnected, often working in concert to mediate emotional regulation and cognition (Thierry et al. 2000). Preclinical studies in rodents have not only shown parallel changes but have provided more precise information on morphological changes in these areas.

Loss of prefrontal cortex (PFC) function can impair working memory and cognitive flexibility as well as other higher-order cognitive functions (Chudasama and Robbins 2006; Dalley et al. 2004; Robbins 2007). In rodents, the prelimbic and infralimbic cortices are subregions of the PFC thought to mediate fear expression and fear extinction, respectively (Corcoran and Quirk 2007; Myers and Davis 2007; Quirk and Mueller 2008). Four weeks of CUS result in neuronal atrophy, specifically reduced volume, neuron number and apical dendritic length in layers I–III in both these regions (Cerqueira et al. 2005; Dias-Ferreira et al. 2009). This stress effect is replicated pharmacologically with four weeks of corticosterone or dexamethasone, a corticosterone-receptor agonist, injections (Cerqueira et al. 2005). In addition to neuronal atrophy, decreases in spine density and spine surface area have been observed in the medial PFC following exposure to three weeks of restraint stress (Liston et al. 2006; Radley et al. 2004, 2006, 2008). Interestingly, while restraint stress reduced apical dendritic length in the medial PFC, there was a corresponding increase in the orbitofrontal cortex, suggesting that subregions of the PFC may respond differently to stress (Liston et al. 2006). Shorter stress exposures have highlighted the potential sensitivity of the PFC to stress. A reduced restraint stress paradigm, ten days of two-hour restraint stress, reduced apical dendritic length in rat infralimbic cortex (Shansky and Morrison 2009). Seven days of only ten-minute daily restraint stress was effective in reducing apical branch number and length in layer II/III of the cingulate cortex (Brown et al. 2005). Even more limited, one bout of ten-minute forced swim stress was suffient to reduce apical dendritic length in layer II/III of the mouse infralimbic cortex though no change was seen in the prelimbic cortex (Izquierdo et al. 2006).

Also key in the fear learning circuitry are the hippocampus and amygdala (Maren and Quirk 2004). The hippocampus has a well-established role in learning and memory and is particularly important for spatial learning, as in the MWM or with contextual fear learning (Bird and Burgess 2008; Ji and Maren 2007; Maren and Holt 2000). Stress has a similar effect on morphology in the hippocampus as it does in the PFC (Joels et al. 2004; McEwen 2001). Five weeks of social conflict decreased cell proliferation and survival in the rat dentate gyrus, an effect rescued by concomitant fluoxetine treatment (Czeh et al. 2007). Rats exposed to either a brief (two days) or chronic (21 days) social stressor induced morphology changes, though changes were more robust following chronic stress (Kole et al. 2004). Four weeks of CUS or pharmacological stress by corticosterone or dexamethasone injections resulted in reduced CA3 and DG volume and decreased dendritic length of granule cells, CA3 and CA1 pyramidal cells (Cerqueira et al. 2007; Sousa et al. 2000). Shorter durations of stress, 21 days of restraint stress or ten days of CUS, also results in hippocampal dendritic retraction which was reversed when tianeptine was given in conjunction with stress (Vyas et al. 2002; Watanabe et al. 1992b).

In contrast to the PFC and hippocampus, stress induces neuronal hypertrophy in the amygdala, a region essential for the acquisition and consolidation of aversive memories (Maren and Quirk 2004; Pare et al. 2004; Roozendaal et al. 2009). While 21 days of restraint stress decreased hippocampal dendritic length, this same stressor increased dendritic length and branching in neurons of the basolateral amygdala (BLA) as well as the bed nucleus of stria terminalis (Vyas et al. 2002, 2003). This effect has been replicated with either CUS or corticosterone injections which lead to increases in dendritic length and spine density in the BLA and bed nucleus of stria terminalis (Pego et al. 2008). Stress-induced changes appear to be longer lasting in the amygdala than in the hippocampus or PFC. Following 21 days of recovery from stress, BLA pyramidal cells of stressed rats were still longer than controls, whereas CA3 hippocampal neurons were no different and overextension was seen in proximal dendritic arbors of IL neurons (Goldwater et al. 2009; Vyas et al. 2004).

As with behavior, stress results in different changes in neuronal structure in the PFC, hippocampus and amygdala. Generally, stress-induced dendritic retraction in the PFC and hippocampus but induced extension in the amygdala. This appears to align with the differing effects of stress on behavior, with impairments in largely PFC and hippocampus-associated behaviors and enhancements in amygdala-associated behaviors. However, it is unlikely that structural changes alone are the cause of stress-induced behavioral changes. More likely stress results in a cascade of changes on a molecular level of which modification to structure is an observable consequence.

4 Molecular Mechanisms of Structural Changes: Focus on Glutamate

It is clear that exposure to stress can alter neuronal morphology in regions associated with the emotional and cognitive changes seen with psychiatric disorders. Stress can alter multiple molecular signaling cascades that could explain the observed structural changes (a subset of these are reviewed in Pittenger and Duman (2008). Here, we focus on evidence for the role of the glutamatergic system in mediating stress-induced changes because of its known role in mechanisms of plasticity, cognition, and increasingly, emotional and psychiatric disorders.

Glutamate is the main excitatory neurotransmitter in the CNS and binds to multiple metabotropic receptors and three families of ionotropic receptors: alpha-amino-3-hydroxy-5-methyl-4-isoxazolepropionic acid (AMPA) receptors, kainate receptors and *N*-methyl-D-aspartate (NMDA) receptors. Activity-dependent alterations in glutamate transmission are a key mechanism of synaptic plasticity that in turn influences multiple downstream signaling pathways, including those activating neurotrophic factors involved in cell growth. Activation of synaptic NMDA receptors appears to be crucial for activity-dependent changes in synapses and even neuronal survival (Hardingham and Bading 2003). Changes in synaptic plasticity have been shown to influence spine morphology, as LTP-inducing protocols can lead to new dendritic spines (Engert and Bonhoeffer 1999; Matsuzaki et al. 2004). Conversely, elevated glutamate levels and excessive activation of NMDA receptors can lead to cell damage and even death in many neuropathological conditions, (Dirnagl et al. 1999; Lipton and Rosenberg 1994; Olney et al. 1986; Sapolsky 2000a, 2000b). The opposing effects of glutamate transmission most likely depend on the pattern of NMDA receptor activation and have been reviewed in (Hardingham and Bading 2003). Variations in the pattern of glutamate receptor activation could in part explain a shift in the activation of cell growth versus cell death pathways that influence the observed morphological changes.

Stress is known to lead to changes in glutamate transmission in the same regions showing stress-induced morphological changes. Various forms of stress as well as acute injections of corticosterone lead to a rapid and transient increase in extracellular glutamate in the hippocampus and PFC (Moghaddam 1993; Venero and Borrell 1999). Interference with glutamate signaling using glutamatergic receptor antagonists can attenuate the dendritic atrophy observed after chronic restraint stress. The anti-epileptic drug phenytoin is known to reduce excitatory amino acid release including glutamate and injections of phenytoin block stress-induced atrophy of the CA3 hippocampal neurons (Watanabe et al. 1992a). More specifically, blocking NMDA but not AMPA receptors attenuates CA3 dendritic atrophy suggesting that the NMDA receptor has a particular role in mediating the morphological changes seen with stress (Magarinos and McEwen 1995).

In the PFC, less is understood about how NMDA receptor antagonists alter stress-induced atrophy. However, lesions resulting in cholinergic deafferentation lead to dendritic remodeling and increased spine density in the PFC, an effect shown to be dependent on NMDA receptors (Garrett et al. 2006). Interestingly, systemic administration of the NMDAR antagonist CPP during chronic restraint stress not only blocks dendritic atrophy in the mPFC, but actually induces dendritic hypertrophy in this region, an effect not observed in unstressed control rats (Martin and Wellman 2011). Most recently, the NMDA receptor antagonist ketamine was found to block the loss of synaptic proteins in the PFC after stress induced by the learned helplessness paradigm (Li et al. 2010). NMDA receptor antagonists not only block stress-induced atrophy but have also been found to increase spine density in medial PFC pyramidal cells 24 h after administration in the absence of stress (Li et al. 2010). While additional data directly studying the role of glutamate transmission in each region showing stress-induced structural changes is sparse, much is known about the effects of glutamate transmission in cognitive and emotional behavioral paradigms mediated by these regions.

5 Glutamate in Rodent Behavioral Correlates of Emotion and Cognition

The NMDA receptors are heteromeric assemblies composed of an obligatory NR1 subunit and one or more NR2 (NR2A-NR2D) (Rosenmund et al. 1998) or NR3 subunits (Ciabarra et al. 1995). *Postmortem* studies of the brains of depressed patients have found reduced NR1 mRNA in the hippocampus (Law and Deakin 2001) and decreased NR2A and NR2B protein levels in the PFC (Feyissa et al. 2009). In rodent models, a single exposure to forced swim stress leads to increases in surface (but not total) NR1, NR2A, NR2B and AMPA receptor subunits in the PFC (Yuen et al. 2009). As mentioned earlier, NMDA, but not AMPA receptor antagonism, attenuates dendritic atrophy resulting from chronic stress or glucocorticoid injections (Magarinos and McEwen 1995). Similarly, NMDA receptor antagonists block stress-induced alterations in long-term potentiation and long-term depression in hippocampal CA1 neurons (Kim et al. 1996). Together, NMDA receptors appear to be modulated by stress and necessary for some stress-induced morphological changes. Further evidence has demonstrated that NMDA receptors are also involved in the cognitive and emotion-related paradigms affected by stress.

Genetic manipulations in rodent models have allowed for a subunit-specific analysis of the NMDA receptor's role in behaviors shown to be altered by stress. While there are no NR2A subunit-specific antagonists, a mutant mouse line lacking this subunit has been developed and displays decreased anxiety- and depression-related behaviors (Boyce-Rustay and Holmes 2006). While wild-type mice normally show restraint stress-induced changes in the light/dark exploration task, this effect was blocked or even reversed in NR2A knockout mice, highlighting the importance of NR2A in behavioral changes consequential to stress (Mozhui et al. 2010). Supporting the importance of NMDA receptors in morphology, these knockout mice also demonstrated decreased spine density on pyramidal neurons of the BLA (Mozhui et al. 2010). As mentioned earlier, stress-induced increases in anxiety-like behavior are often paralleled by increases in BLA spine density. The loss of spine density in the BLA NR2A KO mice could in part explain their reduced anxiety-like behavior. Future studies could evaluate if the increase of spines after stress is also blocked in these mice, and if local NR2A knockdown in the BLA still blocks the stress-induced behavioral changes in the light/dark exploration task.

Prenatal deletion of the NR1 or NR2B subunit is lethal. However, several genetic models with reduced NR1 or NR2B levels have been successfully used. With the development of a floxed NR1 mouse line, it has been possible to study the effects of NR1 deletion in specific regions of the brain. One such model limited NR1 deletion to hippocampal CA1 pyramidal cells and found a deficit in trace, but not delay, fear conditioning and impaired spatial (MWM, T-maze), but not non-spatial memory (Huerta et al. 2000; Tonegawa et al. 1996; Tsien et al. 1996a, 1996b). Mice with a deletion of NR1 restricted to hippocampal CA3 pyramidal cells are impaired in retention of one-trial context discrimination at three h, but not 24 h after avoidance training (Cravens et al. 2006). Similar results were found in mice with the NR2B subunit deleted in CA1 and cortical pyramidal cells (Brigman et al. 2010). Mice lacking the NR2B subunit on pyramidal cells of the cortex and CA1 also showed multiple learning and memory deficits, including in the MWM, T-maze and trace, but not delay fear conditioning. These same mice additionally showed decreased spine density in pyramidal cells of the CA1. While they demonstrated normal depressive-like behavior in the six-minute forced swim test, these mice developed less depressive-like behavior over the course of a novel 10 day swim stress paradigm, indicating a phenotype specific to repeated stress (Kiselycznyk et al. 2011).

Ketamine and other NMDA receptor antagonists have acute behavioral effects in humans and rodent models that appear to resemble many of the symptoms of schizophrenia. This has in part led to the NMDA receptor hypofunction hypothesis, suggesting that NMDA receptor hypofunction on inhibitory neurons in the PFC is responsible for excessive neuronal excitability in this region thus disrupting executive functions. Mice that are viable with only 5–10% of the normal NR1 levels have been used to test the relationship between NR1 subunits and schizophrenia. These mice have decreased prepulse inhibition, sociability and anxiety-like behavior but increased locomotion, a constellation of behaviors similar to those seen in patients (Halene et al. 2009). A recent study found that deletion of NR1 on interneurons during early postnatal development resulted in schizophrenia-like phenotypes in mice. This included novelty-induced hyperlocomotion, impaired nest building and mating, increased anxiety- and anhedonia-like behaviors, behaviors that were exacerbated by social isolation stress (Belforte et al. 2010). The schizophrenia-like behavior was selective to mice with early postnatal deletion (approximately equivalent to late gestation to age 2 in humans) and was not observed in mice with post-adolescence NR1 deletion. This is consistent with hypotheses proposing a developmental cause in schizophrenia, with stress exacerbating deficits in cortical development.

An additional post-synaptic receptor of glutamate, the AMPA receptor, is also essential for synaptic plasticity. The AMPA receptor is a heteromeric receptor composed of a combination of four subunits: GluR1-GluR4. Various antidepressants with diverse structures have been found to alter the phosphorylation of GluR1 receptors, therefore altering glutamatergic synaptic transmission

(Svenningsson et al. 2007). In schizophrenia, studies have found decreased GluR1 expression in the hippocampus, striatum and PFC that is reversed by treatment with neuroleptics (Sokolov 1998). Mice lacking the GluR1 subunit display novelty-induced hyperlocomotion that is reversed by the antipsychotic haloperidol (Wiedholz et al. 2008). These mice also demonstrated altered social behavior and deficits in prepulse inhibition and impairments in a spatial reversal learning task. In additional studies, GluR1 KO mice were found to demonstrate behaviors indicative of mania-related phenotypes as they also showed stress-induced hyperactivity, reduced immobility in the forced swim test, and alterations in approach/avoid conflict tests. Treatment with lithium reversed the KO's anxiety-like phenotype and partially reversed their stress-induced hyperlocomotion (Fitzgerald et al. 2010). However, deletion of GluR1 did not affect the stress-induced changes in light/dark exploration, as was observed in NR2A knockout mice (Mozhui et al. 2010). As mentioned earlier, AMPA receptor antagonists alone do not block stress-induced changes in morphology as with the NMDA antagonists (Magarinos and McEwen 1995). While this appears contradictory, it is likely that activation of these two types of glutamatergic receptors activate a different collection of downstream signaling pathways (such as calcium-activated pathways), and there are some pathways specific to NMDA receptors that explain stress-induced changes.

The amount of glutamate available to NMDA and AMPA receptors is partially regulated by reuptake with glutamate transporters (EAAT1-5 in humans) located both on neurons and neighboring glial cells (Anderson and Swanson 2000). Postmortem studies of schizophrenic patients have found elevated expression of mRNA but decreased protein expression of EAAT1 in the dorsolateral PFC and anterior cingulate cortex (Bauer et al. 2008) and increased thalamic EAAT1 and EAAT2 (Smith et al. 2001). Similarly, variants of the EAAT1 gene (SLC1A3) have been linked with schizophrenia (Walsh et al. 2008). Conversely, microarrays of postmortem samples of depressed patients found a decrease of EAAT1 and EAAT2 mRNA in the frontal cortex (Choudary et al. 2005) and decreases in EAAT3 and EAAT4 mRNA in the striatum (McCullumsmith and Meador-Woodruff 2002). The antidepressant riluzole has been found to enhance glutamate transporter activity (Fumagalli et al. 2008), and the mood stabilizer valproate has been shown to increase EAAT1 but decrease EAAT2 levels in the hippocampus (Hassel et al. 2001; Ueda and Willmore 2000).

Human EAAT1-4 corresponds to the rodent GLAST, GLT-1, EAAC1 and EAAT4, respectively, (Arriza et al. 1994), with GLT-1 (human EAAT2) responsible for the majority of extracellular glutamate regulation (Arriza et al. 1994; Shigeri et al. 2004; Zarate et al. 2002). Complete GLT-1 knockout results in hippocampal damage and spontaneous epileptic seizures and is often lethal. To avoid these effects, antisense knockdown of GLAST, GLT-1 and EAAC1 have been used and revealed that GLAST and GLT-1 knockdown cause an elevation in extracellular glutamate, neurodegeneration and paralysis while EAAC1 knockdown does not elevate extracellular glutamate and produces only mild neurotoxicity (Rao et al. 2001a, 2001b; Rothstein et al. 1996). GLAST knockout mice are viable and have a phenotype resembling the positive and negative symptoms of schizophrenia, including increased novelty-induced hyperlocomotion that is reversed by the antipsychotic haloperidol, abnormal sociability, reduced acoustic startle response and impaired visual discrimination (Karlsson et al. 2008, 2009). It is not known, however, how a deficit in glutamate transporters affects depression-related behaviors or response to stress. Future studies could investigate the role of glutamate transporters in buffering pyramidal cells to the increased glutamate release during stress.

Because of their role in synaptic plasticity, these glutamatergic receptors have been extensively studied in animal models of cognition. Animals lacking these receptors often show deficits in cognitive-related tasks that overlap those seen in models of neuropsychiatric disorders. Their role in emotional-related behaviors and stress response, however, has not been systematically studied and still remains untested in many of these mutant models. There have been examples of baseline alterations in emotionality (as with the NR2A mice) and schizophrenia-related behaviors, suggesting that dysfunction of the glutamatergic system alone can result in pathology. A more nuanced approach, however, is seen in those studies analyzing changes in the normal stress response in animals with alterations in the glutamatergic system. As mentioned earlier, stress can act as a predisposing factor for many neuropsychiatric illnesses and it is important to see how stress can lead to a disruption in cognition and emotional regulation. Here, we briefly discussed the dysregulated stress response in NMDA, but not AMPA, receptor-related manipulations. In the future, it will be interesting to see studies of stress response applied to more models of glutamatergic function.

6 Conclusions

Stressful experiences are often reported in the life histories of neuropsychiatric patients and it is well established that stress can be a risk factor for the development of neuropsychiatric disorders. Typical to this clinical population are deficits in emotional regulation and cognitive function. Preclinical data showing parallel behavioral and structural changes in the PFC, hippocampus and amygdala following stress support a direct role for stress modifying these emotional and cognitive functions. Multiple molecular mechanisms could explain these stress evoked changes. Here, we have shown how the glutamate system can be linked to the structural and ultimately behavioral changes seen in neuropsychiatric disorders.

Glutamatergic transmission is integral to multiple levels of neural function. At one level, alterations in glutamate transmission can influence downstream signaling pathways involved in cell death or cell growth. Changes in glutamate transmission during stress could lead to differential activation of these signaling pathways, possibly explaining the morphological changes observed. One such signaling pathway of interest involves brain-derived neurotrophic factor (BDNF).

In its mature form, BDNF promotes neuronal growth and survival and its release is dependent on glutamatergic synaptic activity (Carvalho et al. 2008; Lu 2003). BDNF itself also influences glutamatergic transmission, altering the release of glutamate, glutamate receptor composition and synaptic plasticity (Carvalho et al. 2008; Kuczewski et al. 2009; Lu 2003). BDNF expression is reduced with stress, an effect reversed by antidepressants, and has itself been shown to be reduced in depression-like behaviors (Gourley et al. 2009; Nibuya et al. 1995; Shirayama et al. 2002; Siuciak et al. 1997). Thus far few studies have collectively examined the effects of stress, morphology and behavior in relation to BDNF. Preliminary work has shown that genetically induced BDNF overexpression can prevent stress-induced hippocampal dendritic retraction and reduced learned helplessness, making BDNF a particularly relevant molecule in the study of stress and cognition (Govindarajan et al. 2006). The contributions of BDNF and other downstream signaling pathways will be important targets elucidating the mechanisms of stress-induced morphological changes.

In addition, many studies have used genetic techniques to understand the contribution of glutamate receptors in emotional and cognitive behavioral tasks. These studies have shown that changes in activity and composition of glutamate receptors can lead to changes in emotional regulation and cognitive deficits. However, few studies have used these models to directly investigate the role of receptors in the context of stress. These receptors are integrally involved in synaptic plasticity mechanisms, thus examining the effects of stress on these receptors could help explain why stress is a precursor to the development of neuropsychiatric disorders.

Furthering our understanding of how stress affects glutamate receptors and alters downstream signaling pathways will ultimately contribute to the development of improved therapeutics for individuals suffering from neuropsychiatric disorders.

References

Anderson CM, Swanson RA (2000) Astrocyte glutamate transport: review of properties, regulation, and physiological functions. Glia 32:1–14

Arnsten AF (2009) Stress signalling pathways that impair prefrontal cortex structure and function. Nat Rev Neurosci 10:410–422

Arriza JL, Fairman WA, Wadiche JI, Murdoch GH, Kavanaugh MP, Amara SG (1994) Functional comparisons of three glutamate transporter subtypes cloned from human motor cortex. J Neurosci 14:5559–5569

Baran SE, Armstrong CE, Niren DC, Hanna JJ, Conrad CD (2009) Chronic stress and sex differences on the recall of fear conditioning and extinction. Neurobiol Learn Mem 91:323–332

Bauer D, Gupta D, Harotunian V, Meador-Woodruff JH, McCullumsmith RE (2008) Abnormal expression of glutamate transporter and transporter interacting molecules in prefrontal cortex in elderly patients with schizophrenia. Schizophr Res 104:108–120

Belforte JE, Zsiros V, Sklar ER, Jiang Z, Yu G, Li Y, Quinlan EM, Nakazawa K (2010) Postnatal NMDA receptor ablation in corticolimbic interneurons confers schizophrenia-like phenotypes. Nat Neurosci 13:76–83

Berton O, McClung CA, Dileone RJ, Krishnan V, Renthal W, Russo SJ, Graham D, Tsankova NM, Bolanos CA, Rios M, Monteggia LM, Self DW, Nestler EJ (2006) Essential role of BDNF in the mesolimbic dopamine pathway in social defeat stress. Science 311:864–868

Bird CM, Burgess N (2008) The hippocampus and memory: insights from spatial processing. Nat Rev Neurosci 9:182–194

Blechert J, Michael T, Vriends N, Margraf J, Wilhelm FH (2007) Fear conditioning in posttraumatic stress disorder: evidence for delayed extinction of autonomic, experiential, and behavioural responses. Behav Res Ther 45:2019–2033

Bondi CO, Rodriguez G, Gould GG, Frazer A, Morilak DA (2008) Chronic unpredictable stress induces a cognitive deficit and anxiety-like behavior in rats that is prevented by chronic antidepressant drug treatment. Neuropsychopharmacology 33:320–331

Boyce-Rustay JM, Holmes A (2006) Genetic inactivation of the NMDA receptor NR2A subunit has anxiolytic- and antidepressant-like effects in mice. Neuropsychopharmacology 31:2405–2414

Brigman JL, Wright T, Talani G, Prasad-Mulcare S, Jinde S, Seabold GK, Mathur P, Davis MI, Bock R, Gustin RM, Colbran RJ, Alvarez VA, Nakazawa K, Delpire E, Lovinger DM, Holmes A (2010) Loss of GluN2B-containing NMDA receptors in CA1 hippocampus and cortex impairs long-term depression, reduces dendritic spine density, and disrupts learning. J Neurosci 30:4590–4600

Brown SM, Henning S, Wellman CL (2005) Mild, short-term stress alters dendritic morphology in rat medial prefrontal cortex. Cereb Cortex 15:1714–1722

Campbell AM, Park CR, Zoladz PR, Munoz C, Fleshner M, Diamond DM (2008) Pre-training administration of tianeptine, but not propranolol, protects hippocampus-dependent memory from being impaired by predator stress. Eur Neuropsychopharmacol 18:87–98

Carvalho AL, Caldeira MV, Santos SD, Duarte CB (2008) Role of the brain-derived neurotrophic factor at glutamatergic synapses. Br J Pharmacol 153(1):S310–S324

Cerqueira JJ, Mailliet F, Almeida OF, Jay TM, Sousa N (2007) The prefrontal cortex as a key target of the maladaptive response to stress. J Neurosci 27:2781–2787

Cerqueira JJ, Pego JM, Taipa R, Bessa JM, Almeida OF, Sousa N (2005) Morphological correlates of corticosteroid-induced changes in prefrontal cortex-dependent behaviors. J Neurosci 25:7792–7800

Choudary PV, Molnar M, Evans SJ, Tomita H, Li JZ, Vawter MP, Myers RM, Bunney WE Jr, Akil H, Watson SJ, Jones EG (2005) Altered cortical glutamatergic and GABAergic signal transmission with glial involvement in depression. Proc Natl Acad Sci USA 102:15653–15658

Chudasama Y, Robbins TW (2006) Functions of frontostriatal systems in cognition: comparative neuropsychopharmacological studies in rats, monkeys and humans. Biol Psychol 73:19–38

Ciabarra AM, Sullivan JM, Gahn LG, Pecht G, Heinemann S, Sevarino KA (1995) Cloning and characterization of chi-1: a developmentally regulated member of a novel class of the ionotropic glutamate receptor family. J Neurosci 15:6498–6508

Conboy L, Tanrikut C, Zoladz PR, Campbell AM, Park CR, Gabriel C, Mocaer E, Sandi C, Diamond DM (2009) The antidepressant agomelatine blocks the adverse effects of stress on memory and enables spatial learning to rapidly increase neural cell adhesion molecule (NCAM) expression in the hippocampus of rats. Int J Neuropsychopharmacol 12:329–341

Corcoran KA, Quirk GJ (2007) Activity in prelimbic cortex is necessary for the expression of learned, but not innate, fears. J Neurosci 27:840–844

Cravens CJ, Vargas-Pinto N, Christian KM, Nakazawa K (2006) CA3 NMDA receptors are crucial for rapid and automatic representation of context memory. Eur J Neurosci 24:1771–1780

Czeh B, Muller-Keuker JI, Rygula R, Abumaria N, Hiemke C, Domenici E, Fuchs E (2007) Chronic social stress inhibits cell proliferation in the adult medial prefrontal cortex:

hemispheric asymmetry and reversal by fluoxetine treatment. Neuropsychopharmacology 32:1490–1503

Dalley JW, Cardinal RN, Robbins TW (2004) Prefrontal executive and cognitive functions in rodents: neural and neurochemical substrates. Neurosci Biobehav Rev 28:771–784

de Kloet ER, Joels M, Holsboer F (2005) Stress and the brain: from adaptation to disease. Nat Rev Neurosci 6:463–475

Diamond DM, Park CR, Heman KL, Rose GM (1999) Exposing rats to a predator impairs spatial working memory in the radial arm water maze. Hippocampus 9:542–552

Dias-Ferreira E, Sousa JC, Melo I, Morgado P, Mesquita AR, Cerqueira JJ, Costa RM, Sousa N (2009) Chronic stress causes frontostriatal reorganization and affects decision-making. Science 325:621–625

Dirnagl U, Iadecola C, Moskowitz MA (1999) Pathobiology of ischaemic stroke: an integrated view. Trends Neurosci 22:391–397

Eiland L, McEwen BS (2012) Early life stress followed by subsequent adult chronic stress potentiates anxiety and blunts hippocampal structural remodeling. Hippocampus 22(1):82–91

Engert F, Bonhoeffer T (1999) Dendritic spine changes associated with hippocampal long-term synaptic plasticity. Nature 399:66–70

Ferreri F, Lapp LK, Peretti CS (2011) Current research on cognitive aspects of anxiety disorders. Curr Opin Psychiatry 24(1):49–54

Feyissa AM, Chandran A, Stockmeier CA, Karolewicz B (2009) Reduced levels of NR2A and NR2B subunits of NMDA receptor and PSD-95 in the prefrontal cortex in major depression. Prog Neuropsychopharmacol Biol Psychiatry 33:70–75

Fitzgerald PJ, Barkus C, Feyder M, Wiedholz LM, Chen YC, Karlsson RM, Machado-Vieira R, Graybeal C, Sharp T, Zarate C, Harvey-White J, Du J, Sprengel R, Gass P, Bannerman D, Holmes A (2010) Does gene deletion of AMPA GluA1 phenocopy features of schizoaffective disorder? Neurobiol Dis 40(3):608–621

Fumagalli E, Funicello M, Rauen T, Gobbi M, Mennini T (2008) Riluzole enhances the activity of glutamate transporters GLAST, GLT1 and EAAC1. Eur J Pharmacol 578:171–176

Garrett JE, Kim I, Wilson RE, Wellman CL (2006) Effect of N-methyl-d-aspartate receptor blockade on plasticity of frontal cortex after cholinergic deafferentation in rat. Neuroscience 140:57–66

Goldwater DS, Pavlides C, Hunter RG, Bloss EB, Hof PR, McEwen BS, Morrison JH (2009) Structural and functional alterations to rat medial prefrontal cortex following chronic restraint stress and recovery. Neuroscience 164:798–808

Gourley SL, Kedves AT, Olausson P, Taylor JR (2009) A history of corticosterone exposure regulates fear extinction and cortical NR2B, GluR2/3, and BDNF. Neuropsychopharmacology 34(3):707–716

Gourley SL, Kiraly DD, Howell JL, Olausson P, Taylor JR (2008) Acute Hippocampal Brain-Derived Neurotrophic Factor Restores Motivational and Forced Swim Performance After Corticosterone. Biol Psychiatry 64(10):884–890

Govindarajan A, Rao BS, Nair D, Trinh M, Mawjee N, Tonegawa S, Chattarji S (2006) Transgenic brain-derived neurotrophic factor expression causes both anxiogenic and antidepressant effects. Proc Natl Acad Sci USA 103:13208–13213

Halene TB, Ehrlichman RS, Liang Y, Christian EP, Jonak GJ, Gur TL, Blendy JA, Dow HC, Brodkin ES, Schneider F, Gur RC, Siegel SJ (2009) Assessment of NMDA receptor NR1 subunit hypofunction in mice as a model for schizophrenia. Genes Brain Behav 8:661–675

Hammen C (2005) Stress and depression. Annu Rev Clin Psychol 1:293–319

Hardingham GE, Bading H (2003) The Yin and Yang of NMDA receptor signalling. Trends Neurosci 26(2):81–89

Hassel B, Iversen EG, Gjerstad L, Tauboll E (2001) Up-regulation of hippocampal glutamate transport during chronic treatment with sodium valproate. J Neurochem 77:1285–1292

Holmes A, le Guisquet AM, Vogel E, Millstein RA, Leman S, Belzung C (2005) Early life genetic, epigenetic and environmental factors shaping emotionality in rodents. Neurosci Biobehav Rev 29:1335–1346

Holmes A, Wellman CL (2009) Stress-induced prefrontal reorganization and executive dysfunction in rodents. Neurosci Biobehav Rev 33:773–783
Huerta PT, Sun LD, Wilson MA, Tonegawa S (2000) Formation of temporal memory requires NMDA receptors within CA1 pyramidal neurons. Neuron 25:473–480
Huot RL, Thrivikraman KV, Meaney MJ, Plotsky PM (2001) Development of adult ethanol preference and anxiety as a consequence of neonatal maternal separation in Long Evans rats and reversal with antidepressant treatment. Psychopharmacology (Berl) 158:366–373
Izquierdo A, Wellman CL, Holmes A (2006) Brief uncontrollable stress causes dendritic retraction in infralimbic cortex and resistance to fear extinction in mice. J Neurosci 26:5733–5738
Ji J, Maren S (2007) Hippocampal involvement in contextual modulation of fear extinction. Hippocampus 17:749–758
Joels M, Karst H, Alfarez D, Heine VM, Qin Y, van Riel E, Verkuyl M, Lucassen PJ, Krugers HJ (2004) Effects of chronic stress on structure and cell function in rat hippocampus and hypothalamus. Stress 7:221–231
Joels M, Karst H, Krugers HJ, Lucassen PJ (2007) Chronic stress: implications for neuronal morphology, function and neurogenesis. Front Neuroendocrinol 28:72–96
Judo C, Matsumoto M, Yamazaki D, Hiraide S, Yanagawa Y, Kimura S, Shimamura K, Togashi H (2010) Early stress exposure impairs synaptic potentiation in the rat medial prefrontal cortex underlying contextual fear extinction. Neuroscience 169:1705–1714
Karlsson RM, Heilig M, Holmes A (2008) Loss of Glial Glutamate and Aspartate Transporter (Excitatory Amino Acid Transporter 1) Causes Locomotor Hyperactivity and Exaggerated Responses to Psychotomimetics: Rescue by Haloperidol and Metabotropic Glutamate 2/3 Agonist. Biol Psychiatry 64(9):810–814
Karlsson RM, Tanaka K, Saksida LM, Bussey TJ, Heilig M, Holmes A (2009) Assessment of glutamate transporter GLAST (EAAT1)-deficient mice for phenotypes relevant to the negative and executive/cognitive symptoms of schizophrenia. Neuropsychopharmacology 34(6):1578–1589
Kim JJ, Foy MR, Thompson RF (1996) Behavioral stress modifies hippocampal plasticity through N-methyl-D-aspartate receptor activation. Proc Natl Acad Sci USA 93:4750–4753
Kim JJ, Lee HJ, Han JS, Packard MG (2001) Amygdala is critical for stress-induced modulation of hippocampal long-term potentiation and learning. J Neurosci 21:5222–5228
Kiselycznyk C, Svenningsson P, Delpire E, Holmes A (2011) Genetic, pharmacological, and lesion analyses reveal a selective role for corticohippocampal GLUN2B in a novel repeated swim stress paradigm. Neuroscience 193:259–268
Kole MH, Costoli T, Koolhaas JM, Fuchs E (2004) Bidirectional shift in the cornu ammonis 3 pyramidal dendritic organization following brief stress. Neuroscience 125:337–347
Kuczewski N, Porcher C, Lessmann V, Medina I, Gaiarsa JL (2009) Activity-dependent dendritic release of BDNF and biological consequences. Mol Neurobiol 39:37–49
Lapiz-Bluhm MD, Soto-Pina AE, Hensler JG, Morilak DA (2009) Chronic intermittent cold stress and serotonin depletion induce deficits of reversal learning in an attentional set-shifting test in rats. Psychopharmacology (Berl) 202:329–341
Law AJ, Deakin JF (2001) Asymmetrical reductions of hippocampal NMDAR1 glutamate receptor mRNA in the psychoses. Neuroreport 12:2971–2974
Li N, Lee B, Liu RJ, Banasr M, Dwyer JM, Iwata M, Li XY, Aghajanian G, Duman RS (2010) mTOR-dependent synapse formation underlies the rapid antidepressant effects of NMDA antagonists. Science 329:959–964
Lipton SA, Rosenberg PA (1994) Excitatory amino acids as a final common pathway for neurologic disorders. N Engl J Med 330:613–622
Liston C, Miller MM, Goldwater DS, Radley JJ, Rocher AB, Hof PR, Morrison JH, McEwen BS (2006) Stress-induced alterations in prefrontal cortical dendritic morphology predict selective impairments in perceptual attentional set-shifting. J Neurosci 26:7870–7874
Lu B (2003) BDNF and activity-dependent synaptic modulation. Learn Mem 10:86–98

Lupien SJ, McEwen BS, Gunnar MR, Heim C (2009) Effects of stress throughout the lifespan on the brain, behaviour and cognition. Nat Rev Neurosci 10:434–445

Magarinos AM, McEwen BS (1995) Stress-induced atrophy of apical dendrites of hippocampal CA3c neurons: comparison of stressors. Neuroscience 69:83–88

Maren S, Holt W (2000) The hippocampus and contextual memory retrieval in Pavlovian conditioning. Behav Brain Res 110:97–108

Maren S, Quirk GJ (2004) Neuronal signalling of fear memory. Nat Rev Neurosci 5:844–852

Martin KP, Wellman CL (2011). NMDA receptor blockade alters stress-induced dendritic remodeling in medial prefrontal cortex. Cereb Cortex 21(10):2366–2373

Matsuzaki M, Honkura N, Ellis-Davies GC, Kasai H (2004) Structural basis of long-term potentiation in single dendritic spines. Nature 429:761–766

McCullumsmith RE, Meador-Woodruff JH (2002) Striatal excitatory amino acid transporter transcript expression in schizophrenia, bipolar disorder, and major depressive disorder. Neuropsychopharmacology 26:368–375

McEwen BS (2000a) Allostasis and allostatic load: implications for neuropsychopharmacology. Neuropsychopharmacology 22:108–124

McEwen BS (2000b) The neurobiology of stress: from serendipity to clinical relevance. Brain Res 886:172–189

McEwen BS (2001) Plasticity of the hippocampus: adaptation to chronic stress and allostatic load. Ann N Y Acad Sci 933:265–277

McLean A, Dowson J, Toone B, Young S, Bazanis E, Robbins TW, Sahakian BJ (2004) Characteristic neurocognitive profile associated with adult attention-deficit/hyperactivity disorder. Psychol Med 34:681–692

Michael T, Blechert J, Vriends N, Margraf J, Wilhelm FH (2007) Fear conditioning in panic disorder: Enhanced resistance to extinction. J Abnorm Psychol 116:612–617

Miracle AD, Brace MF, Huyck KD, Singler SA, Wellman CL (2006) Chronic stress impairs recall of extinction of conditioned fear. Neurobiol Learn Mem 85:213–218

Mizoguchi K, Yuzurihara M, Ishige A, Sasaki H, Chui DH, Tabira T (2000) Chronic stress induces impairment of spatial working memory because of prefrontal dopaminergic dysfunction. J Neurosci 20:1568–1574

Moghaddam B (1993) Stress preferentially increases extraneuronal levels of excitatory amino acids in the prefrontal cortex: comparison to hippocampus and basal ganglia. J Neurochem 60:1650–1657

Mozhui K, Karlsson RM, Kash TL, Ihne J, Norcross M, Patel S, Farrell MR, Hill EE, Graybeal C, Martin KP, Camp M, Fitzgerald PJ, Ciobanu DC, Sprengel R, Mishina M, Wellman CL, Winder DG, Williams RW, Holmes A (2010) Strain differences in stress responsivity are associated with divergent amygdala gene expression and glutamate-mediated neuronal excitability. J Neurosci 30:5357–5367

Murphy FC, Michael A, Robbins TW, Sahakian BJ (2003) Neuropsychological impairment in patients with major depressive disorder: the effects of feedback on task performance. Psychol Med 33:455–467

Murray GK, Cheng F, Clark L, Barnett JH, Blackwell AD, Fletcher PC, Robbins TW, Bullmore ET, Jones PB (2008) Reinforcement and reversal learning in first-episode psychosis. Schizophr Bull 34:848–855

Myers KM, Davis M (2007) Mechanisms of fear extinction. Mol Psychiatry 12:120–150

Nibuya M, Morinobu S, Duman RS (1995) Regulation of BDNF and trkB mRNA in rat brain by chronic electroconvulsive seizure and antidepressant drug treatments. J Neurosci 15:7539–7547

Olney JW, Collins RC, Sloviter RS (1986) Excitotoxic mechanisms of epileptic brain damage. Adv Neurol 44:857–877

Pare D, Quirk GJ, Ledoux JE (2004) New vistas on amygdala networks in conditioned fear. J Neurophysiol 92:1–9

Park CR, Zoladz PR, Conrad CD, Fleshner M, Diamond DM (2008) Acute predator stress impairs the consolidation and retrieval of hippocampus-dependent memory in male and female rats. Learn Mem 15:271–280

Pego JM, Morgado P, Pinto LG, Cerqueira JJ, Almeida OF, Sousa N (2008) Dissociation of the morphological correlates of stress-induced anxiety and fear. Eur J Neurosci 27:1503–1516

Pittenger C, Duman RS (2008) Stress, depression, and neuroplasticity: a convergence of mechanisms. Neuropsychopharmacology 33:88–109

Quirk GJ, Mueller D (2008) Neural mechanisms of extinction learning and retrieval. Neuropsychopharmacology 33:56–72

Radley JJ, Rocher AB, Miller M, Janssen WG, Liston C, Hof PR, McEwen BS, Morrison JH (2006) Repeated stress induces dendritic spine loss in the rat medial prefrontal cortex. Cereb Cortex 16:313–320

Radley JJ, Rocher AB, Rodriguez A, Ehlenberger DB, Dammann M, McEwen BS, Morrison JH, Wearne SL, Hof PR (2008) Repeated stress alters dendritic spine morphology in the rat medial prefrontal cortex. J Comp Neurol 507:1141–1150

Radley JJ, Sisti HM, Hao J, Rocher AB, McCall T, Hof PR, McEwen BS, Morrison JH (2004) Chronic behavioral stress induces apical dendritic reorganization in pyramidal neurons of the medial prefrontal cortex. Neuroscience 125:1–6

Rao VL, Dogan A, Bowen KK, Todd KG, Dempsey RJ (2001a) Antisense knockdown of the glial glutamate transporter GLT-1 exacerbates hippocampal neuronal damage following traumatic injury to rat brain. Eur J Neurosci 13:119–128

Rao VL, Dogan A, Todd KG, Bowen KK, Kim BT, Rothstein JD, Dempsey RJ (2001b) Antisense knockdown of the glial glutamate transporter GLT-1, but not the neuronal glutamate transporter EAAC1, exacerbates transient focal cerebral ischemia-induced neuronal damage in rat brain. J Neurosci 21:1876–1883

Rau V, DeCola JP, Fanselow MS (2005) Stress-induced enhancement of fear learning: an animal model of posttraumatic stress disorder. Neurosci Biobehav Rev 29:1207–1223

Robbins TW (2007) Shifting and stopping: fronto-striatal substrates, neurochemical modulation and clinical implications. Philos Trans R Soc Lond B Biol Sci 362:917–932

Rodrigues SM, LeDoux JE, Sapolsky RM (2009) The influence of stress hormones on fear circuitry. Annu Rev Neurosci 32:289–313

Rogers MA, Kasai K, Koji M, Fukuda R, Iwanami A, Nakagome K, Fukuda M, Kato N (2004) Executive and prefrontal dysfunction in unipolar depression: a review of neuropsychological and imaging evidence. Neurosci Res 50:1–11

Roozendaal B, McEwen BS, Chattarji S (2009) Stress, memory and the amygdala. Nat Rev Neurosci 10(6):423–433

Rosenmund C, Stern-Bach Y, Stevens CF (1998) The tetrameric structure of a glutamate receptor channel. Science 280.1596–1599

Rothstein JD, Dykes-Hoberg M, Pardo CA, Bristol LA, Jin L, Kuncl RW, Kanai Y, Hediger MA, Wang Y, Schielke JP, Welty DF (1996) Knockout of glutamate transporters reveals a major role for astroglial transport in excitotoxicity and clearance of glutamate. Neuron 16:675–686

Sandi C, Merino JJ, Cordero MI, Touyarot K, Venero C (2001) Effects of chronic stress on contextual fear conditioning and the hippocampal expression of the neural cell adhesion molecule, its polysialylation, and L1. Neuroscience 102:329–339

Sapolsky RM (2000a) Glucocorticoids and hippocampal atrophy in neuropsychiatric disorders. Arch Gen Psychiatry 57:925–935

Sapolsky RM (2000b) The possibility of neurotoxicity in the hippocampus in major depression: a primer on neuron death. Biol Psychiatry 48:755–765

Schneiderman N, Ironson G, Siegel SD (2005) Stress and health: psychological, behavioral, and biological determinants. Annu Rev Clin Psychol 1:607–628

Schwabe L, Schachinger H, de Kloet ER, Oitzl MS (2010) Stress impairs spatial but not early stimulus–response learning. Behav Brain Res 213:50–55

Shansky RM, Morrison JH (2009) Stress-induced dendritic remodeling in the medial prefrontal cortex: effects of circuit, hormones and rest. Brain Res 1293:108–113

Shigeri Y, Seal RP, Shimamoto K (2004) Molecular pharmacology of glutamate transporters, EAATs and VGLUTs. Brain Res Brain Res Rev 45:250–265

Shirayama Y, Chen AC, Nakagawa S, Russell DS, Duman RS (2002) Brain-derived neurotrophic factor produces antidepressant effects in behavioral models of depression. J Neurosci 22:3251–3261

Sinha R (2008) Chronic stress, drug use, and vulnerability to addiction. Ann N Y Acad Sci 1141:105–130

Siuciak JA, Lewis DR, Wiegand SJ, Lindsay RM (1997) Antidepressant-like effect of brain-derived neurotrophic factor (BDNF). Pharmacol Biochem Behav 56:131–137

Smith RE, Haroutunian V, Davis KL, Meador-Woodruff JH (2001) Vesicular glutamate transporter transcript expression in the thalamus in schizophrenia. Neuroreport 12:2885–2887

Sokolov BP (1998) Expression of NMDAR1, GluR1, GluR7, and KA1 glutamate receptor mRNAs is decreased in frontal cortex of "neuroleptic-free" schizophrenics: evidence on reversible up-regulation by typical neuroleptics. J Neurochem 71:2454–2464

Sousa N, Lukoyanov NV, Madeira MD, Almeida OF, Paula-Barbosa MM (2000) Reorganization of the morphology of hippocampal neurites and synapses after stress-induced damage correlates with behavioral improvement. Neuroscience 97:253–266

Sterner EY, Kalynchuk LE (2010) Behavioral and neurobiological consequences of prolonged glucocorticoid exposure in rats: relevance to depression. Prog Neuropsychopharmacol Biol Psychiatry 34:777–790

Svenningsson P, Bateup H, Qi H, Takamiya K, Huganir RL, Spedding M, Roth BL, McEwen BS, Greengard P (2007) Involvement of AMPA receptor phosphorylation in antidepressant actions with special reference to tianeptine. Eur J Neurosci 26:3509–3517

Thierry AM, Gioanni Y, Degenetais E, Glowinski J (2000) Hippocampo-prefrontal cortex pathway: anatomical and electrophysiological characteristics. Hippocampus 10:411–419

Tonegawa S, Tsien JZ, McHugh TJ, Huerta P, Blum KI, Wilson MA (1996) Hippocampal CA1-region-restricted knockout of NMDAR1 gene disrupts synaptic plasticity, place fields, and spatial learning. Cold Spring Harb Symp Quant Biol 61:225–238

Tsien JZ, Chen DF, Gerber D, Tom C, Mercer EH, Anderson DJ, Mayford M, Kandel ER, Tonegawa S (1996a) Subregion- and cell type-restricted gene knockout in mouse brain. Cell 87:1317–1326

Tsien JZ, Huerta PT, Tonegawa S (1996b) The essential role of hippocampal CA1 NMDA receptor-dependent synaptic plasticity in spatial memory. Cell 87:1327–1338

Ueda Y, Willmore LJ (2000) Molecular regulation of glutamate and GABA transporter proteins by valproic acid in rat hippocampus during epileptogenesis. Exp Brain Res 133:334–339

van Harmelen AL, van Tol MJ, van der Wee NJ, Veltman DJ, Aleman A, Spinhoven P, van Buchem MA, Zitman FG, Penninx BW, Elzinga BM (2010) Reduced Medial Prefrontal Cortex Volume in Adults Reporting Childhood Emotional Maltreatment. Biol Psychiatry 68(9):832–838

Venero C, Borrell J (1999) Rapid glucocorticoid effects on excitatory amino acid levels in the hippocampus: a microdialysis study in freely moving rats. Eur J Neurosci 11:2465–2473

Vyas A, Bernal S, Chattarji S (2003) Effects of chronic stress on dendritic arborization in the central and extended amygdala. Brain Res 965:290–294

Vyas A, Mitra R, Shankaranarayana Rao BS, Chattarji S (2002) Chronic stress induces contrasting patterns of dendritic remodeling in hippocampal and amygdaloid neurons. J Neurosci 22:6810–6818

Vyas A, Pillai AG, Chattarji S (2004) Recovery after chronic stress fails to reverse amygdaloid neuronal hypertrophy and enhanced anxiety-like behavior. Neuroscience 128:667–673

Walsh T, McClellan JM, McCarthy SE, Addington AM, Pierce SB, Cooper GM, Nord AS, Kusenda M, Malhotra D, Bhandari A, Stray SM, Rippey CF, Roccanova P, Makarov V, Lakshmi B, Findling RL, Sikich L, Stromberg T, Merriman B, Gogtay N, Butler P, Eckstrand K, Noory L, Gochman P, Long R, Chen Z, Davis S, Baker C, Eichler EE, Meltzer PS, Nelson SF, Singleton AB, Lee MK, Rapoport JL, King MC, Sebat J (2008) Rare structural variants disrupt multiple genes in neurodevelopmental pathways in schizophrenia. Science 320:539–543

Waltz JA, Gold JM (2007) Probabilistic reversal learning impairments in schizophrenia: further evidence of orbitofrontal dysfunction. Schizophr Res 93:296–303

Watanabe Y, Gould E, Cameron HA, Daniels DC, McEwen BS (1992a) Phenytoin prevents stress- and corticosterone-induced atrophy of CA3 pyramidal neurons. Hippocampus 2:431–435

Watanabe Y, Gould E, McEwen BS (1992b) Stress induces atrophy of apical dendrites of hippocampal CA3 pyramidal neurons. Brain Res 588:341–345

Wiedholz LM, Owens WA, Horton RE, Feyder M, Karlsson RM, Hefner K, Sprengel R, Celikel T, Daws LC, Holmes A (2008) Mice lacking the AMPA GluR1 receptor exhibit striatal hyperdopaminergia and 'schizophrenia-related' behaviors. Mol Psychiatry 13:631–640

Wood GE, Norris EH, Waters E, Stoldt JT, McEwen BS (2008) Chronic immobilization stress alters aspects of emotionality and associative learning in the rat. Behav Neurosci 122:282–292

Yamamoto S, Morinobu S, Takei S, Fuchikami M, Matsuki A, Yamawaki S, Liberzon I (2009) Single prolonged stress: toward an animal model of posttraumatic stress disorder. Depress Anxiety 26:1110–1117

Yuen EY, Liu W, Karatsoreos IN, Feng J, McEwen BS, Yan Z (2009) Acute stress enhances glutamatergic transmission in prefrontal cortex and facilitates working memory. Proc Natl Acad Sci USA 106:14075–14079

Zarate CA, Quiroz J, Payne J, Manji HK (2002) Modulators of the glutamatergic system: implications for the development of improved therapeutics in mood disorders. Psychopharmacol Bull 36:35–83

Susceptibility Genes for Schizophrenia: Mutant Models, Endophenotypes and Psychobiology

Colm M. P. O'Tuathaigh, Lieve Desbonnet, Paula M. Moran and John L. Waddington

Abstract Schizophrenia is characterised by a multifactorial aetiology that involves genetic liability interacting with epigenetic and environmental factors to increase risk for developing the disorder. A consensus view is that the genetic component involves several common risk alleles of small effect and/or rare but penetrant copy number variations. Furthermore, there is increasing evidence for broader, overlapping genetic-phenotypic relationships in psychosis; for example, the same susceptibility genes also confer risk for bipolar disorder. Phenotypic characterisation of genetic models of candidate risk genes and/or putative pathophysiological processes implicated in schizophrenia, as well as examination of epidemiologically relevant gene × environment interactions in these models, can illuminate molecular and pathobiological mechanisms involved in schizophrenia. The present chapter outlines both the evidence from phenotypic studies in mutant mouse models related to schizophrenia and recently described mutant models addressing such gene × environment interactions. Emphasis is placed on evaluating the extent to which mutant phenotypes recapitulate the totality of the disease

C. M. P. O'Tuathaigh · L. Desbonnet · J. L. Waddington
Molecular and Cellular Therapeutics,
Royal College of Surgeons in Ireland,
Dublin 2, Ireland

C. M. P. O'Tuathaigh (✉)
School of Medicine, University College Cork,
Western Gateway Building, Cork, Ireland
e-mail: c.otuathaigh@ucc.ie

P. M. Moran
School of Psychology, University of Nottingham,
Nottingham, NG7 2RD, UK

Curr Topics Behav Neurosci (2012) 12: 209–250
DOI: 10.1007/7854_2011_194

Published Online: 26 February 2012

phenotype or model selective endophenotypes. We also discuss new developments and trends in relation to the functional genomics of psychosis which might help to inform on the construct validity of mutant models of schizophrenia and highlight methodological challenges in phenotypic evaluation that relate to such models.

Keywords Schizophrenia · Psychotic illness · Susceptibility gene · Mutant model · Phenotype · Gene × environment interaction

Contents

1 Introduction 211
1.1 'Missing Heritability'? Evaluating Genetic Evidence for Schizophrenia Risk Genes 211
1.2 The Influence of Technological Progress on Mutant Modelling of Psychosis 212
1.3 Assessing Construct Validity of Mutant Models of Schizophrenia 213
1.4 Gene × Environment Interactions in Schizophrenia: Implications for Modelling 214
2 Genetic Models Relating to Dopaminergic Dysfunction 215
2.1 Dopamine Receptors 215
2.2 Dopamine Transporter 216
2.3 Catechol-*O*-Methyltransferase 217
2.4 AKT1/GSK-3 218
3 Genetic Models Relating to Glutamatergic Dysfunction 219
3.1 NMDAR-Mediated Neurotransmission 219
3.2 Glycine 220
3.3 D-serine 221
3.4 NMDAR Signalling Complex 222
3.5 Metabotropic Glutamate Receptors 222
4 Genetic Models Relating to Candidate Risk Genes and Copy Number Variations 223
4.1 Disrupted-in-schizophrenia-1 223
4.2 Dysbindin 225
4.3 Neuregulin 226
4.4 Copy Number Variations 228
5 Evaluation of G × E Hypotheses in Mutant Models of Schizophrenia 229
5.1 DISC1 × Immune Challenge 229
5.2 NRG1/COMT × Exposure to Drugs of Abuse 230
5.3 Genes Related to the Pathophysiology of Schizophrenia × Exposure to Prenatal Stress 231
5.4 Overview of G × E studies 232
6 Discussion 233
6.1 Divining gene-Phenotype Associations in Schizophrenia Risk Gene Models 233
6.2 Integration of Novel Gene Manipulation Strategies 234
6.3 Examination of Gene × Gene Interactions in Schizophrenia 238
7 Conclusions 239
References 239

1 Introduction

1.1 'Missing Heritability'? Evaluating Genetic Evidence for Schizophrenia Risk Genes

Schizophrenia is a highly heritable disorder [having a heritability estimate of approximately 0.8] that is characterised by a multifaceted psychopathology of psychotic symptoms, negative symptoms and cognitive deficits (Tandon et al. 2009; Waddington et al. 2011). The genetic basis of schizophrenia has proven complex and difficult to capture, with the expected breakthrough from genome-wide association studies (GWAS) failing to support the more prominent common risk alleles identified to date (Gill et al. 2009; Owen et al. 2010). Lack of consistent supportive data for specific genes or loci associated with schizophrenia has lead investigators to speculate regarding the underlying causes of 'missing heritability' (Manolio et al. 2007; International Schizophrenia Consortium 2009) in schizophrenia.

Two main hypotheses of schizophrenia genetics have suggested that such 'missing heritability' reflects either or both of multiple common risk alleles, each of small effect, or the impact of rare but highly penetrant alleles (Allen et al. 2008; Gill et al. 2009; Bassett et al. 2010; Owen et al. 2010). The latter hypothesis is supported by recent microarray findings which have highlighted the involvement of copy number variations (CNVs) in susceptibility to schizophrenia (Xu et al. 2008, 2009; International Schizophrenia Consortium 2008; Walsh et al. 2008). These are not mutually incompatible hypotheses and it is likely that both contribute to risk. It has been noted that genetic studies of schizophrenia might be handicapped by several factors, including genetic and phenotypic heterogeneity, epistatic gene interactions and the role that environmental events play in the development and expression of psychiatric illness (Burmeister 1999; Burmeister et al. 2008). An additional factor is uncertainty as to the genetic-phenotypic boundaries of schizophrenia, with both common risk alleles and rare CNVs associated also with risk for bipolar disorder (Gill et al. 2009; International Schizophrenia Consortium 2009; Grozeva et al. 2010; Steinberg et al. 2011), in accordance with a dimensional, as opposed to a categorical, model of psychotic illness (Waddington et al. 2011). Thus, henceforth, our use of the term 'schizophrenia' should be interpreted as shorthand for a breadth of psychotic illness, the boundaries of which remain to be defined.

Evidence to date indicates that while a number of genetic variants have been associated with multiple diagnostic domains, it is clear that some contribute to the expression of a limited set of endophenotypes (van Os et al. 2010). This places considerable emphasis on dissection of the schizophrenia phenotype into distinct and accessible endophenotypes that may relate more closely to underlying pathobiology. Endophenotypes are defined as measurable, intermediate disease features that bridge the gap between the overt manifestations of schizophrenia and underlying risk genes (Gottesman and Gould 2003; Braff et al. 2007). We have previously noted the utility of endophenotypes in relation to mutant phenotypes, being relatively closer to the genetic basis of the disorder at issue and, therefore, less complex to model; it is assumed that while a single gene can affect multiple behavioural endpoints, an intermediate biological or behavioural endophenotype is less susceptible to confounding influences and is therefore more amenable to investigation (O'Tuathaigh et al. 2007a; Desbonnet et al. 2009; Kirby et al. 2010). An endophenotype-based approach to schizophrenia risk genes may explain how a gene with modest association support may still have a function of potential relevance to the disorder, insofar as it may be associated with one or more putative endophenotype in humans, or their equivalents in mice. Among the most prominent candidate genes implicated in risk for schizophrenia by association or CNV studies of risk for schizophrenia, those for which the evidence is most strong and replicable are Disrupted-in-Schizophrenia-1 (DISC1), dystrobrevin-binding protein 1 (DTNBP1; dysbindin), and neuregulin 1 (NRG1) (Waddington et al. 2007; Allen et al. 2008; Bertram 2008; Gill et al. 2009; Owen et al. 2010).

1.2 The Influence of Technological Progress on Mutant Modelling of Psychosis

Constitutive mutant models involving the construction of mice with gene disruption, either by deletion [i.e. knockout (KO)] or insertion/over-expression [i.e. transgenic/knockin] have been used routinely to study the effect of developmental, whole-body loss-of-target gene function. These techniques have not been without drawbacks, with extrapolation of gene-phenotype relationships potentially confounded by: putative compensatory and redundancy mechanisms; the possibility of embryonic lethality where the gene is functionally pleiotropic; potentially, whole-body ablation or overexpression of a given gene may result in phenotypes that involve both diverse brain regions and extracerebral mechanisms and thus complicate interepretation of phenotypic data. However, each technique presents its own profile of advantage and disadvantage.

Progress in mouse genetic technology over the last decade has facilitated the development of conditional genetic models that are differentiated on the basis of cell-type specificity, regional brain selectivity and temporal activation of the mutation. Unsurprisingly, the evidence indicates that each of these factors can be critical determinants of whether a specific gene disruption can give rise to a given phenotypic outcome or otherwise; indeed, contradictory data have also been reported for the same gene mutation depending on cell-, regional- and possibly temporal specificity (e.g. Singer et al. 2011).

1.3 Assessing Construct Validity of Mutant Models of Schizophrenia

Criteria employed to evaluate a mutant model of schizophrenia involve several forms of validity: face validity [whether the model produces a behavioural phenotype which resemble human disease features]; predictive validity [the degree to which deficits or behavioural signs observed in the model are responsive to agents which either ameliorate or worsen analogous symptoms in patients]; construct validity [whether the experimental/genetic manipulation can be directly related to the known aetiology or pathophysiological features of the illness]. Assessing predictive validity is problematic, due to the lack of efficacy of existing antispsychotic drugs against negative symptoms and cognitive deficits in schizophrenia, both of which have been most strongly associated with prognosis and functional outcome.

The symptoms of schizophrenia, which emerge during adolescence, include positive symptoms [i.e. hallucinations, delusions and thought disorder], negative symptoms [e.g. avolition, anhedonia, blunted affect, poverty of speech and social withdrawal], and cognitive dysfunction [e.g. impairment in working memory, executive function and attention]. Phenotypic modelling of positive symptoms at the level of behaviour is restricted to 'proxy' indices of positive symptoms, such as hyperactivity in response to a novel stimulus and to psychotomimetics such as amphetamine or phencyclidine (PCP); prepulse inhibition (PPI) and latent inhibition (LI), involving sensory gating, selective attention to relevant over irrelevant stimuli and attribution of salience, are at the interface of psychotic and cognitive processes (Amann et al. 2010; Desbonnet et al. 2009; Arguello and Gogos 2010; Kirby et al. 2010; O'Tuathaigh and Waddington 2010; van den Buuse 2010). Modelling of negative symptoms has focused primarily on a restricted range of social, emotional, and motivational behaviours that apply to, and are accessible in, both humans and animals, i.e. asociality and, to a lesser extent, anhedonia (Arguello and Gogos 2006; O'Tuathaigh et al. 2010a).

Although the issue of which cognitive domains are disrupted-in-schizophrenia has proven controversial (Harvey et al. 2010), there is some agreement that impairments in working memory, executive function and attention are core features. Disruption to these domains of cognition [including PPI and LI, which are at the interface of cognitive dysfunction and positive symptoms] is amenable to investigation in animals. Particular emphasis has been placed on spatial working memory and executive function, processes which can be accessed using the maze-based and several operant reward-based tasks (Arguello and Gogos 2006, 2010; Amann et al. 2010).

Evaluating the construct validity of a genetic model of psychosis places the focus on (a) how accurately the genetic model mimics the human risk polymorphism, (b) the status of the clinical data linking the risk variant with the entirety of the disorder and/or limited to disease-relevant endophenotypes, and (c) the availability of a valid and reliable phenotypic armamentarium capable of exploring hypothesised gene-phenotype relationships (Arguello and Gogos 2006; Arguello and Gogos 2010; Desbonnet et al. 2009; Kirby et al. 2010; Low and Hardy 2007).

1.4 Gene × Environment Interactions in Schizophrenia: Implications for Modelling

With ongoing progress in the field of psychiatric genetics and the modifying influence of environmental factors, current conceptualisations of schizophrenia have addressed the complex interaction of genes (G) and environment (E) (van Os et al. 2010). Focusing on studies examining the interaction of epidemiologically relevant risk factors and candidate gene mutant models, it has been noted that such an approach has the potential to produce G × E models with increased construct validity (Desbonnet et al. 2009; Kirby et al. 2010). Consistent with the well-considered 'stress-vulnerability' aetiological model, which proposes an interaction between inherited genetic vulnerability and subsequent exposure to adverse environmental risk factors, there is a growing body of evidence linking early pre-l and postnatal life trauma (e.g. maternal infection, childhood abuse), societal factors (e.g. urbanicity, migration), and drug use (e.g. cannabis exposure) with risk for schizophrenia (van Os et al. 2010). While clinical studies linking specific genetic factors with the psychosis-inducing or precipitating effects of environmental factors have been slow to emerge, there has been some supportive evidence sensitive enough to detect the interaction of environmental adversity with a putative risk gene within a young population followed prospectively (Caspi et al. 2005).

Recent years have seen a movement towards the generation of animal models of psychosis based on the interaction of genetic mutations and well-characterised environmental factors (Ayhan et al. 2009; Gray and Hannan 2007). As outlined elsewhere (Burrows et al. 2011), the utility of such G × E models depends upon the degree to which the environmental models at a preclinical level possess construct validity with respect to environmental factors which have been epidemiologically linked with increased risk for schizophrenia; thus, although issues of validity have most frequently been explored in relation to the target gene mutation employed in a given model, whether epidemiologically relevant environmental factors have been successfully translated into current preclinical G × E models constituting a particular challenge (Burrows et al. 2011).

The present chapter will discuss the evidence on (a) mutant models relating to the putative pathophysiology of schizophrenia, primarily dopaminergic and glutamatergic dysfunction, (b) mutant models involving candidate risk genes for schizophrenia, with an emphasis on evaluating the extent to which mutant phenotypes recapitulate the totality of the disease phenotype or model selective endophenotypes, and (c) mutant models addressing G × E interactions relevant to schizophrenia.

2 Genetic Models Relating to Dopaminergic Dysfunction

2.1 Dopamine Receptors

Constitutive KO of dopamine (DA) receptor subtypes in mice has been an important experimental tool in understanding their roles in behaviours considered as endophenotypes for schizophrenia (O'Sullivan et al. 2010). Amphetamine has been shown to disrupt sensory-motor gating, considered to be a behavioural proxy for the impaired ability of schizophrenia patients to gate information flux appropriately. The indirect DA agonist d-amphetamine disrupts sensory gating in D1, D3, and D4 KO mice, but not D2 KO mice, suggesting that the D2 receptor is required for its disruption by amphetamine (Ralph et al. 1999; Ralph-Williams et al. 2002). The KO approach has recently suggested dissociable roles for D1 and D2 receptors in habituation and sensitisation to acoustic startle (Halberstadt and Geyer 2009), which may be of relevance to subgroups of patients that display habituation abnormalities. Antipsychotic drugs enhance low levels of LI. This behavioural index of the ability to ignore irrelevant information is abnormal in patients with schizophrenia and is enhanced in D2 KO mice (Bay-Richter et al. 2009), indicating the importance of the D2 receptor in the processes by which antipsychotic drugs exert their behavioural effects. We have recently shown that enhancement of LI by the antipsychotic drugs haloperidol and clozapine is prevented in D2 KO mice, confirming directly that they act via D2 receptors to restore the ability to ignore irrelevant information. However, we have also shown that amphetamine impairment of LI and its restoration by haloperidol and clozapine is intact in D2 KO mice, suggesting that the role of the D2 receptor differs depending on how disruption to LI is produced. New therapeutic approaches are seeking to circumvent direct effects at the D2 receptor, given their association with undesirable motor and other side effects side effects, hence amphetamine disruption of LI in D2 KO mice may prove to be a useful new application of the KO approach to produce animal models for the identification and evaluation of new treatments (Bay-Richter et al., unpublished data).

The D2 receptor has two isoforms, long (D2L) and short (D2S), derived from alternative splicing. Studies in isoform-specific KOs have indicated the D2L isoform to be crucial for antipsychotic drug action (Xu et al. 2002) and may be of greater importance in extrapyramidal side effects (Wang et al. 2000); however, effects on complex cognitive behaviours more specifically related to schizophrenia have yet to receive systematic evaluation.

Cognitive functions such as working memory and cognitive flexibility are known to be impaired in schizophrenia and are underpinned by prefrontal [PFC] cortical brain regions. There is a high density of D1 receptors in these regions, hence it is not surprising that D1 KO mice show deficits in spatial working memory and reversal learning (Holmes et al. 2004; El-Ghundi et al. 2007). The D2 receptor may also be important in set shifting and reversal learning. D2 KO mice show impairment in adjusting responding to previously reinforced stimuli when unexpected outcomes are encountered and both D2 KO and antipsychotic drugs have recently been shown to produce similar reversal learning deficits in a set shifting paradigm (Glickstein et al. 2002; Kruzich and Grandy 2004; DeSteno and Schmauss 2009). This raises the possibility that the D2 receptor is important in producing drug-induced impairments in these functions. D3 KO mice also show deficits in spatial working memory in some (Glickstein et al. 2002) but not all studies (Xing et al. 2010). Mutants with selective overexpression of subcortical D2 receptors show behavioural changes that include reduced incentive motivation, as indexed by reduced lever pressing for food reward in operant tasks, and abnormal timing behaviour, which may be relevant to perturbations in timing seen in a number of psychiatric diseases, including schizophrenia (Drew et al. 2009).

2.2 Dopamine Transporter

Extracellular DA concentration is regulated via a plasma membrane DA transporter (DAT). Deletion of this entity dramatically alters both presynaptic homeostasis and extracellular dynamics of DA. These models have been particularly useful in understanding the mechanism of action of psychostimulant drugs (Gainetdinov 2008). While most studies have used DAT KO and heterozygous mice (Giros et al. 1996), DAT knockdown (Zhuang et al. 2001) mice with moderately increased DA levels have also been investigated (Gainetdinov 2008). As a consequence of permanently enhanced DAergic tone, DAT KO mice show locomotor hyperactivity and stereotypy in a novel environment, as well as memory and sensory gating deficits (Gainetdinov 2008). These mice also show social interaction deficits related to behavioural inflexibility (Giros et al. 1996; Rodriguiz et al. 2004), a more positive bias towards a hedonically positive tastant (Costa et al. 2007) and enhanced resistance to extinction of food-reinforced operant behaviour (Hironaka et al. 2004), indicating a role for DA in updating rewarding values, habit learning and memory.

A variety of behavioural responses to psychostimulants and other drugs of abuse have been reported in DAT KO mice and have been reviewed previously (Gainetdinov 2008). Sex-dependent effects on ethanol preference and consumption has been reported in DAT KO mice (Savelieva et al. 2002; Hall et al. 2003; Mathews et al. 2006; Gainetdinov 2008) and have been interpreted as reflecting altered hedonic mechanisms; KO mice showed greater motivation for the task than WT (i.e. 'wanting'), without affecting their responsivity for sucrose reward (i.e. 'liking'). These data suggest that chronically elevated tonic DAergic activity produces changes in incentive motivation; however, these are in the opposite direction to those that might form part of the negative symptom profile of schizophrenia (Cagniard et al. 2006). Rodriguiz et al. (2004) found several patterns of abnormal social responsivity in DAT KO mice, including disruption of the stability of social hierarchies. This suggests that chronic DAergic hyperfunction may play an important role in abnormal social interactions.

2.3 *Catechol-O-Methyltransferase*

The enzyme catechol-*O*-methyltransferase (COMT) is involved in the catabolism of catecholeamine neurotransmitters. COMT is expressed in the pyramidal neurons of the prefrontal cortex and hippocampus and plays a specific role in the catabolism of cortical DA but not noradrenaline (Papaleo et al. 2008). This specificity, together with the COMT gene lying within a chromosomal region of interest for psychosis [22q11] and showing polymorphism that alters activity of the enzyme, has made COMT the subject of extensive study in schizophrenia (Craddock et al. 2006). Several studies suggest that functional polymorphisms of the COMT gene are associated with performance on frontal cortical tasks such as the Wisconsin card sort task in patients with schizophrenia and controls and confer enhanced vulnerability to environmental triggers for schizophrenia such as cannabis (Caspi et al. 2005).

Studies in COMT KO mice have investigated motor activity, anxiety, aggression, and sensorimotor gating (Gogos et al. 1998, Huotari et al. 2002, Haasio et al. 2003, Tammimäki et al. 2008). The findings include impaired emotional reactivity in female but not male KO and increased aggression and ethanol consumption in male but not female KO; these sex-specific effects support an emerging body of evidence that the influence of COMT on a variety of phenotypes may be different in males and females (Harrison and Tunbridge 2008). Babovic et al. (2007) have studied the phenotypic ethogram of COMT mutants and found that heterozygous but not homozygous, mice show abnormal exploration of and habituation to a novel environment. It is likely that behavioural effects are influenced by DA state according to the Yerkes-Dodson arousal curve, which indicates an inverted U-shaped relationship between arousal and behavioural performance; this has been evoked previously to explain paradoxical effects of prefrontal DAergic manipulations on behavioural performance (Seamans and Yang 2004). Male COMT KO show improvement in spatial working memory (Babovic et al. 2008; Papaleo et al. 2008), whereas a transgenic COMT mutant involving overexpression of the human COMT-VAL polymorphism shows deficits in attentional set shifting, recognition memory, and working memory performance (Papaleo et al. 2008); low-dose amphetamine restored recognition memory, confirming DAergic involvement and providing further support for an inverted U-shaped relationship between cognitive function and cortical DAergic activity (Seamans and Yang 2004; Tunbridge et al. 2006).

Recent studies across commonly used inbred strains have shown that a polymorphism in COMT, a B2 SINE insertion, is associated with behavioural differences related to exploration (Kember et al. 2010), altered expression of genes implicated in synaptic function and intracellular signalling, as well as genes linked with DAergic, glutamatergic, and GABAergic systems (Li et al. 2010).

2.4 AKT1/GSK-3

Striatal D2 receptors have been shown to exert their action in a cAMP-independent manner by promoting the formation of a signalling complex composed of AKT, protein phosphatase-2A (PP2A), and ß-arrestin 2 (Beaulieu et al. 2004, 2005). Formation of this complex leads to the inactivation of AKT after the dephosphorylation of its regulatory threonine 308 (Thr-308) residue by PP2A (Beaulieu et al. 2005). Inactivation of AKT in response to DA results in the activation of glycogen synthase kinase 3 (GSK-3; see below). Antipsychotic drugs such as haloperidol have been shown to activate AKT phosphorylation in the mouse brain (Emamian et al. 2004). A number of studies have indicated a role for ß-arrestin, AKT, and GSK-3 in DAergic regulation of behaviour. ß-arrestin KO mice show reduced climbing and reduced sensitivity to the effects of amphetamine on locomotor activity, while AKT1 KO mice show enhanced sensitivity to the disruptive effects of amphetamine on PPI (Beaulieu et al. 2009). Recently, deficits in PPI in females only has been demonstrated in AKT1 KO mice (Chen and Lai 2011).

Several studies have identified SNPs in the DISC1 gene to be associated with schizophrenia and DISC1 has been shown to regulate the phosphorylation of certain GSK-3 substrates and neuronal progenitor proliferation via modulation of GSK-3 signalling (Mao et al. 2009). Takashima (2009) have shown that GSK-3 KO mice show deficits in memory reconsolidation, but not acquisition deficits, in water maze and contextual fear conditioning paradigms. While direct studies in negative symptom models are awaited, the association between neuroplasticity, neurodevelopment, and DA signalling makes this a very promising candidate for understanding negative symptoms.

GSK-3 is also a central component of the developmentally important Wnt signalling pathway, which has been suggested to be important for certain aspects of neuronal functioning that may relate to negative symptoms, including synaptogenesis (Hur and Zhou 2010). Mice lacking the homologue of the Drosophila segment polarity gene Dvl-1, a Wnt signalling partner, are reported as having deficiencies in PPI and social interactions (Lijam et al. 1997). The same group replicated their study six years later and included some additional behavioural tasks (Long et al. 2004). In common with the earlier study, DVL-1 mice were less likely to huddle together and less likely to build nests, although no PPI deficits were seen. Other behavioural tests showed that mice lacking Dvl-1 had other impairments in behaviours related to group dynamics. It is not clear why, several generations later, mice lost their deficit in PPI; an environmental effect was considered most likely but an interaction with the background strain cannot be excluded. It should be noted that there are very few research groups that have performed follow-up studies of this type within colonies and it is possible that many phenotypic differences reported for a variety of mutant mice could turn out to show similar environmental sensitivity.

3 Genetic Models Relating to Glutamatergic Dysfunction

Hypoglutamatergic function may play an important pathobiological role in schizophrenia (Coyle 2006; Balu and Coyle 2011). The supporting arguments are based on: (a) the properties of NMDA receptor (NMDAR) antagonists to be psychotomimetic in healthy humans (Adler et al. 1999; Kegeles et al. 2000; Coyle 2006) and to exacerbate or precipitate psychotic symptoms in patients with schizophrenia (Malhotra et al. 1997a, 1997b); (b) reports of NMDAR abnormalities in schizophrenia, both in postmortem brain (Stone et al. 2007) and in living patients using SPECT imaging (Pilowsky et al. 2007); (c) limited evidence to suggest that pharmacological enhancement of NMDAR activity may be therapeutically useful against the symptoms of schizophrenia (Javitt et al. 2004; Coyle 2006; Pinard et al. 2010). Clinical genetic evidence linking gene variants associated with NMDAR activity to schizophrenia has proven inconsistent (Allen et al. 2008).

3.1 NMDAR-Mediated Neurotransmission

Mice deficient in the NMDAR NR1 subunit exhibit deficiencies across several schizophrenia-related phenotypes: decreased responsivity to the NMDAR antagonists PCP and MK-801, hyperactivity in a novel environment and deficits in PPI that were reversible particularly by second-generation antipsychotics, and social deficits that were less responsive to antipsychotics (Mohn et al. 1999; Fradley et al. 2005; Duncan et al. 2004, 2006). Mice containing a targeted mutation in NR1 which prevented NR1 phosphorylation at serine 897, shown to be dramatically reduced in patients with schizophrenia (Emamian et al. 2004), induced a marked disruption of NMDAR-mediated synaptic transmission; these mice displayed deficits in both PPI and social interaction, as indexed by a modified sociability and preference for social novelty paradigm, but no genotypic changes in locomotor activity (Li et al. 2009). These mutants also displayed decreased AMPA receptor-mediated transmission and decreased AMPAR GluR1 subunit expression in the synapse.

A recent study further refined our understanding of the role NMDAR signalling in modulation of schizophrenia endophenotypes by examining the effect of targeted postnatal or post-adolescent ablation of the NR1 subunit in cortical and hippocampal GABAergic neurons, based on evidence supporting corticolimbic GABAergic interneurons as a site of NMDA receptor hypofunction (Belforte et al. 2010). Several schizophrenia-related phenotypes, including novelty-induced hyperlocomotion and nest-building behaviour, were found to be present in postnatally targeted NR1 mutants when assessed in adulthood but not before adolescence. These NR1 mutants also displayed social recognition deficits and a mild anhedonic phenotype, measured in the sucrose preference test, as well as deficits in spatial working memory, assessed by spontaneous alternation behaviour. Deficits in spontaneous alternation and PPI were reversed by risperidone; other schizophrenia-related behaviours which were not reversed by risperidone included nest building. Interestingly, the same schizophrenia-related phenotypic profile was not observed in mice containing post-adolescent ablation of NR1, supporting the neurodevelopmental hypothesis of schizophrenia and highlighting the importance of developmental dysregulation of GABAergic neurons.

An alternative NMDAR mutant line, involving deletion of the NR1-associated NR2A (GluR1) subunit, was associated with hyperactivity that was ameliorated by haloperidol and risperidone, together with deficits in spatial and latent learning and augmented DA metabolism in striatum and frontal cortex (Miyamoto and Nabeshima 2002). Additionally, mice with partial reduction of the gene SP4 (a member of the SP1 family of transcription factors) evidence a reduction in the expression of the NR1 subunit, accompanied by a specific learning deficit in the Barnes maze; no genotypic differences were found in tests of social interaction, anxiety-related behaviour or other measures of cognition (Zhou et al. 2010).

3.2 Glycine

NMDAR activation involves postsynaptic depolarisation and the binding of two agonists, glutamate, and either glycine or D-serine at the glycine modulatory site (Tsien 2000). Glycine acts at an accessory site necessary for NMDAR function to facilitate NMDA-mediated transmission; thus, glycine agonists promote NMDA-mediated transmission and, controversially, may evidence some efficacy against the negative symptoms of schizophrenia (Tuominen et al. 2005; Patil et al. 2007; Pinard et al. 2010). A mutant line carrying a point mutation in the NMDAR glycine binding site, Grin1 (D481 N), has been shown to display hyperactivity in a novel environment that was not reversed by antipsychotics, reduced sociability that was reversed by D-serine and partially by clozapine, as well as abnormalities in spatial learning and memory (Ballard et al. 2002; Labrie et al. 2008).

As the availability of glycine is partially modulated by glycine transporters GlyT1 and GlyT2, with GlyT1 known to be co-expressed with NMDAR in glutamatergic synapses in the forebrain, mutant studies have reported enhanced NMDAR activity in mice with conditional GlyT1 deletion in forebrain neurons (Yee et al. 2006). Forebrain GlyT1 deletion has been associated with enhancement of associative learning (Yee et al. 2006), object recognition memory (Singer et al. 2007), selective attention, as measured by LI (Yee et al. 2006), and reversal learning in a water maze task (Singer et al. 2009); no effect of the mutation was observed on spatial working memory (Singer et al. 2009). Mice with constitutive heterozygous GlyT1 KO (Tsai et al. 2004) or lacking GlyT1 in the hippocampus and cerebellum (Singer et al. 2011) evidence unchanged PPI and non-responsivity to the PPI-disruptive effects of the NMDAR antagonist MK-801. However, conditional forebrain-specific GlyT1 mutants evidenced impaired PPI and intact responsivity to the PPI-disruptive effects of MK-801 (Singer et al. 2011). Overall, the diverse precognitive profile observed following enhancement of NMDA transmission via GlyT1 inactivation supports a role for GlyT1 inhibitors in the treatment of cognitive symptoms in schizophrenia.

3.3 D-serine

D-Serine is a highly selective endogenous activator of NMDAR signalling. A mouse strain lacking activity of D-amino acid oxidase (DAO, an enzyme involved in metabolism of D-serine), due to a spontaneous point mutation in the gene encoding DAO, is accompanied by reduced behavioural sensitivity to MK-801 (Hashimoto et al. 2005) and PCP (Almond et al. 2006); these changes were accompanied by an increase in NMDA receptor-mediated neurotransmission (Maekawa et al. 2005). DAO mutants demonstrated no disruption to PPI but reduced exploration in a novel environment, with attenuation of stereotype and an increase in locomotor response to methamphetamine (Almond et al. 2006). More recently, using a different procedure to that employed by Almond and colleagues, DAO mutants showed enhanced PPI, increased sensitivity to the PPI-disruptive effect of the NMDA receptor antagonist SDZ 220-581 and a subtle performance deficit in the Morris water maze, but no change in anxiety-related behaviours (Zhang et al. 2011).

Grin1 (D481 N) mutants were crossed with the same DAO mutant line in order to examine whether genetic inactivation of DAO would increase availability of D-serine and ameliorate schizophrenia-related phenotypes in a mouse model characterised by deficient NMDA glycine site activation (Labrie and Roder 2010). In agreement with their hypothesis, mice with both DAO inactivation and Grin (D481 N) mutation showed normalisation of social affiliative behaviour, as measured in the sociability and preference for social novelty test, and of LI, with partial reversal of spatial recognition and sensorimotor gating deficits. These data would suggest that inhibition of DAO function may, in a domain-selective manner, partially improve schizophrenia-related behavioural deficits caused by decreased occupancy of glycine sites asociated with NMDA receptors.

Mutants unable to produce D-serine due to disruption of its synthetic enzyme serine racemase display mild hyperactivity (Basu et al. 2009). While social interaction and PCP- or amphetamine-induced hyperactivity and disruption of PPI were found to be unaffected in serine racemase mutants, PCP-enhanced startle reactivity was found to be disrupted in these mutants (DeVito et al. 2011; Benneyworth et al. 2011). Serine racemase KO mice also demonstrate a subtle cognitive phenotype characterised by spatial learning deficits (Basu et al. 2009) and disrupted episodic memory performance in an object recognition paradigm, while other aspects of task performance which measured novelty detection, recognition memory, and relational memory were intact (DeVito et al. 2011); these deficits were accompanied by dendritic morphological abnormalities of pyramidal neurons in the medial prefrontal cortex. In overview, while the endpoint associated with many of these lines is NMDA receptor hypofunction, the diverse modes for achieving this outcome carry distinct phenotypic consequences compensatory mechanisms may account for some of these differences. However, phenotypic diversity may reflect (a) differential availability of D-serine and glycine at the glycine modulatory site and (b) whether the mutation affects receptor function directly or indirectly via limited availability in a subset of glycine modulatory site ligands.

3.4 NMDAR Signalling Complex

Abnormalities in NMDAR-interacting protein function have been implicated in schizophrenia; these include the glial glutamate and aspartate transporter (GLAST), which has been shown to be expressed differentially in the dorsolateral prefrontal cortex, anterior cingulate cortex and thalamus in postmortem brains from patients with schizophrenia (Smith et al. 2001; Bauer et al. 2008). GLAST KO exhibit antipsychotic-sensitive hyperactivity in a novel environment, increased locomotor responsivity to the NMDAR antagonist MK-801 and impaired sociability (Karlsson et al. 2008, 2009). Knockout of the NMDAR signalling molecule SynGAP is associated with increased novelty-induced hyperlocomotion and reduced behavioural sensitivity to MK-801 administration (Guo et al. 2009). SynGAP mutants evidence intact sociability but impaired social novelty preference, disrupted PPI, enhanced startle reactivity, and deficits in spatial working memory (Guo et al. 2009).

3.5 Metabotropic Glutamate Receptors

Metabotropic glutamate receptor (mGluR) involvement in the pathophysiology of schizophrenia, together with the diverse range of phenotypes associated with mutation for members of this receptor family (mGluR1-8), are discussed in detail elsewhere (Krivoy et al. 2008). Mutants with AMPA GluR1 KO display antipsychotic-sensitive hyperactivity in a novel environment but not in the home cage, reduction of MK-801-induced hyperactivity and impaired PPI, in association with reduced clearance of DA (Wiedholz et al. 2008). Examination of the behavioural phenotype of mGluR4 KO mice revealed impaired PPI and enhanced sensitivity to the locomotor stimulatory effects of the NMDAR antagonist MK-801; no differences, however, were observed in relation to social behaviour and spatial working memory (Sagata et al. 2010). mGluR5 KO mutants show hyperactivity and impaired PPI that are reversible by antipsychotics; hyperactivity induced by the NMDA antagonist MK-801 was also increased following deletion of mGluR5 (Brody et al. 2004; Gray et al. 2009).

In summary, the evidence from metabotropic glutamate receptor mutants would implicate this receptor family in PPI impairment. Other studies have documented depression- and anxiety-related behavioural phenotypes across a variety of assays (Feyder et al. 2007; Chourbaji et al. 2008). GluR1 KO also exhibited increased novelty- or stress (injection, restraint or forced stress)-induced hyperactivity, as well as reduced immobility in the forced swim test (Fitzgerald et al. 2010). These deficits were accompanied by a reduced anxiety phenotype in the elevated plus maze that was normalised following chronic (2 week) treatment with lithium, used as a moods stabiliser in bipolar disorder. The authors suggested that the combination of schizophrenia-related phenotypes and affective/manic behavioural signs in GluR1 KO mice, particularly under conditions of stress, may provide a novel model of schizoaffective disorder.

4 Genetic Models Relating to Candidate Risk Genes and Copy Number Variations

4.1 Disrupted-in-schizophrenia-1

Familial mutation in the disrupted-in-schizophrenia-1 (DISC1) gene, due to a balanced chromosomal translocation at 1q42.1-1q42.3, has been associated with schizophrenia and other psychiatric disorders across diverse populations (Chubb et al. 2008; Johnstone et al. 2011); while there are initial reports of associations between DISC1 and positive symptoms, these are inconsistent as to whether the relationship is with delusions (DeRosse et al. 2007) or hallucinations (Szeszko et al. 2008). Genetic lineage and association studies have further suggested DISC1 as a general risk factor for schizophrenia, schizoaffective disorder, bipolar disorder, major depression, autism, and Asperger syndrome (Chubb et al. 2008). The mechanism through which DISC1 dysfunction may contribute to a wide spectrum of psychiatric disorders remains unknown (Hennah and Porteous 2009). During embryonic development, DISC1 appears to play an important role in neurodevelopment and structural plasticity via interaction with several proteins, including phosphodiesterase-4B, Fez1, NDEL1 and LIS1 (Duan et al. 2007; Chubb et al. 2008; Muir et al. 2008).

While generation of a DISC1 KO model has not been achieved, largely due to the complexity of exon usage of the DISC1 gene (Wang et al. 2008), several mutant models of DISC1 gene function have been described in the literature: (1) A naturally occurring DISC1 mutation (25 base pair deletion at exon 6) in the commonly employed 129S6 Sv/Ev mouse strain, which additionally carries a termination codon in exon 8 and a polyadenylation signal in the adjacent intron, shows impairment in spatial working memory using a delayed-non-match-to-place task, but not using the Morris water maze, tests of associative learning-contextual conditioning or other alternative measures of memory-related processes

(Koike et al. 2006; Kvajo et al. 2008); (2) ENU mutagenesis in exon 2 of DISC1 is characterised by hyperactivity, with antipsychotic-sensitive deficits in PPI and LI and disrupted working memory in a T-maze task (Clapcote et al. 2007); (3) A transgenic line with inducible expression of a DISC1 C-terminal fragment evidenced impaired spatial working memory using the delayed-non-matched-to-place task, as well as reduced sociability (Li et al. 2007); (4) A transgenic line with expression of a dominant-negative truncated form of DISC1 under the CaMKII promoter, mutants showed hyperactivity but no changes in PPI, social interaction or cognition (Hikida et al. 2007); (5) An inducible transgenic line with forebrain-specific expression of mutant human DISC1 (hDISC1) under the CaMKII promoter showed hyperactivity without material disruption to PPI, sex-specific impairment of working memory using the Morris water maze, with a deficit in dyadic social interactions but not sociability (Pletnikov et al. 2008); (6) A transgenic line with expression of truncated DISC1 exhibited normal exploratory activity but disruption to LI (Shen et al. 2008).

Interestingly, lentiviral silencing of DISC1 expression in the adult dentate gyrus was accompanied by an increase in novelty-induced hyperlocomotion; this behavioural effect was reversed following treatment with the GSK-3β inhibitor SB-216763, suggesting that increased GSK-3β activity secondary to DISC1 loss-of-function might be associated with schizophrenia-like behaviours (Mao et al. 2009). Further support for this hypothesis has come from a recent report that pharmacological or genetic inactivation of GSK-3 in the ENU-generated DISC1 mutant (L100P) reversed phenotypic deficits in PPI and LI, and normalised their hyperactivity profile (Lipina et al. 2010). This further implicates impaired DISC1-GSK-3 interplay in schizophrenia-relevant behaviours.

Employing the mutant hDISC1 transgenic model (Pletnikov et al. 2008), a follow-up study assessed the effect of timing of the DISC1 mutation on the expression of schizophrenia-related phenotypes (Ayhan et al. 2011). In this study, the effect of mutant hDISC1 expression was compared during prenatal, postnatal, or both periods. Regardless of timepoint of expression, hDISC1 mutants exhibited fewer cortical parvalbumin-positive cells, as well as cortical hypodopaminergia relative to controls. Combined pre- and postnatal mutant hDISC1 expression was associated with the most prominent schizophrenia-related abnormalities; these included decreased social interaction in a test of dyadic interaction, increased sensitivity to psychostimulants and increased immobility in the tail suspension test, with lateral ventricular enlargement and morphological abnormalities in granule cells of the dentate gyrus. In contrast, neither pre- nor postnatal expression alone fully recapitulated the phenotype associated with combined timepoint expression.

Extending these findings and introducing regional specificity, Niwa and colleagues (2010), using in *utero* electroporation to produce selective knockdown of DISC1 in pyramidal neurons in prefrontal cortex of the foetal mouse, reported post-pubertal emergence of disrupted PPI, impaired object recognition memory and delayed T-maze alternation; there was disturbed maturation of mesocortical DAergic projections to and interneurons in the medial prefrontal cortex. Several of these phenotypes, including impaired PPI and increased extracellular DA levels in the prefrontal cortex, were normalised following administration of clozapine.

In summary, the diversity of DISC1 models available indicates that the variable phenotypic effects of different DISC1 mutations appears to be highly dependent upon time of expression, environmental factors (see 6.1) and, potentially, interactions with other risk genes (Ross et al. 2006; Ayhan et al. 2011; see 6.3).

4.2 Dysbindin

Systematic review and meta-analysis has indicated dystrobrevin-binding protein 1 (DTNBP1; dysbindin) to be a replicable risk gene for schizophrenia (Allen et al. 2008; Gill et al. 2009). Initial data indicated that dysbindin mRNA expression was decreased in schizophrenia in PFC and hippocampus (Weickert et al. 2004, 2008) but subsequent studies have failed to replicate this finding in the dorsolateral prefrontal cortex (Tang et al. 2009). In vitro work indicates that reduction in dysbindin can lower glutamate release, while overexpression of dysbindin elevates glutamate release, suggesting a modulatory role for dysbindin in glutamate neurotransmission in cortical neuronal cultures (Numakawa et al. 2004).

The 'sandy'(*sdy*) mouse, a spontaneous mutation identified in the DBA/2J strain, carries a naturally occurring deletion that includes the DTNBP1 gene. Behavioural phenotyping of *sdy* mutants has produced inconsistent findings, there being reports of unaltered (Feng et al. 2008; Bhardwaj et al. 2009; Li et al. 2003), decreased (Takao et al. 2007; Hattori et al. 2008) or increased (Cox et al. 2009; Ji et al. 2009) spontaneous exploratory activity in the open field. Sensitisation to amphetamine is enhanced, with both increased turnover (Murotani et al. 2007) and decreased concentration (Hattori et al. 2008) of DA having been reported. While the DBA/2J strain does not show robust PPI, precluding assessment therein, backcrossing of the *sdy* mutation onto a C57BL6 line revealed either reduced PPI (Talbot 2009) or enhanced PPI and startle reactivity (Papaleo et al. 2012).

S*dy* mutants evidence reduction in social contacts and social contact time in a dyadic paradigm (Feng et al. 2008; Hattori et al. 2008); genotypic disruption of spatial learning and working memory using the Barnes maze, the Morris water maze and the T-maze forced alternation task (Takao et al. 2007; Cox et al. 2009), as well as deficits in recognition memory using the novel object recognition paradigm, have also been observed (Feng et al. 2008; Hattori et al. 2008).

It has been suggested that reduction in DTNBP1 expression, associated with decreased glutamatergic neurotransmission (Numakawa et al. 2004), may contribute to aberrant prefrontal cortex function and consequent working memory deficits in schizophrenia (Jentsch et al. 2009). These authors found modest disruption in working memory in *Sdy* mutants using the choice accuracy measure in a delayed-non-match-to-position task; this was accompanied by disturbance of excitatory neurotransmission in the PFC, as indexed by reduction in amplitude of eEPSCs and frequency of miniature EPSCs, with abolition of paired-pulse facilitation. However, abundant dysbindin expression in the hippocampus may indicate an alternative mechanism for such dysfunction via synaptic plasticity and associated memory functions (Takao et al. 2008; Cox et al. 2009). Mutation of DTNBP1 also caused deficits in working memory, as measured by the delayed-non-match-to-position task; this deficit was accompanied by a decrease in cortical NR1 mRNA expression that was correlated with working memory performance (Karlsgodt et al. 2011).

In a separate study, *Sdy* mice displayed enhanced learning in the acquisition phase of a modified T-maze working memory task; with introduction of delay intervals, DTNBP1 mutants then displayed worse performance (Papaleo et al. 2012). This genotype-specific and delay-dependent disruption was exacerbated by exposure to a mild stressor during testing, where mice were housed in a new, clean cage rather than the familiar home cage; these data are consistent with the recognised ability of mild stressors to impair PFC mediated working memory function. Cellular findings suggest that the interaction between CaMKII and enhanced signalling at cortical D2 receptors may contribute, at least partially, to observed working memory phenotypes (Ji et al. 2009; Papaleo et al. 2012).

4.3 Neuregulin

Neuregulin-1 (NRG1) has been associated functionally with numerous neurodevelopmental processes, including neuronal migration, neurotransmitter receptor expression/activation, myelination, and synaptic plasticity (Mei and Xiong 2008). Multiple NRG1 isoforms have been described, the diversity of which has been attributed to alternative splicing and the existence of multiple 5′ flanking regulatory elements. NRG1 I–III share the EGF–like signalling domain; interaction of these EGF-like domains with membrane-associated tyrosine kinases (ErbB receptors) activates intracellular signalling pathways implicated in the pathobiology of schizophrenia (Harrison and Law 2006; Mei and Xiong 2008; O'Tuathaigh et al. 2009). Initially associated with increased risk for schizophrenia in an Icelandic sample (Stefansson et al. 2002), NRG1 has been confirmed on meta-analysis to be a replicable risk gene for schizophrenia (Bertram 2008). Further evidence indicating disruption to NRG1-ErbB signalling in the pathobiology of schizophrenia derives from studies in human postmortem brain tissue and cell lines (Harrison and Law 2006; Mei and Xiong 2008).

A missense mutation in exon 11, which codes for the transmembrane region of NRG1, has been associated with schizophrenia (Walss-Bass et al. 2006). The phenotype of mutants with heterozygous deletion of the transmembrane (TM)-domain of NRG1 [homozygous KO being lethal] is characterised by hyperactivity, including sex-specific effects among individual topographies of the murine repertoire, that is sensitive to amelioration by clozapine (O'Tuathaigh et al. 2006, 2007a; Karl et al. 2007); there is modest disruption to PPI (Stefansson et al. 2002; however, see Van den Buuse et al. 2009; O'Tuathaigh et al. 2008, for conflicting data). Although initial studies indicated no effect of TM-domain NRG1 loss-of-function on various learning and memory measures such as Y maze alternation and the Barnes maze (O'Tuathaigh et al. 2007b), a separate investigation revealed mild domain-specific cognitive abnormalities in TM-domain NRG1 mutants in novel object recognition memory and in contextual but not cue-directed fear learning (Duffy et al. 2010). Consistent with earlier reports (O'Tuathaigh et al. 2007b), no differences in spatial learning or working memory were observed in this study.

Heterozygous KO of TM-domain NRG1 likely affects several NRG1 isoforms. In terms of phenotypic comparisons with more isoform-specific NRG1 deletions, type III NRG1 mutants exhibit greater deficits in PPI relative to TM-NRG1 mutants (Chen et al. 2007). However, neither type III NRG1 mutants nor those with targeted disruption of type I/type II NRG1 display the hyperactivity that is characteristic of TM-NRG1 mutants (Rimer et al. 2005; Chen et al. 2007). In contrast, mutants with heterozygous deletion in the EGF-like domain [that might be expected to impact on all NRG1 isoforms] exhibit an initial increase in locomotor activity and more rapid habituation of exploration in a novel environment; in the same mutant line, showing baseline PPI to be unaltered, PPI was disrupted by MK-801 but not by amphetamine (Duffy et al. 2008).

To assess the impact of over-expression of NRG1, mice carrying the transgene of mouse type-1 NRG1 cDNA were examined for the presence of schizophrenia-related abnormalities (Kato et al. 2010). Each of two separate transgenic lines displayed: (a) increased novelty-induced hyperactivity, although habituation levels did not differ across genotypes; (b) intact PPI and startle responsivity; (c) impairment in contextual learning; (d) increased aggression and decreased social behaviours in the resident-intruder paradigm. HPLC analysis in various brain regions found decreased DA and DOPAC levels in the hippocampus of over-expressor mutants. Paradoxically, both hypermorphic and hypomorphic expression of NRG1 produce a common behavioural phenotypic profile. A different type-1 NRG1 over-expressor line, driven by a Thy-1 promoter in brain projection neurons, displayed marked impairment of PPI; however, a tremorous phenotype likely confounds behavioural assessment (Deakin et al. 2009). Phenotypic differences between both NRG1 Type-1 over-expressor lines may be attributable to the use of distinct gene promoters to regulate transgene expression (Kato et al. 2010).

Although it was initially shown that NMDAR expression is reduced in TM-domain NRG1 mutants (Stefansson et al. 2002), recent evidence has questioned the role of NRG1 in NMDAR function, with brain-specific mutation of ErbB2 and ErbB4 having no effect on the expression of NMDAR subtypes (Barros et al. 2009; Gajendran et al. 2009). However, phosphorylation of NR2B receptor subunits appears reduced in the TM-domain NRG1 mutant (Bjarnadottir et al. 2007). Interestingly, ablation of the ErbB4 receptor in parvalbumin-positive interneurons in the prefrontal cortex prevented NRG1-induced stimulation of GABA and produced a schizophrenia-related phenotype comparable to that observed in heterozygous NRG1 or ErbB4 KO models; these included PPI deficits, impaired working memory and hyperactivity (Wen et al. 2010). This conditional line also showed impaired working memory in the 4- or 8-arm radial maze; a similar profile has not been observed in NRG1 and ErbB4 mutants (O'Tuathaigh et al. 2007a, b). These data would implicate parvalbumin-positive neurons as a target for NRG1/ErbB4 modulation of behaviours relevant to schizophrenia. Recent experiments demonstrate that ErbB4 also controls the formation and/or maintenance of excitatory synapses on specific populations of GABAergic interneurons and that postsynaptic ErbB4 function is probably required in this process (Fazzari et al. 2010).

A potential role for NRG1 in G × E interactions related to schizophrenia (see 5.2) is highlighted by a recent study which examined HPA axis activity and stress responsivity in a line of NRG1 mutant rats [generated using the sleeping beauty transposon] with disruption of the NRG1 Type II isoform (Taylor et al. 2011). NRG1 mutant rats displayed increased basal corticosterone levels and decreased corticosterone levels after 30 min restraint stress relative to basal levels, with increased expression of glucocorticoid receptor expression in the pituitary and hippocampus and decreased expression in the paraventricular nucleus. These mutants also displayed slower habituation in a novel environment, consistent with that observed in NRG1 hypomorphic mice (O'Tuathaigh et al. 2006), which the authors interpreted as increased reactivity to environmental stimuli.

4.4 Copy Number Variations

Studies now indicate the importance of rare structural mutations in schizophrenia, although the size of their contribution or possible associations with symptom subtypes remains unclear (Bassett et al. 2010; Owen et al. 2010). These mutations usually consist of DNA sequences that vary between individuals due to either duplication or deletion of chromosomal material. Initial support for CNVs came from patients with velocardiofacial syndrome, caused by a microdeletion at chromosome 22q11 and associated with substantive increase in risk for psychosis (Karayiorgou and Gogos 2006). Mice containing heterozygous disruption in a subset of 22q11.2 genes (Gnb1l, Dgcr8 and Tbx1) demonstrate significantly reduced PPI (Long et al. 2006; Paylor et al. 2006; Stark et al. 2008). Other schizophrenia-related phenotypes of mutants carrying a multigene deletion across the 22q11 region include neuronal migratory defects and disruption of cortical neurogenesis (Meechan et al. 2007a, 2007b; Stark et al. 2008).

There is now considerable evidence for an increased number of low-frequency CNVs, across numerous loci, in individuals with schizophrenia (International Schizophrenia Consortium 2008; Walsh et al. 2008; Kirov et al. 2009a; Levinson et al. 2011). Several studies have identified common CNV loci which map to chromosomes 1q21.1 and 15q13.2 (International Schizophrenia Consortium 2008; Stefansson et al. 2002; Levinson et al. 2011). The 15q13.2 region contains the $\alpha7$ nicotinic cholinergic receptor gene (CHRNA7); mice containing partial knockout of the CHRNA7 gene evidence attentional deficits (Young et al. 2007), while agonists at the α7 nicotinic cholinergic receptor may have some efficacy against both the positive and negative symptoms of schizophrenia (Hajós and Rogers 2010).

In recent years, several studies have supported a link between increased susceptibility for schizophrenia and microdeletions affecting the gene neurexin 1 (NRXN1; Kirov et al. 2008, 2009b; Ikeda et al. 2010; Rujescu et al. 2009; Levinson et al. 2011). Neurexins are presynaptic proteins that function as synaptic recognition molecules and may contribute to various aspects of synaptic function via binding to neuroligins (Ichtchenko et al. 1995). Mice containing deletion of NRXN1α display behavioural and electrophysiological deficits relevant to schizophrenia, including disruption to synaptic transmission and PPI (Etherton et al. 2009). Interestingly, dysregulation of NRXN1 expression throughout development has been observed in mice containing mutation of the DISC1 gene (the ENU-generated L100P mutation), providing a measure of functional association between two prominent risk genes.

5 Evaluation of G × E Hypotheses in Mutant Models of Schizophrenia

According to the neurodevelopmental model of schizophrenia, genetic (G) and environmental (E) influences combine to influence early CNS development and function, producing pathophysiological deficits which precede the onset of psychotic symptoms (Fatemi and Folsom 2009; Waddington et al. 2011). A variety of early insults, ranging from viral infection during pregnancy, through psychosocial adversities during childhood, to psychosocial stressors and substance abuse over adolescence/young adulthood, have been associated with schizophrenia (van Os et al. 2010) and shown to produce behavioural and neurobiological changes reminiscent of schizophrenia in adult rodents and non-human primates (O'Tuathaigh et al. 2007a, 2010a, O'Tuathaigh et al. 2010b). One important consideration in the development of G × E models is the timing of the environmental insult. The majority of preclinical G × E studies have sought to employ environmental manipulations at specific periods of brain development, from early stages of pregnancy through to adolescence, which have been identified as critical to the pathogenesis of schizophrenia (Cannon et al. 2003; Waddington et al. 2011).

5.1 DISC1 × Immune Challenge

A number of recent studies have examined the behavioural consequences of manipulation of the pre- or post-natal environment in mice harbouring genetic mutation. Epidemiological findings have found an association between elevated cytokines in maternal serum and increased risk for schizophrenia in offspring

(Brown et al. 2004), leading some authors to suggest that G × E interaction in schizophrenia may involve genetic vulnerability to abnormal immune responses following environmental challenge (Ayhan et al. 2009). A recent study examined the interaction of the schizophrenia risk gene DISC1 with neonatal exposure to the immune system activating agent Poly I:C (Ibi et al. 2010). The authors proposed that the neonatal period in mice corresponds to post-conception days 128–158 for cortical events and 93–115 for limbic development in humans (Clancy et al. 2007; Ibi et al. 2010). Transgenic mice expressing a dominant-negative form of DISC1 demonstrated a selective and more pronounced schizophrenia-related phenotype across various cognitive measures (i.e. spontaneous alternation, object recognition memory, contextual fear memory) following neonatal immune challenge; social interaction and MK-801-induced hyperactivity were also selectively altered in Poly I: C-treated DISC1 mutants. These behavioural deficits were accompanied by a decrease in parvalbumin-positive interneurons in the medical prefrontal cortex.

Another model of G × E interaction related to schizophrenia has been described recently in mice with inducible expression of mutant hDISC1 in forebrain neurons (Pletnikov et al. 2008; Abazyan et al. 2010). Rather than examining neonatal immune challenge, this study employed the well-characterised prenatal Poly I: C immune challenge paradigm (Meyer et al. 2009). These authors demonstrated in male mice expressing mutant hDISC1 that treatment with Poly I:C at gestational day (GD) 9, corresponding to the end of the first trimester of human pregnancy, produced a behavioural profile that included elevated anxiety, depression-like behaviours and disrupted social interaction; these behavioural deficits were accompanied by altered 5-HT neurotransmission in the hippocampus, decreased HPA axis reactivity and differential modulation of secretion of inflammatory cytokines, as well as attenuation of genotypic enlargement of the lateral ventricles. In contrast with the earlier study (Ibi et al. 2010), no changes in cognition were observed. These data may suggest an environmental mechanism by which the same DISC1 mutation can produce diverse neuropsychiatric phenotypes, in a manner comparable to that observed with the original Scottish chromosomal dislocation (Chubb et al. 2008; Hennah et al. 2009). With respect to potential mechanisms underlying putative interactions between DISC1 dysfunction and immune activation, it has been suggested that a key target may be disturbance of neuronal maturation, which may be effected via proliferation, migration, early dendritic development, and axonal outgrowth, as well as synapse formation and maturation (Ayhan et al. 2009).

5.2 NRG1/COMT × Exposure to Drugs of Abuse

Epidemiological data have indicated an association between substance abuse and risk for psychosis. Although clinical data have not indicated a relationship between NRG1 genotype and behavioural responsivity to drugs of abuse, studies conducted in mutant models of NRG1 function have indicated changes in behavioural and neural sensitivity to cannabis and psychostimulant drugs in heterozygous NRG1 TM-domain mutants and COMT mutants (Boucher et al. 2007a, b; O'Tuathaigh et al. 2010c).

NRG1 male mutants demonstrated increased sensitivity to the hypoactivity-inducing and PPI-disruptive effects of an acute dose of Δ^9-THC, the psychoactive constituent of cannabis (Boucher et al. 2007a). In a separate follow-up study, female NRG1 mutants were shown to be insensitive to Δ^9-THC -induced social withdrawal, although its hypoactivity-inducing effects were observed in both genotypes (Long et al. 2010). Increased sensitivity to Δ^9-THC-induced activation of *c-fos*, a marker of neuronal activation, was also found in NRG1 male mutants across several brain regions (Boucher et al. 2007b). Subchronic exposure to the synthetic cannabinoid CP55, 940 for 15 days in NRG1 TM-domain mutants was associated with more rapid tolerance to its hypoactivity-inducing and hypothermic effects (Boucher et al. 2010); in contrast, NRG1 mutants were insensitive to its anxiety-modulating effects. More recently, it was shown that the effects of subchronic administration of NMDA receptor antagonists, a pharmacological regimen known to produce schizophrenia-related deficits in rodents, were altered in a paradoxical manner in NRG1 TM-domain mutants; specifically, the hyper-activity and social interaction phenotypes reported in vehicle-treated mutants were absent following NMDA receptor antagonist treatment (O'Tuathaigh et al. 2010b).

Cannabis consumption, particularly when used in adolescence, has been associated with a doubling of risk for schizophrenia (Arseneault et al. 2004; Henquet et al. 2005; Moore et al. 2007). A longitudinal birth cohort study has shown that cannabis use was more likely to be followed by psychosis among those exposed during adolescence and homozygous for the COMT Val108Met allele (Caspi et al. 2005). We have recently shown that in COMT KO, genotype exerted specific modulation of responsivity to chronic Δ^9-THC administration during adolescence in terms of exploratory activity, spatial working memory and anxiety; this illuminates the interaction between genes and adverse environmental exposures over a particular stage of development in the expression of the psychosis phenotype (O'Tuathaigh et al. 2010c).

5.3 Genes Related to the Pathophysiology of Schizophrenia × Exposure to Prenatal Stress

Mice containing either knockdown or a point mutation in the synaptosomal-associated protein of 25 kDa (SNAP-25), a gene associated with synaptic transmission and linked with schizophrenia in meta-analysis of genomic association data (Lewis et al. 2003), were particularly vulnerable to the effects of prenatal stressors on the emergence of schizophrenia-releveant behavioural phenotypes (Oliver and Davies 2009). Specifically, during embryonic day 11.5–17.5, SNAP-25 mutants exposed to a combination of physical (restraint, swim), situational (open field), and social stress, demonstrated significantly enhanced PPI deficits relative to stressed WT; these deficits were reversed by haloperidol or clozapine. Interestingly, social interaction deficits, in terms of both sociability and preference for social novelty, were observed only in SNAP-25 point mutants that had been exposed to prenatal stress; this indicates that combining genetic vulnerability with exposure to an environmental stressor both modifies and produces novel phenotypes relevant to schizophrenia. Prenatal stress increased anxiety-related behaviour (elevated plus maze, open field) in all mice regardless of genotype.

Reelin is a protein that has been functionally linked with the control of synaptic functions and a dysfunctional Reelin-mediated signalling pathway has been suggested to play a role in the pathophysiology of schizophrenia (Knuesel 2010). Several studies have examined the effect of early life adversity on the heterozygous Reeler mouse phenotype. In Reeler mutants who had been prenatally exposed over GD 14-16 to the neurotoxin chlopyrifos, development was altered in a paradoxical fashion, reversing genotypic abnormalities across a number of behavioural domains: ultrasound vocalisation, increased amphetamine-induced locomotion, and stereotype (Laviola et al. 2006). Interestingly, a similar effect was observed in Reeler mutants who were exposed to early maternal separation (PND2-6); maternally separated Reeler mutants failed to show abnormalities in ultrasonic vocalisations and exploratory behaviour, as well as social interaction (Ognibene et al. 2007). These data illustrate the emergence of unexpected and novel phenotypes in risk gene mutants as a result of gene × environment interactions.

5.4 Overview of G × E studies

Going forward, systematic examination of G × E risk factor manipulations in genetic models will be necessary to identify common and distinct G × E risk profiles at the level of endophenotypes and underlying neurobiology. With respect to diagnostic categories which may be differentially susceptible to environmental versus genetic influences, a recent review of clinical evidence for G × E interactions in schizophrenia hypothesised that environmental factors associated with schizophrenia may have a greater impact on affective and psychotic dimensions, while developmental influences may have a greater impact on negative and cognitive dimensions (van Os et al. 2010). One of the criticisms of the genetic model data to date has focused on the translational success of modelling certain environmental factors or stressors (e.g. psychosocial stressors) between human and small animal species (van Os et al. 2010). Additionally, for logistical reasons, the feasibility of studies combining exposure to several environmental factors with any given genetic vulnerability factor(s) may be difficult to realise.

In teasing out the biological mechanisms underlying synergistic effects of environmental and genetic factors on risk for disease, it is known from preclinical studies that exposure to several environmental stressors implicated in schizophrenia (prenatal immune challenge, early life adversity, psychosocial stressors, adolescent cannabis/psychostimulant exposure) may be associated with disruption to DAergic transmission. Given interactions between DAergic function and GABAergic, glutamatergic and endocannabinoid signalling, future research on the biological impact of these environmental insults should also examine neurodevelopmental processes and functions implicated in schizophrenia, including changes in myelination, synapse formation, and immune system function.

6 Discussion

6.1 Divining gene-Phenotype Associations in Schizophrenia Risk Gene Models

6.1.1 Sex Differences

Despite considerable epidemiological evidence for sex differences in the incidence, psychopathology and course of schizophrenia, the notable absence of mutant studies which have systematically compared male and female murine phenotypes has been noted (O'Tuathaigh et al. 2010a; Chen and Lai 2011). Additionally, it has been noted that aside from the potential explanatory role of sexually dimorphic biology in relation to the epidemiological findings, sex differences in the nature of environmental interactions may also help to explain sex differences in the expression of psychiatric symptoms (Accortt et al. 2008; Burrows et al. 2011). The presence of sexually dimorphic effects of susceptibility gene mutation on the expression of schizophrenia-related endophenotypes, which have been noted throughout our review of the literature, might suggest the involvement of sex hormonal regulation of AKT1, DISC1, NRG1 and other risk gene function in a manner which may contribute to the pathogenesis and/or expression of schizophrenia.

6.1.2 Improved Characterisation of Behavioural Models of Schizophrenia

As others have pointed out, progress in molecular and genomic techniques, and the implications for genetic modelling in schizophrenia, has not been paralleled by comparable advances in the generation of murine behavioural paradigms capable of measuring specific forms of cognition and social and motivational behaviours which are disrupted-in-schizophrenia; furthermore, these relate to clinical features that are largely insensitive to existing antipsychotic treatments (Brigman et al. 2010). While the field has made incremental progress in generating novel murine paradigms to assess these domains, including translating across species from rat to mouse, significant progress have been made in some areas such as the development of tasks employing touchscreen technology to measure executive function processes (Brigman et al. 2010).

Despite cognitive deficits being considered a core feature of schizophrenia, several aspects of cognition have not been addressed in existing behavioural models. Recent studies have described novel assessment methods for problem-solving behaviour, a form of executive functioning (Ben Abdallah et al. 2011). In this paradigm, which is dependent upon a set of ethologically appropriate escape behaviours within a limited time period, the greatest executive function deficits were observed in mice deficient in the GluA1 subunit of AMPA receptors or mice which had received hippocampal lesions. Mice with lesions of the medial prefrontal cortex showed more subtle and specific performance deficits under one of the task conditions, while similarly subtle difficulties in problem-solving behaviour were found in mice containing overexpression of the D2 receptor or in mice which were chronically treated with an NMDA receptor antagonist (Ben Abdallah et al. 2011). Although variable performance in control groups across each of these experimental conditions indicates that further work is required in relation to standardisation of the procedure, the relatively simple box design and experimental procedure indicates a promising new tool for assessment of cognitive processes, including executive functions, relating to schizophrenia.

Stress responsivity has been proposed as a modulating trait which might increase risk for psychotic disorders (Goodwin et al. 2004; Myin-Germeys et al. 2009). Individual differences in affective response to stressful stimuli, alongside the associated physiological response, may play a role in induction of psychosis. It has been suggested that inherited liability to psychotic disorder may be expressed as altered responsivity to everyday stressful situations (Myin-Germeys et al. 2005).

6.1.3 Strain Differences

Even where a single gene mutation has been introduced, evidence indicates that a mutant model bred onto different background genetic strains can show very different phenotypes, suggesting gene $\times$ gene interactions and/or genetic modifier effects (Burrows et al. 2011; see Waddington et al. 2005 for further discussion of implications of strain differences in CNS KO models).

6.2 Integration of Novel Gene Manipulation Strategies

Review of single gene mutant and G $\times$ E studies over the past few years, which have increasingly involved the application of conditional transgenic technologies and alternative gene manipulation strategies, demonstrates the power of examining the effect of dynamic genetic manipulations throughout development, followed by multidomain behavioural, cellular, and molecular phenotyping. Further integration of new strategies to modulate gene expression in a phenotypic and regional manner will facilitate the development of more accurate in vivo disease models based on available pathophysiological data for disease states (Garbett et al. 2010).

Table 1 Mutant mouse phenotypes relating to dopaminergic and glutamatergic mechanisms implicated in schizophrenia

	Anxiety	Novelty-induced activity	Cognition			Sociability			Psychopharmacology		
			Working memory	Attention	PPI	Social memory[a]	Social aggression	Social interaction[b]	Increased sensitivity to psychostimulants[c]	Increased sensitivity to stress	Responsivity to antipsychotics
Dopamine mutation											
D1 Receptor KO	–	↑	↓	–	=	–	–	–	=	–	–
D2 Receptor KO	–	↓	↓	–	=	–	–	–	↑/↓	–	–
D2L KO	–	=	–	–	=	–	–	–	=	–	↓
D2 Receptor OE	–	–	–	–	–	–	–	–	–	–	–
D3 Receptor KO	–	–	↓	–	–	–	–	–	–	–	–
DAT KO	–	↑	↓	–	↓	–	↑	↓	–	–	–
COMT Het	↑	=	=/↑	–	=	=	↑	=	–	–	–
COMT KO	↑/↓	=	↑	–	=	=	=	=	↓	↑	–
COMT-Val Tg	↓	=	↓	↓	=	–	–	–	↑	↓	–
AKT1 KO	–	=	–	–	↓	–	–	–	↑	–	–
GSK-3 KO	–	–	–	–	–	–	–	–	–	–	–
DVL-1 KO	–	=	–	–	↓/=	=	–	↓	–	–	–
Glutamate mutation											
NR1 KO	–	↑	–	–	↓	↓	↓	↓	↓	–	=/↓
NR1 S897	–	–	–	–	↓	–	–	↓	–	–	–
NR1 PND	↑	↑	↓	–	↓	↓	–	–	–	–	–
GluRε1 KO	–	↑	–	–	–	–	–	–	–	–	↑
GlyT1 cKO	=	–	=	↑	=	–	–	–	↓/=	–	–

(continued)

Table 1 (continued)

	Anxiety	Novelty-induced activity	Cognition			Sociability			Psychopharmacology		
			Working memory	Attention	PPI	Social memory[a]	Social aggression	Social interaction[b]	Increased sensitivity to psychostimulants[c]	Increased sensitivity to stress	Responsivity to antipsychotics
DAO KO	↓	↓/↑	–	–	=/↑	–	–	–	↓/↑	–	–
DSR KO	–	↑	–	–	=	=	–	=	–	–	–
GLAST KO	–	↑	–	–	–	–	–	↓	↑	–	↑
SynGAP KO	–	↑	↓	–	↓	↓	–	=	↓	–	–
AMPA GluR1 KO	↓	↑	–	–	↓	–	–	–	↓	↑	↑
mGluR4 KO	–	–	=	–	↓	–	–	↓	↑	–	–
mGluR5 KO	–	↑	–	–	↓	–	–	–	↑	–	↑
VGluT1 Het	–	–	–	–	–	–	–	–	–	↑	–
VGluT2 Het	–	–	–	–	–	–	↓	↑	–	–	–

↑, increased relative to wildtype; ↓, decreased relative to wildtype; =, no difference;/, separates different findings; -, not reported. Abbreviations: *PPI* prepulse inhibition, *KO* knockout mutants, *Het* heterozygous mutants, *OE* over-expression, *DAT* dopamine transporter, *COMT* Catechol-*O*-methyltransferase, *Tg* transgenic mice, *GSK-3* glycogen synthase kinase 3, *NR1* NMDA receptor 1, *NR1 PND* NMDA receptor postnatal deletion, *GluRε1* NR1-associated NR2A subunit, *GlyT1 cKO* Glycine transporter 1 conditional knockout, *DAO* D-amino acid oxidase, *DSR* D-serine, *GLAST* glutamate and aspartate transporter, *mGluR* metabotopic glutamate receptor, *VGluT* vesicular glutamate transporter. [a] Includes social novelty preference and social recognition tests. [b] Includes investigation of stationary mouse within a cage and free social interaction tests. [c] Includes dopamine-enhancing agents, NMDA receptor antagonists

Table 2 Mutant mouse phenotypes relating to genes associated with risk for schizophrenia

	Anxiety	Novelty-induced activity	Cognition			Sociability			Psychopharmacology		
			Working memory	Attention	PPI	Social novelty preference	Social aggression	Social interaction[a]	Increased sensitivity to psychostimulants[b]	Increased sensitivity to stress	Responsivity to antipsychotics
Dysbindin sdy	=/↑	↓/=/↑	↓	–	↓/=/↑	–	–	↓	–	↑	–
DISC1 deletion	↓/=/↑	↑	–	↓/=/↑	–	–	↓	–	↑	–	↓
DISC1 Q31 L	–	=	–	–	–	–	–	↓	–	↑	–
DISC1 L100P	–	↑	↓	↓	↓	–	–	–	–	=	↑
DISC1-C Tg	–	–	↓	–	–	–	–	↓	–	↑	–
DISC1-DN	=	↑	=	–	=	–	–	=	↓	↑	↓
DISC1-hu Tg	=	↑	↓	–	=	=	↓	↓	–	–	–
DISC1-(tr) Tg	–	–	–	↓	–	–	–	–	–	–	–
NRG1–TM Het	=	↑	=	–	↓	↓	↑	=	↑	↑	–
NRG1 III Het	–	=	↓	–	↓	–	–	–	–	–	–
NRG1 Ig Het	–	=	–	↓	–	–	–	–	–	–	↑
NRG1 EGF Het	–	↑	–	–	=	–	–	–	–	–	–
NRG1 (I) Tg	–	↑	–	–	=	–	↑	↓	–	–	–
NRG1 (I-Thy1) Tg	=	=	–	–	↓	–	–	–	–	–	–
ErbB4 Het	–	=/↑	↓	–	–	–	–	–	–	–	–
ErbB4 PV	–	↑	↓	–	↓	–	–	–	–	–	–

↑, increased relative to wildtype; ↓, decreased relative to wildtype; =, no difference; /, separates different findings; -, not reported; *PPI* prepulse inhibition, *DAO* D-amino acid oxidase, *DISC1* disrupted-in-schizophrenia-1, *DISC1 C Tg* DISC1 C-terminal fragment transgenic mice, *DISC1-DN* DISC1 dominant-negative mice, *DISC1-hu Tg* DISC1 human form transgenic mice, *DISC1-(tr)* truncated DISC1 transgenic mice, *NRG1-TM* neuregulin 1 transmembrane domain, *Het* heterozygous mutants, *NRG1-EGF* neuregulin1 epidermal growth factor domain, *NRG1 (I)* neuregulin1 type I over-expressing mice, *NRG1 (I-Thy1) Tg* neuregulin1 type 1 (Thy1 promoter driven) over-expressing mice, *ErbB4 PV* mice with ablation of parvalbumin-positive ErbB4. [a] Includes investigation of stationary mouse within a cage and free social interaction tests. [b] includes dopamine-enhancing agents, NMDA receptor antagonists

In addition to inducible and conditional systems, the in utero gene transfer technique, with expression and/or shRNA constructs for RNAi, may prove a promising methodological innovation by enabling the selection of target cell populations for genetic modulation, while the introduction of inducible expression constructs will allow better temporal control. Furthermore, experimental designs for future studies on mutant models of psychosis would be improved, where appropriate, by the incorporation of a rescue strategy aimed at restoring or compensating for the loss-of-gene function; this would assist in providing mechanistic insight into the role of the gene in mediating schizophrenia-related phenotypes and the development of novel antipsychotic targets. By using the technical advantage of in utero transfer, in which expression of more than one gene can be modified at one time by cotransfection of expression and/or RNAi constructs (Shu et al. 2004; Young-Pearse et al. 2007), it should prove possible to test the synergistic or other epistatic effects of multiple genetic factors.

Evidence from the emerging field of molecular epigenomics suggests a putative deficit in microRNA processing in schizophrenia. MicroRNAs are small, non-coding RNAs that are believed to target at least a third of protein coding genes and are associated with the regulation of mRNA levels for multiple genes in the human cerebral cortex (Farh et al. 2005; Zhang and Su 2008). Dysregulation of miRNA expression may account for some of the neurodevelopmental aspects of psychiatric disorders and the difficulty in identifying individual causative genes (Dinan 2010; Miller and Wahlestedt 2010). For example, disruption of miRNA biogenesis, as a result of conditional deletion of DiGeorge syndrome critical region gene 8 (*Dgcr8*), has been associated with the expression of behavioural and structural brain endophenotypes related to schizophrenia (Stark et al. 2008). Increased study of the role of miRNAs in regulating phenotypic expression in existing mutant models has the potential to provide further insight into the mechanisms by which genes may contribute to specific aspects of the schizophrenia phenotype.

6.3 Examination of Gene × Gene Interactions in Schizophrenia

Assuming a polygenic basis for schizophrenia, neither partial nor complete loss-of-function or over-expression of any single gene in mice is likely to produce an animal model with construct validity for the disease; simultaneous dysregulation of several susceptibility genes likely better reflects the genetic contribution to risk for developing the disorder, emphasising the importance of generating and phenotypically characterising compound heterozygous mutants. In these animals, partial loss- or gain-of-function in several components could perhaps mimic more precisely the etiopathologic mechanisms, as well as the pathophysiological features of schizophrenia. Evidence from cellular and molecular studies conducted till date provides an important framework to understand how several common susceptibility genes and/or multiple, rare CNVs might converge functionally onto common pathways that may reflect the etiopathobiology of schizophrenia.

For example, there is evidence to connect NRG1, AKT, and DISC1 in a common pathway that may regulate neurodevelopment and contribute to susceptibility to schizophrenia (Kim et al. 2009) Future characterisation of mice with simultaneous dysregulation of several risk genes may provide novel insights into whether those risk genes may contribute to unique phenotypes and/or whether they might contribute in a synergistic manner with other risk genes to the expression of schizophrenia-related endophenotypes.

7 Conclusions

Increasingly, the evidence from mutant studies is showing that targeted and transient ontogenetic challenges of certain spatial and temporal specificity may be sufficient to invoke a disease process (Thompson and Levitt 2010). Additionally, recent technical advances allowing manipulation of a single gene in a regionally or temporally specific manner has provide the opportunity to assess how specific risk genes might interact with environmental factors to produced schizophrenia-relevant phenotypes. Clearly, the construct validity of any genetic model is determined by the degree to which the mouse gene mutation successfully mimics the disease-associated gene variant or mutation, as well as the available arsenal of accurate and reliable phenotyping instruments. This is particularly the case for a heterogenous and/or polygenic disorder such as schizophrenia; however, single and compound gene mutant models have enduring potential to inform on the contribution of a given risk gene or networks of genes to the expression of schizophrenia-related endophenotypes (Tables 1, 2).

Acknowledgments The authors' studies are supported by Science Foundation Ireland Principal Investigator grant 07/IN.1/B960 and Health Research Board of Ireland Postdoctoral Fellowship PD/2007/20.

References

Abazyan B, Nomura J, Kannan G et al (2010) Prenatal interaction of mutant DISC1 and immune activation produces adult psychopathology. Biol Psychiatry 68:1172–1181
Accortt EE, Freeman MP, Allen JJ (2008) Women and major depressive disorder: clinical perspectives on causal pathways. J Womens Health 17:1583–1590
Adler CM, Malhotra AK, Elman I et al (1999) Comparison of ketamine-induced thought disorder in healthy volunteers and thought disorder in schizophrenia. Am J Psychiatry 156:1646–1649
Allen NC, Bagade S, McQueen MB et al (2008) Systematic meta-analyses and field synopsis of genetic association studies in schizophrenia: the SzGene database. Nat Genet 40:827–834
Almond SL, Fradley RL, Armstrong EJ et al (2006) Behavioral and biochemical characterization of a mutant mouse strain lacking D-amino acid oxidase activity and its implications for schizophrenia. Mol Cell Neurosci 32:324–334
Amann LC, Gandal MJ, Halene TB et al (2010) Mouse behavioral endophenotypes for schizophrenia. Brain Res Bull 83:147–161

Arguello PA, Gogos JA (2006) Modeling madness in mice: one piece at a time. Neuron 52:179–196

Arguello PA, Gogos JA (2010) Cognition in mouse models of schizophrenia susceptibility genes. Schizophr Bull 36:289–300

Arseneault L, Cannon M, Witton J et al (2004) Causal association between cannabis and psychosis: examination of the evidence. Br J Psychiatry 184:110–117

Ayhan Y, Sawa A, Ross CA (2009) Animal models of gene-environment interactions in schizophrenia. Behav Brain Res 204:274–281

Ayhan Y, Abazyan B, Nomura J et al (2011) Differential effects of prenatal and postnatal expressions of mutant human DISC1 on neurobehavioral phenotypes in transgenic mice: evidence for neurodevelopmental origin of major psychiatric disorders. Mol Psychiatry 16:293–306

Babovic D, O'Tuathaigh CM, O'Connor AM et al (2008) Phenotypic characterization of cognition and social behavior in mice with heterozygous versus homozygous deletion of catechol-O-methyltransferase. Neuroscience 155:1021–1029

Babovic D, O'Tuathaigh CM, O'Sullivan GJ et al (2007) Exploratory and habituation phenotype of heterozygous and homozygous COMT knockout mice. Behav Brain Res 183:236–239

Ballard TM, Pauly-Evers M, Higgins GA et al (2002) Severe impairment of NMDA receptor function in mice carrying targeted point mutations in the glycine binding site results in drug-resistant nonhabituating hyperactivity. J Neurosci 22:6713–6723

Balu DT, Coyle JT (2011) Neuroplasticity signaling pathways linked to the pathophysiology of schizophrenia. Neurosci Biobehav Rev 35:848–870

Barros CS, Calabrese B, Chamero P et al (2009) Impaired maturation of dendritic spines without disorganization of cortical cell layers in mice lacking NRG1/ErbB signaling in the central nervous system. Proc Natl Acad Sci USA 106:4507–4512

Bassett AS, Scherer SW, Brzustowicz LM (2010) Copy number variations in schizophrenia: critical review and new perspectives on concepts of genetics and disease. Am J Psychiatry 167:899–914

Basu AC, Tsai GE, Ma CL et al (2009) Targeted disruption of serine racemase affects glutamatergic neurotransmission and behavior. Mol Psychiatry 14:719–727

Bauer D, Gupta D, Harotunian V et al (2008) Abnormal expression of glutamate transporter and transporter interacting molecules in prefrontal cortex in elderly patients with schizophrenia. Schizophr Res 104:108–120

Bay-Richter C, O'Tuathaigh CM, O'Sullivan G et al (2009) Enhanced latent inhibition in dopamine receptor-deficient mice is sex-specific for the D1 but not D2 receptor subtype: implications for antipsychotic drug action. Int J Neuropsychopharmacol 17:1–12

Beaulieu JM, Sotnikova TD, Yao WD et al (2004) Lithium antagonizes dopamine-dependent behaviors mediated by an AKT/glycogen synthase kinase 3 signaling cascade. Proc Natl Acad Sci U S A 101:5099–5104

Beaulieu JM, Sotnikova TD, Marion S (2005) An Akt/beta-arrestin 2/PP2A signaling complex mediates dopaminergic neurotransmission and behavior. Cell 122:261–273

Beaulieu JM, Gainetdinov RR, Caron MG (2009) Akt/GSK3 signaling in the action of psychotropic drugs. Annu Rev Pharmacol Toxicol 49:327–347

Belforte JE, Zsiros V, Sklar ER et al (2010) Postnatal NMDA receptor ablation in corticolimbic interneurons confers schizophrenia-like phenotypes. Nat Neurosci 13:76–83

Ben Abdallah NM, Fuss J, Trusel M et al (2011) The puzzle box as a simple and efficient behavioral test for exploring impairments of general cognition and executive functions in mouse models of schizophrenia. Exp Neurol 227:42–52

Benneyworth MA, Basu AC, Coyle JT (2011) Discordant behavioral effects of psychotomimetic drugs in mice with altered NMDA receptor function. Psychopharmacology 213:143–153

Bertram L (2008) Genetic research in schizophrenia: new tools and future perspectives. Schizophr Bull 34:806–812

Bhardwaj SK, Baharnoori M, Sharif-Askari B et al (2009) Behavioral characterization of dysbindin-deficient sandy mice. Behav Brain Res 197:435–441

Bjarnadottir M, Misner DL, Haverfield-Gross S et al (2007) Neuregulin1 (NRG1) signaling through Fyn modulates NMDA receptor phosphorylation: differential synaptic function in NRG1 ± knock-outs compared with wild-type mice. J Neurosci 27:4519–4529
Boucher AA, Arnold JC, Duffy L et al (2007a) Heterozygous neuregulin 1 mice are more sensitive to the behavioural effects of Delta9-tetrahydrocannabinol. Psychopharmacology 192:325–336
Boucher AA, Hunt GE, Karl T et al (2007b) Heterozygous neuregulin 1 mice display greater baseline and Delta(9)-tetrahydrocannabinol-induced c-Fos expression. Neuroscience 149:861–70
Boucher AA, Hunt GE, Micheau J et al (2010) The schizophrenia susceptibility gene neuregulin 1 modulates tolerance to the effects of cannabinoids. Int J Neuropsychopharmacol 12:1–13
Braff D, Schork NJ, Gottesman II (2007) Endophenotyping schizophrenia. Am J Psychiatry 164:705–707
Brigman JL, Graybeal C, Holmes A (2010) Predictably irrational: assaying cognitive inflexibility in mouse models of schizophrenia. Front Neurosci 4:13
Brody SA, Dulawa SC, Conquet F (2004) Assessment of a prepulse inhibition deficit in a mutant mouse lacking mGlu5 receptors. Mol Psychiatry 9:35–41
Brown AS, Begg MD, Gravenstein S et al (2004) Serologic evidence of prenatal influenza in the etiology of schizophrenia. Arch Gen Psychiatry 61:774–780
Burmeister M (1999) Basic concepts in the study of diseases with complex genetics. Biol Psychiatry 45:522–532
Burmeister M, McInnis MG, Zöllner S (2008) Psychiatric genetics: progress amid controversy. Nat Rev Genet 9:527–540
Burrows EL, McOmish CE, Hannan AJ (2011) Gene-environment interactions and construct validity in preclinical models of psychiatric disorders. Prog Neuropsychopharmacol Biol Psychiatry 35:1376–1382
Cagniard B, Balsam PD, Brunner D et al (2006) Mice with chronically elevated dopamine exhibit enhanced motivation, but not learning, for a food reward. Neuropsychopharmacology 31:1362–1370
Cannon TD, van Erp TG, Bearden CE et al (2003) Early and late neurodevelopmental influences in the prodrome to schizophrenia: contributions of genes, environment, and their interactions. Schizophr Bull 29:653–669
Caspi A, Moffitt TE, Cannon M et al (2005) Moderation of the effect of adolescent-onset cannabis use on adult psychosis by a functional polymorphism in the catechol-O-methyltransferase gene: longitudinal evidence of a gene X environment interaction. Biol Psychiatry 57:1117–1127
Chen YW, Lai WS (2011) Behavioral phenotyping of v-akt murine thymoma viral oncogene homolog 1-deficient mice reveals a sex-specific prepulse inhibition deficit in females that can be partially alleviated by glycogen synthase kinase-3 inhibitors but not by antipsychotics. Neuroscience 174:178–189
Chen YJ, Johnson MA, Lieberman MD et al (2007) Type III neuregulin-1 is required for normal sensorimotor gating, memory-related behaviors, and corticostriatal circuit components. J Neurosci 28:6872–6883
Chourbaji S, Vogt MA, Fumagalli F et al (2008) AMPA receptor subunit 1 (GluR-A) knockout mice model the glutamate hypothesis of depression. FASEB J 22:3129–3134
Chubb JE, Bradshaw NJ, Soares DC (2008) The DISC locus in psychiatric illness. Mol Psychiatry 13:36–64
Clancy B, Finlay BL, Darlington RB et al (2007) Extrapolating brain development from experimental species to humans. Neurotoxicology 28:931–937
Clapcote SJ, Lipina TV, Millar JK et al (2007) Behavioral phenotypes of Disc1 missense mutations in mice. Neuron 54:387–402
Costa RM, Gutierrez R, de Araujo IE et al (2007) Dopamine levels modulate the updating of tastant values. Genes Brain Behav 6:314–320

Cox MM, Tucker AM, Tang J et al (2009) Neurobehavioral abnormalities in the dysbindin-1 mutant, sandy, on a C57BL/6 J genetic background. Genes Brain Behav 8:390–397
Coyle JT (2006) Glutamate and schizophrenia: beyond the dopamine hypothesis. Cell Mol Neurobiol 26:365–384
Craddock N, Owen MJ, O'Donovan MC (2006) The catechol-O-methyl transferase (COMT) gene as a candidate for psychiatric phenotypes: evidence and lessons. Mol Psychiatry 11:446–458
Deakin IH, Law AJ, Oliver PL et al (2009) Behavioural characterization of neuregulin 1 type I overexpressing transgenic mice. Neuroreport 20:1523–1528
DeRosse P, Hodgkinson CA, Lencz T et al (2007) Disrupted in schizophrenia 1 genotype and positive symptoms in schizophrenia. Biol Psychiatry 61:1208–1210
Desbonnet L, Waddington JL, O'Tuathaigh CM (2009) Mice mutant for genes associated with schizophrenia: common phenotype or distinct endophenotypes? Behav Brain Res 204:258–273
DeSteno DA, Schmauss C (2009) A role for dopamine D2 receptors in reversal learning. Neuroscience 162:118–127
Devito LM, Balu DT, Kanter BR et al (2011) Serine racemase deletion disrupts memory for order and alters cortical dendritic morphology. Genes Brain Behav 10:210–222
Dinan TG (2010) MicroRNAs as a target for novel antipsychotics: a systematic review of an emerging field. Int J Neuropsychopharmacol 23:1–10
Drew MR, Simpson EH, Kellendonk C et al (2009) Transient overexpression of striatal D2 receptors impairs operant motivation and interval timing. J Neurosci 27:7731–7739
Duan X, Chang JH, Ge S et al (2007) Disrupted-In-Schizophrenia 1 regulates integration of newly generated neurons in the adult brain. Cell 130:1146–1158
Duffy L, Cappas E, Scimone A et al (2008) Behavioral profile of a heterozygous mutant mouse model for EGF-like domain neuregulin 1. Behav Neurosci 122:748–759
Duffy L, Cappas E, Lai D et al (2010) Cognition in transmembrane domain neuregulin 1 mutant mice. Neuroscience 170:800–807
Duncan GE, Moy SS, Perez A et al (2004) Deficits in senosrimotor gating and tests of social behavior in a genetic model of reduced NMDA receptor function. Behav Brain Res 153:507–519
Duncan GE, Moy SS, Lieberman JA et al (2006) Effects of haloperidol, clozapine, and quetiapine on sensorimotor gating in a genetic model of reduced NMDA receptor function. Psychopharmacology 184:190–200
El-Ghundi M, O'Dowd BF, George SR (2007) Insights into the role of dopamine receptor systems in learning and memory. Rev Neurosci 18:37–66
Emamian ES, Hall D, Birnbaum MJ (2004) Convergent evidence for impaired AKT1-GSK3beta signaling in schizophrenia. Nat Genet 36:131–137
Etherton MR, Blaiss CA, Powell CM et al (2009) Mouse neurexin-1 alpha deletion causes correlated electrophysiological and behavioral changes consistent with cognitive impairments. Proc Natl Acad Sci USA 106:17998–18003
Farh KK, Grimson A, Jan C et al (2005) The widespread impact of mammalian MicroRNAs on mRNA repression and evolution. Science 310:1817–1821
Fatemi SH, Folsom TD (2009) The neurodevelopmental hypothesis of schizophrenia, revisited. Schizophr Bull 35:528–548
Fazzari P, Paternain AV, Valiente M et al (2010) Control of cortical GABA circuitry development by Nrg1 and ErbB4 signalling. Nature 464:1376–1380
Feng YQ, Zhou ZY, He X et al (2008) Dysbindin deficiency in sandy mice causes reduction of snapin and displays behaviors related to schizophrenia. Schizophr Res 106:218–228
Feyder M, Wiedholz L, Sprengel R et al (2007) Impaired associative fear learning in mice with complete loss or haploinsufficiency of AMPA GluR1 receptors. Front Behav Neurosci 1:4
Fitzgerald PJ, Barkus C, Feyder M et al (2010) Does gene deletion of AMPA GluA1 phenocopy features of schizoaffective disorder? Neurobiol Dis 40:608–621

Fradley RL, O'Meara GF, Newman RJ et al (2005) STOP knockout and NMDA NR1 hypomorphic mice exhibit deficits in sensorimotor gating. Behav Brain Res 163:257–264
Gainetdinov RR (2008) Dopamine transporter mutant mice in experimental neuropharmacology. Naunyn Schmiedebergs Arch Pharmacol 377:301–313
Gajendran N, Kapfhammer JP, Lain E et al (2009) Neuregulin signaling is dispensable for NMDA- and GABA(A)-receptor expression in the cerebellum in vivo. J Neurosci 29:2404–2413
Garbett KA, Horváth S, Ebert PJ et al (2010) Novel animal models for studying complex brain disorders: BAC-driven miRNA-mediated in vivo silencing of gene expression. Mol Psychiatry 15:987–995
Gill M, Donohoe G, Corvin A (2009) What have the genomics ever done for psychoses? Psychol Med 40:529–540
Giros B, Jaber M, Jones SR et al (1996) Hyperlocomotion and indifference to cocaine and amphetamine in mice lacking the dopamine transporter. Nature 379:606–612
Glickstein SB, Hof PR, Schmauss C (2002) Mice lacking dopamine D2 and D3 receptors have spatial working memory deficits. J Neurosci 22:5619–5629
Gogos JA, Morgan M, Luine V et al (1998) Catechol-O-methyltransferase-deficient mice exhibit sexually dimorphic changes in catecholamine levels and behavior. Proc Natl Acad Sci USA 95:9991–9996
Goodwin RD, Fergusson DM, Horwood LJ (2004) Panic attacks and psychoticism. Am J Psychiatry 161:88–92
Gottesman II, Gould TD (2003) The endophenotype concept in psychiatry: etymology and stategic intentions. Am J Psychiatry 160:636–645
Gray L, Hannan AJ (2007) Dissecting cause and effect in the pathogenesis of psychiatric disorders: genes, environment and behaviour. Curr Mol Med 7:470–478
Gray L, van den Buuse M, Scarr E et al (2009) Clozapine reverses schizophrenia-related behaviours in the metabotropic glutamate receptor 5 knockout mouse: association with N-methyl-D-aspartic acid receptor up-regulation. Int J Neuropsychopharmacol 12:45–60
Grozeva D, Kirov G, Ivanov D et al (2010) Rare copy number variants: a point of rarity in genetic risk for bipolar disorder and schizophrenia. Arch Gen Psychiatry 67:318–327
Guo X, Hamilton PJ, Reish NJ et al (2009) Reduced expression of the NMDA receptor-interacting protein SynGAP causes behavioral abnormalities that model symptoms of schizophrenia. Neuropsychopharmacology 34:1658–1672
Haasio K, Huotari M, Nissinen E (2003) Tissue histopathology, clinical chemistry and behaviour of adult Comt-gene-disrupted mice. J Appl Toxicol 23:213–219
Hajós M, Rogers BN (2010) Targeting alpha7 nicotinic acetylcholine receptors in the treatment of schizophrenia. Curr Pharm Des 16:538–554
Halberstadt AL, Geyer MA (2009) Habituation and sensitization of acoustic startle: opposite influences of dopamine D1 and D2 family receptors. Neurobiol Learn Mem 92:243–248
Hall FS, Sora I, Uhl GR (2003) Sex-dependent modulation of ethanol consumption in vesicular monoamine transporter 2 (VMAT2) and dopamine transporter (DAT) knockout mice. Neuropsychopharmacology 28:620–628
Harrison PJ, Law AJ (2006) Neuregulin 1 and schizophrenia: genetics, gene expression, and neurobiology. Biol Psychiatry 60:132–140
Harrison PJ, Tunbridge EM (2008) Catechol-O-methyltransferase (COMT): a gene contributing to sex differences in brain function, and to sexual dimorphism in the predisposition to psychiatric disorders. Neuropsychopharmacology 33:3037–3045
Harvey PD, Wingo AP, Burdick KE, Baldesarini RJ (2010) Cognition and disability in bipolar disorder: lessons from schizophrenia research. Bipolar Disord 12:364–375
Hashimoto A, Yoshikawa M, Niwa A et al (2005) Mice lacking D-amino acid oxidase activity display marked attenuation of stereotypy and ataxia induced by MK-801. Brain Res 1033:210–215

Hattori S, Murotani T, Matsuzaki S et al (2008) Behavioral abnormalities and dopamine reductions in sdy mutant mice with a deletion in Dtnbp1, a susceptibility gene for schizophrenia. Biochem Biophys Res Commun 373:298–302

Hennah W, Porteous D (2009) The DISC1 pathway modulates expression of neurodevelopmental, synaptogenic and sensory perception genes. PLoS One 4:e4906

Hennah W, Thomson P, McQuillin A et al (2009) DISC1 association, heterogeneity and interplay in schizophrenia and bipolar disorder. Mol Psychiatry 14:865–873

Henquet C, Murray R, Linszen D, van Os J (2005) The environment and schizophrenia: the role of cannabis use. Schizophr Bull 31:608–612

Hikida T, Jaaro-Peled H, Seshadri S et al (2007) Dominant-negative DISC1 transgenic mice display schizophrenia-associated phenotypes detected by measures translatable to humans. Proc Natl Acad Sci USA 104:14501–14506

Hironaka N, Ikeda K, Sora I et al (2004) Food-reinforced operant behavior in dopamine transporter knockout mice: enhanced resistance to extinction. Ann N Y Acad Sci 1025:140–145

Holmes A, Lachowicz JE, Sibley DR (2004) Phenotypic analysis of dopamine receptor knockout mice; recent insights into the functional specificity of dopamine receptor subtypes. Neuropharmacology 47:1117–1134

Huotari M, Santha M, Lucas LR et al (2002) Effect of dopamine uptake inhibition on brain catecholamine levels and locomotion in catechol-O-methyltransferase-disrupted mice. J Pharmacol Exp Ther 303:1309–1316

Hur EM, Zhou FQ (2010) GSK3 signalling in neural development. Nat Rev Neurosci 11:539–551

Ibi D, Nagai T, Koike H et al (2010) Combined effect of neonatal immune activation and mutant DISC1 on phenotypic changes in adulthood. Behav Brain Res 206:32–37

Ichtchenko K, Hata Y, Nguyen T et al (1995) Neuroligin 1: a splice site-specific ligand for beta-neurexins. Cell 81:435–443

Ikeda M, Aleksic B, Kirov G et al (2010) Copy number variation in schizophrenia in the Japanese population. Biol Psychiatry 67:283–286

International Schizophrenia Consortium (2008) Rare chromosomal deletions and duplications increase risk of schizophrenia. Nature 455:237–241

International Schizophrenia Consortium, Purcell SM, Wray NR et al (2009) Common polygenic variation contributes to risk of schizophrenia and bipolar disorder. Nature 460:748–52

Javitt DC, Balla A, Burch S et al (2004) Reversal of phencyclidine-induced dopaminergic dysregulation by N-methyl-D-aspartate receptor/glycine-site agonists. Neuropsychopharmacology 29:300–307

Jentsch JD, Trantham-Davidson H, Jairl C et al (2009) Dysbindin modulates prefrontal cortical glutamatergic circuits and working memory function in mice. Neuropsychopharmacology 34:2601–2608

Ji Y, Yang F, Papaleo F et al (2009) Role of dysbindin in dopamine receptor trafficking and cortical GABA function. Proc Natl Acad Sci USA 106:19593–19598

Johnstone M, Thomson PA, Hall J et al (2011) DISC1 in schizophrenia: genetic mouse models and human genomic imaging. Schizophr Bull 37:14–20

Karayiorgou M, Gogos JA (2006) Schizophrenia genetics: uncovering positional candidate genes. Eur J Hum Genet 14:512–519

Karl T, Duffy L, Scimone A et al (2007) Altered motor activity, exploration and anxiety in heterozygous neuregulin 1 mutant mice: implications for understanding schizophrenia. Genes Brain Behav 6:677–687

Karlsgodt KH, Robleto K, Trantham-Davidson H et al (2011) Reduced dysbindin expression mediates N-methyl-D-aspartate receptor hypofunction and impaired working memory performance. Biol Psychiatry 69:28–34

Karlsson RM, Tanaka K, Heilig M et al (2008) Loss of glial glutamate and aspartate transporter (excitatory amino acid transporter 1) causes locomotor hyperactivity and exaggerated responses to psychotomimetics: rescue by haloperidol and metabotropic glutamate 2/3 agonist. Biol Psychiatry 64:810–814

Karlsson RM, Tanaka K, Saksida LM et al (2009) Assessment of glutamate transporter GLAST (EAAT1)-deficient mice for phenotypes relevant to the negative and executive/cognitive symptoms of schizophrenia. Neuropsychopharmacology 34:1578–1589

Kato T, Kasai A, Mizuno M et al (2010) Phenotypic characterization of transgenic mice overexpressing neuregulin-1. PLoS One 5:e14185

Kegeles LS, Abi-Dargham A, Zea-Ponce Y et al (2000) Modulation of amphetamine-induced striatal dopamine release by ketamine in humans: implications for schizophrenia. Biol Psychiatry 48:627–640

Kember RL, Fernandes C, Tunbridge EM et al (2010) A B2 SINE insertion in the Comtl gene (Comtl(B2i)) results in an overexpressing, behavior modifying allele present in classical inbred mouse strains. Genes Brain Behav 9:925–932

Kim JY, Duan X, Liu CY et al (2009) DISC1 regulates new neuron development in the adult brain via modulation of AKT-mTOR signaling through KIAA1212. Neuron 63:761–773

Kirby B, Waddington JL, O'Tuathaigh CMP (2010) Advancing a functional genomics for schizophrenia: psychopathological and cognitive phenotypes in mutants with gene disruption. Brain Res Bull 83:162–176

Kirov G, Gumus D, Chen W et al (2008) Comparative genome hybridization suggests a role for NRXN1 and APBA2 in schizophrenia. Hum Mol Genet 17:458–465

Kirov G, Grozeva D, Norton N et al (2009a) Support for the involvement of large copy number variants in the pathogenesis of schizophrenia. Hum Mol Genet 18:1497–1503

Kirov G, Rujescu D, Ingason A et al (2009b) Neurexin 1 (NRXN1) deletions in schizophrenia. Schizophr Bull 35:851–4

Koike H, Arguello PA, Kvajo M et al (2006) Disc1 is mutated in the 129S6/SvEv strain and modulates working memory in mice. Proc Natl Acad Sci USA 103:3693–3697

Knuesel I (2010) Reelin-mediated signaling in neuropsychiatric and neurodegenerative diseases. Prog Neurobiol 91:257–274

Krivoy A, Fischel T, Weizman A (2008) The possible involvement of metabotropic glutamate receptors in schizophrenia. Eur Neuropsychopharmacol 18:395–405

Kruzich PJ, Grandy DK (2004) Dopamine D2 receptors mediate two-odor discrimination and reversal learning in C57BL/6 mice. BMC Neurosci 5:12

Kvajo M, McKellar H, Arguello PA et al (2008) A mutation in mouse Disc 1 that models a schizophrenia risk allele leads to specific alterations in neuronal architecture and cognition. Proc Natl Acad Sci USA 105:7076–7081

Labrie V, Lipina T, Roder JC (2008) Mice with reduced NMDA receptor glycine affinity model some of the negative and cognitive symptoms of schizophrenia. Psychopharmacology 200:217–230

Labrie V, Roder JC (2010) The involvement of the NMDA receptor D-serine/glycine site in the pathophysiology and treatment of schizophrenia. Neurosci Biobehav Rev 34:351–372

Laviola G, Adriani W, Gaudino C et al (2006) Paradoxical effects of prenatal acetylcholinesterase blockade on neuro-behavioral development and drug-induced stereotypies in reeler mutant mice. Psychopharmacology 187:331–344

Levinson DF, Duan J, Oh S et al (2011) Copy number variants in schizophrenia: confirmation of five previous findings and new evidence for 3q29 microdeletions and VIPR2 duplications. Am J Psychiatry 168:302–316

Lewis CM, Levinson DF, Wise LH et al (2003) Genome scan meta-analysis of schizophrenia and bipolar disorder, part II: schizophrenia. Am J Hum Genet 73:34–48

Li W, Zhang Q, Oiso N et al (2003) Hermansky-Pudlak syndrome type 7 (HPS-7) results from mutant dysbindin, a member of the biogenesis of lysosome-related organelles complex 1 (BLOC-1). Nat Genet 35:84–89

Li W, Zhou Y, Jentsch JD et al (2007) Specific developmental disruption of disrupted-in-schizophrenia-1 function results in schizophrenia-related phenotypes in mice. Proc Natl Acad Sci USA 104:18280–18285

Li B, Devidze N, Barengolts D et al (2009) NMDA receptor phosphorylation at a site affected in schizophrenia controls synaptic and behavioral plasticity. J Neurosci 29:11965–11972

Li Z, Mulligan MK, Wang X (2010) A transposon in comt generates mRNA variants and causes widespread expression and behavioral differences among mice. PLoS One 5:e12181

Lipina TV, Niwa M, Jaaro-Peled H et al (2010) Enhanced dopamine function in DISC1-L100P mutant mice: implications for schizophrenia. Genes Brain Behav 9:777–789

Lijam N, Paylor R, McDonald MP et al (1997) Social interaction and sensorimotor gating abnormalities in mice lacking Dvl1. Cell 90:895–905

Long JM, LaPorte P, Paylor R et al (2004) Expanded characterization of the social interaction abnormalities in mice lacking Dvl1. Genes Brain Behav 3:51–62

Long JM, LaPorte P, Merscher S et al (2006) Behavior of mice with mutations in the conserved region deleted in velocardiofacial/DiGeorge syndrome. Neurogenetics 7:247–257

Long LE, Chesworth R, Arnold JC et al (2010) A follow-up study: acute behavioural effects of Delta (9)-THC in female heterozygous neuregulin 1 transmembrane domain mutant mice. Psychopharmacology 211:277–289

Low NC, Hardy J (2007) What is a schizophrenic mouse? Neuron 54:348–349

Maekawa M, Okamura T, Kasai N et al (2005) D-amino-acid oxidase is involved in D-serine-induced nephrotoxicity. Chem Res Toxicol 18:1678–1682

Malhotra AK, Adler CM, Kennison SD et al (1997a) Clozapine blunts N-methyl-D-aspartate antagonist-induced psychosis: a study with ketamine. Biol Psychiatry 42:664–668

Meechan DW, Maynard TM, Gopalakrishna D et al (2007a) When half is not enough: gene expression and dosage in the 22q11 deletion syndrome. Gene Expr 13:299–310

Manolio TA, Rodriguez LL, Brooks L et al (2007) New models of collaboration in genome-wide association studies: the Genetic Association Information Network. Nat Genet 39:1045–1051

Malhotra AK, Pinals DA, Adler CM et al (1997b) Ketamine-induced exacerbation of psychotic symptoms and cognitive impairment in neuroleptic-free schizophrenics. Neuropsychopharmacology 17:141–150

Mao Y, Ge X, Frank CL et al (2009) Disrupted in schizophrenia 1 regulates neuronal progenitor proliferation via modulation of GSK3beta/beta-catenin signaling. Cell 136:1017–1031

Mathews TA, John CE, Lapa GB (2006) No role of the dopamine transporter in acute ethanol effects on striatal dopamine dynamics. Synapse 60:288–294

Meechan DW, Maynard TM, Gopalakrishna D et al (2007b) When half is not enough: gene expression and dosage in the 22q11 deletion syndrome. Gene Expr 13:299–310

Mei L, Xiong WC (2008) Neuregulin 1 in neural development, synaptic plasticity and schizophrenia. Nat Rev Neurosci 9:437–452

Meyer U, Feldon J, Fatemi SH (2009) In vivo rodent models for the experimental investigation of prenatal immune activation effects in neurodevelopmental brain disorders. Neurosci Biobehav Rev 33:1061–1079

Miller BH, Wahlestedt C (2010) MicroRNA dysregulation in psychiatric disease. Brain Res 1338:89–99

Miyamoto Y, Nabeshima T (2002) Analysis of neuronal functions in mice lacking the NMDA receptor epsilon 1 subunit. Nippon Yakurigaku Zasshi 119:327–335

Mohn AR, Gainetdinov RR, Caron MG et al (1999) Mice with reduced NMDA receptor expression display behaviors related to schizophrenia. Cell 98:427–436

Moore TH, Zammit S, Lingford-Hughes A (2007) Cannabis use and risk of psychotic or affective mental health outcomes: a systematic review. Lancet 370:319–328

Muir WJ, Pickard BS, Blackwood DH (2008) Disrupted-in-Schizophrenia-1. Curr Psychiatry Rep 10:140–147

Murotani T, Ishizuka T, Hattori S et al (2007) High dopamine turnover in the brains of Sandy mice. Neurosci Lett 421:47–51

Myin-Germeys I, Delespaul P, van Os J (2005) Behavioural sensitization to daily life stress in psychosis. Psychol Med 35:733–741

Myin-Germeys I, Oorschot M, Collip D et al (2009) Experience sampling research in psychopathology: opening the black box of daily life. Psychol Med 39:1533–1547

Niwa M, Kamiya A, Murai R et al (2010) Knockdown of DISC1 by in utero gene transfer disturbs postnatal dopaminergic maturation in the frontal cortex and leads to adult behavioral deficits. Neuron 65:480–489

Numakawa T, Yagasaki Y, Ishimoto T (2004) Evidence of novel neuronal functions of dysbindin, a susceptibility gene for schizophrenia. Hum Mol Genet 13:2699–2708

Ognibene E, Adriani W, Macrì S et al (2007) Neurobehavioural disorders in the infant reeler mouse model: interaction of genetic vulnerability and consequences of maternal separation. Behav Brain Res 177:142–149

Oliver PL, Davies KE (2009) Interaction between environmental and genetic factors modulates schizophrenic endophenotypes in the Snap-25 mouse mutant blind-drunk. Hum Mol Genet 18:4576–4589

O'Sullivan, GJ, O'Tuathaigh C, Tomiyama K, Koshikawa N, Waddington JL (2010) Dopamine receptor subtypes and behaviour: from psychopharmacology to mutant models. In: Neve K (ed) The dopamine receptors, pp 323–371. Humana Press, Totowa

O'Tuathaigh CM, O'Sullivan GJ, Kinsella A et al (2006) Sexually dimorphic changes in the exploratory and habituation profiles of heterozygous neuregulin-1 knockout mice. Neuroreport 17:79–83

O'Tuathaigh CMP, Babovic D, O'Meara G et al (2007a) Susceptibility genes for schizophrenia: phenotypic characterisation of mutant models. Neurosci Biobehav Rev 31:60–78

O'Tuathaigh CM, Babovic D, O'Sullivan GJ et al (2007b) Phenotypic characterization of spatial cognition and social behavior in mice with 'knockout' of the schizophrenia risk gene neuregulin 1. Neuroscience 147:18–27

O'Tuathaigh CM, O'Connor AM, O'Sullivan GJ et al (2008) Disruption to social dyadic interactions but not emotional/anxiety-related behaviour in mice with heterozygous 'knockout' of the schizophrenia risk gene neuregulin-1. Prog Neuropsychopharmacol Biol Psychiatry 32:462–466

O'Tuathaigh CM, Desbonnet L, Waddington JL (2009) Neuregulin-1 signalling in schizophrenia: 'Jack of all trades' or master of some? Expert Rev Neurother 9:1–3

O'Tuathaigh CMP, Waddington JL (2010) Mutant mouse models: phenotypic relationships to domains of psychopathology and pathobiology in schizophrenia. Schizophr Bull 36:243–245

O'Tuathaigh CMP, Kirby BP, Moran PM et al (2010a) Mutant mouse models: genotype-phenotype relationships to negative symptoms in schizophrenia. Schizophr Bull 36:271–288

O'Tuathaigh CMP, Harte M, Tighe O et al (2010b) Schizophrenia-related endophenotypes in heterozygous neuregulin-1 'knockout' mice: NMDA-receptor antagonist effects, neurochemistry and brain structure. Eur J Neurosci 31:349–58

O'Tuathaigh CM, Hryniewiecka M, Behan A et al (2010c) Chronic adolescent exposure to delta 9-tetrahydrocannabinol in COMT knockout mice: Impact on phenotypes relevant to psychosis. Neuropsychopharmacology 35:2262–2273

Owen MJ, Craddock N, O'Donovan MC (2010) Suggestion of roles for both common and rare risk variants in genome-wide studies of schizophrenia. Arch Gen Psychiatry 67:667–673

Papaleo F, Crawley JN, Song J et al (2008) Genetic dissection of the role of catechol-O-methyltransferase in cognition and stress reactivity in mice. J Neurosci 28:8709–8723

Papaleo F, Yang F, Garcia S et al (2012) Dysbindin-1 modulates prefrontal cortical activity and schizophrenia-like behaviors via dopamine/D2 pathways. Mol Psychiatry 17:85–98

Patil ST, Zhang L, Martenyi F et al (2007) Activation of mGlu2/3 receptors as a new approach to treat schizophrenia: a randomized Phase 2 clinical trial. Nat Med 13:1102–1107

Paylor R, Glaser B, Mupo A et al (2006) Tbx1 haploinsufficiency is linked to behavioral disorders in mice and humans: implications for 22q11 deletion syndrome. Proc Natl Acad Sci USA 103:7729–7734

Pilowsky LS, Bressan RA, Stone JM et al (2007) First in vivo evidence of an NMDA receptor deficit in medication-free schizophrenic patients. Mol Psychiatry 11:118–119

Pinard E, Alanine A, Alberati D et al (2010) Selective GlyT1 inhibitors: discovery of [4-(3-fluoro-5-trifluoromethylpyridin-2-yl)piperazin-1-yl][5-methanesulfonyl-2-((S)-2,2,2-trifluoro-1-methylethoxy)phenyl]methanone (RG1678), a promising novel medicine to treat schizophrenia. J Med Chem 53:4603–4614

Pletnikov MV, Ayhan Y, Nikolskaia O et al (2008) Inducible expression of mutant human DISC1 in mice is associated with brain and behavioural abnormalities reminiscent of schizophrenia. Mol Psychiatry 13:173–186

Ralph RJ, Varty GB, Kelly MA et al (1999) The dopamine D2, but not D3 or D4, receptor subtype is essential for the disruption of prepulse inhibition produced by amphetamine in mice. J Neurosci 19:4627–4633

Ralph-Williams RJ, Lehmann-Masten V, Otero-Corchon V et al (2002) Differential effects of direct and indirect dopamine agonists on prepulse inhibition: a study in D1 and D2 receptor knockout mice. J Neurosci 22:9604–9611

Rimer M, Barrett DW, Maldonado MA, Vock VM et al (2005) Neuregulin-1 immunoglobulin-like domain mutant mice: clozapine sensitivity and impaired latent inhibition. Neuroreport 16:271–275

Rodriguiz RM, Chu R, Caron MG et al (2004) Aberrant responses in social interaction of dopamine transporter knockout mice. Behav Brain Res 148:185–198

Ross CA, Margolis RL, Reading SA et al (2006) Neurobiology of schizophrenia. Neuron 52:139–153

Rujescu D, Ingason A, Cichon S et al (2009) Disruption of the neurexin 1 gene is associated with schizophrenia. Hum Mol Genet 18:988–996

Sagata N, Iwaki A, Aramaki T et al (2010) Comprehensive behavioural study of GluR4 knockout mice: implication in cognitive function. Genes Brain Behav 9:899–909

Savelieva KV, Caudle WM, Findlay GS (2002) Decreased ethanol preference and consumption in dopamine transporter female knock-out mice. Alcohol Clin Exp Res 26:758–764

Seamans JK, Yang CR (2004) The principal features and mechanisms of dopamine modulation in the prefrontal cortex. Prog Neurobiol 74:1–58

Shen S, Lang B, Nakamoto C et al (2008) Schizophrenia-related neural and behavioural phenotypes in transgenic mice expressing truncated DISC1. J Neurosci 28:10893–10904

Shu T, Ayala R, Nguyen MD et al (2004) Ndel1 operates in a common pathway with LIS1 and cytoplasmic dynein to regulate cortical neuronal positioning. Neuron 44:263–277

Singer P, Boison D, Möhler H (2007) Enhanced recognition memory following glycine transporter 1 deletion in forebrain neurons. Behav Neurosci 121:815–825

Singer P, Feldon J, Yee BK et al (2009) Are DBA/2 mice associated with schizophrenia-like endophenotypes? A behavioural contrast with C57BL/6 mice. Psychopharmacology 206:677–698

Singer P, Boison D, Möhler H et al (2011) Modulation of sensorimotor gating in prepulse inhibition by conditional brain glycine transporter 1 deletion in mice. Eur Neuropsychopharmacol 21:401–413

Smith RE, Haroutunian V, Davis KL et al (2001) Expression of excitatory amino acid transporter transcripts in the thalamus of subjects with schizophrenia. Am J Psychiatry 158:1393–1399

Stark KL, Xu B, Bagchi A et al (2008) Altered brain microRNA biogenesis contributes to phenotypic deficits in a 22q11-deletion mouse model. Nat Genet 40:751–760

Stefansson H, Sigurdsson E, Steinthorsdottir V et al (2002) Neuregulin 1 and susceptibility to schizophrenia. Am J Hum Genet 71:877–892

Steinberg S, Mors O, Børglum AD et al (2011) Expanding the range of ZNF804A variants conferring risk of psychosis. Mol Psychiatry 16:59–66

Stone JM, Morrison PD, Pilowsky LS (2007) Glutamate and dopamine dysregulation in schizophrenia-a synthesis and selective review. J Psychopharmacol 21:440–452

Szeszko PR, Hodgkinson CA, Robinson DG et al (2008) DISC1 is associated with prefrontal cortical gray matter and positive symptoms in schizophrenia. Biol Psychol 79:103–110

Takao K, Yamasaki N, Miyakawa T (2007) Impact of brain-behavior phenotyping of genetically-engineered mice on research of neuropsychiatric disorders. Neurosci Res 58:124–132

Takao K, Toyama K, Nakanishi K et al (2008) Impaired long-term memory retention and working memory in sdy mutant mice with a deletion in Dtnbp1, a susceptibility gene for schizophrenia. Mol Brain 1:11

Takashima A (2009) Drug development targeting the glycogen synthase kinase-3beta (GSK-3beta)-mediated signal transduction pathway: role of GSK-3beta in adult brain. J Pharmacol Sci 109:174–178
Talbot K (2009) The sandy (sdy) mouse: a dysbindin-1 mutant relevant to schizophrenia research. Prog Brain Res 179:87–94
Tammimäki A, Forsberg MM, Karayiorgou M et al (2008) Increase in free choice oral ethanol self-administration in catechol-o-methyltransferase gene-disrupted male mice. Basic Clin Pharmacol Toxicol 103:297–304
Tandon R, Nasrallah HA, Keshavan MS (2009) Schizophrenia, "just the facts" 4 Clinical features and conceptualization. Schizophr Res 110:1–23
Tang J, LeGros RP, Louneva N et al (2009) Dysbindin-1 in dorsolateral prefrontal cortex of schizophrenia cases is reduced in an isoform-specific manner unrelated to dysbindin-1 mRNA expression. Hum Mol Genet 18:3851–3863
Taylor SB, Taylor AR, Markham JA et al (2011) Disruption of the neuregulin 1 gene in the rat alters HPA axis activity and behavioral responses to environmental stimuli. Physiol Behav 104:205–214
Thompson BL, Levitt P (2010) The clinical-basic interface in defining pathogenesis in disorders of neurodevelopmental origin. Neuron 67:702–712
Tsai G, Ralph-Williams RJ, Martina M et al (2004) Gene knockout of glycine transporter 1: characterization of the behavioral phenotype. Proc Natl Acad Sci U S A 101:8485–8490
Tsien JZ (2000) Linking Hebb's coincidence-detection to memory formation. Curr Opin Neurobiol 10:266–273
Tunbridge EM, Harrison PJ, Weinberger DR (2006) Catechol-o-methyltransferase, cognition, and psychosis: Val158Met and beyond. Biol Psychiatry 60:141–151
Tuominen HJ, Tiihonen J, Wahlbeck K (2005) Glutamatergic drugs for schizophrenia: a systematic review and meta-analysis. Schizophr Res 72:225–234
van den Buuse M, Wischhof L, Lee RX et al (2009) Neuregulin 1 hypomorphic mutant mice: enhanced baseline locomotor activity but normal psychotropic drug-induced hyperlocomotion and prepulse inhibition regulation. Int J Neuropsychopharmacol 12:1383–1393
Van den Buuse M (2010) Modelling the positive symptoms of schizophrenia in genetically-modified mice: pharmacology and methodology aspects. Schizophr Bull 36:246–270
van Os J, Kenis G, Rutten BP (2010) The environment and schizophrenia. Nature 468:203–212
Waddington JL, O'Tuathaigh C, O'Sullivan G et al (2005) Phenotypic studies on dopamine receptor subtype and associated signal transduction mutants: insights and challenges from 10 years at the psychopharmacology-molecular biology interface. Psychopharmacology 181:611–638
Waddington JL, Corvin AP, Donohoe G, O'Tuathaigh CMP, Mitchell KJ, Gill M (2007) Functional genomics and schizophrenia: endophenotypes and mutant models. Psychiat Clin N Amer 30:365–399
Waddington JL, Hennessy RJ, O'Tuathaigh CMP, Owoeye O, Russell V (2011) Schizophrenia and the lifetime trajectory of psychotic illness: developmental neuroscience and pathobiology, redux. In: Brown AS, Patterson PH (eds.) The Origins of Schizophrenia. Columbia University Press, New York (in press)
Walsh T, McClellan JM, McCarthy SE et al (2008) Rare structural variants disrupt multiple genes in neurodevelopmental pathways in schizophrenia. Science 320:539–543
Walss-Bass C, Liu W, Lew DF et al (2006) A novel missense mutation in the transmembrane domain of neuregulin 1 is associated with schizophrenia. Biol Psychiatry 60:548–553
Wang Y, Xu R, Sasaoka T et al (2000) Dopamine D2 long receptor-deficient mice display alterations in striatum-dependent functions. J Neurosci 20:8305–8314
Wang Q, Jaaro-Peled H, Sawa A (2008) How has DISC1 enabled drug discovery? Mol Cell Neurosci 37:187–195
Wen L, Lu YS, Zhu XH et al (2010) Neuregulin 1 regulates pyramidal neuron activity via ErbB4 in parvalbumin-positive interneurons. Proc Natl Acad Sci USA 107:1211–1216

Weickert CS, Straub RE, McClintock BW et al (2004) Human dysbindin (DTNBP1) gene expression in normal brain and in schizophrenic prefrontal cortex and midbrain. Arch Gen Psychiatry 61:544–555

Wiedholz LM, Owens WA, Horton RE et al (2008) Mice lacking the AMPA GluR1 receptor exhibit striatal hyperdopaminergia and 'schizophrenia-related' behaviors. Mol Psychiatry 13:631–640

Xing B, Kong H, Meng X et al (2010) Dopamine D1 but not D3 receptor is critical for spatial learning and related signaling in the hippocampus. Neuroscience 169:1511–1519

Xu R, Hranilovic D, Fetsko LA et al (2002) Dopamine D2S and D2L receptors may differentially contribute to the actions of antipsychotic and psychotic agents in mice. Mol Psychiatry 7:1075–1082

Xu B, Roos JL, Levy S et al (2008) Strong association of de novo copy number mutations with sporadic schizophrenia. Nat Genet 40:880–885

Xu B, Woodroffe A, Rodriguez-Murillo L (2009) Elucidating the genetic architecture of familial schizophrenia using rare copy number variant and linkage scans. Proc Natl Acad Sci USA 106:16746–16751

Yee BK, Balic E, Singer P et al (2006) Disruption of glycine transporter 1 restricted to forebrain neurons is associated with a procognitive and antipsychotic phenotypic profile. J Neurosci 26:3169–3181

Young JW, Crawford N, Kelly JS et al (2007) Impaired attention is central to the cognitive deficits observed in alpha 7 deficient mice. Eur Neuropsychopharmacol 17:145–155

Young-Pearse TL, Bai J, Chang R et al (2007) A critical function for beta-amyloid precursor protein in neuronal migration revealed by in utero RNA interference. J Neurosci 27:14459–14469

Zhang R, Su B (2008) MicroRNA regulation and the variability of human cortical gene expression. Nucleic Acids Res 36:4621–4628

Zhang M, Ballard ME, Basso AM et al (2011) Behavioral characterization of a mutant mouse strain lacking D-amino acid oxidase activity. Behav Brain Res 217:81–87

Zhou X, Nie Z, Roberts A et al (2010) Reduced NMDAR1 expression in the Sp4 hypomorphic mouse may contribute to endophenotypes of human psychiatric disorders. Hum Mol Genet 19:3797–3805

Zhuang X, Oosting RS, Jones SR et al (2001) Hyperactivity and impaired response habituation in hyperdopaminergic mice. Proc Natl Acad Sci U S A 98:1982–1987

Genetic Models of Sensorimotor Gating: Relevance to Neuropsychiatric Disorders

Susan B. Powell, Martin Weber and Mark A. Geyer

Abstract Sensorimotor gating, or the ability of a sensory event to suppress a motor response, can be measured operationally via prepulse inhibition (PPI) of the startle response. PPI is deficient in schizophrenia patients as well as other neuropsychiatric disorders, can be measured across species, and has been used widely as a translational tool in preclinical neuropharmacological and genetic research. First developed to assess drug effects in pharmacological and developmental models, PPI has become one of the standard behavioral measures in genetic models of schizophrenia and other neuropsychiatric disorders that exhibit PPI deficits. In this chapter we review the literature on genetic models of sensorimotor gating and discuss the utility of PPI as a tool in phenotyping mutant mouse models. We highlight the approaches to genetic mouse models of neuropsychiatric disease, discuss some of the important caveats to these approaches, and provide a comprehensive table covering the more recent genetic models that have evaluated PPI.

Keywords Prepulse inhibition · Startle · Mouse models · Schizophrenia · Genetic · Mutant

S. B. Powell (✉) · M. A. Geyer
Department of Psychiatry, University of California San Diego,
9500 Gilman Dr, La Jolla, CA 92093-0804, USA
e-mail: sbpowell@ucsd.edu

M. A. Geyer
e-mail: mgeyer@ucsd.edu

S. B. Powell · M. A. Geyer
Research Service, VA San Diego Healthcare System,
San Diego, CA, USA

M. Weber
Department of Neuroscience, Genentech Inc, 1 DNA Way,
South San Francisco, CA 94080-4990, USA
e-mail: weber.martin@gene.com

Curr Topics Behav Neurosci (2012) 12: 251–318
DOI: 10.1007/7854_2011_195

Published Online: 26 February 2012

Contents

1 Introduction: Definitions and Measures of Sensorimotor Gating 252
2 Utility of Prepulse Inhibition Measures in Genetic Models of Schizophrenia 254
  2.1 Relationship of PPI to Psychiatric Symptoms, Neurocognitive Measures, and Overall Function 254
  2.2 PPI as an Indicator of Neural Processes and a Pharmacological Screen 255
  2.3 PPI as a Tool to Evaluate Gene–Environment Interactions 256
  2.4 Evaluating the Role of a Susceptibility Gene in Pathology 262
  2.5 Methodological Considerations with Specific Relevance to Genetic Models of Sensorimotor Gating 275
3 Approaches to Genetic Models of Sensorimotor Gating 276
  3.1 Overview of Molecular Genetic Techniques 276
  3.2 Models Based on Hypotheses Regarding the Pathophysiology of Disease 277
  3.3 Genetic Mouse Models as Pharmacological Tools 295
  3.4 Genetic Models of Candidate Genes for Neuropsychiatric Diseases 296
  3.5 Phenotype-Based Models Revealing the Function of Genes 298
4 Discussion 301
References 303

1 Introduction: Definitions and Measures of Sensorimotor Gating

Sensorimotor gating refers to the regulation of sensory information as it is transmitted to motor output systems. When sensory information is processed centrally, it requires some degree of filtering or "gating" prior to accessing and impinging upon motor output. While this process has long been observed at multiple levels of biology, the synaptic and cellular mechanism of sensorimotor gating (Frost et al. 2003; Nusbaum and Contreras 2004; Rose and Scott 2003) as well as the functional implications (Braff 2010, 2011; Swerdlow et al. 2008) are starting to be elucidated. One form of sensorimotor gating that is widely studied in humans and animals is prepulse inhibition (PPI) of startle. PPI is a form of startle plasticity in which presentation of a weak stimulus (prepulse) preceding an intense startling stimulus (pulse) by 30–500 ms inhibits the startle response (Graham 1975; Hoffman and Ison 1980). The fundamental mechanism underlying this inhibition is thought to resemble the normal process of filtering incoming sensory stimuli (Geyer and Braff 1987). PPI levels may indicate the current integrity of sensorimotor gating mechanisms, providing an operational measure of sensorimotor gating. In humans, startle is measured from the eye blink response through electromyographic recordings of the orbicularis oculi muscle (Fridlund and Cacioppo 1986) and typically involves startle to acoustic stimuli, although tactile stimuli have been used as well (Braff et al. 1992; Kumari et al. 2003; Neuner et al. 2010; Swerdlow et al. 2001b). PPI deficits are observed in schizophrenia patients [for review see (Braff et al. 2001; Swerdlow et al. 2008)], their unaffected first degree relatives (Cadenhead et al. 2000), and patients with schizotypal personality disorder (Cadenhead et al. 1993). Additionally, several other neuropsychiatric disorders are associated with decreased PPI, including Obsessive–Compulsive Disorder (Swerdlow et al. 1993), Tourette's syndrome (Swerdlow et al. 2001b), Huntington's disease (Swerdlow et al. 1995), manic bipolar patients (Perry et al. 2001), Panic Disorder (Ludewig et al. 2002), Fragile X syndrome (Frankland et al. 2004; Hessl et al. 2009), and adults with autism (Perry et al. 2007). Although the core symptoms of these disorders are diverse, a feature common to all of them is deficient gating, with a gating deficit predominating in the cognitive sphere in some disorders and in the sensory or motor domains in others. Thus, deficient gating has been reported across a variety of neuropsychiatric disorders, with PPI deficits in schizophrenia patients being the best characterized and the most widely replicated (Braff et al. 2001; Kumari et al. 2008; Ludewig et al. 2003; Mackeprang et al. 2002; Swerdlow et al. 2008).

Sensorimotor gating abnormalities, as measured by PPI, are being used as an endophenotype in genetic studies of schizophrenia (Braff et al. 2007; Greenwood et al. 2011), and meet the criteria outlined for a viable endophenotype (Turetsky et al. 2007). With an increased focus on observable and measurable behaviors as rational approaches to genetic studies of heterogeneous neuropsychiatric diseases such as schizophrenia (Gottesman and Gould 2003), large-scale genetic studies are examining neurophysiological measures such as PPI, P50 auditory evoked suppression, antisaccade eye movement, mismatch negativity, and P300 event-related potential (Turetsky et al. 2007). Recent studies suggest that PPI is heritable (Hasenkamp et al. 2010) and associated with polymorphisms in the CHRNA3 gene (Petrovsky et al. 2010), neuregulin 1 (Roussos et al. 2011), and COMT (Giakoumaki et al. 2008; Quednow et al. 2008; Roussos et al. 2008). The use of PPI as an endophenotype in genetic studies of schizophrenia, combined with the observation that PPI has a strong genetic component in mice (Francis et al. 2003), indicates that PPI may be a useful behavioral phenotype to consider in genetic mouse models related to neuropsychiatric disease, particularly schizophrenia. While there are certainly many other symptoms, behavioral traits, and neurophysiological deficits observed across the heterogeneous group of patients with schizophrenia, PPI appears to be a viable endophenotype for human genetic studies and thus a reasonable approach to investigate in genetic animal models. Additionally, considering that PPI measures a fundamental component of information processing and is observable across many species, it is a useful endpoint with which to understand the more general impact of specific genes on neurobehavioral function and on the neural substrates underlying this function.

Mutant mouse models related to schizophrenia have been based primarily on the pathophysiology of schizophrenia, the known effects of antipsychotic drugs, and candidate genes for schizophrenia. In this review, we provide an overview of PPI in genetic mouse models, concentrating on the time period since the Swerdlow et al. (2008) review. We discuss the contribution and usefulness of PPI as a phenotype in the context of genetic mouse models and speculate about the significance of PPI deficits and future directions for the field. There have been previous reviews on candidate genes for schizophrenia (Arguello and Gogos 2010; O'Tuathaigh et al. 2007), and reviews focusing specifically on PPI, which summarized studies of strain differences, genetic mutants, and the pharmacology of PPI in mice (Geyer et al. 2002; Powell et al. 2009; Swerdlow et al. 2008). In this review, we highlight the approaches to genetic mouse models of neuropsychiatric disease, discuss some of the important caveats to these approaches, and provide a comprehensive table covering the genetic models that have evaluated PPI since the Swerdlow et al. (2008) review. To that end, we discuss PPI as a phenotype in genetic mouse models generated to address hypotheses regarding the pathophysiology of neuropsychiatric disease, candidate genes, and basic proteins involved in neural development or synaptic function. We also discuss the usefulness of PPI in phenotype-driven approaches in which a PPI phenotype could lead to "bottom up" approaches of identifying novel genes of relevance to PPI and/or neuropsychiatric disease [i.e. hypothesis-generating; or phenotype-genotype approaches (Jacobson and Cryan 2010)].

2 Utility of Prepulse Inhibition Measures in Genetic Models of Schizophrenia

Mutant mouse models of neuropsychiatric disease have targeted genes involved in the pathophysiology of the disease, candidate or susceptibility genes thought to be involved in the etiology of the disease, or genes involved in basic physiological processes. Because many of the target genes overlap at the functional level across different disorders (e.g. proteins involved in basic processes of neurodevelopment), manipulation of these genes may have relevance to many neuropsychiatric disorders. Thus, it is important to determine the relevant behavioral tests to best understand the underlying brain abnormalities resulting from genetic alterations of proteins. Second, it is critical to develop animal models to screen novel therapeutics for specific domains of function. In this context, PPI has emerged as a useful behavior with which to assess the integrity of basic neural circuits in genetic mouse models and to screen for pharmacological agents, particularly in the context of identifying treatments for schizophrenia.

2.1 Relationship of PPI to Psychiatric Symptoms, Neurocognitive Measures, and Overall Function

Attempts to relate PPI deficits to the positive, negative, and cognitive symptoms of schizophrenia have yielded mixed results (Thaker 2007). PPI negatively correlates with thought disorder (Meincke et al. 2004; Perry and Braff 1994; Perry et al. 1999) and distractibility (Karper et al. 1996) in schizophrenia. PPI increases observed with risperidone treatment correlated with improvements in PANSS general psychopathology subscale scores (Martinez-Gras et al. 2009) in schizophrenia patients. In a well-powered study with over 300 subjects, PPI did not correlate with cognitive measures using traditional "pen and paper" tests [i.e. Wisconsin card sorting task (WCST), California verbal learning task, etc.], but did correlate positively with global assessment of function (GAF) and Independent Living scales (Swerdlow et al. 2006). There have been some studies demonstrating a relationship between cognitive constructs and PPI levels. For example, converging evidence indicates that PPI is correlated with strategy formation and execution time in the Cambridge Neuropsychological Test Automated Battery (CANTAB) in healthy controls (Bitsios et al. 2006; Csomor et al. 2008; Giakoumaki et al. 2006), a finding that should be examined further in patient populations. The cognitive neuroscience treatment research to improve cognition in Schizophrenia (CNTRICS) program funded by the National institute of mental health considered PPI to provide a measure of the cognitive construct of "gain control" as a specific aspect of the perceptual abnormalities seen in patients with schizophrenia (Green et al. 2009). The series of CNTRICS workshops concluded that PPI may have utility as a biomarker for use in proof of concept studies of potential treatments for the cognitive deficits in schizophrenia that are not ameliorated by existing antipsychotic drugs. For the purpose of evaluating genetic mouse models of neuropsychiatric disease, the more useful comparison to make is not between PPI and specific symptoms of schizophrenia, but rather the relationship between a gene and the observable dependent measure, i.e. PPI. The approach of using endophenotypes in genetic studies has greatly strengthened the ability to conduct cross-species translational studies by providing specific observables for study in experimental animals [reviewed in (Geyer and Markou 2002; Gould and Gottesman 2006) Waddington, chapter in this book]. Useful endophenotypes in this context are measures that are observed in humans and can be measured in animals.

2.2 PPI as an Indicator of Neural Processes and a Pharmacological Screen

A PPI deficit in a genetic mutant could indicate that the gene may be involved in the neural circuitry known to modulate PPI [e.g. cortical, limbic, striatal (Swerdlow et al. 2001a)]; in other words it could function as a "surrogate measure for neural processes" as Swerdlow et al. (2008) suggest. While a PPI deficit per se is not indicative of altered striatal or limbic circuitry, the presence of the deficit may suggest that these brain regions are affected by the genetic manipulation and provide a reasonable starting place for further hypothesis testing regarding the neurobiological implications of the genetic manipulation. Additionally, mutant mouse models offer the opportunity to screen putative antipsychotics that may involve a novel target and avoid the problems of receptor tautology inherent in many pharmacological studies [e.g. dopamine agonist-induced disruption blocked by a dopamine antagonist; as discussed in Powell et al. (2009). Using mutant mice to screen for putative antipsychotics may provide a means to develop novel drug targets not achieved with current pharmacological disruptions of PPI (see Table 1 for examples).

2.3 PPI as a Tool to Evaluate Gene–Environment Interactions

Studies of gene–environment interactions may be particularly informative for neuropsychiatric diseases, most of which likely involve a genetic susceptibility combined with environmental factors (e.g. stress) to observe the full manifestation of the disease (Gottesman 1991) (see also Sen and Karg, Gross and Carola chapters in this book). Three ways in which genetics and environmental manipulations have been utilized in genetic mouse models are: (1) using a mutant [e.g. knockout (KO)] to delineate the physiological mechanism of an environmental manipulation; (2) rescuing a phenotype in a mutant with an environmental manipulation; or (3) potentiating or unmasking a phenotype in a genetic mutant with an environmental manipulation. There are a few examples in which PPI has been a useful endpoint with which to assess gene–environment interactions in mouse models. For example, PPI deficits associated with maternal immune activation (MIA) with PolyI:C during mid-gestation, which typically leads to deficits in PPI in adult offspring (Meyer et al. 2005; Shi et al. 2003), are blocked in interleukin (IL)-6 KO dams (Smith et al. 2007). Thus, PPI in a genetic mutant (IL-6 KO mice) was used to determine the mechanism for the effects of an environmental manipulation (immune activation) on brain development. An example of a PPI phenotype being "rescued" in a KO mouse comes from studies in Phospholipase C $-\beta 1$ KO mice, in which PPI deficits and locomotor hyperactivity were attenuated in KO mice by environmental enrichment or clozapine (McOmish et al. 2008) (Table 1). More recently, and perhaps most important to etiological models of neuropsychiatric disease, there have been several studies examining the "two-hit" approach (Eells et al. 2006; Ibi et al. 2010). For example, nuclear receptor null Nurr1 heterozygous mice, which display reduced mesocortical and mesolimbic dopamine (Eells et al. 2002), showed reduced PPI after postnatal isolation rearing, an effect that was not observed with either isolation rearing or genotype alone (Eells et al. 2006). This study provides a good example of the utility of PPI in gene–environment models relevant to schizophrenia, specifically those designed to test the "two-hit" hypothesis for the etiology of schizophrenia. It should be kept in mind, however, that many studies assessing gene–environment effects are evaluating additive effects of two manipulations and must be interpreted with caution.

Table 1 Genetically engineered model organisms based on genes related to: *vulnerability for schizophrenia*

References	Mouse strain, sex	Model description/background/ rationale	Basal PPI	Effects of drugs or manipulations typically used to induce PPI-deficits	Effects of (presumed) antipsychotics/other treatments
		DISC1			
Hikida et al. (2007)	DN-DISC1 TG, M	DISC1 is a potential SZ susceptibility gene	↓PPI in TG versus WT at lowest PP intensity only		
Ibi et al. (2010)	DN-DISC1 TG, combined with neonatal PolyI:C 5 mg/kg or saline (PND2-6), M & F	Two-hit model of SZ, genetic susceptibility combined with environmental insult (immune activation)	∅PPI & ∅PA in TG versus WT		
Niwa et al. (2010)	DISC1 KD via *in utero* gene transfer (E14), M & F	DISC1 plays important role in brain development. In utero gene transfer allows for analysis of DISC1 function during development and impact on adult brain function	↓PPI & ∅PA in KD versus WT at P56, ∅PPI at P28, effect slightly less in F compared to M		*CLO*: ↑PPI & ∅PA in KD; ∅PPI & ∅PA in WT
Lipina et al. (2010)	DISC1L100P point mutation	Missense mutation in exon 2 of DISC1 at L100P in mice	↓PPI & ↓PA in mutant versus WT	AMP (0.5 mg/kg) ↓PPI in mutant but ∅PPI in WT	HAL: ↑PPI in mutant (↑PPI in WT at PP16); ↓PA in WT, ÆPA in mutant
Lipina et al. (2011)	DISC1L100P point mutation (combined with GSK3 mutant)	Missense mutation in exon 2 of DISC1 at L100P in mice; DISC1L100P mice crossed with Glycogen synthase kinase-3 (GSK-3), a serine/threonine protein kinase that interacts with the N-terminal region of DISC1	↓PPI in DISC1L100P mutant vs. WT; ∅PPI in DISC1L100P x GSK-3α HET		TDZD-8 (GSK-3 inhibitor): ↑PPI & ∅PA in DISC1L100P mutant; ∅PPI & ∅PA in WT

(continued)

Table 1 (continued)

References	Mouse strain, sex	Model description/background/ rationale	Basal PPI	Effects of drugs or manipulations typically used to induce PPI-deficits	Effects of (presumed) antipsychotics/other treatments
		Reelin			
Barr et al. (2008)	Reelin KO, M	Reelin levels in brains of SZ are reduced	∅(unimodal acoustic) PPI, but ↓crossmodal PPI at 100 ms PP intervals, PPP at PP intervals ≤ 60 ms, ↓PA habituation & ↑PA at 100–110 dB(A) intensity for +/− versus WT		
		Neuregulin			
Chen et al. (2008)	Type III NRG1 mutants, M	NRG1 may represent a SZ susceptibility gene. Involved in cell-cell interaction; signals via erbB TK	↓PPI & ∅PA in +/− versus WT		Chronic NIC (↑PPI in +/−; ∅PPI in WT)
Duffy et al. (2008)	EGF-like domain NRG1 mutants, M	The EGF-like domain of NRG1 is sufficient for NRG's activity	∅PPI in −/− versus WT	MK801 (↓PPI in −/− versus WT); AMPH (↓PPI in −/− *and* WT; no significant differences between −/− & WT)	
Deakin et al. (2009)	NRG1$^{Type\ 1\text{-}Tg}$	Type I NRG1 increased in Hpc and PFC of SZ patients	↓PPI & ↑PA in Tg versus WT		
Barros et al. (2009)	ERBB2/B4-CNS-specific KO, M & F	NRG1 and ERBB4 are candidate genes for SZ	↓PPI in KO versus WT (M only; ∅PPI in F); PA data not shown		CLO ↑PPI in KO; ∅PPI in WT CLO ↑ spine density

(continued)

Table 1 (continued)

References	Mouse strain, sex	Model description/background/ rationale	Basal PPI	Effects of drugs or manipulations typically used to induce PPI-deficits	Effects of (presumed) antipsychotics/other treatments
Wen et al. (2010)	PV-Cre; ErbB4 KO	In adulthood NRG1 and ErbB4 may regulate neurotransmission. Evidence that ErbB4 is located at the postsynaptic density of excitatory neurons, likely on GABA interneurons	↓PPI & ∅PA in −/− versus WT		*DIAZ*: ↑PPI in −/− mice; ∅PPI in WT; PA data not shown
Kato et al. (2010)	NRG1 TG5, TG7, M & F	Postmortem studies indicate increased NRG1 in SZ brain. Transgenic mutants to examine hyper NRG1 signaling	∅PPI & ∅PA in TG versus WT		
		COMT			
Papaleo et al. (2008)	COMT KO, COMT Val-Tg	Polymorphisms in COMT gene reported in SZ; mice overexpress human COMT-Val variant	∅PPI in −/− or TG versus WT; ↑PA in −/− versus WT, ↓PA in TG versus WT		
Tammimäki et al. (2010)	S-COMT KO, M & F	Two isoforms of COMT, S-COMT and MB-COMT, differentially contribute to DA function in the frontal cortex	∅PPI & ↑PA in −/− versus WT (although lack of ↑PA with ↑ dB in WT mice)		
		G-alpha			
Kelly et al. (2009)	Forebrain-specific Galpha OE, M, F	↑mRNA G-protein subunit Galpha levels, genetically linked to SZ	↓PPI & ∅PA in +/− versus WT; developmental or adult expression ∅PPI; ↓PPI correlates with ↓CTX & HPC cAMP levels		*HAL*: ↑PPI in OE; ∅PPI in WT *Rolipram* (↑*cAMP levels*): ↑PPI in OE; ∅PPI in WT

(continued)

Table 1 (continued)

References	Mouse strain, sex	Model description/background/rationale	Basal PPI	Effects of drugs or manipulations typically used to induce PPI-deficits	Effects of (presumed) antipsychotics/other treatments
		G72/G30			
Otte et al. (2009)	G72Tg (G72/G30 "humanized" BAC transgenic)	G72/G30 gene locus implicated in SZ, BD, and panic disorder	↓PPI & ∅PA in Tg versus WT		HAL ↑PPI in Tg; ∅PPI in WT (slight ↑PPI at low PP intensity in WT); PA data not reported
		SMARCA			
Koga et al. (2009)	SMARCA2 KO, M & F	SMARCA encodes BRM, a chromatin remodeling multi-protein complex; highly expressed in differentiating neurons; SNP in gene found in SZ (this paper)	↓PPI in −/− versus WT; PA data not shown		
		Neurexin			
Etherton et al. (2009)	Neurexin-1α KO, M & F	CNVs for neurexin-1 α identified in SZ & Autism. Neurexins bind to neuroligins and function as synaptic recognition molecules	↓PPI in −/− versus WT; ↑PA to stimuli 80–100 dB in −/− versus WT		
		TCF4			
Brzózka et al. (2010)	TCF4 Tg, M	TCF4, a helix-loop-helix transcription factor, identified in several GWAS studies as susceptibility gene for SZ	↓PPI & ∅PA in Tg versus WT		

(continued)

Table 1 (continued)

References	Mouse strain, sex	Model description/background/rationale	Basal PPI	Effects of drugs or manipulations typically used to induce PPI-deficits	Effects of (presumed) antipsychotics/other treatments
		AKT1			
Chen and Lai (2011)	AKT1 KO, M & F	AKT1 has been associated with schizophrenia in genetic studies as well as postmortem brain analyses AKT1 signal transduction pathway important in diverse biological functions	*F*: ↓PPI & ∅PA in −/− versus WT; PA *M*: ∅PPI & ∅PA in −/− versus WT		*RAC*: ∅PPI & ∅PA in −/− or WT (F only) *CLOZ*: ∅PPI & ∅PA in −/− or WT (F only) *8-OH-DPAT*: ∅PPI in −/− versus WT (attenuation weak; no genotype × drug interaction) *SB216763*: ∅PPI in −/− versus WT (attenuation weak; no genotype × drug interaction)

Abbreviations: *5-HT* serotonin, *AAV* adeno-associated virus, *Aβ* amyloid β- peptide, *ACE* angiotensin converting enzyme, *ACH* acetylcholine (receptor), *AD* Alzheimer's disease, *ADX* adrenalectomy, *AIL* advanced intercross line, *AMP* amphetamine, *AMY* amygdala, *APD* antipsychotic drug, *APP* amyloid precursor protein, *ARIP* aripiprazole, *APO* apomorphine, *AT* angiotensin, *BAC* bacterial artificial chromosome, *BACE β*-site APP cleaving enzyme, *BD* bipolar disorder, *EG* background, *CaMKIV* calcium-calmodulin-dependent protein kinease IV, *cAMP* cyclic adenosine monophosphate, *Ckr* chakragati, *CLO* clozapine, *CNS* central nervous system, *COC* cocaine, *COMT* catechol-O-methyltransferase, *CORT* corticosterone, *CRF* corticotropin releasing factor, *CTR* controls, *CTX* cortex, *DA* dopamine (receptor), *DAT* dopamine transporter, *dB* decibel, *DCC* deleted in colorectal cancer, *DIAZ* diazepam, *DIZ* dizocilpine, *DN* dominant-negative, *DDO* D-aspartate oxidase, *E* embryonic day, *EE* environmental enrichment, *EGF* epidermal growth factor, *ENU* N-ethyl-N-nitrosourea, *f* frontal, *F* female, *FABP* fatty acid binding protein, *FGF* fibroblast growth factor, *Fmr1* fragile × mental retardation 1 gene, *FMRP* fragile × mental retardation protein, *FXS* fragile × Syndrome, *GLAST* glutamate and aspartate transporter, *GLU* glutamate, *GluR* glutamate receptor, *GLUT* glucose transporter, *GR* glucocorticoid receptor, *GRIP* glutamate receptor interacting protein, *GSK* glycogen synthase kinase, *GWAS* genome wide association studies, *HAL* haloperidol, *HD* Huntington's disease, *HPC* hippocampus, *IC* imprinted cluster, *IL* interleukin, *ISI* interstimulus interval, *KI* knock-in, *KO* knock-out, *M* male, *m* metabotropic, *MDB* methyl-CpG binding protein, *mo* month, *NIC* nicotine, *MPEP* 2-methyl-6-(phenylethylyn)-pyridine hydrocholoride, *METH* metamphetamine, *Mgat* N-acetylglucosaminyltransferase, *NAA* N-acetyl-aspartate, *NAAG* N-acetylalpha L-aspartyl-L-glutamate, *NAC* nucleus accumbens, *NCAM* neural cell adhesion molecule, *NIS* nisoxetine, *nNOS* neuronal nitric oxide synthase, *NO* nitric oxide, *NPS* neuropeptide S (receptor), *NPY* neuropeptide Y, *NR* NMDA receptor subunit, *NRG* neuregulin, *NRL* neuroligin, *ns* not significant(ly), *NSE* neuron-specific enolase, *NT* neurotensin, *OE* overexpressor, *OXO* oxotremorine, *PA* magnitude of response to pulse alone, *PACAP* pituitary adenylate-cyclase-activating polypeptide, *PD* Parkinson's disease, *PDE* phosphodiesterase, *PET-1* plasmocytoma expressed transcript-1, *PND* postnatal day, *PLC* phospholipase C, *PNS* prenatal stress, poly I:C polyinosinic: polycytidylic acid, *PP* prepulse, *PPI* prepulse inhibition of startle, *PPP* prepulse potentiation, *PS1* presinilin1, *PWS* Prader–Willi syndrome, *QTL* quantitative trait locus, *QUET* quetiapine, *RAC* raclopride, *RAGE* receptor for advanced glycation end-products, *RasGAP* Ras GTPase-activating protein, *RE* renin-enhancer, *RIS* risperidone, *SCOP* scopolamine, *SI* social isolation, *SNP* single nucleotide polymorphism, *SOCS* suppressor of cytokine signaling; *SREB* superconserved receptor expressed in brain; *STR* striatum, *SYN* synapsin, *SynGAP* synaptic GTP-ase-activating protein, *SZ* schizophrenia, *TAAR* trace amine-associated receptor, *TG* transgenic, TGF-*β* transforming growth factor beta, *TK* typosine kinase, *TMS* transcranial magnetic stimulation, *V1b* vasopressin receptor 1b, *WT* wild-type

↓ decreased, ↑ increased, ∅ unchanged, −/− homozygous mice, +/− heterozygous mice

2.4 Evaluating the Role of a Susceptibility Gene in Pathology

When evaluating the role of a susceptibility gene implicated in any neuropsychiatric disease, it is important to consider what criteria should be placed on a genetic/etiological model. In the present context, it is relevant to consider whether or not deficient PPI is a necessary phenotype with which to evaluate the usefulness of a targeted gene deletion of potential relevance to the given neuropsychiatric condition under study (e.g. schizophrenia). Using schizophrenia as an example, the failure to see a PPI deficit in a mouse model may indicate a "false negative" particularly if other key behaviors relevant to schizophrenia are observed (e.g. deficient social interaction, disruptions in attentional set shifting). The lack of a PPI deficit in this case does not indicate that the genetic model is not of relevance to schizophrenia. There are several examples provided in Table 1 in which a PPI deficit was not present in a mutant mouse but other behavioral differences such as deficits in memory, social interaction, or set shifting were apparent in the mice. The likelihood of being able to reproduce all aspects of a heterogeneous disease in another species with a genetic mutation (most often a single gene deletion) is very rare if not impossible (see Jones et al. (2008); Powell et al. (2009) for discussion). For that matter, it is not the case that all aspects of schizophrenia are observed in each patient carrying the diagnosis. Rather, support for a model should be based on the convergence of data from multiple sources [e.g. many animal models, human genetic studies, etc. (Jones et al. 2008)]. Thus, no one phenotype should be considered as being either necessary or sufficient to support a model for a neuropsychiatric disease, particularly since the distributions of the behavioral measure often overlap between healthy volunteers and patients, as is the case with PPI and schizophrenia (Swerdlow et al. 2008). Thus an animal model should not be rejected based on "normal" PPI. For example, GLAST KO mice lacking the glutamate and aspartate transporter do not show any deficits in PPI but do show deficits in social approach, nest building, and pairwise discrimination learning (Karlsson et al. 2009). Along the same lines, there is the possibility that a PPI deficit in a mutant mouse model could represent a "false positive", in which a PPI phenotype may be suggestive of an association between that gene or pathway and disease and no such association is found. As we have argued previously, the PPI phenotype should be interpreted as meaning that the given genetic manipulation may be involved in the regulation of PPI expression and caution that PPI phenotypes should not be automatically associated with a specific disease, e.g. schizophrenia (Powell et al. 2009). Interestingly, there are a few examples where KO or transgenic mice actually had increased PPI relative to wild-type controls. One such example is the FMR1 KO mice (Paylor et al. 2008; Thomas et al. 2011a, b; but see de Vrij et al. 2008; Table 2). It remains unclear how such findings should be interpreted, but the possibility that the loss of the gene may lead to gain of function should be considered.

Table 2 Genetically engineered model organisms based on genes related to: *neurotransmitters, neuropeptides, and orphan receptors*

References	Mouse Strain, Sex	Model description/ background/ rationale	Basal PPI	Effects of drugs or manipulations typically used to induce PPI-deficits	Effects of (presumed) antipsychotics/other treatments
			A. DOPAMINE		
		DA receptors			
Doherty et al. (2008)	DA D1, D2 & D3 KO, M, F	Abnormalities in DA signaling in SZ patients in part have led to the DA hypothesis for SZ	∅PPI in D1, D2 & D3 -/- versus WT; ↑PA in D1 -/- versus WT; ∅PA in D2 (M only); ↓PA in D2 -/- (F only) & D3 -/- versus WT & in D3 (F) versus D3 (M) -/-	COC (∅PPI in D1 -/-; ↓PPI in D2 -/-, D3 -/- & WT mice; but: attenuated ↓PPI in D2-/- versus WT and more pronounced ↓PPI in D3 -/- versus WT; ∅PA in WT	RAC; SCH23390 (a D1 antagonist) both ↓COC-induced PPI deficits in WT; ∅PA in WT)
Young et al. (2011b)	DA D4 KO, M	Alterations in D4R may contribute to cognitive and attentional deficits in neuropsychiatric disorders.	∅PPI & ∅PA in -/- or +/- versus WT		
		DAT			
Powell et al. (2008)	DAT KO, (F) M	A hyper-DA state and DAT abnormalities have been implicated in SZ	M: ↓PPI & ∅PA in -/- versus WT; F: inconsistent ↓PPI in -/- versus WT		M: CLO (PPI in -/- normalized to WT levels, but ∅PPI in -/- & WT versus VEH) ↓PA (main effect); QUET (↑PPI & ∅PA in -/- & ∅PPI & ↓PA in WT)
		Other Dopamine related			
Grant et al. (2007)	DCC KO, M (PPI studies)	DCC mice are netrin 1 receptor KO mice. Netrin 1 is strongly expressed on and may affect development of DA neurons	↓PPI (yet ns) and ↑PA (trend only) in +/- versus +/+	∅AMP (∅PPI in +/-, but ↓PPI in +/+)	

(continued)

Table 2 (continued)

References	Mouse Strain, Sex	Model description/ background/ rationale	Basal PPI	Effects of drugs or manipulations typically used to induce PPI-deficits	Effects of (presumed) antipsychotics/other treatments
			B. GLUTAMATE		
		NR1			
Duncan et al. (2010)	NR1 hypomorph (NR1 $^{neo/neo}$),	Developmental effects of NR1 hypomorph on other Glu receptors could contribute to the SZ-like phenotype. Evidence for alterations in kainic acid receptors in SZ.	↓PPI & ↑PA in hypomorph versus WT		LY382884, kainic acid antagonist, ↓PA in hypomorphs & ∅PA in WT; ↑PPI in hypomorphs & WT
Belforte et al. (2010)	Early postnatal NR1 reduction in GABA interneurons *(Ppp1r2-cre +/−; NR1*loxP/loxP-line A [KO]), M	NMDAR hypofunction, specifically in GABA interneurons, may contribute to SZ pathophysiology. NR1 is an obligatory subunit of NMDAR.	Early postnatal NR1 reduction: ↓PPI & ∅PA in KO versus Ctrl lines (Flox-A & Cre), in isolated & group-housed mice Adult NR1 reduction: ∅PPI & ∅PA		
		AMPA			
Sagata et al. (2010)	GluR4 KO, M & F	GluR4 subunit of AMPA receptor involved in synaptic plasticity; genetic association between GluR4 and SZ	↓PPI & ↓PA in -/- versus WT		
		mGLU			
Gray et al. (2009)	mGlu5 KO, M, F	mGlu5 receptor interacts with NMDA receptor and has been implicated in behavioral endophenotypes of SZ	↓PPI in -/- versus WT; ∅PA in -/- versus WT		Chronic (8 wks) CLO (↑PPI in -/- & ∅PPI in WT); ↑NMDA receptors with CLO treatment

(continued)

Table 2 (continued)

References	Mouse Strain, Sex	Model description/ background/ rationale	Basal PPI	Effects of drugs or manipulations typically used to induce PPI-deficits	Effects of (presumed) antipsychotics/other treatments
Fendt et al. (2010)	mGlu8 KO, M	mGlu8 is a presynaptic receptor which regulates Glu release; may be important for neuropsychiatric disorders with altered Glu function (e.g. SZ).	∅PPI and ↓PA in -/- versus WT	PCP: ↓PPI in -/- and WT; ∅PA in -/- or WT	
Chen et al. (2010)	mGlu5 KO, M & F	Evidence of NMDA hypofunction in mGlu5 KO mice. mGlu5 KO mice may be useful as novel screen of putative glutamatergic antipsychotics.	↓PPI & ∅PA in -/- versus WT		Sarcosine: ↑PPI in -/-; ∅PPI in WT; ∅PA in -/- or WT LY379268: ∅PPI & ∅PA in -/- or WT N-acetylcysteine: ↑PPI in -/- & WT; ↑PPI in -/- not blocked with LY341495
		Vesicular glutamate transporter			
Wallén-Mackenzie et al. (2009)	VGLUT2$^{f/f;CKII/Cre}$ (preadolescent forebrain-specific KO), M & F	Forebrain glutamate plays an important role in many sensory and cognitive functions relevant to psychiatric disease; downstream changes in DA (this study)	↓PPI & ↓PA in VGLUT2$^{f/f;CKII/Cre}$ versus WT	PCP: ↓PPI in WT; ∅PPI in VGLUT2$^{f/f;CKII/Cre}$ (no drug versus post-PCP); PA (no data shown)	ARIP (no-drug versus post-ARIP): ↑PPI in KO; ∅PPI in WT; PA (no data shown)

(continued)

Table 2 (continued)

References	Mouse Strain, Sex	Model description/ background/ rationale	Basal PPI	Effects of drugs or manipulations typically used to induce PPI-deficits	Effects of (presumed) antipsychotics/other treatments
		Glycine			
Benneyworth et al. (2011)	SR KO, M GCP2 HET, M	D-serine is an agonist and NAG an antagonist at the glycine modulatory site on NMDA receptors. Serine racemate (SR) and glutamic acid decarboxylase 2 (GCP2) regulate D-serine and NAG, respectively.	∅PPI & ∅PA in SR -/- versus WT; ∅PPI & ∅PA in GCP2 +/- versus WT	PCP: ↓PPI in SR -/-, GCP2 +/-, & WT; inverted U-shaped dose response function on PA: high dose (6 mg/kg) ∅PA in WT, ↑PA in SR -/-; 3 mg/kg dose ↑PA in WT, ∅PA in GCP2 +/- AMPH: ↓PPI in Sr-/-, GCP2 +/-, & WT; U-shaped dose response function on PA; no difference in SR -/-, GCP2 +/-, & WT	
		Other glutamate related			
Karlsson et al. (2009)	GLAST (EAAT1) KO, M, F	Glutamate dysfunction associated with SZ; glutamate transporter may be useful pharmacological target for SZ.	∅PPI in -/- or +/- versus WT; ↓PA in -/- versus +/- & WT		
Guo et al. (2009)	SynGAP +/- gene deletion	SynGAP is a RasGAP and NMDAR-associated protein; interacts with NMDA receptors	↓PPI and ↑PA in +/- versus WT		CLO: ∅PPI in +/- or WT; ↓PA in +/- and WT
Han et al. (2009)	GCPII +/-, M & F	Glutamate carboxypeptidase II (GCPII) hydrolyzes NAAG into glutamate and NAA	∅PPI & ∅PA in +/- versus WT		
Wang et al. (2009)	Norbin KO	Norbin interacts with mGluR5 receptors and positively regulates mGluR5 signaling.	↓PPI & ∅PA in -/- versus WT		

(continued)

Table 2 (continued)

References	Mouse Strain, Sex	Model description/ background/ rationale	Basal PPI	Effects of drugs or manipulations typically used to induce PPI-deficits	Effects of (presumed) antipsychotics/other treatments
		C. SEROTONIN (5-HT)			
Semenova et al. (2008)	5-HT7 KO, M	Antipsychotics have high affinity for 5-HT7 receptors.	∅PPI across PP intensities and ISI, ∅PA	PCP ↓PPI in WT, attenuated in -/-; Apo and Amp ↓PPI in WT & -/-; ∅PA with Apo, Amp, PCP	
Shanahan et al. (2009)	5-HTT KO, F	OCD patients have PPI deficits, gain of function 5-HTT alleles associated with OCD	∅PPI in -/-, +/- versus WT	RU24969 (5HT1B agonist) ↓PPI in WT, ∅PPI in -/-, intermediate effect in +/-	
Schaefer et al. (2009)	Pet-1 KO, M &F	Pet-1 is a transcription factor restricted to 5-HT neurons. Gene deletion results in loss of 5-HT.	↑PA & ∅PPI in -/- versus WT		
Hill et al. (2011)	5HT2C KO	5HT and BDNF interact. ↑BDNF levels 5HT2C KO mice (this study)	∅PPI & ∅PA in -/- versus WT		
Groenink et al. (2011)	5HT1A KO, M (129S6, Swiss Webster [SW] backgrounds)	5HT1A gene may interact with adverse life events.	∅PPI & ∅PA in -/- versus WT	129S6: Maternal separation ↓PPI & ↑PA in -/- & WT SW: Maternal separation ∅PPI & ∅PA in -/- & WT	
van den Buuse et al. (2011a)	5HT1A KO, M & F	5HT1A receptor may mediate effects of MDMA on PPI	∅PPI & ∅PA in -/- versus WT	MDMA: ↓PPI in -/- at lower doses than WT at 100 ms ISI (M only); ↓PA in -/- & WT	
van den Buuse et al. (2011b)	5HT1A KO, M & F	5HT1A receptor may interact with neurotransmitter systems involved in SZ	∅PPI & ∅PA in -/- versus WT	APO: ↓PPI & ∅PA in -/- & WT	

(continued)

Table 2 (continued)

References	Mouse Strain, Sex	Model description/ background/ rationale	Basal PPI	Effects of drugs or manipulations typically used to induce PPI-deficits	Effects of (presumed) antipsychotics/other treatments
		D. ACETHYLCHOLINE (ACh)			
		Nicotinic			
Darvas et al. (2009)	LYPD6 Tg	LYPD6 modulates nicotinic ACh receptors; LYPD6 Tg show enhanced nicotinic tone	↑PPI in Tg versus WT; ∅PA in Tg versus WT		
Young et al. (2011a)	α7-nAChR KO, M & F	α7-nAChR is being investigated as a pro-cognitive target for SZ; comparison of KO mice across several cognitive tests	∅PPI in -/-, +/-, and WT; ↑PA in -/- versus +/- & WT;		
		Muscarinic			
Turner et al. (2010)	M2, M4 KO, M	M2 and M4 muscarinic acetylcholine receptors differ in tissue distribution and physiology; may play a role in sleep and arousal	∅PPI in M2 or M4 -/- versus C57BL/6 N mice; ∅PA in M2 -/- versus M4 -/-; ↑PA in M2 -/- versus C57; ∅PA in M4 -/- versus C57 (store bought mice; not littermates)		

(continued)

Table 2 (continued)

References	Mouse Strain, Sex	Model description/ background/ rationale	Basal PPI	Effects of drugs or manipulations typically used to induce PPI-deficits	Effects of (presumed) antipsychotics/other treatments
Thomsen et al. (2010)	M1, M4 KO, F M1KO/ M4KO,M,F	Muscarinic receptors may be related to pathophysiology of SZ, and muscarinic agonists have been proposed as targets for antipsychotics and cognition enhancers	F: ↓PPI & ∅PA in M1-/-/ M4-/- versus WT; ∅PPI & ∅PA in M1 -/- versus WT; ∅PPI & ↑PA in M4 -/-WT; M: ∅PPI & ∅PA in M1-/-/M4-/- versus WT	SCOP: ↓PPI in WT & M4-/- mice (F) & slight ↓PPI in M1-/-/ M4-/- (M); ∅PPI in M1-/- (F) or M1-/-/M4-/- (F) ↑PA in all groups; M1-/-/ M4-/- (F) and M4-/- (F) more sensitive to ↑PA	Xanomeline: blocked SCOP-induce ↓PPI in WT & M1-/-, but not in M4-/- mice; data in M1-/-/ M4-/- difficult to interpret because ∅PPI with SCOP; blocked SCOP-induced ↑PA in M1-/- & M1-/-/M4-/-; ∅PA in SCOP-treated M4-/- OXO (versus SCOP): blocked SCOP-induced ↓PPI in all genotypes (although SCOP effect not significant in M1-/-/M4-/-); trend for reversal of SCOP-induced ↑PA OXO (alone): ↓PPI in WT & ↑PPI in M1-/-/ M4-/-; ↓PA in WT & M1-/-M4-/- (which may lead to variability in PPI) CLO: ↑PPI in M1-/-/M4-/- at doses with ∅PPI in WT; ↓PPI in WT & M1-/-; ∅PPI in M4-/-; ↓PA in WT & M1-/-; ∅PA in M1-/-/ M4-/- & M4-/- HAL: ↑PPI in WT & M1-/-/M4-/-; ∅PA in WT or M1-/-/M4-/-
		Vesicular ACh Transporter (VAChT)			
Schmid et al. (2011)	VAChT KD, M	Cholinergic transmission associated with cognitive function and startle behavior. VAChT KD mice have reduced cholinergic tone.	∅PPI & ∅PA in -/- versus WT; ↓ between day HAB in -/- versus WT		

(continued)

Table 2 (continued)

References	Mouse Strain, Sex	Model description/ background/ rationale	Basal PPI	Effects of drugs or manipulations typically used to induce PPI-deficits	Effects of (presumed) antipsychotics/other treatments
		E. (NEURO)PEPTIDES			
		Angiotensin/Renin			
Martin et al. (2008)	RE KO, NSE-AT1 TG, M	AT/Renin may be involved in brain (DA) signaling. ACE levels were reported to be altered in SZ patients. REKO have reduced renin signaling; NSE-AT1 TG have over-expression of AT1 receptors.	∅PPI (RE -/- & NSE-AT1 TG vs. WT); ↓PA (RE KO (trend)) & ∅PA in NSE-AT1 versus WT	APO, AMP & DIZ (all ↓PA (RE KO vs. WT; main effect across treatment conditions); ↓PPI in RE -/-, NSE-AT1 TG & WT (no genotype x drug interactions); APO (↓PA in RE -/-, NSE-AT1 TG & WT); AMP (↑PA in RE -/- & WT (trend), ∅PA in NSE-AT1 TG & WT;) DIZ (↑PA in RE -/- & WT; more pronounced in RE -/-, ↑PA (trend) in NSE-AT1 TG & WT;	
		Neurotensin			
Feifel et al. (2010a)	NT1 KO, M & F	Neurotensin (NT) may play a role in SZ pathophysiology. NT levels decreased in CSF from SZ patients, and APDs increase NT levels.	∅PPI in -/- versus WT; ∅PA in -/- versus WT	AMP ↓PPI & ∅PA in -/- & WT; MK801 ↓PPI; MK801 ↑PA (low dose only; more pronounced in WT than -/-)	PD149163, NT1 receptor agonist, ↑PPI in WT, ∅PPI in -/-;
Feifel et al. (2010b)	NT2 KO, M & F	NT2 receptors highly expressed in brain but not well studied	↑PPI & ↓PA in -/- versus WT (both sexes)		

(continued)

Table 2 (continued)

References	Mouse Strain, Sex	Model description/ background/ rationale	Basal PPI	Effects of drugs or manipulations typically used to induce PPI-deficits	Effects of (presumed) antipsychotics/other treatments
Oliveros et al. (2010)	NT1 KO & NT2 KO, M	NT receptors may play a role in psychotomimetic and antipsychotic effects of drugs	∅PPI & ∅PA in NT1 -/-, NT2 -/-, & WT	AMP: ↓PPI in NT2 -/- & WT, ∅PPI in NT1 -/-; ↓PA in WT, ∅PA in NT1 -/- or NT2 -/- DIZ: ↓PPI in NT2 -/- & WT, ∅PPI in NT1 -/-	CLO: blocks AMP-induced ↓PPI in WT & NT2 -/-; blocks DIZ-induced ↓PPI in WT, does not block DIZ-induced ↓PPI in NT2 -/- NT69L: blocks AMP-induced ↓PPI in WT & NT2 -/-; blocks DIZ-induced ↓PPI in NT2 -/-, does not block DIZ-induced ↓PPI in WT; ↓PA in WT & NT2-/-
		CRF			
Groenink et al. (2008)	CRF-OE, M	Stress physiology, particularly CRF and glucocorticoids, implicated in neuropsychiatric disorders	↓PPI in CRF-OE versus WT at lower PP intensities; ∅PA in CRF-OE versus WT	CORT implant: ∅PPI in CRF-OE or WT mice.	CRF1 antagonists CP154,526 & DMP695 normalized PPI deficit in CRF-OE mice; GR antagonists and ADX did not; CP154,526 ↓PA
		Oxytocin			
Caldwell et al. (2009)	Oxt KO, M, F	Oxytocin has shown antipsychotic potential in animal models of SZ and in some preliminary clinical trials	∅PPI & ∅PA in -/- versus WT	PCP: Exacerbated ↓PPI in -/- versus WT, ∅PA; Apo & Amp: ↓PPI in -/- & WT; ∅PA	
		Arginine Vasopressin			
Egashira et al. (2009)	V1aR KO, V1bR KO	Arginine vasopressin (AVP), a diuretic peptide with receptors widely expressed in the CNS, has been implicated in neuropsychiatric disorders.	↓PPI in V1aR & V1bR -/- versus WT; ↑PA in V1aR & V1bR -/- versus WT		
		Neuropeptide Y			
Karl et al. (2010a)	Y2 KO, M & F	NPY levels altered in SZ patients. NPY Y2 receptor expressed in HPC and AMY	M: ↑PPI & ∅PA in -/- versus WT F: ∅PPI in -/- versus WT	MK801: ↓PPI & ↑PA in -/- & WT AMP: ↓PPI & ↑PA in -/- & WT	

(continued)

Table 2 (continued)

References	Mouse Strain, Sex	Model description/ background/ rationale	Basal PPI	Effects of drugs or manipulations typically used to induce PPI-deficits	Effects of (presumed) antipsychotics/other treatments
Karl et al. (2010b)	Y1 KO, M	Decreased Y1 receptor expression in lymphocytes of SZ patients	∅PPI, ↓PA, & ↓HAB in -/- versus WT	MK801: ↓PPI & ↑PA in -/- & WT AMP: ↓PPI & ∅PA in -/- & WT (effects are ISI & PP dependent)	
		Neuropeptide S			
Zhu et al. (2010)	NPSR1 KO, M & F	NPSR1 is a G protein coupled receptor located in brain regions involved in anxiety and learning and memory, whose endogenous ligand is NPS. NPSR1 SNPs identified in panic disorder.	M: ↓PA & ∅PPI in -/- versus WT; F: ∅PA & ∅PPI in -/- versus WT		
Fendt et al. (2011)	NPSR KO, M	NPS implicated in regulation of emotional behavior	∅PPI in -/- or +/- versus WT; ↓PA in -/- & +/- versus WT		HAL: ↑PPI & ↓PA in -/-, +/-, & WT
		PACAP			
Hashimoto et al. (2007)	PACAP KO, M	PACAP affects neurotransmission; potential SZ susceptibility gene	↓PPI in -/- versus WT		RIS (↑PPI in -/-; ∅ in WT)
Ishihama et al. (2010)	PACAP KO, M	Genetic susceptibility plus environmental factors may contribute to SZ	↓PPI in -/- versus WT (PA data not shown)	SI: ↓PPI & ∅PA in -/- & WT	EE: ↓PPI in -/- & WT

(continued)

Table 2 (continued)

References	Mouse Strain, Sex	Model description/ background/ rationale	Basal PPI	Effects of drugs or manipulations typically used to induce PPI-deficits	Effects of (presumed) antipsychotics/other treatments
		Relaxin-3			
Smith et al. (2009)	Relaxin-3 KO, M & F	Relaxin-3 modulates feeding, stress response, exploratory behavior, and arousal	∅PPI in -/- versus WT; PA data not reported		
		F. ORPHAN RECEPTORS			
Matsumoto et al. (2008)	SREB2 KO, SREB2 TG, M	SREB (= GPR85) highly conserved GPCR; expressed in brain regions of high plasticity; overtransmission of 2 SNPs in SZ (this study)	↓PPI & ∅PA in TG versus WT; ∅PPI & ∅PA in -/- versus WT		
Logue et al. (2009)	GPR88 KO, M & F	GPR88 is highly expressed in brain regions implicated in SZ and regulates DA function (this paper).	↓PPI in -/- versus WT; ∅PA in -/- versus WT		HAL: ↑PPI in -/-, ∅PPI in WT; ↓PA in -/- & WT RISP: ↑PPI in -/- & WT; ↓PA in -/- & WT
Kim et al. (2010)	TAK1 KO, M	TAK1 is a nuclear orphan receptor that regulates transcription; TAK1 deletion delays migration of cerebellar granular neurons	↓PPI, ↓PA, ↓HAB in -/- versus WT		

(continued)

Table 2 (continued)

References	Mouse Strain, Sex	Model description/ background/ rationale	Basal PPI	Effects of drugs or manipulations typically used to induce PPI-deficits	Effects of (presumed) antipsychotics/other treatments
		G. TRACE AMINES			
Karmacharya et al. (2011)	TAAR1 KO, M	Trace amines are very similar to biogenic amines and are implicated in neuropsychiatric disease; trace amine receptors may mediate effects of antipsychotic drugs	∅PPI in -/- versus WT (PA data not shown)		CLO: ↑PPI in WT, ∅PPI in -/-

Abbreviations: *5-HT* serotonin, *AAV* adeno-associated virus, *Aβ* amyloid β- peptide, *ACE* angiotensin converting enzyme, *ACH* acetylcholine (receptor), *AD* Alzheimer's disease, *ADX* adrenalectomy, *AIL* advanced intercross line, *AMP* amphetamine, *AMY* amygdala, *APD* antipsychotic drug, *APP* amyloid precursor protein, *ARIP* aripiprazole, *APO* apomorphine, *AT* angiotensin, *BAC* bacterial artificial chromosome, *BACE* β-site APP cleaving enzyme, *BD* bipolar disorder, *BG* background, *CaMKIV* calcium-calmodulin-dependent protein kinease IV, *cAMP* cyclic adenosine monophosphate, *Ckr* chakragati, *CLO* clozapine, *CNS* central nervous system, *COC* cocaine, *COMT* catechol-O-methyltransferase, *CORT* corticosterone, *CRF* corticotropin releasing factor, *CTR* controls, *CTX* cortex, *DA* dopamine (receptor), *DAT* dopamine transporter, *dB* decibel, *DCC* deleted in colorectal cancer, *DIAZ* diazepam, *DIZ* dizocilpine, *DN* dominant-negative, *DDO* D-aspartate oxidase, *E* embryonic day, *EE* environmental enrichment, *EGF* epidermal growth factor, *ENU* N-ethyl-N-nitrosourea, *f* frontal, *F* female, *FABP* fatty acid binding protein, *FGF* fibroblast growth factor, *Fmr1* fragile × mental retardation 1 gene, *FMRP* fragile × mental retardation protein, *FXS* fragile × Syndrome, *GLAST* glutamate and aspartate transporter, *GLU* glutamate, *GluR* glutamate receptor, *GLUT* glucose transporter, *GR* glucocorticoid receptor, *GRIP* glutamate receptor interacting protein, *GSK* glycogen synthase kinase, *GWAS* genome wide association studies, *HAL* haloperidol, *HD* Huntington's disease, *HPC* hippocampus, *IC* imprinted cluster, *IL* interleukin, *ISI* interstimulus interval, *KI* knock-in, *KO* knock-out, *M* male, *m* metabotropic, *MDB* methyl-CpG binding protein, *mo* month, *NIC* nicotine, *MPEP* 2-methyl-6-(phenylethylyn)-pyridine hydrocholoride, *METH* metamphetamine, *Mgat* N-acetylglucosaminyltransferase, *NAA* N-acetyl-aspartate, *NAAG* N-acetylalpha L-aspartyl-L-glutamate, *NAC* nucleus accumbens, *NCAM* neural cell adhesion molecule, *NIS* nisoxetine, *nNOS* neuronal nitric oxide synthase, *NO* nitric oxide, *NPS* neuropeptide S (receptor), *NPY* neuropeptide Y, *NR* NMDA receptor subunit, *NRG* neuregulin, *NRL* neuroligin, *ns* not significant(ly), *NSE* neuron-specific enolase, *NT* neurotensin, *OE* overexpressor, *OXO* oxotremorine, *PA* magnitude of response to pulse alone, *PACAP* pituitary adenylate-cyclase-activating polypeptide, *PD* Parkinson's disease, *PDE* phosphodiesterase, *PET-1* plasmocytoma expressed transcript-1, *PND* postnatal day, *PLC* phospholipase C, *PNS* prenatal stress, poly I:C polyinosinic: polycytidylic acid, *PP* prepulse, *PPI* prepulse inhibition of startle, *PPP* prepulse potentiation, *PS1* presinilin1, *PWS* Prader–Willi syndrome, *QTL* quantitative trait locus, *QUET* quetiapine, *RAC* raclopride, *RAGE* receptor for advanced glycation end-products, *RasGAP* Ras GTPase-activating protein, *RE* renin-enhancer, *RIS* risperidone, *SCOP* scopolamine, *SI* social isolation, *SNP* single nucleotide polymorphism, *SOCS* suppressor of cytokine signaling; *SREB* superconserved receptor expressed in brain; *STR* striatum, *SYN* synapsin, *SynGAP* synaptic GTP-ase-activating protein, *SZ* schizophrenia, *TAAR* trace amine-associated receptor, *TG* transgenic, TGF-β transforming growth factor beta, *TK* typosine kinase, *TMS* transcranial magnetic stimulation, *V1b* vasopressin receptor 1b, *WT* wild-type

↓ decreased, ↑ increased, ∅ unchanged, −/− homozygous mice, +/− heterozygous mice

2.5 Methodological Considerations with Specific Relevance to Genetic Models of Sensorimotor Gating

For a detailed description of the methods of acoustic startle and PPI in mice see Geyer and Dulawa (2003). When conducting an initial analysis of startle and PPI in a genetic mutant, several considerations should be made when evaluating PPI. Typically, startle sessions involve variable prepulse intensities (e.g. 3, 6, 12 dB above background), which should produce an incremental increase in PPI with increasing prepulse intensities. Of course any evaluation of a PPI phenotype should be considered in the context of a thorough assessment of physical and sensory abnormalities (e.g. hearing loss), as pointed out in Geyer et al. (2002). In a study examining strain differences in hearing and PPI, Willot et al. (1994) showed a relationship between PPI levels and hearing impairments. Hearing loss is also exacerbated by age, particularly in specific strains of mice such as C57s, which experience high frequency hearing loss with age (Willott et al. 1994). One way to avoid relying solely on auditory processing is to assess multimodal PPI using light as a prepulse and/or airpuffs as a startling stimulus (Brody et al. 2004; Young et al. 2010). This multisensory approach can avoid potential confounds of hearing loss and still enable the evaluation of PPI. It should be noted, however, that some degree of noise is often produced with the delivery of airpuff stimuli, therefore, data on airpuff-tactile startle should be interpreted with this consideration in mind. Other tactile stimuli, such as mild footshock could provide an alternative approach to this issue. Another way to provide a gross measure of hearing in genetic mutants is to incorporate a "startle threshold" block into the session, in which increasing decibels of acoustic stimuli are presented to measure startle magnitude across these intensities (Brody et al. 2004). The threshold at which the mouse begins to startle can then be evaluated. Additionally, differences in baseline startle magnitude can confound effects on PPI. Although dissociations between startle and PPI have been reported in mice (Brody et al. 2004) and rats (Sipes and Geyer 1994), large differences in baseline startle magnitude can complicate interpretations of changes in PPI. If large differences in baseline startle are observed, startle magnitude can be "matched" either by using a range of startle stimuli (e.g. 110, 120 dB) paired with the prepulse stimuli or doing a post hoc analysis of only those animals across genotypes that have comparable levels of startle. We have included results on startle magnitude [pulse alone (PA)] in Tables 1–5 when the data were provided and indicated studies that did not report data on startle magnitude. A good example of testing over a wide range of startle amplitudes can be found in Savonenko et al. (2009) in their studies of EP2 KO mice. The background strain on which genetic mutants are made should be considered as well. For a discussion of the issue surrounding background strain the reader is referred to Crawley (2007) and Geyer et al. (2002).

3 Approaches to Genetic Models of Sensorimotor Gating

3.1 Overview of Molecular Genetic Techniques

Broadly defined, behavioral genetic approaches can be divided into two main approaches either beginning with the gene of interest or the behavior of interest (Jacobson and Cryan 2010). Gene-based or "reverse genetics" approaches begin with the targeted gene being manipulated in the animal and the resultant behavior evaluated. Phenotype-based or "forward genetics" approaches begin with the targeted behavioral phenotype (trait) and then involve subsequent genetic analysis of the trait. Most genetic studies of sensorimotor gating have focused on gene-based approaches, and as such, these approaches are discussed in more detail than phenotype-based approaches, which are discussed more extensively in Tarantino and Eisener-Doman (2011). The goals for any genetic approach can be very different—some approaches seek to understand the more global role of a specific gene in a particular phenotype related to a disease, while others may be designed to elucidate the role of a particular gene in a cellular process, in neural circuit abnormalities, or in brain development. Studies representing all of these goals are well represented in the literature on genetic models of sensorimotor gating (see Tables 1–5). The techniques used to generate molecular genetic models using the gene-based (reverse genetic) approach include constitutive gene deletion (knockout), insertion of exogenous DNA into the genome (transgenic), and insertion of gene at a particular locus via homologous recombination (knockin) [reviewed in (Crawley 2007; Tecott and Wehner 2001)]. Conditional genetic manipulations, which restrict the expression of a targeted gene either temporally or regionally, are useful in avoiding complications of in utero lethality or compensatory brain changes, or when specific hypotheses exist regarding the developmental expression of a gene or the specific neuroanatomical function of the gene. Examples of conditional genetic manipulations include the Cre-LoxP system in which the targeted gene is floxed with lox-p sites. When combined with Cre recombinase, the floxed gene is expressed, giving regional, cell-specific, or temporal control over expression or deletion of targeted gene (van der Neut 1997; Wang 2009). Restriction to specific brain regions can be achieved through the use of specific promoters used to drive expression. For example, the CaMKIIα promoter restricts expression of a genetic mutation to the forebrain (Mayford et al. 1996). Additional examples of tools to achieve temporal control over gene function involve the Tet-On and Tet-Off systems. With these systems, the target gene is linked to either the tetracycline-controlled transactivator (tTA) or the reverse tTA (rtTA) and can be either turned "on" with doxycycline administered (tTA; Tet-On) or turned "off" in the presence of doxycycline (rtTA; Tet-Off) (Mansuy and Bujard 2000). Thus, by combining some of these approaches, conditional and inducible gene expression in mouse brain can be achieved. Additionally, insertion of large segments of DNA into the genome can be achieved through the use of bacterial artificial chromosomes (BAC transgenics) (Heintz 2001). The BAC transgenic technique is a useful approach when a large segment of DNA has been implicated in a disease (e.g. 22q11.2 microdeletion or microduplications). For example, BAC transgenic technology was used to generate two lines of transgenic mice expressing different genes located on the 22q11.2 segment of the chromosome. Interestingly, mice overexpressing Prodh and Vpreb2 showed increased PPI, while mice overexpressing Zdhhc8, Ranbp1, Htf9c, T10, Arvcf, COMT showed no differences in PPI (Stark et al. 2009). More recent techniques, such as in utero gene transfer, are being developed to modulate the expression of genes during specific stages of embryonic development (Niwa et al. 2010).

"Phenotype-based" approaches to behavioral genetics include "forward genetic screens" such as mutagenesis via radiation or chemical means [e.g. ENU mutagenesis; (Tarantino and Bucan 2000) and quantitative trait loci (QTL) studies on crosses of inbred mouse strains with distinct phenotypes or on mice or rats selectively bred for PPI levels (Hitzemann et al. 2008; Tarantino and Eisener-Doman 2011). Thus far, most of the PPI mutants identified through ENU mutagenesis screens have had some amount of hearing loss, and thus the specificity of the PPI "phenotype" was most likely confounded by deafness (Lisa Tarantino, personal communication). Other molecular genetic tools such as siRNA, in which double stranded RNA homologous to the targeted gene is made and then inserted into the cell to block expression of that gene, and viral transfection, in which genes are inserted into viral vectors and injected into the nucleus of cells, are additional techniques used in neuroscience research assessing sensorimotor gating, but will not be discussed in this chapter. Due to the increasingly large number of genetic models generated in which PPI was evaluated, we will provide some examples in the text of different genetic approaches undertaken to study sensorimotor gating, but the reader is referred to Tables 1–5 for a comprehensive review of the recent genetic models assessing PPI.

3.2 Models Based on Hypotheses Regarding the Pathophysiology of Disease

Genetic models of sensorimotor gating deficits were based initially on the hypothesized pathophysiology of schizophrenia or the known mechanism of action of both antipsychotic and psychotomimetic drugs [e.g. amphetamine, PCP]. Thus, early genetic mutants focused primarily on dopamine (Ralph et al. 2001) and glutamate (Brody and Geyer 2004; Duncan et al. 2004) neurotransmitters and receptors (e.g. dopamine-related genes, glutamate-related genes). Many of these observed effects [e.g. PPI deficits in dopamine transporter KO mice (Ralph et al. 2001)] were not surprising based on the extensive pharmacology of PPI deficits in rodents, but nevertheless the genetic models provided proof of concept that constitutive gene deletion could produce dramatic functional effects on sensorimotor gating in adult animals. PPI deficits in dopamine transporter knockout mice strengthened the converging evidence that dopamine hyperfunction played an important role in sensorimotor gating and potentially the pathophysiology of schizophrenia. Similarly, PPI deficits in metabotropic glutamate mutants (Brody et al. 2003; Brody and Geyer 2004; Gray et al. 2009; Kinney et al. 2003) have stimulated interest in these receptors as targets for drug development (Cleva and Olive 2011). Genetic approaches to pathophysiology of neuropsychiatric disease are becoming increasingly sophisticated and many of these models incorporate assessments of PPI as a relevant behavioral phenotype (Tables 3, 4). Jumping off from the original observation that N-methyl-D-aspartate (NMDA) NR1 hypomorphs show PPI deficits in addition to other behavioral abnormalities (Duncan et al. 2004), several genetic mutants have been made based on reduced NMDA receptor expression, including NMDA subtype-specific mutants (reviewed in Powell et al. (2009)) and downstream signaling proteins (Table 3).

The converging GABA-glutamate theory of schizophrenia stems from two lines of evidence (Coyle 2004). First, the glutamate hypothesis of schizophrenia is derived from evidence that acute administration of phencyclidine (PCP), a non-competitive NMDA antagonist, produces schizophrenia-like symptoms in healthy humans (Javitt 2004; Javitt and Zukin 1991). Extending such observations, several experimental studies have utilized another NMDA antagonist, ketamine, to induce a model psychosis in normal volunteers (Abel et al. 2003; Krystal et al. 1994; Oranje et al. 2002; van Berckel et al. 1998) and to exacerbate symptoms in patients with schizophrenia (Malhotra et al. 1997a, b; Krystal et al. 1994, 2003). Additionally, studies on postmortem brain tissue from schizophrenia patients show altered GABAergic interneuron function, particularly in parvalbumin (PV) interneurons (Benes and Berretta 2001; Lewis et al. 2005). Evidence that NMDA hypofunction specifically on GABA interneurons plays a role in the pathogenesis of schizophrenia comes from several lines of preclinical research. First, NMDA antagonists increase the firing rate of pyramidal cells and increase prefrontal glutamate release (Jackson et al. 2004; Moghaddam et al. 1997), suggesting that NMDA antagonists are preferentially blocking inhibitory interneurons. Second, preclinical studies suggest that GABA interneurons are more sensitive to NMDA antagonists than other neuronal subtypes such as pyramidal neurons (Grunze et al. 1996). Third, repeated exposure to NMDA antagonists such as ketamine decreases PV and GAD67 expression (Behrens et al. 2007). Based on these converging lines of evidence, Belforte et al. (2010) used the Cre-LoxP system to engineer mice with deletion of NR1 specifically in GABAergic interneurons (Table 1). Their elegant set of studies showed that reduction in NR1 early in postnatal development produced deficits in PPI, but that adult NR1 reduction had no effect on PPI. These studies corroborate previous studies reporting PPI deficits with constitutive NR1 reduction throughout the brain (Duncan et al. 2004). There have been several other genetic mutants created to test hypotheses regarding glutamate signaling (see Table 4). For example, mice with pre-adolescent forebrain-specific deletions of the vesicular glutamate transporter VGLUT2$^{f/f;CKII/Cre}$ show PPI deficits, which are attenuated with the antipsychotic drug aripiprazole (Wallén-Mackenzie et al. 2009). Genetic mutants of specific proteins involved in NMDA receptor signaling such as mice heterozygous for SynGAP gene deletion also show PPI deficits in

Table 3 Genetically engineered model organisms based on genes related to: *miscellaneous biological processes*

References	Mouse strain, sex	Model description/background/rationale	Basal PPI	Effects of drugs or manipulations typically used to induce PPI-deficits	Effects of (presumed) antipsychotics/other treatments
		Second messengers			
Siuciak et al. (2008)	PDE4B KO, M	PDE4 catalyzes the degradation of cAMP. Widely expressed in brain	↓PPI & ↑PA in −/− versus WT (↓PPI not dependent on ↑PA)		
McOmish et al. (2008)	PLC-β1 KO	PLC-β1 is a rate-limiting enzyme involved in the intracellular signaling cascade linked to several metabotropic receptors implicated in SZ; PLC-β1 reduced in SZ brain	↓PPI & PA (−/− vs. WT)		*EE (main effects)*: ↑PPI & ↓PA (trend) versus standard housing); EE × genotype × PP intensity effect on PPI (↑PPI in −/−); EE × genotoype effect on PA (↓PA in WT); HAL (no rescue of PPI deficit of −/− vs. VEH treated WT; ∅PA); CLO (main effect ↑PPI; genotype × CLO × PP intensity effect: rescue of PPI deficit of −/− vs. VEH treated WT; main effect: ↓PA vs. VEH; genotype × CLO; more pronounced ↓PA in WT)
Koh et al. (2008)	PLC-β1 KO		↓PPI & trend for ↓PA (−/− vs. WT)		*HAL*: normalized PPI in −/−, no effect in WT; ∅PA
		Growth factors			
Ransome et al. (2008)	SOCS-2 TG	SOCS-2 affects growth hormone signaling & potentially hippocampal neurogenesis	∅PPI in TG versus WT		
Scearce-Levie et al. (2008)	Fgf17 KO, M & F	Fgf17 is involved in region-specific development of the rodent fCTX	∅PPI & ∅PA in −/− versus +/− or WT		

(continued)

Table 3 (continued)

References	Mouse strain, sex	Model description/background/rationale	Basal PPI	Effects of drugs or manipulations typically used to induce PPI-deficits	Effects of (presumed) antipsychotics/other treatments
Ohgake et al. (2009)	Mdk KO, M & F	Midkine involved in neurodevelopment and adult neuroplasticity; abnormal serum levels in SZ and AD; KO produces ↓DA and DAR in striatum (this paper)	↓PPI in −/− versus WT, +/− intermediate (adult); ∅PPI in −/− versus WT (4 wk old); ∅PA in −/−, +/−, WT (adult, 4wk)		*HAL*: ↑PPI in −/−; ∅PPI in WT *CLO*: ↑PPI in −/−; ∅PPI in WT Overall ↓PA with APD; CLO (3 mg/kg) ↓PA in WT
Sun et al. (2010)	Forebrain-specific SMAD4 KO, M	SMAD4 is an intracellular transducer of TGF-β signaling; TGF-β signaling important for neurodevelopment and shown to be abnormal in SZ	↓PPI & ∅PA in −/− versus WT		
Nguyen et al. (2011)	PDGFR-$\beta^{ff:Nestin\ Cre+}$ KO, M (neuron-specific PDGFR KO)	Platelet derived growth factor (PDGF) is important for embryogenesis and CNS development. PDFGR-β has been associated with SZ and ASD	↓PPI & ∅PA in −/− versus WT		
		Protein kinase			
Bersudsky et al. (2008)	GSK3β KO, M	GSK3β levels were reported to be reduced in SZ patients. Li inhibits GSK3β *in vitro*	∅PPI & ∅PA in +/− versus WT		
Kaidanovich-Beilin et al. (2009)	GSK3α KO, M & F	GSK3 implicated in several neuronal signaling pathways	↑PPI & ∅PA in −/− versus WT		
Takao et al. (2010)	CaMKIV KO, M	CaMKIV is a protein kinase that activates transcription factor CREB and thus may play a role in synaptic plasticity	∅PPI & ∅PA in −/− versus WT		

(continued)

Table 3 (continued)

References	Mouse strain, sex	Model description/background/rationale	Basal PPI	Effects of drugs or manipulations typically used to induce PPI-deficits	Effects of (presumed) antipsychotics/other treatments
Siuta et al. (2010)	Rictor $^{ff:Nestin\ Cre+}$ KO (neuron-specific Rictor KO)	Rictor is a component of mTOR2C, the kinase responsible for phosphorylation of Akt. Akt deficiencies associated with SZ	↓PPI & ∅PA in −/− versus WT		*NIS*: (↑PPI in −/−; no data in WT; PA data not shown) *CLO*: (∅PPI in −/−; PA data not shown)
		Proteases			
Horii et al. (2008)	Neuropsin KO, M	The serine protease neuropsin may be involved in neuronal plasticity/degeneration	∅PPI & PA in −/− versus WT		
Savonenko et al. (2008)	BACE1 KO, M	BACE1 is critical for APP cleavage/amyloid β production. May also play a role in SZ via proteolysis of NRG & axon myelination	↓PPI in −/− & ∅PPI in +/− versus WT; ∅PA in −/− & +/− versus WT; ↑PA latency in −/− versus +/− & WT		*CLO:* (amelioration of PPI deficits in −/−; ∅PPI in +/− & WT)
		Synaptic proteins			
Dyck et al. (2007)	SYN II KO	SYN II, a vesicle-linked phosphoprotein plays a role in neuronal development & transmitter release; hypothesized to contribute to the etiology of SZ	↓PPI (−/− relative to +/− & WT)		
Dyck et al. (2009)	SYN II KO, M & F	↓SYNII in mPFC of SZ patients; ↑SYNII levels with chronic APD	↓PPI in −/− versus +/− and WT; ↓PA in −/− versus +/− and WT; lack of HAB in −/−		

(continued)

Table 3 (continued)

References	Mouse strain, sex	Model description/background/rationale	Basal PPI	Effects of drugs or manipulations typically used to induce PPI-deficits	Effects of (presumed) antipsychotics/other treatments
Savonenko et al. (2009)	EP2 KO, F	EP2 is a prostaglandin receptor involved in activity-dependent synaptic plasticity	↓PPI in −/− versus WT; ∅PA in −/− versus WT		
Oliver and Davies (2009)	SNAP-25 *Bdr* (*Blind drunk* mutant), SNAP-25 +/−, combined with prenatal stress (PNS), M	SNARE protein SNAP-25 important for synaptic function and neurotransmitter release; SNAP-25 genomic region linked to SZ	↓PPI & ∅PA in *Bdr* mutant versus WT, ↓PPI exacerbated by PNS; ∅PPI & ∅PA in +/− versus WT, no effect of PNS in +/−		*CLO*: ↑PPI in BDR/NS & BDR/PNS; ∅PPI in WT/NS or WT/PNS *HAL*: ↑PPI in Bdr/PNS at 86dB PP only; ∅PPI in WT/NS, Bdr/NS, or WT/PNS
Blundell et al. (2010)	Rim1α KO, M Rab3A KO, M Syt1R233Q (point mutation) KI, M&F Rim1αS413A KI, M	Presynaptic proteins involved in neurotransmitter release and associated with SZ in genetic studies	*Rim1α*: ↓PPI &∅PA in −/− versus WT *Rab3A*: ↓PPI & ∅PA in −/−versus WT *Syt1R233Q*: ↓PPI & ∅PA in KI versus WT *Rim1αS413A*: ∅PPI & ∅PA in KI versus WT		
		Calcium Signaling			
Matsuo et al. (2009)	RyR3 KO	Ryanodine receptor 3 is an intracellular calcium release channel preferentially expressed in HPC and STR and involved in synaptic transmission and plasticity. RyR3 forms a signaling complex with calcineurin, a potential susceptibility gene for SZ and BD	↓PPI in −/− versus WT at 78 dB PP + 110 dB startle pulse only; ∅PA in −/− versus WT		

(continued)

Table 3 (continued)

References	Mouse strain, sex	Model description/background/rationale	Basal PPI	Effects of drugs or manipulations typically used to induce PPI-deficits	Effects of (presumed) antipsychotics/other treatments
Nakagawasai et al. (2010)	$Ca_v2.2$ KO, M	N-type calcium channels are important for neurotransmitter release, brain development, and activity-depdendent plasticity	↓PPI & ∅PA in −/− versus WT		
		Potassium channel signaling			
Kapfhamer et al. (2010)	Ppp2r5δ KD, M & F GSK3β KD, M KCNQ2 KD, M & F KD created via gene trap (GT), reduced expression by 50%	KCNQ2 gene encodes an M-type potassium channel subunit and may be a substrate for GSK3β; PP2A regulates GSK3β; GSK3β and M-type potassium channels may be linked to SZ	*Ppp2r5δGt*: ↓PPI & ∅PA in GT/+ versus WT *GSK3β*: ↓PPI & ∅PA in +/− versus WT *KCNQ2*: ↓PPI & ∅PA in +/− versus WT		
		Immune activation			
Smith et al. (2007)	IL6 KO	Maternal immune activation (here simulated via poly I:C injection) has been implied in the pathogenesis of SZ		*Maternal poly I:C injection.* *Effects in offspring*: ∅PPI in −/− but ↓PPI in WT	
Asp et al. (2010)	Tap1 KO mice	Prenatal influenza infection associated with increased SZ risk. Transporter associated with antigen processing 1 mice have reduced expression of MHC class I	No direct comparisons between −/− and WT mice. C57BL6/J WT mice purchased from Jackson Laboratories	Influenza infection on PND 3 or 4: ↓PPI in adult −/− mice (during ISI block), trend for ↑PA in infected mice in startle threshold block; ∅PPI & ∅PA in infected versus non-infected WT C57 mice	

(continued)

Table 3 (continued)

References	Mouse strain, sex	Model description/background/rationale	Basal PPI	Effects of drugs or manipulations typically used to induce PPI-deficits	Effects of (presumed) antipsychotics/other treatments
López-Ramos et al. (2010)	TgIE96, M	Mice expressing pseudorabies virus immediate-early protein IE180 (TgIE96) have cerebellar abnormalities	↓PPI & ∅PA in Tg versus WT (C57BL/6J WT mice purchased from Jackson Laboratories)		
		Glucose metabolism			
Schmidt et al. (2008)	GLUT3 (Slc2a3) KO, M	GLUT is widely expressed in brain. Role in neuronal glucose homeostasis and likely role in diverse behaviors	↓PPI at lower PP intensities & ↑PA in +/− versus WT		
		Neuromodulatory functions			
Errico et al. (2008)	DDO KO, M	The enzyme DDO degrades D-aspartate, an amino acid with elusive function in the CNS. A neuromodulatory function at NMDA receptors has been implied	∅PPI & ∅PA in −/− versus WT	AMP (∅PPI in DDO −/−; ↓PPI in WT) DIZ (↓PPI in −/− & WT, more pronounced in WT)	Chronic D-aspartate (∅PPI & PA) antagonized AMP & DIZ-induced ↓PPI in WT (not tested in −/−)
Tanda et al. (2009)	nNOS KO, M	nNOS synthesizes NO from L-arginine; involved in many intracellular signaling processes; ↓nNOS expression in STR and HPC of SZ; polymorphisms of nNOS associated with SZ	∅PPI & ∅PA in −/− versus WT	*SKF81297 (DA D1 agonist)*: ↓PPI & ∅PA in −/−; ∅PPI & ↓PA in WT	
		Altered myelination			
Tanaka et al. (2009)	$PLP1^{(tg/-)}$	Mylein and oligodendrocyte dysfunction may contribute to schizophrenia pathogenesis. Myelin proteolipid protein (plp1) decreased in brains of SZ patients	↓PPI in Tg versus WT at higher PP intensity (78dB), PA data not shown		

(continued)

Table 3 (continued)

References	Mouse strain, sex	Model description/background/rationale	Basal PPI	Effects of drugs or manipulations typically used to induce PPI-deficits	Effects of (presumed) antipsychotics/other treatments
		Spine morphogenesis			
Cahill et al. (2009)	KALRN KO	Kalrinin, a guanine-nucleotide exchange factor (GEF) for Rac-like GTPases, regulates spine morphogenesis. KALRN −/− have decreased spine density (this paper)	↓PPI in −/− versus WT; PA data not shown		
		Oxidative stress			
Benoit et al. (2010)	QR2 KO, M	QR2 enhances production of quinone and reactive oxygen species (RCS) and is upregulated in cognitively impaired aged rats; polymorphism in QR2 promoter region weakly associated with SZ	∅PPI in −/− versus WT; PA data not shown		
Cole et al. (2011)	GCLM KO, M	GCLM is rate-limiting enzyme in glutathione (GSH) synthesis; GSH defends against oxidative stress	∅PPI & ∅PA in −/− versus WT		
		Metallothionines			
Koumura et al. (2009)	MT3 KO, M	Metallothionines (MTs) bind zinc and other metals and are neuroprotective in some models	↓PPI & ∅PA in −/− versus WT		
		Neurodevelopmental genes (other)			
Willi et al. (2010)	Nogo-A KO, M	Nogo-A inhibits growth during development and in adulthood and thus may be involved in neurodev and neural plasticity; some evidence of pathology in SZ	↓PPI & ∅PA in −/− versus WT		

(continued)

Table 3 (continued)

References	Mouse strain, sex	Model description/background/rationale	Basal PPI	Effects of drugs or manipulations typically used to induce PPI-deficits	Effects of (presumed) antipsychotics/other treatments
Takata et al. (2010)	XBP1 KO, M	XBP1 gene encodes transcription factor in the ER unfolded protein response; involved in brain development and potentially pathogenesis of bipolar disorder and depression	↑PPI & ∅PA in +/− versus WT		
		Glycoproteins			
Sakatani et al. (2009)	RAGE KO, M	RAGE is a receptor for multiple ligands including Aβ, HMGB1, and S100B, which contribute to the pathology of AD, epilepsy, and ischemia	↑PPI in −/− versus WT ↑PA in −/− versus WT at low sound pressure levels (70–100 dB); ↓PA in −/− versus WT at 120 dB		
Fukuda et al. (2011)	Fut8 KO, M	α1,6 fucosyltransferase (Fut8)[2] transfers fucose from a GDP-fucose to position 6 of N-acetylglucosamine to form N-linked oligosaccharides of glycoproteins. Fucosylated glycoproteins widely distributed in brain	↓PPI & ∅PA in −/− versus +/− & WT	*After restraint stress*: ↓PPI & ∅PA in +/− versus WT	

(continued)

Table 3 (continued)

References	Mouse strain, sex	Model description/background/rationale	Basal PPI	Effects of drugs or manipulations typically used to induce PPI-deficits	Effects of (presumed) antipsychotics/other treatments
		Sex hormones			
Chavez et al. (2009)	ArKO, M, F	Gender differences in the age of onset, severity, and response to treatment in SZ. Second increase in SZ incidence in older age for women when estrogen levels decline. Thus, estrogen may be neuroprotective against development of SZ. Aromatase (Ar) converts androgens to estrogens	∅PPI in −/− versus WT	AMP ↓PPI in −/− & WT (↓PPI slightly attenuated in F −/−), ↓PA in −/− & WT *Apo*: ↓PPI in −/− & WT (↓PPI slightly attenuated in F −/−), ↓ PA in −/− & WT *MK801*: ↓PPI & ↑PA in −/− & WT	

Abbreviations: *5-HT* serotonin, *AAV* adeno-associated virus, *Aβ* amyloid β- peptide, *ACE* angiotensin converting enzyme, *ACH* acetylcholine (receptor), *AD* Alzheimer's disease, *ADX* adrenalectomy, *AIL* advanced intercross line, *AMP* amphetamine, *AMY* amygdala, *APD* antipsychotic drug, *APP* amyloid precursor protein, *ARIP* aripiprazole, *APO* apomorphine, *AT* angiotensin, *BAC* bacterial artificial chromosome, *BACE* β-site APP cleaving enzyme, *BD* bipolar disorder, *EG* background, *CaMKIV* calcium-calmodulin-dependent protein kinease IV, *cAMP* cyclic adenosine monophosphate, *Ckr* chakragati, *CLO* clozapine, *CNS* central nervous system, *COC* cocaine, *COMT* catechol-O-methyltransferase, *CORT* corticosterone, *CRF* corticotropin releasing factor, *CTR* controls, *CTX* cortex, *DA* dopamine (receptor), *DAT* dopamine transporter, *dB* decibel, *DCC* deleted in colorectal cancer, *DIAZ* diazepam, *DIZ* dizocilpine, *DN* dominant-negative, *DDO* D-aspartate oxidase, *E* embryonic day, *EE* environmental enrichment, *EGF* epidermal growth factor, *ENU* N-ethyl-N-nitrosourea, *f* frontal, *F* female, *FABP* fatty acid binding protein, *FGF* fibroblast growth factor, *Fmr1* fragile × mental retardation 1 gene, *FMRP* fragile × mental retardation protein, *FXS* fragile × Syndrome, *GLAST* glutamate and aspartate transporter, *GLU* glutamate, *GluR* glutamate receptor, *GLUT* glucose transporter, *GR* glucocorticoid receptor, *GRIP* glutamate receptor interacting protein, *GSK* glycogen synthase kinase, *GWAS* genome wide association studies, *HAL* haloperidol, *HD* Huntington's disease, *HPC* hippocampus, *IC* imprinted cluster, *IL* interleukin, *ISI* interstimulus interval, *KI* knock-in, *KO* knock-out, *M* male, *m* metabotropic, *MDB* methyl-CpG binding protein, *mo* month, *NIC* nicotine, *MPEP* 2-methyl-6-(phenylethylyn)-pyridine hydrocholoride, *METH* metamphetamine, *Mgat* N-acetylglucosaminyltransferase, *NAA* N-acetyl-aspartate, *NAAG* N-acetylalpha L-aspartyl-L-glutamate, *NAC* nucleus accumbens, *NCAM* neural cell adhesion molecule, *NIS* nisoxetine, *nNOS* neuronal nitric oxide synthase, *NO* nitric oxide, *NPS* neuropeptide S (receptor), *NPY* neuropeptide Y, *NR* NMDA receptor subunit, *NRG* neuregulin, *NRL* neuroligin, *ns* not significant(ly), *NSE* neuron-specific enolase, *NT* neurotensin, *OE* overexpressor, *OXO* oxotremorine, *PA* magnitude of response to pulse alone, *PACAP* pituitary adenylate-cyclase-activating polypeptide, *PD* Parkinson's disease, *PDE* phosphodiesterase, *PET-1* plasmocytoma expressed transcript-1, *PND* postnatal day, *PLC* phospholipase C, *PNS* prenatal stress, poly I:C polyinosinic: polycytidylic acid, *PP* prepulse, *PPI* prepulse inhibition of startle, *PPP* prepulse potentiation, *PS1* presinilin1, *PWS* Prader–Willi syndrome, *QTL* quantitative trait locus, *QUET* quetiapine, *RAC* raclopride, *RAGE* receptor for advanced glycation end-products, *RasGAP* Ras GTPase-activating protein, *RE* renin-enhancer, *RIS* risperidone, *SCOP* scopolamine, *SI* social isolation, *SNP* single nucleotide polymorphism, *SOCS* suppressor of cytokine signaling; *SREB* superconserved receptor expressed in brain; *STR* striatum, *SYN* synapsin, *SynGAP* synaptic GTP-ase-activating protein, *SZ* schizophrenia, *TAAR* trace amine-associated receptor, *TG* transgenic, TGF-β transforming growth factor beta, *TK* typosine kinase, *TMS* transcranial magnetic stimulation, *V1b* vasopressin receptor 1b, *WT* wild-type

↓ decreased, ↑ increased, ∅ unchanged, −/− homozygous mice, +/− heterozygous mice

Table 4 Genetically engineered model organisms (ca. 07/01/2007-06/28/2011) based on genes related to: *models for specific disorders*

References	Mouse strain, sex	Model description/background/rationale	Basal PPI	Effects of drugs or manipulations typically used to induce PPI-deficits	Effects of (presumed) antipsychotics/other treatments
		Alzheimer's disease			
Dejaegere et al. (2008)	Alpha 1B/C γ-secretase KO	γ-secretases have been implicated in the cleavage of membrane proteins incl. APP, & NRG (a SZ vulnerability gene) & in NOTCH signaling	↓PPI (most pronounced at high PA intensities) & ∅PA in −/− versus WT		*HAL:* (↑PPI & ∅PA in −/− & WT) *CLO:* (↑PPI in −/−; ∅PPI in WT; ∅PA in −/− & WT)
Gruart et al. (2008)	APP, PS1, APP+PS1 TG, M	Neuropathological changes in brain regions regulating PPI may occur in AD models & may affect PPI	↓PPI in 18 mo old WT, APP KO, PS1 KO & APP+PS1 TG versus 3 mo old WT. Most pronounced ↓PPI in 18 mo APP+PS1 TG. PA data not shown		
Tsujimura et al. (2008)	KF-1 KO, M	KF-1 gene increased in AD patients; localized in hippocampus, cerebellum; modulates protein levels as a ubiquitin ligase; increased in frontal ctx after chronic antidepressant treatment, electroconvulsive therapy, and TMS	↑PPI in −/− versus WT; ↓PA in −/− versus WT		
Takeuchi et al. (2011)	P301S Tg, M	Tau, a microtubule-associated protein, can cause cell death and is associated with AD and other neurodegenerative diseases. FTDP17 (frontotemporal dementia and PD linked to chromosome 17) is linked to a point mutation in tau (P301S)	↑PPI & ↓PA at 120 dB only in Tg versus WT		

(continued)

Table 4 (continued)

References	Mouse strain, sex	Model description/background/rationale	Basal PPI	Effects of drugs or manipulations typically used to induce PPI-deficits	Effects of (presumed) antipsychotics/other treatments
		Autism			
Allan et al. (2008)	MDB1 KO, M & F	Via DNA methylation, MDB mediate epigenetic gene regulation, which has been linked to neurodevelopmental disorders, including autism	↓PPI & ∅PA in −/− versus WT		
Chadman et al. (2008)	NL3 KI, M & F	Neuroligin-3 (NL3) is a cell adhesion molecule involved in synapse development & implicated in autism; "knock in" of point mutation identified in autism	∅PPI in KI versus WT; ↓PA in KI versus WT		
Radyushkin et al. (2009)	NL3 KO, M	Point mutations in neuroligin-3 are associated with autism	∅PPI & ∅PA in −/− versus WT		
Moy et al. (2009)	NRCAM KO, M & F	NRCAM identified as a candidate gene for autism; involved in cell-cell interactions during brain development and thus affect axonal guidance and neural circuit development	↓PPI in −/− versus WT (M only; ∅PPI in F); ∅PA in −/− versus WT		
DeLorey et al. (2011)	GABRB3 KO, M & F	GABRB3 gene associated with autism	∅PPI (F) & ↑PPI (M) in +/− versus WT; ↓PA (M & F) in +/− versus WT		
Mejias et al. (2011)	GRIP1/2 DKO (double KO, Grip1 flox/KO, Grip2 KO/KO)	GRIP1 scaffolding protein that interacts with GLU2/3 via PDZ domains 4-6. SNP in PDZ4-6 domain and 5 missense variants identified in ASD (this paper)	↓PPI in −/− versus WT mice		

(continued)

Table 4 (continued)

References	Mouse strain, sex	Model description/background/rationale	Basal PPI	Effects of drugs or manipulations typically used to induce PPI-deficits	Effects of (presumed) antipsychotics/other treatments
		Fragile X-syndrome (FXS)			
deVrij et al. (2008)	FMR1 KO	FXS involves mental retardation & is caused by the absence of FMRP. PPI is reduced in FXS patients	↓PPI at most PP intensities in −/− versus WT		MPEP (mGluR5 antagonist; ↑PPI in −/− & WT)
Paylor et al. (2008)	FMR1 KO, FMR1 KO with a yeast artificial chromosome containing human FMR1 gene, M		↑PPI & ↓PA (trend) in −/− versus WT		Addition of human FMR1 gene normalized, i.e. ↓PPI & ↑PA in FMR1 KO to WT levels, & ↓PPI & ∅PA in WT
Baker et al. (2010)	FMR1 KO, M & F	Background strain may affect behavioral phenotype in FMR1 KO mice; bred to albino C57BL/6J-Tyr(c-Brd) background	↑PPI & ↓PA in −/− versus WT (both sexes)		
Levenga et al. (2011)	FMR1 KO, M	FMRP involved in translation of mRNA at the synapse, downstream from mGlu5 signaling	↓PPI in −/− versus WT; PA data not reported (startle measured via eyeblink)		*AFQ056, mGluR5 antagonist*: ↑PPI in −/−; ∅PPI in WT
Veeraragavan et al. (2011)	FMR1 KO, M	Increased M1 muscarinic signaling in FMR1 KO mice	∅PPI & ∅PA in −/− versus WT		*Dicyclomine*: ∅PPI & ∅PA in −/− or WT
Thomas et al. (2011a)	mG1.FMR1 (mGlu1 HZ + FMR1 KO); mG5.FMR1 (mGlu5 HZ + FMR1 KO); FMR1 KO	FMR1 may increase signaling through Group1 mGluRs (mGlu1, mGlu5)	↑PPI & ↓PA in FMR1−/− & mG5.FMR1 versus WT; ↑PPI & ↓PA in FMR1−/− & mG1.FMR1 versus WT		

(continued)

Table 4 (continued)

References	Mouse strain, sex	Model description/background/rationale	Basal PPI	Effects of drugs or manipulations typically used to induce PPI-deficits	Effects of (presumed) antipsychotics/other treatments
Thomas et al. (2011b)	FMR1 KO, M & F	Group 1 mGluRs may be pharmacological target in Fragile X	↓PA & ∅PPI (trend for predicted ↑) in −/− versus WT	*JNJ (mGlu1 antagonist)*: ↓PPI in WT, ∅PPI in −/−; ∅PA in −/− & WT *MPEP (mGlu5 antagonist)*: ∅PPI & ∅PA in −/− & WT	
		Huntington's disease			
Menalled et al. (2009)	R6/2 TG, R6/2B TG, YAC128 TG, YAC128B TG, BACHD TG, HDH^{Q111} KI, M & F	HD results from a known genetic CAG repeat mutation; ↓PPI in HD patients; HD mutant mice show varying degrees of neuropathology and behavioral abnormalities; comparison controlling for many factors including sex, background strain, husbandry, and handling	*R6/2*: ↓PPI in TG versus WT beginning at 12 wks; ↓PA *R6/2B*: ↓PPI in TG versus WT *BACHD*: ↓PPI beginning at 24 wks; ∅PA *YAC128B*: ↓PPI at 54 wks; ∅PA *YAC128*: ∅PPI; ∅PA HDH^{Q111} KI: ∅PPI; ∅PA		
Brooks et al. (2010)	$Hdh^{(CAG)150}$ KI, M & F	Longitudinal characterization of KI mouse carrying 150 CAG repeats on the mouse Htt locus	↓PPI & ↓PA in KI versus WT		
Brooks et al. (2011)	R6/1 TG, M & F	R6/1 TG mouse contains ~115 CAG repeats on exon 1 of the HD gene. Characterization of R6/1 TG mice on C57BL/6 background is needed	↓PPI & ↓PA in TG versus WT		

(continued)

Table 4 (continued)

References	Mouse strain, sex	Model description/background/rationale	Basal PPI	Effects of drugs or manipulations typically used to induce PPI-deficits	Effects of (presumed) antipsychotics/other treatments
Pietropaolo et al. (2011)	R6/1 TG, M & F	Investigate neuropsychiatric symptoms in pre-motor stage of pathology	∅PPI & ∅PA in TG versus WT		
		Hearing disorders			
Allen et al. (2008)	Kv1.1 KO	PPI in rodents is a powerful means to identify hearing deficits. The role of Kv1.1 channels in PPI gap detection paradigms was assessed	↓PPI (for long gaps only)		
		22q11 deletion & duplication syndromes			
Suzuki et al. (2009b)	SEPT5 (Septin5) KO, M	22q11.2 microdeletions result in cognitive & behavioral abnormalities and are associated with SZ and autism; SEPT5 gene located on 200kb region spanning the deletion	↑PPI & ∅PA in −/− versus WT		
Stark et al. (2009)	*BAC Tg-1*: Overexpress Prodh & Vpreb2 *BAC Tg-2*: overexpress Zdhhc8, Ranbp1, Htf9c, T10, Arvcf, & COMT	22q11.2 microduplications are also associated with behavioral problems and learning disabilities; gain of function mutations can also be informative for the deletion syndrome	*Tg-1*: ↑PPI & ∅PA in Tg-1 versus WT *Tg-2*: ∅PPI & ∅PA in Tg-2 versus WT		

(continued)

Table 4 (continued)

References	Mouse strain, sex	Model description/background/rationale	Basal PPI	Effects of drugs or manipulations typically used to induce PPI-deficits	Effects of (presumed) antipsychotics/other treatments
Suzuki et al. (2009a)	*BAC Tg, M*: overexpress 190 kb segment of 22q11.2 including TXNRD2, COMT, ARVCF	22q11.2 is a hotspot for CNVs. 97% of children with 22q11.2 duplications show cognitive deficits	∅PPI & ∅PA in Tg versus WT		
		Down syndrome			
Ortiz-Abalia et al. (2008)	TG DyrK1A overexpressors, M	DyrK1A overexpression has been implicated in Down Syndrome	↓PPI at higher PP intensities in TG versus WT (trend only); PA not shown		*Intrastriatal infusion of AAV iRNA against Dyrk1A*: PP intensity-dependent ↑PPI in TG treated with functional versus scrambled viral vectors & pre versus post-treatment with functional vector; PA not shown
		Glycosilation IIa congenital disorders			
Soleimani et al. (2008)	Mgat5 KO, M & F	Mgat5 is involed in glycosylation of cell surface glycoproteins, including receptors & transporters. Abundant in the CNS	∅PPI & PA in −/− versus WT; ↑PA in M versus F (main effect)		

(continued)

Table 4 (continued)

References	Mouse strain, sex	Model description/background/rationale	Basal PPI	Effects of drugs or manipulations typically used to induce PPI-deficits	Effects of (presumed) antipsychotics/other treatments
		Prader-Willi syndrome			
Relkovic et al. (2010)	PWS-IC (+/−), M & F	Prader-Willi Syndrome (PWS) neurodevelopmental disorder resulting from deletion of paternally expressed imprinted genes of chromosome 15q11-q13, a region also implicated in autism and SZ	↓PPI & ↑PA in +/− versus WT in M & F (only at P105 in M)		

Abbreviations: *5-HT* serotonin, *AAV* adeno-associated virus, *Aβ* amyloid *β*- peptide, *ACE* angiotensin converting enzyme, *ACH* acetylcholine (receptor), *AD* Alzheimer's disease, *ADX* adrenalectomy, *AIL* advanced intercross line, *AMP* amphetamine, *AMY* amygdala, *APD* antipsychotic drug, *APP* amyloid precursor protein, *ARIP* aripiprazole, *APO* apomorphine, *AT* angiotensin, *BAC* bacterial artificial chromosome, *BACE β*-site APP cleaving enzyme, *BD* bipolar disorder, *BG* background, *CaMKIV* calcium-calmodulin-dependent protein kinease IV, *cAMP* cyclic adenosine monophosphate, *Ckr* chakragati, *CLO* clozapine, *CNS* central nervous system, *COC* cocaine, *COMT* catechol-O-methyltransferase, *CORT* corticosterone, *CRF* corticotropin releasing factor, *CTR* controls, *CTX* cortex, *DA* dopamine (receptor), *DAT* dopamine transporter, *dB* decibel, *DCC* deleted in colorectal cancer, *DIAZ* diazepam, *DIZ* dizocilpine, *DN* dominant-negative, *DDO* D-aspartate oxidase, *E* embryonic day, *EE* environmental enrichment, *EGF* epidermal growth factor, *ENU* N-ethyl-N-nitrosourea, *f* frontal, *F* female, *FABP* fatty acid binding protein, *FGF* fibroblast growth factor, *Fmr1* fragile × mental retardation 1 gene, *FMRP* fragile × mental retardation protein, *FXS* fragile × Syndrome, *GLAST* glutamate and aspartate transporter, *GLU* glutamate, *GluR* glutamate receptor, *GLUT* glucose transporter, *GR* glucocorticoid receptor, *GRIP* glutamate receptor interacting protein, *GSK* glycogen synthase kinase, *GWAS* genome wide association studies, *HAL* haloperidol, *HD* Huntington's disease, *HPC* hippocampus, *IC* imprinted cluster, *IL* interleukin, *ISI* interstimulus interval, *KI* knock-in, *KO* knock-out, *M* male, *m* metabotropic, *MDB* methyl-CpG binding protein, *mo* month, *NIC* nicotine, *MPEP* 2-methyl-6-(phenylethylyn)-pyridine hydrocholoride, *METH* metamphetamine, *Mgat* N-acetylglucosaminyltransferase, *NAA* N-acetyl-aspartate, *NAAG* N-acetylalpha L-aspartyl-L-glutamate, *NAC* nucleus accumbens, *NCAM* neural cell adhesion molecule, *NIS* nisoxetine, *nNOS* neuronal nitric oxide synthase, *NO* nitric oxide, *NPS* neuropeptide S (receptor), *NPY* neuropeptide Y, *NR* NMDA receptor subunit, *NRG* neuregulin, *NRL* neuroligin, *ns* not significant(ly), *NSE* neuron-specific enolase, *NT* neurotensin, *OE* overexpressor, *OXO* oxotremorine, *PA* magnitude of response to pulse alone, *PACAP* pituitary adenylate-cyclase-activating polypeptide, *PD* Parkinson's disease, *PDE* phosphodiesterase, *PET-1* plasmocytoma expressed transcript-1, *PND* postnatal day, *PLC* phospholipase C, *PNS* prenatal stress, poly I:C polyinosinic: polycytidylic acid, *PP* prepulse, *PPI* prepulse inhibition of startle, *PPP* prepulse potentiation, *PS1* presinilin1, *PWS* Prader–Willi syndrome, *QTL* quantitative trait locus, *QUET* quetiapine, *RAC* raclopride, *RAGE* receptor for advanced glycation end-products, *RasGAP* Ras GTPase-activating protein, *RE* renin-enhancer, *RIS* risperidone, *SCOP* scopolamine, *SI* social isolation, *SNP* single nucleotide polymorphism, *SOCS* suppressor of cytokine signaling; *SREB* superconserved receptor expressed in brain; *STR* striatum, *SYN* synapsin, *SynGAP* synaptic GTP-ase-activating protein, *SZ* schizophrenia, *TAAR* trace amine-associated receptor, *TG* transgenic, TGF-*β* transforming growth factor beta, *TK* typosine kinase, *TMS* transcranial magnetic stimulation, *V1b* vasopressin receptor 1b, *WT* wild-type

↓ decreased, ↑ increased, ∅ unchanged, −/− homozygous mice, +/− heterozygous mice

addition to other behavioral abnormalities (Guo et al. 2009). For a review of other neurotransmitter-related genetic mutants [e.g. acetylcholine, serotonin, dopamine (see Table 4)].

In addition to classic neurotransmitters, the important role of neuropeptides in neuropsychiatric disease is becoming more widely appreciated. PPI has been assessed in genetic mutants for many neuropeptides including angiotensin, neurotensin, corticotropin releasing factor (CRF) overexpression, oxytocin, arginine vasopressin, neuropeptide Y, PACAP, and Relaxin-3 (Table 4). Stress physiology, particularly CRF and glucocorticoids, has been implicated in neuropsychiatric disease. Mice overexpressing a CRF transgene exhibit deficits in PPI (at low prepulse intensities), which are normalized by CRF1 antagonists, whereas glucocorticoid receptor antagonists did not reverse the PPI deficit (Groenink et al. 2008). Moreover, several genetic mutants have been created to determine the role of specific proteins in brain processes, including second messenger signaling, synaptic proteins, synaptic vesicles, protein kinases, etc. (Table 3). While the direct link for many of these proteins has not necessarily been established for specific neuropsychiatric diseases, a thorough behavioral evaluation of these mice including an assessment of PPI can help elucidate the functional implications of the target protein. For example, gene deletion of PDE4B, which catalyzes the degradation of cAMP and is widely expressed in brain, is associated with PPI deficits (Siuciak et al. 2008). Decreased presynaptic proteins such as synaptophysin, SNAP-25, and complexin II have been observed in postmortem brains of schizophrenia patients (Harrison and Weinberger 2005). These data, combined with evidence of problems with neuronal migration and abnormal neuronal processes, have led investigators to conceptualize schizophrenia as a disease of functional "dysconnectivity" (Friston and Frith 1995; McGlashan and Hoffman 2000; Weinberger et al. 1992) or a "disorder of the synapse" (Frankle et al. 2003; Mirnics et al. 2001) reviewed in (Harrison and Weinberger 2005). Thus many genetic mutants have been created for specific synaptic proteins (Table 3). For example, SYNII, a vesicle-linked phosphoprotein that plays a role in neuronal development and neurotransmitter release (Cesca et al. 2010), is genetically associated with schizophrenia, decreased in brain of schizophrenia patients, and increased with chronic antipsychotic drug treatment (Chen et al. 2004; Chong et al. 2002). SYNII KO mice exhibit decreased PPI, providing more evidence for the essential role of SYNII in synaptic function and behavior (Dyck et al. 2007, 2009). Similarly, a mutation of SNAP-25, a SNARE protein that is integral to synaptic function and neurotransmitter release, is associated with PPI deficits (Oliver and Davies 2009).

3.3 Genetic Mouse Models as Pharmacological Tools

Because many of the early molecular genetic approaches focused on neurotransmitter receptor genes, several of these mutants have also been used to delineate the receptor mechanisms of drugs that disrupt PPI in order to expand our understanding of the neurotransmitter systems and neural circuitry underlying deficient PPI (Doherty et al. 2008; Dulawa et al. 1997; Ralph et al. 1999; van den Buuse et al. 2011a) (Table 4). Since most of these genetic mutants were built on rat pharmacology, we must keep in mind several important species differences between the pharmacology of mouse versus rat PPI (e.g. differential effects of dopamine D1 and D2 receptors, 5-HT agonists, etc.; Powell et al. 2009). Pharmacological disruptions of PPI and their reversal with antipsychotics have been better characterized in rats than in mice, although the literature on PPI pharmacology in mice is rapidly increasing (Swerdlow et al. 2008). In order to use mutant mouse models to examine the receptor mechanisms for alterations in PPI, more complete dose response studies are warranted in mice. Nevertheless, drugs that both impair and improve PPI continue to be investigated in genetic mutants. This approach is particularly useful in determining the functional role of a receptor (e.g. does the gene deletion alter the response to a psychotomimetic drug) and in determining the receptor mechanisms for a drug effect when either receptor-selective drugs are not available or when novel drugs are being evaluated. For example, amphetamine disrupts PPI in WT mice, but not in mice with gene deletions of DDO, an enzyme that degrades D-aspartate (Errico et al. 2008), suggesting a functional role of DDO in PPI disruptions. Interestingly, oxytocin KO mice show an increased sensitivity to the PPI-disruptive effects of PCP (Caldwell et al. 2009), suggesting that endogenous oxytocin may protect against PCP-induced disruptions of PPI, a finding that supports the clinical data showing putative antipsychotic properties of oxytocin (Feifel et al. 2010a). A similar approach was used to show protective properties of nNOS, which synthesizes NO from L-arginine, against PPI-induced disruptions with the dopamine D1 agonist SKF81297 (Tanda et al. 2009).

3.4 Genetic Models of Candidate Genes for Neuropsychiatric Diseases

As the field of neuropsychiatric genetics has grown, several gene targets have been identified consistently for particular disorders and have thus been modified in mouse models through gene deletion, the addition of a transgene, etc. When evaluating candidate genes for mouse models, one should consider that many single nucleotide polymorphisms (SNPs) identified in genetic screens for neuropsychiatric disease involve mutations in introns of genes. Thus, it is very important to consider (1) whether or not functional mutations in the gene have been identified before embarking on mutant mouse models, and (2) the impact of species–specific alterations in gene function in the mutant mouse (Low and Hardy 2007). Table 1 summarizes genetic mouse models of vulnerability genes for schizophrenia, and Table 2 summarizes genetic models of other neuropsychiatric disorders in which PPI deficits have been observed. Two approaches in human studies have increased our understanding of the genetics of sensorimotor gating (1) assessing the relationship between polymorphisms in a target gene (e.g. NRG1, COMT) and PPI levels in healthy volunteers and (2) endophenotype-based genetic studies of schizophrenia (e.g. Consortium on the Genetics of Schizophrenia; COGS). For example, recent studies suggest that PPI is associated with polymorphisms in CHRNA3 (Petrovsky et al. 2010), neuregulin 1 (Roussos et al. 2011), and COMT (Giakoumaki et al. 2008; Quednow et al. 2008; Roussos et al. 2008) genes. Some of these same genes (e.g. NRG1, COMT) are also associated with prepulse inhibition in the COGS data set (Greenwood et al. 2011).

Several different mutants for neuregulin 1 isoforms have been created with varying effects on behavior and neuroanatomy (reviewed in Duffy et al. 2008; O'Tuathaigh et al. 2007). Overall, the PPI phenotype in NRG1 mutants has been inconsistent (O'Tuathaigh et al. 2011, 2009). The Type III NRG1 heterozygote mouse has shown abnormalities in multiple neuroanatomical and behavioral endpoints relevant to schizophrenia, including PPI deficits that are improved with chronic nicotine administration (Chen et al. 2008). CNS-specific ERBB2/B4 KO mice exhibit PPI deficits, which are attenuated by the antipsychotic drug clozapine (Barros et al. 2009). Conditional knockouts with better temporal and regional specificity have allowed for evaluation of more specific roles of NRG1 and its receptors. For example, deletion of ErbB4 selectively in PV interneurons produces PPI deficits, suggesting a functional interaction between NRG1 and the integrity of PV interneurons (Wen et al. 2010).

The COMT Val allele is associated with reduced P300 and P50 ERPs (Gallinat et al. 2003; Golimbet et al. 2006; Lu et al. 2007) and reduced PPI (Quednow et al. 2008; Roussos et al. 2008), presumably due to decreased dopamine function in the prefrontal cortex. There have been several mice created to examine the effects of COMT gene on schizophrenia relevant phenotypes, COMT KO mice (Gogos et al. 1998; Papaleo et al. 2008; Yavich et al. 2007), COMT Tg mice that overexpress the human COMT-Val polymorphism (Papaleo et al. 2008), and S-COMT isoform KO (Tammimäki et al. 2010) Thus far, none of the COMT mouse models have shown robust deficits in PPI (Table 1), although other behaviors probing prefrontal cortex such as the attentional set shifting task (ASST) are altered in the COMT-Val TG mice (Papaleo et al. 2008).

The disrupted in Schizophrenia 1 (DISC1) gene has shown association with schizophrenia and behavioral and neuroanatomical biomarkers for schizophrenia (for review see Mackie et al. 2007; Porteous et al. 2006). DISC1 is associated with schizophrenia neuropathology (Cannon et al. 2005; Callicott et al. 2005), abnormal P300 ERP (Blackwood et al. 2001), and neurocognitive function (e.g. learning and memory; Burdick et al. 2005; Cannon et al. 2005; Hennah et al. 2005). The data on the dominant negative DISC1 models have reported both decreased PPI (Hikida et al. 2007) and no change in PPI (Ibi et al. 2010); whereas, the DISC1L100P mutations have shown more consistent decreases in PPI (Lipina et al. 2011; Lipina et al. 2010). The new technology of in utero gene transfer, allows for the knockdown of DISC1 into the mouse brain during early embryonic development.

Mice with in utero knockdown of DISC1 show reduced PPI when tested in adulthood (Niwa et al. 2010). As we have discussed previously, modeling the human DISC1 mutation in mice is difficult because most of the models have assumed that the mutation in neuropsychiatric disease is due to the formation of a truncated DISC1 protein (Powell et al. 2009). Another gene, Boymaw, is also disrupted on chromosome 11 in the Scottish family. Two fusion transcripts are generated between DISC1 and Boymaw genes in the translocation carriers in the Scottish schizophrenia family (Zhou et al. 2008). Thus, expressing the two fusion transcripts may be a better strategy for creating mutant DISC1 mice for the study of neuropsychiatric disorders. The increasing identification of copy number variants (CNVs) associated with schizophrenia and large-scale GWAS studies identifying novel candidate genes will lead to even greater potential to create new mouse models (see Tables for examples). Additionally, PPI is being used more widely to characterize mutant mouse models of candidate genes for other neuropsychiatric disorders such as autism and Huntington's Disease (Table 2).

3.5 Phenotype-Based Models Revealing the Function of Genes

PPI deficits have also been used in phenotype-to-genotype approaches, or "forward genetic" screens such as ENU mutagenesis, QTL on strain crosses, and QTL with selective breeding. For example, using F2 mice from a C57BL/6 x C3h/HE cross, QTL identified 6 loci for PPI, including the gene FABP7, which has been linked to NMDA receptor function (Watanabe et al. 2007). In a study comparing A/J, C57Bl6/J, and congenic crosses thereof (recombinant congenic mouse strains) (Torkamanzehi et al. 2008) found some common markers for startle but not PPI compared to previous studies (Joober et al. 2002). Selective breeding for high and low levels of PPI has identified QTLs on chromosomes 11 and 16 in low PPI versus high PPI mouse lines (Hitzemann et al. 2008; Schwabe et al. 2007). An ENU mutagenesis screen produced 2 PPI mutants (Cook et al. 2007). One major problem with these approaches is that many of the PPI mutants, particularly with ENU mutagenesis, will likely be deaf or have some degree of hearing loss. In the (Torkamanzehi et al. 2008) study on crosses of A/J, C57Bl6/J and congenic crosses, light prepulses and tactile startle pulses were used to measure PPI to avoid this potential confound (but airpuff startle also has an acoustic component to it).

The field of molecular genetics continues to produce new mutant mouse models with unknown effects on the central nervous system. Many of these mutants have behavioral abnormalities that have been observed anecdotally. Phenotypic characterization of mice with mutations of genes heretofore not known to be relevant to CNS disease (some examples found in Tables 1–5) may reveal novel genes for further study of neuropsychiatric genetics. Although these approaches were gene-based, some have the capacity to identify novel genes for a particular disease, which is why we have included these examples here. Two examples of the way a novel gene of relevance to psychiatric conditions can be discovered through the

Table 5 Phenotype-based approaches

References	Mouse strain, sex	Model description/ background/rationale	Basal PPI	Effects of drugs or manipulations typically used to induce PPI-deficits	Effects of (presumed) antipsychotics/other treatments
		QTL approaches			
Cook et al. (2007)	ENU-mutagenesis-induced variants	Genome-wide screening for neurobehavioral phenotypes, including PPI	2 PPI mutants with ↓PPI		
Watanabe et al. (2007)	F2 mice from C57BL/6 × C3H/HE crosses, M, & F	Screening for PPI and startle phenotypes	6 loci for PPI, 6 for startle and four for startle latency were identified, incl. FABP7 (linked to NMDAR function (↓PPI & ↓PA latency)		
Hitzemann et al. (2008)	C57BL/6J, DBA/2J, BALBc/J and LP/J crosses, M & F	Selective breeding for high and low PPI	∅PPI & PA in M versus F and no correlation between PPI & PA initially. Breeding line × generation effects on PPI (low vs. high PPI line) after five generations. No line × generation effects on PA, but ↑PA in low PPI line versus ↓PA in high PPI line (main effect only). QTL on chromosomes 11 & 16 in low PPI versus high PPI line	METH (Line × dose interaction. ↓PPI in high PPI line, but ∅PPI in low PPI line at highest METH dose). DIZ (↓PPI; main effect of dose only, no line × dose interaction)	HAL(↑PPI; main effect of dose only, no line × dose interaction)
Torkamanzehi et al. (2008)	A/J, C57Bl6/J and congenic crosses thereof, M	Screening for PPI and PA phenotypes using light prepulses and tactile startle pulses	Comparable PPI, but not PA in parental strains. Significant variation for congenic strains for PA for both BG and for PPI in C57Bl6 BG strains. Common markers with acoustic PA, but not acoustic PPI from previous study (Joober et al. 2002)		
Samocha et al. (2010)	LG/J x SM/J F_2 cross & F_{34} advanced intercross line (AIL), M, F	PPI and PA phenotyping followed by QTL in F_2 generation and genotyped F_{34} line at 3000 SNP markers	QTLs found at chromosome 12 for all PP intensities; QTLs for PP6 & PP12 on chromosome 7; QTLs on chromosomes 7 & 17 for startle response; QTLs on chromosome 4 for startle habituation; AIL mice reduced size of some QTLs to less than 5 cm		

(continued)

Table 5 (continued)

References	Mouse strain, sex	Model description/ background/rationale	Basal PPI	Effects of drugs or manipulations typically used to induce PPI-deficits	Effects of (presumed) antipsychotics/other treatments
		Insertional mutagenesis			
Verma et al. (2008)	Ckr mouse, M, F	*Ckr* serendipidous mutation resulted in mouse with abnormal circling behavior	↓PPI (modest) at all PP intensities in Ckr versus (+/− & WT); ∅PA in Ckr versus −/+ or WT		CLO (↑PPI in Ckr; ∅PPI in WT; RIS (∅PPI in Ckr & WT), HAL (∅PPI in Ckr & WT); PA data not shown

Abbreviations: *5-HT* serotonin, *AAV* adeno-associated virus, *Aβ* amyloid β- peptide, *ACE* angiotensin converting enzyme, *ACH* acetylcholine (receptor), *AD* Alzheimer's disease, *ADX* adrenalectomy, *AIL* advanced intercross line, *AMP* amphetamine, *AMY* amygdala, *APD* antipsychotic drug, *APP* amyloid precursor protein, *ARIP* aripiprazole, *APO* apomorphine, *AT* angiotensin, *BAC* bacterial artificial chromosome, *BACE* β-site APP cleaving enzyme, *BD* bipolar disorder, *BG* background, *CaMKIV* calcium-calmodulin-dependent protein kinease IV, *cAMP* cyclic adenosine monophosphate, *Ckr* chakragati, *CLO* clozapine, *CNS* central nervous system, *COC* cocaine, *COMT* catechol-O-methyltransferase, *CORT* corticosterone, *CRF* corticotropin releasing factor, *CTR* controls, *CTX* cortex, *DA* dopamine (receptor), *DAT* dopamine transporter, *dB* decibel, *DCC* deleted in colorectal cancer, *DIAZ* diazepam, *DIZ* dizocilpine, *DN* dominant-negative, *DDO* D-aspartate oxidase, *E* embryonic day, *EE* environmental enrichment, *EGF* epidermal growth factor, *ENU* N-ethyl-N-nitrosourea, *f* frontal, *F* female, *FABP* fatty acid binding protein, *FGF* fibroblast growth factor, *Fmr1* fragile × mental retardation 1 gene, *FMRP* fragile × mental retardation protein, *FXS* fragile × Syndrome, *GLAST* glutamate and aspartate transporter, *GLU* glutamate, *GluR* glutamate receptor, *GLUT* glucose transporter, *GR* glucocorticoid receptor, *GRIP* glutamate receptor interacting protein, *GSK* glycogen synthase kinase, *GWAS* genome wide association studies, *HAL* haloperidol, *HD* Huntington's disease, *HPC* hippocampus, *IC* imprinted cluster, *IL* interleukin, *ISI* interstimulus interval, *KI* knock-in, *KO* knock-out, *M* male, *m* metabotropic, *MDB* methyl-CpG binding protein, *mo* month, *NIC* nicotine, *MPEP* 2-methyl-6-(phenylethylyn)-pyridine hydrocholoride, *METH* metamphetamine, *Mgat* N-acetylglucosaminyltransferase, *NAA* N-acetyl-aspartate, *NAAG* N-acetylalpha L-aspartyl-L-glutamate, *NAC* nucleus accumbens, *NCAM* neural cell adhesion molecule, *NIS* nisoxetine, *nNOS* neuronal nitric oxide synthase, *NO* nitric oxide, *NPS* neuropeptide S (receptor), *NPY* neuropeptide Y, *NR* NMDA receptor subunit, *NRG* neuregulin, *NRL* neuroligin, *ns* not significant(ly), *NSE* neuron-specific enolase, *NT* neurotensin, *OE* overexpressor, *OXO* oxotremorine, *PA* magnitude of response to pulse alone, *PACAP* pituitary adenylate-cyclase-activating polypeptide, *PD* Parkinson's disease, *PDE* phosphodiesterase, *PET-1* plasmocytoma expressed transcript-1, *PND* postnatal day, *PLC* phospholipase C, *PNS* prenatal stress, poly I:C polyinosinic: polycytidylic acid, *PP* prepulse, *PPI* prepulse inhibition of startle, *PPP* prepulse potentiation, *PS1* presinilin1, *PWS* Prader–Willi syndrome, *QTL* quantitative trait locus, *QUET* quetiapine, *RAC* raclopride, *RAGE* receptor for advanced glycation end-products, *RasGAP* Ras GTPase-activating protein, *RE* renin-enhancer, *RIS* risperidone, *SCOP* scopolamine, *SI* social isolation, *SNP* single nucleotide polymorphism, *SOCS* suppressor of cytokine signaling; *SREB* superconserved receptor expressed in brain; *STR* striatum, *SYN* synapsin, *SynGAP* synaptic GTP-ase-activating protein, *SZ* schizophrenia, *TAAR* trace amine-associated receptor, *TG* transgenic, TGF-β transforming growth factor beta, *TK* typosine kinase, *TMS* transcranial magnetic stimulation, *V1b* vasopressin receptor 1b, *WT* wild-type

↓ decreased, ↑ increased, ∅ unchanged, −/− homozygous mice, +/− heterozygous mice

creation of a mutant mouse are the SP4 and the SREB2 genes. SP4 is a member of the Sp1 family of transcription factors. Hypomorphic Sp4 mice showed vacuolization in the hippocampus, age-dependent decrease in neurotrophin-3 expression in the dentate granule cells, and robust deficits in both PPI and contextual memory (Zhou et al. 2004). These studies revealed a novel Sp4 pathway that is important for hippocampal development and essential to many behaviors, including PPI, relevant to schizophrenia (Zhou et al. 2004). Zhou et al. (2009) have gone on to examine the role of the human SP4 gene in schizophrenia and bipolar disorder. Several SNPs from the human SP4 gene are found to associate with both bipolar disorder and schizophrenia in both Caucasian and Chinese samples. This work represents an example in which a PPI phenotype, in combination with other behavioral abnormalities, suggested the association of this gene with neuropsychiatric disorders. A similar finding was also observed with SREB2 Tg mice, which display PPI deficits. Follow-up genetic studies in schizophrenia found a genetic association between SREB2 and schizophrenia (Matsumoto et al. 2008).

4 Discussion

Mutant mouse models of schizophrenia provide a unique way to assess the function of a susceptibility gene, test hypotheses about the pathophysiology of disease, address receptor mechanisms of drugs, and generate hypotheses about the function of relatively unknown genes. In this review, we provided a comprehensive overview of PPI deficits in genetic models since July, 2007 and elaborate on specific examples where appropriate. As illustrated in Tables 1–5, perhaps the most notable advances in genetic approaches to sensorimotor gating over the last few years have come from candidate genes for schizophrenia and other neuropsychiatric diseases exhibiting PPI deficits, genes involved in basic synaptic processes and receptor signaling, and conditional genetic approaches that help in understanding the dynamic function of specific genes both regionally and temporally. In particular, since our last reviews (Powell et al. 2009; Swerdlow et al. 2008), many genetic mutants of proteins involved in synaptic plasticity, second messenger systems, calcium signaling, etc. have been developed. In this case, PPI is used as a functional measure of sensorimotor processing in models of basic brain development. Whether or not these cellular processes turn out to be relevant for a specific disease such as schizophrenia remains to be determined. Thus far, most models are still in the "characterization" phase—testing multiple behavioral/cognitive constructs that are deficient across neuropsychiatric disorders (e.g. sensorimotor gating, attention, social interaction, learning, and memory, etc.). Few models, however, have shown any predictive power for drug development. Perhaps this new wave of mutants based on susceptibility genes, synaptic function, or brain development, as opposed to genetic manipulation based on mechanism of action of antipsychotics or psychotomimetics, will be better models to lead the field forward in medication development for mental illness. As Moore (2010) argues, too many models that are not based on either "etiologic or pathogenic" theories, may result in too many "false positives" in drug screens. Rather, refined models that more closely mimic the etiological risk factors and/or neuropathology of disease (e.g. schizophrenia) may generate more predictive models for drug development (Moore 2010). While this approach might be very useful if the goal is to use the mutant mice for drug development, we have illustrated the benefits of careful behavioral characterization in genetic models of basic brain processes as well. As mentioned above, it is unlikely that all aspects of a heterogeneous disease will be recapitulated in another species with a genetic mutation. Investigators must rely on convergence of behavioral and neuroanatomical or neurochemical data in a given mutant mouse to support clinical data on the link between the candidate gene and disease. No single phenotype such as PPI should be considered as being either necessary or sufficient to substantiate a model as having relevance to neuropsychiatric disease.

Animal studies addressing neuropsychiatric disease would benefit greatly from more neurobiologically based biomarkers for these disorders. Such an approach has been taken in the CNTRICS initiative, in an explicit attempt to incorporate more cognitive neuroscience-based testing into treatment trials of putative cognitive therapies for schizophrenia. Along the same lines, genetic studies of schizophrenia have focused on psychophysiological endophenotypes such as PPI instead of the broader, more heterogeneous diagnosis of schizophrenia. In fact, many laboratories are now using PPI as an endophenotype in genetic studies of schizophrenia (Braff et al. 2007; Greenwood et al. 2007; Greenwood et al. 2011; Hokyo et al. 2010). Some of these genetic studies have generated further support for a genetic contribution to PPI (Greenwood et al. 2007, 2011), and other studies have suggested that genetic variants in COMT (Quednow et al. 2008; Roussos et al. 2008), CHRNA3 (Petrovsky et al. 2010), and neuregulin 1 (Roussos et al. 2011) directly affect PPI levels (as reviewed above). Thus, as human studies materialize with more neurobiologically defined behavioral measures, the ability to translate these measures or "endophenotypes" into animal models should improve dramatically. PPI in genetic models offers a behavioral endpoint that has shown predictive validity in rat pharmacological models, cross-species homology with the same measure in humans, and alterations in response to genetic manipulations implicated in the pathogenesis of schizophrenia. While these mutant mouse models are not without shortcomings, they offer some of the best attempts at etiological models that are possible in rodents. Merging the genetic etiological models with a second hit approach may strengthen some of the mutant mouse models. Hence, consideration for other factors, such as the importance of environmental risk factors (e.g. prenatal infection) and the role of epigenetics (e.g. DNA methylation) in the etiology of schizophrenia, should also be incorporated with genetic models.

Acknowledgments This work was supported by grants from the National Institute of Mental Health (R01MH042228, R01MH052885, R01MH091407) and by the Veterans Affairs VISN 22 Mental Illness Research, Education, and Clinical Center.

References

Abel KM, Allin MP, Hemsley DR, Geyer MA (2003) Low dose ketamine increases prepulse inhibition in healthy men. Neuropharmacol 44:729–737

Allan AM, Liang X, Luo Y, Pak C, Li X, Szulwach KE, Chen D, Jin P, Zhao X (2008) The loss of methyl-CpG binding protein 1 leads to autism-like behavioral deficits. Hum Mol Genet 17: 2047–2057

Allen PD, Schmuck N, Ison JR, Walton JP (2008) Kv1.1 channel subunits are not necessary for high temporal acuity in behavioral and electrophysiological gap detection. Hear Res 246:52–58

Arguello PA, Gogos JA (2010) Cognition in mouse models of schizophrenia susceptibility genes. Schizophr Bull 36:289–300

Asp L, Holtze M, Powell SB, Karlsson H, Erhardt S (2010) Neonatal infection with neurotropic influenza A virus induces the kynurenine pathway in early life and disrupts sensorimotor gating in adult Tap −/− mice. Int J Neuropsychopharmacol 13:475–485

Baker KB, Wray SP, Ritter R, Mason S, Lanthorn TH, Savelieva KV (2010) Male and female Fmr1 knockout mice on C57 albino background exhibit spatial learning and memory impairments. Genes Brain Behav 9:562–574

Barr AM, Fish KN, Markou A, Honer WG (2008) Heterozygous reeler mice exhibit alterations in sensorimotor gating but not presynaptic proteins. Eur J Neurosci 27:2568–2574

Barros CS, Calabrese B, Chamero P, Roberts AJ, Korzus E, Lloyd K, Stowers L, Mayford M, Halpain S, Muller U (2009) Impaired maturation of dendritic spines without disorganization of cortical cell layers in mice lacking NRG1/ErbB signaling in the central nervous system. Proc Natl Acad Sci U S A 106:4507–4512

Behrens MM, Ali SS, Dao DN, Lucero J, Shekhtman G, Quick KL, Dugan LL (2007) Ketamine-induced loss of phenotype of fast-spiking interneurons is mediated by NADPH-oxidase. Science 318:1645–1647

Belforte JE, Zsiros V, Sklar ER, Jiang Z, Yu G, Li Y, Quinlan EM, Nakazawa K (2010) Postnatal NMDA receptor ablation in corticolimbic interneurons confers schizophrenia-like phenotypes. Nat Neurosci 13:76–83

Benes FM, Berretta S (2001) GABAergic interneurons: implications for understanding schizophrenia and bipolar disorder. Neuropsychopharmacology 25:1–27

Benneyworth MA, Basu AC, Coyle JT (2011) Discordant behavioral effects of psychotomimetic drugs in mice with altered NMDA receptor function. Psychopharmacology (Berl) 213: 143–153

Benoit CE, Bastianetto S, Brouillette J, Tse Y, Boutin JA, Delagrange P, Wong T, Sarret P, Quirion R (2010) Loss of quinone reductase 2 function selectively facilitates learning behaviors. J Neurosci 30:12690–12700

Bersudsky Y, Shaldubina A, Kozlovsky N, Woodgett JR, Agam G, Belmaker RH (2008) Glycogen synthase kinase-3beta heterozygote knockout mice as a model of findings in postmortem schizophrenia brain or as a model of behaviors mimicking lithium action: negative results. Behav Pharmacol 19:217–224

Bitsios P, Giakoumaki SG, Theou K, Frangou S (2006) Increased prepulse inhibition of the acoustic startle response is associated with better strategy formation and execution times in healthy males. Neuropsychologia 44:2494–2499

Blackwood DH, Fordyce A, Walker MT, Clair DM, Porteous DJ, Muir WJ (2001) Schizophrenia and affective disorders—cosegregation with a translocation at chromosome 1q42 that directly disrupts brain-expressed genes: clinical and P300 findings in a family. Am J Hum Genet 69:428–433

Blundell J, Kaeser PS, Sudhof TC, Powell CM (2010) RIM1alpha and interacting proteins involved in presynaptic plasticity mediate prepulse inhibition and additional behaviors linked to schizophrenia. J Neurosci 30:5326–5333

Braff DL (2010) Prepulse inhibition of the startle reflex: a window on the brain in schizophrenia. Curr Top Behav Neurosci 4:349–371

Braff DL (2011) Gating in schizophrenia: from genes to cognition (to real world function?). Biol Psychiatry 69:395–396

Braff DL, Freedman R, Schork NJ, Gottesman II (2007) Deconstructing schizophrenia: an overview of the use of endophenotypes in order to understand a complex disorder. Schizophr Bull 33:21–32

Braff DL, Geyer MA, Swerdlow NR (2001) Human studies of prepulse inhibition of startle: normal subjects, patient groups, and pharmacological studies. Psychopharmacology (Berl) 156:234–258

Braff DL, Grillon C, Geyer MA (1992) Gating and habituation of the startle reflex in schizophrenic patients. Arch Gen Psychiatry 49:206–215

Brody SA, Conquet F, Geyer MA (2003) Disruption of prepulse inhibition in mice lacking mGluR1. Eur J Neurosci 18:3361–3366

Brody SA, Dulawa SC, Conquet F, Geyer MA (2004) Assessment of a prepulse inhibition deficit in a mutant mouse lacking mGlu5 receptors. Mol Psychiatry 9:35–41

Brody SA, Geyer MA (2004) Interactions of the mGluR5 gene with breeding and maternal factors on startle and prepulse inhibition in mice. Neurotoxic Res 6:79–90

Brooks S, Higgs G, Jones L, Dunnett SB (2010) Longitudinal analysis of the behavioural phenotype in Hdh ((CAG)150) Huntington's disease knock-in mice. Brain Res Bull. May 8. [Epub ahead of print]

Brooks SP, Janghra N, Workman VL, Bayram-Weston Z, Jones L, Dunnett SB (2011) Longitudinal analysis of the behavioural phenotype in R6/1 (C57BL/6J) Huntington's disease transgenic mice. Brain Res Bull. Jan 25. [Epub ahead of print]

Brzózka MM, Radyushkin K, Wichert SP, Ehrenreich H, Rossner MJ (2010) Cognitive and sensorimotor gating impairments in transgenic mice overexpressing the schizophrenia susceptibility gene Tcf4 in the brain. Biol Psychiatry 68:33–40

Burdick KE, Hodgkinson CA, Szeszko PR, Lencz T, Ekholm JM, Kane JM, Goldman D, Malhotra AK (2005) DISC1 and neurocognitive function in schizophrenia. Neuroreport 16:1399–1402

Cadenhead KS, Geyer MA, Braff DL (1993) Impaired startle prepulse inhibition and habituation in patients with schizotypal personality disorder. Am J Psychiatry 150:1862–1867

Cadenhead KS, Swerdlow NR, Shafer KM, Diaz M, Braff DL (2000) Modulation of the startle response and startle laterality in relatives of schizophrenic patients and in subjects with schizotypal personality disorder: evidence of inhibitory deficits. Am J Psychiatry 157: 1660–1668

Cahill ME, Xie Z, Day M, Photowala H, Barbolina MV, Miller CA, Weiss C, Radulovic J, Sweatt JD, Disterhoft JF, Surmeier DJ, Penzes P (2009) Kalirin regulates cortical spine morphogenesis and disease-related behavioral phenotypes. Proc Natl Acad Sci U S A 106: 13058–13063

Caldwell HK, Stephens SL, Young WS 3rd (2009) Oxytocin as a natural antipsychotic: a study using oxytocin knockout mice. Mol Psychiatry 14:190–196

Callicott JH, Straub RE, Pezawas L, Egan MF, Mattay VS, Hariri AR, Verchinski BA, Meyer-Lindenberg A, Balkissoon R, Kolachana B, Goldberg TE, Weinberger DR (2005) Variation in DISC1 affects hippocampal structure and function and increases risk for schizophrenia. Proc Natl Acad Sci U S A 102:8627–8632

Cannon TD, Hennah W, van Erp TG, Thompson PM, Lonnqvist J, Huttunen M, Gasperoni T, Tuulio-Henriksson A, Pirkola T, Toga AW, Kaprio J, Mazziotta J, Peltonen L (2005) Association of DISC1/TRAX haplotypes with schizophrenia, reduced prefrontal gray matter, and impaired short- and long-term memory. Arch Gen Psychiatry 62:1205–1213

Cesca F, Baldelli P, Valtorta F, Benfenati F (2010) The synapsins: key actors of synapse function and plasticity. Prog Neurobiol 91:313–348

Chadman KK, Gong S, Scattoni ML, Boltuck SE, Gandhy SU, Heintz N, Crawley JN (2008) Minimal aberrant behavioral phenotypes of neuroligin-3 R451C knockin mice. Autism Res 1:147–158

Chavez C, Gogos A, Jones ME, van den Buuse M (2009) Psychotropic drug-induced locomotor hyperactivity and prepulse inhibition regulation in male and female aromatase knockout (ArKO) mice: role of dopamine D1 and D2 receptors and dopamine transporters. Psychopharmacology (Berl) 206:267–279

Chen HH, Stoker A, Markou A (2010) The glutamatergic compounds sarcosine and N-acetylcysteine ameliorate prepulse inhibition deficits in metabotropic glutamate 5 receptor knockout mice. Psychopharmacology (Berl) 209:343–350

Chen Q, He G, Qin W, Chen QY, Zhao XZ, Duan SW, Liu XM, Feng GY, Xu YF, Clair D, Li M, Wang JH, Xing YL, Shi JG, He L (2004) Family-based association study of synapsin II and schizophrenia. Am J Hum Genet 75:873–877

Chen Y-JJ, Johnson MA, Lieberman MD, Goodchild RE, Schobel S, Lewandowski N, Rosoklija G, Liu R-C, Gingrich JA, Small S, Moore H, Dwork AJ, Talmage DA, Role LW (2008) Type III Neuregulin-1 is required for normal sensorimotor gating, memory-related behaviors, and corticostriatal circuit components. J Neurosci 28:6872–6883

Chen YW, Lai WS (2011) Behavioral phenotyping of v-akt murine thymoma viral oncogene homolog 1-deficient mice reveals a sex-specific prepulse inhibition deficit in females that can be partially alleviated by glycogen synthase kinase-3 inhibitors but not by antipsychotics. Neuroscience 174:178–189

Chong VZ, Young LT, Mishra RK (2002) cDNA array reveals differential gene expression following chronic neuroleptic administration: implications of synapsin II in haloperidol treatment. J Neurochem 82:1533–1539

Cleva RM, Olive MF (2011) Positive allosteric modulators of type 5 metabotropic glutamate receptors (mGluR5) and their therapeutic potential for the treatment of CNS disorders. Molecules 16:2097–2106

Cole TB, Giordano G, Co AL, Mohar I, Kavanagh TJ, Costa LG (2011) Behavioral characterization of GCLM-knockout mice, a model for enhanced susceptibility to oxidative stress. J Toxicol 2011:157687

Cook MN, Dunning JP, Wiley RG, Chesler EJ, Johnson DK, Miller DR, Goldowitz D (2007) Neurobehavioral mutants identified in an ENU-mutagenesis project. Mamm Genome 18:559–572

Coyle JT (2004) The GABA-glutamate connection in schizophrenia: which is the proximate cause? Biochem Pharmacol 68:1507–1514

Crawley JN (2007) What's wrong with my mouse? Behavioral phenotyping of transgenic and knockout mice, Wiley

Csomor PA, Stadler RR, Feldon J, Yee BK, Geyer MA, Vollenweider FX (2008) Haloperidol differentially modulates prepulse inhibition and p50 suppression in healthy humans stratified for low and high gating levels. Neuropsychopharmacology 33:497–512

Darvas M, Morsch M, Racz I, Ahmadi S, Swandulla D, Zimmer A (2009) Modulation of the Ca2+ conductance of nicotinic acetylcholine receptors by Lypd6. Eur Neuropsychopharmacol 19:670–681

de Vrij FM, Levenga J, van der Linde HC, Koekkoek SK, De Zeeuw CI, Nelson DL, Oostra BA, Willemsen R (2008) Rescue of behavioral phenotype and neuronal protrusion morphology in Fmr1 KO mice. Neurobiol Dis 31:127–132

Deakin IH, Law AJ, Oliver PL, Schwab MH, Nave KA, Harrison PJ, Bannerman DM (2009) Behavioural characterization of neuregulin 1 type I overexpressing transgenic mice. Neuroreport 20:1523–1528

Dejaegere T, Serneels L, Schafer MK, Van Biervliet J, Horre K, Depboylu C, Alvarez-Fischer D, Herreman A, Willem M, Haass C, Hoglinger GU, D'Hooge R, De Strooper B (2008) Deficiency of Aph1B/C-gamma-secretase disturbs Nrg1 cleavage and sensorimotor gating that can be reversed with antipsychotic treatment. Proc Natl Acad Sci U S A 105:9775–9780

DeLorey TM, Sahbaie P, Hashemi E, Li WW, Salehi A, Clark DJ (2011) Somatosensory and sensorimotor consequences associated with the heterozygous disruption of the autism candidate gene, Gabrb3. Behav Brain Res 216(1):36–45

Doherty JM, Masten VL, Powell SB, Ralph RJ, Klamer D, Low MJ, Geyer MA (2008) Contributions of dopamine D1, D2, and D3 receptor subtypes to the disruptive effects of cocaine on prepulse inhibition in mice. Neuropsychopharmacology 33:2648–2656

Duffy L, Cappas E, Scimone A, Schofield PR, Karl T (2008) Behavioral profile of a heterozygous mutant mouse model for EGF-like domain neuregulin 1. Behav Neurosci 122:748–759

Dulawa SC, Hen R, Scearce-Levie K, Geyer MA (1997) Serotonin1B receptor modulation of startle reactivity, habituation, and prepulse inhibition in wild-type and serotonin1B knockout mice. Psychopharmacology (Berl) 132:125–134

Duncan GE, Inada K, Koller BH, Moy SS (2010) Increased sensitivity to kainic acid in a genetic model of reduced NMDA receptor function. Brain Res 1307:166–176

Duncan GE, Moy SS, Perez A, Eddy DM, Zinzow WM, Lieberman JA, Snouwaert JN, Koller BH (2004) Deficits in sensorimotor gating and tests of social behavior in a genetic model of reduced NMDA receptor function. Behav Brain Res 153:507–519

Dyck BA, Skoblenick KJ, Castellano JM, Ki K, Thomas N, Mishra RK (2007) Synapsin II knockout mice show sensorimotor gating and behavioural abnormalities similar to those in the phencyclidine-induced preclinical animal model of schizophrenia. Schizophr Res 97:292–293

Dyck BA, Skoblenick KJ, Castellano JM, Ki K, Thomas N, Mishra RK (2009) Behavioral abnormalities in synapsin II knockout mice implicate a causal factor in schizophrenia. Synapse 63:662–672

Eells JB, Lipska BK, Yeung SK, Misler JA, Nikodem VM (2002) Nurr1-null heterozygous mice have reduced mesolimbic and mesocortical dopamine levels and increased stress-induced locomotor activity. Behav Brain Res 136:267–275

Eells JB, Misler JA, Nikodem VM (2006) Early postnatal isolation reduces dopamine levels, elevates dopamine turnover and specifically disrupts prepulse inhibition in Nurr1-null heterozygous mice. Neuroscience 140:1117–1126

Egashira N, Mishima K, Iwasaki K, Oishi R, Fujiwara M (2009) New topics in vasopressin receptors and approach to novel drugs: role of the vasopressin receptor in psychological and cognitive functions. J Pharmacol Sci 109:44–49

Errico F, Rossi S, Napolitano F, Catuogno V, Topo E, Fisone G, D'Aniello A, Centonze D, Usiello A (2008) D-aspartate prevents corticostriatal long-term depression and attenuates schizophrenia-like symptoms induced by amphetamine and MK-801. J Neurosci 28: 10404–10414

Etherton MR, Blaiss CA, Powell CM, Sudhof TC (2009) Mouse neurexin-1alpha deletion causes correlated electrophysiological and behavioral changes consistent with cognitive impairments. Proc Natl Acad Sci U S A 106:17998–18003

Feifel D, Macdonald K, Nguyen A, Cobb P, Warlan H, Galangue B, Minassian A, Becker O, Cooper J, Perry W, Lefebvre M, Gonzales J, Hadley A (2010a) Adjunctive intranasal oxytocin reduces symptoms in schizophrenia patients. Biol Psychiatry 68:678–680

Feifel D, Pang Z, Shilling PD, Melendez G, Schreiber R, Button D (2010b) Effects of neurotensin-2 receptor deletion on sensorimotor gating and locomotor activity. Behav Brain Res 212:174–178

Feifel D, Pang Z, Shilling PD, Melendez G, Schreiber R, Button D (2010c) Sensorimotor gating in neurotensin-1 receptor null mice. Neuropharmacology 58:173–178

Fendt M, Buchi M, Burki H, Imobersteg S, Ricoux B, Suply T, Sailer AW (2011) Neuropeptide S receptor deficiency modulates spontaneous locomotor activity and the acoustic startle response. Behav Brain Res 217:1–9

Fendt M, Burki H, Imobersteg S, van der Putten H, McAllister K, Leslie JC, Shaw D, Holscher C (2010) The effect of mGlu8 deficiency in animal models of psychiatric diseases. Genes Brain Behav 9:33–44

Francis DD, Szegda K, Campbell G, Martin WD, Insel TR (2003) Epigenetic sources of behavioral differences in mice. Nat Neurosci 6:445–446

Frankland PW, Wang Y, Rosner B, Shimizu T, Balleine BW, Dykens EM, Ornitz EM, Silva AJ (2004) Sensorimotor gating abnormalities in young males with fragile X syndrome and Fmr1-knockout mice. Mol Psychiatry 9:417–425

Frankle WG, Lerma J, Laruelle M (2003) The synaptic hypothesis of schizophrenia. Neuron 39:205–216

Fridlund AJ, Cacioppo JT (1986) Guidelines for human electromyographic research. Psychophysiology 23(5):567–589

Friston KJ, Frith CD (1995) Schizophrenia: a disconnection syndrome? Clin Neurosci 3:89–97

Frost WN, Tian LM, Hoppe TA, Mongeluzi DL, Wang J (2003) A cellular mechanism for prepulse inhibition. Neuron 40:991–1001

Fukuda T, Hashimoto H, Okayasu N, Kameyama A, Onogi H, Nakagawasai O, Nakazawa T, Kurosawa T, Hao Y, Isaji T, Tadano T, Narimatsu H, Taniguchi N, Gu J (2011) {alpha}1, 6-Fucosyltransferase-deficient mice exhibit multiple behavioral abnormalities associated with a schizophrenia-like phenotype: importance of the balance between the dopamine and serotonin systems. J Biol Chem 286:18434–18443

Galici R, Boggs JD, Miller KL, Bonaventure P, Atack JR (2008) Effects of SB-269970, a 5-HT7 receptor antagonist, in mouse models predictive of antipsychotic-like activity. Behav Pharmacol 19:153–159

Gallinat J, Bajbouj M, Sander T, Schlattmann P, Xu K, Ferro EF, Goldman D, Winterer G (2003) Association of the G1947A COMT (Val(108/158)Met) gene polymorphism with prefrontal P300 during information processing. Biol Psychiatry 54:40–48

Geyer MA, Braff DL (1987) Startle habituation and sensorimotor gating in schizophrenia and related animal models. Schizophr Bull 13:643–668

Geyer MA, Dulawa SC (2003) Assessment of murine startle reactivity, prepulse inhibition, and habituation. Curr Protoc Neurosci Chapter 8: Unit 8, p 17

Geyer MA, Markou A (2002) The role of preclinical models in the development of psychotropic drugs. In: Davis KL, Charney D, Coyle JT, Nemeroff CB (eds) Neuropsychopharmacology: the fifth generation of progress. Lippincott, Williams Wilkins, pp 445–455

Geyer MA, McIlwain KL, Paylor R (2002) Mouse genetic models for prepulse inhibition: an early review. Mol Psychiatry 7:1039–1053

Giakoumaki SG, Bitsios P, Frangou S (2006) The level of prepulse inhibition in healthy individuals may index cortical modulation of early information processing. Brain Res 1078: 168–170

Giakoumaki SG, Roussos P, Bitsios P (2008) Improvement of prepulse inhibition and executive function by the COMT inhibitor tolcapone depends on COMT Val158Met polymorphism. Neuropsychopharmacology 33:3058–3068

Gogos JA, Morgan M, Luine V, Santha M, Ogawa S, Pfaff D, Karayiorgou M (1998) Catechol-O-methyltransferase-deficient mice exhibit sexually dimorphic changes in catecholamine levels and behavior. Proc Natl Acad Sci U S A 95:9991–9996

Golimbet V, Gritsenko I, Alfimova M, Lebedeva I, Lezheiko T, Abramova L, Kaleda V, Ebstein R (2006) Association study of COMT gene Val158Met polymorphism with auditory P300 and performance on neurocognitive tests in patients with schizophrenia and their relatives. World J Biol Psychiatry 7:238–245

Gottesman II, Gould TD (2003) The endophenotype concept in psychiatry: etymology and strategic intentions. Am J Psychiatry 160:636–645

Gottesman II (1991) Schizophrenia Genesis. WH Freeman

Gould TD, Gottesman II (2006) Psychiatric endophenotypes and the development of valid animal models. Genes Brain Behav 5:113–119

Graham FK (1975) Presidential address, 1974. The more or less startling effects of weak prestimulation. Psychophysiology 12:238–248

Grant A, Hoops D, Labelle-Dumais C, Prevost M, Rajabi H, Kolb B, Stewart J, Arvanitogiannis A, Flores C (2007) Netrin-1 receptor-deficient mice show enhanced mesocortical dopamine transmission and blunted behavioural responses to amphetamine. Eur J Neurosci 26: 3215–3228

Gray L, van den Buuse M, Scarr E, Dean B, Hannan AJ (2009) Clozapine reverses schizophrenia-related behaviours in the metabotropic glutamate receptor 5 knockout mouse: association with N-methyl-D-aspartic acid receptor up-regulation. Int J Neuropsychopharmacol 12:45–60

Green MF, Butler PD, Chen Y, Geyer MA, Silverstein S, Wynn JK, Yoon JH, Zemon V (2009) Perception measurement in clinical trials of schizophrenia: promising paradigms from CNTRICS. Schizophr Bull 35:163–181

Greenwood TA, Braff DL, Light GA, Cadenhead KS, Calkins ME, Dobie DJ, Freedman R, Green MF, Gur RE, Gur RC, Mintz J, Nuechterlein KH, Olincy A, Radant AD, Seidman LJ, Siever LJ, Silverman JM, Stone WS, Swerdlow NR, Tsuang DW, Tsuang MT, Turetsky BI, Schork NJ (2007) Initial heritability analyses of endophenotypic measures for schizophrenia: the consortium on the genetics of schizophrenia. Arch Gen Psychiatry 64:1242–1250

Greenwood TA, Lazzeroni LC, Murray SS, Cadenhead KS, Calkins ME, Dobie DJ, Green MF, Gur RE, Gur RC, Hardiman G, Kelsoe JR, Leonard S, Light GA, Nuechterlein KH, Olincy A, Radant AD, Schork NJ, Seidman LJ, Siever LJ, Silverman JM, Stone WS, Swerdlow NR, Tsuang DW, Tsuang MT, Turetsky BI, Freedman R, Braff DL (2011) Analysis of 94 candidate genes and 12 endophenotypes for schizophrenia from the consortium on the genetics of schizophrenia. Am J Psychiatry

Groenink L, Bijlsma EY, van Bogaert MJ, Oosting RS, Olivier B (2011) Serotonin1A receptor deletion does not interact with maternal separation-induced increases in startle reactivity and prepulse inhibition deficits. Psychopharmacology (Berl) 214:353–365

Groenink L, Dirks A, Verdouw PM, de Graaff M, Peeters BW, Millan MJ, Olivier B (2008) CRF1 not glucocorticoid receptors mediate prepulse inhibition deficits in mice overexpressing CRF. Biol Psychiatry 63:360–368

Gruart A, Lopez-Ramos JC, Munoz MD, Delgado-Garcia JM (2008) Aged wild-type and APP, PS1, and APP+PS1 mice present similar deficits in associative learning and synaptic plasticity independent of amyloid load. Neurobiol Dis 30:439–450

Grunze HC, Rainnie DG, Hasselmo ME, Barkai E, Hearn EF, McCarley RW, Greene RW (1996) NMDA-dependent modulation of CA1 local circuit inhibition. J Neurosci 16:2034–2043

Guo X, Hamilton PJ, Reish NJ, Sweatt JD, Miller CA, Rumbaugh G (2009) Reduced expression of the NMDA receptor-interacting protein SynGAP causes behavioral abnormalities that model symptoms of Schizophrenia. Neuropsychopharmacology 34:1659–1672

Han L, Picker JD, Schaevitz LR, Tsai G, Feng J, Jiang Z, Chu HC, Basu AC, Berger-Sweeney J, Coyle JT (2009) Phenotypic characterization of mice heterozygous for a null mutation of glutamate carboxypeptidase II. Synapse 63:625–635

Harrison PJ, Weinberger DR (2005) Schizophrenia genes, gene expression, and neuropathology: on the matter of their convergence. Mol Psychiatry 10: 40–68, image 5

Hasenkamp W, Epstein MP, Green A, Wilcox L, Boshoven W, Lewison B, Duncan E (2010) Heritability of acoustic startle magnitude, prepulse inhibition, and startle latency in schizophrenia and control families. Psychiatry Res 178:236–243

Hashimoto R, Hashimoto H, Shintani N, Chiba S, Hattori S, Okada T, Nakajima M, Tanaka K, Kawagishi N, Nemoto K, Mori T, Ohnishi T, Noguchi H, Hori H, Suzuki T, Iwata N, Ozaki N, Nakabayashi T, Saitoh O, Kosuga A, Tatsumi M, Kamijima K, Weinberger DR, Kunugi H, Baba A (2007) Pituitary adenylate cyclase-activating polypeptide is associated with schizophrenia. Mol Psychiatry 12:1026–1032

Heintz N (2001) BAC to the future: the use of BAC transgenic mice for neuroscience research. Nat Rev Neurosci 2:861–870

Hennah W, Tuulio-Henriksson A, Paunio T, Ekelund J, Varilo T, Partonen T, Cannon TD, Lonnqvist J, Peltonen L (2005) A haplotype within the DISC1 gene is associated with visual memory functions in families with a high density of schizophrenia. Mol Psychiatry 10: 1097–1103

Hessl D, Berry-Kravis E, Cordeiro L, Yuhas J, Ornitz EM, Campbell A, Chruscinski E, Hervey C, Long JM, Hagerman RJ (2009) Prepulse inhibition in fragile X syndrome: feasibility, reliability, and implications for treatment. Am J Med Genet B Neuropsychiatr Genet 150B:545–553

Hikida T, Jaaro-Peled H, Seshadri S, Oishi K, Hookway C, Kong S, Wu D, Xue R, Andrade M, Tankou S, Mori S, Gallagher M, Ishizuka K, Pletnikov M, Kida S, Sawa A (2007) Dominant-negative DISC1 transgenic mice display schizophrenia-associated phenotypes detected by measures translatable to humans. Proc Natl Acad Sci U S A 104:14501–14506

Hill RA, Murray SS, Halley PG, Binder MD, Martin SJ, van den Buuse M (2011) Brain-derived neurotrophic factor expression is increased in the hippocampus of 5-HT(2C) receptor knockout mice. Hippocampus 21:434–445

Hitzemann R, Malmanger B, Belknap J, Darakjian P, McWeeney S (2008) Short-term selective breeding for high and low prepulse inhibition of the acoustic startle response; pharmacological characterization and QTL mapping in the selected lines. Pharmacol Biochem Behav 90:525–533

Hoffman HS, Ison JR (1980) Reflex modification in the domain of startle: I. Some empirical findings and their implications for how the nervous system processes sensory input. Psychol Rev 87:175–189

Hokyo A, Kanazawa T, Uenishi H, Tsutsumi A, Kawashige S, Kikuyama H, Glatt SJ, Koh J, Nishimoto Y, Matsumura H, Motomura N, Yoneda H (2010) Habituation in prepulse inhibition is affected by a polymorphism on the NMDA receptor 2B subunit gene (GRIN2B). Psychiatr Genet 20:191–198

Horii Y, Yamasaki N, Miyakawa T, Shiosaka S (2008) Increased anxiety-like behavior in neuropsin (kallikrein-related peptidase 8) gene-deficient mice. Behav Neurosci 122(3):498–504

Ibi D, Nagai T, Koike H, Kitahara Y, Mizoguchi H, Niwa M, Jaaro-Peled H, Nitta A, Yoneda Y, Nabeshima T, Sawa A, Yamada K (2010) Combined effect of neonatal immune activation and mutant DISC1 on phenotypic changes in adulthood. Behav Brain Res 206:32–37

Ishihama T, Ago Y, Shintani N, Hashimoto H, Baba A, Takuma K, Matsuda T (2010) Environmental factors during early developmental period influence psychobehavioral abnormalities in adult PACAP-deficient mice. Behav Brain Res 209:274–280

Jackson ME, Homayoun H, Moghaddam B (2004) NMDA receptor hypofunction produces concomitant firing rate potentiation and burst activity reduction in the prefrontal cortex. Proc Natl Acad Sci U S A 101:8467–8472

Jacobson LH, Cryan JF (2010) Genetic approaches to modeling anxiety in animals. Curr Top Behav Neurosci 2:161–201

Javitt DC (2004) Glutamate as a therapeutic target in psychiatric disorders. Mol Psychiatry 9:984–997

Javitt DC, Zukin SR (1991) Recent advances in the phencyclidine model of schizophrenia. Am J Psychiatry 148:1301–1308

Jones DNC, Gartlon JE, Minassian A, Perry W, Geyer MA (2008) Developing new drugs for schizophrenia: from animals to the clinic. In: Borsini R, McArthur R (eds) Animal and translational models for CNS drug discovery. Psychiatric Disorders Elsevier, New York, pp 199–262

Joober R, Zarate JM, Rouleau GA, Skamene E, Boksa P (2002) Provisional mapping of quantitative trait loci modulating the acoustic startle response and prepulse inhibition of acoustic startle. Neuropsychopharmacol 27:765–781

Kaidanovich-Beilin O, Lipina TV, Takao K, van Eede M, Hattori S, Laliberte C, Khan M, Okamoto K, Chambers JW, Fletcher PJ, Macaulay K, Doble BW, Henkelman M, Miyakawa T, Roder J, Woodgett JR (2009) Abnormalities in brain structure and behavior in GSK-3alpha mutant mice. Mol Brain 2:35

Kapfhamer D, Berger KH, Hopf FW, Seif T, Kharazia V, Bonci A, Heberlein U (2010) Protein Phosphatase 2a and glycogen synthase kinase 3 signaling modulate prepulse inhibition of the acoustic startle response by altering cortical M-Type potassium channel activity. J Neurosci 30:8830–8840

Karl T, Chesworth R, Duffy L, Herzog H (2010a) Acoustic startle response and sensorimotor gating in a genetic mouse model for the Y1 receptor. Neuropeptides 44:233–239

Karl T, Chesworth R, Duffy L, Herzog H (2010b) Schizophrenia-relevant behaviours in a genetic mouse model for Y2 deficiency. Behav Brain Res 207:434–440

Karlsson RM, Tanaka K, Saksida LM, Bussey TJ, Heilig M, Holmes A (2009) Assessment of glutamate transporter GLAST (EAAT1)-deficient mice for phenotypes relevant to the negative and executive/cognitive symptoms of schizophrenia. Neuropsychopharmacol 34: 1578–1589

Karmacharya R, Lynn SK, Demarco S, Ortiz A, Wang X, Lundy MY, Xie Z, Cohen BM, Miller GM, Buttner EA (2011) Behavioral effects of clozapine: involvement of trace amine pathways in *C. elegans* and *M. musculus*. Brain Res 1393:91–99

Karper LP, Freeman GK, Grillon C, Morgan CA 3rd, Charney DS, Krystal JH (1996) Preliminary evidence of an association between sensorimotor gating and distractibility in psychosis. J Neuropsychiatry Clin Neurosci 8:60–66

Kato T, Kasai A, Mizuno M, Fengyi L, Shintani N, Maeda S, Yokoyama M, Ozaki M, Nawa H (2010) Phenotypic characterization of transgenic mice overexpressing neuregulin-1. PLoS One 5(12):e14185

Kelly MP, Stein JM, Vecsey CG, Favilla C, Yang X, Bizily SF, Esposito MF, Wand G, Kanes SJ, Abel T (2009) Developmental etiology for neuroanatomical and cognitive deficits in mice overexpressing Galphas, a G-protein subunit genetically linked to schizophrenia. Mol Psychiatry 14(398–415):347

Kim YS, Harry GJ, Kang HS, Goulding D, Wine RN, Kissling GE, Liao G, Jetten AM (2010) Altered cerebellar development in nuclear receptor TAK1/TR4 null mice is associated with deficits in GLAST(+) glia, alterations in social behavior, motor learning, startle reactivity, and microglia. Cerebellum

Kinney GG, Burno M, Campbell UC, Hernandez LM, Rodriguez D, Bristow LJ, Conn PJ (2003) Metabotropic glutamate subtype 5 receptors modulate locomotor activity and sensorimotor gating in rodents. J Pharmacol Exp Ther 306:116–123

Koga M, Ishiguro H, Yazaki S, Horiuchi Y, Arai M, Niizato K, Iritani S, Itokawa M, Inada T, Iwata N, Ozaki N, Ujike H, Kunugi H, Sasaki T, Takahashi M, Watanabe Y, Someya T, Kakita A, Takahashi H, Nawa H, Muchardt C, Yaniv M, Arinami T (2009) Involvement of SMARCA2/BRM in the SWI/SNF chromatin-remodeling complex in schizophrenia. Hum Mol Genet 18:2483–2494

Koh HY, Kim D, Lee J, Lee S, Shin HS (2008) Deficits in social behavior and sensorimotor gating in mice lacking phospholipase Cbeta1. Genes Brain Behav 7:120–128

Koumura A, Kakefuda K, Honda A, Ito Y, Tsuruma K, Shimazawa M, Uchida Y, Hozumi I, Satoh M, Inuzuka T, Hara H (2009) Metallothionein-3 deficient mice exhibit abnormalities of psychological behaviors. Neurosci Lett 467:11–14

Krystal J, D'Souza DC, Mathalon D, Perry E, Belger A, Hoffman R (2003) NMDA receptor antagonist effects, cortical glutamatergic function, and schizophrenia: toward a paradigm shift in medication development. Psychopharmacol 169:215–233

Krystal JH, Karper LP, Seibyl JP, Freeman GK, Delaney R, Bremner JD, Heninger GR, Bowers MB Jr, Charney DS (1994) Subanesthetic effects of the noncompetitive NMDA antagonist, ketamine, in humans. Psychotomimetic, perceptual, cognitive, and neuroendocrine responses. Arch Gen Psychiatry 51:199–214

Kumari V, Fannon D, Geyer MA, Premkumar P, Antonova E, Simmons A, Kuipers E (2008) Cortical grey matter volume and sensorimotor gating in schizophrenia. Cortex 44:1206–1214

Kumari V, Gray JA, Geyer MA, ffytche D, Soni W, Mitterschiffthaler MT, Vythelingum GN, Simmons A, Williams SC, Sharma T (2003) Neural correlates of tactile prepulse inhibition: a functional MRI study in normal and schizophrenic subjects. Psychiatry Res 122:99–113

Levenga J, Hayashi S, de Vrij FM, Koekkoek SK, van der Linde HC, Nieuwenhuizen I, Song C, Buijsen RA, Pop AS, Gomezmancilla B, Nelson DL, Willemsen R, Gasparini F, Oostra BA (2011) AFQ056, a new mGluR5 antagonist for treatment of fragile X syndrome. Neurobiol Dis 42(3):311–317

Lewis DA, Hashimoto T, Volk DW (2005) Cortical inhibitory neurons and schizophrenia. Nat Rev Neurosci 6:312–324

Lipina TV, Kaidanovich-Beilin O, Patel S, Wang M, Clapcote SJ, Liu F, Woodgett JR, Roder JC (2011) Genetic and pharmacological evidence for schizophrenia-related DISC1 interaction with GSK-3. Synapse 65:234–248

Lipina TV, Niwa M, Jaaro-Peled H, Fletcher PJ, Seeman P, Sawa A, Roder JC (2010) Enhanced dopamine function in DISC1-L100P mutant mice: implications for schizophrenia. Genes Brain Behav 9:777–789

Logue SF, Grauer SM, Paulsen J, Graf R, Taylor N, Sung MA, Zhang L, Hughes Z, Pulito VL, Liu F, Rosenzweig-Lipson S, Brandon NJ, Marquis KL, Bates B, Pausch M (2009) The orphan GPCR, GPR88, modulates function of the striatal dopamine system: a possible therapeutic target for psychiatric disorders? Mol Cell Neurosci 42:438–447

López-Ramos JC, Tomioka Y, Morimatsu M, Yamamoto S, Ozaki K, Ono E, Delgado-García JM (2010) Motor-coordination-dependent learning, more than others, is impaired in transgenic mice expressing pseudorabies virus immediate-early protein IE180. PLoS One 5(8):e12123

Low NC, Hardy J (2007) What is a schizophrenic mouse? Neuron 54:348–349

Lu BY, Martin KE, Edgar JC, Smith AK, Lewis SF, Escamilla MA, Miller GA, Canive JM (2007) Effect of catechol O-methyltransferase val(158)met polymorphism on the p50 gating endophenotype in schizophrenia. Biol Psychiatry 62:822–825

Ludewig K, Geyer MA, Vollenweider FX (2003) Deficits in prepulse inhibition and habituation in never-medicated, first-episode schizophrenia. Biol Psychiatry 54:121–128

Ludewig S, Ludewig K, Geyer MA, Hell D, Vollenweider FX (2002) Prepulse inhibition deficits in patients with panic disorder. Depress Anxiety 15:55–60

Mackeprang T, Kristiansen KT, Glenthoj BY (2002) Effects of antipsychotics on prepulse inhibition of the startle response in drug-naive schizophrenic patients. Biol Psychiatry 52: 863–873

Mackie S, Millar JK, Porteous DJ (2007) Role of DISC1 in neural development and schizophrenia. Curr Opin Neurobiol 17:95–102

Malhotra AK, Adler CM, Kennison SD, Elman I, Pickar D, Breier A (1997a) Clozapine blunts N-methyl-D-aspartate antagonist-induced psychosis: a study with ketamine. Biol Psychiatry 42:664–668

Malhotra AK, Pinals DA, Adler CM, Elman I, Clifton A, Pickar D, Breier A (1997b) Ketamine-induced exacerbation of psychotic symptoms and cognitive impairment in neuroleptic-free schizophrenics. Neuropsychopharmacol 17:141–150

Mansuy IM, Bujard H (2000) Tetracycline-regulated gene expression in the brain. Curr Opin Neurobiol 10:593–596

Martin S, Markus MA, Morris BJ, Davisson RL, Lawrence AJ, van den Buuse M (2008) Does angiotensin interact with dopaminergic mechanisms in the brain to modulate prepulse inhibition in mice? Neuropharmacol 54:399–404

Martinez-Gras I, Rubio G, del Manzano BA, Rodriguez-Jimenez R, Garcia-Sanchez F, Bagney A, Leza JC, Borrell J (2009) The relationship between prepulse inhibition and general psychopathology in patients with schizophrenia treated with long-acting risperidone. Schizophr Res 115(2–3):215–221

Matsumoto M, Straub RE, Marenco S, Nicodemus KK, Matsumoto S, Fujikawa A, Miyoshi S, Shobo M, Takahashi S, Yarimizu J, Yuri M, Hiramoto M, Morita S, Yokota H, Sasayama T, Terai K, Yoshino M, Miyake A, Callicott JH, Egan MF, Meyer-Lindenberg A, Kempf L, Honea R, Vakkalanka RK, Takasaki J, Kamohara M, Soga T, Hiyama H, Ishii H, Matsuo A, Nishimura S, Matsuoka N, Kobori M, Matsushime H, Katoh M, Furuichi K, Weinberger DR (2008) The evolutionarily conserved G protein-coupled receptor SREB2/GPR85 influences brain size, behavior, and vulnerability to schizophrenia. Proc Natl Acad Sci U S A 105: 6133–6138

Matsuo N, Tanda K, Nakanishi K, Yamasaki N, Toyama K, Takao K, Takeshima H, Miyakawa T (2009) Comprehensive behavioral phenotyping of ryanodine receptor type 3 (RyR3) knockout mice: decreased social contact duration in two social interaction tests. Front Behav Neurosci 3:3

Mayford M, Bach ME, Huang YY, Wang L, Hawkins RD, Kandel ER (1996) Control of memory formation through regulated expression of a CaMKII transgene. Science 274:1678–1683

McGlashan TH, Hoffman RE (2000) Schizophrenia as a disorder of developmentally reduced synaptic connectivity. Arch Gen Psychiatry 57:637–648

McOmish CE, Burrows E, Howard M, Scarr E, Kim D, Shin HS, Dean B, van den Buuse M, Hannan AJ (2008) Phospholipase C-beta1 knockout mice exhibit endophenotypes modeling schizophrenia which are rescued by environmental enrichment and clozapine administration. Mol Psychiatry 13:661–672

Meincke U, Morth D, Voss T, Thelen B, Geyer MA, Gouzoulis-Mayfrank E (2004) Prepulse inhibition of the acoustically evoked startle reflex in patients with an acute schizophrenic psychosis–a longitudinal study. Eur Arch Psychiatry Clin Neurosci 254:415–421

Mejias R, Adamczyk A, Anggono V, Niranjan T, Thomas GM, Sharma K, Skinner C, Schwartz CE, Stevenson RE, Fallin MD, Kaufmann W, Pletnikov M, Valle D, Huganir RL, Wang T (2011) Gain-of-function glutamate receptor interacting protein 1 variants alter GluA2 recycling and surface distribution in patients with autism. Proc Natl Acad Sci U S A 108:4920–4925

Menalled L, El-Khodor BF, Patry M, Suarez-Farinas M, Orenstein SJ, Zahasky B, Leahy C, Wheeler V, Yang XW, MacDonald M, Morton AJ, Bates G, Leeds J, Park L, Howland D, Signer E, Tobin A, Brunner D (2009) Systematic behavioral evaluation of Huntington's disease transgenic and knock-in mouse models. Neurobiol Dis 35:319–336

Meyer U, Feldon J, Schedlowski M, Yee BK (2005) Towards an immuno-precipitated neurodevelopmental animal model of schizophrenia. Neurosci Biobehav Rev 29:913–947

Mirnics K, Middleton FA, Lewis DA, Levitt P (2001) Analysis of complex brain disorders with gene expression microarrays: schizophrenia as a disease of the synapse. Trends Neurosci 24:479–486

Moghaddam B, Adams B, Verma A, Daly D (1997) Activation of glutamatergic neurotransmission by ketamine: a novel step in the pathway from NMDA receptor blockade to dopaminergic and cognitive disruptions associated with the prefrontal cortex. J Neurosci 17:2921–2927

Moore H (2010) The role of rodent models in the discovery of new treatments for schizophrenia: updating our strategy. Schizophr Bull 36:1066–1072

Moy SS, Nonneman RJ, Young NB, Demyanenko GP, Maness PF (2009) Impaired sociability and cognitive function in Nrcam-null mice. Behav Brain Res 205:123–131

Nakagawasai O, Onogi H, Mitazaki S, Sato A, Watanabe K, Saito H, Murai S, Nakaya K, Murakami M, Takahashi E, Tan-No K, Tadano T (2010) Behavioral and neurochemical characterization of mice deficient in the N-type Ca2+ channel alpha1B subunit. Behav Brain Res 208:224–230

Neuner I, Stocker T, Kellermann T, Ermer V, Wegener HP, Eickhoff SB, Schneider F, Shah NJ (2010) Electrophysiology meets fMRI: neural correlates of the startle reflex assessed by simultaneous EMG-fMRI data acquisition. Hum Brain Mapp 31:1675–1685

Nguyen PT, Nakamura T, Hori E, Urakawa S, Uwano T, Zhao J, Li R, Bac ND, Hamashima T, Ishii Y, Matsushima T, Ono T, Sasahara M, Nishijo H (2011) Cognitive and socio-emotional deficits in platelet-derived growth factor receptor-beta gene knockout mice. PLoS One 6:e18004

Niwa M, Kamiya A, Murai R, Kubo K, Gruber AJ, Tomita K, Lu L, Tomisato S, Jaaro-Peled H, Seshadri S, Hiyama H, Huang B, Kohda K, Noda Y, O'Donnell P, Nakajima K, Sawa A, Nabeshima T (2010) Knockdown of DISC1 by in utero gene transfer disturbs postnatal dopaminergic maturation in the frontal cortex and leads to adult behavioral deficits. Neuron 65:480–489

Nusbaum MP, Contreras D (2004) Sensorimotor gating: startle submits to presynaptic inhibition. Curr Biol 14:R247–R249

O'Tuathaigh CM, Babovic D, O'Meara G, Clifford JJ, Croke DT, Waddington JL (2007) Susceptibility genes for schizophrenia: characterisation of mutant mouse models at the level of phenotypic behaviour. Neurosci Biobehav Rev 31:60–78

O'Tuathaigh CM, Desbonnet L, Moran PM, Kirby BP, Waddington JL (2011) Molecular genetic models related to schizophrenia and psychotic illness: heuristics and challenges. Curr Top Behav Neurosci 7:87–119

O'Tuathaigh CM, Desbonnet L, Waddington JL (2009) Neuregulin-1 signaling in schizophrenia: 'Jack of all trades' or master of some? Expert Rev Neurother 9:1–3

Ohgake S, Shimizu E, Hashimoto K, Okamura N, Koike K, Koizumi H, Fujisaki M, Kanahara N, Matsuda S, Sutoh C, Matsuzawa D, Muramatsu H, Muramatsu T, Iyo M (2009) Dopaminergic hypofunctions and prepulse inhibition deficits in mice lacking midkine. Prog Neuropsychopharmacol Biol Psychiatry 33:541–546

Oliver PL, Davies KE (2009) Interaction between environmental and genetic factors modulates schizophrenic endophenotypes in the Snap-25 mouse mutant blind-drunk. Hum Mol Genet 18:4576–4589

Oliveros A, Heckman MG, Corena-McLeod DPM, Williams K, Boules M, Richelson E (2010) Sensorimotor gating in NTS1 and NTS2 null mice: effects of d-amphetamine, dizocilpine, clozapine and NT69L. J Exp Biol 213:4232–4239

Oranje B, Gispen-de Wied CC, Verbaten MN, Kahn RS (2002) Modulating sensory gating in healthy volunteers: the effects of ketamine and haloperidol. Biol Psychiatry 52:887–895

Ortiz-Abalia J, Sahun I, Altafaj X, Andreu N, Estivill X, Dierssen M, Fillat C (2008) Targeting Dyrk1A with AAVshRNA attenuates motor alterations in TgDyrk1A, a mouse model of down syndrome. Am J Hum Genet 83:479–488

Otte DM, Bilkei-Gorzo A, Filiou MD, Turck CW, Yilmaz O, Holst MI, Schilling K, Abou-Jamra R, Schumacher J, Benzel I, Kunz WS, Beck H, Zimmer A (2009) Behavioral changes in G72/G30 transgenic mice. Eur Neuropsychopharmacol 19:339–348

Papaleo F, Crawley JN, Song J, Lipska BK, Pickel J, Weinberger DR, Chen J (2008) Genetic dissection of the role of catechol-O-methyltransferase in cognition and stress reactivity in mice. J Neurosci 28:8709–8723

Paylor R, Yuva-Paylor LA, Nelson DL, Spencer CM (2008) Reversal of sensorimotor gating abnormalities in Fmr1 knockout mice carrying a human Fmr1 transgene. Behav Neurosci 122: 1371–1377

Perry W, Braff DL (1994) Information-processing deficits and thought disorder in schizophrenia. Am J Psychiatry 151:363–367

Perry W, Geyer MA, Braff DL (1999) Sensorimotor gating and thought disturbance measured in close temporal proximity in schizophrenic patients. Arch Gen Psychiatry 56:277–281

Perry W, Minassian A, Feifel D, Braff DL (2001) Sensorimotor gating deficits in bipolar disorder patients with acute psychotic mania. Biol Psychiatry 50:418–424

Perry W, Minassian A, Lopez B, Maron L, Lincoln A (2007) Sensorimotor gating deficits in adults with autism. Biol Psychiatry 61:482–486

Petrovsky N, Quednow BB, Ettinger U, Schmechtig A, Mossner R, Collier DA, Kuhn KU, Maier W, Wagner M, Kumari V (2010) Sensorimotor gating is associated with CHRNA3 polymorphisms in schizophrenia and healthy volunteers. Neuropsychopharmacology 35: 1429–1439

Pietropaolo S, Delage P, Cayzac S, Crusio WE, Cho YH (2011) Sex-dependent changes in social behaviors in motor pre-symptomatic r6/1 mice. PLoS One 6:e19965

Porteous DJ, Thomson P, Brandon NJ, Millar JK (2006) The genetics and biology of Disc1—an emerging role in psychosis and cognition. Biol Psychiatry 60:123–131

Powell SB, Young JW, Ong JC, Caron MG, Geyer MA (2008) Atypical antipsychotics clozapine and quetiapine attenuate prepulse inhibition deficits in dopamine transporter knockout mice. Behav Pharmacol 19:562–565

Powell SB, Zhou X, Geyer MA (2009) Prepulse inhibition and genetic mouse models of schizophrenia. Behav Brain Res 204:282–294

Quednow BB, Wagner M, Mossner R, Maier W, Kuhn K-U (2008) Sensorimotor gating of schizophrenia patients depends on catechol O-Methyltransferase Val158Met polymorphism. Schizophr Bull: sbn088

Radyushkin K, Hammerschmidt K, Boretius S, Varoqueaux F, El-Kordi A, Ronnenberg A, Winter D, Frahm J, Fischer J, Brose N, Ehrenreich H (2009) Neuroligin-3-deficient mice: model of a monogenic heritable form of autism with an olfactory deficit. Genes Brain Behav 8:416–425

Ralph RJ, Paulus MP, Fumagalli F, Caron MG, Geyer MA (2001) Prepulse inhibition deficits and perseverative motor patterns in dopamine transporter knock-out mice: differential effects of D1 and D2 receptor antagonists. J Neurosci 21:305–313

Ralph RJ, Varty GB, Kelly MA, Wang YM, Caron MG, Rubinstein M, Grandy DK, Low MJ, Geyer MA (1999) The dopamine D2, but not D3 or D4, receptor subtype is essential for the disruption of prepulse inhibition produced by amphetamine in mice. J Neurosci 19:4627–4633

Ransome MI, Turnley AM (2008) Growth hormone signaling and hippocampal neurogenesis: insights from genetic models. Hippocampus 18:1034–1050

Relkovic D, Doe CM, Humby T, Johnstone KA, Resnick JL, Holland AJ, Hagan JJ, Wilkinson LS, Isles AR (2010) Behavioural and cognitive abnormalities in an imprinting centre deletion mouse model for Prader–Willi syndrome. Eur J Neurosci 31:156–164

Rose PK, Scott SH (2003) Sensory-motor control: a long-awaited behavioral correlate of presynaptic inhibition. Nat Neurosci 6:1243–1245

Roussos P, Giakoumaki SG, Adamaki E, Bitsios P (2011) The influence of schizophrenia-related neuregulin-1 polymorphisms on sensorimotor gating in healthy males. Biol Psychiatry 69: 479–486

Roussos P, Giakoumaki SG, Rogdaki M, Pavlakis S, Frangou S, Bitsios P (2008) Prepulse inhibition of the startle reflex depends on the catechol O-methyltransferase Val158Met gene polymorphism. Psychol Med :1–8

Sagata NA, Iwaki A, Aramaki T, Takao K, Kura S, Tsuzuki T, Kawakami R, Ito I, Kitamura T, Sugiyama H, Miyakawa T, Fukumaki Y (2010) Comprehensive behavioral study of GluR4 knockout mice, implication in cognitive function. Genes Brain Behav

Sakatani S, Yamada K, Homma C, Munesue S, Yamamoto Y, Yamamoto H, Hirase H (2009) Deletion of RAGE causes hyperactivity and increased sensitivity to auditory stimuli in mice. PLoS One 4:e8309

Samocha KE, Lim JE, Cheng R, Sokoloff G, Palmer AA (2010) Fine mapping of QTL for prepulse inhibition in LG/J and SM/J mice using F2 and advanced intercross lines. Genes Brain Behav 9:759–767

Savonenko A, Munoz P, Melnikova T, Wang Q, Liang X, Breyer RM, Montine TJ, Kirkwood A, Andreasson K (2009) Impaired cognition, sensorimotor gating, and hippocampal long-term depression in mice lacking the prostaglandin E2 EP2 receptor. Exp Neurol 217:63–73

Savonenko AV, Melnikova T, Laird FM, Stewart KA, Price DL, Wong PC (2008) Alteration of BACE1-dependent NRG1/ErbB4 signaling and schizophrenia-like phenotypes in BACE1-null mice. Proc Natl Acad Sci U S A 105:5585–5590

Scearce-Levie K, Roberson ED, Gerstein H, Cholfin JA, Mandiyan VS, Shah NM, Rubenstein JL, Mucke L (2008) Abnormal social behaviors in mice lacking Fgf17. Genes Brain Behav 7:344–354

Schaefer TL, Vorhees CV, Williams MT (2009) Mouse plasmacytoma-expressed transcript 1 knock out induced 5-HT disruption results in a lack of cognitive deficits and an anxiety phenotype complicated by hypoactivity and defensiveness. Neuroscience 164:1431–1443

Schmid S, Azzopardi E, De Jaeger X, Prado MA, Prado VF (2011) VAChT knock-down mice show normal prepulse inhibition but disrupted long-term habituation. Genes Brain Behav

Schmidt S, Richter M, Montag D, Sartorius T, Gawlik V, Hennige AM, Scherneck S, Himmelbauer H, Lutz SZ, Augustin R, Kluge R, Ruth P, Joost HG, Schürmann A (2008) Neuronal functions, feeding behavior, and energy balance in Slc2a3+/- mice. Am J Physiol Endocrinol Metab 295(5):E1084–E1094

Schneider M, Spanagel R, Zhang SJ, Bading H, Klugmann M (2007) Adeno-associated virus (AAV)-mediated suppression of Ca2+/calmodulin kinase IV activity in the nucleus accumbens modulates emotional behaviour in mice. BMC Neurosci 8:105

Schwabe K, Freudenberg F, Koch M (2007) Selective breeding of reduced sensorimotor gating in wistar rats. Behav Genet 37:706–712

Semenova S, Geyer MA, Sutcliffe JG, Markou A, Hedlund PB (2008) Inactivation of the 5-HT(7) receptor partially blocks phencyclidine-induced disruption of prepulse inhibition. Biol Psychiatry 63:98–105

Shanahan NA, Holick Pierz KA, Masten VL, Waeber C, Ansorge M, Gingrich JA, Geyer MA, Hen R, Dulawa SC (2009) Chronic reductions in serotonin transporter function prevent 5-HT1B-induced behavioral effects in mice. Biol Psychiatry 65:401–408
Shi L, Fatemi SH, Sidwell RW, Patterson PH (2003) Maternal influenza infection causes marked behavioral and pharmacological changes in the offspring. J Neurosci 23:297–302
Sipes TA, Geyer MA (1994) Multiple serotonin receptor subtypes modulate prepulse inhibition of the startle response in rats. Neuropharmacology 33:441–448
Siuciak JA, McCarthy SA, Chapin DS, Martin AN (2008) Behavioral and neurochemical characterization of mice deficient in the phosphodiesterase-4B (PDE4B) enzyme. Psychopharmacology (Berl) 197:115–126
Siuta MA, Robertson SD, Kocalis H, Saunders C, Gresch PJ, Khatri V, Shiota C, Kennedy JP, Lindsley CW, Daws LC, Polley DB, Veenstra-Vanderweele J, Stanwood GD, Magnuson MA, Niswender KD, Galli A (2010) Dysregulation of the norepinephrine transporter sustains cortical hypodopaminergia and schizophrenia-like behaviors in neuronal rictor null mice. PLoS Biol 8:e1000393
Smith CM, Lawrence AJ, Sutton SW, Gundlach AL (2009) Behavioral phenotyping of mixed background (129S5:B6) relaxin-3 knockout mice. Ann N Y Acad Sci 1160:236–241
Smith SE, Li J, Garbett K, Mirnics K, Patterson PH (2007) Maternal immune activation alters fetal brain development through interleukin-6. J Neurosci 27:10695–10702
Soleimani L, Roder JC, Dennis JW, Lipina T (2008) Beta N-acetylglucosaminyltransferase V (Mgat5) deficiency reduces the depression-like phenotype in mice. Genes Brain Behav 7:334–343
Stark KL, Burt RA, Gogos JA, Karayiorgou M (2009) Analysis of prepulse inhibition in mouse lines overexpressing 22q11.2 orthologues. Int J Neuropsychopharmacol: 1–7
Sun M, Gewirtz JC, Bofenkamp L, Wickham RJ, Ge H, O'Connor MB (2010) Canonical TGF-beta signaling is required for the balance of excitatory/inhibitory transmission within the hippocampus and prepulse inhibition of acoustic startle. J Neurosci 30:6025–6035
Suzuki G, Harper KM, Hiramoto T, Sawamura T, Lee M, Kang G, Tanigaki K, Buell M, Geyer MA, Trimble WS, Agatsuma S, Hiroi N (2009a) Sept5 deficiency exerts pleiotropic influence on affective behaviors and cognitive functions in mice. Hum Mol Genet 18:1652–1660
Suzuki G, Harper KM, Hiramoto T, Funke B, Lee M, Kang G, Buell M, Geyer MA, Kucherlapati R, Morrow B, Mannisto PT, Agatsuma S, Hiroi N (2009b) Over-expression of a human chromosome 22q11.2 segment including TXNRD2, COMT and ARVCF developmentally affects incentive learning and working memory in mice. Hum Mol Genet 18:3914–3925
Swerdlow NR, Benbow CH, Zisook S, Geyer MA, Braff DL (1993) A preliminary assessment of sensorimotor gating in patients with obsessive compulsive disorder. Biol Psychiatry 33:298–301
Swerdlow NR, Geyer MA, Braff DL (2001a) Neural circuit regulation of prepulse inhibition of startle in the rat: current knowledge and future challenges. Psychopharmacology (Berl) 156:194–215
Swerdlow NR, Karban B, Ploum Y, Sharp R, Geyer MA, Eastvold A (2001b) Tactile prepuff inhibition of startle in children with Tourette's syndrome: in search of an "fMRI-friendly" startle paradigm. Biol Psychiatry 50:578–585
Swerdlow NR, Light GA, Cadenhead KS, Sprock J, Hsieh MH, Braff DL (2006) Startle gating deficits in a large cohort of patients with schizophrenia: relationship to medications, symptoms, neurocognition, and level of function. Arch Gen Psychiatry 63:1325–1335
Swerdlow NR, Paulsen J, Braff DL, Butters N, Geyer MA, Swenson MR (1995) Impaired prepulse inhibition of acoustic and tactile startle response in patients with Huntington's disease. J Neurol Neurosurg Psychiatry 58:192–200
Swerdlow NR, Weber M, Qu Y, Light GA, Braff DL (2008) Realistic expectations of prepulse inhibition in translational models for schizophrenia research. Psychopharmacology (Berl) 199:331–388

Takao K, Tanda K, Nakamura K, Kasahara J, Nakao K, Katsuki M, Nakanishi K, Yamasaki N, Toyama K, Adachi M, Umeda M, Araki T, Fukunaga K, Kondo H, Sakagami H, Miyakawa T (2010) Comprehensive behavioral analysis of calcium/calmodulin-dependent protein kinase IV knockout mice. PLoS One 5:e9460

Takata A, Kakiuchi C, Ishiwata M, Kanba S, Kato T (2010) Behavioral and gene expression analyses in heterozygous XBP1 knockout mice: possible contribution of chromosome 11qA1 locus to prepulse inhibition. Neurosci Res

Takeuchi H, Iba M, Inoue H, Higuchi M, Takao K, Tsukita K, Karatsu Y, Iwamoto Y, Miyakawa T, Suhara T, Trojanowski JQ, Lee VM, Takahashi R (2011) P301S mutant human tau transgenic mice manifest early symptoms of human tauopathies with dementia and altered sensorimotor gating. PLoS One 6:e21050

Tammimäki A, Kaenmaki M, Kambur O, Kulesskaya N, Keisala T, Karvonen E, Garcia-Horsman JA, Rauvala H, Mannisto PT (2010) Effect of S-COMT deficiency on behavior and extracellular brain dopamine concentrations in mice. Psychopharmacology (Berl) 211:389–401

Tanaka H, Ma J, Tanaka KF, Takao K, Komada M, Tanda K, Suzuki A, Ishibashi T, Baba H, Isa T, Shigemoto R, Ono K, Miyakawa T, Ikenaka K (2009) Mice with altered myelin proteolipid protein gene expression display cognitive deficits accompanied by abnormal neuron-glia interactions and decreased conduction velocities. J Neurosci 29:8363–8371

Tanda K, Nishi A, Matsuo N, Nakanishi K, Yamasaki N, Sugimoto T, Toyama K, Takao K, Miyakawa T (2009) Abnormal social behavior, hyperactivity, impaired remote spatial memory, and increased D1-mediated dopaminergic signaling in neuronal nitric oxide synthase knockout mice. Mol Brain 2:19

Tarantino M, Eisener-Dorman AF (2011) Forward genetic approachesto understanding complex behaviors (in this volume)

Tarantino LM, Bucan M (2000) Dissection of behavior and psychiatric disorders using the mouse as a model. Hum Mol Genet 9:953–965

Tecott LH, Wehner JM (2001) Mouse molecular genetic technologies: promise for psychiatric research. Arch Gen Psychiatry 58:995–1004

Thaker GK (2007) Schizophrenia endophenotypes as treatment targets. Expert Opin Ther Targets 11:1189–1206

Thomas AM, Bui N, Graham D, Perkins JR, Yuva-Paylor LA, Paylor R (2011a) Genetic reduction of group 1 metabotropic glutamate receptors alters select behaviors in a mouse model for fragile X syndrome. Behav Brain Res 223:310–321

Thomas AM, Bui N, Perkins JR, Yuva-Paylor LA, Paylor R (2011b) Group I metabotropic glutamate receptor antagonists alter select behaviors in a mouse model for fragile X syndrome. Psychopharmacology (Berl)

Thomsen M, Wess J, Fulton BS, Fink-Jensen A, Caine SB (2010) Modulation of prepulse inhibition through both M(1) and M (4) muscarinic receptors in mice. Psychopharmacology (Berl) 208:401–416

Torkamanzehi A, Boksa P, Joober R (2008) Prepulse inhibition (PPI) of tactile startle response in recombinant congenic strains of mice: QTL mapping and comparison with acoustic PPI. J Genet Genomics 35:139–151

Tsujimura A, Matsuki M, Takao K, Yamanishi K, Miyakawa T, Hashimoto-Gotoh T (2008) Mice lacking the kf-1 gene exhibit increased anxiety- but not despair-like behavior. Front Behav Neurosci 2:4

Turetsky BI, Calkins ME, Light GA, Olincy A, Radant AD, Swerdlow NR (2007) Neurophysiological endophenotypes of schizophrenia: the viability of selected candidate measures. Schizophr Bull 33:69–94

Turner J, Hughes LF, Toth LA (2010) Sleep, activity, temperature and arousal responses of mice deficient for muscarinic receptor M2 or M4. Life Sci 86:158–169

van Berckel BN, Oranje B, van Ree JM, Verbaten MN, Kahn RS (1998) The effects of low dose ketamine on sensory gating, neuroendocrine secretion and behavior in healthy human subjects. Psychopharmacology (Berl) 137:271–281 van den Buuse M, Becker T, Kwek P, Martin S, Ruimschotel E, Risbrough V (2011a) Disruption of prepulse inhibition by 3,4-methylenedioxymethamphetamine (MDMA): comparison between male and female wild-type and 5-HT1A receptor knockout mice. Int J Neuropsychopharmacol: 1–6

van den Buuse M, Ruimschotel E, Martin S, Risbrough VB, Halberstadt AL (2011b) Enhanced effects of amphetamine but reduced effects of the hallucinogen, 5-MeO-DMT, on locomotor activity in 5-HT(1A) receptor knockout mice: implications for schizophrenia. Neuropharmacology

van der Neut R (1997) Targeted gene disruption: applications in neurobiology. J Neurosci Methods 71:19–27

Veeraragavan S, Bui N, Perkins JR, Yuva-Paylor LA, Carpenter RL, Paylor R (2011) Modulation of behavioral phenotypes by a muscarinic M1 antagonist in a mouse model of fragile X syndrome. Psychopharmacology (Berl)

Verma V, Tan CH, Ong WY, Grigoryan GA, Jones CA, Stolzberg D, Salvi R, Gross KW, Ratty AK, Dawe GS (2008) The chakragati mouse shows deficits in prepulse inhibition of acoustic startle and latent inhibition. Neurosci Res 60:281–288

Wallén-Mackenzie A, Nordenankar K, Fejgin K, Lagerstrom MC, Emilsson L, Fredriksson R, Wass C, Andersson D, Egecioglu E, Andersson M, Strandberg J, Lindhe O, Schioth HB, Chergui K, Hanse E, Langstrom B, Fredriksson A, Svensson L, Roman E, Kullander K (2009) Restricted cortical and amygdaloid removal of vesicular glutamate transporter 2 in preadolescent mice impacts dopaminergic activity and neuronal circuitry of higher brain function. J Neurosci 29:2238–2251

Wang H, Westin L, Nong Y, Birnbaum S, Bendor J, Brismar H, Nestler E, Aperia A, Flajolet M, Greengard P (2009) Norbin is an endogenous regulator of metabotropic glutamate receptor 5 signaling. Science 326:1554–1557

Wang X (2009) Cre transgenic mouse lines. Methods Mol Biol 561:265–273

Watanabe A, Toyota T, Owada Y, Hayashi T, Iwayama Y, Matsumata M, Ishitsuka Y, Nakaya A, Maekawa M, Ohnishi T, Arai R, Sakurai K, Yamada K, Kondo H, Hashimoto K, Osumi N, Yoshikawa T (2007) Fabp7 maps to a quantitative trait locus for a schizophrenia endophenotype. PLoS Biol 5:e297

Weinberger DR, Berman KF, Suddath R, Torrey EF (1992) Evidence of dysfunction of a prefrontal-limbic network in schizophrenia: a magnetic resonance imaging and regional cerebral blood flow study of discordant monozygotic twins. Am J Psychiatry 149:890–897

Wen L, Lu YS, Zhu XH, Li XM, Woo RS, Chen YJ, Yin DM, Lai C, Terry AV Jr, Vazdarjanova A, Xiong WC, Mei L (2010) Neuregulin 1 regulates pyramidal neuron activity via ErbB4 in parvalbumin-positive interneurons. Proc Natl Acad Sci U S A 107:1211–1216

Willi R, Weinmann O, Winter C, Klein J, Sohr R, Schnell L, Yee BK, Feldon J, Schwab ME (2010) Constitutive genetic deletion of the growth regulator Nogo-A induces schizophrenia-related endophenotypes. J Neurosci 30:556–567

Willott JF, Carlson S, Chen H (1994) Prepulse inhibition of the startle response in mice: relationship to hearing loss and auditory system plasticity. Behav Neurosci 108:6

Yavich L, Forsberg MM, Karayiorgou M, Gogos JA, Mannisto PT (2007) Site-specific role of catechol-O-methyltransferase in dopamine overflow within prefrontal cortex and dorsal striatum. J Neurosci 27:10196–10209

Young JW, Meves JM, Tarantino IS, Caldwell S, Geyer MA (2011a) Delayed procedural learning in alpha7-nicotinic acetylcholine receptor knockout mice. Genes Brain Behav

Young JW, Powell SB, Scott CN, Zhou X, Geyer MA (2011b) The effect of reduced dopamine D4 receptor expression in the 5-choice continuous performance task: Separating response inhibition from premature responding. Behav Brain Res 222:183–192

Young JW, Wallace CK, Geyer MA, Risbrough VB (2010) Age-associated improvements in cross-modal prepulse inhibition in mice. Behav Neurosci 124:133–140

Zhou X, Geyer MA, Kelsoe JR (2008) Does disrupted-in-schizophrenia (DISC1) generate fusion transcripts? Mol Psychiatry 13:361–363

Zhou X, Long JM, Geyer MA, Masliah E, Kelsoe JR, Wynshaw-Boris A, Chien KR (2004) Reduced expression of the Sp4 gene in mice causes deficits in sensorimotor gating and memory associated with hippocampal vacuolization. Mol Psychiatry 10:393–406

Zhou X, Tang W, Greenwood TA, Guo S, He3 L, Geyer MA, Kelsoe JR (2009) Transcription factor SP4 is a susceptibility gene for bipolar disorder. PlosOne In Press

Zhu H, Mingler MK, McBride ML, Murphy AJ, Valenzuela DM, Yancopoulos GD, Williams MT, Vorhees CV, Rothenberg ME (2010) Abnormal response to stress and impaired NPS-induced hyperlocomotion, anxiolytic effect and corticosterone increase in mice lacking NPSR1. Psychoneuroendocrinology 35:1119–1132

Part II
Clinical Behavioral Genetics

Applications of Second Generation Sequencing Technologies in Complex Disorders

Mònica Bayés, Simon Heath and Ivo Glynne Gut

Abstract Second generation sequencing (2ndGS) technologies generate unprecedented amounts of sequence data very rapidly and at relatively limited costs, allowing the sequence of a human genome to be completed in a few weeks. The principle is on the basis of generating millions of relatively short reads from amplified single DNA fragments using iterative cycles of nucleotide extensions. However, the data generated on this scale present new challenges in interpretation, data analysis and data management. 2ndGS technologies are becoming widespread and are profoundly impacting biomedical research. Common applications include whole-genome sequencing, target resequencing, characterization of structural and copy number variation, profiling epigenetic modifications, transcriptome sequencing and identification of infectious agents. New methodologies and instruments that will enable to sequence the complete human genome in less than a day at a cost of less than $1,000 are currently in development.

Keywords Sequencing technologies · Bioinformatics · Personalized genomics

Contents

1 Sequencing Technologies 322
1.1 Pyrosequencing (454-Roche-IonTorrent) 324
1.2 Sequencing with Reversible Terminators (Illumina) 325
1.3 Sequencing by Ligation (Applied Biosystems) 325
1.4 Single-Molecule Sequencing (Pacific Biosciences) 326

M. Bayés (✉) · S. Heath · I. G. Gut
Centro Nacional de Análisis Genómico, C/Baldiri Reixac 4,
08028 Barcelona, Spain
e-mail: mbayes@pcb.ub.es

Curr Topics Behav Neurosci (2012) 12: 321–343
DOI: 10.1007/7854_2011_196

Published Online: 14 February 2012

1.5 Novel Technologies in Development 326
1.6 Comparison of Cost, Throughput and Accuracy 327
2 Informatics and Bioinformatics for Second Generation Sequencing 328
2.1 Data Storage 329
2.2 Checks of Sequence Quality and Integrity 329
2.3 Sequence Alignment to the Reference Genome 330
2.4 Single-Nucleotide Variant Calling 331
3 Application of Second Generation Sequencing Technologies 332
3.1 Whole-Genome Sequencing 332
3.2 Targeted Sequencing 334
3.3 Characterizing Structural and Copy Number Variation 337
3.4 Profiling Epigenetic Modifications 338
3.5 Transcriptome Sequencing 339
3.6 Identification of Infectious Agents 340
References 340

1 Sequencing Technologies

In 1977 two very different methods were presented for directly sequencing DNA: the Maxam–Gilbert chemical cleavage method (Maxam and Gilbert 1977) and the Sanger chain-termination method (Sanger et al. 1977). The Sanger sequencing method relies on base-specific termination of the growing chain during DNA synthesis caused by the incorporation of dideoxy-nucleotides (ddNTPs) that lack the hydroxyl group necessary for DNA chain elongation. The labeled DNA fragments are then separated according to their length on a gel matrix. Sanger and Gilbert shared the Nobel prize in 1980. Sanger's method became widely used first in slab-gel implementations with radioactive detection and later in capillary electrophoretic systems using fluorescent-labeled ddNTPs. It revolutionized the genetics field and eventually led to the sequencing of the human (Lander et al. 2001; Venter et al. 2001) and many other genomes. However, the effort to sequence the human genome required many hundreds of automated capillary sequencers over a ten-year period and cost over \$10 M. At the present time, automated capillary sequencing instruments with up to 384 capillaries can generate around 1 Mbase of sequence every 24 h.

In 2005 there was a paradigm shift in DNA sequencing with the introduction of the first second generation sequencing (2ndGS) technologies (for a review see Ansorge 2009). While Sanger sequencing relies on the individual analysis of DNA fragments (one fragment per capillary), second generation sequencers are able to analyze millions of DNA templates in parallel (Fig. 1, Table 1). This is achieved first by placing single DNA molecules into individual and miniaturized reaction vessels for amplification. In emulsion PCR, template DNA molecules are individually captured on beads, and compartmentalized and clonally amplified in droplets within an oil emulsion (Diehl et al. 2006). An alternative approach is to spread out the DNA fragments on a planar surface and generate clonally amplified

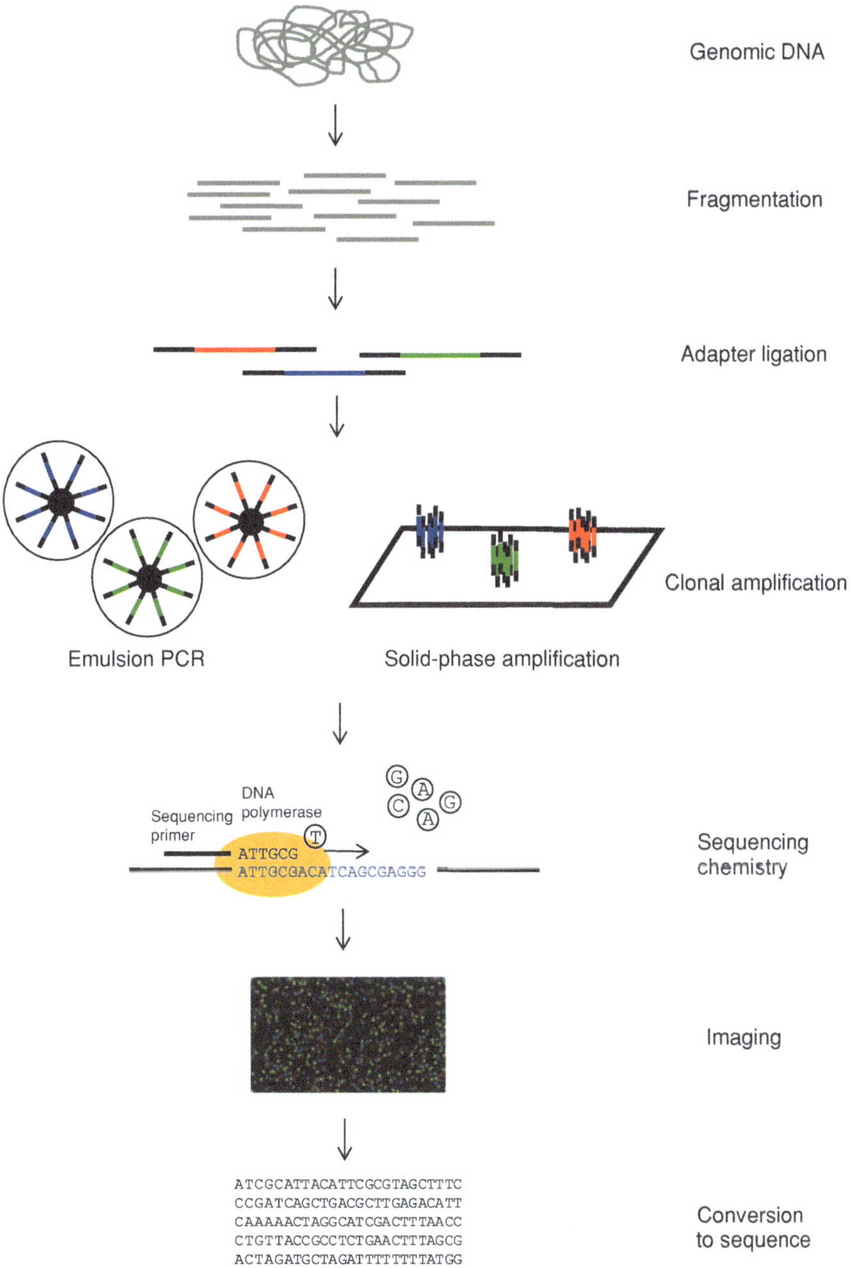

Fig. 1 Schematic representation of the principles of 2ndGS technologies. Genomic DNA (*gray*) is randomly sheared and ligated to universal adapter sequences (*black*). Clonal amplification of the fragments is achieved either in droplets within an oil emulsion (emulsion PCR; *left*) or on the surface of a solid substrate (solid-phase amplification; *right*). The sequencing process consists of iterative cycles of enzyme-driven nucleotide extension and imaging-based data acquisition

Table 1 Comparison of sequencing platforms

	ABI3730xl (Applied Byosistems)	GS FLX (454, Roche)	HiSeq2000 (Solexa, Illumina)	5500xl SOLiD (Life Technologies)	PacBio RS (Pacific Biosciences)
Amplification approach	PCR/ cloning	Emulsion PCR	Solid-phase amplification	Emulsion PCR	None
Sequencing chemistry	Sanger method	Pyrosequencing	Reversible dye terminators	Ligation based	
		sequencing	Single-molecule real-time sequencing		
Reads per run	1,536	1 M	3,000 M[a]	2,400 M[a]	NA
Bases per read	850 bp	700 bp	2 × 100 bp	2 × 60 bp	1,000 bp
Output per run	1.3 Mb	700 Mb	600 Gb[a]	300 Gb[a]	NA
Time per run	1.5 days	1 day	11 days	7 days	<1 day

[a] From two flowcells or slides

templates in situ (Bentley et al. 2008). Once the clonal amplification of the original templates is completed, millions of DNA "polymerase clones" or "polonies" are subjected to a sequencing chemistry in a microfluidic device coupled with a high-resolution imaging system. For each DNA "polony", the resulting images are converted into sequence reads.

1.1 Pyrosequencing (454-Roche-IonTorrent)

The Roche/454 system (www.454.com) was the first 2ndGS system released to the market. The templates are amplified by emulsion PCR and the sequences are obtained by iterative pyrosequencing (Magulies et al. 2005). Pyrosequencing (Ronaghi et al. 1996) is a sequencing-by-synthesis method that relies on an enzymatic cascade in which the release of pyrophosphate during the incorporation of a nucleotide drives a chemiluminescent reaction that can be detected with a high-resolution charge-coupled device (CCD) camera. It uses regular deoxy-nucleotides that are flowed sequentially in a known order. The amount of light emitted is proportional to the number of base incorporations. Because dNTPs are used, homopolymers in the template DNA are the main source of errors due to the difficulty to unambiguously assign the number of bases incorporated. The current 454/Roche platform provides reads that are several hundred bases, more than any other 2ndGS technology. Roche markets two different systems, the Genome Sequencer (GS) FLX instrument that can generate about 0.7 Gbase of sequence per run, and the GS Junior that produces about 35 Mbases (Table 1).

Recently Ion Torrent introduced an instrument that uses the same underlying methods as the Roche system for clonal amplification and parallel sequencing but with a different detection method (www.iontorrent.com). Rather than detecting chemiluminescence generated through an enzymatic cascade, the Ion Torrent system measures directly the pH change caused by the release of a positively charged hydrogen ion during each base incorporation. The system uses semiconductor technology that converts chemical information into digital, without any optics. The specification that currently is being provided moves this instrument into the realm of the Roche FLX junior.

1.2 Sequencing with Reversible Terminators (Illumina)

The first Solexa sequencing platform (Genome Analyzer 1G) was commercialized in late 2006 and Illumina acquired the company soon after the launch of this instrument (www.illumina.com). In this technology templates are amplified in near vicinity by solid-surface bridge amplification using a microfluidic cluster station (Bentley et al. 2008) (Fig. 1). Sequencing proceeds on the same surface using four reversible terminator nucleotides each labeled with a different fluorescent dye. The identity of each incorporated nucleotide is determined by recording the fluorescence with a CCD camera. After each imaging cycle the fluorescent moiety is cleaved from the base and the 3′end is regenerated to enable next nucleotide incorporation. The technology has undergone several technical and chemical upgrades to give read lengths over 100 bases. The new version of the Illumina instrument, the HiSeq2000, can run two slides simultaneously and generate over 600 Gbases of sequence in 11 days of operation (Table 1). This equates to 30x coverage of six human genomes in a single run.

1.3 Sequencing by Ligation (Applied Biosystems)

Applied Biosystems introduced the SOLiD system in late 2007 (http://www.appliedbiosystems.com/absite/us/en/home/applications-technologies/solid-next-generation-sequencing.html). As for the Roche/454 system, this system uses emulsion PCR for clonal amplification. The key distinguishing feature of this platform is the sequencing chemistry that uses a DNA ligase rather than a DNA polymerase (for a review see Ansorge 2009). A set of four fluorescently labeled octamers whose fourth and fifth bases are encoded by the attached fluorescent group are hybridized to the template. The probes that match the templates are then ligated to a primer oligonucleotide in a template-directed reaction. The fluorescent readout identifies the fixed bases. A cleavage reaction removes the last four bases with the fluorescent tag and hybridization and ligation cycles are repeated to determine in successive cycles bases 9 and 10, 14 and 15 and so on. After completion of a defined number of cycles, a new primer is annealed that is offset from the first primer by one base. The ligation procedure is repeated to give the sequence of bases 3 and 4, 8 and 9, and 13 and 14. The same procedure is repeated for the two-base, three-base and four-base offset primers, which completes the sequence. Because in the whole process each base is interrogated twice, the SOLiD technology shows the greatest accuracy. The latest version of the SOLiD instrument, 5500xl, can generate 60 base reads and process two slides simultaneously to produce more than 300 Gbases of sequence in a 7-day run (Table 1).

All of the three 2ndGS systems described above can be used with paired-end protocols that enable the sequence to be obtained from the two extremities of the templates (for a review see Holt and Jones 2008). Paired-end protocols partially compensate for the shortcoming of short read lengths of these systems compared to Sanger sequencing by providing linking information. Illumina offers a paired-end protocol to read 2 $\times$ 100 bases at a distance of up to 600 bases and a mate-pair procedure in which the insert size can be up to 5 kbases.

1.4 Single-Molecule Sequencing (Pacific Biosciences)

A new system from Pacific Biosciences was introduced in 2011 (www.pacificbiosciences.com). It is considered the first of a third generation of DNA sequencing instruments because it can sequence single DNA fragments without the need for PCR amplification. Single molecules of DNA polymerase are immobilized at the bottom of nanometer scale aperture chambers called zero-mode waveguides (ZMWs) (Korlach et al. 2010). The template-directed primer extension reaction with nucleotides labeled with fluorescence at the end of the phosphate moiety is monitored in real time. The zero-mode wave guides restrict the observation to the volume containing the DNA polymerase. This dramatically reduces the perturbation due to background fluorescence of the labeled nucleotides that are not involved in the nucleotide incorporation reaction. Because the individual base incorporation error rate is quite high, DNA molecules are circularized and read several times to generate a consensus sequence. Reads of several kbases in length have been achieved by the developers, although total throughput of the system is still in the order of Mbases.

1.5 Novel Technologies in Development

There are several companies and academic groups working in the development of new methodologies and instruments to enable sequencing of the complete human genome in less than a day at a cost of less than $1,000 (for a review see Schadt et al. 2010). Oxford Nanopore Technologies is developing a nanopore-based sequencing platform (www.nanoporetech.com). The principle of the method is the detection of nucleotides as they are driven by an electrophoretic field pass through an α–hemolysin membrane nanopore that is coupled with an exonuclease.

The change in the membrane potential at the nanopore is used to determine the identity of the nucleotide passing through. The system offers the potential to identify individual nucleotide modifications including 5-methylcytosine.

The unprecedented resolution of the new generation of sequencing technologies will allow high-throughput single cell DNA and RNA sequencing in the near future. The underlying concept is that a target sequence of DNA is generated in all of the cells of a histological section simultaneously. Several studies have already shown the feasibility of sequencing the transcriptome of individual cells with current instruments using whole-genome or transcriptome amplification procedures (Tang et al. 2010).

Current sequencing methods do not provide long-range haplotype information than can be very important for applications such as disease mapping or population history analysis. Molecular procedures for the determination of haplotypes will become more important when the methods for the determination of simple sequence motifs are well established. There are several approaches to stretch and immobilize long lengths of single genomic DNA molecules on a surface, allowing molecular sequencing tests to be carried out in situ that will provide haplotype information.

1.6 Comparison of Cost, Throughput and Accuracy

The three second generation platforms described in this chapter have their own advantages and disadvantages. Currently the HiSeq2000 and 5500xl SOLiD platforms have the higher sequencing capacity, 25–50 Gbases per day, while the GS FLX can generate up to 1 Gbase per day (Table 1). In addition, the reagent cost per Gbase on the GS FLX is more than two orders of magnitude higher than on the other platforms. However, these numbers have to be taken with caution because the rapid increase in throughput at constant reagent consumption has raised the importance of other cost components such as instrument depreciation, labor and data management that need to be considered.

The GS FLX system provides intermediate read length, with an average of 700 bases. In contrast, the 5500xl SOLiD and HiSeq2000 platforms read many more DNA molecules in parallel (up to 3000 million) but have shorter read lengths (up to 100 bp) (Table 1). This would be a severe limitation if the companies were not offering paired-end reads, which facilitate their mapping onto the reference sequence.

Error rates largely dictate the read length limits of different sequencing platforms. However, the different systems are difficult to compare in terms of accuracy because base quality values are not comparable across platforms. Raw accuracies of >99.94, >99 and >98.5% are reported in 5500xl SOLiD, GS FLX and HiSeq2000 specifications, respectively. In the GS FLX platform, the most common errors are insertions and deletions in homopolymer segments. By contrast, the 5500xl SOLiD and HiSeq2000 instruments have a very distinct error profile, with the vast majority of the sequencing errors being base substitutions. The Life-Technologies two-base encoding system shows the highest individual base call accuracy, although it needs to be pointed out that the method uses the reference sequence to make corrections to the raw base calls. At present, the error distribution of third generation technologies based on single-molecule sequencing is much higher than that of PCR-based methods.

As with genotyping technologies, it is essential to use the right 2ndGS instrument for the application in hand. For applications requiring long reads, such as metagenomic sequencing, and de novo sequencing of genomes, GS FLX is the technology of choice. The long reads guarantee best possible assembly of the data. The 5500xl SOLiD and HiSeq2000 systems work better for tag counting applications such as mRNA and smallRNA profiling and identification of DNA–protein binding sites, because they provide very high number of individual reads with just enough sequence to map them to the reference genome. Because of the lower reagent cost, HiSeq 2000 and 5500xl SOLiD are also the ideal solution for resequencing of genomes or exomes where reference sequences are available. The use of the SOLiD platform with the very accurate two-base encoding system is recommended for low coverage projects and for detecting rare variants and somatic mutations. In some cases, the use of various 2ndGS platforms is recommended to take the advantages of each technology: error profile, read length, paired-end reads, number of tags, etc.

2 Informatics and Bioinformatics for Second Generation Sequencing

The use of 2ndGS technologies that produce hundreds of millions of short reads brings with it a collection of informatics and bioinformatics challenges. The most immediate issue is that of data storage, with even small projects needing tens of terabytes of storage, and sequencing centers having storage requirements in the petabyte range. The next concerns the assessment of the quality of the produced sequence data. Each sequencing platform produces its own quality metrics, but it is often advisable to add additional quality checks to check the consistency of the results. The final challenge that will be addressed in this chapter is that of the primary data analysis, which will generally include assembly and/or alignment of the individual sequence reads followed by subsequent analyses that will depend on the particular sequencing experiment being performed. A survey of all the analyses used for the different types of sequencing experiments that have been described using 2ndGS technologies is beyond the scope of this chapter, but a short description of variant calling for the case of the resequencing of genomic data will be given. A list of available 2ndGS software has not been included as such lists become quickly out of date with the development of new packages and the improvement of existing packages. Up-to-date information on analysis packages can be found from online forums such as SEQanswers (http://www.seqanswers.com), while the focus of this chapter is on the important features that such packages should possess, allowing the reader to make their own selection.

2.1 Data Storage

The principal data produced by the sequencers are the sequence reads and associated quality measures, which alone can require a large amount of storage space. For example, WGS of a human genome at 30× coverage will require almost 100 Gbases of sequence. The common data formats used for this type of data require 2 bytes per base (1 for the base and 1 for the quality) with some extra to store the read identity, so 100 Gbases of sequence will require >200 Gbytes of storage, although this can be reduced to <50 Gbytes using data compression techniques. The sequence reads, however, are only a part of the story, and to get full picture of the storage requirements it is necessary to look at the complete analysis pipeline from the initial data produced by the sequencers (depending on the platform these could be the sequence reads or could be unprocessed data such as images) to the final products of the pipeline (i.e., called variants). Each stage of the pipeline will require space to calculate and store the results of the analysis, and a working space of at last several Tbytes will typically be required to work on WGS data from higher organisms.

An important decision to be made that has a large impact on the total storage requirements is how long the results files from different stages of the pipeline should be kept. Many of the stages in the analysis pipeline (particularly the early stages) involve data reduction and some consequent data loss, so that it is often not possible to take the results of a particular analysis stage and recreate the results of an earlier stage. A balance must therefore be struck between the cost of recreating the data from a particular stage if necessary, and the costs of storing them.

2.2 Checks of Sequence Quality and Integrity

Current 2ndGS platforms provide measures of sequencing quality for each sequenced base. Base calling is generally done by calculating the probability that the true base at a given read position is A, C, G or T, and selecting the most likely base at each position. A common measure of quality is based on the negative of the log ratio of the probability of the most likely base to that of the next most likely base. A low quality score indicates that the probability of the called base being incorrect is high, whereas a high quality score indicates that the probability of a false call is low (e.g., an Illumina quality score of 30 means that the probability that this base is called incorrect is 0.01%).

It is important to keep account of the quality scores, both as a measure of the overall quality of the sequencing run, and for downstream analyses.

For example, a useful metric to assess the overall quality of a sequencing run is the proportion of sequenced bases that have quality scores that exceed a certain threshold. For downstream analyses there are two basic approaches to use quality scores: (a) reads with low quality bases can be trimmed so that all remaining bases are high quality and (b) the analyses explicitly take account of the quality scores by weighing bases according to their qualities, with low quality bases being down weighted and so having less effect on the analyses.

In addition to errors in the base calling, another potential source of problems is sample mix-ups, leading to the sample that is being analyzed not being what is expected. Such mix-ups can occur at many points in the process, for example at the collection facility, during library preparation, at the moment the samples are placed on the sequencer or during the analyses themselves. It is difficult to completely eliminate the possibility of sample mix-ups, but a combination of good laboratory practice and analysis checks can both greatly reduce the chances of mix-ups and detect the majority of mix-ups that do occur. For example, a common practice is to perform systematically SNP genotyping on the same samples that are sequenced using one of the genome-wide SNP array platforms. This provides a set of reference genotypes that can be checked against genotypes predicted from the sequence data (Koboldt et al. 2010).

2.3 Sequence Alignment to the Reference Genome

For species for which a good reference sequence exists, the first analysis to be performed after base calling and initial quality control is to align the short sequence reads to the reference. When 2ndGS technologies were first developed, existing sequence aligners were not adapted to handle efficiently the very large number of short reads. In recent years a large number of aligners have been developed explicitly for short sequence reads, and now aligning hundreds of Gbases of sequence for WGS studies is a matter of hours rather than days or weeks (Li and Homer 2010; Koboldt et al. 2010). The aim of this section is not to give an exhaustive survey of available alignment software but rather to discuss the features that a good aligner should have, that may itself depend on the type of analysis that is required. All aligners report for each read the number of matches that have been found with an indication of the differences between the read and the reference at the match position. To reduce the number of possible matches, aligners only consider matches that are close to the reference, although how 'close' is defined can vary between aligners, and can often be adjusted by the user. A typical definition of close would be where the read differs from the reference at n or less positions, possibly only considering high quality bases in the comparison with the reference. Important considerations when choosing an aligner are whether the search for matches is exhaustive (i.e., is there a guarantee that all close matches will be found), what the behavior is with reads that map to repetitive regions (are all matches returned in these cases, or a subset of matches or a single match selected using some criterion), whether the aligner can find gapped or split alignments (discussed below), and the computational requirements. For variant calling (discussed below) it is important that only uniquely mapping reads are used otherwise there is a high risk of false calls being made, but for different analyses (i.e. of CNVs), nonuniquely mapping reads give important information and should be made available by the aligner.

The alignment of reads from paired-end libraries requires a post-processing step after alignment where the potential matches from both ends of the same fragment are examined to see if a consistent pair of matches exists where the orientation and distance of the two ends is as expected given the library construction method. In this way a nonuniquely mapping read can be 'rescued' if the corresponding pair is itself uniquely mapped. Paired-end reads are useful for the detection of structural variation as large deletions/insertions or translocations and inversions will create sets of paired-reads with incoherent matches (i.e., with the wrong distance between the pairs or with the two ends mapping in the wrong orientation or on different chromosomes).

This approach is powerful for larger structural variants in the order of several hundred base pairs or larger, but for smaller variants the distortion of the distance between paired-reads is too small to allow detection by this method. Smaller insertion/deletion variants can be detected using gapped and split alignments, if the aligner can generate these. Gapped alignments allow a short insertion/deletion (typically <30 bp to be present) within the reads, while split alignments are a generalization of gapped alignments where an arbitrary gap can exist in a read, with the two extremities of the same read being potentially mapped to different chromosomes. The use of gapped/split alignments along with paired-end reads allows the detection of deletions across the entire scale of the genome from single bases up to entire chromosomes. Other types of structural rearrangements can also be detected across most of the range, although with current technologies there is a gap between very short features (<30 bp) and features $> \sim 200$ bp where rearrangements cannot be reliably detected, although increasing read lengths should reduce this gap in the near future. It should also be mentioned here that gapped/split alignments are also very useful in the mapping of reads from the transcriptome on to the genomic reference, allowing the mapping of exon/intron boundaries (Trapnell et al. 2009; Bryant et al. 2010).

2.4 Single-Nucleotide Variant Calling

The principle of Single-Nucleotide Variant (SNV) calling is straightforward; at each position the number of reads of each base is counted, and the genotype at the position is called using the base counts. All SNV callers work very similarly; if only one base is seen then the most likely genotype is homozygous for the observed base, and if a more or less equal number of two different bases are seen then a heterozygous call will be the most likely. However, different callers do differ in the details of how ambiguous situations are dealt with, and how the confidence for the call is calculated. Simple callers use heuristic filters to determine thresholds for calling different genotypes, while more sophisticated callers use explicit statistical models that allow statements to be made about the precision of the estimation, and that permit the incorporation of prior information such as the base quality scores and knowledge about the frequency and location of known variants (Li et al. 2009; Koboldt et al. 2009).

3 Application of Second Generation Sequencing Technologies

Current high-throughput sequencing technologies are becoming widespread and allow even individual research groups to generate unprecedented amounts of sequence data very rapidly and at relatively limited costs. They enable new lines of experimental investigation in basic and clinical research (Voelkerding et al. 2009; Mardis 2009). In the near future, the cost of genome sequencing will be greatly reduced and will make personalized genomics for medical purposes a close reality.

The next sections describe current applications of 2ndGS technologies in the medical field. With the development of 2ndGS technologies, searching for changes in DNA sequence, copy number, structure, methylation status and expression profile can be accomplished with a single platform.

3.1 Whole-Genome Sequencing

Today, 2ndGS technologies are being applied to sequence complete human genomes in order to unravel the complexity of the normal and disease genomes in terms of single-nucleotide variants (SNVs), small insertions and deletions (indels), structural variants (SVs) and copy number variations (CNVs). The number of individual genomes that has been publicly announced to be sequenced is rapidly increasing. However, it is difficult to identify variants contributing to disease from individual sequences (Fig. 2b). As for the Genome-Wide Association Studies (GWAS) performed with SNP arrays, sample size needs to be very large to establish genotype-phenotype relationships for most disorders.

There are several ongoing large-scale projects that aim to sequence the whole genome of several hundreds or even thousands of individuals. The 1,000 Genome Project (www.1000genomes.org) aims to sequence the genomes of a large number of people to provide a comprehensive catalog of human genetic variation, including rare variants (present in <5% in the population) that are thought to play a major role in the genetic basis of complex diseases and that are not readily screened by current genotyping technologies. The International Cancer Genome Consortium (ICGC, www.icgc.org) has been organized as an international effort to obtain a comprehensive description of genomic, transcriptomic and epigenomic changes in 50 different tumor types (The International Cancer Genome Consortium 2010).

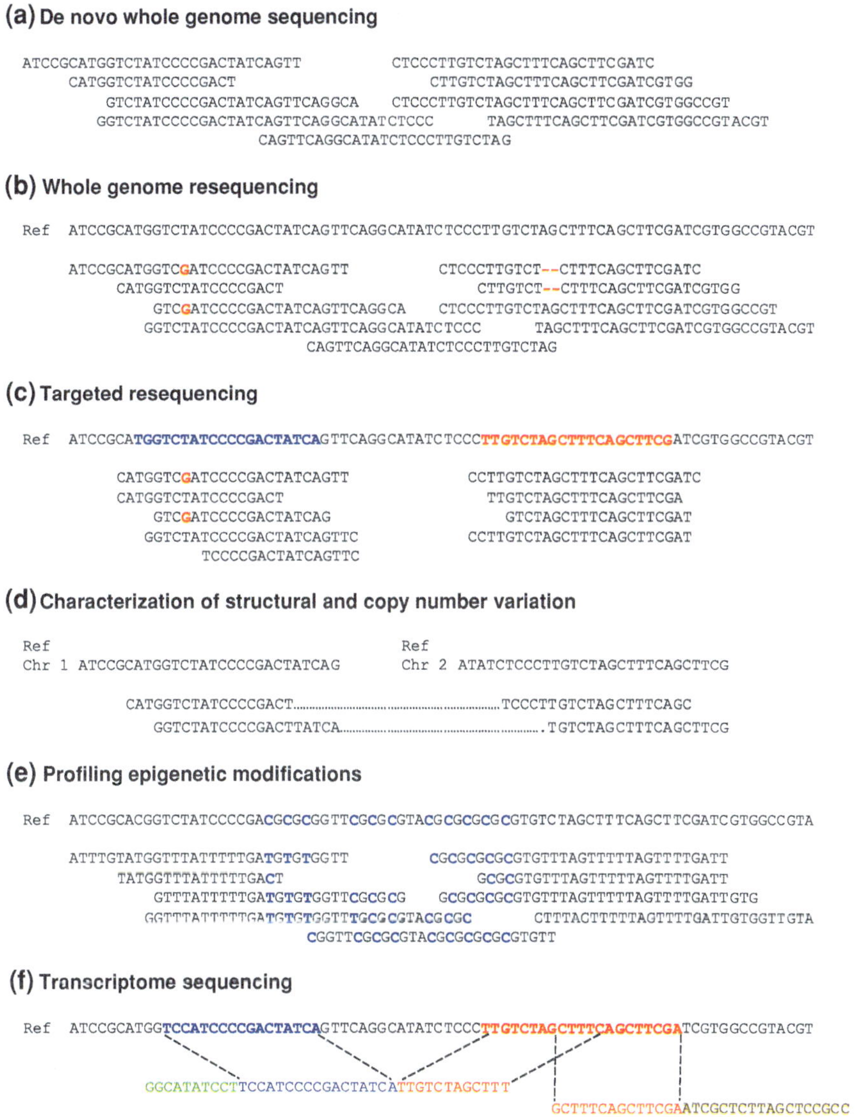

Fig. 2 Examples of the sequencing reads obtained in different applications. When appropriate the genome reference sequence (Ref) is shown above, with exons marked with different colors. Sequence variants in the reads (SNVs and indels) are shown in *red* (**b**, **c**). Paired-end reads mapping to different chromosomal locations are shown in **d**. Positions displaying complete methylation (C), no methylation (T) and partial methylation (T or C) after bisulfite treatment are depicted in *blue* (**e**)

Application of 2ndGS technologies to tumor samples has already yielded significant insights. Several landmark studies have determined the complete DNA sequences of clinical tumor samples or cell lines, and compared these to normal tissues from the same individual. In this way, genome-wide molecular profiles of cancer have been obtained for individuals with acute myeloid leukemia, lung cancer, glioblastoma, malignant melanoma and breast cancer revealing a large range of mutation loads, ranging from 10 somatic missense mutations in leukemia genomes to more than 100 mutations in tumors subjected to substantial exogenous mutagenic exposures such as lung and skin cancer (Robison 2010). Results obtained so far support the hypothesis that there is extensive genetic heterogeneity within a given tumor type, with a large number of infrequently mutated genes and a few genes with mutations of high incidence (Stratton et al. 2009). In addition, resequencing complete cancer genomes from multiple samples from the same patient allows the identification of changes that occur during cancer progression and will possibly lead to improved drug therapies (Ding et al. 2010).

Whole genome sequencing (WGS) has the potential to identify mutations located outside the protein-coding regions of genes, including putative deep intronic mutations or pathogenic variants in regulatory regions that are not being addressed by other methods. All the genomes sequenced so far have yielded similar results in terms of SNVs: around 3 million SNVs per genome of which around 0.5 million are novel SNVs not present in dbSNP. However, because of the difficulties of establishing the pathological relevance of these variants, the analysis of WGS data has been focused on the identification of a subset of variants located in protein-coding regions and large SVs. Comprehensive mutation databases such as COSMIC (Catalog of Somatic Mutations in Cancer, www.sanger.ac.uk/genetics/CGP/cosmic) will substantially facilitate the interpretation of findings.

Reduced costs are causing a rise in WGS. Two recent studies have applied WGS for genetic diagnosis in two patients with Charcot-Marie-Tooth disease and metachondromatosis (Lupski et al. 2010; Sobreira et al. 2010). Some companies and Institutions, such as Complete Genomics (www.completegenomics.com), Illumina (www.illumina.com) and the Beijing Genomics Institute (www.genomics.cn), offer individual genome sequencing services to researchers and also directly to consumers. However, WGS is still not affordable for large genetic epidemiologic studies that require large sample sizes to find out genome-wide significant associations. WGS of carefully selected individuals that are at the extreme ends of a trait distribution sequencing selected families with multiple affected individuals or the use of pooled samples are appealing alternatives.

3.2 Targeted Sequencing

Because whole-genome sequencing is still an expensive endeavor, many researchers need to focus their analysis on specific genomic regions, specific genes or the whole exome to approach disease genetics (Fig. 2c). 2ndGS can be coupled

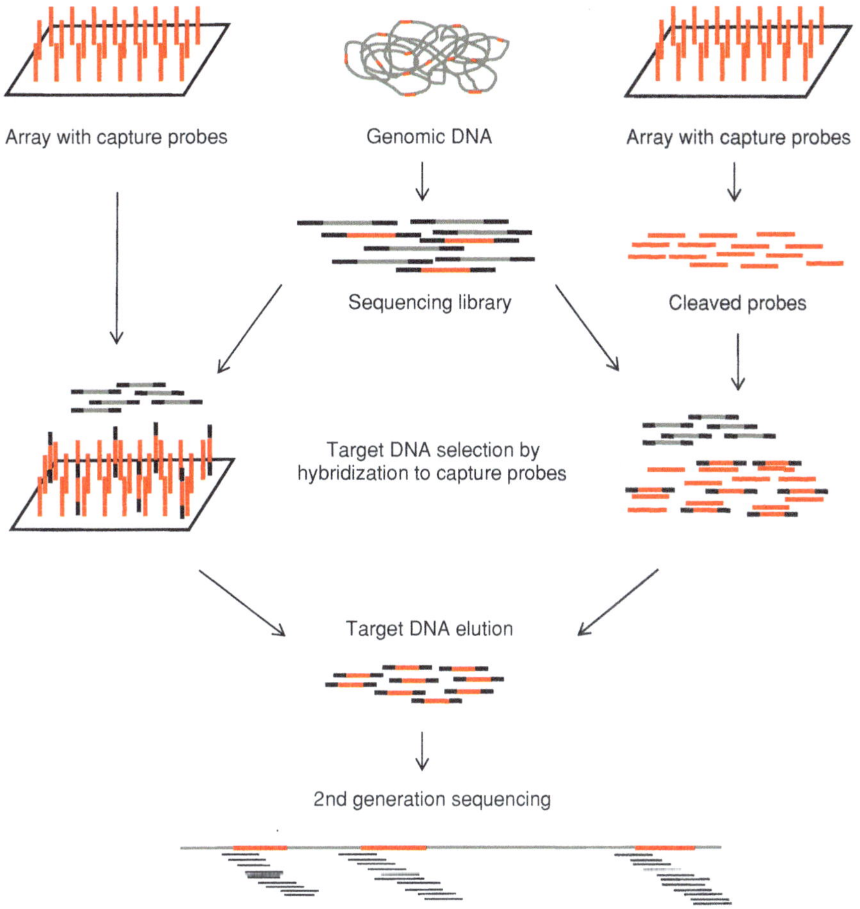

Fig. 3 Strategies for selecting specific regions of interest. Genomic DNA (*gray*, with regions of interest in *red*) is randomly sheared and ligated to universal adapter sequences (*black*). The fragments of interest are then captured by hybridization to target-specific probes either on a microarray surface (*left*) or in solution (*right*). The hybrid-selected enriched output library is eluted and sequenced

with DNA capturing or enrichment methods to resequence targeted regions of interest (Fig. 3).

Hybridization-based methods have been developed to capture specific genomic regions using a set of long oligonucleotide probes that are complementary to the regions of interest. Randomly fragmented, denatured genomic DNA is hybridized to the oligonucleotide probes that are immobilized on a solid surface or used in solution. The bound DNA is recovered and processed for sequencing. After sequencing, the percentage of mappable reads that target to the region of interest is typically >60%, although this is highly variable between regions

(Mamanova et al. 2010). Robust and reproducible protocols are currently being worked out in order to minimize variations in capture uniformity across regions.

3.2.1 Sequencing Specific Genes

Some Mendelian disorders are known to be caused by mutations in tens or hundreds of different genes. For these diseases, it is desirable to sequence a panel of candidate genes at high depth of coverage to identify the causal mutations in a particular family. PCR coupled with 2ndGS can be used to screen for mutations in disease-causing genes but it is unpractical if the number of exons to analyze is very large. High-throughput PCR systems that are based on microfluidics, such as the ones developed by RainDance Technologies (www.raindancetechnologies.com) or Fluidigm (www.fluidigm.com), partially circumvent the limited multiplex capability and high costs of traditional PCR. Alternatively, hybridization-based capturing procedures can be used to enrich the sample for the genes of interest. As an example, Vasta et al. (2009) developed an assay for the diagnosis of mitochondrial disorders that allows the simultaneous capture and sequencing of the entire mitochondrial genome and the exons of 362 nuclear gene encoding mitochondrial proteins. Other groups have designed similar multigene-resequencing assays for disorders that exhibit a high degree of locus and allelic heterogeneity such as breast/ovarian cancer (Morgan et al. 2010; Walsh et al. 2010) and familial hypertrophic cardiomyopathy (Dames et al. 2010).

At some point in the future these methods are likely to replace some of the genetic disease diagnostic products that are on the market today. In addition to reductions in cost and time, the digital nature of 2ndGS-based methods will allow for the detection of somatic variants that are present in a subpopulation of cells and that may indicate the presence of a preclinical disease.

3.2.2 Sequencing Exomes

Most Mendelian disorders are caused by mutations in protein-coding regions that represent a small part (up to 2%) of the human genome. Given the elevated cost of WGS at high coverage, exome sequencing (sequencing the collection of all human exons) has emerged as an alternative screening strategy to find variants underlying a mendelian disease. Companies such as Agilent (www.agilent.com) and Nimblegen (www.nimblegen.com) have developed user-ready products that enable to capture the human exome very cost-effectively (Mamanova et al. 2010).

These exome-capture methods coupled with 2ndGS technologies are extremely valuable for finding mutations that cause rare Mendelian disorders using small family pedigrees that are not tractable to linkage analysis. The approach has successfully been applied to familial pancreatic cancer (Jones et al. 2009), Miller syndrome (Ng et al. 2010) and other rare monogenic diseases. A typical project of this type involves sequencing a few individuals with highly selected clinical phenotypes, identifying all variants with respect to the reference sequence, filtering out common variation present in public SNV databases and selecting and validating the putative pathogenic variants shared by the affected individuals.

In the near future, exome sequencing will probably trigger the discovery of non-recurring variants with large effects on complex phenotypes such as schizophrenia, autism, amyotrophic lateral sclerosis and X-linked mental retardation (McCarthy 2009).

3.2.3 Sequencing Large Genomic Regions

Several GWAS in complex disorders have identified regions of the genome that are significantly associated with the risk of the disease (www.genome.gov/26525384). However, the majority of these reported associations involve variants that lie outside known genes, with no obvious functional consequences, and it remains a challenge to pinpoint the true causative variants.

Using targeted capture and 2ndGS methods, large contiguous stretches of genome (up to few Mbases) identified through GWAS can be sequenced to help refine association signals and find the biologically causative mutation among these correlated variants. Following this approach, Yeager et al. (2008) resequenced the gene-poor 8q24 region that is associated to breast, prostate and colorectal cancer and generated a detailed map of common and rare genetic variation across the region. Hopefully, this kind of data together with the development of assays that test directly the effect of the newly discovered variants will shed some light on the biologically relevant variants that predispose to complex disorders.

In a similar way, other groups have used targeted capture methods to quickly and cost-effectively sequence the genomic intervals identified by linkage analysis in rare Mendelian disorders (Brkanac et al. 2009; Volpi et al. 2010 and Nikopoulos et al. 2010). In both applications, a major limitation of this approach is that in all hybridization-based capture methods, repetitive regions are masked during the probe design process leaving only the unique fraction of the genomes represented.

3.3 Characterizing Structural and Copy Number Variation

Like SNVs, CNVs and SVs may account for the modulation of many complex phenotypic traits and disease susceptibility in humans. 2ndGS has been quickly applied to obtain single-base resolution data on CNVs and SVs partially supplanting array Comparative Genome Hybridization (aCGH) and SNP genotyping arrays.

An adequate capture of SVs can be obtained with the analysis of sequence reads from both ends of contiguous DNA fragments that are few hundred base pairs long (paired-reads) or from captured distal ends of larger fragments (mate pairs). Long mate paired-reads are more sensitive for detecting large events, although they have a broader size distribution and therefore have less resolution for breakpoint localization. The analysis of the separation distance of the paired-reads with respect to the reference genome enables the identification of genomic insertions and deletions. In addition and unlike microarrays, 2ndGS can also provide data on copy neutral variations such as translocations and inversions by searching for read pairs with ends mapping to different genomic locations or in opposite orientation, respectively (Fig. 2). The formation of chimeric clones during the library preparation and the sequencing errors resulting in incorrectly mapped reads are the major limitations (Medvedev et al. 2009).

In a massively parallel sequencing experiment with several millions of reads, the number of sequences aligning to a genomic region is proportional to the number of times this region appears in the genome. Therefore, CNVs can be detected at high sensitivity by determining the depth of coverage across the genome after correcting for differences in GC content and uniqueness across the genome. Reports on WGS of individual genomes have identified several thousands of CNVs which is in several fold more than the numbers that were reported in array-based studies (Ku et al. 2010).

However, paired-end WGS may be not meaningful for those patients in which cytogenetic techniques have previously localized the disease-associated chromosome rearrangement. In these cases, sequencing of sorted derivative chromosomes or microdissected chromosomal regions has proven to be a feasible approach for the characterization of chromosomal breakpoints (Chen et al. 2008; Weise et al. 2010).

3.4 *Profiling Epigenetic Modifications*

Eukaryote genomes carry relatively stable chemical marks that are added to either DNA or its packing histones. These epigenetic changes are known to play an important role in the regulation of gene expression and to contribute to the understanding of some cancers, congenital anomaly syndromes and other complex non-neoplastic disorders, including some metabolic and cardiovascular diseases. 2ndGS using slightly modified sample preparation protocols allows for an unbiased exploration of DNA methylation and histone modifications and has substantially accelerated epigenetic research.

Standard methodology for the detection of cytosine methylation, the most commonly studied epigenetic modification, is based on bisulfate treatment that converts the cytosine residues to uracils while leaving 5′-methylcytosines unchanged. Bisulfite Sequencing (BS-Seq) combines bisulfite treatment with 2ndGS technology, providing a digital measure of the frequency that a cytosine is methylated (Lister and Ecker 2009) (Fig. 2e). Other methods such as reduced representation BS sequencing or MeDIP sequencing allow saving some costs by targeting only CpG-rich or highly methylated regions, respectively. In the near future, novel single-molecule sequencing approaches, such as the ones being developed by Pacific Biosciences (www.pacificbiosciences.com) and Oxford

Nanopore Technologies (www.nanoporetech.com), will directly detect 5-methylcytosine (mC) and 5-hydroxymethylcytosine (hmC) without prior bisulfite treatment thus enabling comprehensive epigenome mapping.

Post-translational covalent modifications of histone tails, including methylation, acetylation and phosphorylation, can be characterized by chromatin immunoprecipitation with an antibody that specifically recognizes the histone variant and characterization of the bound DNA by massive parallel sequencing (ChIP-Seq) (Park 2008). Follow-on bioinformatic analysis enables the genome-wide characterization of the binding sites of the histone of interest with high precision and provides new insights into how eukaryotic genomes are organized.

Several consortium projects such as the Human Epigenome Project (HEP, www.epigenome.org) and the Encyclopedia of DNA Elements (ENCODE) aim to produce reference epigenome maps for a variety of human cells. Lister et al. (2009) provided the first complete methylation map, also called the methylome, of two human cell types and interpreted this data in relation to mRNA expression levels. Several other groups are currently using 2ndGS technologies to study the influence of cytosine methylation and histone modifications in developmental stages and disease states.

3.5 *Transcriptome Sequencing*

2ndGS technologies have been proven to be very useful for sequencing RNA (RNA-Seq) and are currently replacing microarray-based methods. They not only provide a digital overview of gene expression profiles but also enable the characterization of splice variants, antisense transcription, fusion genes, allele-specific expression, RNA editing and other forms of sequence variation in the transcribed portion of genes (Fig. 2f).

In RNA-Seq, the expression level of a gene is deduced from the total number of reads each gene sequence appears and normalized by the length of the gene. Low abundant transcripts that would escape detection by microarray-based methods can be detected by increasing sequencing depth. In this regard, short read sequencing technologies provide much deeper coverage at lower cost although they may be problematic when characterizing novel exon junctions (Morozova et al. 2009).

Aberrant transcriptional events resulting from chromosomal rearrangements play a causative role in many human cancers. In recent years several studies have successfully used paired-end RNA-Seq to detect fusion transcripts in cancer at single-nucleotide resolution (Maher et al. 2009; Zhao et al. 2009). In addition, sequencing the transcriptome is an alternative to exome sequencing for identifying expressed genetic variants when performed in the appropriate tissue type (Cirulli et al. 2010). For example, a recurrent somatic mutation in *FOXL2* gene has been found in granulosa-cell tumors using this approach (Shah et al. 2009).

A related application of 2ndGS technologies is the discovery and profiling of non-coding RNAs (ncRNA) in order to elucidate their role in health and disease. ncRNAs are small RNA molecules involved in post-transcriptional regulation of gene expression that have been suggested to modulate parameters such as disease onset, severity or progression. The short length of ncRNAs makes 2ndGS platforms ideal for ncRNA characterization on a genome-wide basis, as illustrated by a recent study by Uziel et al. (2009) that demonstrated how miRNA overexpression collaborates with the Sonic Hedgehog pathway in medulloblastoma. Similarly, other groups have reported the involvement of specific microRNAs in the etiology of ovarian and breast cancer that could be used as biomarkers for tumor classification and prognosis (Wyman et al. 2009; Nygaard et al. 2009).

3.6 Identification of Infectious Agents

2ndGS has also been used for the rapid identification of pathogens in chronic as well as acute disease replacing conventional Sanger sequencing methods. This is typically accomplished by extracting all nucleic acids from the sample of interest, sequencing the complex DNA or RNA mixture and then assembling the non-human sequence reads derived from the infectious agents. In 2008, a new arena virus and a new polyomavirus were associated with febrile illness after visceral organ transplantation and Merkel cell carcinoma, respectively, by sequencing RNA from affected tissues (Palacios et al. 2008; Feng et al. 2008).

For known pathogens, the high sensitivity of 2ndGS has proven to be ideal to detect rare variants that might confer to the pathogen some resistance to treatment or other virulence markers. Given the high throughput of second GS technologies, hundreds of samples individually tagged with a molecular barcode can be analyzed simultaneously. This approach has been used to identify drug-resistant HIV strains in patients at the earliest stages (Hofmann et al. 2007).

The Human Microbiome project (http://nihroadmap.nih.gov/hmp/) aims to analyze the entire collection of microbes (microbiome) that are present in the inner and outer surfaces of our body, including the skin, the oral cavity, the vagina and the intestine, and analyze its role in human health and disease. It uses a metagenomic approach that allows the study of native microbial or viral communities with no initial cloning steps combined with the pyrosequencing technology that provides with long (>300 bp) and highly accurate reads (>99.5%) (MacLean et al. 2009). In the near future, sequencing the microbiome at intermediate stages from health to disease will greatly increase our understanding of some human conditions such as obesity and cancer. In addition, WGS of the host genomes will allow us to investigate the genetic basis of susceptibility to microbial infection.

References

Ansorge WJ (2009) Next-generation DNA sequencing techniques. N Biotechnol 25(4):195–203
Bentley DR, Balasubramanian S, Swerdlow HP et al (2008) Accurate whole human genome sequencing using reversible terminator chemistry. Nature 456:53–59

Brkanac Z, Spencer D, Shendure J et al (2009) IFRD1 is a candidate gene for SMNA on chromosome 7q22-q23. Am J Hum Genet 84(5):692–697

Bryant DW Jr, Shen R, Priest HD, Wong WK, Mockler TC (2010) Supersplat–spliced RNA-seq alignment. Bioinformatics 26(12):1500–1505

Cirulli ET, Singh A, Shianna KV et al (2010) Screening the human exome: a comparison of whole genome and whole transcriptome sequencing. Genome Biol 11(5):R57

Chen W, Kalscheuer V, Tzschach A et al (2008) Mapping translocation breakpoints by next-generation sequencing. Genome Res 18(7):1143–1149

Dames S, Durtschi J, Geiersbach K, Stephens J, Voelkerding KV (2010) Comparison of the Illumina genome analyzer and roche 454 GS FLX for resequencing of hypertrophic cardiomyopathy-associated genes. J Biomol Tech 21(2):73–80

Diehl F, Li M, He Y, Kinzler KW, Vogelstein B, Dressman D (2006) BEAMing: single-molecule PCR on microparticles in water-in-oil emulsions. Nat Methods 3(7):551–559

Ding L, Ellis MJ, Li S, Larson DE et al (2010) Genome remodelling in a basal-like breast cancer metastasis and xenograft. Nature 464(7291):999–1005

Feng H, Shuda M, Chang Y, Moore PS (2008) Clonal integration of a polyomavirus in human Merkel cell carcinoma. Science 319(5866):1096–1100

Hoffmann C, Minkah N, Leipzig J et al (2007) DNA bar coding and pyrosequencing to identify rare HIV drug resistance mutations. Nucleic Acids Res 35(13):e91

Holt RA, Jones SJ (2008) The new paradigm of flow cell sequencing. Genome Res 18(6):839–846

Jones S, Hruban RH, Kamiyama M et al (2009) Exomic sequencing identifies PALB2 as a pancreatic cancer susceptibility gene. Science 324(5924):217

Koboldt DC, Chen K, Wylie T et al (2009) VarScan: variant detection in massively parallel sequencing of individual and pooled samples. Bioinformatics 25(17):2283–2285

Koboldt DC, Ding L, Mardis ER, Wilson RK (2010) Challenges of sequencing human genomes. Brief Bioinform 11(5):484–498

Korlach J, Bjornson KP, Chaudhuri BP et al (2010) Real-time DNA sequencing from single polymerase molecules. Methods Enzymol 472:431–455

Ku CS, Loy EY, Salim A, Pawitan Y, Chia KS (2010) The discovery of human genetic variations and their use as disease markers: past, present and future. J Hum Genet 55(7):403–415

Li H, Handsaker B, Wysoker A et al (2009) The sequence alignment/map format and SAMtools. Bioinformatics 25(16):2078–2079

Li H, Homer N (2010) A survey of sequence alignment algorithms for next-generation sequencing. Brief Bioinform 11(5):473–483

Lander ES, Linton LM, Birren B et al (2001) Initial sequencing and analysis of the human genome. Nature 409(6822):860–921

Lister R, Ecker JR (2009) Finding the fifth base: genome-wide sequencing of cytosine methylation. Genome Res 19(6):959–966

Lister R, Pelizzola M, Dowen RH et al (2009) Human DNA methylomes at base resolution show widespread epigenomic differences. Nature 462(7271):315–322

Lupski JR, Reid JG, Gonzaga-Jauregui C et al (2010) Whole-genome sequencing in a patient with Charcot-Marie-Tooth neuropathy. N Engl J Med 362(13):1181–1191

MacLean D, Jones JD, Studholme DJ (2009) Application of 'next-generation' sequencing technologies to microbial genetics. Nat Rev Microbiol 7(4):287–296

Maher CA, Palanisamy N, Brenner JC et al (2009) Chimeric transcript discovery by paired-end transcriptome sequencing. Proc Natl Acad Sci U S A 106(30):12353–12358

Mamanova L, Coffey AJ, Scott CE et al (2010) Target-enrichment strategies for next-generation sequencing. Nat Methods 7(2):111–118

Mardis ER (2009) New strategies and emerging technologies for massively parallel sequencing: applications in medical research. Genome Med 1(4):40

Margulies M, Egholm M, Altman WE et al (2005) Genome sequencing in microfabricated high-density picolitre reactors. Nature 437(7057):376–380

Maxam AM, Gilbert W (1977) A new method for sequencing DNA. Proc Natl Acad Sci 74(2):560–564

Medvedev P, Stanciu M, Brudno M (2009) Computational methods for discovering structural variation with next-generation sequencing. Nat Methods 6(11 Suppl):S13–S20

McCarthy MI (2009) Exploring the unknown: assumptions about allelic architecture and strategies for susceptibility variant discovery. Genome Med 1(7):66

Morgan JE, Carr IM, Sheridan E et al (2010) Genetic diagnosis of familial breast cancer using clonal sequencing. Hum Mutat 31(4):484–491

Morozova O, Hirst M, Marra MA (2009) Applications of new sequencing technologies for transcriptome analysis. Annu Rev Genomics Hum Genet 10:135–151

Ng SB, Buckingham KJ, Lee C et al (2010) Exome sequencing identifies the cause of a mendelian disorder. Nat Genet 42(1):30–35

Nikopoulos K, Gilissen C, Hoischen A et al (2010) Next-generation sequencing of a 40 Mb linkage interval reveals TSPAN12 mutations in patients with familial exudative vitreoretinopathy. Am J Hum Genet 86(2):240–247

Nygaard S, Jacobsen A, Lindow M et al (2009) Identification and analysis of miRNAs in human breast cancer and teratoma samples using deep sequencing. BMC Med Genomics 2:35

Palacios G, Druce J, Du L et al (2008) A new arenavirus in a cluster of fatal transplant-associated diseases. N Engl J Med 358(10):991–998

Park PJ (2008) Epigenetics meets next-generation sequencing. Epigenetics 3(6):318–321

Robison K (2010) Application of second-generation sequencing to cancer genomics. Brief Bioinform 11:524–534 (Epub ahead of print)

Ronaghi M, Karamohamed S, Pettersson B et al (1996) Real-time DNA sequencing using detection of pyrophosphate release. Anal Biochem 242:84–89

Sanger F, Nicklen S, Coulson AR (1977) DNA Sequencing with chain-terminating inhibitors. Proc Natl Acad Sci 74:546–567

Schadt EE, Turner S, Kasarskis A (2010) A window into third-generation sequencing. Hum Mol Genet 19(R2):R227–R240

Shah SP, Köbel M, Senz J et al (2009) Mutation of FOXL2 in granulosa-cell tumors of the ovary. N Engl J Med 360(26):2719–2729

Sobreira NL, Cirulli ET, Avramopoulos D et al (2010) Whole-genome sequencing of a single proband together with linkage analysis identifies a Mendelian disease gene. PLoS Genet 6(6):e1000991

Stratton MR, Campbell PJ, Futreal PA (2009) The cancer genome. Nature 458(7239):719–724

Tang F, Barbacioru C, Nordman E et al (2010) RNA-Seq analysis to capture the transcriptome landscape of a single cell. Nat Protoc 5(3):516–535

The international cancer genome consortium. Hudson TJ, Anderson W, Artez A et al (2010) International network of cancer genome projects. Nature 464(7291):993–998

Trapnell C, Pachter L, Salzberg SL (2009) Tophat: discovering splice junctions with RNA-Seq. Bioinformatics 25(9):1105–1111

Uziel T, Karginov FV, Xie S et al (2009) The MIR-17 ~ 92 cluster collaborates with the Sonic Hedgehog pathway in medulloblastoma. Proc Natl Acad Sci U S A 106(8):2812–2817

Vasta V, Ng SB, Turner EH, Shendure J, Hahn SH (2009) Next generation sequence analysis for mitochondrial disorders. Genome Med 1(10):100

Venter JC, Adams MD, Myers EW et al (2001) The sequence of the human genome. Science 291(5507):1304–1351

Voelkerding KV, Dames SA, Durtschi JD (2009) Next-generation sequencing: from basic research to diagnostics. Clin Chem 55(4):641–658

Volpi L, Roversi G, Colombo EA et al (2010) Targeted next-generation sequencing appoints c16orf57 as clericuzio-type poikiloderma with neutropenia gene. Am J Hum Genet 86(1):72–76

Walsh T, Lee MK, Casadei S et al (2010) Detection of inherited mutations for breast and ovarian cancer using genomic capture and massively parallel sequencing. Proc Natl Acad Sci 107(28):12629–12633

Weise A, Timmermann B, Grabherr M et al (2010) High-throughput sequencing of microdissected chromosomal regions. Eur J Hum Genet 18(4):457–462

Wyman SK, Parkin RK, Mitchell PS et al (2009) Repertoire of microRNAs in epithelial ovarian cancer as determined by next generation sequencing of small RNA cDNA libraries. PLoS One 4(4):e5311

Yeager M, Xiao N, Hayes RB et al (2008) Comprehensive resequence analysis of a 136 kb region of human chromosome 8q24 associated with prostate and colon cancers. Hum Genet 124(2):161–170

Zhao Q, Caballero OL, Levy S et al (2009) Transcriptome-guided characterization of genomic rearrangements in a breast cancer cell line. Proc Natl Acad Sci U S A 106(6):1886–1891

Rare Genomic Deletions and Duplications and their Role in Neurodevelopmental Disorders

Joseph T. Glessner, John J.M. Connolly and Hakon Hakonarson

Abstract Copy number variations (CNVs) are deletions and duplications of DNA sequences that vary in length from a few base pairs to several million. While these structural variations are often benign, they can disrupt vital biological functions and result in disease. CNVs have been identified as causal in a number of neurodevelopmental disorders (NDs), including but not limited to, autism, attention-deficit/hyperactivity disorder (ADHD), and schizophrenia. Here, we examine CNV research into these disorders, and discuss relevant methodological considerations. By identifying specific rare deletions and duplications, we may be better able to determine the etiology of neurodevelopmental disorders and identify appropriate treatments.

Keywords Copy number variation · Neurodevelopmental disorder · Autism · Schizophrenia · Attention deficit hyperactivity disorder

J. T. Glessner · J. J.M. Connolly · H. Hakonarson (✉)
Center for Applied Genomics,
The Children's Hospital of Philadelphia,
Philadelphia, PA 19104, USA
e-mail: hakonarson@email.chop.edu

H. Hakonarson
Division of Genetics,
The Children's Hospital of Philadelphia,
Philadelphia, PA 19104, USA

H. Hakonarson
Department of Pediatrics,
University of Pennsylvania School of Medicine,
Philadelphia, PA 19104, USA

Curr Topics Behav Neurosci (2012) 12: 345–360
DOI: 10.1007/7854_2011_179

Published Online: 13 January 2012

Contents

1 Introduction 346
2 Methodological Considerations 346
 2.1 CNV Detection and Validation 346
 2.2 Interpretation 350
3 Large Constitutional Deletions and Duplications in Neurodevelopmental Disease 351
4 Common Neurodevelopmental Disorders 353
 4.1 Autism 354
 4.2 Schizophrenia 354
 4.3 ADHD 355
5 Conclusion 356
References 357

1 Introduction

NDs are present in approximately 6% of children (Polanczyk et al. 2007; Newschaffer et al. 2007; Arajärvi et al. 2005), require lengthy evaluation, can be difficult to treat, and contribute significantly to overall disease burden and healthcare costs. While the causes of NDs remain largely unknown, recent research efforts have identified CNVs as potentially important pathogenic factors. In the past several years, techniques used to identify CNVs have advanced considerably, making it easier to identify structural variations. Two common techniques for detecting CNVs include 1) using microarray assays from single nucleotide polymorphism (SNP) genotypes and 2) using comparative genomic hybridization (CGH) arrays based on intensity data to identify contiguous probes that signal CNV events. The improved sophistication of whole-genome discovery approaches necessitates a renewed emphasis on rigorous statistical oversight and validation methods. In the section below, we discuss some of the challenges present in studying CNVs, while later we focus on these structural variations in relation to specific NDs.

2 Methodological Considerations

2.1 CNV Detection and Validation

Identifying CNVs using high-density microarrays requires high-quality genotype signals. Factors that help validate the genome signal include increasing the proportion of SNPs in genotype regions, minimizing the standard deviations of intensity signals, monitoring the number of CNVs putatively associated with given loci (a high number of CNVs may indicate erroneous identification), controlling population substructures, and validating calls by using an alternate method (e.g. quantitative PCR or multiplex ligation-dependent probe amplification (MLPA)).

Table 1 Deletions and duplications associated with neurodevelopmental disorders

Locus	CNV Type	Disease	Gene
22q11.2	Del/Dup	DiGeorge syndrome and velocardiofacial syndrome	*TBX1+*
17p13.3	Del	Miller–Dieker syndrome	*LIS1 +*
17q11.2	Del/Dup	Neurofibromatosis type 1	*NF1 +*
Xq22	Del/Dup	Pelizaeus–Merzbacher disease	*PLP1*
15q11–q13	Del	Prader–Willi syndrome/Angelman syndrome	*UBE3A +*
15q11–q13	Dup	Autism	*UBE3A +*
17p11.2	Del	Smith–Magenis syndrome	*RAI1 +*
17p11.2	Dup	Potocki–Lupski syndrome	*RAI1 +*
7q11.23	Del/Dup	Williams syndrome (WS)	*many*
11q14.2–14.3	Del	Attention-Deficit/Hyperactivity Disorder	*GRM5*
3p26.1	Del	Attention-Deficit/Hyperactivity Disorder	*GRM7*
22q13	Del	Autism	*SHANK3*
Xp22.33	Del	Autism	*NLGN4*
16p11.2	Del	Autism	*many*
4q28.3	Del	Autism	*PCDH10*
3q24	Del	Autism	*NHE9*
15q11–q13	Dup	Autism	*many*
2p16.3	Del	Autism	*NRXN1*
3p26.3	Del/Dup	Autism	*CNTN4*
22q11.21	Dup	Autism	*TBX1+*
6q26	Del	Autism	*PARK2*
1q25.2	Dup	Autism	*RFWD2, PAPPA2*
7p14.3	Dup	Autism	*AK057321*
3q13.33	Dup	Autism	*FBXO40, GOLGB1*
2p24.3	Dup	Autism	*AK123120*
3p26.2	Del	Autism	*UNQ3037*
10q23.2	Del	Autism	*GRID1*
3q26.31	Dup	Autism	*NLGN1*
4q31.21	Dup	Autism	*GYPELOC441046*
3q13.3	Dup	Bipolar disorder	*GSK3B*
Xq28	Del	Rett syndrome	*MECP2*
2q34	Del	Schizophrenia	*ERBB4*
5p13.3	Del	Schizophrenia	*SLC1A3*
2q31.2	Del	Schizophrenia	*RAPGEF4*
12q24	Dup	Schizophrenia	*CIT*
16q22.1	Del	Schizophrenia	*PDPR*
16p11.2	Dup	Schizophrenia	*QPRT, DOC2A, TBX6+*
22q11.21	Del	Schizophrenia	*COMT, TBX1+*
9q34.3	Del	Schizophrenia	*CACNA1B*
10q11.21	Del	Schizophrenia	*RET*
4p16.1	Del	Schizophrenia	*WDR1*
18q12.3	Del	Schizophrenia	*RIT2, PIK3C3*
3p26.2	Del	Schizophrenia	*SUMF1*
Xp22.33	Del	Tourette's syndrome	*NLGN4*

Another important validation measure is to limit signal intensity waves—genome-wide artifacts of signal intensities that alternately yield inflated or deflated signals (Marioni et al. 2007). Typically, these correlate with genomic content (total guanines (Gs) and cytosines (Cs) per total of the four bases) and require the imposition of wave-correction models. In a study of Illumina®'s BeadStudio software, we previously demonstrated that approximately 10% of DNA samples from commercial cell line repositories show visually discernable wavy patterns (Diskin et al. 2008). Because CNVs are typically inferred from gains and losses of intensity signals, the imposition of wave-correction models are critical for accurate detection.

CNV detection algorithms, including the methods and options they implement, are still under development and debate—although PennCNV (Wang et al. 2007) and BirdSuite (Korn et al. 2008) have gained popularity. Similarly, while Illumina® (San Diego, CA) (Steemers and Gunderson, 2007) and Affymetrix® (Santa Clara, CA) (Fodor et al. 1991) genome-wide arrays are widely used, the actual array content is constantly evolving, and new versions and probe designs make the task of integrating datasets a difficult one (Fig. 1). The CNV signal can vary depending on the type of source sample used (e.g. cell lines or saliva), batch effects, equipment variation, reagent variation, and the approach taken to signal normalization. All of these factors can introduce disparities that need to be modeled and statistically corrected.

Some individuals with associated CNVs may not show any notable phenotypic variation. This indicates a threshold effect or a possible second-hit modifying mutation. The second hit is extremely hard to determine given the rarity of the initial CNV, which limits the sample size. At rare CNV frequencies, small numbers of controls with a given CNV quickly dilute significance tests *versus* CNVs not detected in a large control cohort. It should also be noted that differences in the number of statistical tests being performed on CNV regions mean that classical multiple-test statistical corrections or Bonferroni corrections are difficult to adapt to rare CNV association. A more practical alternative is to use the significance observed in control-excess as a model for the null distribution. This model allows us to accept CNV regions where case-excess is of greater significance as surpassing the null distribution.

Related to this are issues of pleiotropy and/or variable expressivity of the underlying genes. More dense continuous coverage is needed to confidently call CNVs, culminating in genome-sequencing technologies. Individual SNPs and their underlying probe sequences have different efficiencies in differentiating modes of CNVs. Enrichment for clearly differentiating probes is essential for SNP arrays with strong CNV detection utility. Nevertheless, rare CNVs may be so rare that individual labs evaluating their own dataset may lack the requisite power to reach confident association. Collaboration and data sharing are important to boosting marginally significant signals. Collaboration is becoming more widespread based on collegial agreements and a trend among funding agencies to stipulate that raw data be deposited in the public domain.

The development of public datasets such as the rapidly growing dbGaP has the added benefit of facilitating replication studies, which are critical to validation. Existing CNV data can also be integrated into the major public databases.

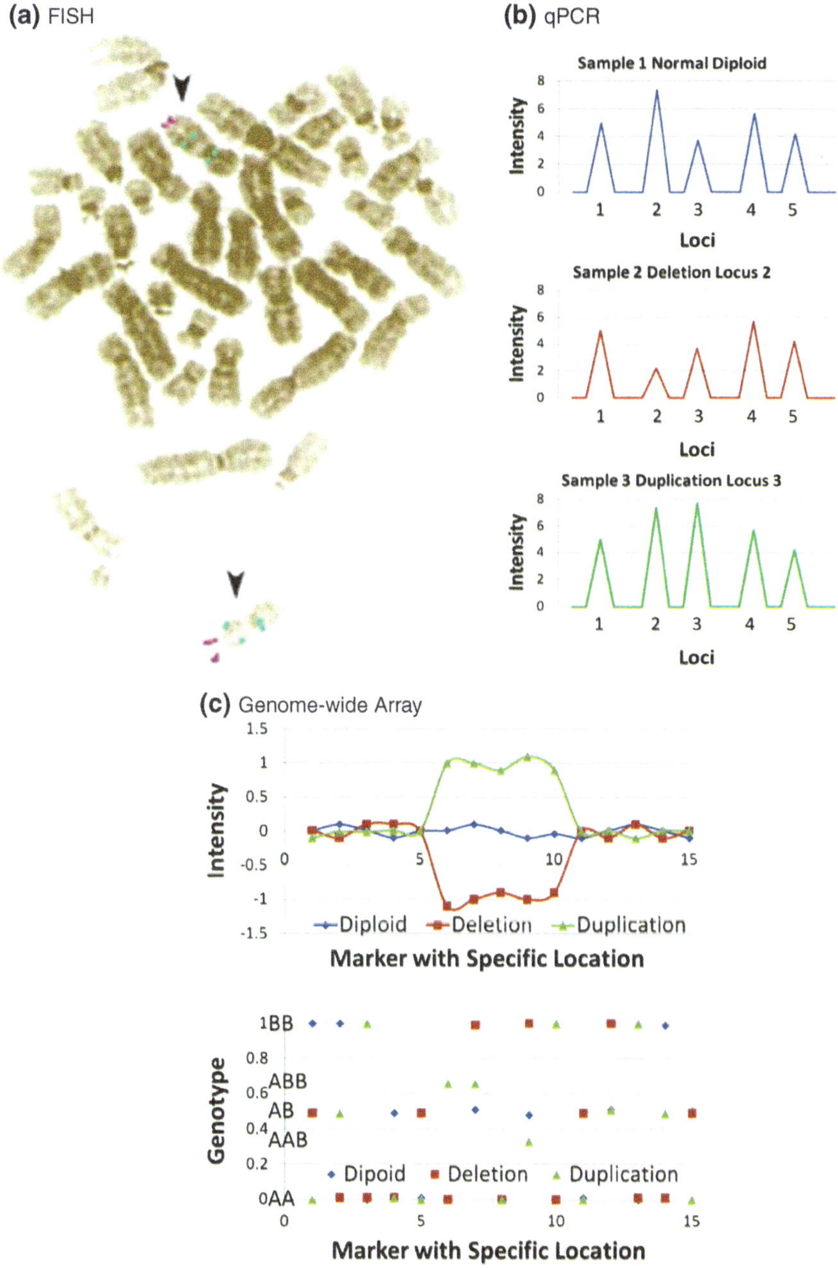

Fig. 1 CNV detection methods and data. **A** Two-color FISH shows large CNV, deletion shown. **B** Based on a control value, a ratio of observed to control for specific loci signal CNV. **C** SNP arrays or aCGH assay contiguous SNPs or CN probes are bound to the genome allowing for detection of variations from normal diploid

Genome-wide association remains focused on SNP genotypes by the current definition, but the underlying principles can be incorporated into copy number variations. The central databases for disease association results such as Online Mendelian Inheritance in Man (OMIM) and NHGRI: A Catalog of Published Genome-Wide Association Studies largely exclude CNV association results. While the CNV event is certainly more difficult to represent, a central consensus region SNP ID could be very analogous to reporting significant SNPs. Single loci reporting is complicated by factors that include varying confidence of specific calls, association, and marginal success in replication (due to rarity of individual variants). Because of the lack of CNV entries in disease association resources, such a query becomes a large primary literature search. An alternative is to present a gene network of related function or interaction of around 100 genes. Further abstract is the idea that psychiatric patients have a CNV burden in genic regions, meaning there are more large rare CNVs in psychiatric patients than controls.

The primary catalog for CNVs is currently the Database of Genomic Variants (*projects.tcag.ca/variation/*), which is also integrated into the UCSC browser. There are key caveats for interpreting this data. Firstly, a variety of array platforms and CNV-calling algorithms are used. Secondly, the subjects are part of studies which typically focus on a specific phenotype. By virtue of being de-identified for research purposes, they do not have a complete phenotype to apply to other disease studies. Thirdly, acquiring the source sample to validate the CNV with the array and calling algorithm from an alternate laboratory is not straightforward and is often unsuccessful. Since 2004, papers describing genome-wide maps of copy number variation have revealed that CNVs and segmental duplications are very abundant features of the human genome.

Given the massive amounts of CNVs that can be detected and specifically assayed, collections of CNV-specific markers such as copy number polymorphisms (CNPs) can be developed. Similar to SNPs, CNPs are assumed to be common in above 1% population frequency. Although the hypothesis of common disease explained by common variants makes intuitive sense, mounting support for common disease rare variants in the form of CNVs are being found associated with NDs.

2.2 Interpretation

When CNVs are observed in a parent but not in a child, this may indicate an error in the CNV call. The converse is not necessarily the case, and de novo CNVs have been strongly associated with a number of NDs (e.g. Sebat et al. 2007). De novo CNVs may be particularly informative for pronounced NDs such as brain malformations, where the parent is often annotated as healthy. De novo mutations may also be of special relevance to heterogeneous NDs such as autism and ADHD, where etiologies are complex and a specific disease label can have very different phenotypes and combinations (Geschwind and Levitt 2007). Further, CNVs may have prospective diagnostic value in differentiating behaviorally similar NDs. In certain cases, the parent may have an attenuated ND or an additional protective variant, which has obvious implications for treatment. Rare private mutations in certain families are expected from a population genetics perspective without detrimental effect. This necessitates observation of recurrent rare CNVs in unrelated individuals from diverse geographic and ethnic populations, which can be detected by different array technologies.

Proximal to CNV regions, highly homologous low-copy-number repeats (LCRs) are frequently observed, which can predispose to recombination between non-allelic LCRs (non-allelic homologous recombination) (Britten and Davidson, 1976; Lupski 1998). Evolution and genome condensation occur through various mechanisms, including chromosome splicing of highly similar sequences known as homologous recombination (HR). In somatic cells, HR is needed to repair extreme DNA damage such as double strand breaks (DSB). If spliced incorrectly, CNVs and genomic instability can result. An intermediate state is formed between two DNA strands, which proceeds by crossover (two-way sharing, meiosis, and DSB) or gene conversion (one-way sharing, DSB). The human genome has much segmental duplication, which provides similar sequences for HR to occur. HR may expand gene families with diverse functions but also leads to more mistakes. Segmental duplications can masquerade as allelic sequences during meiosis and lead to erroneous splicing, termed non-allelic homologous recombination. Gene conversion can result in inserts of non-expressed poor function elements into homologous expressed genes. Segmental duplications are also referred to as low-copy repeats. Non-homologous end-joining (NHEJ) (Moore and Haber, 1996), fork stalling, and template switching (FoSTeS) (Lee et al. 2007) can result in less typical CNV generation.

The widespread and, in some regions, high prevalence of CNVs is somewhat counterintuitive for such an aberration from the classical genetics of a maternal and paternal diploid chromosome set. However, it is likely that CNVs may contribute to evolutionary mechanisms, whereby gene families with homologous sequences are extended into complementary functions over time. Perhaps CNVs involving genes responsible for behavior, cognition, and psychological state are most active in humans representing a successful expansion of gene functions based on ancestral duplications.

3 Large Constitutional Deletions and Duplications in Neurodevelopmental Disease

Early studies with chromosomal karyotypes and fluorescent in situ hybridization (FISH) were limited in their ability to detect only the largest CNVs, but increased resolution of SNP microarrays and DNA sequencing technologies have made smaller CNVs far easier to identify. Large CNVs inevitably have a greater likelihood of disrupting crucial functional elements but examples exist of small targeted CNV that can cause phenotype variation. Certain CNVs can give rise to very specific phenotypes while others may have variable expressivity. Some genes may have complementary function to other genes such that deletion of one gene homolog does not result in significant phenotypic variation.

Perhaps the most familiar CNV linked with ND involvement is trisomy 21 or Down syndrome, which results from the duplication of an entire chromosome 21 (Lejeune et al. 1959). Symptoms include distinct facial features, as well as delayed mental and social development. The CNV variant is highly penetrant since the duplication is directly associated with a specific phenotype.

Williams syndrome is associated with a deletion of 7q11.23 and is characterized by a distinctive facial appearance, visuospatial deficits, learning disorders, ADHD, and hypersociability (Francke 1999; Fishman et al. 2010) (Table 1).

Deletion and duplication of 15q11–q13 have been described with variable expressivity based on maternal or paternal inheritance. Maternal 15q11–q13 deletion results in Angelman syndrome (AS), which is characterized by developmental delay, sleep disturbance, seizures, jerky movements (especially hand-flapping), frequent laughter or smiling, and a happy demeanor (Greenberg et al. 1991). AS is caused by the under-expression of *UBE3A*, whose protein product E6-AP is involved in the transfer of ubiquitin molecules for degradation (Kishino et al. 1997; Matsuura et al. 1997). Paternal deletion of 15q11–13 causes Prader–Willi syndrome (PWS), which is characterized by hypotonia in infancy, hypogonadotrophic hypogonadism, facial dysmorphism, obesity, short stature, and cognitive and behavioral impairments (Cassidy, Dykens, and Williams, 2000). The microdeletions associated with PWS all share a 4.3 kb deleted region, which includes the promoter region, first exon, and part of the first intron of the SNRPN gene (Rodriguez-Jato et al. 2005). A number of phenotypes have also been associated with duplication in the 15q11–13 region. Hypotonia, language and motor delay, epilepsy, facial dysmorphism, and cognitive and learning impairments are key features of 15q duplication syndrome. Maternal 15q11–q13 duplication results in autism, while paternal duplication often is of no effect or slight developmental delay (Gillberg et al. 1991). These same features are associated with 15q trisomy (Chamberlain and Lalande, 2010).

Miller–Dieker syndrome (MLS) is associated with deletion of 17p13.3 impacting the *LIS1* gene (Dobyns et al. 1993). Features of MLS include lissencephal (smooth brain), mental retardation, and facial dysmorphism (Schiff et al. 2010). Other than brain malformations, many patients also present epilepsy, spastic behavior, and shortened life expectancy. Larger deletions incorporating proximal genes result in more severe phenotypes, indicating the involvement of long range regulation or epistatic interaction between genes.

Another example is the 22q11.2 deletion that causes a multi-systemic disorder. Many patients with the deletion have a learning disorder, congenital heart defects, a malformed palate, mild facial abnormalities, schizophrenia, and sensitivity to infections (Liu et al. 2002). The 22q11.2 deletion syndrome (also known as velocardiofacial syndrome or DiGeorge syndrome) is caused by a microdeletion in chromosome 22. The phenotype is variable and can include facial dysmorphism, cardiovascular abnormalities, short stature, cognitive and behavioral impairments, and a high risk of schizophrenia (Karayiorgou et al. 2010). Genes in the 22q11.2 region include *COMT* (Shashi et al. 2006) and *TBX1* (Paylor et al. 2006). Duplication of 22q11.2 also has variable clinical presentations, including developmental delay, dysmorphic facial features, autism, and cognitive and behavioral impairments (Portnoï 2009). Additional structural variation of this locus includes (11;22) translocation causing the der(22) syndrome and (2) the formation of a supernumerary chromosome resulting in Cat-Eye syndrome. Eight different LCRs are located in proximal 22q, and are presumably susceptible to homologous recombination events and promote nonallelic homologous recombination (Shaikh et al. 2001). The 22q13 deletion causes Phelan–McDermid syndrome, characterized by pervasive developmental disorders, speech delay, and hypotonia (McDermid and Morrow, 2002).

Rett syndrome (RTT), which is found almost exclusively in females, is associated with deletion of Xq28 including the *MECP2* gene. Duplications of *MECP2* were found in males exhibiting severe intellectual disability, and autistic features (Ramocki et al. 2010). *MECP2* appears to be the singularly important gene that is impacted by the Xq28 CNV.

Pelizaeus–Merzbacher disease is associated with *PLP1* duplication and disease is characterized by coordination, motor abilities, and intellectual function deterioration, which were described as early as 1885 (Pelizaeus 1885). Xq26.2–q27.1 duplication impacting the *SOX3* gene was identified in females in a family with language impairments that included stuttering and dyslalia, short height, and dysmorphic features. Deletions on chromosome X may be due to submicroscopic X-to-autosome translocation, Alu–Alu recombination, and non-homologous end-joining (Inoue et al. 2002).

Thus, large CNVs are high confidence calls that can be detected on different resolution arrays. These large CNVs often encompass many genes and in turn cause multi-systemic syndromes rather than singular features. As such, they are straightforward to detect with available technology but in the majority of cases the actual critical gene(s) or variant(s) within the CNV, or the micro level instability, remains to be elucidated.

4 Common Neurodevelopmental Disorders

Taken together, multiple rare CNVs have been implicated in the pathogenesis of NDs. However, a few common variant models have also been implicated in predisposing to several of the NDs, including autism and schizophrenia (Wang et al. 2009; International Schizophrenia Consortium 2009). While common SNP genotypes have been implicated to a greater degree in inflammatory and metabolic disorders, such as type 1 diabetes, inflammatory bowel disease and type 2 diabetes, it is conceivable that with greater sample sizes that many more common SNP variants will be uncovered that predispose to ND susceptibility. In addition, common CNVs are also hypothetically possible but current methods often under call their true frequency. Thus, rare CNVs have been most successful in NDs lending support to a common disease-rare variant model that is disparate from the more typical common disease-common variant model, previously implicated by many (Schork et al. 2009).

4.1 Autism

A potentially efficient approach to identifying CNVs in neurodevelopmental disorders is to target functional gene clusters of associated loci. Previously reported ASD candidate genes, such as *NRXN1* (Kim et al. 2008) and *CNTN4* (Roohi et al. 2008; Fernandez et al. 2008) have been independently replicated by Glessner et al., Glessner et al. 2009. We also identified several new susceptibility genes encoding neuronal cell-adhesion molecules, including *NLGN1* and *ASTN2*, which were enriched with CNVs in ASD cases. Moreover, CNVs within or surrounding genes in ubiquitin pathways—*UBE3A*, *PARK2*, *RFWD2*, and *FBXO40*—were observed in cases but not in controls. Duplications 55 kilobases upstream of complementary DNA *AK123120* were identified (Glessner et al. 2009). Using a ~550,000 SNP array, we observed an average, 15.5 CNV calls for each individual, with a similar frequency observed in cases and controls.

Recently, Pinto et al. (2010) identified inherited and de novo CNVs associated with 25 genes that were enriched in a large ASD sample. These include *SHANK2*, *SYNGAP1*, *DLGAP2*, and the *DDV53-PTCHD1* locus. Interestingly, the group also found that individuals with ASDs have neither significantly more, nor significantly larger, CNVs than matched controls. However, they are more likely to have CNVs in genic regions, particularly in loci previously associated with ASDs and/or intellectual disability. This is important because it replicates our previous finding that CNV frequencies are comparable for cases and controls. This underscores the point that individuals with NDs may not be any more prone to duplications or deletions *per se*, but more importantly to CNVs in specific coding regions. The localization of CNVs at genes linked with intellect is also theoretically significant in that it suggests a putative link between the two (Skuse 2007). CNVs impacting *DPP6, DPP10,* and *PCDH9* were also associated with ASDs (Marshall et al. 2008). CNVs on 16p11.2 and genotype association on 5p15 (*SEMA5A/TAS2R1*) were also found in autism (Weiss et al. 2008; Weiss and Arking 2009).

4.2 Schizophrenia

We recently used 1.7 million probes to perform a whole-genome CNV analysis on a cohort of 977 schizophrenia cases and 2,000 controls, validating positive findings in an independent cohort of 758 schizophrenia cases and 1,485 controls (Glessner et al. 2010). The gene ontology synaptic transmission family of genes was notably enriched for CNVs in the cases ($P = 1.5 \times 10^{-7}$). Among these, *CACNA1B* and *DOC2A*, both calcium-signaling genes responsible for neuronal excitation, were deleted in 16 cases and duplicated in 10 cases, respectively. In addition, *RET* and *RIT2*, both ras-related genes important for neural crest development, were significantly affected by CNVs. *RET* deletion was exclusive to seven cases, and *RIT2* deletions were overrepresented in the schizophrenia cases. Similar to the ASD studies above, rare CNVs were not overrepresented in cases *versus* controls. This is largely consistent with the literature, though (Walsh et al. 2008) found novel large rare deletions and duplications of genes observed in 15% of cases versus 5% of controls. A study of CNVs in Chinese schizophrenia patients detected no significant difference in rare CNVs between cases and controls (Shi et al. 2008). Another study of 1,013 cases and 1,084 controls of European ancestry also failed to find more rare CNVs >100 kb in cases or enrichment for neurodevelopmental pathways (Need et al. 2009).

Previous studies have associated various CNVs with schizophrenia, including deletions of 22q11.2 (Liu et al. 2002), *NRXN1* (Kirov et al. 2007), *APBA2* (Kirov et al. 2007), and *CNTNAP2* (Friedman et al. 2008). However, each of these CNVs is rare and they account for a relatively small proportion of the overall genetic risk in schizophrenia. Large rare CNVs impacting many different genes enriched in neurodevelopmental pathways have been reported elsewhere (Walsh et al. 2008; International Schizophrenia Consortium 2008; Stefansson et al. 2008). Specific loci exhibiting runs of homozygosity (ROHs) have also been associated with schizophrenia (Lencz et al. 2007), and de novo CNVs ($P = 7.8 \times 10^{-4}$) were recently observed in sporadic schizophrenia cases in comparison with controls (Xu et al. 2008). Comparison of genomic findings in schizophrenia and autism has suggested a diametric etiology (Crespi et al. 2010).

4.3 ADHD

Attention-deficit/hyperactivity disorder (ADHD), the most common neuropsychiatric disorder in children, with a 5.2% worldwide prevalence, causes significant academic, behavioral, and social impairment throughout the life span (Polanczyk et al. 2007). The phenotype consists of extreme manifestations of continuous traits, including hyperactivity, impulsivity, and dysfunction in executive and self-regulatory skills that involve working memory, temporal organization, planning, organizing, maintaining focus, effort, and motivation (Elia and Devoto 2007). Genome-wide association studies of ADHD including CNVs have been few despite high prevalence.

Specific genetic factors underlying risk for ADHD remain elusive. To assess the role of structural variation in ADHD, we previously identified 222 inherited copy number variations (CNVs) within 335 ADHD patients and their parents that were not detected in 2,026 unrelated healthy individuals. (Elia et al. 2010). Again, no excess CNVs, either deletions or duplications, were found in the ADHD cohort relative to controls. Inherited rare CNVs impacted candidate genes of autism, schizophrenia, and Tourette syndrome, including *A2BP1*, *AUTS2*, *CNTNAP2*, and *IMMP2L*. The ADHD CNVs enriched in cases were related to psychological and neurological functions, including learning, behavior, synaptic transmission, and central nervous system development. A deletion within the glutamate receptor gene, *GRM5*, was found in an affected parent and all three affected offspring.

GRM7 deletion was also observed in cases and not in controls (Elia et al., 2010). Rare inherited structural variations play an important role in ADHD development and indicate a set of putative candidate genes for further study in the etiology of ADHD.

5 Conclusion

In summary, both common and rare variants have been implicated in NDs. While considerable progress has been made over the past few years in unveiling the genetic causes of NDs, genome-wide studies explain roughly 10–15% of the heritability of NDs. A recent simulation study suggests that some of the ‘missing’ heritability might be accountable for synthetic associations—a term used to describe the relationship between a (common) causal variant and a (rarer) non-causal variant. Dickson et al. (2010) simulated output from a genome-wide association study (GWAS), where some individuals had rare variants (0.5–2% of the population) that likely caused a disease. They found that the GWAS technique erroneously identified the synthetically associated variant as causal, essentially obscuring the signal from the rarer “at risk” variant. In other words, synthetic associations infer that many people share a variant that confers some small risk of disease. In fact, only a few of these individuals share a rarer variant that carries a much higher risk.

CNV studies, which are sensitive to specific rare deletions and duplications, provide a potential route around this problem. However, as noted above, this approach is relatively insensitive to second-hit variants because of the limits of sample size. A combination of the two approaches is arguably the most attractive. The explosion of GWAS publications in the past several years has generated numerous loci putatively associated with NDs. Perhaps the most important replication that can be made without the need for genotyping multitudes of additional samples is in functional gene clusters of associated loci. This also boosts confidence in association signals of rare CNVs, which singularly may have arisen by chance. The CNVs were recently identified at CNTN4 and NRXN1 have well-established associations with autism (Glessner et al. 2009).

Some of the missing heritability should also disappear as arrays with increasingly powerful resolutions are developed. Although much progress has been made to this end, further resolution of data is needed to gain confidence in CNV calls, and to establish clear breakpoints. Forty-two million probes have already been achieved with genome-tiling set of multiple arrays (Conrad et al. 2010). The best resolution, of course, is sequencing the genome for every contiguous base, which has also been investigated (Kidd et al. 2008).

Many of the issues allied to structural studies will doubtless be resolved by technological advances. Nevertheless, additional complexity in the CNV theme exists in mosaicism, translocations, and inversions. Neither translocations nor inversions can be detected with a SNP array, and clone paired-end sequencing (fosmid) and sequence-assembly comparison are the only genome-wide technologies that can detect these structural variants (excepting targeted methods such as FISH and Southern blotting). Other challenges are organizational, and these include issues of data-integration, sharing, and management, all of which have been touched-on above.

With the ability to recognize sequence-level variation already a diagnostic reality, the necessity of quickly dissociating benign and disease-causing changes will become more pressing. Rett syndrome, a ND on the autism spectrum, is one example of a specific microdeletion associated with a specific phenotype. Similar associations are likely to emerge from the CNV field over the coming years. Similarly, family based studies may help to differentiate de novo and inherited forms, and may ultimately help identify protective variants.

References

Arajärvi R et al (2005) Prevalence and diagnosis of schizophrenia based on register, case record and interview data in an isolated Finnish birth cohort born 1940–1969. Soc Psychiatry Psychiatr Epidemiol 40:808–816

Britten RJ, Davidson EH (1976) Studies on nucleic acid reassociation kinetics: empirical equations describing DNA reassociation. Proc Natl Acad Sci USA 73:415–419

Chamberlain SJ, Lalande M (2010) Neurodevelopmental disorders involving genomic imprinting at human chromosome 15q11-q13. Neurobiol Dis 39(1):13–20

Conrad DF, Pinto D, Redon R, Feuk L, Gokcumen O, Zhang Y, Aerts J, Andrews TD, Barnes C, Campbell P, Fitzgerald T, Hu M, Ihm CH, Kristiansson K, Macarthur DG, Macdonald JR, Onyiah I, Pang AW, Robson S, Stirrups K, Valsesia A, Walter K, Wei J; Wellcome Trust Case Control Consortium, Tyler-Smith C, Carter NP, Lee C, Scherer SW, Hurles ME (2010) Origins and functional impact of copy number variation in the human genome. Nature, 464(7289):704–12. Epub 2009 Oct 7

Crespi B, Stead P, Elliot M (2010) Evolution in health and medicine Sackler colloquium: Comparative genomics of autism and schizophrenia. Proc Natl Acad Sci USA, 107 Suppl 1:1736–41. Epub 2009 Dec 1

Diskin SJ, Li M, Hou C, Yang S, Glessner J, Hakonarson H, Bucan M, Maris JM, Wang K (2008) Adjustment of genomic waves in signal intensities from whole-genome SNP genotyping platforms. Nucleic Acids Res, 36(19):e126. Epub 2008 Sep 10

Dobyns WB, Reiner O, Carrozzo R, Ledbetter DH (1993) Lissencephaly. A human brain malformation associated with deletion of the LIS1 gene located at chromosome 17p13. JAMA 270:2838–2842

Elia J, Devoto M (2007) ADHD genetics: 2007 update. Curr Psychiatry Rep. 9:434–439

Elia J, Gai X, Xie HM, Perin JC, Geiger E, Glessner JT, D'arcy M, deBerardinis R, Frackelton E, Kim C, Lantieri F, Muganga BM, Wang L, Takeda T, Rappaport EF, Grant SF, Berrettini W, Devoto M, Shaikh TH, Hakonarson H, White PS (2010) Rare structural variants found in attention-deficit hyperactivity disorder are preferentially associated with neurodevelopmental genes. Mol Psychiatry. 15(6):637–46. Epub 2009 Jun 23

Fernandez T et al (2008) Disruption of Contactin 4 (CNTN4) results in developmental delay and other features of 3p deletion syndrome. Am. J. Hum. Genet. 82:1385

Fishman I, Yam A, Bellugi U, Lincoln A, Mills D (2010) Contrasting patterns of language-associated brain activity in autism and Williams syndrome. Social Cognitive and Affective Neuroscience, Aug 27. [Epub ahead of print, doi:10.1093/scan/nsq075]

Fodor SP, Read JL, Pirrung MC, Stryer L, Lu AT, Solas D (1991) Light-directed, spatially addressable parallel chemical synthesis. Science 251:767–773

Francke U (1999) Williams syndrome: genes and mechanisms. Hum Mol Genet 8:1947–1954

Friedman JI, Vrijenhoek T, Markx S, Janssen IM, van der Vliet WA et al (2008) CNTNAP2 gene dosage variation is associated with schizophrenia and epilepsy. Mol. Psychiatry 13:261–266

Garbern JY (2006) Pelizaeus–Merzbacher disease: Genetic and cellular pathogenesis. Cell Mol Life Sci: CMLS 64(1):50–65

Geschwind DH, Levitt P (2007) Autism spectrum disorders: developmental disconnection syndromes. Current Opin Neurobiol 17(1):103–111

Gillberg C, Steffenburg S, Wahlström J, Gillberg IC, Sjöstedt A, Martinsson T, Liedgren S, Eeg-Olofsson O (1991) Autism associated with marker chromosome. J Am Acad Child Adolesc Psychiatry 30(3):489–494

Glessner JT, Hakonarson H (2009) Common variants in polygenic schizophrenia. Genome Biol, 10(9):236. Epub 2009 Sep 29

Glessner JT, Reilly MP, Kim CE, Takahashi N, Albano A, Hou C, Bradfield JP, Zhang H, Sleiman PM, Flory JH, Imielinski M, Frackelton EC, Chiavacci R, Thomas KA, Garris M, Otieno FG, Davidson M, Weiser M, Reichenberg A, Davis KL, Friedman JI, Cappola TP, Margulies KB, Rader DJ, Grant SF, Buxbaum JD, Gur RE, Hakonarson H (2010) Strong synaptic transmission impact by copy number variations in schizophrenia. Proc Natl Acad Sci USA, 107(23):10584–10589. Epub 2010 May 20

Glessner JT, Wang K, Cai G, Korvatska O, Kim CE et al (2009) Autism genomewide copy number variation reveals ubiquitin and neuronal genes. Nature 459:569–573

Greenberg F, Guzzetta V, Montes de Oca-Luna R, Magenis RE, Smith AC, Richter SF, Kondo I, Dobyns WB, Patel PI, Lupski JR (1991) Molecular analysis of the Smith-Magenis syndrome: a possible contiguous-gene syndrome associated with del(17)(p11.2). Am. J. Hum. Genet, 49:1207–1218

Inoue K, Osaka H, Thurston VC, Clarke JTR, Yoneyama A, Rosenbarker L, Bird TD, Hodes ME, Shaffer LG, Lupski JR (2002) Genomic rearrangements resulting in PLP1 deletion occur by nonhomologous end joining and cause different dysmyelinating phenotypes in males and females. Am. J. Hum. Genet 71:838–853

International Schizophrenia Consortium, Purcell SM, Wray NR, Stone JL, Visscher PM, O'Donovan MC, Sullivan PF, Sklar P (2009) Common polygenic variation contributes to risk of schizophrenia and bipolar disorder. Nature, 460(7256):748–752. Epub 2009 Jul 1

International Schizophrenia Consortium, Stone JL, O'Donovan MC, Gurling H, Kirov GK et al (2008) Rare chromosomal deletions and duplications increase risk of schizophrenia. Nature 455:237–241

Korn JM et al (2008) Integrated genotype calling and association analysis of SNPs, common copy number polymorphisms and rare CNVs. Nat Genet 40:1253–1260

Karayiorgou M, Simon TJ, Gogos JA (2010) 22q11.2 microdeletions: linking DNA structural variation to brain dysfunction and schizophrenia. Nat Rev Neurosci 11(6):402–416

Kidd JM, Cooper GM, Donahue WF, Hayden HS, Sampas N, Graves T, Hansen N, Teague B, Alkan C, Antonacci F, Haugen E, Zerr T, Yamada NA, Tsang P, Newman TL, Tüzün E, Cheng Z, Ebling HM, Tusneem N, David R, Gillett W, Phelps KA, Weaver M, Saranga D, Brand A, Tao W, Gustafson E, McKernan K, Chen L, Malig M, Smith JD, Korn JM, McCarroll SA, Altshuler DA, Peiffer DA, Dorschner M, Stamatoyannopoulos J, Schwartz D, Nickerson DA, Mullikin JC, Wilson RK, Bruhn L, Olson MV, Kaul R, Smith DR, Eichler EE (2008) Mapping and sequencing of structural variation from eight human genomes. Nature 453(7191):56–64

Kim HG et al (2008) Disruption of neurexin 1 associated with autism spectrum disorder. Am. J. Hum. Genet 82:199–207

Kirov G, Gumus D, Chen W, Norton N, Georgieva L et al (2007) Comparative genome hybridization suggests a role for NRXN1 and APBA2 in schizophrenia. Hum. Mol. Genet. 17(3):458–465

Kishino T, Lalande M, Wagstaff J (1997) UBE3A/E6-AP mutations cause Angelman syndrome. Nat Genet 15(1):70–73

Lee JA, Carvalho CM, Lupski JR (2007) A DNA replication mechanism for generating nonrecurrent rearrangements associated with genomic disorders. Cell 131(7):1235–1247

Lejeune J, Turpin R, Gautier M (1959) Le mongolisme: premier exemple d'aberration autosomique humaine. Ann Genet 1:41

Lencz T, Lambert C, DeRosse P, Burdick KE, Morgan TV et al (2007) Runs of homozygosity reveal highly penetrant recessive loci in schizophrenia. PNAS 104:19942–19947

Liu H, Abecasis GR, Heath SC, Knowles A, Demars S et al (2002) Genetic variation in the 22q11 locus and susceptibility to schizophrenia. Proc Natl Acad Sci USA 99:16859–16864

Lupski JR (1998) Genomic disorders: structural features of the genome can lead to DNA rearrangements and human disease traits. Trends Genet 14:417–422

Marioni JC, Thorne NP, Valsesia A, Fitzgerald T, Redon R, Fiegler H, Andrews TD, Stranger BE, Lynch AG, Dermitzakis ET, Carter NP, Tavaré S, Hurles ME (2007) Breaking the waves: improved detection of copy number variation from microarray-based comparative genomic hybridization. Genome Biol 8(10):R228

Marshall CR, Noor A, Vincent JB, Lionel AC, Feuk L, Skaug J, Shago M, Moessner R, Pinto D, Ren Y, Thiruvahindrapduram B, Fiebig A, Schreiber S, Friedman J, Ketelaars CE, Vos YJ, Ficicioglu C, Kirkpatrick S, Nicolson R, Sloman L, Summers A, Gibbons CA, Teebi A, Chitayat D, Weksberg R, Thompson A, Vardy C, Crosbie V, Luscombe S, Baatjes R, Zwaigenbaum L, Roberts W, Fernandez B, Szatmari P, Scherer SW (2008) Structural variation of chromosomes in autism spectrum disorder. Am J Hum Genet 82(2):477–488. Epub 2008 Jan 17

Matsuura T, Sutcliffe JS, Fang P, Galjaard RJ, Jiang YH, Benton CS, Rommens JM, Beaudet AL (1997) De novo truncating mutations in E6-AP ubiquitin-protein ligase gene (UBE3A) in Angelman syndrome. Nat Genet 15(1):74–77

McDermid HE, Morrow BE (2002) Genomic disorders on 22q11. Am J Hum Genet 70: 1077–1088

Moore JK, Haber JE (1996) Cell cycle and genetic requirements of two pathways of nonhomologous end-joining repair of double-strand breaks in Saccharomyces cerevisiae. Mol Cell Biol 16(5): 2164–2173

Need AC, Ge D, Weale ME, Maia J, Feng S et al (2009) A genome-wide investigation of SNPs and CNVs in Schizophrenia. PLoS Genet 5(2):e1000373

Newschaffer CJ et al (2007) The epidemiology of autism spectrum disorders. Annu Rev Public Health 28:235–258

Paylor R, Glaser B, Mupo A, Ataliotis P, Spencer C, Sobotka A, Sparks C, Choi CH, Oghalai J, Curran S et al (2006) Tbx1 haploinsufficiency is linked to behavioral disorders in mice and humans: implications for 22q11 deletion syndrome. Proc Natl Acad Sci USA 103:7729–7734

Pelizaeus F (1885) Über eine eigenthümliche Form spastischer Lähmung mit Cerebralerschinungen auf hereditärer Grundlage (Multiple Sklerose). Arch. Psychiatr. Nervenkr,16:698–710

Pinto D, Pagnamenta AT, Klei L, Anney R, Merico D, Regan R, Conroy J, Magalhaes TR, Correia C, Abrahams BS, Almeida J, Bacchelli E, Bader GD, Bailey AJ, Baird G, Battaglia A, Berney T, Bolshakova N, Bölte S, Bolton PF, et al (2010) Functional impact of global rare copy number variation in autism spectrum disorders. Nature, 466(7304):368–372. Epub 2010 Jun 9

Polanczyk G, de Lima MS, Horta BL, Biederman J, Rohde LA (2007) The worldwide prevalence of ADHD: a systematic review and metaregression analysis. Am J Psychiatry 164(6):942–948

Portnoï MF (2009) Microduplication 22q11.2: a new chromosomal syndrome. Eur J Med Genet 52(2–3):88–93

Ramocki MB, Tavyev YJ, Peters SU (2010) The MECP2 duplication syndrome. Am J Med Genet Part A 152A(5):1079–1088

Rodriguez-Jato S, Nicholls RD, Driscoll DJ, Yang TP (2005) Characterization of cis- and trans-acting elements in the imprinted human SNURF-SNRPN locus. Nucleic Acids Res 33(15):4740–4753

Roohi J et al (2008) Disruption of contactin 4 in three subjects with autism spectrum disorder. J Med Genet 46:176–182

Schiff M, Delahaye A, Andrieux J, Sanlaville D et al (2010) Further delineation of the 17p13.3 microdeletion involving YWHAE but distal to PAFAH1B1: Four additional patients. Eur J Med Genet 53(5):303–308

Schork NJ, Murray SS, Frazer KA, Topol EJ (2009) Common vs. rare allele hypotheses for complex diseases. Curr Opin Genet Dev 19(3):212–219

Sebat J, Lakshmi B, Malhotra D, Troge J, Lese-Martin C, Walsh T, Yamrom B, Yoon S, Krasnitz A, Kendall J, Leotta A, Pai D, Zhang R, Lee YH, Hicks J, Spence SJ, Lee AT, Puura K, Lehtimaki T, Ledbetter D, Gregersen PK, Bregman J, Sutcliffe JS, Jobanputra V, Chung W, Warburton D, King MC, Skuse D, Geschwind DH, Gilliam TC, Ye K, Wigler M (2007) Strong association of de novo copy number mutations with autism. Sci 316:445–449

Shaikh TH, Kurahashi H, Emanuel BS (2001) Evolutionarily conserved low copy repeats (LCRs) in 22q11 mediate deletions, duplications, translocations, and genomic instability: an update and literature review. Genet Med 3(1):6–13

Shashi V, Keshavan MS, Howard TD, Berry MN, Basehore MJ, Lewandowski E, Kwapil TR (2006) Cognitive correlates of a functional COMT polymorphism in children with 22q11.2 deletion syndrome. Clin Genet 69:234–238

Shi YY, He G, Zhang Z, Tang W, Zhang J Jr et al (2008) A study of rare structural variants in schizophrenia patients and normal controls from Chinese Han population. Mol Psychiatry 13:911–913

Skuse DH (2007) Rethinking the nature of genetic vulnerability to autistic spectrum disorders. Trends in Genetics 23(8):387–395

Steemers FJ, Gunderson KL (2007) Whole genome genotyping technologies on the BeadArray platform. Biotechnol J 2:41–49

Stefansson H, Rujescu D, Cichon S, Pietiläinen OP, Ingason A et al (2008) Large recurrent microdeletions associated with Schizophrenia. Nature 455:232–236

Walsh T, McClellan JM, McCarthy SE, Addington AM, Pierce SB et al (2008) Rare Structural Variants Disrupt Multiple Genes in Neurodevelopmental Pathways in Schizophrenia. Science 320:539–543

Wang K, Zhang H, Ma D, Bucan M, Glessner JT et al (2009) Common genetic variants on 5p14.1 associate with autism spectrum disorders. Nature 459:528–533

Wang K et al (2007) PennCNV: an integrated hidden Markov model designed for high-resolution copy number variation detection in whole-genome SNP genotyping data. Genome Res 17:1665–1674

Weiss LA, Arking DE (2009) Gene Discovery Project of Johns Hopkins & the Autism Consortium, Daly MJ, Chakravarti A. A genome-wide linkage and association scan reveals novel loci for autism. Nature, 461(7265):802–808

Weiss LA et al (2008) Association between microdeletion and microduplication at 16p11.2 and autism. N Engl J Med 358:667–675

Xu B, Roos JL, Levy S, van Rensburg EJ, Gogos JA et al (2008) Strong association of de novo copy number mutations with sporadic schizophrenia. Nat Genet 40:880–885

Toward a Mechanistic Understanding of How Variability in Neurobiology Shapes Individual Differences in Behavior

Ryan Bogdan, Justin M. Carré and Ahmad R. Hariri

Abstract Research has begun to identify how variability in brain function contributes to individual differences in complex behavioral traits. Examining variability in molecular signaling pathways with emerging and established methodologies such as pharmacologic fMRI, multimodal PET/fMRI, and hormonal assays are beginning to provide a mechanistic understanding of how individual differences in brain function arise. Against this background, functional genetic polymorphisms are being utilized to understand the origins of variability in signaling pathways as well as to efficiently model how such emergent variability impacts behaviorally relevant brain function and health outcomes. This chapter provides an overview of a research strategy that integrates these complimentary levels of analysis; existing empirical data is used to illustrate the effectiveness of this approach in illuminating the mechanistic neurobiology of individual differences in complex behavioral traits. This chapter also discusses how such efforts can contribute to the identification of predictive risk markers that interact with unique environmental factors to precipitate psychopathology.

R. Bogdan (✉) · J. M. Carré · A. R. Hariri
Laboratory of NeuroGenetics, Department of Psychology & Neuroscience, Institute for Genome Sciences & Policy, Duke University, 417 Chapel Drive, Durham, NC 27708, USA
e-mail: bogdan.ryan@gmail.com

J. M. Carré
Laboratory of Social Neuroendocrinology, Department of Psychology, Wayne State University, Detroit, MI 48202, USA
e-mail: justin@carrelab.com

A. R. Hariri
e-mail: ahmad@haririlab.com

Curr Topics Behav Neurosci (2012) 12: 361–393
DOI: 10.1007/7854_2011_182

Published Online: 23 March 2012

Keywords Neurogenetics · Amygdala · Striatum · Threat · Reward · Aggression · Stress

Contents

1 Introduction 362
2 Trait Anxiety, the Amygdala and Serotonin 364
3 Aggression, the Amygdala and Testosterone 368
4 Impulsivity, the Ventral Striatum and Dopamine 371
5 Endocannabinoids, Threat- and Reward-Related Brain Functions 373
6 Stress, the HPA Axis and the Mineralocorticoid Receptor 377
7 Summary and Future Directions 380
References 384

1 Introduction

Individual differences in trait affect, personality and temperament critically shape complex human behaviors, successfully navigating social interactions and overcoming challenges from our ever changing environments. Such individual differences may also serve as important predictors of vulnerability to psychopathology including depression, anxiety, addiction, and antisocial personality disorder, especially upon exposure to environmental adversity. Accordingly, identifying the biological mechanisms which give rise to trait individual differences affords unique opportunity to develop a deeper understanding of complex human behaviors, disease liability and treatment. Having established multiple modal neural processes supporting specific aspects of complex social behavior, research has now begun to reveal the neural substrates of inter-individual variability in these and related constructs. Moreover, recent studies have established that these neural substrates represent temporally stable and reliable indices of brain function. Thus, much like their behavioral counterparts, brain function represents an enduring, trait-like phenomenon, which in and of themselves may serve as important markers of individual differences as well as disease liability and pathophysiology.

As research continues to illustrate the predictive relationship between brain activation and trait-like behaviors (e.g., increased amygdala reactivity predicts trait anxiety), an important next step is to systematically identify the underlying mechanisms driving variability in brain circuit function. In this regard, neuroimaging studies employing pharmacologic challenge paradigms, principally targeting monoamine neurotransmission and hormonal systems, have revealed that even subtle alterations in dopaminergic, noradrenergic, serotonergic and neuroendocrine signaling can have profound impact on the functional response of brain circuitries supporting affect, personality and temperament. Similarly, multimodal neuroimaging approaches have provided evidence for directionally specific relationships between key components of monoaminergic signaling cascades, assessed with radiotracer positron emission tomography (PET), and brain function, assessed with BOLD fMRI. Collectively, pharmacological challenge neuroimaging and multimodal PET/fMRI are revealing how variability in behaviorally relevant brain activation emerges as a function of underlying variability in key brain signaling pathways (e.g., increased serotonin signaling predicting increased amygdala reactivity). The next logical step is to identify the sources of inter-individual variability in these key neurochemical signaling mechanisms.

In the modern era of human molecular genetics, one step is firmly planted in the direction of identifying common variation in the genes that influence the functioning or availability of components in these pathways. As DNA sequence variation across individuals represents the ultimate wellspring of variability in emergent molecular, neurobiological and related behavioral processes, understanding the relationships between genes, brain and behavior is important for establishing a mechanistic foundation for individual differences in behavior and related psychiatric disease. Moreover, such genetic polymorphisms can be readily identified from DNA collected via cells from individual blood or even saliva samples using relatively well-tolerated, inexpensive and standardized laboratory protocols. Once collected and isolated, an individual's DNA can be amplified repeatedly providing an almost endless reservoir of material for genotyping of additional candidate polymorphisms as they are identified. When a precise cascade of related neurobiological and behavioral effects are clearly established, common polymorphisms can represent incredibly powerful predictive markers of such emergent properties that are more readily accessible (e.g., samples can be collected in doctor's offices), applicable (e.g., even newborns can be genotyped) and economical (e.g., costing only tens of dollars per sample in comparison to the hundreds and even thousands required for fMRI and PET) than their technological counterparts in neuroimaging and neuropharmacology. Of course, arriving at this ultimate reduction requires intensive and expansive efforts wherein all these technologies as well as epidemiological and clinical studies are first brought to bear on explicating the detailed biological mechanisms mediating individual differences in trait behaviors and related risk for neuropsychiatric disease.

In the last 5 years, significant progress has been made in describing the contributions of multiple common genetic polymorphisms to individual differences in complex behavioral phenotypes and disease liability—in particular, by identifying effects of functional genetic variation on the neural processes that mediate behavioral responses to environmental challenge (Caspi and Moffitt 2006; Hariri and Holmes 2006). The current chapter will review how the integration of psychology, neuroimaging, neuroendocrinology, neuropharmacology and molecular genetics can work toward the ultimate goal of understanding the detailed mechanisms mediating individual differences in human behavior and, in turn, establish predictive markers of disease vulnerability. The vast potential of such an integrated approach will be highlighted by reviewing recent studies whose collective results demonstrate that common sequence variation in human genes that bias key components of molecular signaling cascades results in altered brain circuit function that mediates individual differences in complex behavioral traits such as temperamental anxiety, aggression, stress responsiveness and impulsivity. With their increased utilization and continued expansion each level of analysis in this integrative strategy—brain circuit function, neural signaling cascades and molecular genetics—also has the potential to uniquely illuminate clinically relevant information that can be used in efforts to devise individually tailored treatment regimes and establish predictive disease markers. In lieu of further describing a general framework, five specific examples will be used to illustrate the effectiveness of this integrated strategy to parse biological mechanisms mediating individual differences in complex behaviors.

Multiple mechanisms involving de novo biosynthesis, vesicular release, active reuptake, metabolic degradation as well as a myriad of both pre- and post-synaptic receptors contribute to the regulation of neurotransmission and its subsequent modulation of brain function. To illustrate the powerful capacity of functional genetic polymorphisms to model emergent variability in signaling pathways, the five exemplars below focus on different critical node in regulating the magnitude of neurotransmission, namely autoregulatory negative feedback, active synaptic reuptake, post-synaptic receptor binding, intracellular receptor binding, and enzymatic degradation. In the first example, individual differences in trait anxiety will be mapped onto threat-related amygdala reactivity. Variability in amygdala reactivity will, in turn, be mapped to serotonin signaling. Finally, variability in serotonin signaling will be mapped to a common functional polymorphism impacting the capacity for negative feedback inhibition of serotonergic neurons in the midbrain. In the second example, a similar relationship will be described between variability in aggression, amygdala reactivity, testosterone signaling and a variable number of tandem repeats in the androgen receptor. The third example describes variability in impulsivity, reward-related ventral striatum reactivity, dopamine signaling and a polymorphism impacting synaptic clearance of striatal dopamine. In the fourth example, a common polymorphism affecting the enzymatic degradation of endocannabinoids will be linked to divergent effects on threat-related amygdala reactivity and reward-related ventral striatum reactivity. In the fifth and last example, variability in stress-responsiveness and hypothalamic–pituitary–adrenal axis function will be linked to a missense polymorphism affecting the mechanistic action of the mineralocorticoid receptor.

2 Trait Anxiety, the Amygdala and Serotonin

The experience of anxiety is commonplace amongst both human and non-human primates as well as other highly social animals. In the context of social interactions, especially within delimited social hierarchies consisting of dominant and subordinate individuals, anxiety serves to shape appropriate and often opposing responses to precipitating events such as competition for limited resources (e.g., food, water, reproductive partners). Sensitivity to potentially threatening social cues (e.g., affective facial expressions) varies considerably between individuals and represents a core component of commonly employed constructs representing trait anxiety. Individuals with high trait anxiety exhibit a propensity to more frequently appraise situations as more threatening than do others and are generally more sensitive to social cues including those representing both explicit and implicit threat (e.g., angry and fearful facial expressions). In turn, these individuals are at increased risk for developing psychopathology characterized by abnormal social and emotional behaviors such as depression and often precipitated by exposure to chronic or severe stressors. Examining the neural correlates of individual variability in dispositional temperament such as trait anxiety represents an important step in understanding key socioemotional behaviors as well as an effective means of elucidating pathophysiological processes contributing to related disordered states.

Converging evidence from animal and human studies clearly demonstrates that the amygdala is centrally involved in mediating both physiological (e.g., autonomic reactivity) and behavioral (e.g., reallocation of attentional resources) effects that allow an individual to respond adaptively to varied environmental and social challenges (LeDoux 2000). A large corpus of human neuroimaging research reveals that the amygdala is robustly engaged by varied biologically salient stimuli, most notably emotional facial expressions especially those representing threat. However, individuals differ appreciably in the magnitude of amygdala activation on exposure to emotionally expressive facial expressions, and these individual differences appear to be stable over time (Johnstone et al. 2005; Manuck et al. 2007). Thus, they may contribute to the emergence of stable differences in temperament such as trait anxiety.

Recent neuroimaging studies have reported positive relationships between the magnitude of amygdala reactivity to affective, especially threatening, stimuli and inter-individual variability in indices of trait (Dickie and Armony 2008; Etkin et al. 2004; Haas et al. 2007; Killgore and Yurgelun-Todd 2005; Most et al. 2006; Ray et al. 2005) and also state anxiety (Bishop et al. 2004; Somerville et al. 2004). In one study, Stein et al. (2007) report that high trait anxiety is associated with greater amygdala reactivity not only to angry and fearful but also happy facial expressions. Consistent with this pattern of normal variability, various mood and anxiety disorders (e.g., unipolar and bipolar depression, generalized anxiety disorder, social phobia) have been linked with greater amygdala responses to facial expressions depicting fear and anger, as well as sadness and disgust, and, more variably, to emotionally neutral facial expressions (Cooney et al. 2006; Evans et al. 2008; Phan et al. 2006; Phillips et al. 2003; Stein et al. 2002; Whalen et al. 2002). Such findings demonstrate that anxiety-related psychopathology is associated with a heightened amygdala response to diverse affective stimuli. More importantly, in the absence of such disorders, variability in the magnitude of threat-related amygdala reactivity is an important predictor of individual differences in trait anxiety.

Having first established a predictive link between amygdala reactivity and trait anxiety, factors that drive such behaviorally relevant variability in brain function can be now be identified in the broader context of detailing the biological mechanisms mediating individual differences in temperamental anxiety. Converging preclinical and clinical evidence indicates that amygdala functioning is sensitive to the effects of central serotonin (Sadikot and Parent 1990), whose principle forebrain innervation is provided by the midbrain dorsal raphe nuclei (DRN). Available data from animal studies indicate that relative increases in local 5-HT result in potentiation of amygdala activation and associated behavioral phenomenon, such as fear conditioning (Amat et al. 1998, 2004; Burghardt et al. 2004, 2007; Forster et al. 2006; Maier and Watkins 2005). As advanced in the introduction of this chapter, recent neuroimaging studies using multimodal PET/fMRI or pharmacological challenge BOLD fMRI have provided direct evidence for parallel effects of 5-HT in humans. Specifically, in vivo PET has revealed that decreased endogenous capacity for local 5-HT reuptake (Rhodes et al. 2007) is associated with relatively increased amygdala reactivity. Acute IV administration of a selective serotonin reuptake inhibitor, which reduces capacity for 5-HT reuptake, during BOLD fMRI is likewise associated with not only increased amygdala reactivity but also decreased habituation of amygdala reactivity over time (Bigos et al. 2008). These data clearly indicate that variability in the regulation of 5-HT signaling is an important source of individual differences in amygdala reactivity.

Crucial among components regulating 5-HT neurotransmission and its subsequent modulation of brain function is activation of somatodendritic 5-HT$_{1A}$ autoreceptors, which mediate negative feedback on DRN neurons resulting in decreased 5-HT release at postsynaptic targets in the forebrain (Sharp et al. 2007). Using multimodal PET/fMRI, we previously reported that the density of 5-HT$_{1A}$ autoreceptors accounts for 30–44% of variability in amygdala reactivity in healthy adults (Fisher et al. 2006), confirming the important role of 5-HT$_{1A}$ autoreceptors in modulating the activity of serotonergic target regions. Given the critical role of 5-HT$_{1A}$ autoreceptors in regulating 5-HT signaling and its resulting influence on the functioning of major brain targets, such as the amygdala, as well as complex behavioral processes (Cowen et al. 1994; Hansenne et al. 2002; Lesch and Gutknecht 2004), it is important to identify sources of emergent variability in 5-HT$_{1A}$ function.

Common sequence variation in the human 5-HT$_{1A}$ gene (*HTR1A*) represents one potential source of such inter-individual variability. Recently, a relatively frequent single nucleotide polymorphism, C(-1019)G, in the promoter region of *HTR1A* was demonstrated to impact transcriptional regulation of the gene through altered binding of the transcription factors. Specifically, the -1019G allele abolishes or impairs transcriptional repression of the promoter and, as a consequence, is associated with increased 5-HT$_{1A}$ expression (Lemonde et al. 2003), a phenomenon that appears to be specific to autoreceptors (Czesak et al. 2006). Consistent with this finding, in vivo human PET has revealed specifically increased 5-HT$_{1A}$ autoreceptor density in both healthy adults and depressed patients carrying the -1019G allele (Parsey et al. 2006). However, a similar effect was not observed in an earlier PET study (David et al. 2005). Regardless, the in vitro effects of the *HTR1A* -1019G allele and the more general relationship

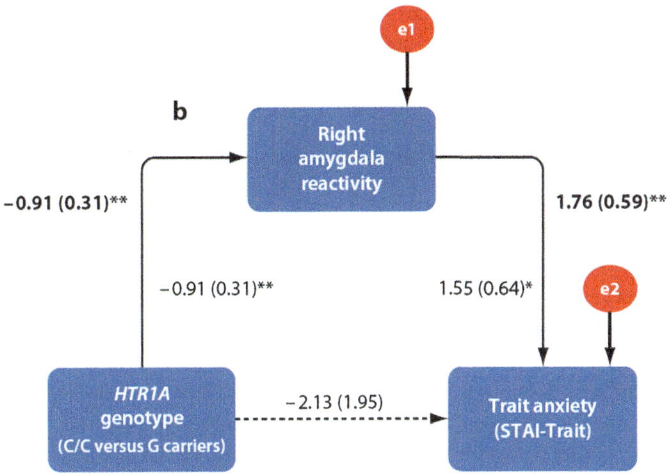

Fig. 1 Trait anxiety is indirectly predicted by *HTR1A* genotype (rs6295) through amygdala reactivity (adapted from Fakra et al 2009; Hariri 2009). Lines are labeled with unstandardized path coefficients and standard errors in parentheses. Bolded coefficients outside of the lines represent values from the trimmed model. Unbolded coefficients presented internally represent values from the full model with all paths included. Significant indirect effects of *HTR1A* genotype on trait anxiety were observed ($\alpha\beta = -1.60$, SE = 0.73, $P < 0.05$) while direct effects were nonsignificant and dropped from the model. E1 and e2 represent the residual variances not explained by variables included in the model. $*P < 0.05$, $**P < 0.01$

documented between increased 5-HT_{1A} autoreceptor density and decreased amygdala reactivity (Fisher et al. 2006) suggest that this common functional genetic variation may contribute significantly to the emergence of inter-individual variability in serotonin signaling which, in turn, biases amygdala reactivity.

Consistent with the existing data (i.e., increased 5-HT_{1A} autoreceptors leading to increased negative feedback inhibition of DRN and decreased 5-HT release), we recently demonstrated that the *HTR1A* -1019G allele is associated with significantly decreased threat-related amygdala reactivity (Fakra et al. 2009). In addition, we found that *HTR1A* genotype effects on trait anxiety were mediated through its impact on threat-related amygdala reactivity, which presumably reflects the genotypes modulation of postsynaptic 5-HT release. Specifically, while path models revealed no significant direct genotype effect on trait anxiety they demonstrated that *HTR1A* C(-1019)G and amygdala reactivity indirectly predicted a significant proportion (9.2%) of individual differences in trait anxiety through their respective indirect and direct paths (Fig. 1). The data from this study is remarkably consistent with that reported for other common functional polymorphisms also associated with relatively increased 5-HT signaling, most notably the 5-HTTLPR short allele (Hariri et al. 2002b; Munafo et al. 2008) and *MAOA* low-activity alleles (Meyer-Lindenberg et al. 2006). More importantly, these findings represent an important step in this avenue of research by providing empirical documentation for the basic premise that genetic variation in neural signaling cascades indirectly impact emergent behavioral processes by biasing the response of underlying neural circuitries (Hariri et al. 2006b; Hariri and Weinberger 2003).

3 Aggression, the Amygdala and Testosterone

Aggression, defined as any behavior directed toward the goal of harming or injuring another living being (Baron and Richardson 1994), has a major negative impact on society. For instance, the World Health Organization estimated that over 500 individuals between the ages of 10 and 29 die every day as a result of inter-personal conflict while countless more suffer psychologically and physically from aggression-related events. Nevertheless, despite these negative consequences, the use (or threat) of aggression can be beneficial under certain conditions (e.g., athletic competition, self-defense and establishment of status hierarchies).

Two factors contributing to the expression of aggressive behavior are the pursuit of a desired goal (e.g., money, territory, status, mates) and interpersonal provocation. Accordingly, researchers have typically classified aggressive behavior as either proactive or reactive. Proactive aggression, also referred to as instrumental aggression, occurs in the absence of direct provocation, does not involve physiological arousal, and is a goal-oriented behavior aimed at the acquisition of a valued resource (Dodge and Coi 1987). On the other hand, reactive aggression is a defensive response to perceived or actual provocation and is characterized by anger, impulsivity, affective instability and high levels of physiological arousal (Dodge and Coi 1987). Many cases of proactive aggression are highlighted in the media (e.g., assassinations, serial murders); however, reactive aggression likely accounts for most societal problems (Nelson and Trainor 2007). Thus, in the current section, we will focus primarily on putative neurobiological mechanisms underlying reactive aggression.

Non-human animal and human neuroscience research indicates that aggressive behavior is regulated by several inter-connected nodes of a 'social behavior network' including the hypothalamus, amygdala, bed nucleus of the stria terminalis, lateral septum, periaquaductal gray, and orbitofrontal cortex (see Nelson and Trainor 2007; Davidson et al. 2000 for reviews). In humans, a number of studies converge to implicate amygdala hyper-reactivity and/or reduced amygdala-orbitofrontal cortex (OFC) coupling in response to social threat among individuals prone to anger and reactive aggression (see Siever 2008 for review). For example, in a PET study with criminal offenders, Raine et al. (1997) reported that affective murderers (i.e., reactively aggressive inmates) demonstrated increased glucose metabolism in subcortical structures (including the amygdala) and decreased glucose metabolism in the prefrontal cortex. Subsequent imaging studies found differences in neural responses to threat among individuals characterized by reactive aggressive behavior. For instance, Coccaro et al. (2007) reported that adults diagnosed with intermittent explosive disorder displayed amygdala hyper-reactivity and decreased amygdala-OFC coupling to angry facial expressions.

Patients with borderline personality disorder, who are prone to engage in reactive aggression, demonstrate relatively increased amygdala reactivity to facial expressions depicting threat (Mauchnik and Schmahl 2010) and decreased amygdala-PFC coupling (New et al. 2007). Finally, spousal abusers characterized by reactive (but not proactive) aggression display heightened amygdala reactivity and also demonstrate attentional biases for aggressive words (Lee et al. 2008; Chan et al. 2010).

Studies in non-clinical samples indicate that even normal variation in constructs linked to aggressive behavior maps onto variability in threat-related neural responses. For instance, Beaver et al. (2008) reported that individual differences in approach motivation, a construct linked to reactive aggression (Harmon-Jones 2003), were positively correlated with amygdala reactivity to angry facial expressions. Other research has found that individual differences in approach motivation were associated with decreased ventral ACC-amygdala coupling during processing of angry facial expressions (Passamonti et al. 2008). Given the important role of highly interconnected prefrontal regions (e.g., ventral ACC, OFC) in mediating top-down regulation of amygdala driven emotional reactivity (see Davidson et al. 2000 for review), such decreased functional coupling may, in part, explain the positive link between approach motivation and aggressive behavior. Also, individual differences in trait anger are positively correlated with amygdala reactivity to angry faces, but only among men with relatively elevated trait anxiety scores (Carré et al. in press-a). Finally, in a more direct test of the hypothesis that amygdala reactivity to facial cues of threat represents a neurobiological marker for reactive aggression, we have found that variation in self-reported physical aggression is positively correlated with amygdala reactivity to neutral and angry facial expression in two independent samples of healthy men (Carré et al. in press-b). Together, these findings converge to suggest that amygdala hyper-reactivity and/or decreased amygdala-OFC coupling during processing of threat-related stimuli may represent a distinct neural signature for one's propensity to engage in reactive aggression (Siever 2008). It is important to note that individuals characterized by callous-unemotional traits and proactive aggression (e.g., conduct disorder, psychopathy) display amygdala and OFC *hypo*-reactivity to facial signals of threat (see Blair 2010 for review).

As described above, it is important to consider the underlying molecular substrates that give rise to individual variation in threat-related amygdala reactivity and amygdala-OFC coupling. Testosterone, the end-product of the hypothalamic pituitary gonadal axis, is one prime candidate. The physiological effects of testosterone occur mainly through binding to intra-cellular steroid hormone receptors (i.e., androgen and estrogen receptors) to ultimately influence gene transcription. Importantly, androgen (and estrogen) receptors are abundantly located in the amygdala and interconnected limbic structures involved in mediating aggressive behavior (see Newman 1999; Simon 2002 for reviews). Thus, through stimulation of steroid hormone receptors and subsequent modulation of cell function (see Adkins-Regan 2005 for review) testosterone can influence the functioning of neural circuits implicated in the expression of human aggression.

Recent functional neuroimaging studies have detailed some of the neural structures that are sensitive to individual differences in testosterone concentrations. In particular, individual differences in baseline testosterone concentrations are positively correlated with amygdala reactivity to facial expressions signaling threat (Derntl et al. 2009; Manuck et al. 2010) and negatively correlated with OFC reactivity to provocation (Mehta and Beer 2010). Consistent with these correlational studies, pharmacologic challenge experiments indicate that acutely raising testosterone concentrations causes an increase in amygdala reactivity and a decrease in amygdala-OFC connectivity in response to facial signals of threat (Hermans et al. 2008; van Wingen et al. 2008, 2010). These findings suggest that the association between acute fluctuations in testosterone and reactive aggression in men (Carré and McCormick 2008; Carré et al. 2009, 2010) may be due to the influence of testosterone on neural processing of threat (e.g., angry faces or provocation), which may ultimately bias aggressive behavior during social challenges.

As discussed above, many of the physiological effects associated with testosterone are mediated by activation of intra-cellular androgen receptors. Specifically, when activated by testosterone, androgen receptors (AR) migrate to the cell nucleus where they regulate gene transcription by activating hormone response elements (HRE) located within gene regulatory sequences. Importantly, the transcription potential of the androgen receptor varies with the expansion of a polyglutamine stretch in the N-terminal domain of the AR protein, as encoded by a trinucleotide (CAG) repeat polymorphism in exon 1 of the X chromosome-linked *AR* gene (Zitzmann and Nieschlag 2003). Specifically, in vitro work indicates that the transactivation potential of the androgen receptor declines in relation to an increase in the number of CAG repeats (Chamberlain et al. 1994), and that androgen receptor concentrations decline with an increasing number of CAG repeats (Choong et al. 1998).

Recent evidence indicates that the number of CAG repeats correlate negatively with testosterone responses to social interactions with attractive women (Roney et al. 2009). In other words, men with fewer *AR* CAG repeats demonstrate a more robust neuroendocrine response to potential mates, suggesting that this androgen receptor polymorphism may influence the efficiency with which an individual may mount an endocrine response to social interactions. Furthermore, genetic studies have found that high testosterone men with fewer CAG repeats are more aggressive (Rajender et al. 2008; Vermeesch et al. 2010). Finally, Manuck et al. (2010) found that individual differences in CAG repeat length were negatively correlated with ventral amygdala reactivity to facial expressions depicting threat, particularly among men with relatively high baseline testosterone concentrations. Thus, variation in the number of *AR* CAG repeats modulates testosterone responses to social interactions and amygdala reactivity to facial signals of threat.

Collectively, these findings provide support for the idea that heightened amygdala reactivity to social threat (e.g., provocation and/or facial signals of threat) may represent a neurobiological mechanism through which androgens modulate human aggressive behavior. Pharmacologic challenge experiments in which aggressive behavior is measured directly during fMRI are needed to confirm the causal role of testosterone in modulating human aggressive and threat-related neural responses.

4 Impulsivity, the Ventral Striatum and Dopamine

Discounting future outcomes underlies much of human decision making and figures prominently in several overlapping psychological constructs such as self-regulation, impulse-control, delay of gratification and intertemporal choice (Manuck et al. 2003). Moreover, individuals who strongly prefer immediate over deferred rewards of larger nominal value are often generally impulsive or lacking in self-control and at risk for addictive disorders such as pathological gambling, cigarette smoking and drug and alcohol abuse (Alessi and Petry 2003; Bickel et al. 1999; Kirby et al. 1999; Madden et al. 1997). In experimental research on intertemporal choice, discounting of future rewards or delay discounting (DD) is a well-characterized behavioral measure of preference for immediate over delayed rewards and provides an index of impulsive tendencies in humans (Green and Myerson 2004). Behavioral tests used to derive estimates of DD commonly ask participants to choose between multiple immediate rewards that vary in value and a constant, larger reward available after varying intervals of delay. In such tasks, rates of discounting often differ appreciably and consistently among individuals (Simpson and Vuchinich 2000). Thus, DD represents a potentially important psychometric index of individual differences in present versus future-oriented tendencies.

Similar to the research on trait anxiety and amygdala reactivity explication of the underlying neural processes that give rise to such inter-individual variability has the potential to allow for a more comprehensive understanding of the mechanisms leading not only to normal variability in such behaviors but also the pathophysiology of addiction and related disorders. Through reciprocal cortical and subcortical connections, the nucleus accumbens (NAcc) and, more broadly, the ventral striatum (VS), contribute to the motivational salience of stimuli and abet appetitive or reward-dependent behaviors (Berridge and Robinson 2003). Activity of the VS increases in response to both the anticipation and receipt of rewarding stimuli including primary (e.g., food) and secondary (e.g., money) reinforcers (O'Doherty 2004). Moreover, in addiction, craving and compulsive drug seeking as well as sensitivity to drug cues are associated with dysregulated increases in VS activity (Kalivas and Volkow 2005). Because the response of the VS involves an immediate response to rewards, the magnitude of VS activity may contribute to individual differences in a relative preference for immediate, compared to delayed, rewards.

Using BOLD fMRI, we have demonstrated that the magnitude of VS reactivity predicts individual differences in a simple laboratory measure of DD (Hariri et al. 2006a). Specifically, analyses revealed that individual differences in DD correlate positively with magnitude of VS activation in response to both positive and negative feedback as well as with differential reward-related VS activation in response to positive compared with negative feedback. Consistent with the strong general correlation between DD and traditional self-report measures of impulsivity (De Wit et al. 2004, 2007), we have also found that reward-related VS reactivity is positively correlated with scores from the Barratt Impulsiveness Scale (Forbes et al. 2009). Collectively, our results suggest that increased self-reported impulsivity as well as the preference for smaller immediate over larger delayed rewards reflect both a relatively indiscriminate and hyper-reactive VS circuitry. Similar variability in VS function has also been associated with more complex measures of incentive-based decision making (Knutson et al. 2007). Moreover, dysregulation of the VS contributes to addiction, perhaps by affecting impulsive decision making (Kalivas and Volkow 2005). As such, inter-individual variability in VS reactivity to reward-related stimuli likely contributes to the emergence of differences in the intermediate behavioral risk factors for, as well as the clinical expression of, addiction. Identifying variability in neural signaling pathways that contributes to individual differences in VS function offers additional traction in the search for underlying biological mechanisms.

Dopamine modulation of neuronal activity, especially in the VS (i.e., mesolimbic system), serves as a nexus for the expression of DA signaling at the level of reward-related behaviors (Cardinal et al. 2004; Kelley 2004). Functioning of the DA system has been linked to normal individual differences in reward-related traits (Depue et al. 1994), and disorders involving enhanced reward-seeking, such as addiction, have been hypothesized to reflect maladaptive alterations of this mesolimbic reward system (Hyman et al. 2006; Volkow et al. 1999). Multimodal and pharmacological neuroimaging studies of DA effects on brain function again offer a unique opportunity to more directly evaluate underlying molecular mechanisms regulating this circuitry. A recent in vivo human study reported a direct relationship between striatal DA synthesis, assessed with PET and brain activity, assessed with BOLD fMRI (Siessmeier et al. 2006). Acute increase of DA release via oral amphetamine has also been linked with relatively increased extent of BOLD fMRI assessed VS activity (Menon et al. 2007). More generally, acute pharmacologic increase of DA in both healthy volunteers (Hariri et al. 2002a) and patients with Parkinson's disease (Tessitore et al. 2002) results in relatively increased BOLD fMRI assessed activity in closely related limbic brain regions, namely the amygdala. Given the importance of DA in modulating this behaviorally relevant neural circuitry, identifying factors that determine inter-individual variability in DA signaling and its related impact on the reactivity of the VS will facilitate our understanding of the neurobiological mechanisms governing reward-related behaviors and augment efforts to improve the treatment and even prevention of pathological behaviors such as drug abuse and addiction.

We have explored the role of altered DA signaling, resulting from a common functional polymorphism impacting active synaptic reuptake in the striatum, in determining inter-individual variability in reward-related VS reactivity and correlated variability in behavioral impulsivity. Consistent with the research on serotonin signaling, amygdala reactivity and trait anxiety, the selection of our candidate polymorphism was driven by available in vitro and/or in vivo assays demonstrating significant impact of the variant on aspects of biological function related to DA neurotransmission and not on available data from association studies with behavioral (e.g., impulsivity) or clinical (e.g., alcoholism) phenotypes. While association studies are necessary for understanding the ultimate contribution of genetic polymorphisms to variability in behavioral and clinical phenomena, they do not readily allow for inferences regarding polymorphic effects on gene or protein function. Such inferences are instrumental for the development of biologically plausible and tractable hypotheses regarding the impact of genetic variation on inter-individual variability in brain function and associated behaviors such as those pursued in our current work (Hariri et al. 2006b; Hariri and Weinberger 2003).

The dopamine transporter is responsible for the active clearance of synaptic DA and, thus, plays a critical role in regulating the duration of postsynaptic DA signaling, especially in the striatum (Sesack et al. 1998). Accumulating evidence indicates that a 40-base pair variable number of tandem repeats (VNTR) polymorphism in the 3′ untranslated region of the DAT gene (*SLC6A3*) impacts the expression and availability of DAT (Bannon et al. 2001). Although a genotype effect has not been consistently observed across all studies (Martinez et al. 2001; Michelhaugh et al. 2001; Mill et al. 2005; van Dyck et al. 2005), several suggest that in comparison to the 9-repeat allele, the 10-repeat is associated with relatively increased levels of DAT both in vivo (Cheon et al. 2005; Heinz et al. 2000) and in vitro (Mill et al. 2002; Van Ness et al. 2005). We hypothesized that there would be relatively greater VS reactivity associated with the 9-repeat allele, which is linked with reduced DAT expression and presumably greater striatal synaptic DA, in comparison with the 10-repeat allele. Consistent with our hypothesis, the DAT1 9-repeat allele was associated with relatively greater VS reactivity and accounted for nearly 12% of the inter-individual variability. In contrast, genetic variation directly affecting DA signaling only in the prefrontal cortex (i.e., COMT Val158Met) was not associated with variability in VS reactivity. These results highlight an important role for a genetic polymorphism affecting striatal DA neurotransmission in mediating inter-individual differences in reward-related VS reactivity. They further suggest that altered VS reactivity may represent a key neurobiological pathway through which these polymorphisms contribute to variability in behavioral impulsivity and related risk for substance use disorders.

5 Endocannabinoids, Threat- and Reward-Related Brain Functions

Modern neuroscience methodologies have greatly advanced our understanding of the intrinsic mechanisms mediating and regulating endogenous cannabinoid or endocannabinoid (eCB) signaling in the CNS (Piomelli 2003). Such eCB signaling has emerged as a potent modulator of neural circuitries mediating both basic physiological (Calignano et al. 1998; Meng et al. 1998) and advanced behavioral responses (Maldonado et al. 2006; Scherma et al. 2008; Viveros et al. 2005). Experimental manipulation of these mechanisms has revealed significant behavioral effects, especially in threat- and reward-related domains, which are generally consistent with the effects of *Cannabis* intoxication, which are largely driven by the constituent chemical Δ^9-tetrahydrocannabinol (Robson 2005). The elucidation of molecular mechanisms regulating eCB signaling, akin to that for serotonin and dopamine, has motivated attempts to understand its possible contribution to the emergence of variability in brain circuit function and related individual differences in behavioral attributes (e.g., anxious or impulsive temperament) associated with increased risk for psychiatric disorders.

After their biosynthesis from arachidonic acid, eCBs such as anandamide (AEA) and 2-arachidonoylglycerol (2-AG) typically modulate synaptic neurotransmission through stimulation of CB1, the principal CNS cannabinoid receptor widely expressed on multiple neuronal subtypes and their distributed circuitries. In turn, the duration and intensity of eCB signaling, especially for AEA, is regulated by two complementary mechanisms: enzymatic degradation via fatty acid amide hydrolase (Cravatt et al. 1996) and active synaptic clearance via the AEA transporter (Piomelli et al. 1999). The psychotropic and THC-like effects of AEA, however, appear to be coupled with fatty acid amide hydrolase (FAAH), but not AEA transporter function (Solinas et al. 2007). Thus, FAAH, an integral membrane enzyme, may uniquely regulate behaviorally relevant eCB signaling by mediating the hydrolytic breakdown of AEA into arachidonic acid and ethanolamine.

Again, common genetic variation (i.e., polymorphisms) affecting the functioning of components involved in eCB neurotransmission (e.g., AEA, CB1, FAAH) may represent a significant potential source of inter-individual variability in eCB signaling that mediates emergent differences in emotion- and reward-related behaviors (Onaivi et al. 2002). Because of its critical role in regulating the signaling duration and intensity of AEA (Cravatt et al. 1996), and its selective contribution to the psychotropic effects of AEA (Solinas et al. 2007), we have recently examined the neurobiological and behavioral effects of a common functional nonsynonymous SNP resulting in the conversion of a conserved proline residue to threonine (P129T) in the amino acid sequence of FAAH (Hariri et al. 2009). In vitro, *FAAH* 385A is associated with normal catalytic properties, but reduced cellular expression of FAAH, possibly through enhanced sensitivity to proteolytic degradation (Chiang et al. 2004; Sipe et al. 2002). Moreover, the C385A is the only common mutation in *FAAH* (Flanagan et al. 2006) and the 385A, which putatively augments AEA signaling via decreased enzymatic degradation, has been associated with reward-related pathologies including street drug use and problem drug/alcohol abuse, as well as being overweight and obese (Flanagan et al. 2006; Sipe et al. 2002).

In animal models, both pharmacologic and genetic disruption of FAAH function result in *decreased* anxiety-like behaviors, as well as *increased* consumption and preference for ethanol (Basavarajappa et al. 2006; Blednov et al. 2007; Kathuria et al. 2003; Moreira et al. 2008; Solinas et al. 2007). Moreover, a recent pharmacologic fMRI study in human subjects has reported that acute oral administration of

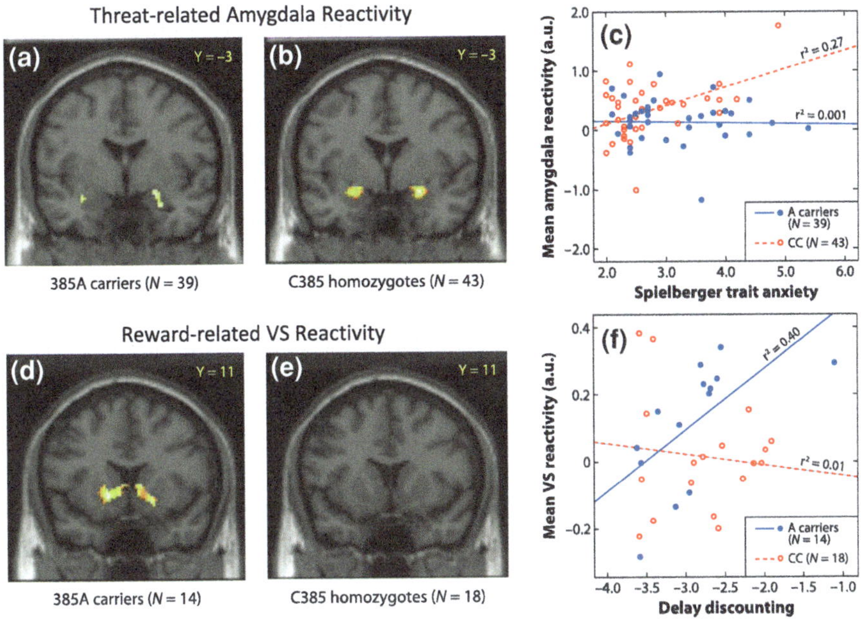

Fig. 2 Effects of *FAAH* genotype (rs324420) on threat- and reward-related brain activation (adapted from Hariri et al. 2009; Hariri 2009). Statistical parametric maps displaying the correlation between threat-related amygdala reactivity and trait anxiety in **a** *FAAH* 385A carriers and **b** C385 homozygotes. **c** Plots of the correlation between threat-related amygdala reactivity and trait anxiety according to *FAAH* C385 genotype. Statistical parametric maps displaying the correlation between reward-related ventral striatal (VS) reactivity and delay discounting in **d** *FAAH* 385A carriers and **e** C385 homozygotes (no significant correlation observed). **f** Plots of the correlation between reward-related VS reactivity and delay discounting according to *FAAH* C385A genotype. FAAH: fatty acid amide hydrolase

THC is associated with reduced amygdala reactivity to threat-related facial expressions of emotion (Phan et al. 2008). Consistent with these effects, we hypothesized that the *FAAH* 385A would be associated with relatively *decreased* threat-related amygdala reactivity, but *increased* reward-related reactivity in the VS. Analyses revealed that carriers of the *FAAH* 385A, associated with reduced enzyme expression and, presumably, increased AEA signaling, have decreased threat-related amygdala reactivity. In contrast, carriers of the FAAH 385A exhibited increased reward-related VS reactivity in comparison to C385 homozygotes. Moreover, divergent effects of *FAAH* C385A genotype on brain function were manifest in a consistent manner at the level of brain-behavior relationships (Fig. 2). Relative to C385 homozygotes, *FAAH* 385A carriers showed a diminished relationship between amygdala reactivity and trait anxiety. In contrast, 385A carriers exhibited a markedly increased relationship between VS reactivity and delay discounting, a behavioral index of impulsivity and reward-sensitivity.

It is important to note that there were no direct associations between *FAAH* genotype and behavioral phenotypes (i.e., anxiety or impulsivity) in this study, a common occurrence when working with relatively small samples, possibly reflecting the minimal effect the proximal biological impact associated with any genotype has on any distal behavioral phenotype (Hariri et al. 2006b; Hariri and Weinberger 2003), as well as the importance of environmental stressors in unmasking genetically driven effects on behavior (Caspi and Moffitt 2006). However, there were robust differences in the relationships between regional brain function and complex behaviors as a function of *FAAH* C385A genotype. These observed brain-behavior patterns may reflect the influence *FAAH* C385A associated differences in endogenous eCB tone on stimulus-driven neural circuit function mediating complex behavioral processes. Relatively higher levels of AEA in the amygdala of *FAAH* 385A carriers may reduce the responsivity of this structure to salient input (possibly through CB1-mediated potentiation of local GABAergic interneurons) and, as a consequence, lead to reduced anxiety-like behaviors predicted by amygdala function. In contrast, higher levels of AEA may increase the responsivity of the VS in *FAAH* 385A carriers (possibly through CB1-mediated increased dopamine release and potentiation of VS neuron activity) leading to increased reward-sensitivity predicted by VS function. Support for this speculation exists in studies reporting a failure of restraint stress to effect changes in amygdala activation in knockouts lacking FAAH or animals treated with FAAH inhibitors (Patel et al. 2005), and increased food-intake as a result of local FAAH inhibition in the nucleus accumbens (Sorice-Gomez et al. 2007). Thus, the endogenous state of eCB signaling associated with either constitutive genetic variation such as the *FAAH* C385A or acute pharmacologic manipulation likely biases the responsivity of neural circuits to behaviorally relevant information and their subsequent regulation of complex behaviors.

Decreased threat-related amygdala reactivity and associated trait anxiety may contribute to the emergence of pathologies such as addiction and obesity, previously associated with the *FAAH* 385A (Flanagan et al. 2006; Sipe et al. 2002; Tyndale et al. 2007), by reducing the sensitivity of these individuals to potential environmental threat or harm. In fact, blunted amygdala reactivity has been reported in individuals at high familial risk for alcoholism and this has been interpreted as possibly contributing to decreased threat-sensitivity and subsequently increased risk-taking behaviors in these genetically predisposed individuals (Glahn et al. 2007). An increase in reward-related VS reactivity and associated impulsivity (e.g., steeper discounting of future, relative to immediate rewards) may likewise contribute to disinhibitory psychopathologies through heightened reward-sensitivity and impulsive decision making. Studies in addicted patients have generally reported a sensitization of the neural circuitry for reward, including the VS (Kalivas and Volkow 2005). And, increased behavioral impulsivity and reward-sensitivity are significant risk factors for addiction (de Wit and Richards 2004). Thus, through divergent effects on both threat- and reward-related brain functions, the influence of *FAAH* C385A on eCB signaling may have a compound and accelerated effect on risk for related pathologies.

6 Stress, the HPA Axis and the Mineralocorticoid Receptor

All organisms strive to maintain homeostasis by regulating physiological states within a dynamic equilibrium. Stress, the perception of inadequate coping resources in the context of environmental demands appraised as threatening, is a common experience that disrupts homeostasis, and triggers a biological-behavioral stress response to promote adaptation/survival. While stress can promote adaptive coping to environmental challenges by recruiting necessary resources, it is also associated with many adverse physical and mental health conditions such as cardiovascular disease, depression, post-traumatic stress disorder (PTSD) and immune system dysfunction (Cohen et al. 2007; McEwen and Gianaros 2010). Importantly, there is tremendous variability in which environmental demands individuals perceive as stressful as well the extent of physiological and psychological response to these demands (Dickerson and Kemeny 2004; Kudielka et al 2009). Because this variability is associated with stress-related physical and mental health outcomes and differences in related neural activation (Marques et al. 2009; Yehuda 2002), identifying the factors that contribute to variability in stress reactivity is an important step for understanding the etiology of stress-related disorders as well as limiting the sequelae of stressful experiences.

A wealth of research has established that stress reactivity is centrally regulated by the hypothalamic–pituitary–adrenal (HPA) axis (for reviews see de Kloet et al. 2005; Ulrich-Lai and Herman 2009). Briefly, pathways from the medial prefrontal cortex, hippocampus, amygdala and brainstem involved in the behavioral, neuroendocrine, autonomic and immune responses to stress converge within the paraventricular nucleus of the hypothalamus to regulate the release of corticotropin releasing hormone (CRH) in response to perceived stress. CRH triggers the release of adrenocorticotropic hormone (ACTH) from the anterior pituitary gland, which upon binding to receptors in the adrenal gland, stimulates cortisol release. As part of a negative feedback loop, increases in cortisol inhibit CRH and ACTH release from the hypothalamus and anterior pituitary gland, respectively. At all levels in the cascade, HPA axis function varies widely across individuals and is relatively stable over time (Fox et al. 2006; Kudielka et al. 2009; Márquez et al. 2005) suggesting that trait-like variation in HPA axis function may contribute to stable differences in responses to stressors.

Rodent knockout and non-human and human pharmacologic challenge studies have demonstrated that disruption at all nodes of the HPA axis (e.g., hypothalamus, pituitary, adrenal gland) can induce anxiety and depressive-like behavior (e.g., Kolber and Muglia 2009; Marques et al. 2009; Müller et al. 2002). Moreover, nearly 50 years of research has strongly linked abnormal function at all levels of the HPA axis to stress-related psychopathology (Gibbons and McHugh 1962; Marquis et al. 2009; Yehuda 2002). For instance, depression is characterized by elevated CRH, ACTH and cortisol as well as a disrupted negative feedback loop whereby cortisol does not effectively inhibit CRH and ACTH (van Praag et al. 2004). PTSD is associated with elevated CRH but diminished ACTH and cortisol as well as an enhanced cortisol-induced inhibition of CRH and ACTH (Yehuda 2002). Furthermore, HPA dysregulation is associated with a number of risk factors for psychopathology such as childhood maltreatment (Tarullo and Gunnar 2006) and low social support (Abercrombie et al. 2004).

More recently, and consistent with earlier non-human animal work (Rodrigues et al. 2009; Ulrich-Lai and Herman 2009), translational neuroimaging research suggests that HPA axis function is associated with structural and functional differences in brain regions, such as the hippocampus, amygdala, basal ganglia and prefrontal cortex, that are relevant to stress-related psychopathology (for review see Pruessner et al. 2010; McEwen and Gianaros 2010). For example, Urry et al. (2006) show that heightened amygdala and reduced ventromedial prefrontal cortex activation during the regulation of negative affect is associated with dysregulated HPA axis function. Moreover, prefrontal glucose metabolism, an index of neuronal activity, predicts variability in HPA axis function (Jahn et al. 2010; Kern et al. 2008), and lesions to these brain regions result in HPA axis dysregulation (Buchanon et al. 2010).

Given the firm association between HPA axis function and stress-related psychopathology and related neural circuitry, factors that shape the responsiveness of this system can be studied with the ultimate goal of identifying mechanisms underlying individual differences in response to stress. As mentioned above, in addition to numerous regulatory mechanisms inside and outside of the HPA axis, cortisol is a major regulator of HPA axis function and related neural systems. In support of a causal relationship between cortisol and stress-related psychopathology, 20% of patients prescribed chronic high doses of hydrocortisone (a synthetic form of cortisol), develop psychopathology including depression, mania and psychosis. In addition, while 75% report some psychiatric symptoms these disappear following treatment cessation (see Marques et al. 2009).

Cortisol operates through a binary corticosteroid receptor system, binding to both the high affinity mineralocorticoid receptor (MR) and low affinity glucocorticoid receptor (GR) which are widely co-expressed in limbic neurons including those in the amygdala (de Kloet et al. 2005; Joëls et al. 2008). These intracellular receptors function as transcriptional regulators whereby binding can alter the expression of 70–100 genes (de Kloet et al. 2005), which likely has widespread consequences for neural activation and stress responsiveness. Recent research provides support for the existence of long-hypothesized membrane-bound MR and GR, although the mechanism(s) through which these effects occur remain unclear (Bartholome et al. 2004; Karst et al. 2006; Joëls et al. 2008).

Because of its low affinity, the GR is only extensively occupied in the wake of large spikes in cortisol such as those associated with stressors or circadian rhythms. One of the primary functions of the GR is to normalize brain activity following the extinction of a stressor and to inhibit continued HPA axis response. On the other hand, because of its high affinity, the MR is almost always occupied and non-human animal research suggests that it is necessary for a stable excitatory tone in the hippocampus which inhibits the HPA axis under basal and stressful circumstances (Reul et al. 2000). In addition to tonically inhibiting HPA axis response, MR binding is relevant to behavioral stress response functions including the appraisal of novel situations as well as the selection of appropriate responses to deal with challenge (de Kloet et al. 2005; Joëls et al. 2008). Thus, it is believed that the MR is prominently involved in basal stress system regulation and the onset of a stress response while the GR primarily functions to terminate an initial stress reaction. As such, the MR is an ideal candidate to contribute to differences in stress reactivity.

In mice, MR knockout or antagonism increases basal and stress-evoked HPA axis activity (Gass et al. 2001) and worsens response to antidepressant medication (DeRijk et al. 2008). Conversely, enhanced MR expression is associated with reduced depressive- and anxiety-like behaviors and corticosterone (the rodent homolog of cortisol) secretion under stressful and basal conditions (Mitra et al. 2009; Rozeboom et al. 2007). Consistent with these mechanisms, antidepressant medications increase MR expression (Brady et al. 1991). Collectively, these findings suggest that reduced MR function may contribute to elevated HPA axis activity at baseline and in dysregulated responses to stressors. Thus, given the prominent role of the MR in stress-responsiveness, stress-related behavior and psychopathology, it is important to identify sources of variability in MR function.

A missense isoleucine(Iso)/valine(Val) polymorphism (rs5522) located in exon 2 of the MR gene (*NR3C2*) occurs in approximately 10% of the Caucasian population. Importantly, in vitro studies show that the Val allele is associated with reduced cortisol, but not aldosterone, binding and reduced cortisol-induced transactivation (De Rijk et al. 2006; Arai et al. 2003). Thus, the MR Val allele may promote elevated HPA axis activity under basal conditions and in response to small stressors due to reduced cortisol-MR binding and subsequently decreased inhibition of the HPA axis. The Val allele may promote the development of HPA axis dysregulation, not only through reduced inhibition of HPA axis responsiveness but also because of reduced MR expression typically elicited by cortisol.

Consistent with these functional associations and speculations, the MR Val allele has been linked to heightened endocrine and autonomic responses to acute stress and stress perception. Specifically, the Val allele is associated with elevated cortisol and heart rate responses to acute social stress (DeRijk et al. 2006), skin conductance responses evoked by acute physical stress (threat-of-shock, Bogdan et al. 2010), perceptions of the aversiveness of shock levels (Bogdan et al. 2010) and waking cortisol levels (van Leeuwen et al. 2010). Furthermore, relative to Iso homozygotes, youth Val carriers have heightened threat-related amygdala reactivity, particularly in the context of low emotional neglect (Bogdan et al. in press). Collectively these finding suggest that the functional effect of the Val allele does result in enhanced stress responsiveness and perception. Furthermore, the Val allele has been associated with enhanced reward learning under basal conditions, but a vulnerability to stress-induced reward learning deficits (Bogdan et al. 2010) as well as depressive symptoms (Kuningas et al. 2007). Thus, it appears that the heightened stress responsiveness characteristic of the Val allele may translate to important behaviorally relevant differences related to psychopathology.

Taken together, individual differences in HPA axis function are predictive of stress-related physical and mental health disturbance in what appears to be a causal relationship. In addition to work showing that environmental experience (e.g., childhood maltreatment) can have long-lasting and even epigenetic effects (McGowan et al. 2009) on HPA axis system function, emerging research has shown that common polymorphisms including that in the MR gene detailed here as well as in the genes for GR (DeRIjk et al. 2008), CRHR1 (Binder 2009; Binder and Nemeroff 2010; Thode et al. unpublished observation) and FKBP5 (REF) influence variability in HPA axis function, stress sensitivity and psychopathology. Given the association of both environmental and genetic factors with individual differences in HPA axis function, this system may be a particularly fruitful area to further pursue gene-by-environment interactions research.

7 Summary and Future Directions

As detailed above, multimodal techniques assessing brain function, have begun to identify how variability in neural substrates associated with processing specific forms of information contribute to emergent individual differences in stable and enduring aspects of human behaviors such as personality and temperament. In parallel, the application of pharmacologic fMRI, multimodal PET/fMRI, and neuroendocrinology is allowing for an understanding of how variability in specific molecular signaling pathways influences individual differences in behaviorally relevant brain function. Moreover, information on DNA sequence variation in humans and related identification of functional genetic polymorphisms is now being utilized to understand the biological origins of variability in component processes of molecular signaling pathways as well as to efficiently model how such emergent variability impacts behaviorally relevant brain function. Such ongoing efforts to understand the detailed mechanisms that mediate individual differences in complex behavioral traits and related psychopathology at the level of brain circuit function, molecular signaling pathways and functional genetic polymorphisms have the potential to inform clinically relevant issues and provide guiding principles for the development of more effective and individually tailored treatment regimes. In addition, the elucidation of such mechanisms, especially those mapped to functional genetic polymorphisms, can lead to identification of predictive risk markers that interact with unique environmental factors to precipitate psychopathology.

While the five examples highlighted in this chapter are evidence for the potential of an informed and integrated research strategy to identify the neurobiology of individual differences in complex behavioral traits and their related clinical endpoints, much work is left to be done. First, to allow for tractable experimental designs and testable hypotheses in existing samples, the studies highlighted above have focused on the effects of a single signaling pathway on behaviorally relevant brain circuitry. Of course, it is very clear that there are numerous complex interactions between signaling pathways and that more than one pathway contributes to the regulation of any brain circuitry. For example, we know that DA plays an important role in modulating amygdala function and anxiety (Hariri et al. 2002a; Tessitore et al. 2002), and that 5-HT can influence reward-related brain circuitry and impulsivity (Manuck et al. 1998). However, existing studies lack the power and sophistication to model such complex interactions while effectively controlling for other important modulatory factors (e.g., age, gender, stress exposure) in the context of BOLD fMRI, pharmacologic fMRI or multimodal PET/fMRI protocols. To do so, we must aggressively expand the scale and scope of our studies to include hundreds and, preferably, thousands of subjects. This will afford opportunities to effectively examine interactions between signaling pathways (e.g., 5-HT and DA) on brain function and behavior through modeling of multiple functional polymorphisms (e.g., *HTR1A* -1019 and DAT1), and examine the effects of genetically driven variation in signaling pathways on multiple behaviorally relevant brain circuitries.

A second important consideration is that existing studies have been largely conducted in ethnically and racially homogenous populations. Thus, the observed effects may not generalize to other populations. This is especially true of studies utilizing functional genetic polymorphisms because the potential effect of any single genetic variant on a complex biological and behavioral phenotype is likely to be small against the background of the approximately 20,000–25,000 human genes and the multitude of other neurobiologically relevant functional variants they likely harbor. In fact, we have already seen that the well-replicated effects of a common functional polymorphism affecting 5-HT signaling on amygdala reactivity in Caucasian subjects may be reversed in those of Asian ancestry (Lee and Ham 2008; Munafo et al. 2008). Importantly, our most recent studies have experimentally controlled for occult genetic stratification independent of self-reported race or ethnicity as well as the independence of the target genotype from other functional polymorphisms impacting the brain functions under study. While such efforts allow for the attribution of emergent variability in brain and behavior to the candidate variant of interest and not to other possible polymorphisms or more general differences between genotype groups in genetic background, it is important to explicitly test the independence of functional polymorphisms through rigorous statistical modeling in larger samples and also to test the validity of any associations derived in one sample population (e.g., Caucasian) to populations with different genetic backgrounds (e.g., Asian or African).

A third important consideration for the future of this research is the need to conduct large-scale prospective studies beginning in childhood to determine any developmental shifts in neurogenetic pathways mediating individual differences in behavior as well as their predictive utility in identifying risk for psychopathology as a function of environmental or other stressors. All of the studies described above and most of the studies available in the literature as a whole have been conducted in adults carefully screened for the absence of psychopathology. Because of this, these findings identify mechanisms contributing to variability in the normative range of behavior only. The utility of these markers of individual differences in behavior be they neural, molecular or genetic in predicting vulnerability to psychopathology is unclear. Such predictive utility is ideally tested through prospective studies beginning with premorbid populations that account for the moderating effects of environmental stress in the emergence of clinical disorder over time (Caspi and Moffitt 2006; Viding et al. 2006).

A fourth issue is the need to further integrate pharmacologic challenge protocols with multimodal PET/fMRI to determine if variability in molecular components of signaling pathways mediate effects of specific neurotransmitters or neuromodulators on individual differences in behaviorally relevant brain circuit function. For example, despite the remarkable convergence of findings implicating variability in eCB signaling in threat- and reward-related brain functions, the exact nature of the downstream signaling pathways through which *FAAH* C385A may modulate neuronal and neural circuit function cannot be determined from the available results. FAAH catalyses the hydrolysis of other biologically active endogenous fatty acid amides (e.g., oleamide and oleoylethanolamide), which impact threat- and reward-related behaviors independently of AEA (Wei et al. 2007; LoVerme et al. 2005). Although, FAAH has high selectivity for AEA (Desarnaud et al. 1995) the effects of *FAAH* C385A cannot be specifically linked to AEA neurotransmission without additional data. If the neural and behavioral effects of *FAAH* C385A are mediated by genotype-driven differential availability of AEA, then these effects should be sensitive to manipulation of CB1 receptors. An interesting test of this putative mechanism would be to examine the impact of CB1 antagonists, such as rimonabant, on neural phenotypes associated with *FAAH* C385A genotype using pharmacologic fMRI. The availability of a PET radiotracer for CB1 (Burns et al. 2007) also allows for the determination of any *FAAH* C385A effects on endogenous receptor concentrations. If this polymorphism biases brain function through AEA stimulation of CB1, then antagonism of the receptor should eliminate the divergent effects on amygdala and VS reactivity documented here. Any genotype related alterations in AEA concentrations may also be reflected in relative up- or down-regulation of CB1 receptors assayed via PET. If CB1 antagonism fails to abolish the differential effects of *FAAH* C385A on brain function or if there are no differences in CB1 concentrations based on the genotype, then the existing effects are likely mediated by non-eCB fatty acid amides. In addition to testing this mechanistic hypothesis with pharmacologic fMRI and multimodal PET/fMRI, future studies with substantially increased sample sizes can model allele load effects of *FAAH* 385A, as well as potential *FAAH* interactions with functional genetic polymorphisms affecting other components of eCB neurotransmission (Chakrabarti et al. 2006).

Fifth, in light of evidence for environmental modulation (Tarullo and Gunnar 2006), gene-by-environment interactions (Caspi and Moffitt 2006) and epigenetic regulation (MacGowan et al. 2009), it is important for future research to not only assess genetic variation, but also to optimize assessments available to assess the objective and subjective impact of positive and negative environmental variables. Examples of this include interview-based methods to assess the objective and subjective impact of stressful life events and prospective study designs to examine the influence of stress over the course of study assessments (e.g., the Life Events and

Difficulties Schedule; Brown and Harris 1978; Monroe 2008; Williamson et al. 2003). Furthermore, in light of theoretical arguments that genotypes associated with adversity may reflect plasticity to the environment and not just vulnerability (e.g., Belsky et al. 2009; Manuck 2010), it is critical for these assessments to also assess positive effects of the environment such as social support (e.g., Hyde et al. 2011).

Finally, there is tremendous potential in developing large databases (again preferably thousands of subjects) with detailed measures of behavioral traits, neuroimaging-based measures of multiple brain circuitries and extensive genotyping. One of the most exciting applications of molecular genetics is in identifying novel biological pathways contributing to the emergence of complex traits (Gibson and Goldstein 2007; McCarthy et al. 2008). The continued refinement of a detailed map of sequence variation across the entire human genome (i.e., SNPs that "tag" every gene) and production of technologies supporting efficient high-throughput identification of such variation in individuals have dramatically accelerated the discovery of genes involved in the emergence of complex disease processes (Fellay et al. 2007; Link et al. 2008) as well as normal variability in continuous traits (Lettre et al. 2008). Many of the genes identified in such studies have illuminated novel pathways not previously implicated in these processes or traits, spurring intensive efforts to understand the potential biological effects of the proteins produced by these genes. As such, these "genome-wide" screens represent an opportunity to leap forward beyond the available pool of candidate molecules and pathways in parsing the mechanisms of complex biological processes. Because neuroimaging-based measures of brain function reveal key mechanisms involved in the emergence of individual differences in behavioral traits and are closer to the biological effects of functional genetic polymorphisms, they are ideal substrates for genome-wide screens. For example, BOLD fMRI estimates of amygdala reactivity predicting variability in trait anxiety can be used as the continuous trait in a genome-wide screen. Any significant associations that emerge between genetic variation and amygdala reactivity may confirm existing relationships (e.g., the importance of genes biasing 5-HT signaling) or, more importantly, reveal unexpected candidate molecules or pathways (e.g., a gene producing a molecule that is expressed in the brain and may function in second-messenger signaling cascades). Once identified and, ideally, replicated in large-scale databases that effectively address confounds common to genome-wide screens (e.g., controlling for multiple comparisons resulting from testing the association of a phenotype with hundreds of thousands or millions of SNPs), the impact of variation in novel genes associated with amygdala reactivity can be explored at each level of the biological cascade leading to trait anxiety (i.e., be fed back into the discovery loop outlined in the introduction). In addition to exponentially improving our understanding of neurobiological pathways leading to individual differences in complex behavioral traits these efforts may lead to the discovery of novel therapeutic strategies targeting related disease processes.

Acknowledgment This manuscript is largely based on an earlier publication in the *Annual Review of Neuroscience* (Hariri 2009).

References

Abercrombie HC, Giese-Davis J, Sephton S, Epel ES, Turner-Cobb JM, Spiegel D (2004) Flattened cortisol rhythms in metastic breast cancer patients. Psychoneuroendocrinology 29: 1082–1092

Adkins-Regan E (2005) Hormones and animal social behavior. Princeton University Press, Princeton

Alessi SM, Petry NM (2003) Pathological gambling severity is associated with impulsivity in a delay discounting procedure. Behav Process 64:345–354

Amat J, Matus-Amat P, Watkins LR, Maier SF (1998) Escapable and inescapable stress differentially alter extracellular levels of 5-HT in the basolateral amygdala of the rat. Brain Res 812:113–120

Amat J, Tamblyn JP, Paul ED, Bland ST, Amat P et al (2004) Microinjection of urocortin 2 into the dorsal raphe nucleus activates serotonergic neurons and increases extracellular serotonin in the basolateral amygdala. Neuroscience 129:509–519

Arai K, Nakagomi Y, Iketani M, Shimura Y, Amemiya S, Ohyama K, Shibasaki T (2003) Functional polymorphisms in the mineralocorticoid receptor and amirolide-sensitive sodium channel genes in a patient with sporadic pseudohypoaldosteronism. Hum Genet 112:91–97

Bannon MJ, Michelhaugh SK, Wang J, Sacchetti P (2001) The human dopamine transporter gene: gene organization, transcriptional regulation, and potential involvement in neuropsychiatric disorders. Eur Neuropsychopharmacol 11:449–455

Baron RA, Richardson D (1994) Human aggression. Plenum, New York

Bartholome B, Spies CM, Gaber T, Schuchmann S, Kunkel D et al (2004) Membrane glucocorticoid receptors (mGCR) are expressed in normal human peripheral blood mononuclea cells and up-regulated after in vitro stimulation and in patients with rheumatoid arthritis. FASEB J 18:70–80

Basavarajappa BS, Yalamanchili R, Cravatt BF, Cooper TB, Hungund BL (2006) Increased ethanol consumption and preference and decreased ethanol sensitivity in female FAAH knockout mice. Neuropharmacology 50:834–844

Beaver JD, Lawrence AD, Passamonti L, Calder AJ (2008) Appetitive motivation predicts the neural response to facial signals of aggression. J Neurosci 28:2719–2725

Belsky J, Jonassaint C, Pluess M, Stanton M, Brummett B, Williams R (2009) Vulnerability genes or plasticity genes? Mol Psychiatry 14:746–754

Berridge KC, Robinson TE (2003) Parsing reward. Trends Neurosci 26:507–513

Bickel WK, Odum AL, Madden GJ (1999) Impulsivity and cigarette smoking: delay discounting in current, never, and ex-smokers. Psychopharmacology (Berl) 146:447–454

Bigos KL, Pollock BG, Aizenstein H, Fisher PM, Bies RR, Hariri AR (2008) Acute 5-HT reuptake blockade potentiates human amygdala reactivity. Neuropsychopharmacology 33:3221–3225

Binder (2009) The role of FKBP5, a co-chaperone of the glucocorticoid receptor in the pathogenesis and therapy of affective and anxiety disorders. Psychoneuroendocrinology, 34(Suppl 1), S186–195

Binder EB, Nemeroff CB (2010) The CRF system, stress, depression and anxiety-insights from human genetic studies. Mol Psychiatry 15:574–588

Bishop SJ, Duncan J, Lawrence AD (2004) State anxiety modulation of the amygdala response to unattended threat-related stimuli. J Neurosci 24:10364–10368

Blair RJ (2010) Neuroimaging of psychopathy and antisocial behavior: a targeted review. Curr Psychiatry Rep 12:76–82

Blednov YA, Cravatt BF, Boehm SL 2nd, Walker D, Harris RA (2007) Role of endocannabinoids in alcohol consumption and intoxication: studies of mice lacking fatty acid amide hydrolase. Neuropsychopharmacology 32:1570–1582

Bogdan R, Perlis RH, Fagerness J, Pizzagalli DA (2010) The impact of mineralocorticoid receptor iso/val genotype (rs5522) and stress on reward learning. Genes Brain Behav 9: 658–667

Bogdan R, Williamson DE, Hariri AR (in press) Mineralocorticoid receptor iso/val genotype (rs5522) moderates the association between priori childhood emotional neglect and amygdala reactivity. Am J Psychiatry

Brady LS, Whitfield HJ, Fox RJ, Gold PW, Herkenham M (1991) Long-term antidepressant administration alters corticotropin-releasing hormone, tyrosine hydroxylase, and mineralocorticoid receptor gene expression in rat brain. Therapeutic implications. J Clin Invest 87:831–837

Brown GW, Harris TO (1978) Social origins of depression: a study of psychiatric disorder in women. Tavistock, London

Buchanan TW, Driscoll D, Mowrer SM, Sollers JJ, Thayer JF et al (2010) Medial prefrontal cortex damage affects physiological and psychological stress responses differently in men and women. Psychoneuroendocrinology 35:56–66

Burghardt NS, Sullivan GM, McEwen BS, Gorman JM, LeDoux JE (2004) The selective serotonin reuptake inhibitor citalopram increases fear after acute treatment but reduces fear with chronic treatment: a comparison with tianeptine. Biol Psychiatry 55:1171–1178

Burghardt NS, Bush DEA, McEwen BS, LeDoux JE (2007) Acute SSRIs increase conditioned fear expression: blockade with a 5-HT2C receptor antagonist. Biol Psychiatry 62:1111–1118

Burns HD, Van Laere K, Sanabria-Bohorquez S, Hamill TG, Bormans G et al (2007) [18F]MK-9470, a positron emission tomography (PET) tracer for in vivo human PET brain imaging of the cannabinoid-1 receptor. Proc Natl Acad Sci U S A 104:9800–9805

Calignano A, La Rana G, Giuffrida A, Piomelli D (1998) Control of pain initiation by endogenous cannabinoids. Nature 394:277–281

Cardinal RN, Winstanley CA, Robbins TW, Everitt BJ (2004) Limbic corticostriatal systems and delayed reinforcement. Ann N Y Acad Sci 1021:33–50

Carré JM, Fisher PM, Manuck SB, Hariri AR (in press-a) Interaction between trait anxiety and trait anger predict predict amygdala reactivity to angry faces in men but not women. Soc Cogn Affect Neurosci

Carré JM, McCormick CM (2008) Aggressive behaviour and change in salivary testosterone concentrations predict willingness to engage in a competitive task. Horm Beh 54:403–409

Carré JM, Murphy KR, Hariri AR (in press-b) What lies beneath the face of aggression. Soc Cogn Affect Neurosci

Carré JM, Putnam SK, McCormick CM (2009) Testosterone responses to competition predict future aggressive behaviour at a cost to reward in men. Psychoneuroendocrinology 343: 561–570

Carré JM, Gilchrist JG, Morrissey MD, McCormick CM (2010) Motivational and situations factors and the relationship between testosterone dynamics and human aggression during competition. Biol Psychol 84:346–353

Caspi A, Moffitt TE (2006) Gene-environment interactions in psychiatry: joining forces with neuroscience. Nat Rev Neurosci 7:583–590

Chakrabarti B, Kent L, Suckling J, Bullmore E, Baron-Cohen S (2006) Variations in the human cannabinoid receptor (CNR1) gene modulate striatal responses to happy faces. Eur J Neurosci 23:1944–1948

Chamberlain NL, Driver ED, Miesfeld RL (1994) The length and location of CAG trinucleotide repeats in the androgen receptor N-terminal dsomain affect transactivation function. Nucleic Acids Res 22:3181–3186

Chan SC, Raine A, Lee TM (2010) Attentional bias toward negative affective stimuli and reactive aggression in male batterers. Psychiatry Res 176:246–249

Cheon KA, Ryu YH, Kim JW, Cho DY (2005) The homozygosity for 10-repeat allele at dopamine transporter gene and dopamine transporter density in Korean children with attention deficit hyperactivity disorder: relating to treatment response to methylphenidate. Eur Neuropsychopharmacol 15:95–101

Chiang KP, Gerber AL, Sipe JC, Cravatt BF (2004) Reduced cellular expression and activity of the P129T mutant of human fatty acid amide hydrolase: evidence for a link between defects in the endocannabinoid system and problem drug use. Hum Mol Genet 13:2113–2119

Choong CS, Wilson EM (1998) Trinucleotide repeats in the human androgen receptor: a molecular basis for disease. J Mol Endocrinol 21:235–257

Coccaro EF, McCloskey MS, Fitzgerald DA, Phan KL (2007) Amygdala and orbitofrontal reactivity to social threat in individuals with impulsive aggression. Biol Psychiatry 62: 168–178

Cohen S, Janicki-Deverts D, Miller GE (2007) Psychological stress and disease. JAMA 298: 1685–1687

Cooney RE, Atlas LY, Joormann J, Eugene F, Gotlib IH (2006) Amygdala activation in the processing of neutral faces in social anxiety disorder: is neutral really neutral? Psychiatry Res 148:55–59

Cowen PJ, Power AC, Ware CJ, Anderson IM (1994) 5-HT1A receptor sensitivity in major depression.A neuroendocrine study with buspirone. Br J Psychiatry 164:372–379

Cravatt BF, Giang DK, Mayfield SP, Boger DL, Lerner RA, Gilula NB (1996) Molecular characterization of an enzyme that degrades neuromodulatory fatty-acid amides. Nature 384:83–87

Czesak M, Lemonde S, Peterson EA, Rogaeva A, Albert PR (2006) Cell-specific repressor or enhancer activities of Deaf-1 at a serotonin 1A receptor gene polymorphism. J Neurosci 26:1864–1871

David SP, Murthy NV, Rabiner EA, Munafo MR, Johnstone EC et al (2005) A functional genetic variation of the serotonin (5-HT) transporter affects 5-HT1A receptor binding in humans. J Neurosci 25:2586–2590

Davidson RJ, Putnam KM, Larson CL (2000) Dysfunction in the neural circuitry of emotion regulation—a possible prelude to violence. Science 289:591–594

de Wit H, Richards JB (2004) Dual determinants of drug use in humans: reward and impulsivity. Nebr Symp Motiv 50:19–55

de Kloet ER, Joëls M, Holsboer F (2005) Stress and the brain: from adaptation to disease. Nat Rev Neurosci 6:463–475

De Wit H, Flory JD, Acheson A, McLoskey M, Manuck SB (2007) IQ and nonplanning impulsivity are independently associated with delay discounting in middle-aged adults. Pers Indiv Differ 42:111–121

Depue RA, Luciana M, Arbisi P, Collins P, Leon A (1994) Dopamine and the structure of personality: relation of agonist-induced dopamine activity to positive emotionality. J Pers Soc Psychol 67:485–498

DeRijk RH, Wüst S, Meijer OC, Zennaro MC, Federenko IS, Hellhammer DH, Giacchetti G, Vreugdenhil E, Zitman FG, de Kloet ER (2006) A common polymorphism in the mineralocorticoid receptor modulates stress responsiveness. J Clin Endocrinol Metab 91:5083–5089

DeRijk RH, van Leeuwen N, Klok MD, Zitman FG (2008) Corticosteroid receptor-gene variants: modulators of the stress-response and implications for mental health. Eur J Pharmacol 585:492–501

Derntl B, Windischberger C, Robinson S, Kryspin-Exner I, Gur RC, Moser E, Habel U (2009) Amygdala activity to fear and anger in healthy young males is associated with testosterone. Psychoneuroendocrinology 34:687–693

Desarnaud F, Cadas H, Piomelli D (1995) Anandamide amidohydrolase activity in rat brain microsomes. J Biol Chem 270:6030–6035

Dickerson SS, Kemeny ME (2004) Acute stressors and cortisol responses: a theoretical integration and synthesis of laboratory research. Psychol Bull 130:355–391

Dickie EW, Armony JL (2008) Amygdala responses to unattended fearful faces: Interaction between sex and trait anxiety. Psychiatry Res 162:51–57

Dodge KA, Coie JD (1987) Social-information processing factors in reactive and proactive aggression in children's peer groups. J Pers Soc Psychol 53:1146–1158

Etkin A, Klemenhagen KC, Dudman JT, Rogan MT, Hen R et al (2004) Individual differences in trait anxiety predict the response of the basolateral amygdala to unconsciously processed fearful faces. Neuron 44:1043–1055

Evans KC, Wright CI, Wedig MM, Gold AL, Pollack MH, Rauch SL (2008) A functional MRI study of amygdala responses to angry schematic faces in social anxiety disorder. Depress Anxiety 25:496–505

Fakra E, Hyde LW, Gorka A, Fisher PM, Munoz KE et al (2009) Effects of HTR1A C(-1019)G on amygdala reactivity and trait anxiety. Arch Gen Psychiatry 66:33–40

Fellay J, Shianna KV, Ge D, Colombo S, Ledergerber B et al (2007) A whole-genome association study of major determinants for host control of HIV-1. Science 317:944–947

Fisher PM, Meltzer CC, Ziolko SK, Price JC, Hariri AR (2006) Capacity for 5-HT_{1A}-mediated autoregulation predicts amygdala reactivity. Nat Neurosci 9:1362–1363

Flanagan JM, Gerber AL, Cadet JL, Beutler E, Sipe JC (2006) The fatty acid amide hydrolase 385 A/A (P129T) variant: haplotype analysis of an ancient missense mutation and validation of risk for drug addiction. Hum Genet 120:581–588

Forbes EE, Brown SM, Kimak M, Ferrell RE, Manuck SB, Hariri AR (2009) Genetic variation in components of dopamine neurotransmission impacts ventral striatal reactivity associated with impulsivity. Mol Psychiatry 14:60–70

Forster GL, Feng N, Watt MJ, Korzan WJ, Mouw NJ et al (2006) Corticotropin-releasing factor in the dorsal raphe elicits temporally distinct serotonergic responses in the limbic system in relation to fear behavior. Neuroscience 141:1047–1055

Fox HC, Wilker EH, Kreek MJ, Sinha R (2006) Reliability of salivary cortisol assessments in cocaine dependent individuals. J Psychopharmacol 20:650–655

Gass P, Reichardt HM, Strekalova T, Henn F, Tronche F (2001) Mice with targetted mutations of glucocorticoid and mineralocorticoid receptors: models for depression and anxiety? Physiol Behav 73:811–825

Gibbons JL, McHugh PR (1962) Plasma cortisol in depressive illness. J Psychiat Res 1:162–171

Gibson G, Goldstein DB (2007) Human genetics: the hidden text of genome-wide associations. Curr Biol 17:R929–R932

Glahn DC, Lovallo WR, Fox PT (2007) Reduced amygdala activation in young adults at high risk of alcoholism: studies from the Oklahoma family health patterns project. Biol Psychiatry 61:1306–1309

Green L, Myerson J (2004) A discounting framework for choice with delayed and probabilistic rewards. Psychol Bull 130:769–792

Haas BW, Omura K, Constable RT, Canli T (2007) Emotional conflict and neuroticism: personality-dependent activation in the amygdala and subgenual anterior cingulate. Behav Neurosci 121:249–256

Hansenne M, Pitchot W, Pinto E, Reggers J, Scantamburlo G et al (2002) 5-HT1A dysfunction in borderline personality disorder. Psychol Med 32:935–941

Hariri AR (2009) The neurobiology of individual differences in complex behavioral traits. Ann Rev Neurosci 32:225–247

Hariri AR, Holmes A (2006) Genetics of emotional regulation: the role of the serotonin transporter in neural function. Trends Cogn Sci 10:182–191

Hariri AR, Weinberger DR (2003) Imaging genomics. Br Med Bull 65:259–270

Hariri AR, Mattay VS, Tessitore A, Fera F, Smith WG, Weinberger DR (2002a) Dextroamphetamine modulates the response of the human amygdala. Neuropsychopharmacology 27:1036–1040

Hariri AR, Mattay VS, Tessitore A, Kolachana B, Fera F et al (2002b) Serotonin transporter genetic variation and the response of the human amygdala. Science 297:400–403

Hariri AR, Brown SM, Williamson DE, Flory JD, de Wit H, Manuck SB (2006a) Preference for immediate over delayed rewards is associated with magnitude of ventral striatal activity. J Neurosci 26:13213–13217

Hariri AR, Drabant EM, Weinberger DR (2006b) Imaging genetics: perspectives from studies of genetically driven variation in serotonin function and corticolimbic affective processing. Biol Psychiatry 59:888–897

Hariri AR, Gorka A, Hyde LW, Kimak M, Halder I et al (2009) Divergent effects of genetic variation in endocannabinoid signaling on human threat- and reward-related brain function. Biol Psychiatry 66:9–16

Harmon-Jones E (2003) Anger and the behavioural approach system. Pers Ind Diff 35:995–1005
Heinz A, Goldman D, Jones DW, Palmour R, Hommer D et al (2000) Genotype influences in vivo dopamine transporter availability in human striatum. Neuropsychopharmacology 22:133–139
Hermans E, Ramsey M, van Honk J (2008) Exogenous testosterone enhances responsiveness to social threat in the neural circuitry of social aggression in humans. Biol Psychiatry 63:263–270
Hyde LW, Gorka A, Manuck SB, Hariri AR (2011) Perceived social support moderates the link between threat-related amygdala reactivity and trait anxiety. Neuropsychologia 49:651–656
Hyman SE, Malenka RC, Nestler EJ (2006) Neural mechanisms of addiction: the role of reward-related learning and memory. Annu Rev Neurosci 29:565–598
Jahn AL, Fox AS, Abercrombie HC, Shelton SE, Oakes TR et al (2010) Subgenual prefrontal cortex activity predicts individual differences in hypothalamic-pituitary-adrenal activity across different contexts. Biol Psychiatry 67:175–181
Joëls M, Karst H, DeRijk R, de Kloet ER (2008) The coming out of the brain mineralocorticoid receptor. Trends Neurosci 31:1–7
Johnstone T, Somerville LH, Alexander AL, Oakes TR, Davidson RJ et al (2005) Stability of amygdala BOLD response to fearful faces over multiple scan sessions. Neuroimage 25: 1112–1123
Kalivas PW, Volkow ND (2005) The neural basis of addiction: a pathology of motivation and choice. Am J Psychiatry 162:1403–1413
Karst H, Berger S, Turiault M, Tronche F, Schütz G, Joëls M (2006) Mineralocorticoid receptors are indispensable for nongenomic modulation of hippocampal glutamate transmission by corticosterone. Proc Natl Acad Sci U S A 102:19204–19207
Kathuria S, Gaetani S, Fegley D, Valino F, Duranti A et al (2003) Modulation of anxiety through blockade of anandamide hydrolysis. Nat Med 9:76–81
Kelley AE (2004) Memory and addiction: shared neural circuitry and molecular mechanisms. Neuron 44:161–179
Kern S, Oakes TR, Stone CK, McAuliff EM, Kirschbaum C, Davidson RJ (2008) Glucose metabolic changes in the prefrontal cortex are associated with HPA axis response to a psychosocial stressor. Psychoneuroendocrinology 33:517–529
Killgore WD, Yurgelun-Todd DA (2005) Social anxiety predicts amygdala activation in adolescents viewing fearful faces. Neuroreport 16:1671–1675
Kirby KN, Petry NM, Bickel WK (1999) Heroin addicts have higher discount rates for delayed rewards than non-drug-using controls. J Exp Psychol Gen 128:78–87
Knutson B, Rick S, Wimmer GE, Prelec D, Loewenstein G (2007) Neural predictors of purchases. Neuron 53:147–156
Kolber BJ, Muglia LJ (2009) Defining brain region-specific glucocorticoid action during stress by conditional gene disruption in mice. Brain Res 1293:85–90
Kudielka BM, Hellhammer DH, Wüst S (2009) Why do we respond so differently? Reviewing determinants of human salivary cortisol responses to challenge. Psychoneuroendocrinology 34:2–18
Kuningas M, de Rijk RH, Westendorp RG, Jolle J, Slagboom PE, van Heemst D (2007) Mental performance in old age dependent on cortisol and genetic variance in the mineralocorticoid and glucocorticoid receptors. Neuropsychopharmacology 32:1295–1301
LeDoux JE (2000) Emotion circuits in the brain. Annu Rev Neurosci 23:155–184
Lee BT, Ham BJ (2008) Serotonergic genes and amygdala activity in response to negative affective facial stimuli in Korean women. Genes Brain Behav 7:899–905
Lee TM, Chan SC, Raine A (2008) Strong limbic and weak frontal activation to aggressive stimuli in spouse abusers. Mol Psychiatry 13:655–656
Lemonde S, Turecki G, Bakish D, Du L, Hrdina PD et al (2003) Impaired repression at a 5-hydroxytryptamine 1A receptor gene polymorphism associated with major depression and suicide. J Neurosci 23:8788–8799
Lesch KP, Gutknecht L (2004) Focus on The 5-HT1A receptor: emerging role of a gene regulatory variant in psychopathology and pharmacogenetics. Int J Neuropsychopharmacol 7:381–385

Lettre G, Jackson AU, Gieger C, Schumacher FR, Berndt SI et al (2008) Identification of ten loci associated with height highlights new biological pathways in human growth. Nat Genet 40: 584–591

Link E, Parish S, Armitage J, Bowman L, Heath S et al (2008) SLCO1B1 variants and statin-induced myopathy—a genomewide study. N Engl J Med 359:789–799

LoVerme J, Gaetani S, Fu J, Oveisi F, Burton K, Piomelli D (2005) Regulation of food intake by oleoylethanolamide. Cell Mol Life Sci 62:708–716

Madden GJ, Petry NM, Badger GJ, Bickel WK (1997) Impulsive and self-control choices in opioid-dependent patients and non-drug-using control participants: drug and monetary rewards. Exp Clin Psychopharmacol 5:256–262

Maier SF, Watkins LR (2005) Stressor controllability and learned helplessness: the roles of the dorsal raphe nucleus, serotonin, and corticotropin-releasing factor. Neurosci Biobehav Rev 29:829–841

Maldonado R, Valverde O, Berrendero F (2006) Involvement of the endocannabinoid system in drug addiction. Trends Neurosci 29:225–232

Manuck SB (2010) The reaction norm in gene x environment interaction. Mol Psychiatry 14:746–754

Manuck SB, Flory JD, McCaffery JM, Matthews KA, Mann JJ, Muldoon MF (1998) Aggression, impulsivity, and central nervous system serotonergic responsivity in a nonpatient sample. Neuropsychopharmacology 19:287–299

Manuck SB, Flory JD, Muldoon MF, Ferrell RE (2003) A neurobiology of intertemporal choice. In: Loewenstein G, Read D, Baumeister RF (eds) Time and decision: economic and psychological perspectives on intertemporal choice. Sage, New York, pp 139–172

Manuck SB, Brown SM, Forbes EE, Hariri AR (2007) Temporal stability of individual differences in amygdala reactivity. Am J Psychiatry 164:1613–1614

Manuck SB, Marsland AL, Flory JD, Gorka A, Ferrell RE, Hariri AR (2010) Salivary testosterone and a trinucleotide (CAG) length polymorphism in the androgen receptor gene predict amygdala reactivity in men. Psychoneuroendocrinology 35:94–104

Marques AH, Silverman MN, Sternberg EM (2009) Glucocorticoid dysregulationss and their clinical correlated: from receptors to therapeutics. Ann N Y Acad Sci 1179:1–18

Márquez C, Nadal R, Armario A (2005) Responsiveness of the hypothalamic-pituitary-adrenal axis to different novel environments is a consistent individual trait in adult male outbred rats. Psychoneuroendocrinology 30:179–188

Martinez D, Gelernter J, Abi-Dargham A, van Dyck CH, Kegeles L et al (2001) The variable number of tandem repeats polymorphism of the dopamine transporter gene is not associated with significant change in dopamine transporter phenotype in humans. Neuropsychopharmacology 24:553–560

Mauchnik J, Schmahl C (2010) The latest neuroimaging findings in borderline personality disorder. Curr Psychiatry Rep 12:46–55

McCarthy MI, Abecasis GR, Cardon LR, Goldstein DB, Little J et al (2008) Genome-wide association studies for complex traits: consensus, uncertainty and challenges. Nat Rev Genet 9:356–369

McEwen BS, Gianaros PJ (2010) Central role of the brain in stress and adaptation: links to socioeconomic status, health, and disease. Ann N Y Acad Sci 1186:190–222

McGowan PO, Sasaki A, D'Alessio AC, Gymov S, Labonté B et al (2009) Epigenetic regulation of the glucocorticoid receptor in human brain associates with childhood abuse. Nat Neurosci 12:342–348

Mehta PH, Beer J (2010) Neural mechanisms of the testosterone-aggression relation: the role of the orbito-frontal cortex. J Cog Neuro 22:2357–2368

Meng ID, Manning BH, Martin WJ, Fields HL (1998) An analgesia circuit activated by cannabinoids. Nature 395:381–383

Menon M, Jensen J, Vitcu I, Graff-Guerrero A, Crawley A et al (2007) Temporal difference modeling of the blood-oxygen level dependent response during aversive conditioning in humans: effects of dopaminergic modulation. Biol Psychiatry 62:765–772

Meyer-Lindenberg A, Buckholtz JW, Kolachana B, Hariri AR, Pezawas L et al (2006) Neural mechanisms of genetic risk for impulsivity and violence in humans. Proc Natl Acad Sci U S A 103:6269–6274
Michelhaugh SK, Fiskerstrand C, Lovejoy E, Bannon MJ, Quinn JP (2001) The dopamine transporter gene (SLC6A3) variable number of tandem repeats domain enhances transcription in dopamine neurons. J Neurochem 79:1033–1038
Mill J, Asherson P, Browes C, D'Souza U, Craig I (2002) Expression of the dopamine transporter gene is regulated by the 3′ UTR VNTR: evidence from brain and lymphocytes using quantitative RT-PCR. Am J Med Genet 114:975–979
Mill J, Asherson P, Craig I, D'Souza UM (2005) Transient expression analysis of allelic variants of a VNTR in the dopamine transporter gene (DAT1). BMC Genet 6:3
Mitra R, Ferguson D, Sapolsky RM (2009) Mineralocorticoid receptor overexpression in basolateral amygdala reduces corticosterone secretion and anxiety. Biol Psychiatry 66: 686–690
Monroe SM (2008) Modern approaches to conceptualizing and measuring human life stress. Annu Rev Clin Psychol 4:33–52
Moreira FA, Kaiser N, Monory K, Lutz B (2008) Reduced anxiety-like behaviour induced by genetic and pharmacological inhibition of the endocannabinoid-degrading enzyme fatty acid amide hydrolase (FAAH) is mediated by CB1 receptors. Neuropharmacology 54:141–150
Most SB, Chun MM, Johnson MR, Kiehl KA (2006) Attentional modulation of the amygdala varies with personality. Neuroimage 31:934–944
Müller M, Holsboer F, Keck ME (2002) Genetic modification of corticosteroid receptor signaling: novel insights into pathophysiology and treatment strategies of human affective disorders. Neuropeptides 36:117–131
Munafo MR, Brown SM, Hariri AR (2008) Serotonin transporter (5-HTTLPR) genotype and amygdala activation: a meta-analysis. Biol Psychiatry 63:852–857
Nelson RJ, Trainor BC (2007) Neural mechanisms of aggression. Nat Rev Neurosci 8:536–546
New AS, Hazlett EA, Buchsbaum MS, Goodman M, Mitelman SA, Newmark R et al (2007) Amygdala-prefrontal disconnection in borderline personality disorder. Neuropsychopharmacology 32:1629–1640
Newman S (1999) The medial extended amygdala in male reproductive behavior. A node in the mammalian social behavior network. Ann N Y Acad Sci 877:242–257
O'Doherty JP (2004) Reward representations and reward-related learning in the human brain: insights from neuroimaging. Curr Opin Neurobiol 14:769–776
Onaivi ES, Leonard CM, Ishiguro H, Zhang PW, Lin Z et al (2002) Endocannabinoids and cannabinoid receptor genetics. Prog Neurobiol 66:307–344
Parsey RV, Oquendo MA, Ogden RT, Olvet DM, Simpson N et al (2006) Altered serotonin 1A binding in major depression: a [carbonyl-C-11]WAY100635 positron emission tomography study. Biol Psychiatry 59:106–111
Passamonti L, Rowe JB, Ewbank M, Hampshire A, Keane J, Calder AJ (2008) Connectivity from the ventral anterior cingulate to the amygdala is modulated by appetitive motivation in response to facial signals of aggression. Neuroimage 43:562–570
Patel S, Cravatt BF, Hillard CJ (2005) Synergistic interactions between cannabinoids and environmental stress in the activation of the central amygdala. Neuropsychopharmacology 30:497–507
Phan KL, Fitzgerald DA, Nathan PJ, Tancer ME (2006) Association between amygdala hyperactivity to harsh faces and severity of social anxiety in generalized social phobia. Biol Psychiatry 59:424–429
Phan KL, Angstadt M, Golden J, Onyewuenyi I, Popovska A, de Wit H (2008) Cannabinoid modulation of amygdala reactivity to social signals of threat in humans. J Neurosci 28: 2313–2319
Phillips ML, Drevets WC, Rauch SL, Lane R (2003) Neurobiology of emotion perception II: implications for major psychiatric disorders. Biol Psychiatry 54:515–528

Piomelli D (2003) The molecular logic of endocannabinoid signalling. Nat Rev Neurosci 4:873–884
Piomelli D, Beltramo M, Glasnapp S, Lin SY, Goutopoulos A et al (1999) Structural determinants for recognition and translocation by the anandamide transporter. Proc Natl Acad Sci U S A 96:5802–5807
Pruessner JC, Dedovic K, Pressner M, Lord C, Buss C et al (2010) Stress regulation in the central nervous system: evidence from structural and functional neuroimaging studies in human populations. Psychoneuroendocrinology 35:179–191
Raine A, Buchsbaum M, LaCasse L (1997) Brain abnormalities in murderers indicated by positron emission tomography. Biol Psychiatry 42:495–508
Rajender S, Pandu G, Sharma JD, Gandhi KP, Singh L, Thangaraj K (2008) Reduced CAG repeats length in androgen receptor gene is associated with violent criminal behavior. Int J Legal Med 122:367–372
Ray RD, Ochsner KN, Cooper JC, Robertson ER, Gabrieli JD, Gross JJ (2005) Individual differences in trait rumination and the neural systems supporting cognitive reappraisal. Cogn Affect Behav Neurosci 5:156–168
Reul JM, Gesing A, Droste S, Stec IS, Weber A et al (2000) The brain mineralocorticoid receptor: greedy for ligand, mysterious in function. Euro J Pharmacol 405:235–249
Rhodes RA, Murthy NV, Dresner MA, Selvaraj S, Stavrakakis N et al (2007) Human 5-HT transporter availability predicts amygdala reactivity in vivo. J Neurosci 27:9233–9237
Robson P (2005) Human studies of cannabinoids and medicinal cannabis. Handb Exp Pharmacol 168:719–756
Rodrigues SM, LeDoux JE, Sapolsky RM (2009) The influence of stress hormones on fear circuitry. Annu Rev Neurosci 32:289–313
Roney JR, Simmons ZL, Lukaszewski AW (2009) Androgen receptor gene sequence and basal cortisol concentrations predict men's hormonal responses to potential mates. Proc Soc Biol 277:57–63
Rozeboom AM, Akil H, Seasholtz AF (2007) Mineralocorticoid receptor overexpression in forebrain decreases anxiety-like behavior and alters the stress response in mice. Proc Natl Acad of Sci U S A 104:4688–4693
Sadikot AF, Parent A (1990) The monoaminergic innervation of the amygdala in the squirrel monkey: an immunohistochemical study. Neuroscience 36:431–447
Scherma M, Medalie J, Fratta W, Vadivel SK, Makriyannis A et al (2008) The endogenous cannabinoid anandamide has effects on motivation and anxiety that are revealed by fatty acid amide hydrolase (FAAH) inhibition. Neuropharmacology 54:129–140
Sesack SR, Hawrylak VA, Guido MA, Levey AI (1998) Cellular and subcellular localization of the dopamine transporter in rat cortex. Adv Pharmacol 42:171–174
Sharp T, Boothman L, Raley J, Queree P (2007) Important messages in the 'post': recent discoveries in 5-HT neurone feedback control. Trends Pharmacol Sci 28:629–636
Siessmeier T, Kienast T, Wrase J, Larsen JL, Braus DF et al (2006) Net influx of plasma 6-[18F]fluoro-l-DOPA (FDOPA) to the ventral striatum correlates with prefrontal processing of affective stimuli. Eur J Neurosci 24:305–313
Siever LJ (2008) Neurobiology of aggression and violence. Am J Psychiatry 165:429–442
Simon NG (2002) Hormonal processes in the development and expression of aggressive behavior. In: Pfaff DW et al (eds) Hormones, brain and behavior. Academic Press, San Diego
Simpson CA, Vuchinich RE (2000) Reliability of a measure of temporal discounting. Psychol Rec 50:3–16
Sipe JC, Chiang K, Gerber AL, Beutler E, Cravatt BF (2002) A missense mutation in human fatty acid amide hydrolase associated with problem drug use. Proc Natl Acad Sci U S A 99: 8394–8399
Solinas M, Tanda G, Justinova Z, Wertheim CE, Yasar S et al (2007) The endogenous cannabinoid anandamide produces delta-9-tetrahydrocannabinol-like discriminative and neurochemical effects that are enhanced by inhibition of fatty acid amide hydrolase but not by inhibition of anandamide transport. J Pharmacol Exp Ther 321:370–380

Somerville LH, Kim H, Johnstone T, Alexander AL, Whalen PJ (2004) Human amygdala responses during presentation of happy and neutral faces: correlations with state anxiety. Biol Psychiatry 55:897–903

Sorice-Gomez E, Matias I, Rueda-Orozco PE, Cisneros M, Petrosino S et al (2007) Pharmacological enhancement of the endocannabinoid system in the nucleus accumbens shell stimulates food intake and increases c-Fos expression in the hypothalamus. Br J Pharmacol 151:1109–1116

Stein MB, Goldin PR, Sareen J, Zorrilla LT, Brown GG (2002) Increased amygdala activation to angry and contemptuous faces in generalized social phobia. Arch Gen Psychiatry 59: 1027–1034

Stein MB, Simmons AN, Feinstein JS, Paulus MP (2007) Increased amygdala and insula activation during emotion processing in anxiety-prone subjects. Am J Psychiatry 164:318–327

Tarullo AR, Gunnar MR (2006) Child maltreatment and the developing HPA axis. Hormones Behav 50:632–639

Tessitore A, Hariri AR, Fera F, Smith WG, Chase TN et al (2002) Dopamine modulates the response of the human amygdala: a study in Parkinson's disease. J Neurosci 22:9099–9103

Thode K, Walss-Bass C, Hariri AR, Olvera R, Munoz K, et al Functional evidence implicating the role of the corticotropin-releasing hormone receptor 1 in the development of stress-related disorders. Am J Psychiatry (unpublished observation)

Tyndale RF, Payne JI, Gerber AL, Sipe JC (2007) The fatty acid amide hydrolase C385A (P129T) missense variant in cannabis users: studies of drug use and dependence in Caucasians. Am J Med Genet B Neuropsychiatr Genet 144:660–666

Ulrich-Lai YM, Herman JP (2009) Neural regulation of endocrine and autonomic stress responses. Nat Rev Neurosci 10:397–409

Urry HL, van Reekum CM, Johnstone T, Kalin NH, Thurow ME et al (2006) Amygdala and ventromedial prefrontal cortex are inversely coupled during regulation of negative affect and predict diurnal pattern of cortisol secretion among older adults. J Neurosci 26:4415–4425

van Dyck CH, Malison RT, Jacobsen LK, Seibyl JP, Staley JK et al (2005) Increased dopamine transporter availability associated with the 9-repeat allele of the SLC6A3 gene. J Nucl Med 46:745–751

Van Leeuwen N, Kumsta R, Entringer S, de Kloet ER, Zitman FG, DeRijk RH, Wust S (2010) Functional mineralocorticoid receptor (MR) gene variation influences the cortisol awakening response after dexamethasone. Psychoneuroendocrinology 35:339–349

Van Ness SH, Owens MJ, Kilts CD (2005) The variable number of tandem repeats element in DAT1 regulates in vitro dopamine transporter density. BMC Genet 6:55

van Praag HM, de Kloet R, van Os J (2004) Stress the brain and depression. Cambridge University Press, New York

van Wingen GA, Zylick SA, Pieters S, Mattern C, Verkes RJ, Buitelaar JK, Fernandez G (2008) Testosterone increases amygdala reactivity in middle-aged women to a young adulthood level. Neuropsychopharmacology 34:539–547

van Wingen GA, Mattern C, Verkes RJ, Buitelaar J, Fernandez G (2010) Testosterone reduces amygdala-orbitalfrontal cortex coupling. Psychoneuroendocrinology 35:105–113

Vermeersch H, T'Sjoen G, Kaufman JM, Vincke J, Van Houtte M (2010) Testosterone, androgen receptor gene CAG repeat length, mood and behaviour in adolescent males. Eur J Endocrinol 163:319–328

Viding E, Williamson DE, Hariri AR (2006) Developmental imaging genetics: challenges and promises for translational research. Dev Psychopathol 18:877–892

Viveros MP, Marco EM, File SE (2005) Endocannabinoid system and stress and anxiety responses. Pharmacol Biochem Behav 81:331–342

Volkow ND, Fowler JS, Wang GJ (1999) Imaging studies on the role of dopamine in cocaine reinforcement and addiction in humans. J Psychopharmacol 13:337–345

Wei XY, Yang JY, Dong YX, Wu CF (2007) Anxiolytic-like effects of oleamide in group-housed and socially isolated mice. Prog Neuropsychopharmacolo Biol Psychiatry 31:1189–1195

Whalen PJ, Shin LM, Somerville LH, McLean AA, Kim H (2002) Functional neuroimaging studies of the amygdala in depression. Semin Clin Neuropsychiatry 7:234–242

Williamson DE, Birmaher B, Ryan ND, Shiffrin TP, Lusky JA et al (2003) The stressful life events schedule for children and adolescents: development and validation. Psychiatry Res 119:225–241

Yehuda R (2002) Post-traumatic stress disorder. N Engl J Med 346:108–114

Zitzmann M, Nieschlag E (2003) The CAG repeat polymorphism within the androgen receptor gene and maleness. Int J Androl 26:76–83

Behavioural Genetics of Childhood Disorders

Christine M. Freitag, Philip Asherson and Johannes Hebebrand

Abstract After a general introduction into genetic risk factors for child psychiatric disorders, four specific child psychiatric disorders with a strong genetic component, namely, Autism Spectrum Disorders, Attention Deficit / Hyperactivity Disorder, Nocturnal Enuresis, and obesity, are discussed in detail. Recent evidence of linkage, candidate gene, and genome-wide association studies are presented. This chapter ends with a prospectus on further research needs.

Keywords Autism spectrum disorders · Attention deficit/hyperactivity disorder · Nocturnal enuresis · Obesity · BMI · Linkage · Association · Genetic

C. M. Freitag (✉)
Department of Child and Adolescent Psychiatry,
Psychosomatics and Psychotherapy, Goethe-Universität Frankfurt am Main,
Deutschordenstraße 50, 60528 Frankfurt am Main, Germany
e-mail: C.Freitag@em.uni-frankfurt.de

P. Asherson
MRC Social, Genetic and Developmental Psychiatry Centre,
Institute of Psychiatry, London, UK
e-mail: philip.asherson@kcl.ac.uk

J. Hebebrand
Department of Child and Adolescent Psychiatry,
University of Duisburg-Essen, Essen, Germany
e-mail: johannes.hebebrand@uni-due.de

Curr Topics Behav Neurosci (2012) 12: 395–428
DOI: 10.1007/7854_2011_178

Published Online: 1 March 2012

Contents

1 Introduction ... 396
2 Selected Child Psychiatric Disorders with a Strong Genetic Component ... 401
2.1 Autism Spectrum Disorders ... 401
2.2 Attention-Deficit/Hyperactivity Disorder ... 405
2.3 Nocturnal Enuresis ... 411
2.4 Obesity ... 412
3 Prospectus ... 415
References ... 415

1 Introduction

Over the past 50 years, substantial progress has been made in linking specific behavioural and psychiatric phenotypes to chromosomal aberrations or genetic variation at the DNA level. Prerequisites of this development were significant conceptual, methodological and technical advances in both molecular and statistical genetics as well as in phenotypic assessment.

Advances in *cytogenetics* allowed the identification of specific syndromes based on quantitative chromosomal imbalances of complete chromosomes, as in trisomy 21 in 1959, and of chromosomal regions, as in the cri du chat syndrome due to monosomy 5p, in 1963. Approximately five megabases represent the resolution limits of banding techniques; fluorescent in situ hybridisation (FISH) allows detection of deletions as small as 1.5 Mb. Recently, microarray-based comparative genomic hybridisation (array CGH) has been introduced as a complementary method with even higher resolution (Miller et al. 2008).

Micro-deletions can result in syndromes with distinct behavioural features such as the Williams-Beuren Syndrome (7q11.23), Prader-Willi syndrome (paternal 15q11), Angelman syndrome (maternal 15q11), 16p11.2 deletion, the velocardiofacial/DiGeorge syndrome (22q11) and 22q13.3 deletion syndrome (Xiang et al. 2010). Because a single gene on average encompasses 10–15 (kb), the phenotypes of micro-deletion syndromes result from the loss of a number of genes (partial monosomy). Corresponding to micro-deletions, several *micro-duplication syndromes* exist, e.g., corresponding to the Williams-Beuren Syndrome region at 7q11.23, and the chromosome 16p11.2-duplication (Shen et al. 2010). The numerous findings of copy number variations (CNVs) in recent whole genome studies on child psychiatric disorders, as Attention-Deficit/Hyperactivity Disorder (ADHD) or Autism Spectrum Disorders (ASD) (see below) emphasises the relevance of small cytogenetic findings for psychiatric disorders.

Quantitative imbalances can functionally result in over expression (e.g., trisomy or partial trisomy) or under expression (monosomy or partial monosomy) of those genes located on a chromosome or within a specific chromosomal region. Due to the high proportion of genes expressed in the central nervous system, e.g., $\sim$80% in mice (Lein et al. 2007), brain function is almost always perturbed in such disorders.

Findings in Down syndrome have revealed that of the approximately 200 genes on chromosome 21, only a subset is responsible for the characteristic phenotype. One well-known example is the early onset of dementia in subjects with Down syndrome, which can partially be attributed to the over expression of the amyloid-ß precursor protein gene (APP) located at 21q21-22. In addition to the quantitative imbalance, the Down syndrome phenotype is affected by allelic variation; for example, differences in length of a tetranucleotide repeat in intron 7 of the APP locus explain substantial variation in age at onset of dementia in subjects with trisomy 21 (Margallo-Lana et al. 2004).

The first large scale and systematic genotype–phenotype studies with a primary behavioural, psychological and psychiatric focus centred on sex chromosome (gonosome) disorders. Long-term follow-up of their development into adulthood revealed that these disorders are characterised by subtle neuropsychiatric and neuropsychological symptoms such as an IQ distribution shifted slightly to the left and elevated rates of attention problems, speech and reading difficulties and reduced impulse-control. Nevertheless, the dissection of a highly specific behavioural phenotype associated with any one of these sex chromosome disorders was not possible (Hebebrand 1990; Walzer 1985).

The ability to detect variation at the DNA level, e.g., mutations, formed the basis for the successful elucidation of the molecular mechanisms underlying many monogenetic disorders, which can entail more or less specific behavioural phenotypes. Examples are shown in Table 1. Frequently, such variation is detectable in exons of genes underlying monogenetic disorders and simplistically either entails that the respective gene product is structurally altered or not formed at all. For example, missense mutations entail the substitution of the regular amino acid at a specific position of the protein with another; this alteration of the amino acid sequence of the respective protein can have functional implications at the levels of the cell, tissue and organism. Two monogenetic disorders, which can cause autism, tuberous sclerosis (TSC1-/TSC2-gene) and fragile X syndrome, have been studied thoroughly, and due to the elucidation of the cellular pathways affected by a lack of hamartin or tuberin (tuberous sclerosis) or the FMR1 proteine (fragile X syndrome), new treatment options have been developed which are currently tested in randomised controlled trials (Ehninger et al. 2009; Hagerman et al. 2009).

The majority of DSM-IV TR psychiatric disorders are most likely complex implying that they only infrequently result from single gene mutations. Instead, it is assumed that several gene variants interact in a complex manner with environmental factors to produce the phenotype (Kendler 2005). Complex disorders typically show higher concordance rates in monozygotic than dizygotic twins; concordance rates in monozygotic twins are typically below 1 implying that environmental factors play a role in the manifestation of the disorder. Family studies have shown that complex psychiatric disorders are characterised by elevated recurrence risks in first and second degree relatives which are below those expected for classical monogenetic dominant or recessive traits. Moreover, for the complex psychiatric disorders a steep decline in recurrence risks is observable between first and second degree family members. In third degree relatives recurrence risks are usually only minimally elevated above population based rates of the respective disorder (Hill et al. 2006; Knoblauch 2007).

Table 1 Examples of the influence of single gene disorders on cognition and behaviour

Monogenic disorder	Mode of inheritance	Gene	Cognitive phenotype	Behavioural phenotype
Lesch-Nyhan syndrome	X-chromosomal recessive	Hypoxanthine guanine phosphoribosyltransferase (*HPRT*)	Mental retardation	Self-mutilative biting of fingers and lips
Tuberous sclerosis	Autosomal dominant	Tuberous sclerosis-1 and -2 (*TSC1/-2*)	Learning difficulties, low IQ, mental retardation	Behavioural problems, autism, autism spectrum disorder
Fragile X syndrome	X-chromosomal, triplet repeat expansion	Fragile X mental retardation protein (*FMR1*)	Mental retardation,	Autism spectrum disorder, social phobia, ADHD
Leptin deficiency	Autosomal recessive	Leptin (*lep*)		Diminished perception of food reward, decreased response to satiety signals, hyperphagia
Chorea Huntington	Autosomal dominant, triplet repeat expansion	Huntingtin	Dementia	Depression, schizophrenia, personality change, obsessive–compulsive behaviour
Phenylketonuria	Autosomal recessive	Phenylalanine hydroxylase (*PAH*)	Mental retardation	Autism
Wilson disease	Autosomal recessive	Membrane copper-transport protein (*ATP7B*)		Psychiatric symptoms (personality changes, depression, psychosis)
Neurofibromatosis	Autosomal dominant	*Neurofibromin*	Mental retardation	ADHD

Table 2 Heritability estimates of selected behaviours and developmental milestones

	Assessments	Heritability estimates	
Behaviours			
Dieting	EAT	42%	
Body dissatisfaction	EDI	52%	
Drive for thinness	EDI	44%	
Disinhibition of eating	TFEQ	40%	
Restrained eating	TFEQ	28%	
Hunger	TFEQ	28%	
Obsessive compulsive behaviour	CBCL	45–58%	
Developmental milestones			
Motor	Crawling, sitting	0–90%	
Development	Standing, walking		
		Female	*Male*
Expressive language "vocabulary"	MCD-I-R	8%	20%
"Two-word-combination-use"	MCD-I-R	28%	10%

EAT= Eating Attitudes Test; *EDI*= Eating Disorder Inventory; *MCD-I-R*= MacArthur Communicative Development Inventories-Short Form; *CBCL*= Child Behaviour Checklist; *TFEQ*= Three Factor Eating
Table adapted from Hebebrand et al. (2010)

Within child and adolescent psychiatry, most twin, family and adoption studies have been carried out for attention-deficit/hyperactivity disorder (ADHD) (Freitag et al. 2010a). ADHD is also noteworthy because large population-based twin studies have analysed quantitative dimensions of the disorder; complex gene-environment twin studies have also been performed. For many behavioural traits and developmental milestones heritability estimates based on categorical or dimensional (quantitative) data indicate that overall approximately half of the variance is explained by genetic factors (Plomin 2005), the other half by the environment (Table 2). For many behavioural traits the environmental influences are predominantly unique effects that explain differences for familial traits; with some exceptions, such as conduct problems, where the familial environment also plays an important role. It is worth noting that in the absence of direct evidence for the impact of the familial environment on a trait, there may still be a role for such influences through gene–environment interactions, which are part of the heritable component in genetic model fitting studies.

For most psychiatric disorders, which are usually assessed categorically, genetic factors have also been shown to play an important role (Table 3). Heritability estimates typically exceed 0.5. ADHD and autistic disorders have been shown to be one of the most highly heritable child and adolescent psychiatric disorders (Asherson et al. 2005; Freitag et al. 2010a, b; Hebebrand et al. 2010). Knowledge of the magnitude of the genetic basis of a particular disorder is valuable for interpreting psychiatric findings within a patient's family and for probing for specific disorders in relatives of the index patient.

Until recently, candidate gene and linkage studies dominated the attempts to uncover genetic variation underlying such complex disorders. However, viewed in

Table 3 Heritability estimates of selected psychiatric disorders

Disorder	Heritability estimates	Reference
PDD	70–90%	Freitag (2007)
Enuresis	67–70%	Von Gontard et al. (2001)
Conduct Disorder	53%	Gelhorn et al. (2005)
OCD	47%	Bolton et al. (2007)
Anxiety Disorders	30–40%	Eley et al. (2003)
ADHD	70–80%	Faraone et al. (2005)
Anorexia Nervosa	48–88%	Hinney et al. (2010)
Bulimia Nervosa	28–83%	Hinney et al. (2010)
Schizophrenia	73–90%	Sullivan et al. (2003)
Bipolar Disorder	85%	McGuffin et al. (2003)
Major Depression	31–42%	Sullivan et al. (2000)

Table adapted from Hebebrand et al. (2010)
OCD= Obsessive Compulsive Disorder; *PDD*= Pervasive Developmental Disorders (including autistic disorder, Asperger disorder, disintegrative disorder and PDD Not Otherwise Specified); *ADHD*= Attention-Deficit and Hyperactivity Disorder

retrospect it can be concluded that these large scaled efforts were largely unsuccessful; progress in the molecular dissection of complex psychiatric phenotypes proved to be exceedingly slow for a period of over 20 years (Burmeister et al. 2008). Candidate gene studies could only be performed for those genes for which an a priori hypothesis existed as to their relevance for the respective disorder; obviously for each disorder this represented only a very limited number in light of the totally known number of human genes. In addition, candidate gene studies in different psychiatric disorders frequently focussed on the same set of genes of a particular neurotransmitter system, such as dopamine and serotonin transporters and receptors. In other words, the candidate gene studies reflected the paucity of hypotheses as to the underlying pathways involved in complex psychiatric disorders.

In 2006, the first genome-wide association studies (GWAS) based on DNA chip technology were introduced (Hardy et al. 2009). Whereas it is still too early to judge the total insight that this novel technology will provide into the pathogenesis of complex disorders, we can nevertheless already conclude that GWAS have entailed a paradigm shift, thus justifying the nomination as 'breakthrough of the year' by Science magazine in 2007 (Pennisi 2007). For many complex somatic and neuropsychiatric disorders, novel genes have been detected which provide initial insights into frequently unknown pathways involved in the respective disorders. For many disorders, different groups pooled their data to come up with several thousand cases and controls, such numbers had almost never been analysed in the pre-GWAS era. A major finding has been that the effect sizes of validated trait or disease-related SNPs are in most cases extremely small. According to a recent synopsis (Hindorff et al. 2009) the median odds ratio was 1.33 with an interquartile range of 1.2–1.61, although more recent evidence for very large-scale studies of traits such as height and body mass index in samples of 100,000–200,000 individuals, indicate effect sizes equivalent to odds ratios of 1.1 or less, with hundreds or risk alleles involved (Speliotes et al. 2010).

Developmental aspects represent a key feature of child and adolescent psychiatry. The unfolding of gene expression provides virtually all of the information necessary to guide the orderly succession of events underlying the development of any organism and the central nervous system in particular. A behavioural trait or the symptoms of any given mental disorder are more uniform for a specific developmental stage than across all of infancy, childhood and adolescence; in addition, comorbidity is dependent on developmental stage. Many traits and disorders must be viewed in the context of brain development, elucidation of the underlying molecular mechanisms will contribute to the identification of genes involved in normal development of the central nervous system and its function. Dyslexia genes, which are involved in global brain-development processes such as neural migration and axonal guidance represent just one such example (Scerri et al. 2010).

Specific disorders run their course during specific developmental phases. Examples include nocturnal enuresis which at age 7 affects approximately 10% and at age 18 only 1% (von Gontard et al. 2001). The reduction of hyperactivity in ADHD during adolescence is another example (Kessler et al. 2005). Both anorexia and bulimia nervosa rarely start in childhood and only infrequently persist beyond 30 years of age (Holtkamp et al. 2005); and in Tourette's disorder both comorbid disorders and the development of the tics show age-related patterns. It appears probable that alterations in expression levels of specific genes partially account for disorder specific manifestation ages and the symptom development over time.

2 Selected Child Psychiatric Disorders with a Strong Genetic Component

Genetic findings in four exemplary child psychiatric disorders, ASD, ADHD, nocturnal enuresis (NE) and obesity will be presented in more detail in the second part of this chapter.

2.1 Autism Spectrum Disorders

ASD are characterised by social interaction and communication difficulties as well as stereotyped behaviour and special interests. ASD show a heritability of around 80–90% (Freitag 2007; Lichtenstein et al. 2010). *Linkage* has been found in at least two independent studies in regions 2q21–33, 3q25–27, 3p25, 4q32, 6q14–21, 7q22, 7q31–36, 11p12–13, 17q11–21 (Alarcon et al. 2008; Duvall et al. 2007; Freitag 2007; Ma et al. 2007; Schellenberg et al. 2006; Szatmari et al. 2007). A meta-analysis confirmed the region 7q22–32, and reported suggestive evidence for linkage to 10p12–q11.1 and 17p11.2–q12 (Trikalinos et al. 2006).

Following linkage studies, genes in replicated linkage regions have been screened for *mutations or common variants* increasing the risk for ASDs. In the following overview on *candidate gene association studies*, only replicated findings of genes in linked regions will be presented.

Chromosome 2

Different common variants in the mitochondrial aspartate/glutamate carrier (*SLC25A12*) gene located on 2q24 were associated with ASD in several studies (Freitag 2007; Silverman et al. 2008).

Chromosome 3

Oxytocin plays a crucial role in social cognition and behaviour (Donaldson et al. 2008) and the oxytocin receptor gene (*OXTR*) is located under the linkage peak on 3p25. Several single nucleotide polymorphism (SNP) alleles and haplotypes of *OXTR* were associated with ASD (Jacob et al. 2007; Lerer et al. 2008; Wu et al. 2005).

Chromosome 6

Three studies found evidence for association of different SNP genotypes or alleles in the Glutamate receptor 6 (*GluR6*) gene under the linkage peak on chromosome 6 (Freitag 2007).

Chromosome 7

Most candidate genes assessed in ASD are located on chromosome 7q22–36, as this area is best replicated from linkage studies. Mutations or variants in the following candidate genes could not clearly be replicated as risk factors for ASD (Freitag 2007): *FOXP2* (a gene which was mutated in a severe monogenic form of speech and language impairment in one family) (Fisher et al. 1998), *RELN* (neuronal migration, formation of cortical layers, synaptogenesis), *PTPRZ1* (highly expressed during embryogenesis), *NRCAM*, *WNT2* and *HOXA1* (hindbrain development in mouse).

The *RELN* (Reelin) gene codes for a signalling protein that plays a crucial role in neuronal migration, formation of cortical layers and synaptogenesis. Three studies supported the involvement of a trinucleotide repeat polymorphism in the 5′UTR region in Reelin in ASD (Persico et al. 2001; Serajee et al. 2006; Skaar et al. 2005), whereas five other, at least similarly powered, studies did not (Bonora et al. 2003; Devlin et al. 2004; Krebs et al. 2002; Li et al. 2004; Zhang et al. 2002).

Different rare and common, possibly functional variants in *LAMB1* were associated with ASD in two studies (Bonora et al. 2005; Freitag 2007; Hutcheson et al. 2004). *LAMB1* encodes the $\beta 1$ chain of laminin, which is an important glycoprotein promoting neuronal migration and neurite outgrowth in the developing nervous system.

Different alleles of two SNPs in the Engrailed 2 (*EN2*)-gene on chromosome 7q36 were associated with ASD in several independent samples (Brune et al. 2008; Freitag 2007; Wang et al. 2008; Yang et al. 2008). *EN2* is a homeobox transcription factor which plays a role during the development of cerebellar and brainstem functions.

Common variants as well as rare mutations in the contactin-associated protein-like 2 (*CNTNAP2*), a member of the neurexin superfamily involved in cell-adhesion and neuronal migration (Alarcon et al. 2008; Arking et al. 2008; Bakkaloglu et al. 2008) also increased the risk for ASD in several independent samples. Gene-expression analyses in the developing human brain identified *CNTNAP2* as enriched in circuits important for language development.

The gene encoding the pleiotropic MET receptor tyrosine kinase plays a role in brain development and gastrointestinal repair. As some individuals with ASD suffer from gastrointestinal symptoms (GIS), this gene was specifically assessed in individuals with ASD and GIS. A functional promotor variant, several other SNP genotypes or alleles as well as rare mutations were associated with ASD in several independent samples, predominantly in individuals with GIS (Campbell et al. 2006, 2009; Sousa et al. 2009). In a post-mortem brain protein expression analysis, ASD individuals showed lower levels of the MET protein compared to controls (Campbell et al. 2007).

Chromosome 17

Due to findings of platelet hyperserotonemia in children with autism and their first-degree relatives, common variants in the Serotonin-transporter gene (*SLC6A4*) were assessed by several studies (Freitag 2007). A recent meta-analysis, however, did not report an effect of 5-HTTLPR and STin2 alleles on ASD risk (Huang et al. 2008).

In addition to association of common variants, co-occurrence of ASD and *single gene disorders* has been observed for a long time. The most prevalent single gene disorders in ASD are tuberous sclerosis (*TSC1/TSC2*; around 1%) and fragile X syndrome (around 3–5%). More rare ($\ll$1%), but medically treatable single gene disorders are phenylketonuria (PKU) and Smith-Lemli-Opitz (SLO) syndrome caused by mutations in 7-dehydrocholestrol reductase (*DHCR7*) (Baieli et al. 2003; Freitag 2007; Sikora et al. 2006). The rate of ASD in these disorders also is increased, but ASD is not observed in all individuals carrying the mutation: fragile X syndrome ca. 25% (males), tuberous sclerosis ca. 20%; SLO ca. 50%; PKU ca. 10% (Abrahams et al. 2008). Mutation screening and subsequent association studies have elucidated other rare causes of (most likely) single gene disorders in individuals with ASD. Despite rare positive linkage findings for loci on the X-chromosome, several variants in genes on the X-chromosome were assessed for association with ASD, as the sex distribution in ASD is markedly skewed (male:female = 4:1). Two neuroligin genes on Xq13 and Xp22 were screened for mutations in several studies. Neuroligins are essential components of synaptogenesis. Despite the findings of several non-conservative mutations in single families in *NLGN3* and *NLGN4* these could not be replicated in larger samples of individuals with ASD (Freitag 2007). Similarly, two mutations identified in the ribosomal protein gene *RPL10* on Xq28 were not replicated in a subsequent study (Gong et al. 2009; Klauck et al. 2006).

For a long-time suspected causes of heterogeneity, ASD are *chromosomal abnormalities* which can be observed by standard karyotyping (Vorstman et al. 2006b). With a rate of approximately 1%, the most prevalent cytogenetic abnormalities are observed on chromosome 15q11-13 (Depienne et al. 2009; Freitag 2007) in most cases maternally, but also paternally inherited duplications. Further, relatively frequent findings are deletions of chromosome 2q37, chromosome 7q31 and deletions or duplications of chromosome 22q13. In addition, Klinefelter Syndrome (XXY) as well as duplications of the Williams-Beuren Syndrome region 7q11.23 and deletions of 22q11 (Velo-cardio-facial syndrome) are associated with increased autistic traits (Berg et al. 2007; van Rijn et al. 2008; Vorstman et al. 2006a).

Due to technological advances, it is now possible to also assess small cytogenetic abnormalities ('copy number variations' CNVs) not detected by standard karyotyping (Henrichsen et al. 2009). Recent publications show that some–especially rare–CNVs are observed more frequently in ASD patients compared to control subjects (Bucan et al. 2009; Glessner et al. 2009; Kumar et al. 2008; Marshall et al. 2008; Mefford et al. 2008; Pinto et al. 2010; Sebat et al. 2007; Weiss et al. 2008). Similar to the cytogenetic findings obtained by standard karyotyping, most CNVs represent rare, unique events rather than representing recurrent deletions or duplications. *Replicated rare CNVs from genome-wide studies*, which were observed more frequently in ASD compared to control individuals, are located on the following chromosomes: 1q21, 2p16.3 (*NRXN1*), 3p25–26 (*CNTN4*), 7q36.2 (*DPP6*), 15q11–13 (*UBE3A*, *OR4M2*, *OR4N4*); 16p11.2 (*MAPK3*, *MAZ*, *DOC2A*, *SEZ6L2*, *HIRIP3*, *IL6*); 22q11.2; X (*DDX51-PTCHD1*; maternally inherited). Some of these CNVs were observed also more frequently in individuals with mental retardation or schizophrenia than in controls. ASD specific CNVs were not exclusively observed in ASD individuals with specific dysmorphic features or mental retardation but were also present in high functioning patients with autism with only minor dysmorphology (van der Zwaag et al. 2009).

Molecular genetic studies in ASD have come a long way from the early linkage studies, which aimed at describing a few loci and subsequently finding one or a few genes of major effect relevant for all cases of ASD. It has now become clear that ASD are heterogeneous disorders, caused by several rare—most likely—monogenetic disorders (as fragile X syndrome, mutations in *TSC1/TSC2*, *LAMB1*, *CNTNAP2*, *PTEN*, *DHCR7*, *SHANK3*, *NLGN3/4* or *RPL10*). In addition, 'contiguous gene syndromes' are likely causes of ASD, as the overall rate of rare CNVs and large chromosomal deletions, duplications and translocations is increased in individuals with ASD compared to controls.

Common variants, on the other hand, may shape the phenotype or eventually may lead to the disorder by interacting with rare mutations or rare CNVs. A mechanism like this has been shown for *PTEN* haploinsufficient individuals.

The serotonin-transporter gene *SLC6A4* has been discussed as both an ASD susceptibility gene and a second-site modifier in ASD (Bartlett et al. 2005; Hessl et al. 2008). A study in *PTEN* haploinsufficient mice (Page et al. 2009) demonstrated that the phenotypes of these mice were modified in an additive fashion by *SLC6A4* haploinsufficiency. In addition, the role of *PTEN* in the maintenance of genomic stability (Shen et al. 2007; Stiles 2009) makes it likely that *PTEN* haploinsufficiency may increase the probability of a secondary modifying event, such as a copy number variation in a chromosomal region relevant to ASD. Common variants also might increase the risk for autistic traits in the general population as well as for less severe autistic disorders as Asperger Syndrome or PDD-NOS.

From results of current genetic findings in ASD, it is likely that mutations or common variants in genes coding for gene products involved in (1) cell–cell interaction and synaptic function, including development of dendritic spines, (2) neuronal migration and growth or (3) excitatory and inhibitory neurotransmission are causes of ASD. The pathway influencing cell–cell interaction and synaptic function includes *NRXNs*, *NLNGs*, *CNTN3/4*, *CNTNAP2* and *SHANK3*. In addition, the FMR protein, which is missing in fragile X syndrome, modulates dendritic spine formation and synaptic plasticity by inhibiting mGluR1/5-mediated dendritic protein synthesis (Hagerman et al. 2009). Neuronal migration and growth are influenced by gene products of *LAMB1*, *EN2* or the MET receptor tyrosine kinase gene. The mTor/PI3-kinase (PI3 K) pathway involves *PTEN*, *TSC1/2* and several other genes, which were observed in rare CNVs in individuals with ASD (Cusco et al. 2009). It strongly influences (neuronal) cell growth. Gene products influencing the regulation of excitatory and inhibitory neurotransmission are GABA and glutamate receptors. In addition, disbalance of excitatory and inhibitory neurotransmission was also observed in fragile X syndrome. Clearly, this list of possibly involved pathways is not exhaustive, and other mechanisms or pathways may emerge as results of further studies will be published.

2.2 Attention-Deficit/Hyperactivity Disorder

ADHD is characterised by age-inappropriate hyperactivity, impulsivity and attention problems (American Psychiatric Association 2000). The disorder usually starts in early childhood and persists into adults in around two-thirds of cases (Faraone et al. 2006). It is thought to be caused by the interplay of genetic and environmental risk factors and shows a heritability of 60–80% (Freitag et al. 2010a). A recent meta-analysis on twin studies in ADHD has reported additive and dominant genetic as well as non-shared environmental effects in ADHD (Burt 2009), a finding, which was discussed and challenged by a comment, emphasising the role of appropriate data transformation and analysis as well as the limitations of models obtained on twins reared together, possibly underestimating common environmental effects obtained from twin studies on ADHD (Wood 2010). Nevertheless, based on current data common (familial) environment is not thought to play an important role in ADHD.

The genetic effects on the development of ADHD throughout the life span are not well understood. The heritability of self-rated ADHD symptoms in adults is far lower than that reported for children and adolescents, based on parent and teacher reports; being around 30% (Boomsma et al. 2010) compared to the average heritability in children of around 76% (Faraone et al. 2005). This may be a measurement artefact, based on self versus informant ratings, and confounding with adult onset disorder that might mimic some of the symptoms of ADHD, or might reflect a growing impact of unique environmental effects on the long-term course of ADHD into adulthood. Further work is needed to clarify this important question.

In terms of the genetic architecture, a segregation study resulted in a model implicating non-Mendelian major gene effects with low penetrance in ADHD (Maher et al. 1999). However, these findings are partly in contrast to the observation of twin studies, which point to a model implying multiple common genetic variants of small effect in the aetiology of the disorder (Levy et al. 1997).

Results of hypothesis free molecular genetic studies are currently supportive of both models, too, indicating clear genetic heterogeneity in ADHD. Linkage studies elicited several loci likely containing rare genetic variants of major effect only present in single large families, but also common variants of smaller effect present across several families (Romanos et al. 2008). In addition, genome-wide association studies indicate a very limited number of common variants of small effect to date (Franke et al. 2009; McCarthy et al. 2008) and rare copy number variants that confer greater (moderate) levels of risk (Elia et al. 2010; Williams et al. 2010). We therefore conclude that on the basis of current data, both types of genetic effects likely play a role with evidence for both common risk alleles of minor effect, and variants that are individually rare in the population but have a larger impact on risk for the disorder. As we shall see much more work is needed to clarify the balance between these two types of genetic effects and it remains feasible that at least some of the risk for ADHD is conferred by many individually rare alleles of minor effect; which would be extremely difficult to detect.

Another question that can be addressed by quantitative genetic approaches is the aetiological relationship between inattention and hyperactivity-impulsivity. Twin studies that have investigated this have found that these are separable traits with only partially overlapping genetic influences. While the correlation between the two domains is largely accounted for by shared genetic effects there are in additional unique genetic effects acting on each of the two domains (McLoughlin et al. 2007). Genetic investigation of the two domains would therefore be expected to provide some evidence for unique as well as shared aetiological pathways.

In support of this general conclusion we see differential relationships between each of the two domains and comorbid traits. One example is reading disability (RD) and inattention where the correlation between ADHD and RD was found to be largely driven by genetic factors not shared with hyperactivity-impulsivity (Paloyelis et al. 2010). A contrasting example is a recent study of ADHD and oppositional behaviour that found very highly phenotypic and genetic correlations with hyperactivity-impulsivity but much lower with inattention

(Wood et al. 2009). These data suggested that in middle childhood ADHD and conduct problems may represent the same underlying liability, distinct from the inattentive domain.

Twin studies have found that there are many other examples of overlapping genetic influences between ADHD symptoms and comorbid disorders and traits including autism spectrum disorder (Ronald et al. 2008), motor coordination (Francks et al. 2003), conduct disorder (Thapar et al. 2001) and depression (Cole et al. 2009) among others. These studies demonstrate the complexity of shared and unique genetic and other aetiological influences among many mental health disorders and comorbid traits. At a basic level further work using longitudinal twin designs can further delineate the causal relationships between ADHD and co-occurring traits which include pleiotropic effects (multiple effects of individual sets of genes), mediating effects (brain functions or behaviours that mediate genetic effects on ADHD) and risk models (one disorder leading to another).

Dopaminergic Candidate Gene Studies

Molecular genetic studies on ADHD started with candidate gene association studies in the mid 1990s with the first two reported associations between repeat length polymorphisms in the dopamine D4 receptor (*DRD4*) (LaHoste et al. 1996; Swanson et al. 1998a, b) and dopamine transporter (*DAT1*) genes (Cook et al. 1995). Subsequently association was reported with a microsatellite marker near to the dopamine D5 receptor gene (*DRD5*) (Lowe et al. 2004). Since then there have been numerous further studies with relatively few independent replications. However, meta-analysis of available data reports strong evidence for the association of dopamine system genes, in particular with the 7-repeat allele of a variable number tandem repeat (VNTR) polymorphism within *DRD4* and the microsatellite which lies upstream of *DRD5*. These two associations are significant because they are as yet the only genetic association findings to reach genome-wide levels of significance, in the meta-analytic study of Li et al. (2006), which is generally accepted as the gold standard for determining that a genetic association is true. The recommended level of significance, in the region of 5×10^{-8} (Dudbridge et al. 2008) is important because it means that in a systematic screen of the genome for association, these findings would occur less than 5% of the time by chance alone; whereas conventional levels of significance such as 0.05 would be detected by chance every 20th selection of a random independent genetic polymorphism, of which there are hundreds of thousands across the genome. We can therefore be confident that these two findings are truly associated with ADHD.

The other genetic association that is often cited in the previous literature is between the 10-repeat allele of a VNTR within the 3′-UTR of the dopamine transporter gene (*DAT1*). However, the evidence for this association from meta-analytic studies is far from convincing with three negative reports (Li et al. 2006; Maher et al. 2002; Purper-Ouakil et al. 2005) while three other reports found only weak evidence of association (Faraone et al. 2005; Gizer et al. 2009; Yang et al. 2007) far below the stringent criteria reached by *DRD4* and *DRD5*. Whether *DAT1* is associated with ADHD therefore remains a contentious question, yet there is evidence for heterogeneity across the datasets included in the meta-analytic studies suggesting that there may be identifiable sources for differences in the strength of the association across the various studies (Gizer et al. 2009; Li et al. 2006).

One potential source of error in the analysis of *DAT1* comes from undetected genotyping errors that cause apparent over transmission of common alleles when using the transmission disequilibrium test (TDT). Mitchell and colleagues found that in studies reporting a positive association using the TDT, 87% identified an association with alleles that were present at a frequency of >50%, whereas in case–control studies, the proportion of common allele associations was 40% (Mitchell et al. 2003). This highly significant difference could only be explained by systematic genotype error related to the method, especially with high frequency alleles such as the *DAT1* 10-repeat which occurs on around 70% of chromosomes from people of European ancestry.

There are, however, more interesting sources of heterogeneity for *DAT1*. First, recent studies of ADHD in adults find evidence for association with the 9-repeat rather than the more common 10-repeat allele, suggesting that the effects of *DAT1* may differ depending on the developmental age, indicating a role for *DAT1* in the modulation of symptoms at different ages, rather than the aetiology of ADHD (Franke et al. 2008). Second, the *DAT1* VNTR has been found to interact with measures of the pre-natal environment including maternal use of tobacco during pregnancy on oppositional and hyperactive-impulsive symptoms (Becker et al. 2008; Kahn et al. 2003) and maternal use of alcohol during pregnancy (Brookes et al. 2006). Interestingly, recent research has shown that when genetic factors are controlled for studies of the association mothers smoking during pregnancy and ADHD, the association appears to be accounted for mainly genetic factors, suggesting that the environmental measure may be correlated with maternal genes (Thapar et al. 2009); suggesting that the observed interactions with *DAT1* may be explained by gene–gene interactions.

Third, there is evidence that two or more haplotypes of *DAT1* may be involved in the association with ADHD, indicating that more than one genetic variant is likely to be involved. Further, the known VNTR may not be the primary functional variant but may tag other genetic variants; and there appears to be more than associated region with the gene (Asherson et al. 2007; Brookes et al. 2006, 2008). Strength of the association would then depend on the relationships between the various functional genetic variants involved, which may differ across sample populations. Fourth, there is evidence that the association with genetic variants across *DAT1* may be specific to non-comorbid ADHD and is not found in children with ADHD plus conduct disorder (Zhou et al. 2008a).

Other Candidate Gene Association Studies

There have now been numerous candidate gene association studies in ADHD and these are best summarised in the recent review and meta-analysis from Gizer et al. (2009). They conducted a systematic review of the association literature with the aim of identifying the most consistent findings in ADHD research to date. They further tested for heterogeneity across studies because other genes, like *DAT1*, might show significant sources of variability due to identifiable moderating factors. Overall they concluded that the following genes were significantly associated with ADHD: *DAT1*, *DRD4*, *DRD5*, *5HTT* (serotonin transporter), *HTR1B* (serotonin 1B receptor) and *SNAP-25* (synaptosomal associated protein); and the following genes showed significant evidence of heterogeneity: *DAT1*, *DRD4*, *DRD5*, *DBH*, *ADRA2A* (adrenergic 2A receptor), *5HTT*, *TPH2* (Tryptophan hydroxylase 2), *MAOA* (monoamine oxidase A) and *SNAP25*. For the candidate genes showing heterogeneity future studies could usefully investigate the potential moderating factors that lead to variability in effect size across studies.

Despite significant progress, the total estimated impact of the most replicated candidate gene findings is around 3.3% of the phenotypic variance of ADHD; accounting for 4.3% of the estimated average heritability of ADHD of 76% (Kuntsi et al. 2006). Further work is clearly needed to explain the rest of the genetic influences on ADHD.

Genome-Wide Association Studies (GWAS)

Other approaches to the identification of novel genes for ADHD include genome-wide association studies which have the potential to detect entirely novel associations where there is no previous a priory hypothesis. Given that we know so little about the function of the brain and how molecular processes lead to susceptibility for ADHD, it makes sense for aetiological research to focus efforts on empirical approaches that systematically screen the entire genome for associations.

The first GWAS study of ADHD investigated 438,784 SNPs in 958 combined type ADHD proband–parent trios. No genes of moderate to large effect were identified and no findings passed genome-wide levels of significance (Neale et al. 2008). However, when a set of 51 candidate genes were investigated, there was significant evidence at the group level that positive association signals were emerging from the selected SNPs, implicating mainly dopamine, noradrenalin and serotonin neurotransmitter genes. Similar findings were subsequently reported in a study that combined genome-wide association data from several studies, with a total sample size of 2,064 trios, 896 cases and 2,455 controls (Neale et al. 2010), confirming the important role of the traditional neurotransmitter systems genes in ADHD.

Overall, the total number of samples analysed in GWAS studies of ADHD to date is far too small and the disappointing findings to date are not unexpected. Further work is required to obtain GWAS information on much larger sets of samples for ADHD. There are, however, potentially interesting new findings that emerge from the GWAS studies of ADHD (Franke et al. 2009). Of particular interest is the Cadherin gene (*CDH13*) which was found to be associated with ADHD in two out of five GWAS studies (Lasky-Su et al. 2008; Lesch et al. 2008) and lies within the only region that reached genome-wide significance in a meta-analysis of linkage studies of ADHD (Zhou et al. 2008b). Furthermore *CDH13* is among the most consistent findings on a wide range variety of phenotypes related to drug abuse and dependence (Uhl et al. 2008). This finding and other hints from GWAS indicate that genes involved in cell division, cell adhesion, neuronal migration and neuronal plasticity may also confer risk for ADHD (Franke et al. 2009).

Overall there is a long way to go to delineate the specific genetic factors that explain the high heritability of the disorder. This is, however, a common phenomenon in common disorders research and several possible explanations for the so-called 'dark-matter' of (missing) heritability has been put forward—dark matter in the sense that we know it exists can detect its influences but simply cannot see it (Manolio et al. 2009). Potential reasons include numerous genes of very small effect, genetic heterogeneity with risk conferred by many different genes and variants within genes, higher order interactions between genes and with environment and aetiological heterogeneity (Manolio et al. 2009).

Copy Number Variations

An important contribution to ADHD risk is rare copy number variants (CNVs) or other types of rare genetic variation. Recent data suggested that in a few cases CNVs may exert a major influence on risk for ADHD and that this risk overlaps with other neurodevelopmental disorders, including schizophrenia and autism (Elia et al. 2010). These initial findings have now been confirmed in one of the most robust studies of ADHD genetics to data. Thapar and colleagues (Williams et al. 2010) investigated a sample of 410 children with ADHD and 1,156 ethically matched controls. The analysis focused on large CNVs greater than 500,000 base pairs in size, that are robust to call from genome-wide SNP arrays, and further validated all identified CNVs using a separate technology. Furthermore, findings were evaluated in a further sample of 825 patients with ADHD from Iceland and 35,243 controls. There were several key findings from this paper. First, the overall burden of risk was increased, with 16% of ADHD cases carrying CNVs, compared to 8% of controls, giving an overall twofold risk. Second, the increased risk was particularly striking for a subgroup of ADHD cases with intellectual disability: 42%, giving an approximate sixfold risk compared to the control population. Third, that the burden of risk remained significant, but much lower, in the group with no intellectual impairment, around 1.7-fold risk. Fourth, there was a clear excess on chromosome 16p13.11, identifying a specific region. Finally, the identified CNVs were significantly enriched for loci previously reported for schizophrenia and autism.

Linkage Analysis and Genes of Major Effect

There is evidence that genes of major effect might lead to ADHD in a few rare families. Furthermore, variation in some of these genes may have more general effects on ADHD. There are two main examples.

The first is a gene called latrophilin 3 (*LPHN3*) that is located within a linkage region on chromosome 4q13.2. This region was first identified in a large multi-generational pedigrees from the Paisa population in Columbia (Arcos-Burgos et al. 2004). Subsequent fine-mapping narrowed the region. Subsequently genetic association analysis in worldwide samples totalling 2,627 ADHD cases and 2,531 controls confirmed the association of *LPHN3* with an average odds ratio of 12, although not yet at genome-wide levels of significance (Arcos-Burgos et al. 2010). Since then further replication was reported in a sample of adults with ADHD, which is interesting because the original extended pedigrees used to identify the linked regions included both children and adults with ADHD (Ribases et al. 2010).

The second example comes may represent a private mutation found in one or a few families in Germany (Lesch et al. 2010). To detect micro-deletions and micro-duplications, they carried out a genome-wide screen for copy number variations in a sample of 99 children and adolescents with severe ADHD. Among several genes containing CNVs in ADHD, they identified the gene encoding neuropeptide Y which was included within a 3 megabase region on chromosome 7p15.2–15.3. Investigation of other family members found evidence that the duplication co-segregated with ADHD.

Conclusion

Overall current molecular genetic studies of ADHD find evidence for both frequent and rare variants that influence the risk for ADHD. The level of evidence is sufficient to confirm that identity of some of the genes and gene regions involved, although the data only explains a very small amount of the total genetic variation that impacts on risk for ADHD. Future studies will need to include very much larger sample sizes and take advantage of the improving technology to screen for both common and rare variants. The specific findings confirm the old neurotransmitter hypotheses of ADHD, but also extend the focus to many other neurobiological processes that impact on brain development and brain function.

2.3 Nocturnal Enuresis

Nocturnal enuresis is characterised by bed wetting after age 5 years old in otherwise healthy children after exclusion of urinary tract infections, malformations, other organic causes or intellectual disability (Von Gontard et al. 1997). Population based twin studies have resulted in heritability estimates of around 70% for NE (Hublin et al. 1998). Despite this relatively large heritability, there is a scarcity of research into NE. Linkage studies have predominantly been performed in large families with autosomal dominant inheritance pattern replicating loci on 4p16.1, 12q24.2, 13q13–q14.3, and 22q11.21 (Arnell et al. 1997; Eiberg et al. 1995, 2001; Eiberg 1995, 1998; Loeys et al. 2002; Von Gontard et al. 1999). Hardly any fine-mapping or candidate gene association studies were performed. Given that nocturnal enuresis can be a persistent disorder more research should be done to elucidate its genetic basis, e.g., by sequencing the replicated segregating loci in the respective families. Interestingly, a recent study reported higher rates of nocturnal enuresis in individuals with schizophrenia, and individuals with enuresis showed a worse performance on two frontal lobe cognitive tests compared to individuals without (Hyde et al. 2008).

2.4 Obesity

Numerous twin, adoption and family studies have allowed for the calculation of heritability estimates for body mass index (BMI), which typically range from 0.4 to 0.7; twin studies have consistently led to higher estimates. Based on complex statistical modelling single family studies have nevertheless also led to estimates in the range of 0.7 (Maes et al. 1997). A recent review of 13 longitudinal studies found a strong genetic continuity in BMI from early childhood to onset of adulthood (Silventoinen and Kaprio 2009). There was also evidence for an effect of a common environment on the tracking of BMI during childhood; in adults the influence of the non-shared environment substantially exceeded that of the shared environment.

Monogenic Obesity

The cloning and initial detection of mutations in the mouse and human (Montague et al. 1997; Zhang et al. 1994) leptin gene marked the beginning of the large-scale elucidation of DNA variation underlying inter-individual differences in body weight. Currently, several types of monogenic forms of obesity have been elucidated. All of the respective mutations are rare, some have only been reported in single cases worldwide; the genes are all expressed in the hypothalamus. The elucidation of such mutations and the successful treatment of leptin deficient extremely obese (Farooqi et al. 1999) individuals has firmly established that the behavioural phenotype hyperphagia/overeating and as a consequence obesity can be caused by mutations in specific genes.

Several of the detected novel monogenic forms of obesity encompass other clinically recognisable features. For instance, mutations in the leptin gene entail hypothalamic hypogonadism, other endocrinological abnormalities and immune system dysfunction (Hinney et al. 2008). In the original report of the clinical features of patients with mutations in the pro-opiomelanocortin gene (*POMC*) adrenal insufficiency and red hair were characteristic features in addition to extreme obesity (Krude et al. 1998). Developmental delay was shown to co-occur with severe obesity in a patient with a mutation in the neurotrophin receptor TrkB (Yeo et al. 2004).

The detection of mutations in the melanocortin-4 receptor gene (*MC4R*) (Hinney et al. 1999; Vaisse et al. 1998; Yeo et al. 1998) marked a turning point in the sense that for the first time the molecular basis of a subgroup of patients with idiopathic obesity was identified. The detection of the association between *MC4R* mutations and obesity is also noteworthy because the combined frequency of all functionally relevant mutations is in the range of 1–6% among extremely obese children and adolescents (Hinney et al. 2003).

Two family-based studies have attempted to quantify the effect of functionally relevant *MC4R* mutations (Dempfle et al. 2004; Stutzmann et al. 2008) by comparing BMI of mutation and wild-type carriers of relatives ascertained via the index cases. According to Dempfle et al. (2004) and Stutzmann et al. (2008) adult male carriers were 4 and 4.3 kg/m^2 heavier than their male relatives with the wildtype genotype. In females the mean effect size of *MC4R* mutations is larger; the corresponding values were 9.5 and 8.7 kg/m^2.

In the pre-GWAS era, the *MC4R* was the first gene shown to harbour a common variant with a small effect size (Geller et al. 2004). The association of the minor allele, coding for Ile103, with leanness was confirmed in subsequent larger meta-analyses with up to almost 124,000 probands (Heid et al. 2008; Speliotes et al. 2010; Wang et al. 2010a; Young et al. 2007). Roughly 3–6% of different populations are heterozygous for this variant, which according to German population-based data for adults entails a 0.5 kg/m^2 reduction in mean BMI. This would qualify this SNP as the one with the strongest effect size of all currently known common variants including the SNPs in intron 1 of the fat mass and obesity associated gene (*FTO*) (Frayling et al. 2007). A second coding *MC4R* variant also protects from obesity (Stutzmann et al. 2007). The *MC4R* is thus currently unique in that it harbours variants of considerably different effect sizes both within and outside the coding region whose minor alleles both increase and decrease BMI.

There are over 90 known coding variants in the *MC4R* that lead to a reduced receptor function seemingly support the common disease—rare variant hypothesis, according to which a substantial proportion of the predisposition to complex disorders is due to rare variants (Bodmer et al. 2008). Indeed, it seems likely to assume that other such loci exist in the genome that cannot readily be picked up via linkage or association analyses due to locus heterogeneity and the low overall combined rate of such mutations at a specific gene locus.

Polygenic Obesity

The results of genome-wide association studies (GWAS) indicate that the common disease—common variant hypothesis is potentially also important. Prior to the most recent meta-analysis of GWAS, ten loci with genome-wide significance defined via a *p*-value $<5 \times 10^{-8}$ had been identified (*FTO*, *MC4R*, *TMEM18*, *GNPDA2*, *BDNF*, *NEGR1*, *SH2B1*, *ETV5*, *MTCH2* and *KCTD15*; Hinney et al. 2009; Willer et al. 2009). The most recent and up-to-date worldwide largest meta-analysis of the GIANT (Genetic Investigation of ANthropmetric Traits) Consortium included 46 studies in total encompassing 123,865 individuals of European ancestry (Speliotes et al. 2010). In a joint analysis of the original data and the confirmatory study encompassing data of up to almost 1,25,000 individuals, a total of 32 SNPs revealed *p*-values $<5 \times 10^{-8}$.

The per allele change in BMI ranged from 0.06 (*MTCH2*, *KCTD15*, *PTBP2*, *RPL27A*, *NUDT3*) to 0.39 (*FTO*) kg/m^2; a total of 10 SNPs showed per allele changes <0.1 kg/m^2 equivalent to less than 324 and 289 g in males and females of average heights (1.8 and 1.7 m). Five SNPs with the lowest per allele change of 0.06 kg/m^2 each account for approximately one half of the aforementioned weights of individuals of average height.

The 32 identified variants explained a mere 1.5% of the BMI variance; this roughly corresponds to 3% of the genetic variance based on an estimated BMI-heritability of 0.5. Speliotes et al. (2010) estimated that there are approximately an additional 200 loci (95% CI: 98–350) with similar effect sizes as the detected 32, which together would account for roughly 3.5% of the variation in BMI or 7% of the genetic variation.

Sub-analyses based on children and adolescents revealed a substantial overlap with the genetic predisposition to adult obesity. Directionally consistent effects were found for 23 of the 32 SNPs. Common variants detected in adults are seemingly important throughout different developmental stages. In accordance with the low heritability estimates for BMI in early life, a significant effect of the *FTO* genotypes was not detectable in children prior to age 4 in two birth cohort studies (Rzehak et al. 2010).

Because most obesity genes are expressed in the central nervous system, the role of the brain in the regulation of body weight is seemingly substantiated. However, the high proportion of all genes expressed centrally needs to be considered prior to drawing firm conclusions. Pathway-based analyses based on the genes identified via the association signals revealed evidence of enrichment for pathways involved in the platelet-derived growth factor signalling, translation elongation, hormone or nuclear hormone receptor binding, homeobox transcription, regulation of cellular metabolism, neurogenesis and neuron differentiation, protein phosphorylation and numerous other pathways related to growth, metabolism, immune and neuronal processes (Speliotes et al. 2010).

The Role of Copy Number Variants in Obesity

Willer et al. (2009) reported genome-wide significant association with BMI of a SNP capturing a common 45-kb deletion near *NEGR1* (neuronal growth regulator 1 precursor gene). Recently, another common CNV (chromosome 10p11.22) was shown to be associated with BMI (Sha et al. 2009) in a Chinese sample (unadjusted *p*-value 0.011). In contrast to the common CNV with relevance for BMI reported above, three recent reports (Bochukova et al. 2010; Walters et al. 2010; Wang et al. 2010b) showed the importance of rare large CNVs for body weight regulation. Two studies (Bochukova et al. 2010; Walters et al. 2010) depicted a genomic region on chromosome 16p11.2 which harbours highly penetrant micro-deletions (about 500 kb) associated to (extreme) obesity. This region had previously been associated with autism and mental retardation (Bochukova et al. 2010; Walters et al. 2010; Weiss et al. 2008); in fact, some of the obese patients analysed by Farooqi and co-workers (Bochukova et al. 2010) additionally had developmental disorders. However, Walters and co-workers found the association of these microdeletions and obesity also in individuals ascertained only for the obese phenotype, suggesting a possible direct association of the deletions at 16p11.2 with obesity, distinct from their implications for cognition.

Wang et al. (2010b) found that CNVs larger than 1 Mb were over-represented in obese cases versus normal weight controls (non-significant odds ratio of 1.5) with a stronger effect for larger CNVs (above 2 Mb). As the effect was more pronounced for CNVs disrupting genes, these genes should be further analysed (Wang et al. 2010b).

3 Prospectus

The genetics of four childhood conditions was elaborated in more detail in this chapter. Other common psychiatric disorders of childhood, adolescence and adulthood also show a heritability >50%, especially anxiety disorders. Gene–environmental effects are likely for this condition, as it is also the case for ADHD and obesity.

Research over the last 10 years in the genetic of child psychiatric disorders has resulted in many promising findings, but also has shown that elucidating the genetic background of psychiatric disorders is challenging and will require new technologies. With sequencing approaches, methylation pattern assessment, and functional studies, new and clinically relevant results are to be expected and will ultimately result in more individualised and targeted treatment options.

References

Abrahams BS, Geschwind DH (2008) Advances in autism genetics: on the threshold of a new neurobiology. Nat Rev Genet 5:341–355

Alarcon M, Abrahams BS, Stone JL, Duvall JA, Perederiy JV, Bomar JM, Sebat J, Wigler M, Martin CL, Ledbetter DH, Nelson SF, Cantor RM, Geschwind DH (2008) Linkage, association, and gene-expression analyses identify CNTNAP2 as an autism-susceptibility gene. Am J Hum Genet 1:150–159

American Psychiatric Association (2000) Diagnostic and statistical manual of mental disorders DSM-IV-TR.

Arcos-Burgos M, Castellanos FX, Pineda D, Lopera F, Palacio JD, Palacio LG, Rapoport JL, Berg K, Bailey-Wilson JE, Muenke M (2004) Attention-deficit/hyperactivity disorder in a population isolate: linkage to loci at 4q13.2, 5q33.3, 11q22, and 17p11. Am J Hum Genet 6:998–1014

Arcos-Burgos M, Jain M, Acosta MT, Shively S, Stanescu H, Wallis D, Domene S, Velez JI, Karkera JD, Balog J, Berg K, Kleta R, Gahl WA, Roessler E, Long R, Lie J, Pineda D, Londono AC, Palacio JD, Arbelaez A, Lopera F, Elia J, Hakonarson H, Johansson S, Knappskog PM, Haavik J, Ribases M, Cormand B, Bayes M, Casas M, Ramos-Quiroga JA, Hervas A, Maher BS, Faraone SV, Seitz C, Freitag CM, Palmason H, Meyer J, Romanos M, Walitza S, Hemminger U, Warnke A, Romanos J, Renner T, Jacob C, Lesch KP, Swanson J,

Vortmeyer A, Bailey-Wilson JE, Castellanos FX, Muenke M (2010) A common variant of the latrophilin 3 gene, LPHN3, confers susceptibility to ADHD and predicts effectiveness of stimulant medication. Mol Psychiatry 11:1053–1066

Arking DE, Cutler DJ, Brune CW, Teslovich TM, West K, Ikeda M, Rea A, Guy M, Lin S, Cook EH, Chakravarti A (2008) A common genetic variant in the neurexin superfamily member CNTNAP2 increases familial risk of autism. Am J Hum Genet 1:160–164

Arnell H, Hjalmas K, Jagervall M, Lackgren G, Stenberg A, Bengtsson B, Wassen C, Emahazion T, Anneren G, Pettersson U, Sundvall M, Dahl N (1997) The genetics of primary nocturnal enuresis: inheritance and suggestion of a second major gene on chromosome 12q. J Med Genet 5:360–365

Asherson P, Kuntsi J, Taylor E (2005) Unravelling the complexity of attention-deficit hyperactivity disorder: a behavioural genomic approach. Br J Psychiatry:103–105

Asherson P, Brookes K, Franke B, Chen W, Gill M, Ebstein RP, Buitelaar J, Banaschewski T, Sonuga-Barke E, Eisenberg J, Manor I, Miranda A, Oades RD, Roeyers H, Rothenberger A, Sergeant J, Steinhausen HC, Faraone SV (2007) Confirmation that a specific haplotype of the dopamine transporter gene is associated with combined-type ADHD. Am J Psychiatry 4:674–677

Baieli S, Pavone L, Meli C, Fiumara A, Coleman M (2003) Autism and phenylketonuria. J Autism Dev Disord 2:201–204

Bakkaloglu B, O'Roak BJ, Louvi A, Gupta AR, Abelson JF, Morgan TM, Chawarska K, Klin A, Ercan-Sencicek AG, Stillman AA, Tanriover G, Abrahams BS, Duvall JA, Robbins EM, Geschwind DH, Biederer T, Gunel M, Lifton RP, State MW (2008) Molecular cytogenetic analysis and resequencing of contactin associated protein-like 2 in autism spectrum disorders. Am J Hum Genet 1:165–173

Bartlett CW, Gharani N, Millonig JH, Brzustowicz LM (2005) Three autism candidate genes: a synthesis of human genetic analysis with other disciplines. Int J Dev Neurosci 2–3:221–234

Becker K, El-Faddagh M, Schmidt MH, Esser G, Laucht M (2008) Interaction of dopamine transporter genotype with prenatal smoke exposure on ADHD symptoms. J Pediatr 2:263–269

Berg JS, Brunetti-Pierri N, Peters SU, Kang SH, Fong CT, Salamone J, Freedenberg D, Hannig VL, Prock LA, Miller DT, Raffalli P, Harris DJ, Erickson RP, Cunniff C, Clark GD, Blazo MA, Peiffer DA, Gunderson KL, Sahoo T, Patel A, Lupski JR, Beaudet AL, Cheung SW (2007) Speech delay and autism spectrum behaviors are frequently associated with duplication of the 7q11.23 Williams-Beuren syndrome region. Genet Med 7:427–441

Bochukova EG, Huang N, Keogh J, Henning E, Purmann C, Blaszczyk K, Saeed S, Hamilton-Shield J, Clayton-Smith J, O'Rahilly S, Hurles ME, Farooqi IS (2010) Large, rare chromosomal deletions associated with severe early-onset obesity. Nature 7281:666–670

Bodmer W, Bonilla C (2008) Common and rare variants in multifactorial susceptibility to common diseases. Nat Genet 6:695–701

Bolton D, Rijsdijk F, O'Connor TG, Perrin S, Eley TC (2007) Obsessive-compulsive disorder, tics and anxiety in 6-year-old twins. Psychol Med 1:39–48

Bonora E, Beyer KS, Lamb JA, Parr JR, Klauck SM, Benner A, Paolucci M, Abbott A, Ragoussis I, Poustka A, Bailey AJ, Monaco AP (2003) Analysis of reelin as a candidate gene for autism. Mol Psychiatry 10:885–892

Bonora E, Lamb JA, Barnby G, Sykes N, Moberly T, Beyer KS, Klauck SM, Poustka F, Bacchelli E, Blasi F, Maestrini E, Battaglia A, Haracopos D, Pedersen L, Isager T, Eriksen G, Viskum B, Sorensen EU, Brondum-Nielsen K, Cotterill R, Engeland H, Jonge M, Kemner C, Steggehuis K, Scherpenisse M, Rutter M, Bolton PF, Parr JR, Poustka A, Bailey AJ, Monaco AP (2005) Mutation screening and association analysis of six candidate genes for autism on chromosome 7q. Eur J Hum Genet 2:198–207

Boomsma DI, Saviouk V, Hottenga JJ, Distel MA, de Moor MH, Vink JM, Geels LM, van Beek JH, Bartels M, de Geus EJ, Willemsen G (2010) Genetic epidemiology of attention deficit hyperactivity disorder (ADHD index) in adults. PLoS One 5:e10621

Brookes KJ, Mill J, Guindalini C, Curran S, Xu X, Knight J, Chen CK, Huang YS, Sethna V, Taylor E, Chen W, Breen G, Asherson P (2006) A common haplotype of the dopamine

transporter gene associated with attention-deficit/hyperactivity disorder and interacting with maternal use of alcohol during pregnancy. Arch Gen Psychiatry 1:74–81
Brookes KJ, Xu X, Anney R, Franke B, Zhou K, Chen W, Banaschewski T, Buitelaar J, Ebstein R, Eisenberg J, Gill M, Miranda A, Oades RD, Roeyers H, Rothenberger A, Sergeant J, Sonuga-Barke E, Steinhausen HC, Taylor E, Faraone SV, Asherson P (2008) Association of ADHD with genetic variants in the 5'-region of the dopamine transporter gene: evidence for allelic heterogeneity. Am J Med Genet B Neuropsychiatr Genet 8:1519–1523
Brune CW, Korvatska E, Allen-Brady K, Cook EH Jr, Dawson G, Devlin B, Estes A, Hennelly M, Hyman SL, McMahon WM, Munson J, Rodier PM, Schellenberg GD, Stodgell CJ, Coon H (2008) Heterogeneous association between engrailed-2 and autism in the CPEA network. Am J Med Genet B Neuropsychiatr Genet 2:187–193
Bucan M, Abrahams BS, Wang K, Glessner JT, Herman EI, Sonnenblick LI, Alvarez Retuerto AI, Imielinski M, Hadley D, Bradfield JP, Kim C, Gidaya NB, Lindquist I, Hutman T, Sigman M, Kustanovich V, Lajonchere CM, Singleton A, Kim J, Wassink TH, McMahon WM, Owley T, Sweeney JA, Coon H, Nurnberger JI, Li M, Cantor RM, Minshew NJ, Sutcliffe JS, Cook EH, Dawson G, Buxbaum JD, Grant SF, Schellenberg GD, Geschwind DH, Hakonarson H (2009) Genome-wide analyses of exonic copy number variants in a family-based study point to novel autism susceptibility genes. PLoS Genet 6:e1000536
Burmeister M, McInnis MG, Zollner S (2008) Psychiatric genetics: progress amid controversy. Nat Rev Genet 7:527–540
Burt SA (2009) Rethinking environmental contributions to child and adolescent psychopathology: a meta-analysis of shared environmental influences. Psychol Bull 4:608–637
Campbell DB, Sutcliffe JS, Ebert PJ, Militerni R, Bravaccio C, Trillo S, Elia M, Schneider C, Melmed R, Sacco R, Persico AM, Levitt P (2006) A genetic variant that disrupts MET transcription is associated with autism. Proc Natl Acad Sci USA 45:16834–16839
Campbell DB, D'Oronzio R, Garbett K, Ebert PJ, Mirnics K, Levitt P, Persico AM (2007) Disruption of cerebral cortex MET signaling in autism spectrum disorder. Ann Neurol 3:243–250
Campbell DB, Buie TM, Winter H, Bauman M, Sutcliffe JS, Perrin JM, Levitt P (2009) Distinct genetic risk based on association of MET in families with co-occurring autism and gastrointestinal conditions. Pediatrics 3:1018–1024
Cole J, Ball HA, Martin NC, Scourfield J, McGuffin P (2009) Genetic overlap between measures of hyperactivity/inattention and mood in children and adolescents. J Am Acad Child Adolesc Psychiatry 11:1094–1101
Cook EH Jr, Stein MA, Krasowski MD, Cox NJ, Olkon DM, Kieffer JE, Leventhal BL (1995) Association of attention-deficit disorder and the dopamine transporter gene. Am J Hum Genet 4:993–998
Cusco I, Medrano A, Gener B, Vilardell M, Gallastegui F, Villa O, Gonzalez E, Rodriguez-Santiago B, Vilella E, Del CM, Perez-Jurado LA (2009) Autism-specific copy number variants further implicate the phosphatidylinositol signaling pathway and the glutamatergic synapse in the etiology of the disorder. Hum Mol Genet 10:1795–1804
Dempfle A, Hinney A, Heinzel-Gutenbrunner M, Raab M, Geller F, Gudermann T, Schafer H, Hebebrand J (2004) Large quantitative effect of melanocortin-4 receptor gene mutations on body mass index. J Med Genet 10:795–800
Depienne C, Moreno-De-Luca D, Heron D, Bouteiller D, Gennetier A, Delorme R, Chaste P, Siffroi JP, Chantot-Bastaraud S, Benyahia B, Trouillard O, Nygren G, Kopp S, Johansson M, Rastam M, Burglen L, Leguern E, Verloes A, Leboyer M, Brice A, Gillberg C, Betancur C (2009) Screening for genomic rearrangements and methylation abnormalities of the 15q11-q13 region in autism spectrum disorders. Biol Psychiatry 4:349–359
Devlin B, Bennett P, Dawson G, Figlewicz DA, Grigorenko EL, McMahon W, Minshew N, Pauls D, Smith M, Spence MA, Rodier PM, Stodgell C, Schellenberg GD (2004) Alleles of a reelin CGG repeat do not convey liability to autism in a sample from the CPEA network. Am J Med Genet B Neuropsychiatr Genet 1:46–50

Donaldson ZR, Young LJ (2008) Oxytocin, vasopressin, and the neurogenetics of sociality. Science 5903:900–904

Dudbridge F, Gusnanto A (2008) Estimation of significance thresholds for genomewide association scans. Genet Epidemiol 3:227–234

Duvall JA, Lu A, Cantor RM, Todd RD, Constantino JN, Geschwind DH (2007) A quantitative trait locus analysis of social responsiveness in multiplex autism families. Am J Psychiatry 4:656–662

Ehninger D, De Vries PJ, Silva AJ (2009) From mTOR to cognition: molecular and cellular mechanisms of cognitive impairments in tuberous sclerosis. J Intellect Disabil Res 10:838–851

Eiberg H (1995) Nocturnal enuresis is linked to a specific gene. Scand J Urol Nephrol (Suppl): 15–16

Eiberg H (1998) Total genome scan analysis in a single extended family for primary nocturnal enuresis: evidence for a new locus (ENUR3) for primary nocturnal enuresis on chromosome 22q11. Eur Urol:34–36

Eiberg H, Berendt I, Mohr J (1995) Assignment of dominant inherited nocturnal enuresis (ENUR1) to chromosome 13q. Nat Genet 3:354–356

Eiberg H, Shaumburg HL, von GA, Rittig S (2001) Linkage study of a large Danish 4-generation family with urge incontinence and nocturnal enuresis. J Urol 6:2401–2403

Eley TC, Bolton D, O'Connor TG, Perrin S, Smith P, Plomin R (2003) A twin study of anxiety-related behaviours in pre-school children. J Child Psychol Psychiatry 7:945–960

Elia J, Gai X, Xie HM, Perin JC, Geiger E, Glessner JT, D'arcy M, deBerardinis R, Frackelton E, Kim C, Lantieri F, Muganga BM, Wang L, Takeda T, Rappaport EF, Grant SF, Berrettini W, Devoto M, Shaikh TH, Hakonarson H, White PS (2010) Rare structural variants found in attention-deficit hyperactivity disorder are preferentially associated with neurodevelopmental genes. Mol Psychiatry 6:637–646

Faraone SV, Perlis RH, Doyle AE, Smoller JW, Goralnick JJ, Holmgren MA, Sklar P (2005) Molecular genetics of attention-deficit/hyperactivity disorder. Biol Psychiatry 11:1313–1323

Faraone SV, Biederman J, Mick E (2006) The age-dependent decline of attention deficit hyperactivity disorder: a meta-analysis of follow-up studies. Psychol Med 2:159–165

Farooqi IS, Jebb SA, Langmack G, Lawrence E, Cheetham CH, Prentice AM, Hughes IA, McCamish MA, O'Rahilly S (1999) Effects of recombinant leptin therapy in a child with congenital leptin deficiency. N Engl J Med 12:879–884

Fisher SE, Vargha-Khadem F, Watkins KE, Monaco AP, Pembrey ME (1998) Localisation of a gene implicated in a severe speech and language disorder. Nat Genet 2:168–170

Francks C, Fisher SE, Marlow AJ, MacPhie IL, Taylor KE, Richardson AJ, Stein JF, Monaco AP (2003) Familial and genetic effects on motor coordination, laterality, and reading-related cognition. Am J Psychiatry 11:1970–1977

Franke B, Hoogman M, Arias VA, Heister JG, Savelkoul PJ, Naber M, Scheffer H, Kiemeney LA, Kan CC, Kooij JJ, Buitelaar JK (2008) Association of the dopamine transporter (SLC6A3/DAT1) gene 9–6 haplotype with adult ADHD. Am J Med Genet B Neuropsychiatr Genet 8:1576–1579

Franke B, Neale BM, Faraone SV (2009) Genome-wide association studies in ADHD. Hum Genet 1:13–50

Frayling TM, Timpson NJ, Weedon MN, Zeggini E, Freathy RM, Lindgren CM, Perry JR, Elliott KS, Lango H, Rayner NW, Shields B, Harries LW, Barrett JC, Ellard S, Groves CJ, Knight B, Patch AM, Ness AR, Ebrahim S, Lawlor DA, Ring SM, Ben-Shlomo Y, Jarvelin MR, Sovio U, Bennett AJ, Melzer D, Ferrucci L, Loos RJ, Barroso I, Wareham NJ, Karpe F, Owen KR, Cardon LR, Walker M, Hitman GA, Palmer CN, Doney AS, Morris AD, Smith GD, Hattersley AT, McCarthy MI (2007) A common variant in the FTO gene is associated with body mass index and predisposes to childhood and adult obesity. Science 5826:889–894

Freitag CM (2007) The genetics of autistic disorders and its clinical relevance: a review of the literature. Mol Psychiatry 1:2–22

Freitag CM, Rohde LA, Lempp T, Romanos M (2010a) Phenotypic and measurement influences on heritability estimates in childhood ADHD. Eur Child Adolesc Psychiatry 3:311–323

Freitag CM, Staal W, Klauck SM, Duketis E, Waltes R (2010b) Genetics of autistic disorders: review and clinical implications. Eur Child Adolesc Psychiatry 3:169–178

Gelhorn HL, Stallings MC, Young SE, Corley RP, Rhee SH, Hewitt JK (2005) Genetic and environmental influences on conduct disorder: symptom, domain and full-scale analyses. J Child Psychol Psychiatry 6:580–591

Geller F, Reichwald K, Dempfle A, Illig T, Vollmert C, Herpertz S, Siffert W, Platzer M, Hess C, Gudermann T, Biebermann H, Wichmann HE, Schafer H, Hinney A, Hebebrand J (2004) Melanocortin-4 receptor gene variant I103 is negatively associated with obesity. Am J Hum Genet 3:572–581

Gizer IR, Ficks C, Waldman ID (2009) Candidate gene studies of ADHD: a meta-analytic review. Hum Genet 1:51–90

Glessner JT, Wang K, Cai G, Korvatska O, Kim CE, Wood S, Zhang H, Estes A, Brune CW, Bradfield JP, Imielinski M, Frackelton EC, Reichert J, Crawford EL, Munson J, Sleiman PM, Chiavacci R, Annaiah K, Thomas K, Hou C, Glaberson W, Flory J, Otieno F, Garris M, Soorya L, Klei L, Piven J, Meyer KJ, Anagnostou E, Sakurai T, Game RM, Rudd DS, Zurawiecki D, McDougle CJ, Davis LK, Miller J, Posey DJ, Michaels S, Kolevzon A, Silverman JM, Bernier R, Levy SE, Schultz RT, Dawson G, Owley T, McMahon WM, Wassink TH, Sweeney JA, Nurnberger JI, Coon H, Sutcliffe JS, Minshew NJ, Grant SF, Bucan M, Cook EH, Buxbaum JD, Devlin B, Schellenberg GD, Hakonarson H (2009) Autism genome-wide copy number variation reveals ubiquitin and neuronal genes. Nature 7246:569–573

Gong X, Delorme R, Fauchereau F, Durand CM, Chaste P, Betancur C, Goubran-Botros H, Nygren G, Anckarsater H, Rastam M, Gillberg IC, Kopp S, Mouren-Simeoni MC, Gillberg C, Leboyer M, Bourgeron T (2009) An investigation of ribosomal protein L10 gene in autism spectrum disorders. BMC Med Genet:7

Hagerman RJ, Berry-Kravis E, Kaufmann WE, Ono MY, Tartaglia N, Lachiewicz A, Kronk R, Delahunty C, Hessl D, Visootsak J, Picker J, Gane L, Tranfaglia M (2009) Advances in the treatment of fragile X syndrome. Pediatrics 1:378–390

Hardy J, Singleton A (2009) Genomewide association studies and human disease. N Engl J Med 17:1759–1768

Hebebrand J, Propping P (1990) Genetic aspects of brain maturation and behavior. In: Rothenberger A (ed) Brain and behavior in child psychiatry, 1st edn. Springer, Berlin

Hebebrand J, Scherag A, Schimmelmann BG, Hinney A (2010) Child and adolescent psychiatric genetics. Eur Child Adolesc Psychiatry 3:259–279

Heid IM, Vollmert C, Kronenberg F, Huth C, Ankerst DP, Luchner A, Hinney A, Bronner G, Wichmann HE, Illig T, Doring A, Hebebrand J (2008) Association of the MC4R V103I polymorphism with the metabolic syndrome: the KORA Study. Obesity (Silver Spring) 2:369–376

Henrichsen CN, Chaignat E, Reymond A (2009) Copy number variants, diseases and gene expression. Hum Mol Genet R1:R1–R8

Hessl D, Tassone F, Cordeiro L, Koldewyn K, McCormick C, Green C, Wegelin J, Yuhas J, Hagerman RJ (2008) Brief report: aggression and stereotypic behavior in males with fragile X syndrome–moderating secondary genes in a "single gene" disorder. J Autism Dev Disord 1:184–189

Hill MK, Sahhar M (2006) Genetic counselling for psychiatric disorders. Med J Aust 9:507–510

Hindorff LA, Sethupathy P, Junkins HA, Ramos EM, Mehta JP, Collins FS, Manolio TA (2009) Potential etiologic and functional implications of genome-wide association loci for human diseases and traits. Proc Natl Acad Sci USA 23:9362–9367

Hinney A, Hebebrand J (2008) Polygenic obesity in humans. Obes Facts 1:35–42

Hinney A, Hebebrand J (2009) Three at one swoop! Obes Facts 1:3–8

Hinney A, Schmidt A, Nottebom K, Heibult O, Becker I, Ziegler A, Gerber G, Sina M, Gorg T, Mayer H, Siegfried W, Fichter M, Remschmidt H, Hebebrand J (1999) Several mutations in the melanocortin-4 receptor gene including a nonsense and a frameshift mutation associated with dominantly inherited obesity in humans. J Clin Endocrinol Metab 4:1483–1486

Hinney A, Hohmann S, Geller F, Vogel C, Hess C, Wermter AK, Brokamp B, Goldschmidt H, Siegfried W, Remschmidt H, Schafer H, Gudermann T, Hebebrand J (2003) Melanocortin-4 receptor gene: case-control study and transmission disequilibrium test confirm that functionally relevant mutations are compatible with a major gene effect for extreme obesity. J Clin Endocrinol Metab 9:4258–4267

Hinney A, Scherag S, Hebebrand J (2010) Genetic findings in anorexia and bulimia nervosa. Prog Mol Biol Transl Sci:241–270

Holtkamp K, Muller B, Heussen N, Remschmidt H, Herpertz-Dahlmann B (2005) Depression, anxiety, and obsessionality in long-term recovered patients with adolescent-onset anorexia nervosa. Eur Child Adolesc Psychiatry 2:106–110

Huang CH, Santangelo SL (2008) Autism and serotonin transporter gene polymorphisms: a systematic review and meta-analysis. Am J Med Genet B Neuropsychiatr Genet 6:903–913

Hublin C, Kaprio J, Partinen M, Koskenvuo M (1998) Nocturnal enuresis in a nationwide twin cohort. Sleep 6:579–585

Hutcheson HB, Olson LM, Bradford Y, Folstein SE, Santangelo SL, Sutcliffe JS, Haines JL (2004) Examination of NRCAM, LRRN3, KIAA0716, and LAMB1 as autism candidate genes. BMC Med Genet:12

Hyde TM, Deep-Soboslay A, Iglesias B, Callicott JH, Gold JM, Meyer-Lindenberg A, Honea RA, Bigelow LB, Egan MF, Emsellem EM, Weinberger DR (2008) Enuresis as a premorbid developmental marker of schizophrenia. Brain (Pt 9):2489–2498

Jacob S, Brune CW, Carter CS, Leventhal BL, Lord C, Cook EH Jr (2007) Association of the oxytocin receptor gene (OXTR) in Caucasian children and adolescents with autism. Neurosci Lett 1:6–9

Kahn RS, Khoury J, Nichols WC, Lanphear BP (2003) Role of dopamine transporter genotype and maternal prenatal smoking in childhood hyperactive-impulsive, inattentive, and oppositional behaviors. J Pediatr 1:104–110

Kendler KS (2005) "A gene for...": the nature of gene action in psychiatric disorders. Am J Psychiatry 7:1243–1252

Kessler RC, Adler LA, Barkley R, Biederman J, Conners CK, Faraone SV, Greenhill LL, Jaeger S, Secnik K, Spencer T, Ustun TB, Zaslavsky AM (2005) Patterns and predictors of attention-deficit/hyperactivity disorder persistence into adulthood: results from the national comorbidity survey replication. Biol Psychiatry 11:1442–1451

Klauck SM, Felder B, Kolb-Kokocinski A, Schuster C, Chiocchetti A, Schupp I, Wellenreuther R, Schmotzer G, Poustka F, Breitenbach-Koller L, Poustka A (2006) Mutations in the ribosomal protein gene RPL10 suggest a novel modulating disease mechanism for autism. Mol Psychiatry 12:1073–1084

Knoblauch H (2007) Schizophrenia, human genetics and genetic counselling. Human genetic counselling as part of the psychiatric/psychotherapeutic treatment concept. Nervenarzt 3(243–8):250

Krebs MO, Betancur C, Leroy S, Bourdel MC, Gillberg C, Leboyer M (2002) Absence of association between a polymorphic GGC repeat in the 5' untranslated region of the reelin gene and autism. Mol Psychiatry 7:801–804

Krude H, Biebermann H, Luck W, Horn R, Brabant G, Gruters A (1998) Severe early-onset obesity, adrenal insufficiency and red hair pigmentation caused by POMC mutations in humans. Nat Genet 2:155–157

Kumar RA, Karamohamed S, Sudi J, Conrad DF, Brune C, Badner JA, Gilliam TC, Nowak NJ, Cook EH Jr, Dobyns WB, Christian SL (2008) Recurrent 16p11.2 microdeletions in autism. Hum Mol Genet 4:628–638

Kuntsi J, Neale BM, Chen W, Faraone SV, Asherson P (2006) The IMAGE project: methodological issues for the molecular genetic analysis of ADHD. Behav Brain Funct:27

LaHoste GJ, Swanson JM, Wigal SB, Glabe C, Wigal T, King N, Kennedy JL (1996) Dopamine D4 receptor gene polymorphism is associated with attention deficit hyperactivity disorder. Mol Psychiatry 2:121–124

Lasky-Su J, Neale BM, Franke B, Anney RJ, Zhou K, Maller JB, Vasquez AA, Chen W, Asherson P, Buitelaar J, Banaschewski T, Ebstein R, Gill M, Miranda A, Mulas F, Oades RD, Roeyers H, Rothenberger A, Sergeant J, Sonuga-Barke E, Steinhausen HC, Taylor E, Daly M, Laird N, Lange C, Faraone SV (2008) Genome-wide association scan of quantitative traits for attention deficit hyperactivity disorder identifies novel associations and confirms candidate gene associations. Am J Med Genet B Neuropsychiatr Genet 8:1345–1354

Lein ES, Hawrylycz MJ, Ao N, Ayres M, Bensinger A, Bernard A, Boe AF, Boguski MS, Brockway KS, Byrnes EJ, Chen L, Chen L, Chen TM, Chin MC, Chong J, Crook BE, Czaplinska A, Dang CN, Datta S, Dee NR, Desaki AL, Desta T, Diep E, Dolbeare TA, Donelan MJ, Dong HW, Dougherty JG, Duncan BJ, Ebbert AJ, Eichele G, Estin LK, Faber C, Facer BA, Fields R, Fischer SR, Fliss TP, Frensley C, Gates SN, Glattfelder KJ, Halverson KR, Hart MR, Hohmann JG, Howell MP, Jeung DP, Johnson RA, Karr PT, Kawal R, Kidney JM, Knapik RH, Kuan CL, Lake JH, Laramee AR, Larsen KD, Lau C, Lemon TA, Liang AJ, Liu Y, Luong LT, Michaels J, Morgan JJ, Morgan RJ, Mortrud MT, Mosqueda NF, Ng LL, Ng R, Orta GJ, Overly CC, Pak TH, Parry SE, Pathak SD, Pearson OC, Puchalski RB, Riley ZL, Rockett HR, Rowland SA, Royall JJ, Ruiz MJ, Sarno NR, Schaffnit K, Shapovalova NV, Sivisay T, Slaughterbeck CR, Smith SC, Smith KA, Smith BI, Sodt AJ, Stewart NN, Stumpf KR, Sunkin SM, Sutram M, Tam A, Teemer CD, Thaller C, Thompson CL, Varnam LR, Visel A, Whitlock RM, Wohnoutka PE, Wolkey CK, Wong VY, Wood M, Yaylaoglu MB, Young RC, Youngstrom BL, Yuan XF, Zhang B, Zwingman TA, Jones AR (2007) Genome-wide atlas of gene expression in the adult mouse brain. Nature 7124:168–176

Lerer E, Levi S, Salomon S, Darvasi A, Yirmiya N, Ebstein RP (2008) Association between the oxytocin receptor (OXTR) gene and autism: relationship to Vineland Adaptive Behavior Scales and cognition. Mol Psychiatry 10:980–988

Lesch KP, Timmesfeld N, Renner TJ, Halperin R, Roser C, Nguyen TT, Craig DW, Romanos J, Heine M, Meyer J, Freitag C, Warnke A, Romanos M, Schafer H, Walitza S, Reif A, Stephan DA, Jacob C (2008) Molecular genetics of adult ADHD: converging evidence from genome-wide association and extended pedigree linkage studies. J Neural Transm 11:1573–1585

Lesch KP, Selch S, Renner TJ, Jacob C, Nguyen TT, Hahn T, Romanos M, Walitza S, Shoichet S, Dempfle A, Heine M, Boreatti-Hummer A, Romanos J, Gross-Lesch S, Zerlaut H, Wultsch T, Heinzel S, Fassnacht M, Fallgatter A, Allolio B, Schafer H, Warnke A, Reif A, Ropers HH, Ullmann R (2010) Genome-wide copy number variation analysis in attention-deficit/hyperactivity disorder: association with neuropeptide Y gene dosage in an extended pedigree. Mol Psychiatry

Levy F, Hay DA, McStephen M, Wood C, Waldman I (1997) Attention-deficit hyperactivity disorder: a category or a continuum? Genetic analysis of a large-scale twin study. J Am Acad Child Adolesc Psychiatry 6:737–744

Li J, Nguyen L, Gleason C, Lotspeich L, Spiker D, Risch N, Myers RM (2004) Lack of evidence for an association between WNT2 and RELN polymorphisms and autism. Am J Med Genet B Neuropsychiatr Genet 1:51–57

Li D, Sham PC, Owen MJ, He L (2006) Meta-analysis shows significant association between dopamine system genes and attention deficit hyperactivity disorder (ADHD). Hum Mol Genet 14:2276–2284

Lichtenstein P, Carlstrom E, Rastam M, Gillberg C, Anckarsater H (2010) The Genetics of Autism Spectrum Disorders and Related Neuropsychiatric Disorders in Childhood. Am J Psychiatry

Loeys B, Hoebeke P, Raes A, Messiaen L, De PA, Vande WJ (2002) Does monosymptomatic enuresis exist? A molecular genetic exploration of 32 families with enuresis/incontinence. BJU Int 1:76–83

Lowe N, Kirley A, Hawi Z, Sham P, Wickham H, Kratochvil CJ, Smith SD, Lee SY, Levy F, Kent L, Middle F, Rohde LA, Roman T, Tahir E, Yazgan Y, Asherson P, Mill J, Thapar A, Payton A, Todd RD, Stephens T, Ebstein RP, Manor I, Barr CL, Wigg KG, Sinke RJ, Buitelaar JK, Smalley SL, Nelson SF, Biederman J, Faraone SV, Gill M (2004) Joint analysis

of the DRD5 marker concludes association with attention-deficit/hyperactivity disorder confined to the predominantly inattentive and combined subtypes. Am J Hum Genet 2:348–356
Ma DQ, Cuccaro ML, Jaworski JM, Haynes CS, Stephan DA, Parod J, Abramson RK, Wright HH, Gilbert JR, Haines JL, Pericak-Vance MA (2007) Dissecting the locus heterogeneity of autism: significant linkage to chromosome 12q14. Mol Psychiatry 4:376–384
Maes HH, Neale MC, Eaves LJ (1997) Genetic and environmental factors in relative body weight and human adiposity. Behav Genet 4:325–351
Maher BS, Marazita ML, Moss HB, Vanyukov MM (1999) Segregation analysis of attention deficit hyperactivity disorder. Am J Med Genet 1:71–78
Maher BS, Marazita ML, Ferrell RE, Vanyukov MM (2002) Dopamine system genes and attention deficit hyperactivity disorder: a meta-analysis. Psychiatr Genet 4:207–215
Manolio TA, Collins FS, Cox NJ, Goldstein DB, Hindorff LA, Hunter DJ, McCarthy MI, Ramos EM, Cardon LR, Chakravarti A, Cho JH, Guttmacher AE, Kong A, Kruglyak L, Mardis E, Rotimi CN, Slatkin M, Valle D, Whittemore AS, Boehnke M, Clark AG, Eichler EE, Gibson G, Haines JL, Mackay TF, McCarroll SA, Visscher PM (2009) Finding the missing heritability of complex diseases. Nature 7265:747–753
Margallo-Lana M, Morris CM, Gibson AM, Tan AL, Kay DW, Tyrer SP, Moore BP, Ballard CG (2004) Influence of the amyloid precursor protein locus on dementia in Down syndrome. Neurology 11:1996–1998
Marshall CR, Noor A, Vincent JB, Lionel AC, Feuk L, Skaug J, Shago M, Moessner R, Pinto D, Ren Y, Thiruvahindrapduram B, Fiebig A, Schreiber S, Friedman J, Ketelaars CE, Vos YJ, Ficicioglu C, Kirkpatrick S, Nicolson R, Sloman L, Summers A, Gibbons CA, Teebi A, Chitayat D, Weksberg R, Thompson A, Vardy C, Crosbie V, Luscombe S, Baatjes R, Zwaigenbaum L, Roberts W, Fernandez B, Szatmari P, Scherer SW (2008) Structural variation of chromosomes in autism spectrum disorder. Am J Hum Genet 2:477–488
McCarthy MI, Abecasis GR, Cardon LR, Goldstein DB, Little J, Ioannidis JP, Hirschhorn JN (2008) Genome-wide association studies for complex traits: consensus, uncertainty and challenges. Nat Rev Genet 5:356–369
McGuffin P, Rijsdijk F, Andrew M, Sham P, Katz R, Cardno A (2003) The heritability of bipolar affective disorder and the genetic relationship to unipolar depression. Arch Gen Psychiatry 5:497–502
McLoughlin G, Ronald A, Kuntsi J, Asherson P, Plomin R (2007) Genetic support for the dual nature of attention deficit hyperactivity disorder: substantial genetic overlap between the inattentive and hyperactive-impulsive components. J Abnorm Child Psychol 6:999–1008
Mefford HC, Sharp AJ, Baker C, Itsara A, Jiang Z, Buysse K, Huang S, Maloney VK, Crolla JA, Baralle D, Collins A, Mercer C, Norga K, de RT, Devriendt K, Bongers EM, de LN, Reardon W, Gimelli S, Bena F, Hennekam RC, Male A, Gaunt L, Clayton-Smith J, Simonic I, Park SM, Mehta SG, Nik-Zainal S, Woods CG, Firth HV, Parkin G, Fichera M, Reitano S, Lo GM, Li KE, Casuga I, Broomer A, Conrad B, Schwerzmann M, Raber L, Gallati S, Striano P, Coppola A, Tolmie JL, Tobias ES, Lilley C, Armengol L, Spysschaert Y, Verloo P, De CA, Goossens L, Mortier G, Speleman F, van BE, Nelen MR, Hochstenbach R, Poot M, Gallagher L, Gill M, McClellan J, King MC, Regan R, Skinner C, Stevenson RE, Antonarakis SE, Chen C, Estivill X, Menten B, Gimelli G, Gribble S, Schwartz S, Sutcliffe JS, Walsh T, Knight SJ, Sebat J, Romano C, Schwartz CE, Veltman JA, de Vries BB, Vermeesch JR, Barber JC, Willatt L, Tassabehji M, Eichler EE (2008) Recurrent rearrangements of chromosome 1q21.1 and variable pediatric phenotypes. N Engl J Med 16:1685–1699
Miller DT, Shen Y, Wu BL (2008) Oligonucleotide microarrays for clinical diagnosis of copy number variation. Curr Protoc Hum Genet:Unit
Mitchell AA, Cutler DJ, Chakravarti A (2003) Undetected genotyping errors cause apparent overtransmission of common alleles in the transmission/disequilibrium test. Am J Hum Genet 3:598–610

Montague CT, Farooqi IS, Whitehead JP, Soos MA, Rau H, Wareham NJ, Sewter CP, Digby JE, Mohammed SN, Hurst JA, Cheetham CH, Earley AR, Barnett AH, Prins JB, O'Rahilly S (1997) Congenital leptin deficiency is associated with severe early-onset obesity in humans. Nature 6636:903–908

Neale BM, Lasky-Su J, Anney R, Franke B, Zhou K, Maller JB, Vasquez AA, Asherson P, Chen W, Banaschewski T, Buitelaar J, Ebstein R, Gill M, Miranda A, Oades RD, Roeyers H, Rothenberger A, Sergeant J, Steinhausen HC, Sonuga-Barke E, Mulas F, Taylor E, Laird N, Lange C, Daly M, Faraone SV (2008) Genome-wide association scan of attention deficit hyperactivity disorder. Am J Med Genet B Neuropsychiatr Genet 8:1337–1344

Neale BM, Medland SE, Ripke S, Asherson P, Franke B, Lesch KP, Faraone SV, Nguyen TT, Schafer H, Holmans P, Daly M, Steinhausen HC, Freitag C, Reif A, Renner TJ, Romanos M, Romanos J, Walitza S, Warnke A, Meyer J, Palmason H, Buitelaar J, Vasquez AA, Lambregts-Rommelse N, Gill M, Anney RJ, Langely K, O'Donovan M, Williams N, Owen M, Thapar A, Kent L, Sergeant J, Roeyers H, Mick E, Biederman J, Doyle A, Smalley S, Loo S, Hakonarson H, Elia J, Todorov A, Miranda A, Mulas F, Ebstein RP, Rothenberger A, Banaschewski T, Oades RD, Sonuga-Barke E, McGough J, Nisenbaum L, Middleton F, Hu X, Nelson S (2010) Meta-analysis of genome-wide association studies of attention-deficit/hyperactivity disorder. J Am Acad Child Adolesc Psychiatry 9:884–897

Page DT, Kuti OJ, Prestia C, Sur M (2009) Haploinsufficiency for Pten and Serotonin transporter cooperatively influences brain size and social behavior. Proc Natl Acad Sci USA 6:1989–1994

Paloyelis Y, Rijsdijk F, Wood AC, Asherson P, Kuntsi J (2010) The genetic association between ADHD symptoms and reading difficulties: the role of inattentiveness and IQ. J Abnorm Child Psychol 8:1083–1095

Pennisi E (2007) Breakthrough of the year. Human genetic variation. Science 5858:1842–1843

Persico AM, D'Agruma L, Maiorano N, Totaro A, Militerni R, Bravaccio C, Wassink TH, Schneider C, Melmed R, Trillo S, Montecchi F, Palermo M, Pascucci T, Puglisi-Allegra S, Reichelt KL, Conciatori M, Marino R, Quattrocchi CC, Baldi A, Zelante L, Gasparini P, Keller F (2001) Reelin gene alleles and haplotypes as a factor predisposing to autistic disorder. Mol Psychiatry 2:150–159

Pinto D, Pagnamenta AT, Klei L, Anney R, Merico D, Regan R, Conroy J, Magalhaes TR, Correia C, Abrahams BS, Almeida J, Bacchelli E, Bader GD, Bailey AJ, Baird G, Battaglia A, Berney T, Bolshakova N, Bolte S, Bolton PF, Bourgeron T, Brennan S, Brian J, Bryson SE, Carson AR, Casallo G, Casey J, Chung BH, Cochrane L, Corsello C, Crawford EL, Crossett A, Cytrynbaum C, Dawson G, de JM, Delorme R, Drmic I, Duketis E, Duque F, Estes A, Farrar P, Fernandez BA, Folstein SE, Fombonne E, Freitag CM, Gilbert J, Gillberg C, Glessner JT, Goldberg J, Green A, Green J, Guter SJ, Hakonarson H, Heron EA, Hill M, Holt R, Howe JL, Hughes G, Hus V, Igliozzi R, Kim C, Klauck SM, Kolevzon A, Korvatska O, Kustanovich V, Lajonchere CM, Lamb JA, Laskawiec M, Leboyer M, Le CA, Leventhal BL, Lionel AC, Liu XQ, Lord C, Lotspeich L, Lund SC, Maestrini E, Mahoney W, Mantoulan C, Marshall CR, McConachie H, McDougle CJ, McGrath J, McMahon WM, Merikangas A, Migita O, Minshew NJ, Mirza GK, Munson J, Nelson SF, Noakes C, Noor A, Nygren G, Oliveira G, Papanikolaou K, Parr JR, Parrini B, Paton T, Pickles A, Pilorge M, Piven J, Ponting CP, Posey DJ, Poustka A, Poustka F, Prasad A, Ragoussis J, Renshaw K, Rickaby J, Roberts W, Roeder K, Roge B, Rutter ML, Bierut LJ, Rice JP, Salt J, Sansom K, Sato D, Segurado R, Sequeira AF, Senman L, Shah N, Sheffield VC, Soorya L, Sousa I, Stein O, Sykes N, Stoppioni V, Strawbridge C, Tancredi R, Tansey K, Thiruvahindrapduram B, Thompson AP, Thomson S, Tryfon A, Tsiantis J, van EH, Vincent JB, Volkmar F, Wallace S, Wang K, Wang Z, Wassink TH, Webber C, Weksberg R, Wing K, Wittemeyer K, Wood S, Wu J, Yaspan BL, Zurawiecki D, Zwaigenbaum L, Buxbaum JD, Cantor RM, Cook EH, Coon H, Cuccaro ML, Devlin B, Ennis S, Gallagher L, Geschwind DH, Gill M, Haines JL, Hallmayer J, Miller J, Monaco AP, Nurnberger JI, Jr., Paterson AD, Pericak-Vance MA, Schellenberg GD, Szatmari P, Vicente AM, Vieland VJ, Wijsman EM, Scherer SW, Sutcliffe JS, Betancur C (2010) Functional impact of global rare copy number variation in autism spectrum disorders. Nature 7304:368–372

Plomin R (2005) Finding genes in child psychology and psychiatry: when are we going to be there? J Child Psychol Psychiatry 10:1030–1038

Purper-Ouakil D, Wohl M, Mouren MC, Verpillat P, Ades J, Gorwood P (2005) Meta-analysis of family-based association studies between the dopamine transporter gene and attention deficit hyperactivity disorder. Psychiatr Genet 1:53–59

Ribases M, Ramos-Quiroga JA, Sanchez-Mora C, Bosch R, Richarte V, Palomar G, Gastaminza X, Bielsa A, Arcos-Burgos M, Muenke M, Castellanos FX, Cormand B, Bayes M, Casas M (2010) Contribution of LPHN3 to the genetic susceptibility to ADHD in adulthood: a replication study. Genes Brain Behav

Romanos M, Freitag C, Jacob C, Craig DW, Dempfle A, Nguyen TT, Halperin R, Walitza S, Renner TJ, Seitz C, Romanos J, Palmason H, Reif A, Heine M, Windemuth-Kieselbach C, Vogler C, Sigmund J, Warnke A, Schafer H, Meyer J, Stephan DA, Lesch KP (2008) Genome-wide linkage analysis of ADHD using high-density SNP arrays: novel loci at 5q13.1 and 14q12. Mol Psychiatry 5:522–530

Ronald A, Simonoff E, Kuntsi J, Asherson P, Plomin R (2008) Evidence for overlapping genetic influences on autistic and ADHD behaviours in a community twin sample. J Child Psychol Psychiatry 5:535–542

Rzehak P, Scherag A, Grallert H, Sausenthaler S, Koletzko S, Bauer CP, Schaaf B, von BA, Berdel D, Borte M, Herbarth O, Kramer U, Illig T, Wichmann HE, Hebebrand J, Heinrich J (2010) Associations between BMI and the FTO gene are age dependent: results from the GINI and LISA birth cohort studies up to age 6 years. Obes Facts 3:173–180

Scerri TS, Schulte-Korne G (2010) Genetics of developmental dyslexia. Eur Child Adolesc Psychiatry 3:179–197

Schellenberg GD, Dawson G, Sung YJ, Estes A, Munson J, Rosenthal E, Rothstein J, Flodman P, Smith M, Coon H, Leong L, Yu CE, Stodgell C, Rodier PM, Spence MA, Minshew N, McMahon WM, Wijsman EM (2006) Evidence for multiple loci from a genome scan of autism kindreds. Mol Psychiatry 11(1049–60):979

Sebat J, Lakshmi B, Malhotra D, Troge J, Lese-Martin C, Walsh T, Yamrom B, Yoon S, Krasnitz A, Kendall J, Leotta A, Pai D, Zhang R, Lee YH, Hicks J, Spence SJ, Lee AT, Puura K, Lehtimaki T, Ledbetter D, Gregersen PK, Bregman J, Sutcliffe JS, Jobanputra V, Chung W, Warburton D, King MC, Skuse D, Geschwind DH, Gilliam TC, Ye K, Wigler M (2007) Strong association of de novo copy number mutations with autism. Science 5823:445–449

Serajee FJ, Zhong H, Mahbubul Huq AH (2006) Association of Reelin gene polymorphisms with autism. Genomics 1:75–83

Sha BY, Yang TL, Zhao LJ, Chen XD, Guo Y, Chen Y, Pan F, Zhang ZX, Dong SS, Xu XH, Deng HW (2009) Genome-wide association study suggested copy number variation may be associated with body mass index in the Chinese population. J Hum Genet 4:199–202

Shen WH, Balajee AS, Wang J, Wu H, Eng C, Pandolfi PP, Yin Y (2007) Essential role for nuclear PTEN in maintaining chromosomal integrity. Cell 1:157–170

Shen Y, Dies KA, Holm IA, Bridgemohan C, Sobeih MM, Caronna EB, Miller KJ, Frazier JA, Silverstein I, Picker J, Weissman L, Raffalli P, Jeste S, Demmer LA, Peters HK, Brewster SJ, Kowalczyk SJ, Rosen-Sheidley B, McGowan C, Duda AW III, Lincoln SA, Lowe KR, Schonwald A, Robbins M, Hisama F, Wolff R, Becker R, Nasir R, Urion DK, Milunsky JM, Rappaport L, Gusella JF, Walsh CA, Wu BL, Miller DT (2010) Clinical genetic testing for patients with autism spectrum disorders. Pediatrics 4:e727–e735

Sikora DM, Pettit-Kekel K, Penfield J, Merkens LS, Steiner RD (2006) The near universal presence of autism spectrum disorders in children with Smith-Lemli-Opitz syndrome. Am J Med Genet A 14:1511–1518

Silventoinen K, Kaprio J (2009) Genetics of tracking of body mass index from birth to late middle age: evidence from twin and family studies. Obes Facts 3:196–202

Silverman JM, Buxbaum JD, Ramoz N, Schmeidler J, Reichenberg A, Hollander E, Angelo G, Smith CJ, Kryzak LA (2008) Autism-related routines and rituals associated with a mitochondrial aspartate/glutamate carrier SLC25A12 polymorphism. Am J Med Genet B Neuropsychiatr Genet 3:408–410

Skaar DA, Shao Y, Haines JL, Stenger JE, Jaworski J, Martin ER, DeLong GR, Moore JH, McCauley JL, Sutcliffe JS, Ashley-Koch AE, Cuccaro ML, Folstein SE, Gilbert JR, Pericak-Vance MA (2005) Analysis of the RELN gene as a genetic risk factor for autism. Mol Psychiatry 6:563–571

Sousa I, Clark TG, Toma C, Kobayashi K, Choma M, Holt R, Sykes NH, Lamb JA, Bailey AJ, Battaglia A, Maestrini E, Monaco AP (2009) MET and autism susceptibility: family and case-control studies. Eur J Hum Genet 6:749–758

Speliotes EK, Willer CJ, Berndt SI, Monda KL, Thorleifsson G, Jackson AU, Allen HL, Lindgren CM, Luan J, Magi R, Randall JC, Vedantam S, Winkler TW, Qi L, Workalemahu T, Heid IM, Steinthorsdottir V, Stringham HM, Weedon MN, Wheeler E, Wood AR, Ferreira T, Weyant RJ, Segre AV, Estrada K, Liang L, Nemesh J, Park JH, Gustafsson S, Kilpelainen TO, Yang J, Bouatia-Naji N, Esko T, Feitosa MF, Kutalik Z, Mangino M, Raychaudhuri S, Scherag A, Smith AV, Welch R, Zhao JH, Aben KK, Absher DM, Amin N, Dixon AL, Fisher E, Glazer NL, Goddard ME, Heard-Costa NL, Hoesel V, Hottenga JJ, Johansson A, Johnson T, Ketkar S, Lamina C, Li S, Moffatt MF, Myers RH, Narisu N, Perry JR, Peters MJ, Preuss M, Ripatti S, Rivadeneira F, Sandholt C, Scott LJ, Timpson NJ, Tyrer JP, van WS, Watanabe RM, White CC, Wiklund F, Barlassina C, Chasman DI, Cooper MN, Jansson JO, Lawrence RW, Pellikka N, Prokopenko I, Shi J, Thiering E, Alavere H, Alibrandi MT, Almgren P, Arnold AM, Aspelund T, Atwood LD, Balkau B, Balmforth AJ, Bennett AJ, Ben-Shlomo Y, Bergman RN, Bergmann S, Biebermann H, Blakemore AI, Boes T, Bonnycastle LL, Bornstein SR, Brown MJ, Buchanan TA, Busonero F, Campbell H, Cappuccio FP, Cavalcanti-Proenca C, Chen YD, Chen CM, Chines PS, Clarke R, Coin L, Connell J, Day IN, Heijer M, Duan J, Ebrahim S, Elliott P, Elosua R, Eiriksdottir G, Erdos MR, Eriksson JG, Facheris MF, Felix SB, Fischer-Posovszky P, Folsom AR, Friedrich N, Freimer NB, Fu M, Gaget S, Gejman PV, Geus EJ, Gieger C, Gjesing AP, Goel A, Goyette P, Grallert H, Grassler J, Greenawalt DM, Groves CJ, Gudnason V, Guiducci C, Hartikainen AL, Hassanali N, Hall AS, Havulinna AS, Hayward C, Heath AC, Hengstenberg C, Hicks AA, Hinney A, Hofman A, Homuth G, Hui J, Igl W, Iribarren C, Isomaa B, Jacobs KB, Jarick I, Jewell E, John U, Jorgensen T, Jousilahti P, Jula A, Kaakinen M, Kajantie E, Kaplan LM, Kathiresan S, Kettunen J, Kinnunen L, Knowles JW, Kolcic I, Konig IR, Koskinen S, Kovacs P, Kuusisto J, Kraft P, Kvaloy K, Laitinen J, Lantieri O, Lanzani C, Launer LJ, Lecoeur C, Lehtimaki T, Lettre G, Liu J, Lokki ML, Lorentzon M, Luben RN, Ludwig B, Manunta P, Marek D, Marre M, Martin NG, McArdle WL, McCarthy A, McKnight B, Meitinger T, Melander O, Meyre D, Midthjell K, Montgomery GW, Morken MA, Morris AP, Mulic R, Ngwa JS, Nelis M, Neville MJ, Nyholt DR, O'Donnell CJ, O'Rahilly S, Ong KK, Oostra B, Pare G, Parker AN, Perola M, Pichler I, Pietilainen KH, Platou CG, Polasek O, Pouta A, Rafelt S, Raitakari O, Rayner NW, Ridderstrale M, Rief W, Ruokonen A, Robertson NR, Rzehak P, Salomaa V, Sanders AR, Sandhu MS, Sanna S, Saramies J, Savolainen MJ, Scherag S, Schipf S, Schreiber S, Schunkert H, Silander K, Sinisalo J, Siscovick DS, Smit JH, Soranzo N, Sovio U, Stephens J, Surakka I, Swift AJ, Tammesoo ML, Tardif JC, Teder-Laving M, Teslovich TM, Thompson JR, Thomson B, Tonjes A, Tuomi T, van Meurs JB, van Ommen GJ (2010) Association analyses of 249,796 individuals reveal 18 new loci associated with body mass index. Nat Genet 11:937–948

Stiles BL (2009) Phosphatase and tensin homologue deleted on chromosome 10: extending its PTENtacles. Int J Biochem Cell Biol 4:757–761

Stutzmann F, Vatin V, Cauchi S, Morandi A, Jouret B, Landt O, Tounian P, Levy-Marchal C, Buzzetti R, Pinelli L, Balkau B, Horber F, Bougneres P, Froguel P, Meyre D (2007) Non-synonymous polymorphisms in melanocortin-4 receptor protect against obesity: the two facets of a Janus obesity gene. Hum Mol Genet 15:1837–1844

Stutzmann F, Tan K, Vatin V, Dina C, Jouret B, Tichet J, Balkau B, Potoczna N, Horber F, O'Rahilly S, Farooqi IS, Froguel P, Meyre D (2008) Prevalence of melanocortin-4 receptor deficiency in Europeans and their age-dependent penetrance in multigenerational pedigrees. Diabetes 9:2511–2518

Sullivan PF, Neale MC, Kendler KS (2000) Genetic epidemiology of major depression: review and meta-analysis. Am J Psychiatry 10:1552–1562

Sullivan PF, Kendler KS, Neale MC (2003) Schizophrenia as a complex trait: evidence from a meta-analysis of twin studies. Arch Gen Psychiatry 12:1187–1192

Swanson J, Castellanos FX, Murias M, LaHoste G, Kennedy J (1998a) Cognitive neuroscience of attention deficit hyperactivity disorder and hyperkinetic disorder. Curr Opin Neurobiol 2:263–271

Swanson JM, Sunohara GA, Kennedy JL, Regino R, Fineberg E, Wigal T, Lerner M, Williams L, LaHoste GJ, Wigal S (1998b) Association of the dopamine receptor D4 (DRD4) gene with a refined phenotype of attention deficit hyperactivity disorder (ADHD): a family-based approach. Mol Psychiatry 1:38–41

Szatmari P, Paterson AD, Zwaigenbaum L, Roberts W, Brian J, Liu XQ, Vincent JB, Skaug JL, Thompson AP, Senman L, Feuk L, Qian C, Bryson SE, Jones MB, Marshall CR, Scherer SW, Vieland VJ, Bartlett C, Mangin LV, Goedken R, Segre A, Pericak-Vance MA, Cuccaro ML, Gilbert JR, Wright HH, Abramson RK, Betancur C, Bourgeron T, Gillberg C, Leboyer M, Buxbaum JD, Davis KL, Hollander E, Silverman JM, Hallmayer J, Lotspeich L, Sutcliffe JS, Haines JL, Folstein SE, Piven J, Wassink TH, Sheffield V, Geschwind DH, Bucan M, Brown WT, Cantor RM, Constantino JN, Gilliam TC, Herbert M, Lajonchere C, Ledbetter DH, Lese-Martin C, Miller J, Nelson S, Samango-Sprouse CA, Spence S, State M, Tanzi RE, Coon H, Dawson G, Devlin B, Estes A, Flodman P, Klei L, McMahon WM, Minshew N, Munson J, Korvatska E, Rodier PM, Schellenberg GD, Smith M, Spence MA, Stodgell C, Tepper PG, Wijsman EM, Yu CE, Roge B, Mantoulan C, Wittemeyer K, Poustka A, Felder B, Klauck SM, Schuster C, Poustka F, Bolte S, Feineis-Matthews S, Herbrecht E, Schmotzer G, Tsiantis J, Papanikolaou K, Maestrini E, Bacchelli E, Blasi F, Carone S, Toma C, van Engeland H, de Jonge M, Kemner C, Koop F, Langemeijer M, Hijimans C, Staal WG, Baird G, Bolton PF, Rutter ML, Weisblatt E, Green J, Aldred C, Wilkinson JA, Pickles A, Le Couteur A, Berney T, McConachie H, Bailey AJ, Francis K, Honeyman G, Hutchinson A, Parr JR, Wallace S, Monaco AP, Barnby G, Kobayashi K, Lamb JA, Sousa I, Sykes N, Cook EH, Guter SJ, Leventhal BL, Salt J, Lord C, Corsello C, Hus V, Weeks DE, Volkmar F, Tauber M, Fombonne E, Shih A (2007) Mapping autism risk loci using genetic linkage and chromosomal rearrangements. Nat Genet 3:319–328

Thapar A, Harrington R, McGuffin P (2001) Examining the comorbidity of ADHD-related behaviours and conduct problems using a twin study design. Br J Psychiatry:224–229

Thapar A, Rice F, Hay D, Boivin J, Langley K, van den Bree M, Rutter M, Harold G (2009) Prenatal smoking might not cause attention-deficit/hyperactivity disorder: evidence from a novel design. Biol Psychiatry 8:722–727

Trikalinos TA, Karvouni A, Zintzaras E, Ylisaukko-Oja T, Peltonen L, Jarvela I, Ioannidis JP (2006) A heterogeneity-based genome search meta-analysis for autism-spectrum disorders. Mol Psychiatry 1:29–36

Uhl GR, Drgon T, Liu QR, Johnson C, Walther D, Komiyama T, Harano M, Sekine Y, Inada T, Ozaki N, Iyo M, Iwata N, Yamada M, Sora I, Chen CK, Liu HC, Ujike H, Lin SK (2008) Genome-wide association for methamphetamine dependence: convergent results from 2 samples. Arch Gen Psychiatry 3:345–355

Vaisse C, Clement K, Guy-Grand B, Froguel P (1998) A frameshift mutation in human MC4R is associated with a dominant form of obesity. Nat Genet 2:113–114

van der Zwaag B, Franke L, Poot M, Hochstenbach R, Spierenburg HA, Vorstman JA, van DE, De Jonge MV, Verbeek NE, Brilstra EH, van 't SR, Ophoff RA, van Es MA, Blauw HM, Veldink JH, Buizer-Voskamp JE, Beemer FA, van den Berg LH, Wijmenga C, van Amstel HK, van EH, Burbach JP, Staal WG (2009) Gene-network analysis identifies susceptibility genes related to glycobiology in autism. PLoS ONE 5:e5324

van Rijn S, Swaab H, Aleman A, Kahn RS (2008) Social behavior and autism traits in a sex chromosomal disorder: Klinefelter (47XXY) syndrome. J Autism Dev Disord 9:1634–1641

Von Gontard A, Hollmann E, Eiberg H, Benden B, Rittig S, Lehmkuhl G (1997) Clinical enuresis phenotypes in familial nocturnal enuresis. Scand J Urol Nephrol (Suppl):11–16

Von Gontard A, Eiberg H, Hollmann E, Rittig S, Lehmkuhl G (1999) Molecular genetics of nocturnal enuresis: linkage to a locus on chromosome 22. Scand J Urol Nephrol (Suppl):76–80

Von Gontard A, Schaumburg H, Hollmann E, Eiberg H, Rittig S (2001) The genetics of enuresis: a review. J Urol 6:2438–2443

Vorstman JA, Morcus ME, Duijff SN, Klaassen PW, Heineman-de Boer JA, Beemer FA, Swaab H, Kahn RS, van Engeland H (2006a) The 22q11.2 deletion in children: high rate of autistic disorders and early onset of psychotic symptoms. J Am Acad Child Adolesc Psychiatry 9:1104–1113

Vorstman JA, Staal WG, van Daalen E, van Engeland H, Hochstenbach PF, Franke L (2006b) Identification of novel autism candidate regions through analysis of reported cytogenetic abnormalities associated with autism. Mol Psychiatry 1:1, 18–28

Walters RG, Jacquemont S, Valsesia A, de Smith AJ, Martinet D, Andersson J, Falchi M, Chen F, Andrieux J, Lobbens S, Delobel B, Stutzmann F, El-Sayed Moustafa JS, Chevre JC, Lecoeur C, Vatin V, Bouquillon S, Buxton JL, Boute O, Holder-Espinasse M, Cuisset JM, Lemaitre MP, Ambresin AE, Brioschi A, Gaillard M, Giusti V, Fellmann F, Ferrarini A, Hadjikhani N, Campion D, Guilmatre A, Goldenberg A, Calmels N, Mandel JL, Le CC, David A, Isidor B, Cordier MP, Dupuis-Girod S, Labalme A, Sanlaville D, Beri-Dexheimer M, Jonveaux P, Leheup B, Ounap K, Bochukova EG, Henning E, Keogh J, Ellis RJ, Macdermot KD, van Haelst MM, Vincent-Delorme C, Plessis G, Touraine R, Philippe A, Malan V, Mathieu-Dramard M, Chiesa J, Blaumeiser B, Kooy RF, Caiazzo R, Pigeyre M, Balkau B, Sladek R, Bergmann S, Mooser V, Waterworth D, Reymond A, Vollenweider P, Waeber G, Kurg A, Palta P, Esko T, Metspalu A, Nelis M, Elliott P, Hartikainen AL, McCarthy MI, Peltonen L, Carlsson L, Jacobson P, Sjostrom L, Huang N, Hurles ME, O'Rahilly S, Farooqi IS, Mannik K, Jarvelin MR, Pattou F, Meyre D, Walley AJ, Coin LJ, Blakemore AI, Froguel P, Beckmann JS (2010) A new highly penetrant form of obesity due to deletions on chromosome 16p11.2. Nature 7281:671–675

Walzer S (1985) X chromosome abnormalities and cognitive development: implications for understanding normal human development. J Child Psychol Psychiatry 2:177–184

Wang L, Jia M, Yue W, Tang F, Qu M, Ruan Y, Lu T, Zhang H, Yan H, Liu J, Guo Y, Zhang J, Yang X, Zhang D (2008) Association of the ENGRAILED 2 (EN2) gene with autism in Chinese Han population. Am J Med Genet B Neuropsychiatr Genet 4:434–438

Wang D, Ma J, Zhang S, Hinney A, Hebebrand J, Wang Y, Wang HJ (2010a) Association of the MC4R V103I polymorphism with obesity: a Chinese case-control study and meta-analysis in 55,195 individuals. Obesity (Silver Spring) 3:573–579

Wang K, Li WD, Glessner JT, Grant SF, Hakonarson H, Price RA (2010b) Large copy-number variations are enriched in cases with moderate to extreme obesity. Diabetes 10:2690–2694

Weiss LA, Shen Y, Korn JM, Arking DE, Miller DT, Fossdal R, Saemundsen E, Stefansson H, Ferreira MA, Green T, Platt OS, Ruderfer DM, Walsh CA, Altshuler D, Chakravarti A, Tanzi RE, Stefansson K, Santangelo SL, Gusella JF, Sklar P, Wu BL, Daly MJ (2008) Association between microdeletion and microduplication at 16p11.2 and autism. N Engl J Med 7:667–675

Willer CJ, Speliotes EK, Loos RJ, Li S, Lindgren CM, Heid IM, Berndt SI, Elliott AL, Jackson AU, Lamina C, Lettre G, Lim N, Lyon HN, McCarroll SA, Papadakis K, Qi L, Randall JC, Roccasecca RM, Sanna S, Scheet P, Weedon MN, Wheeler E, Zhao JH, Jacobs LC, Prokopenko I, Soranzo N, Tanaka T, Timpson NJ, Almgren P, Bennett A, Bergman RN, Bingham SA, Bonnycastle LL, Brown M, Burtt NP, Chines P, Coin L, Collins FS, Connell JM, Cooper C, Smith GD, Dennison EM, Deodhar P, Elliott P, Erdos MR, Estrada K, Evans DM, Gianniny L, Gieger C, Gillson CJ, Guiducci C, Hackett R, Hadley D, Hall AS, Havulinna AS, Hebebrand J, Hofman A, Isomaa B, Jacobs KB, Johnson T, Jousilahti P, Jovanovic Z, Khaw KT, Kraft P, Kuokkanen M, Kuusisto J, Laitinen J, Lakatta EG, Luan J, Luben RN, Mangino M, McArdle WL, Meitinger T, Mulas A, Munroe PB, Narisu N, Ness AR, Northstone K, O'Rahilly S, Purmann C, Rees MG, Ridderstrale M, Ring SM, Rivadeneira F, Ruokonen A, Sandhu MS, Saramies J, Scott LJ, Scuteri A, Silander K, Sims MA, Song K, Stephens J, Stevens S, Stringham HM, Tung YC, Valle TT, Van Duijn CM, Vimaleswaran KS, Vollenweider P, Waeber G, Wallace C, Watanabe RM, Waterworth DM, Watkins N, Witteman JC, Zeggini E, Zhai G, Zillikens MC,

Altshuler D, Caulfield MJ, Chanock SJ, Farooqi IS, Ferrucci L, Guralnik JM, Hattersley AT, Hu FB, Jarvelin MR, Laakso M, Mooser V, Ong KK, Ouwehand WH, Salomaa V, Samani NJ, Spector TD, Tuomi T, Tuomilehto J, Uda M, Uitterlinden AG, Wareham NJ, Deloukas P, Frayling TM, Groop LC, Hayes RB, Hunter DJ, Mohlke KL, Peltonen L, Schlessinger D, Strachan DP, Wichmann HE, McCarthy MI, Boehnke M, Barroso I, Abecasis GR, Hirschhorn JN (2009) Six new loci associated with body mass index highlight a neuronal influence on body weight regulation. Nat Genet 1:25–34

Williams NM, Zaharieva I, Martin A, Langley K, Mantripragada K, Fossdal R, Stefansson H, Stefansson K, Magnusson P, Gudmundsson OO, Gustafsson O, Holmans P, Owen MJ, O'Donovan M, Thapar A (2010) Rare chromosomal deletions and duplications in attention-deficit hyperactivity disorder: a genome-wide analysis. Lancet 9750:1401–1408

Wood AC (2010) Rethinking shared environment as a source of variance underlying attention-deficit/hyperactivity disorder symptoms: comment on Burt (2009). Psychol Bull

Wood AC, Rijsdijk F, Asherson P, Kuntsi J (2009) Hyperactive-impulsive symptom scores and oppositional behaviours reflect alternate manifestations of a single liability. Behav Genet 5:447–460

Wu S, Jia M, Ruan Y, Liu J, Guo Y, Shuang M, Gong X, Zhang Y, Yang X, Zhang D (2005) Positive association of the oxytocin receptor gene (OXTR) with autism in the Chinese Han population. Biol Psychiatry 1:74–77

Xiang B, Zhu H, Shen Y, Miller DT, Lu K, Hu X, Andersson HC, Narumanchi TM, Wang Y, Martinez JE, Wu BL, Li P, Li MM, Chen TJ, Fan YS (2010) Genome-wide oligonucleotide array comparative genomic hybridization for etiological diagnosis of mental retardation: a multicenter experience of 1499 clinical cases. J Mol Diagn 2:204–212

Yang B, Chan RC, Jing J, Li T, Sham P, Chen RY (2007) A meta-analysis of association studies between the 10-repeat allele of a VNTR polymorphism in the 3'-UTR of dopamine transporter gene and attention deficit hyperactivity disorder. Am J Med Genet B Neuropsychiatr Genet

Yang P, Lung FW, Jong YJ, Hsieh HY, Liang CL, Juo SH (2008) Association of the homeobox transcription factor gene ENGRAILED 2 with autistic disorder in Chinese children. Neuropsychobiology 1–2:3–8

Yeo GS, Farooqi IS, Aminian S, Halsall DJ, Stanhope RG, O'Rahilly S (1998) A frameshift mutation in MC4R associated with dominantly inherited human obesity. Nat Genet 2:111–112

Yeo GS, Connie Hung CC, Rochford J, Keogh J, Gray J, Sivaramakrishnan S, O'rahilly S, Farooqi IS (2004) A de novo mutation affecting human TrkB associated with severe obesity and developmental delay. Nat Neurosci 11:1187–1189

Young EH, Wareham NJ, Farooqi S, Hinney A, Hebebrand J, Scherag A, O'Rahilly S, Barroso I, Sandhu MS (2007) The V103I polymorphism of the MC4R gene and obesity: population based studies and meta-analysis of 29 563 individuals. Int J Obes (Lond) 9:1437–1441

Zhang Y, Proenca R, Maffei M, Barone M, Leopold L, Friedman JM (1994) Positional cloning of the mouse obese gene and its human homologue. Nature 6505:425–432

Zhang H, Liu X, Zhang C, Mundo E, Macciardi F, Grayson DR, Guidotti AR, Holden JJ (2002) Reelin gene alleles and susceptibility to autism spectrum disorders. Mol Psychiatry 9:1012–1017

Zhou K, Chen W, Buitelaar J, Banaschewski T, Oades RD, Franke B, Sonuga-Barke E, Ebstein R, Eisenberg J, Gill M, Manor I, Miranda A, Mulas F, Roeyers H, Rothenberger A, Sergeant J, Steinhausen HC, Lasky-Su J, Taylor E, Brookes KJ, Xu X, Neale BM, Rijsdijk F, Thompson M, Asherson P, Faraone SV (2008a) Genetic heterogeneity in ADHD: DAT1 gene only affects probands without CD. Am J Med Genet B Neuropsychiatr Genet 8:1481–1487

Zhou K, Dempfle A, Arcos-Burgos M, Bakker SC, Banaschewski T, Biederman J, Buitelaar J, Castellanos FX, Doyle A, Ebstein RP, Ekholm J, Forabosco P, Franke B, Freitag C, Friedel S, Gill M, Hebebrand J, Hinney A, Jacob C, Lesch KP, Loo SK, Lopera F, McCracken JT, McGough JJ, Meyer J, Mick E, Miranda A, Muenke M, Mulas F, Nelson SF, Nguyen TT, Oades RD, Ogdie MN, Palacio JD, Pineda D, Reif A, Renner TJ, Roeyers H, Romanos M, Rothenberger A, Schafer H, Sergeant J, Sinke RJ, Smalley SL, Sonuga-Barke E, Steinhausen HC, van der Meulen E, Walitza S, Warnke A, Lewis CM, Faraone SV, Asherson P (2008b) Meta-analysis of genome-wide linkage scans of attention deficit hyperactivity disorder. Am J Med Genet B Neuropsychiatr Genet 8:1392–1398

Schizophrenia Genes: On the Matter of Their Convergence

Dan Rujescu

Abstract Schizophrenia is a common mental disorder, affecting 0.5–1% of the population. The mode of inheritance is complex and non-Mendelian with a high heritability of ca. 65–80%. Given this complexity, until most recently it was difficult to identify disease genes. But fortunately this has changed. Due to new technologies the last few years have brought highest interest in human genetics of complex diseases. The knowledge resulting from the availability of the complete sequence of the human genome, the systematic identification of single nucleotide polymorphisms (SNPs) throughout the genome, and the development of parallel genotyping technology (microarrays) established the conditions that brought about the current successful time in our ability to probe the genome for identifying disease genes. All these studies showed up new avenues for the biology of common complex diseases and yielded a multitude of genes showing strong association with complex diseases.

Keywords Schizophrenia · Genetics · CNV · Copy number variant · GWA · Polymorphism

Contents

1 Schizophrenia 430
2 Structural Genetic Variants 430
3 Common Genetic Variants 434
References 436

D. Rujescu (✉)
Department of Psychiatry,
Ludwig-Maximilians-University,
Nußbaumstr. 7, 80336 München, Germany
e-mail: Dan.Rujescu@med.uni-muenchen.de

Curr Topics Behav Neurosci (2012) 12: 429–440
DOI: 10.1007/7854_2011_183

Published Online: 14 January 2012

1 Schizophrenia

A major challenge in medicine is to understand genetic, molecular and cellular mechanisms underlying common mental disorders including schizophrenia, which involve complicated genetic and environmental determinants. Schizophrenia is a common mental disorder, affecting 0.5–1% of the population. Schizophrenia mostly presents with several episodes and tends to become chronic. Approximately 30% of patients with schizophrenia require support throughout their lives. Roughly 50% will have lifelong disabilities and social problems. Its direct costs in western countries range between 1.6 and 2.6% of total health care expenditures, and is the seventh most costly medical illness to western societies. Active psychosis has been ranked the third most-disabling condition after quadriplegia and dementia (Ustün et al. 1999), and life expectancy is reduced by ten to twelve years, due to increased physical health problems and a high suicide rate. There is evidence for a strong genetic component in the etiology of schizophrenia, as demonstrated by family, twin and adoption studies. The relative contribution of genetic factors has been estimated to be ca. 80% (Cardno et al. 1999). The mode of inheritance is complex and non-Mendelian (Ross et al. 2006). Given this complexity, together with imprecise and differing definitions of phenotype, until most recently it was notoriously difficult to identify genes involved in this chronic disabling brain disease.

2 Structural Genetic Variants

Several approaches have been used to find causative genetic variants including linkage studies, candidate gene or genome wide association studies as well as studies on structural genetic variants. It is not a new assumption that structural genetic variants including deletions or duplications may play a role in schizophrenia. However, the assumption that rare genetic variants with large effects may account for a significant number of schizophrenia cases has been somehow neglected. This occurred despite the fact that rare variants with large effects on schizophrenia risk are already well established, with the 22q11 deletion syndrome and DISC1 being among the more prominent findings.

David St Clair et al. (1990) identified a balanced 1q43:11q14 translocation associated with major mental illness in a large pedigree showing several psychiatric phenotypes associated with this translocation (St Clair et al. 1990). Within this translocation a disrupted gene (DISC1) was identified which is now an important risk gene for psychosis. Although the mutation is very rare, its discovery boosted a series of studies on the potential pathobiology underlying psychosis. Twenty years of intensive work went by and much research has been undertaken to define the biological function of the DISC1 protein and to further understand how it contributes to the pathogenesis of schizophrenia. There is evidence for a role in multiple cell functions including cAMP signalling, centrosomal and microtubule-based functions, kenesin-mediated intracellular transport and neurite extension which have led to its description as a 'hub protein' for both the developing and adult brain (for review see Gill et al. 2009; Muir et al. 2008). Interestingly, DISC1 expression is regulated by beta-site amyloid precursor protein cleaving enzyme-1-neuregulin cascade (Seshari et al. 2010), implicating another potential schizophrenia risk gene, neuregulin 1 (Stefansson et al. 2003) and its receptors in the same schizophrenia risk pathway as DISC1 (for review see Gill et al. 2010). DISC1 has multiple identified protein interaction partners that highlight pathologically relevant molecular pathways. Amongst these are proteins involved in neuronal migration (e.g. APP, Dixdc1, LIS1, NDE1, NDEL1), neural progenitor proliferation (GSK3β), neurosignalling (Girdin, GSK3β, PDE4) and synaptic function (Kal7, TNIK). Furthermore, emerging evidence of genetic association (NDEL1, PCM1, PDE4B) and copy number variation (NDE1) implicate several DISC1-binding partners as risk factors for schizophrenia (Bradshaw and Porteous 2010).

A microdeletion at 22q11.2 has been repeatedly linked to schizophrenia although it is responsible for the Velo-Cardio-Facial Syndrome (VCFS) also known as DiGeorge syndrome. The first connections starting when Shprintzen et al. (1992) reported schizophrenia-like psychotic symptoms in VCFS affected adolescents and young adults, beside other characteristic symptoms like different physiognomy or cardiac defects. Several follow up studies occurred and provided hope that this deletion could serve as a model for schizophrenia (Karayiorgou and Gogos 2004). A series of animal model and cell culture studies started. One major focus was on the genes lying in this deleted region (for review see Gothelf et al. 2008; Karayiorgou and Gogos 2004). One of the genes within this region is the COMT (Catechol-O-methyl transferase) gene which encodes for the COMT enzyme responsible for the degradation of dopamine, especially in the prefrontal cortex. Although there is evidence for a contribution of COMT to neuropsychological and neurophysiological domains, no clear picture is seen for a major contribution of COMT in schizophrenia *per se* as demonstrated in a recent meta-analysis by Okochi et al. (2009) in 13,088 cases and 16,531 controls. Most probably COMT seems to be more generally involved in neuropsychological and neurophysiological functions including executive functioning, working memory, fluid intelligence and attentional control (for review see Dickinson and Elvevåg 2009).

Other genes which were intensively studied within the 22q11.2 region are e.g. TBX1 and GNB1L. It has been shown that prepulse inhibition deficits in Df1/+ mice are caused by haploinsufficiency of these two genes TBX1 and GNB1L (Paylor et al. 2006). In humans, sequence variation in GNB1L was associated with gene expression and psychosis in VCFS (Williams et al. 2008). Furthermore, mouse models of 22q11.2 deletion syndrome suggest that diminished dosage of certain 22q11 genes disrupts neurogenesis and cortical development (Meechan et al. 2009). Interestingly, Maynard et al. (2008) studied six genes in the 1.5 Mb of the 22q11.2 region which encode mitochondrial proteins (Prodh, Slc25a1, Txnrd2, T10, Zdhhc8). All six genes are expressed in the brain, and maximal expression coincides with peak forebrain synaptogenesis shortly after birth. Furthermore, their protein products are associated with brain mitochondria, including those in synaptic terminals. The authors concluded that 22q11 deletion may alter metabolic properties of cortical mitochondria during early post-natal life, and that several 22q11 mitochondrial genes, particularly during early post-natal cortical development, may disrupt neuronal metabolism or synaptic signalling which seems highly interesting for the pathophysiology of schizophrenia (Maynard et al. 2008).

These two regions were the most prominent large structural variants associated with schizophrenia until recently. However, the rapid technological development which paved the way for genome wide association studies provided new unexpected possibilities for the detection of shorter structural variants, namely for copy number variants (CNVs). The first report on submicroscopic micodeletions or microduplications in schizophrenia was provided by Walsh et al. (2008) and was the starting point for the revival of structural aberrations in schizophrenia. Novel deletions and duplications of genes were present in 5% of controls versus 15% of cases and 20% of young onset cases. These mutations disrupted genes disproportionately more often from signalling networks controlling neurodevelopment, including neuregulin and glutamate pathways. Although the sample size for the detection of these rare events was small, Walsh et al. (2008) suggested that multiple, individually rare mutations altering genes in neurodevelopmental pathways contribute to schizophrenia.

The first well powered study on CNVs associating with schizophrenia was performed by the SGENE+ consortium. A population based sample was used to identify de novo CNVs by analysing 9,878 transmissions from parents to offspring (Stefansson et al. 2008). The 66 de novo CNVs identified were tested for association in a sample of 1,433 schizophrenia cases and 33,250 controls. Three deletions at 1q21.1, 15q11.2 and 15q13.3 showing nominal association with schizophrenia in the first sample were followed up in a second sample of 3,285 cases and 7,951 controls and all three deletions were significantly associated with schizophrenia and related psychoses in the combined sample. Interestingly, two of these regions (1q21.1 and 15q13.3) could be replicated independently by the International Schizophrenia Consortium in 3,391 patients with schizophrenia and 3,181 controls (ISC 2008). The third (15q11.2) association has also been replicated independently by Kirov et al. (2009a). These studies have brought highest interest in human genetics of complex diseases and encouraged other groups to analyse CNVs.

Another main finding in schizophrenia is that deletions within the neurexin 1 gene (NRXN1; 2p16.3) are associated with schizophrenia (Kirov et al. 2008; Walsh et al. 2008). The SGENE+ consortium examined Neurexin 1 for CNVs in 2,977 schizophrenia patients and 33,746 controls from seven European populations using microarray data. The association analysis was restricted to CNVs that disrupt exons. These were significantly associated with a high odds ratio (OR 8.97), showing that Neurexin 1 deletions disrupting exons confer risk of schizophrenia (Rujescu et al. 2009). A meta-analysis which included 8,789 cases and 42,054 controls provided further evidence for the involvement of Neurexin 1 in Schizophrenia.

An excess of deletions >100 kb in patients (0.19 vs. 0.04%, $p = 0.000013$, OR = 4.78) with higher ORs for deletions >100 kb that disrupt exons ($p = 0.000037$, OR = 7.44) was detected (Kirov et al. 2009b).

Interestingly, these microdeletions are also found in patients with a broad overlapping spectrum of other neurodevelopmental phenotypes like autism (Sebat et al. 2007; Abrahams and Geschwind 2008; Weiss et al. 2008), mental retardation (de Vries et al. 2005; Sharp et al. 2006; Murthy et al. 2007; Mefford et al. 2008), epilepsy (Sharp et al. 2008) and further diseases (for review see (Cook and Scherer 2008; Ramocki and Zoghbi 2008; Slavotinek 2008). Therefore, a genomic region on chromosome 16p13.1, which has been implicated in childhood-onset developmental disorders, was studied in schizophrenia by the SGENE consortium (Ingason et al. 2011a). Deletions were detected in 0.12% of cases and 0.04% of controls ($p > 0.05$). Furthermore, also duplications were present in 0.30% of cases versus 0.09% of controls ($p = 0.007$). The region can be divided into three intervals defined by flanking low copy repeats. Duplications spanning intervals I and II showed the most significant ($p = 0.00010$) association with schizophrenia. The age of onset in duplication and deletion carriers among cases ranged from 12 to 35 years, and the majority were males with a family history of psychiatric disorders. In a single Icelandic family, a duplication spanning intervals I and II was present in two cases of schizophrenia, and individual cases of alcoholism, attention deficit hyperactivity disorder and dyslexia (Ingason et al. 2011a).

Given that also duplications seem to be involved in the risk of schizophrenia, another large study on copy number variants was conducted studying maternally derived 15q11–q13 duplication overlapping the Prader-Willi/Angelman syndrome critical region (Ingason et al. 2011b). In a discovery sample of 22 schizophrenia patients with a very early onset of illness (10–15 years of age), one duplication was observed in a patient. Based on this, 7,582 patients with schizophrenia or schizoaffective disorder and 41,370 comparison subjects without known psychiatric illness were screened for copy number variants at 15q11–q13. Duplications were found in further four patients and five comparison subjects. All four patients had maternally derived duplications (0.05%), while only three of the five comparison duplications were maternally derived (0.007%), resulting in a significant excess of maternally derived duplications in case patients (OR = 7.3). This excess is compatible with earlier observations that risk for psychosis in people with Prader–Willi syndrome caused by maternal uniparental disomy is much higher than in those caused by deletion of the paternal chromosome. These findings suggest that the presence of two maternal copies of a fragment of chromosome 15q11.2–q13.1 that overlaps with the Prader–Willi/Angelman syndrome critical region may be a rare risk factor for schizophrenia and other psychoses. Given that maternal duplications of this region are among the most consistent cytogenetic observations in autism, the findings provide further support for a shared genetic etiology between autism and psychosis (Ingason et al. 2011b).

Actually first whole-genome sequencing efforts are under way and raise hope that further rare variants will be found. A recent study based on whole-genome DNA sequencing data from 185 human genomes mapped over 22,000 deletions and 6,000 additional CNVs. Over half of them could be mapped to nucleotide resolution, which facilitates analysing their origin and functional impact. This study shows that CNVs are abundant in humans, differing from other forms of variation in extent, origin and functional impact (Milles et al. 2011). Therefore it is very possible that based on these new findings, rare new genomic disorders will be defined and that our understanding of neurodevelopmental disorders including schizophrenia will hopefully be highly enlarged.

3 Common Genetic Variants

Also common polymorphisms (SNPs) were analysed using genome wide association studies. The first study of this type included 178 cases and 144 controls and was performed by Lencz et al. (2007). The best associated SNP out of 500,000 SNPs was located in CSF2RA (colony stimulating factor, receptor 2 alpha). A second genome wide association (GWA) study did not achieved genome wide significance Sullivan et al. (2008). Both studies were clearly underpowered to provide conclusive result. Therefore, O'Donovan (2008) performed an initial GWAS on a much larger sample including 479 cases and 2,937 controls. Loci surpassing $p < 10\text{-E5}$ were followed up in 6,829 cases and 9,897 controls. Of 12 of these loci, 3 had strong independent support and the overall pattern of replication was unlikely to occur by chance. The evidence for association for the top SNP in the ZNF804A gene strengthened when the affected phenotype included bipolar disorder (O'Donovan et al. 2008). Interestingly, this result was replicated independently in the Irish Case Control Study of Schizophrenia (ICCSS) sample (Riley et al. 2010) as well as in an independent sample of 5,164 schizophrenia cases and 20,709 controls (Steinberg et al. 2011). A fine-mapping, replication and meta-analysis study of 18,945 schizophrenia and schizoaffective disorder patients, 21,274 schizophrenia plus bipolar disorder cases and 38,675 controls showed further evidence for this SNP (Williams et al. 2010). Additionally, there is evidence for the involvement of this SNP in neuropsychological phenotypes. Walters et al. (2010) investigated whether the identified risk allele of the SNP rs1344706 is associated with variations in neuropsychological performance in patients and controls. Patients with DSM-IV diagnosed schizophrenia and healthy controls from independent samples of Irish ($n = 297$ cases and $n = 165$ controls) and German ($n = 251$ cases and $n = 1472$ controls) nationality were included. In the Irish samples ZNF804A genotype was associated with differences in episodic and working memory in patients but not controls. These findings replicated in the same direction in the German sample. Furthermore, in both samples, when patients with lower IQ were excluded the association between ZNF804A and schizophrenia strengthened (Walters et al. 2010). Taken together, the ZNF804A gene is one of the most promising genes for schizophrenia to date.

The largest GWA study to date on schizophrenia was led by Stefansson et al. (2009). 2,663 schizophrenia cases and 13,498 controls from eight European locations were studied within the SGENE+ consortium. Findings from the top 1,500 markers were combined with results for these markers from both the International Schizophrenia Consortium (ISC 2,602 cases/2,885 controls) and the European–American portion of the Molecular Genetics of Schizophrenia Consortium (MGS 2,687 cases/2,656 controls). The best markers were followed up in 5,013 cases and 15,559 controls from four sets of additional samples from Europe. This approach could identify three novel schizophrenia loci: Neurogranin (NRGN), TCF4 and the HLA region (Stefansson et al. 2009). Interestingly, the HLA associations were also found in the large GWAS by the International Schizophrenia Consortium ISC (2009) studying 3,322 European individuals with schizophrenia and 3,587 controls as well as by (Shi et al. 2009) suggesting further evidence for the inflammation theory of schizophrenia. The second gene, TCF4 (transcription cell factor 4), is of high interest too as the Pitt–Hopkins Syndrome's gene was identified by unrelated groups by comparative genomic hybridization (CGH) in 2007 showing that the haploinsufficiency of the TCF4 gene (18q21.2) is due to an autosomal dominant de novo mutation, which is considered to be causative (for review see Taddeucci et al. 2010). Finally, NRGN (neurogranin), the third gene is highly promising. It is a calmodulin-binding protein expressed exclusively in the brain. It is the main postsynaptic protein regulating the availability of calmodulin, binding to it in the absence of calcium (Stefansson et al. 2009). Neurogranin belongs to the calpacitin family and contains an IQ domain that interacts with the Ca^{2+} free form of calmodulin (apo-calmodulin). Depending on the intracellular Ca^{2+} concentration, neurogranin releases calmodulin, so that it can bind Ca^{2+} and activate downstream signalling molecules. Calmodulin furthermore binds different proteins including CaMKII, the calcium/calmodulin-dependent protein kinase II gamma (Hayashi 2009). Zhong et al. (2009) show that neurogranin is concentrated in dendritic spines and that the number of neurogranin molecules available determines the efficiency of calmodulin signalling in the synapse and the strength of AMPA receptor transmission (Zhong et al. 2009). An exogenous overexpression of neurogranin is sufficient to trigger a Ca^{2+} signal that normally induces Long Term Potentiation (LTP) which is crucial for learning and memory. This neurogranin-induced synaptic potentiation and LTP share common signalling mechanisms. Like LTP, neurogranin-induced potentiation requires the activity of NMDA receptors and CaMKII. Furthermore, in such neurons that overexpress neurogranin and where the AMPA receptor response is therefore already potentiated, LTP can no longer be electrophysiologically induced. Taken together, these observations in turn imply the involvement of neurogranin in the signalling pathway of LTP (Hayashi 2009).

Further large genome wide association studies on schizophrenia are currently under way and will hopefully add new relevant knowledge on schizophrenia. Based on this, functional studies including cell and animal models will be necessary. There is new hope that these new avenues will help understanding the neurobiology of schizophrenia in more depth leading to the development of new innovative diagnostic tools and therapies.

References

Abrahams BS, Geschwind DH (2008) Advances in autism genetics: on the threshold of a new neurobiology. Nat Rev Genet 9(5):341–355

Bradshaw NJ, Porteous DJ (2010) DISC1-binding proteins in neural development, signalling and schizophrenia. Neuropharmacology. (Epub ahead of print)

Cardno AG, Marshall EJ, Coid B, Macdonald AM, Ribchester TR, Davies NJ, Venturi P, Jones LA, Lewis SW, Sham PC, Gottesman II, Farmer AE, McGuffin P, Reveley AM, Murray RM (1999) Heritability estimates for psychotic disorders: the Maudsley twin psychosis series. Arch Gen Psychiatry 56:162–168

Cook EH Jr, Scherer SW (2008) Copy-number variations associated with neuropsychiatric conditions. Nature 455(7215):919–923

de Vries BB, Pfundt R, Leisink M, Koolen DA, Vissers LE, Janssen IM, Reijmersdal S, Nillesen WM, Huys EH, Leeuw N, Smeets D, Sistermans EA, Feuth T, Ravenswaaij-Arts CM, van Kessel AG, Schoenmakers EF, Brunner HG, Veltman JA (2005) Diagnostic genome profiling in mental retardation. Am J Hum Genet 77(4):606–616

Dickinson D, Elvevåg B (2009) Genes, cognition and brain through a COMT lens. Neuroscience 164(1):72–87

Gill M, Donohoe G, Corvin A (2010) What have the genomics ever done for the psychoses? Psychol Med Apr 40(4):529–540

Gothelf D, Schaer M, Eliez S (2008) Genes, brain development and psychiatric phenotypes in velo-cardio-facial syndrome. Dev Disabil Res Rev 14(1):59–68

Hayashi Y (2009) Long-term potentiation: two pathways meet at neurogranin. EMBO J 28(19):2859–2860

International Schizophrenia Consortium (ISC) (2008) Rare chromosomal deletions and duplications increase risk of schizophrenia. Nature 455(7210):237–241

International Schizophrenia Consortium (ISC), Purcell SM, Wray NR, Stone JL, Visscher PM, O'Donovan MC, Sullivan PF, Sklar P (2009) Common polygenic variation contributes to risk of schizophrenia and bipolar disorder. Nature 460(7256):748–752

Ingason A, Rujescu D, Cichon S, Sigurdsson E, Sigmundsson T, Pietiläinen OPH, Buizer J, Francks C, Muglia P, Gylfason A, Gustafsson O, Olason PI, Steinberg S, Hansen T, Giegling I, Möller HJ, Hartmann AM, Shianna KV, Ge D, Crombie C, Fraser G, Walker N, Lonnqvist J, Suvisaari J, Tuulio-Henriksson A, Picchioni M, Kiemeney LA, Franke B, Murray R, Vassos E, Ettinger U, Mühleisen TW, Tosato S, Ruggeri M, Djurovic S, Andreasson OA, GROUP, Werge T, Ophoff R, Rietschel M, Goldstein D, Noethen M, Petursson H, Stefansson K, Peltonen L, Collier DA, Stefansson H, St Clair D (2011a) Copy number variations of chromosome 16p13.1 region associated with schizophrenia. Molecular Psychiatry 16(1):17–25

Ingason A, Kirov G, Giegling I, Hansen T, Isles AR, Jakobsen KD, Kristinsson KT, le Roux L, Gustafsson O, Craddock N, Möller HJ, McQuillin A, Muglia P, Cichon S, Rietschel M, Ophoff RA, Djurovic S, Andreassen OA, Pietiläinen OP, Peltonen L, Dempster E, Collier DA, St Clair D, Rasmussen HB, Glenthøj BY, Kiemeney LA, Franke B, Tosato S, Bonetto C, Saemundsen E, Hreidarsson SJ; GROUP Investigators, Nöthen MM, Gurling H, O'Donovan MC, Owen MJ, Sigurdsson E, Petursson H, Stefansson H, Rujescu D, Stefansson K, Werge T. Maternally (2011b) Derived microduplications at 15q11–q13: Implication of imprinted genes in psychotic illness. Am J Psychiatry

Karayiorgou M, Gogos JA (2004) The molecular genetics of the 22q11-associated schizophrenia. Brain Res Mol Brain Res 132(2):95–104

Kirov G, Gumus D, Chen W, Norton N, Georgieva L, Sari M, O'Donovan MC, Erdogan F, Owen MJ, Ropers HH, Ullmann R (2008a) Comparative genome hybridization suggests a role for NRXN1 and APBA2 in schizophrenia. Hum Mol Genet 17(3):458–465

Kirov G, Grozeva D, Norton N, Ivanov D, Mantripragada KK, Holmans P (2009a) International Schizophrenia Consortium; Wellcome Trust Case Control Consortium, Craddock N,

Owen MJ, O'Donovan MC. Support for the involvement of large copy number variants in the pathogenesis of schizophrenia. Hum Mol Genet 18(8):1497–1503

Kirov G, Rujescu D, Ingason A, Collier DA, O'Donovan MC, Owen MJ (2009b) Neurexin 1 (NRXN1) deletions in schizophrenia. Schizophr Bull 35(5):851–854

Lencz T, Morgan TV, Athanasiou M, Dain B, Reed CR, Kane JM, Kucherlapati R, Malhotra AK (2007) Converging evidence for a pseudoautosomal cytokine receptor gene locus in schizophrenia. Mol Psychiatry 12(6):572–580

Maynard TM, Meechan DW, Dudevoir ML, Gopalakrishna D, Peters AZ, Heindel CC, Sugimoto TJ, Wu Y, Lieberman JA, Lamantia AS (2008) Mitochondrial localization and function of a subset of 22q11 deletion syndrome candidate genes. Mol Cell Neurosci 39(3):439–451

Meechan DW, Tucker ES, Maynard TM, LaMantia AS (2009) Diminished dosage of 22q11 genes disrupts neurogenesis and cortical development in a mouse model of 22q11 deletion/ DiGeorge syndrome. Proc Natl Acad Sci U S A 106(38):16434–16445

Mefford HC, Sharp AJ, Baker C, Itsara A, Jiang Z, Buysse K, Huang S, Maloney VK, Crolla JA, Baralle D, Collins A, Mercer C, Norga K, de Ravel T, Devriendt K, Bongers EM, de Leeuw N, Reardon W, Gimelli S, Bena F, Hennekam RC, Male A, Gaunt L, Clayton-Smith J, Simonic I, Park SM, Mehta SG, Nik-Zainal S, Woods CG, Firth HV, Parkin G, Fichera M, Reitano S, Lo Giudice M, Li KE, Casuga I, Broomer A, Conrad B, Schwerzmann M, Räber L, Gallati S, Striano P, Coppola A, Tolmie JL, Tobias ES, Lilley C, Armengol L, Spysschaert Y, Verloo P, De Coene A, Goossens L, Mortier G, Speleman F, van Binsbergen E, Nelen MR, Hochstenbach R, Poot M, Gallagher L, Gill M, McClellan J, King MC, Regan R, Skinner C, Stevenson RE, Antonarakis SE, Chen C, Estivill X, Menten B, Gimelli G, Gribble S, Schwartz S, Sutcliffe JS, Walsh T, Knight SJ, Sebat J, Romano C, Schwartz CE, Veltman JA, de Vries BB, Vermeesch JR, Barber JC, Willatt L, Tassabehji M, Eichler EE (2008) Recurrent rearrangements of chromosome 1q21.1 and variable pediatric phenotypes. N Engl J Med 359(16):1685–1699

Mills RE, Walter K, Stewart C, Handsaker RE, Chen K, Alkan C, Abyzov A, Yoon SC, Ye K, Cheetham RK, Chinwalla A, Conrad DF, Fu Y, Grubert F, Hajirasouliha I, Hormozdiari F, Iakoucheva LM, Iqbal Z, Kang S, Kidd JM, Konkel MK, Korn J, Khurana E, Kural D, Lam HY, Leng J, Li R, Li Y, Lin CY, Luo R, Mu XJ, Nemesh J, Peckham HE, Rausch T, Scally A, Shi X, Stromberg MP, Stütz AM, Urban AE, Walker JA, Wu J, Zhang Y, Zhang ZD, Batzer MA, Ding L, Marth GT, McVean G, Sebat J, Snyder M, Wang J, Ye K, Eichler EE, Gerstein MB, Hurles ME, Lee C, McCarroll SA, Korbel JO (2011) 1000 Genomes Project. Mapping copy number variation by population-scale genome sequencing. Nature 3 470(7332):59–65

Muir WJ, Pickard BS, Blackwood DH (2008) Disrupted-in-Schizophrenia-1. Curr Psychiatry Rep 10(2):140–147

O'Donovan MC, Craddock N, Norton N, Williams H, Peirce T, Moskvina V, Nikolov I, Hamshere M, Carroll L, Georgieva L, Dwyer S, Holmans P, Marchini JL, Spencer CC, Howie B, Leung HT, Hartmann AM, Möller HJ, Morris DW, Shi Y, Feng G, Hoffmann P, Propping P, Vasilescu C, Maier W, Rietschel M, Zammit S, Schumacher J, Quinn EM, Schulze TG, Williams NM, Giegling I, Iwata N, Ikeda M, Darvasi A, Shifman S, He L, Duan J, Sanders AR, Levinson DF, Gejman PV, Cichon S, Nöthen MM, Gill M, Corvin A, Rujescu D, Kirov G, Owen MJ, Buccola NG, Mowry BJ, Freedman R, Amin F, Black DW, Silverman JM, Byerley WF, Cloninger CR (2008) Molecular genetics of schizophrenia collaboration. identification of loci associated with schizophrenia by genome-wide association and follow-up. Nat Genet 40(9):1053–1055

Okochi T, Ikeda M, Kishi T, Kawashima K, Kinoshita Y, Kitajima T, Yamanouchi Y, Tomita M, Inada T, Ozaki N, Iwata N (2009) Meta-analysis of association between genetic variants in COMT and schizophrenia: an update. Schizophr Res 110(1–3):140–148

Paylor R, Glaser B, Mupo A, Ataliotis P, Spencer C, Sobotka A, Sparks C, Choi CH, Oghalai J, Curran S, Murphy KC, Monks S, Williams N, O'Donovan MC, Owen MJ, Scambler PJ, Lindsay E (2006) Tbx1 haploinsufficiency is linked to behavioral disorders in mice and humans: implications for 22q11 deletion syndrome. Proc Natl Acad Sci U S A 103(20): 7729–7734

Ramocki MB, Zoghbi HY (2008) Failure of neuronal homeostasis results in common neuropsychiatric phenotypes. Nature 455(7215):912–918

Riley B, Thiselton D, Maher BS, Bigdeli T, Wormley B, McMichael GO, Fanous AH, Vladimirov V, O'Neill FA, Walsh D, Kendler KS (2010) Replication of association between schizophrenia and ZNF804A in the Irish Case-Control Study of Schizophrenia sample. Mol Psychiatry 15(1):29–37

Ross CA, Margolis RL, Reading SA, Pletnikov M, Coyle JT (2006) Neurobiology of schizophrenia. Neuron 52(1):139–153

Rujescu D, Ingason A, Cichon S, Pietilainen OP, Barnes MR, Toulopoulou T, Picchioni M, Vassos E, Ettinger U, Bramon E, Murray R, Ruggeri M, Tosato S, Bonetto C, Steinberg S, Sigurdsson E, Sigmundsson T, Petursson H, Gylfason A, Olason PI, Hardarsson G, Jonsdottir GA, Gustafsson O, Fossdal R, Giegling I, Moller HJ, Hartmann A, Hoffmann P, Crombie C, Fraser G, Walker N, Lonnqvist J, Suvisaari J, Tuulio-Henriksson A, Andreassen OA, Djurovic S, Hansen T, Werge T, Melle I, Kiemeney LA, Franke B, Buizer-Voskamp JE, Ophoff RA, Rietschel M, Nothen MM, Stefansson K, Peltonen L, St Clair D, Stefansson H, Collier DA (2009) Disruption of the neurexin 1 gene is associated with schizophrenia. Hum Mol Genet 18(5):988–996

Sebat J, Lakshmi B, Malhotra D, Troge J, Lese-Martin C, Walsh T, Yamrom B, Yoon S, Krasnitz A, Kendall J, Leotta A, Pai D, Zhang R, Lee YH, Hicks J, Spence SJ, Lee AT, Puura K, Lehtimaki T, Ledbetter D, Gregersen PK, Bregman J, Sutcliffe JS, Jobanputra V, Chung W, Warburton D, King MC, Skuse D, Geschwind DH, Gilliam TC, Ye K, Wigler M (2007) Strong association of de novo copy number mutations with autism. Science 316(5823): 445–449

Seshadri S, Kamiya A, Yokota Y, Prikulis I, Kano S, Hayashi-Takagi A, Stanco A, Eom TY, Rao S, Ishizuka K, Wong P, Korth C, Anton ES, Sawa A (2010) Disrupted-in-Schizophrenia-1 expression is regulated by beta-site amyloid precursor protein cleaving enzyme-1-neuregulin cascade. Proc Natl Acad Sci U S A 107(12):5622–5627

Sharp AJ, Hansen S, Selzer RR, Cheng Z, Regan R, Hurst JA, Stewart H, Price SM, Blair E, Hennekam RC, Fitzpatrick CA, Segraves R, Richmond TA, Guiver C, Albertson DG, Pinkel D, Eis PS, Schwartz S, Knight SJ, Eichler EE (2006) Discovery of previously unidentified genomic disorders from the duplication architecture of the human genome. Nat Genet 38(9):1038–1042

Sharp AJ, Mefford HC, Li K, Baker C, Skinner C, Stevenson RE, Schroer RJ, Novara F, De Gregori M, Ciccone R, Broomer A, Casuga I, Wang Y, Xiao C, Barbacioru C, Gimelli G, Bernardina BD, Torniero C, Giorda R, Regan R, Murday V, Mansour S, Fichera M, Castiglia L, Failla P, Ventura M, Jiang Z, Cooper GM, Knight SJ, Romano C, Zuffardi O, Chen C, Schwartz CE, Eichler EE (2008) A recurrent 15q13.3 microdeletion syndrome associated with mental retardation and seizures. Nat Genet 40(3):322–328

Shi J, Levinson DF, Duan J, Sanders AR, Zheng Y, Pe'er I, Dudbridge F, Holmans PA, Whittemore AS, Mowry BJ, Olincy A, Amin F, Cloninger CR, Silverman JM, Buccola NG, Byerley WF, Black DW, Crowe RR, Oksenberg JR, Mirel DB, Kendler KS, Freedman R, Gejman PV (2009) Common variants on chromosome 6p22.1 are associated with schizophrenia. Nature 460(7256):753–757

Shprintzen RJ, Goldberg R, Golding-Kushner KJ, Marion RW (1992) Late-onset psychosis in the velo-cardio-facial syndrome. Am J Med Genet 42(1):141–142

Slavotinek AM (2008) Novel microdeletion syndromes detected by chromosome microarrays. Hum Genet 124(1):1–17

St Clair D, Blackwood D, Muir W, Carothers A, Walker M, Spowart G, Gosden C, Evans HJ (1990) Association within a family of a balanced autosomal translocation with major mental illness. Lancet 336(8706):13–16

Stefansson H, Sarginson J, Kong A, Yates P, Steinthorsdottir V, Gudfinnsson E, Gunnarsdottir S, Walker N, Petursson H, Crombie C, Ingason A, Gulcher JR, Stefansson K, St Clair D (2003) Association of neuregulin 1 with schizophrenia confirmed in a Scottish population. Am J Hum Genet 72(1):83–87

Stefansson H, Rujescu D, Cichon S, Pietilainen OP, Ingason A, Steinberg S, Fossdal R, Sigurdsson E, Sigmundsson T, Buizer-Voskamp JE, Hansen T, Jakobsen KD, Muglia P, Francks C, Matthews PM, Gylfason A, Halldorsson BV, Gudbjartsson D, Thorgeirsson TE, Sigurdsson A, Jonasdottir A, Jonasdottir A, Bjornsson A, Mattiasdottir S, Blondal T, Haraldsson M, Magnusdottir BB, Giegling I, Moller HJ, Hartmann A, Shianna KV, Ge D, Need AC, Crombie C, Fraser G, Walker N, Lonnqvist J, Suvisaari J, Tuulio-Henriksson A, Paunio T, Toulopoulou T, Bramon E, Di Forti M, Murray R, Ruggeri M, Vassos E, Tosato S, Walshe M, Li T, Vasilescu C, Muhleisen TW, Wang AG, Ullum H, Djurovic S, Melle I, Olesen J, Kiemeney LA, Franke B, Sabatti C, Freimer NB, Gulcher JR, Thorsteinsdottir U, Kong A, Andreassen OA, Ophoff RA, Georgi A, Rietschel M, Werge T, Petursson H, Goldstein DB, Nothen MM, Peltonen L, Collier DA, St Clair D, Stefansson K (2008) Large recurrent microdeletions associated with schizophrenia. Nature 455(7210):232–236

Stefansson H, Ophoff RA, Steinberg S, Andreassen OA, Cichon S, Rujescu D, Werge T, Pietiläinen OP, Mors O, Mortensen PB, Sigurdsson E, Gustafsson O, Nyegaard M, Tuulio-Henriksson A, Ingason A, Hansen T, Suvisaari J, Lonnqvist J, Paunio T, Børglum AD, Hartmann A, Fink-Jensen A, Nordentoft M, Hougaard D, Norgaard-Pedersen B, Böttcher Y, Olesen J, Breuer R, Möller HJ, Giegling I, Rasmussen HB, Timm S, Mattheisen M, Bitter I, Réthelyi JM, Magnusdottir BB, Sigmundsson T, Olason P, Masson G, Gulcher JR, Haraldsson M, Fossdal R, Thorgeirsson TE, Thorsteinsdottir U, Ruggeri M, Tosato S, Franke B, Strengman E, Kiemeney LA, Genetic Risk and Outcome in Psychosis (GROUP), Melle I, Djurovic S, Abramova L, Kaleda V, Sanjuan J, de Frutos R, Bramon E, Vassos E, Fraser G, Ettinger U, Picchioni M, Walker N, Toulopoulou T, Need AC, Ge D, Yoon JL, Shianna KV, Freimer NB, Cantor RM, Murray R, Kong A, Golimbet V, Carracedo A, Arango C, Costas J, Jönsson EG, Terenius L, Agartz I, Petursson H, Nöthen MM, Rietschel M, Matthews PM, Muglia P, Peltonen L, St Clair D, Goldstein DB, Stefansson K, Collier DA (2009) Common variants conferring risk of schizophrenia. Nature 460(7256): 744–747

Steinberg S, Mors O, Børglum AD, Gustafsson O, Werge T, Mortensen PB, Andreassen OA, Sigurdsson E, Thorgeirsson TE, Böttcher Y, Olason P, Ophoff RA, Cichon S, Gudjonsdottir IH, Pietiläinen OP, Nyegaard M, Tuulio-Henriksson A, Ingason A, Hansen T, Athanasiu L, Suvisaari J, Lonnqvist J, Paunio T, Hartmann A, Jürgens G, Nordentoft M, Hougaard D, Norgaard-Pedersen B, Breuer R, Möller HJ, Giegling I, Glenthøj B, Rasmussen HB, Mattheisen M, Bitter I, Réthelyi JM, Sigmundsson T, Fossdal R, Thorsteinsdottir U, Ruggeri M, Tosato S, Strengman E; GROUP, Kiemeney LA, Melle I, Djurovic S, Abramova L, Kaleda V, Walshe M, Bramon E, Vassos E, Li T, Fraser G, Walker N, Toulopoulou T, Yoon J, Freimer NB, Cantor RM, Murray R, Kong A, Golimbet V, Jönsson EG, Terenius L, Agartz I, Petursson H, Nöthen MM, Rietschel M, Peltonen L, Rujescu D, Collier DA, Stefansson H, St Clair D, Stefansson K (2011) Expanding the range of ZNF804A variants conferring risk of psychosis. Mol Psychiatry 16(1):59–66

Sullivan PF, Lin D, Tzeng JY, van den Oord E, Perkins D, Stroup TS, Wagner M, Lee S, Wright FA, Zou F, Liu W, Downing AM, Lieberman J, Close SL (2008) Genomewide association for schizophrenia in the CATIE study: results of stage 1. Mol Psychiatry 13(6):570–584

Taddeucci G, Bonuccelli A, Mantellassi I, Orsini A, Tarantino E (2010) Pitt-Hopkins syndrome: report of a case with a TCF4 gene mutation. Ital J Pediatr 36(1):12

Ustün TB (1999) The global burden of mental disorders. Am J Public Health. 89(9):1315–1318

Walsh T, McClellan JM, McCarthy SE, Addington AM, Pierce SB, Cooper GM, Nord AS, Kusenda M, Malhotra D, Bhandari A, Stray SM, Rippey CF, Roccanova P, Makarov V, Lakshmi B, Findling RL, Sikich L, Stromberg T, Merriman B, Gogtay N, Butler P, Eckstrand K, Noory L, Gochman P, Long R, Chen Z, Davis S, Baker C, Eichler EE, Meltzer PS, Nelson SF, Singleton AB, Lee MK, Rapoport JL, King MC, Sebat J (2008) Rare structural variants disrupt multiple genes in neurodevelopmental pathways in schizophrenia. Science 320(5875):539–543

Walters JT, Corvin A, Owen MJ, Williams H, Dragovic M, Quinn EM, Judge R, Smith DJ, Norton N, Giegling I, Hartmann AM, Möller HJ, Muglia P, Moskvina V, Dwyer S,

O'Donoghue T, Morar B, Cooper M, Chandler D, Jablensky A, Gill M, Kaladjieva L, Morris DW, O'Donovan MC, Rujescu D, Donohoe G (2010) Psychosis susceptibility gene ZNF804A and cognitive performance in schizophrenia. Arch Gen Psychiatry 67(7):692–700

Weiss LA, Shen Y, Korn JM, Arking DE, Miller DT, Fossdal R, Saemundsen E, Stefansson H, Ferreira MA, Green T, Platt OS, Ruderfer DM, Walsh CA, Altshuler D, Chakravarti A, Tanzi RE, Stefansson K, Santangelo SL, Gusella JF, Sklar P, Wu BL, Daly MJ (2008) Association between microdeletion and microduplication at 16p11.2 and autism. N Engl J Med 358(7):667–675

Williams NM, Glaser B, Norton N, Williams H, Pierce T, Moskvina V, Monks S, Del Favero J, Goossens D, Rujescu D, Giegling I, Kirov G, Craddock N, Murphy KC, O'Donovan MC, Owen MJ (2008) Strong evidence that GNB1L is associated with schizophrenia. Hum Mol Genet 17(4):555–566

Williams HJ, Norton N, Dwyer S, Moskvina V, Nikolov I, Carroll L, Georgieva L, Williams NM, Morris DW, Quinn EM, Giegling I, Ikeda M, Wood J, Lencz T, Hultman C, Lichtenstein P, Thiselton D, Maher BS; Molecular Genetics of Schizophrenia Collaboration (MGS) International Schizophrenia Consortium (ISC), SGENE-plus, GROUP, Malhotra AK, Riley B, Kendler KS, Gill M, Sullivan P, Sklar P, Purcell S, Nimgaonkar VL, Kirov G, Holmans P, Corvin A, Rujescu D, Craddock N, Owen MJ, O'Donovan MC (2010) Fine mapping of ZNF804A and genome-wide significant evidence for its involvement in schizophrenia and bipolar disorder. Mol Psychiatry 2010 (Epub ahead of print)

Zhong L, Cherry T, Bies CE, Florence MA, Gerges NZ (2009) Neurogranin enhances synaptic strength through its interaction with calmodulin. EMBO J 28(19):3027–3039

Gene × Environment Interaction Models in Psychiatric Genetics

Katja Karg and Srijan Sen

Abstract Gene–environment (G × E) interaction research is an emerging area in psychiatry, with the number of G × E studies growing rapidly in the past two decades. This article aims to give a comprehensive introduction to the field, with an emphasis on central theoretical and practical problems that are worth considering before conducting a G × E interaction study. On the theoretical side, we discuss two fundamental, but controversial questions about (1) the validity of statistical models for biological interaction and (2) the utility of G × E research for psychiatric genetics. On the practical side, we focus on study characteristics that potentially influence the outcome of G × E interaction studies and discuss strengths and pitfalls of different study designs, including recent approaches like Genome–Environment Wide Interaction Studies (GEWIS). Finally, we discuss recent developments in G × E interaction research on the most heavily investigated example in psychiatric genetics, the interaction between a serotonin transporter gene promoter variant (5-HTTLPR) and stress on depression.

Keywords Genomic · Stress · Behavior · Serotonin

Contents

1 Introduction 442
1.1 What is a G × E Interaction? 442
1.2 Other Forms of Gene–Environment Co-Action: Gene–Environment Correlations 444
2 Theoretical Considerations for G × E Interaction Studies 444
2.1 Can We Model G × E Interaction in Statistics? 444

K. Karg · S. Sen (✉)
University of Michigan, Ann Arbor, MI, USA
e-mail: srijan@umich.edu

Curr Topics Behav Neurosci (2012) 12: 441–462
DOI: 10.1007/7854_2011_184

Published Online: 13 January 2012

2.2 Is G × E Interaction Research Worth the Effort? 446
3 Practical Considerations for G × E Interaction Studies 447
3.1 Methodological Issues in G × E Interaction Research 447
3.2 Assessment of Environmental Exposure and Disorder Status 449
3.3 Study Designs 450
3.4 Wide Interaction Studies 453
4 Empirical Evidence for G × E Interaction in Psychiatric Genetics 453
5 Future Directions 456
6 Summary 456
References 457

1 Introduction

One of the oldest and most enduring questions in psychiatry is whether mental illness is caused by nature (genes) or nurture (environment). Decades of epidemiology studies have tried to answer this question through twin and adoption studies. These studies have demonstrated a moderate genetic component for some disorders (depression and alcohol dependence) and a high genetic component for others (schizophrenia and autism). The relatively high heritability of psychiatric disorders has prompted investigators to look deeply for direct connections between genes and mental illness. Over the past 20 years, thousands of studies have been performed assessing the direct relationship between genes and mental illness in the form of candidate gene association studies, linkage studies and more recently, genome-wide association studies (GWAS). Despite the intense effort, very few direct genetic effects have been identified (Moffitt et al. 2005; Rutter et al. 2006). Therefore, researchers have increasingly directed their attention to the investigation of interactions between genes and environment, a possibility that has traditionally been understudied in behavioral and psychiatric genetics (Caspi 1998; Scarr 1992). In contrast, G × E interactions have been demonstrated consistently in other branches of medicine (van Os et al. 2008). Hence, G × E interaction research is an emerging discipline in psychiatric genetics with growing numbers of novices in need of a comprehensive introduction to the field. In this chapter we aim to give such an introduction, starting with a detailed definition of G × E interaction. We then discuss two fundamental, but controversial theoretical questions about the validity of statistical models for biological interaction and the utility of G × E interaction research for the field of psychiatric genetics. Finally, we discuss practical aspects of studying G × E interactions, with an emphasis on study designs and assessment methods that may affect the success of G × E interaction studies, and present relevant examples from the field.

1.1 What is a G × E Interaction?

The term "G × E interaction" stems from regression models that seek to divide the population variance for disorder risk into environmental and genetic parts. Effects of these factors that are independent from one another are called

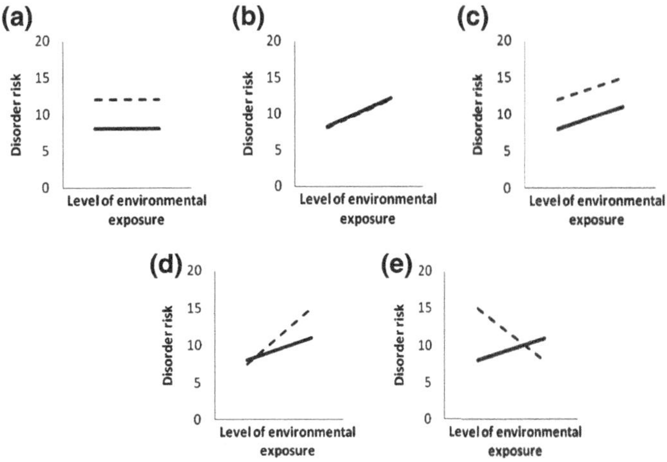

Fig. 1 Illustration of main and interaction effects of genes and environmental exposure on disorder risk. *Solid line* Genotype A, *dashed line* Genotype B. **a** Genetic main effect **b** Environmental main effect **c** Additive effect of genes and environmental exposure **d** Quantitative interaction effect **e** Qualitative interaction effect

main effects. The main effect of either the genetic or the environmental factor can explain the variance for the disorder entirely (Fig. 1a, b) or both factors can coact and explain the variance additively, operating independently alongside each other (Fig. 1c). Consider a child with a retinoblastoma, a malignant tumor of the retina caused by an inherited mutation in one allele of the tumor suppressor gene Rb1. If the patients' unaffected eye gets injured through an accident and the eyesight of this patient becomes worse, the genetic and the environmental factor operate together on the same outcome (eyesight), but are not involved in the same biological pathway and fully independent factors. In contrast, in an interaction effect, the factors depend from each other (Fig. 1d, e). In biological terms, such a G × E interaction effect occurs when *the effect of exposure to an environmental factor on the disorder status depends on the person's genotype* (Moffitt et al. 2006). In other words, a G × E interaction is defined by differences of genotypes in susceptibility to environmental exposure (Kendler and Eaves 1986). For example, our patient with retinoblastoma has an impaired DNA repair system causing her to be markedly more susceptible to UV light compared to an individual without the mutation. By exposure to UV light, tumors develop and worsen the patient's eyesight. Thus, the effect of the exposure to the environmental factor (radiation) on the outcome (eyesight) depends on the person's genotype, constituting an example for G × E interaction. G × E interactions can be quantitative, i. e. the exposure to the environmental pathogen increases the disorder risk for all genotypes, but to different extends (Fig. 1d) or they can be qualitative, i.e. the exposure to the environmental factor increases the risk for one genotype, but decreases it for another (Fig. 1e) (Ottman 1990). With respect to our previous example, a qualitative interaction would occur if UV radiation decreases the risk for retinoblastoma for one genotype, whereas it would increase it for another.

1.2 Other Forms of Gene–Environment Co-Action: Gene–Environment Correlations

Genes and environmental factors can co-act in different ways, and not all of them are G × E interactions [see (Moffitt et al. 2006) for details]. Gene–environment correlations (rGE) are of particular importance, because they can produce false-positive findings in G × E interaction research. rGE can occur when a person's genotype influences her probability of exposure to environmental risks (Plomin et al. 1977; Rutter and Silberg 2002). Several mechanisms have been proposed to drive rGE (Plomin et al. 1977; Jaffee and Price 2007). In active rGE an individual actively selects her environment according to her (genetically influenced) traits and behaviors. For instance, an individual characteristically risk-seeking and impulsive may be much more prone to risk environments than a cautious individual. The presence of rGE has been demonstrated through twin and adoption studies for a wide range of factors, including the occurrence of life events, such as divorce, job loss and serious accidents (Rutter et al. 2006; Rutter and Silberg 2002). The common nature of rGE underscores the danger in the independence assumption of genotype and environment in G × E interaction research. This assumption can be a major problem for some study designs, in particular case-only studies (Jaffee and Price 2007) (further details below).

2 Theoretical Considerations for G × E Interaction Studies

There are two fundamental, theoretical questions about G × E interaction studies that are currently the subject of considerable debate in the literature: (1) Whether the current state of our knowledge about the neurobiology underlying psychiatric disorders allows us to explore G × E interactions in a meaningful way; (2) Whether the expected benefits derived from this research are important enough to justify the considerable resources that these studies require. We address both questions here and try to accurately represent the two opposing camps in the discussion.

2.1 Can We Model G × E Interaction in Statistics?

Although the biological definition of G × E interaction is straightforward, its implementation into statistics is far less clear. Two models are commonly used, the additive and the multiplicative model. The additive model constitutes a G × E interaction when the disorder risk if exposed to both the risk gene (G) and the risk environment (E) differs from the *sum* of the risks if exposed only to G or to E. In biological terms, this is equivalent to the deviation from a simple additive main

Table 1 Illustration of the additive and multiplicative model in statistical G × E interaction testing

	E−	E+
(a)		
G−	2	5
G+	3	10 (6)

	Condition for G × E interaction	Example 1	Example 2
(b)			
Additive model	Risk (G+, E+) $\neq^1$ Risk (G+, E−) + Risk (G−, E+) − Risk (G−, E−)	$10 \neq^1 3+5-2$ (G × E present)	$6 =^1 3+5-2$ (G × E absent)
Multiplicative model	Risk (G+, E+) $\neq^1$ Risk (G+, E−) × Risk (G−, E+)	$10 \neq^1 3 \times 5$ (G × E present)	$6 \neq^1 3 \times 5$ (G × E present)

In Table a, two numerical examples for disorder risk depending on the absence (−) or presence (+) of exposure to the genetic risk factor (G) and environmental risk factor (E) are given, differing only in the (G+, E+) field. Table b illustrates the statistical problem associated with G × E interaction testing: Whereas example 1 leads to the consistent positive result for G × E interaction across the additive and the multiplicative model, the models yield conflicting results for example 2

[1] Statistical significance of the deviation needs to be tested

effects model. This model is used for continuous outcomes, such as depression scores. The multiplicative model constitutes a G × E interaction when the disorder risk if exposed to both G and E differs from the *product* of the risks if exposed only to G or to E. This is used for categorical outcomes, e.g. diagnosis of depression with the two categories "depressed" and "non-depressed". The biological meaning of a multiplicative model is hard to grasp and most researchers argue that the additive model better reflects biological concepts (Rutter et al. 2009). The problem is that in some cases, a study result might deviate significantly from a multiplicative model, but not from an additive model, and vice versa (Kendler and Gardner 2010) (Table 1a, b). This is particularly problematic as continuous outcomes can be converted to categorical outcomes by setting an arbitrary threshold. Given sufficient statistical power, this threshold can be chosen so that either of both models indicate a significant interaction effect. Some researchers argue that this model-dependency renders positive G × E interaction findings arbitrary (Zammit et al. 2010a) and testing for interactions across multiple models is therefore "no different from trawling through many statistical tests looking for those that are significant" (Kendler and Gardner 2010). Therefore, the statistical model to be tested should be carefully selected a priori, based on biological background considerations, and thresholds for categorical data should be set before the analysis. Unfortunately, our current knowledge about neurological pathways is very limited, and, as a result, it is still unclear which statistical model is appropriate (Thompson 1991). This situation has caused some leaders to conclude that we might be unable to move back and forth between statistical and biological interaction models (Kendler and Gardner 2010). The debate remains controversial (Rutter et al. 2006; Zammit et al. 2010a; Caspi and Moffitt 2006; Munafo et al. 2009). One way that investigators have used to circumvent this statistical problem is utilizing new study designs such as case-only or exposed-only designs. These designs do not rely on testing statistical interactions, but directly test differences in exposure rate (case-only design) or in disorder status (exposed-only design) between genotype groups. To date, these designs have mostly been applied in psychiatric G × E interaction research to investigate the interaction between a serotonin transporter gene promoter variant (5-HTTLPR) and stress on the risk of depression (Caspi et al. 2003), with mostly positive results (Karg et al. 2010).

2.2 Is G × E Interaction Research Worth the Effort?

There are three primary arguments for why the identification of G × E interaction effects will substantially advance the field. First, they can help identify new genetic and environmental main effects associated with psychiatric disorders (Kraft et al. 2007). Some risk genes and environments might be masked from detection in scans for direct genes-to-disorder or environment-to-disorder associations because of genotype-specific environmental effects on the disorder status due to G × E interactions. Second, knowledge about the interaction effect of gene and environment on a psychiatric disorder might enhance the identification of the biological pathway underlying the interaction by revealing the genetic and environmental factors involved and thus channel neuroscience studies in a productive direction (Caspi and Moffitt 2006). Third, G × E interaction findings may have clinical relevance and drive the development toward personalized medicine or individual lifestyle recommendations based on the genetic profile (Dempfle et al. 2008; Uher and McGuffin 2007). They could explain differences in response to pharmacological and psychological treatments by differences in the susceptibility of genotypes to environmental factors. Individuals with high-susceptibility genotypes could be identified and prevented from suffering exposure to the relevant environmental pathogens.

Several researchers have criticized this optimistic view, pointing out that the G × E interaction effects identified to date are small, with odds ratios generally between 0.67 and 1.5 (Manolio et al. 2008), limiting the potential influence of G × E interaction on advances in psychiatric genetics and clinical practice (Zammit et al. 2010a, b; Hunter et al. 2008). In particular, the power for finding main effects might only marginally increase by including G × E interaction effects in the statistical model (Munafo et al. 2009). In addition, G × E findings might help identify the underlying biological pathway only through the detection of qualitative G × E interactions, a case known to be rare in epidemiology (Thompson 1991). Thus, there is an ongoing debate about the benefit of G × E interaction research and the considerable amounts of resources spent in the field (Kendler and Gardner 2010; Uher and McGuffin 2007; Zammit et al. 2010b).

3 Practical Considerations for G × E Interaction Studies

Investigating G × E interactions is challenging. For each participating subject, detailed information from three distinct domains is needed: (1) genotype, (2) environmental exposure, (3) psychiatric disorder status. Fortunately, it has become increasingly inexpensive to reliably determine the genotypes of large numbers of subjects due to improved molecular genetic techniques. Gathering valid information in the domains of environmental exposure and disorder status, however, remains expensive and time consuming. This mismatch has led to an increasing number of studies where a huge sample of subjects is genotyped but the quality of phenotype information is comparatively poor. Further, researchers have taken advantage of declined genotyping costs by adding genotype data to studies originally not designed for G × E interaction research (Caspi et al. 2010). Here we give a brief overview on the consequences of these trends and the other methodological issues associated with G × E interaction research. We will present different study design approaches, each with particular advantages and limitations as well as examples from the psychiatric genetics literature (Table 2). For further detailed information on G × E interaction testing see (Caspi and Moffitt 2006; Kendler and Gardner 2010 and Rutter 2002). Complementary research guidelines can be found in (Moffitt et al. 2005, 2006).

3.1 Methodological Issues in G × E Interaction Research

Three major methodological confounding issues are important to consider in planning G × E interaction research: Selection bias, population stratification and recall bias. Selection bias can occur when cases and controls are not drawn from the same underlying population, resulting in erroneous conclusions about associations between genotype, environmental exposure and disorder risk (Hunter 2005). For example, gene–environment correlations can arise in a situation where the presence of a genotype group is correlated with exposure to a particular risk environment. This can result in an overrepresentation of cases with this genotype and therefore steer the study outcome toward false-positive findings regarding differences between cases and controls. Population stratification is the presence of a systematic difference in allele frequencies between subpopulations in a population possibly due to different ancestry (Hunter 2005). Specifically, populations differ with regard to allele frequencies at loci throughout the genome. If these populations also differ in their prevalence of the disorder of interest, spurious associations can be found between this disorder and genetic loci that neither affect the relevant disorder nor are linked to a causative loci. Fortunately, methods have been developed to control for stratification, using unlinked genetic markers to identify and correct for population structure (Cardon and Palmer 2003). These genomic control methods should be utilized in modern day G × E studies.

Table 2 Study designs in G × E interaction

	Design	Description	Advantages	Disadvantages
Family-based designs	Twin study	Comparison of disorder frequency between twin pairs in different environments	No genetic data required; reduced selection and stratification bias	High costs and efforts
	Trio design	Comparison of expected genes in cases to possibly transmitted genes from both parents, stratified by case's environment	Increased power; reduced selection and stratification bias	High costs and efforts
	Sib design	Case-control design with unaffected relative as control	Increased power; reduced selection and stratification bias	High costs and efforts
Traditional population-based designs	Prospective cohort	Comparison of disorder frequency across groups defined by genotype and environment; exposure assessed previous to diagnosis	Reduced selection and stratification bias; reduced recall bias; high-quality measurement for environmental exposure	High costs and efforts; time-consuming; Low, possibly biased follow-up rates
	Cross-sectional	Like prospective cohort, but exposure assessed simultaneously with diagnosis	Reduced selection and stratification bias; more cost-efficient compared to prospective cohort design	Increased recall bias
	Retrospective case-control	Comparison of genotype frequencies and exposure between cases and controls	Increased power and more cost-efficient compared to cohort designs	Increased selection and stratification bias; increased recall bias
Novel population-based designs	Prospective nested case-control	Comparison of cases with matched non-affected cohort members	Combined advantages of prospective cohort and retrospective case-control designs	–
	Case-only	Comparison of exposure across groups defined by genotype	Increased power compared to case-control; cost-efficient	High risk of bias due to confounding with rGE
	Exposed-only	Comparison of genotype frequencies across exposed individuals grouped by genotype	Cost-efficient	Risk of bias due to confounding with rGE; increased selection bias

The third major problem in G × E interaction research is recall bias. Recall bias occurs when subjects cannot accurately recall past events or when particular events become more or less important in retrospect than when they occurred. In particular, patients often overcount potential environmental causes for their disorder, a phenomenon termed mood-congruent memory revision (Joormann et al. 2009; Schwarz and Clore 1983). Recall bias tends to become greater with the greater length of time between the environmental exposure and its report. However, this retrospective forgetting is often selective and its magnitude and character differs between affected and unaffected individuals (Monroe 2008). The difficulties in overcoming the problem of recall bias in retrospective studies provide the impetus for specific novel study designs that we will discuss in later sections.

3.2 Assessment of Environmental Exposure and Disorder Status

An important, but underappreciated factor affecting the power of G × E studies is the assessment method for environmental exposure (Caspi et al. 2010). Poor measurement quality has been correlated with negative findings (Uher and McGuffin 2007, 2010). Simulation studies have demonstrated that in G × E interaction studies, moderate decreases in the measurement accuracy of the environmental variable can result in a 20-fold reduction in statistical power to detect interaction (Moffitt et al. 2005). In line with this simulation result, in a recent meta-analysis on studies investigating the moderating effect of a serotonin transporter gene polymorphism (5-HTTLPR) on the relationship between stressful life events and depression, we found that studies that utilized more intensive stress assessment methods, such as in-person interviews, were more likely to detect an effect than studies that utilized self-report questionnaires (we will discuss the set of 5-HTTLPR-stress studies in more detail in Sect. 4). One reason for these findings is likely that the effect of measurement error, such as recall bias, is more pronounced in self-report questionnaires than in personal interviews because trained interviewers can counteract poor recall by using appropriate techniques such as life event calendars and memory enhancement (Caspi et al. 2010). Self-report event checklists have been shown to result in more imprecise information (Monroe 2008). Objective measurements may also be superior to self-report questionnaires because they minimize the effects of recall bias by focusing objective information. Further, the objective stressor design reduces between-subject heterogeneity by the use of clearly operationalized and objectively identifiable environmental factors, resulting in an increase of internal validity (Caspi et al. 2010). These findings underscore the importance of choosing assessment methods for G × E studies carefully. The use of several independent measurements such as self-report, diagnostic interview or informant reports are excellent possibility to increase the accuracy of assessment (see Caspi et al. (2003), for a good example). A similar set of methodological considerations apply to the assessment of disorder status. In comparison to many systemic disorders, psychiatric disorders are difficult to diagnose, relying on arbitrary thresholds on symptom severity scales (Eaton et al. 2007). For instance, a wide range of threshold scores (12–23) have been suggested for diagnosing depression with the commonly used Beck Depression Inventory (Nuevo et al. 2009). While commonly used diagnostic instruments for many psychiatric disorders (such as depression, alcohol and drug use disorder) have acceptable measurement characteristics, others perform poorly (e.g. panic disorder, obsessive–compulsive disorder, bipolar disorder and schizophrenia), with particularly poor sensitivity (40%) and specificity (89%) for schizophrenia (Eaton et al. 2007).

3.3 Study Designs

G × E interaction study designs can broadly be categorized into family-based designs and population-based designs. Both designs have particular strengths and limitations regarding the methodological issues described above (Table 2). Family-based studies generally assess whether there is a greater than expected transmission of specific alleles to affected family members (Ewens and Spielman 1995). The specific family-based study designs include twin studies (Ottman 1994), trio designs with an affected individual and both parents (Schaid 1999; Witte et al. 1999), and sib designs with one affected and one unaffected sibling or relative (Gauderman et al. 1999). If the frequency of transmission differs between exposed and non-exposed cases, a G × E interaction is present (Schaid 1999). The main advantages of family-based designs is a per subject increase in power compared to population-based designs, and robustness against population stratification. However, family-based designs have some major drawbacks that have limited their use. One is that it is often harder to recruit an adequate number of sibling or twin pairs than unrelated subjects, and the unavailability of living parents can limit the scope of trio studies (Hunter 2005). Further, newer genomic control methods can robustly control for stratification, rendering the primary advantage of family-based methods less useful. Therefore, in most cases of G × E interaction research, population-based designs are used.

In contrast to the family-based design, design studies generally draw from a set of unrelated subjects. These studies differ according to how these subjects are selected. Subjects can be can be drawn from a cohort [cohort study design (Collins 2004)], selected and matched as cases and controls [case-control design (Yang and Khoury 1997)], drawn from affected individuals only [case-only design (Khoury and Flanders 1996)], or from individuals exposed to the environmental risk factor only [exposed-only design (Moffitt et al. 2006)].

Cohort study design. In cohort study designs, the sample studied should accurately represent the target population in terms of genotype, exposure rate and disorder status. Information can be assessed either once (cross-sectional design) or repeatedly over time (prospective/longitudinal design). When analyzing the data, subjects can be assigned to groups according to their genotype and their exposure rate (e.g., genotype A with low environmental exposure vs. genotype B with low environmental exposure), and disorder frequencies can be compared between these groups. If high follow-up rates are obtained, the prospective cohort design can provide high-quality data because it efficiently handles the three major methodological issues facing G × E studies: it minimizes selection bias, because the disorder usually occurs after subjects are selected (Yang and Khoury 1997), it minimizes population stratification by sampling from a defined cohort and it reduces recall bias to a minimum if the baseline information is assessed early in life of the cohort and when it can be followed several times over years (Hunter 2005).

Three of the most important findings in psychiatric G × E interaction research were produced by utilizing through a study a prospective cohort study design, the Dunedin Multidisciplinary Health and Development Study (Dunedin Longitudinal Study) (Caspi et al. 2002, 2003, 2005). The Dunedin Longitudinal Study investigated a large birth cohort of 1,037 children born in 1972–73 in Dunedin, New Zealand. The cohort was first assessed at age three and since then followed up every two years for two decades (Silva 1990). Data from this cohort demonstrated significant G × E interaction effects on violent behavior (Caspi et al. 2002), depression (Caspi et al. 2003) and adult psychosis (Caspi et al. 2005). These landmark studies provide evidence supporting the strength and accuracy of the prospective cohort design.

The downside of this study design is the long time frame necessary to conduct these studies. For instance, the Dunedin Longitudinal Study was started 30 years before the first G × E interaction finding was published. In addition, large samples are needed because the environmental exposure and/or the disorder might be at low prevalence in the cohort (Hunter 2005). As a result, many investigators opt for quicker and less expensive designs. The cross-sectional modification of the cohort study assesses cohort information only once. Although this design loses some of the advantages of a prospective study, the cost and time frame necessary to carry out the study is often more feasible.

Retrospective case-control. Another inexpensive and popular alternative to the prospective cohort design is the retrospective case-control design. Here, affected subjects with the disorder are selected and matched with subjects who do not have the disorder ('controls'). This procedure allows for the controlled sampling of subjects with disorder and/or environmental exposure, yielding the advantage of increased power compared to cohort studies (McClelland and Judd 1993). Information about past exposure is gathered and the exposure rates and genotype frequencies are compared between cases and controls. Due to the selection and matching process, this design is particularly prone to selection bias and population stratification, especially when the source population for controls is hard to define (Hunter 2005). The prospective-nested case-control design is a more sophisticated study design that addresses these methodological problems by selecting cases and controls from a predetermined longitudinal cohort. As cases and controls stem from the same cohort, confounding from selection bias and population stratification is avoided. In addition, recall bias is eliminated because exposure is assessed before the diagnose. Compared to a full cohort approach, this design offers substantial reductions in costs and efforts.

Case-only. Recently, investigators have proposed a study design that eliminates the use of control subjects (Khoury and Flanders 1996; Piegorsch et al. 1994). In the case-only design, affected subjects are selected from the population and grouped according to their genotype and then compared for their exposure rates. In the presence of G × E interaction, some genotypes are more susceptible to the environmental pathogen than others, resulting in an overrepresentation of subjects with environmental exposure in this genotype group. Therefore, differential distributions of exposure rates across genotype groups can be interpreted as a G × E interaction effect. As an example, Mandelli et al. (2006) utilized the case-only design to investigate the interaction effect of 5-HTTLPR and stress on depression. They studied a sample of 686 patients diagnosed for major depression or bipolar disorder and classified them into six groups according to their genotype and the presence or absence of environmental exposure to life stress in the year before depression onset. On comparing the proportion of the sample exposed between each genotype group, they found higher proportions of previously exposed subjects in the genotype groups carrying the short allele. The authors interpreted this finding as evidence for higher stress susceptibility of short allele carriers. However, this conclusion has to be viewed with some caution because the case-only design is prone to confounding. Differential distributions in exposure rates across genotypes can also emerge through G-E correlation, with specific genotypes being more likely to be exposed to the environmental factor than others (Khoury and Flanders 1996). In this study, it is possible that short allele carriers are more prone to experience stressful situations and that this causes their overrepresentation in the exposed group. The only safe way to rule out this potential bias is through the verification of the underlying assumption of gene–environment independency. Therefore, the case-only design should be used only if the independency assumption is verified or for exploratory studies (Albert et al. 2001).

Exposed-only. A related, but subtly different approach that has become increasingly popular is the exposed-only design. Here, subjects exposed to the same environmental factor are selected, grouped according to their genotype and compared for their disorder status. In the presence of G × E interaction, disorder frequencies should be higher in the genotype group with higher susceptibility to the environmental exposure. However, as we discussed concerning the case-only design, this conclusion is only valid in the absence of G-E correlation. An example might illustrate this problem. A recent study utilized an exposed-only design to explore a moderating effect of the FKBP5 (FK506 binding protein 5) gene on the relationship between severe injury and peritraumatic dissociation (Koenen et al. 2005). Peritraumatic dissociation is a evolutionary conserved response to life-threatening events and a risk factor for the development of post-traumatic stress disorder (Ozer et al. 2003). The study sample consisted of 46 severely injured hospitalized children who were genotyped and compared for their peritraumatic dissociation scores with logistic regression analysis. The study revealed a significant G × E interaction effect of FKBP5 genotype and severe injury on the development of peritraumatic dissociation. However, this finding could have arisen through rGE, with one genotype group particularly prone to risk-seeking and therefore more likely to suffer severe injury and corresponding peritraumatic dissociation. This could lead to the erroneous conclusion that this genotype is more susceptible to peritraumatic dissociation than others. In this study, however, injury severity was taken into account in the statistical analysis, rendering a false-positive result due to rGE less likely. Another elegant way to guard against bias due to rGE is exact matching for exposure across participants (Moffitt et al. 2006). This allows investigators to bypass the model-dependency problem. Hence, the problem of rGE in exposed-only designs is much easier to handle than in case-only designs where additional empirical evidence is needed. The exposed-only design is thus an attractive cost-efficient design that can be used to test G × E interaction for candidate genes as well as for the discovery of unknown risk genes (Moffitt et al. 2006).

3.4 Wide Interaction Studies

With the advent of genome-wide association studies (GWAS) it is now possible to genotype up to one million SNPs for each participant, allowing investigators to scan the entire genome for relevant genes without prior hypothesis. While most GWAS to date have explored direct associations, groups have begun to modify GWAS to include assessment of environmental variables in order to conduct Gene–environment wide interaction studies (GEWIS) (Khoury and Wacholder 2009). GEWIS allow us to investigate several candidate pathways at once at relatively low costs and hold the promise to identify new possible G × E interactions. The greatest challenge for GEWIS involves finding a balance between dismissing true findings through stringent correction for multiple testing and reporting false-positive results (Sebastiani et al. 2005). Without any prior hypothesis it is hard to distinguish false from true positives, especially as interaction effects in complex traits such as mental disorders are supposed to be small. However, systematic approaches to the problem are emerging (Onkamo and Toivonen 2006; Wacholder et al. 2004). Despite the great remaining conceptual challenges, GEWIS paired with thorough phenotyping holds promise in producing advances in the field of G × E interaction research.

4 Empirical Evidence for G × E Interaction in Psychiatric Genetics

G × E interactions in psychiatric genetics have been reported for various disorders such as depression, attention deficit/hyperactivity disorder (ADHD), schizophrenia, obesity and substance use disorders (Table 3). The identified environmental pathogens range from prenatal factors such as maternal smoking (Kahn et al. 2003)

Table 3 Selected G × E interaction findings in psychiatric genetics

Gene	Risk environment	Disorder	Original finding
SLC6A4	Stressful life events	Depression	Caspi et al. (2003)
SLC6A4	Childhood maltreatment	Depression	Caspi et al. (2003)
SLC6A4	Mother's expressed emotion	ADHD	Sonuga-Barke et al. (2009)
SLC6A4	Early life stress	Alcohol abuse	Olsson et al. (2005)
MAOA	Childhood maltreatment	Antisocial personality; Conduct disorder	Caspi et al. (2002)
DRD4	Priming alcohol doses	Alcohol craving	Hutchison et al. (2002a)
DRD4	Smoking cues	Tobacco craving	Hutchison et al. (2002b)
DAT1	Prenatal maternal smoking	ADHD	Kahn et al. (2003)
DAT1	Prenatal maternal use of alcohol	ADHD	Brookes et al. (2006)
DAT1	Season of birth	ADHD	Seeger et al. (2004)
DAT1	Psychosocial adversity in childhood	ADHD	Laucht et al. (2007)
DAT1	Mother's expressed emotion	ADHD	Sonuga-Barke et al. (2009)
DAT1	Institutional deprivation	ADHD	Stevens et al. (2009)
COMT	Cannabis use in adolescence	Adult psychosis	Caspi et al. (2005)
COMT	Low birth weight	ADHD	Thapar et al. (2005)
COMT	Stress	Psychosis	van Winkel et al. (2008)
CRHR1	Stress	Alcohol abuse	Blomeyer et al. (2008)
CRHR1	Childhood trauma	Mood and anxiety disorders	Bradley et al. (2008)
FTO	Physical inactivity	Obesity	Andreasen et al. (2008)
FKBP5	Acute injury	Psychological dissociation	Koenen et al. (2005)
FKBP5	Childhood abuse	Mood and anxiety disorders	Binder et al. (2008)

or maternal alcohol use (Brookes et al. 2006) to factors relevant at birth [e.g. season of birth (Seeger et al. 2004), birth weight (Thapar et al. 2005)] and early development [e.g. childhood maltreatment (Caspi et al. 2003), childhood trauma (Bradley et al. 2008)] to factors affecting adolescence [e.g. cannabis use (Caspi et al. 2005)] and adulthood [e.g. stress (Blomeyer et al. 2008), physical inactivity (Andreasen et al. 2008)]. However, track record of replications has often been poor, casting doubt on the validity of these findings (Thomas 2010). Nonreplication can be due to false-negative results, false-positive results or true heterogeneity between studies. False-negative results in psychiatry studies are most often caused by insufficient power, either due to a small sample size or suboptimal phenotyping or genotyping quality. False-positive results can often result from multiple testing and population stratification. True heterogeneity occurs if the interaction exists in some populations studied or with some environmental factors studied but not with others. Here, we present the most heavily investigated example in psychiatric G × E interaction research, a G × E interaction between a polymorphism in the promoter region of the serotonin transporter gene (5-HTTLPR) and both adult stressful life events and childhood maltreatment on the risk of depression (Caspi et al. 2003). We will discuss conflicting results between studies exploring this interaction and potential reasons for the conflict.

The original study exploring this interaction utilized a prospective-longitudinal cohort design with almost 1,000 children and found that individuals homozygous or heterozygous for the low-expressing short variant of 5-HTTLPR are more susceptible to depression after stress than individuals homozygous for the alternate long variant. The same pattern was found for childhood maltreatment. This study caused a great deal of excitement in G × E interaction research and encouraged further research on this issue. To date, there have been 55 follow-up studies with some confirming the original finding, some finding evidence of higher stress susceptibility of individuals with the alternate long allele, and others finding no interaction effect at all (Karg et al. 2010). This inconsistency might be due to the heterogeneity of studies in many relevant aspects. First, studies exploring the relationship between 5-HTTLPR, stress and depression have utilized very different research designs, including longitudinal, cross-sectional, case-control, case-only, exposure-only and family-based designs. Second, studies have measured many different depression phenotypes using diverse assessment strategies, including clinical interviews and self-report checklists, and diverse depression scales, variously yielding both categorical and continuous outcome measures. Third, studies have investigated an extraordinarily varied set of stressors with various assessment methods. For instance, stressors counted in different studies for stressful life events ranged from becoming homeless, and the death of a parent or spouse to growing up in a household with siblings who quarreled or as the child of a father in an unskilled occupation. Other studies used more specific, but highly diverse stressors such as stroke survival, hurricane exposure, bullying victimization or childhood maltreatment. To clarify this confusion, three meta-analyses have been carried out to date. The first two (Uher and McGuffin 2007; Risch et al. 2009) concluded that there was no evidence supporting the presence of an interaction. However, these analyses investigated only small subsamples of all 55 studies due to methodological constraints. The latest meta-analysis (Karg et al. 2010) included all relevant studies and detected stressor type (stressful life events, childhood maltreatment, and specific medical conditions) and stress assessment method (questionnaire, interview, objective) as two critical sources for variability in study outcomes. In particular, studies with childhood maltreatment or specific medical conditions as environmental stressor were more likely to find a significant G × E effect than studies with broader defined stressful life events, as were studies with objective or interview assessment methods for environmental stressors. This again supports the assumption that measurement quality can affect results in G × E research.

Since this original study, further evidence from various fields has emerged (Caspi et al. 2010). First, several empirical studies link the short 5-HTTLPR variant to stress-sensitive phenotypes such as post-traumatic stress disorder (Xie et al. 2009), post-trauma suicide (Roy et al. 2007), stress-related sleep disturbance (Brummett et al. 2007) and anxiety (Stein et al. 2008). Second, a multitude of neuroimaging studies confirmed increased and faster amygdala reactivity following threat in carriers of the short allele e.g. (Furman et al. 2010; Heinz et al. 2005) and linked it to specific brain anatomy characteristics e.g. (Pacheco et al. 2009; Pezawas et al. 2005). Third, Rhesus macaques carrying the short variant exhibit greater anxiety-related behaviors in response to adverse rearing conditions compared to their conspecific with the long alternate (Barr et al. 2004; Spinelli et al. 2007). Fourth, in addition to 5-HTT knockout mice, 5-HTT knockout rats showed increased anxiety levels in response to stress (Homberg et al. 2007). Taken together, these outcomes across a wide variety of techniques, models and species as well as the numerous positive G × E studies robustly demonstrate the interaction effect between stress and 5-HTTLPR genotype on depression and are to date the most intriguing finding of G × E interaction in psychiatric genetics.

5 Future Directions

Although much progress has been made in the past two decades, many questions in G × E interaction research in psychiatric genetics remain open. New, more carefully conducted epidemiological studies could shed light on these questions. Another major step for clarification is the identification of the biological mechanisms underlying interaction effects. Not much is known about how environmental factors can interact with a person's genotype and her nervous system to moderate the disorder risk. Therefore, joining forces with neuroscience is an important step in making progress in the field (Caspi and Moffitt 2006). Many epidemiological studies on G × E interaction in psychiatric genetics were motivated by findings of neuroscience research and positive epidemiological findings, in turn, can stimulate new studies in neuroscience. The interaction between 5-HTTLPR and life stress on depression provides an example where neuroscience studies can illuminate the black box between genes, environment and disorder (Merikangas and Risch 2003) and confirm and explain epidemiological findings. Another fruitful approach for advances in the understanding G × E interaction might be the collaboration with epigenetic research. Many environmental risk factors operate early in development, and fine-tuning of neuronal pathways is known to be affected by environmental factors (Abdolmaleky et al. 2004). If these epigenetic modifications depend on the person's genotype, a plausible mechanism is constituted for G × E interaction in psychiatric genetics. Epigenetic studies for psychiatric disorders are still in their infancy, and new exciting insights in the interplay of genes and environment on the development of mental disorders are to be awaited.

6 Summary

Although the fundamental questions about the validity of statistical models for biological interaction and the utility of G × E interaction findings for advances in psychiatric genetics are still highly debated, novel study designs such as case-only and exposed-only designs can overcome at least some of the statistical concerns. Study designs differ broadly in their strengths and limitations regarding selection bias, population stratification and recall bias. Previously undetermined study characteristics that might additionally affect the outcome of G × E interaction studies are the assessment methods for environmental exposure and disorder status, as shown for the G × E interaction effect between the serotonin transporter promoter variant and stress on depression. New insights into the interplay between genes and environment on the development of mental disorders may emerge through more carefully conducted G × E interaction studies as well as through collaboration with neuroscience and epigenetic research.

References

Abdolmaleky HM, Smith CL, Faraone SV, Shafa R, Stone W, Glatt SJ, Tsuang MT (2004) Methylomics in psychiatry: modulation of Gene-environment interactions may be through DNA methylation. Am J Med Genet B Neuropsychiatr Genet 127B(1):51–59. doi: 10.1002/ajmg.b.20142

Albert PS, Ratnasinghe D, Tangrea J, Wacholder S (2001) Limitations of the case-only design for identifying Gene-environment interactions. Am J Epidemiol 154(8):687–693

Andreasen CH, Stender-Petersen KL, Mogensen MS, Torekov SS, Wegner L, Andersen G, Nielsen AL, Albrechtsen A, Borch-Johnsen K, Rasmussen SS, Clausen JO, Sandbaek A, Lauritzen T, Hansen L, Jorgensen T, Pedersen O, Hansen T (2008) Low physical activity accentuates the effect of the FTO rs9939609 polymorphism on body fat accumulation. Diabetes 57(1):95–101. doi: db07-0910 [pii], 10.2337/db07-0910

Barr CS, Newman TK, Shannon C, Parker C, Dvoskin RL, Becker ML, Schwandt M, Champoux M, Lesch KP, Goldman D, Suomi SJ, Higley JD (2004) Rearing condition and rh5-HTTLPR interact to influence limbic-hypothalamic-pituitary-adrenal axis response to stress in infant macaques. Biol Psychiatry 55(7):733–738. doi: 10.1016/j.biopsych.2003.12.008 S000632230 4000046 [pii]

Binder EB, Bradley RG, Liu W, Epstein MP, Deveau TC, Mercer KB, Tang Y, Gillespie CF, Heim CM, Nemeroff CB, Schwartz AC, Cubells JF, Ressler KJ (2008) Association of fkbp5 polymorphisms and childhood abuse with risk of posttraumatic stress disorder symptoms in adults. JAMA 299(11):1291–1305. doi: 299/11/1291 [pii], 10.1001/jama.299.11.1291

Blomeyer D, Treutlein J, Esser G, Schmidt MH, Schumann G, Laucht M (2008) Interaction between crhr1 gene and stressful life events predicts adolescent heavy alcohol use. Biol Psychiatry 63(2):146–151. doi: S0006-3223(07)00375-7 [pii], 10.1016/j.biopsych.2007.04.026

Bradley RG, Binder EB, Epstein MP, Tang Y, Nair HP, Liu W, Gillespie CF, Berg T, Evces M, Newport DJ, Stowe ZN, Heim CM, Nemeroff CB, Schwartz A, Cubells JF, Ressler KJ (2008) Influence of child abuse on adult depression: moderation by the corticotropin-releasing hormone receptor gene. Arch Gen Psychiatry 65(2):190–200. doi: 10.1001/archgenpsychiatry.2007.26

Brookes KJ, Mill J, Guindalini C, Curran S, Xu X, Knight J, Chen CK, Huang YS, Sethna V, Taylor E, Chen W, Breen G, Asherson P (2006) A common haplotype of the dopamine transporter gene associated with attention-deficit/hyperactivity disorder and interacting with maternal use of alcohol during pregnancy. Arch Gen Psychiatry 63(1):74–81. doi: 63/1/74 [pii] 10.1001/archpsyc.63.1.74

Brummett BH, Krystal AD, Ashley-Koch A, Kuhn CM, Zuchner S, Siegler IC, Barefoot JC, Ballard EL, Gwyther LP, Williams RB (2007) Sleep quality varies as a function of

5-HTTLPR genotype and stress. Psychosom Med 69(7):621–624. doi: PSY.0b013e31814b8de6 [pii], 10.1097/PSY.0b013e31814b8de6

Cardon LR, Palmer LJ (2003) Population stratification and spurious allelic association. Lancet 361(9357):598–604. doi: S0140-6736(03)12520-2 [pii], 10.1016/S0140-6736(03)12520-2

Caspi A (1998) Personality development across the life course. In: Damon W, Eisenberg N (eds) Handbook of child psychology, vol 3. Wiley, New York, pp 311–388

Caspi A, Moffitt TE (2006) Gene-environment interactions in psychiatry: joining forces with neuroscience. Nat Rev Neurosci 7(7):583–590. doi: nrn1925 [pii], 10.1038/nrn1925

Caspi A, McClay J, Moffitt TE, Mill J, Martin J, Craig IW, Taylor A, Poulton R (2002) Role of genotype in the cycle of violence in maltreated children. Science 297(5582):851–854. doi: 10.1126/science.1072290 297/5582/851 [pii]

Caspi A, Sugden K, Moffitt TE, Taylor A, Craig IW, Harrington H, McClay J, Mill J, Martin J, Braithwaite A, Poulton R (2003) Influence of life stress on depression: moderation by a polymorphism in the 5-HTT gene. Science 301(5631):386–389. doi: 10.1126/science.1083968 301/5631/386 [pii]

Caspi A, Moffitt TE, Cannon M, McClay J, Murray R, Harrington H, Taylor A, Arseneault L, Williams B, Braithwaite A, Poulton R, Craig IW (2005) Moderation of the effect of adolescent-onset cannabis use on adult psychosis by a functional polymorphism in the catechol-o-methyltransferase gene: longitudinal evidence of a gene x environment interaction. Biol Psychiatry 57(10):1117–1127. doi: S0006-3223(05)00103-4 [pii], 10.1016/j.biopsych.2005.01.026

Caspi A, Hariri AR, Holmes A, Uher R, Moffitt TE (2010) Genetic sensitivity to the environment: the case of the serotonin transporter gene and its implications for studying complex diseases and traits. Am J Psychiatry 167(5):509–527. doi: 10.1176/appi.ajp. 2010.09101452

Collins FS (2004) The case for a us prospective cohort study of genes and environment. Nature 429(6990):475–477. doi: 10.1038/nature02628 nature02628 [pii]

Dempfle A, Scherag A, Hein R, Beckmann L, Chang-Claude J, Schafer H (2008) Gene-environment interactions for complex traits: definitions, methodological requirements and challenges. Eur J Hum Genet 16(10):1164–1172. doi: ejhg2008106 [pii], 10.1038/ejhg.2008.106

Eaton WW, Hall ALF, MacDonald R, McKibben J (2007) Case identification in psychiatric epidemiology: a review. Int Rev Psych 19(5):497–507. doi: 10.1080/09540260701564906

Ewens WJ, Spielman RS (1995) The transmission/disequilibrium test: history, subdivision, and admixture. Am J Hum Genet 57(2):455–464

Furman DJ, Hamilton JP, Joormann J, Gotlib IH (2010) Altered timing of amygdala activation during sad mood elaboration as a function of 5-HTTLPR. Soc Cogn Affect Neurosci 6(3):270–276. doi: nsq029 [pii], 10.1093/scan/nsq029

Gauderman WJ, Witte JS, Thomas DC (1999) Family-based association studies. J Natl Cancer Inst Monogr 26:31–37

Heinz A, Braus DF, Smolka MN, Wrase J, Puls I, Hermann D, Klein S, Grusser SM, Flor H, Schumann G, Mann K, Buchel C (2005) Amygdala-prefrontal coupling depends on a genetic variation of the serotonin transporter. Nat Neurosci 8(1):20–21. doi: nn1366 [pii], 10.1038/nn1366

Homberg JR, Olivier JD, Smits BM, Mul JD, Mudde J, Verheul M, Nieuwenhuizen OF, Cools AR, Ronken E, Cremers T, Schoffelmeer AN, Ellenbroek BA, Cuppen E (2007) Characterization of the serotonin transporter knockout rat: a selective change in the functioning of the serotonergic system. Neuroscience 146(4):1662–1676. doi: S0306-4522(07)00343-0 [pii], 10.1016/j.neuroscience.2007.03.030

Hunter DJ (2005) Gene-environment interactions in human diseases. Nat Rev Genet 6(4):287–298. doi: 10.1038/nrg1578

Hunter DJ, Altshuler D, Rader DJ (2008) From darwin's finches to canaries in the coal mine–mining the genome for new biology. N Engl J Med 358(26):2760–2763. doi: 358/26/2760 [pii], 10.1056/NEJMp0804318

Hutchison KE, McGeary J, Smolen A, Bryan A, Swift RM (2002a) The DRD4 vntr polymorphism moderates craving after alcohol consumption. Health Psychol 21(2):139–146

Hutchison KE, LaChance H, Niaura R, Bryan A, Smolen A (2002b) The drd4 vntr polymorphism influences reactivity to smoking cues. J Abnorm Psychol 111(1):134–143

Jaffee SR, Price TS (2007) Gene-environment correlations: a review of the evidence and implications for prevention of mental illness. Mol Psychiatry 12(5):432–442. doi: 10.1038/sj.mp. 4001950

Joormann J, Hertel PT, LeMoult J, Gotlib IH (2009) Training forgetting of negative material in depression. J Abnorm Psychol 118(1):34–43. doi: 2009-01738-017 [pii], 10.1037/a0013794

Kahn RS, Khoury J, Nichols WC, Lanphear BP (2003) Role of dopamine transporter genotype and maternal prenatal smoking in childhood hyperactive-impulsive, inattentive, and oppositional behaviors. J Pediatr 143(1):104–110. doi: S0022-3476(03)00208-7 [pii], 10.1016/S0022-3476(03)00208-7

Karg K, Burmeister M, Shedden K, Sen S (2010) The serotonin transporter promoter variant (5-HTTLPR), stress, and depression meta-analysis revisited: evidence of genetic moderation. Arch Gen Psychiatry 12(8):786–792

Kendler KS, Eaves LJ (1986) Models for the joint effect of genotype and environment on liability to psychiatric illness. Am J Psychiatry 143(3):279–289

Kendler KS, Gardner CO (2010) Interpretation of interactions: guide for the perplexed. Br J Psychiatry 197(3):170–171. doi: 197/3/170 [pii], 10.1192/bjp.bp. 110.081331

Khoury MJ, Flanders WD (1996) Nontraditional epidemiologic approaches in the analysis of Gene-environment interaction: case-control studies with no controls! Am J Epidemiol 144(3):207–213

Khoury MJ, Wacholder S (2009) Invited commentary: from genome-wide association studies to Gene-environment-wide interaction studies-challenges and opportunities. Am J Epidemiol 169 (2):227–230; discussion 234–225. doi: kwn351 [pii], 10.1093/aje/kwn351

Koenen KC, Saxe G, Purcell S, Smoller JW, Bartholomew D, Miller A, Hall E, Kaplow J, Bosquet M, Moulton S, Baldwin C (2005) Polymorphisms in FKBP5 are associated with peritraumatic dissociation in medically injured children. Mol Psychiatry 10(12):1058–1059

Kraft P, Yen YC, Stram DO, Morrison J, Gauderman WJ (2007) Exploiting Gene-environment interaction to detect genetic associations. Hum Hered 63(2):111–119. doi: 000099183 [pii], 10.1159/000099183

Laucht M, Skowronek MH, Becker K, Schmidt MH, Esser G, Schulze TG, Rietschel M (2007) Interacting effects of the dopamine transporter gene and psychosocial adversity on attention-deficit/hyperactivity disorder symptoms among 15-year-olds from a high-risk community sample. Arch Gen Psychiatry 64(5):585–590. doi: 64/5/585 [pii], 10.1001/archpsyc.64.5.585

Mandelli L, Serretti A, Marino E, Pirovano A, Calati R, Colombo C (2006) Interaction between serotonin transporter gene, catechol-o-methyltransferase gene and stressful life events in mood disorders. Int J Neuropsychopharmacol 10(04):437–447. doi: 10.1017/s146114570 6006882

Manolio TA, Brooks LD, Collins FS (2008) A hapmap harvest of insights into the genetics of common disease. J Clin Invest 118(5):1590–1605. doi: 10.1172/JCI34772

McClelland GH, Judd CM (1993) Statistical difficulties of detecting interactions and moderator effects. Psychol Bull 114(2):376–390

Merikangas KR, Risch N (2003) Will the genomics revolution revolutionize psychiatry? Am J Psychiatry 160(4):625–635

Moffitt TE, Caspi A, Rutter M (2005) Strategy for investigating interactions between measured genes and measured environments. Arch Gen Psychiatry 62(5):473–481. doi: 62/5/473 [pii], 10.1001/archpsyc.62.5.473

Moffitt TE, Caspi A, Rutter M (2006) Measured Gene-environment interactions in psychopathology: concepts, research strategies, and implications for research, intervention, and public understanding of genetics. Perspect Psychol Sci 1(1):5–27

Monroe SM (2008) Modern approaches to conceptualizing and measuring human life stress. Annu Rev Clin Psychol 4:33–52. doi: 10.1146/annurev.clinpsy.4.022007.141207

Munafo MR, Durrant C, Lewis G, Flint J (2009) Gene $\times$ environment interactions at the serotonin transporter locus. Biol Psychiatry 65(3):211–219. doi: S0006-3223(08)00731-2 [pii], 10.1016/j.biopsych.2008.06.009

Nuevo R, Lehtinen V, Reyna-Liberato PM, Ayuso-Mateos JL (2009) Usefulness of the beck depression inventory as a screening method for depression among the general population of finland. Scand J Public Health 37(1):28–34. doi: 37/1/28 [pii], 10.1177/1403494808097169

Olsson CA, Byrnes GB, Lotfi-Miri M, Collins V, Williamson R, Patton C, Anney RJL (2005) Association between 5-HTTLPR genotypes and persisting patterns of anxiety and alcohol use: results from a 10-year longitudinal study of adolescent mental health. Mol Psychiatry 10(9):868–876

Onkamo P, Toivonen H (2006) A survey of data mining methods for linkage disequilibrium mapping. Hum Genomics 2(5):336–340

Ottman R (1990) An epidemiologic approach to Gene-environment interaction. Genet Epidemiol 7(3):177–185. doi: 10.1002/gepi.1370070302

Ottman R (1994) Epidemiologic analysis of Gene-environment interaction in twins. Genet Epidemiol 11(1):75–86. doi: 10.1002/gepi.1370110108

Ozer EJ, Best SR, Lipsey TL, Weiss DS (2003) Predictors of posttraumatic stress disorder and symptoms in adults: a meta-analysis. Psychol Bull 129(1):52–73

Pacheco J, Beevers CG, Benavides C, McGeary J, Stice E, Schnyer DM (2009) Frontal-limbic white matter pathway associations with the serotonin transporter gene promoter region (5-HTTLPR) polymorphism. J Neurosci 29(19):6229–6233. doi: 29/19/6229 [pii], 10.1523/JNEUROSCI.0896-09.2009

Pezawas L, Meyer-Lindenberg A, Drabant EM, Verchinski BA, Munoz KE, Kolachana BS, Egan MF, Mattay VS, Hariri AR, Weinberger DR (2005) 5-HTTLPR polymorphism impacts human cingulate-amygdala interactions: a genetic susceptibility mechanism for depression. Nat Neurosci 8(6):828–834. doi: nn1463 [pii], 10.1038/nn1463

Piegorsch WW, Weinberg CR, Taylor JA (1994) Non-hierarchical logistic models and case-only designs for assessing susceptibility in population-based case-control studies. Stat Med 13(2):153–162

Plomin R, DeFries JC, Loehlin JC (1977) Genotype-environment interaction and correlation in the analysis of human behavior. Psychol Bull 84(2):309–322

Risch N, Herrell R, Lehner T, Liang KY, Eaves L, Hoh J, Griem A, Kovacs M, Ott J, Merikangas KR (2009) Interaction between the serotonin transporter gene (5-HTTLPR), stressful life events, and risk of depression: a meta-analysis. JAMA 301(23):2462–2471. doi: 10.1001/jama.2009.878

Roy A, Hu XZ, Janal MN, Goldman D (2007) Interaction between childhood trauma and serotonin transporter gene variation in suicide. Neuropsychopharmacology 32(9):2046–2052. doi: 1301331 [pii], 10.1038/sj.npp. 1301331

Rutter M (2002) The interplay of nature, nurture, and developmental influences: the challenge ahead for mental health. Arch Gen Psychiatry 59(11):996–1000. doi: ysa21000 [pii]

Rutter M, Silberg J (2002) Gene-environment interplay in relation to emotional and behavioral disturbance. Annu Rev Psychol 53:463–490. doi: 10.1146/annurev.psych.53.100901.135223 53/1/463 [pii]

Rutter M, Moffitt TE, Caspi A (2006) Gene-environment interplay and psychopathology: multiple varieties but real effects. J Child Psychol Psychiatry 47(3–4):226–261. doi: 10.1002/da.20641

Rutter M, Thapar A, Pickles A (2009) Gene-environment interactions: biologically valid pathway or artifact? Arch Gen Psychiatry 66(12):1287–1289. doi: 66/12/1287 [pii], 10.1001/archgenpsychiatry.2009.167

Scarr S (1992) Developmental theories for the 1990s: development and individual differences. Child Dev 63(1):1–19

Schaid DJ (1999) Case-parents design for Gene-environment interaction. Genet Epidemiol 16(3):261–273. doi: 10.1002/(SICI)1098-2272(1999)16:3<261::AID-GEPI3>3.0.CO;2-M [pii], 10.1002/(SICI)1098-2272(1999)16:3<261:AID-GEPI3>3.0.CO;2-M

Schwarz N, Clore GL (1983) Mood, misattribution, and judgments of well-being: informative and directive functions of affective states. J Pers Soc Psychol 45(3):513–523. doi: 10.1037/0022-3514.45.3.513

Sebastiani P, Ramoni MF, Nolan V, Baldwin CT, Steinberg MH (2005) Genetic dissection and prognostic modeling of overt stroke in sickle cell anemia. Nat Genet 37(4):435–440. doi: ng1533 [pii], 10.1038/ng1533

Seeger G, Schloss P, Schmidt MH, Ruter-Jungfleisch A, Henn FA (2004) Gene-environment interaction in hyperkinetic conduct disorder (HD + CD) as indicated by season of birth variations in dopamine receptor (DRD4) gene polymorphism. Neurosci Lett 366(3):282–286. doi: 10.1016/j.neulet.2004.05.049 S0304394004006305 [pii]

Silva PA (1990) The dunedin multidisciplinary health and development study: a 15 year longitudinal study. Paediatr Perinat Epidemiol 4(1):76–107

Sonuga-Barke EJ, Oades RD, Psychogiou L, Chen W, Franke B, Buitelaar J, Banaschewski T, Ebstein RP, Gil M, Anney R, Miranda A, Roeyers H, Rothenberger A, Sergeant J, Steinhausen HC, Thompson M, Asherson P, Faraone SV (2009) Dopamine and serotonin transporter genotypes moderate sensitivity to maternal expressed emotion: the case of conduct and emotional problems in attention deficit/hyperactivity disorder. J Child Psychol Psychiatry 50(9):1052–1063. doi: JCPP2095 [pii], 10.1111/j.1469-7610.2009.02095.x

Spinelli S, Schwandt ML, Lindell SG, Newman TK, Heilig M, Suomi SJ, Higley JD, Goldman D, Barr CS (2007) Association between the recombinant human serotonin transporter linked promoter region polymorphism and behavior in rhesus macaques during a separation paradigm. Dev Psychopathol 19(4):977–987. doi: S095457940700048X [pii], 10.1017/S095457940700048X

Stein MB, Schork NJ, Gelernter J (2008) Gene-by-environment (serotonin transporter and childhood maltreatment) interaction for anxiety sensitivity, an intermediate phenotype for anxiety disorders. Neuropsychopharmacology 33(2):312–319. doi: 1301422 [pii], 10.1038/sj.npp. 1301422

Stevens SE, Kumsta R, Kreppner JM, Brookes KJ, Rutter M, Sonuga-Barke EJ (2009) Dopamine transporter gene polymorphism moderates the effects of severe deprivation on adhd symptoms: developmental continuities in gene-environment interplay. Am J Med Genet B Neuropsychiatr Genet 150B(6):753–761. doi: 10.1002/ajmg.b.31010

Thapar A, Langley K, Fowler T, Rice F, Turic D, Whittinger N, Aggleton J, Van den Bree M, Owen M, O'Donovan M (2005) Catechol o-methyltransferase gene variant and birth weight predict early-onset antisocial behavior in children with attention-deficit/hyperactivity disorder. Arch Gen Psychiatry 62(11):1275–1278. doi: 62/11/1275 [pii], 10.1001/archpsyc.62.11.1275

Thomas D (2010) Gene-environment-wide association studies: emerging approaches. Nat Rev Genet 11(4):259–272. doi: nrg2764 [pii], 10.1038/nrg2764

Thompson WD (1991) Effect modification and the limits of biological inference from epidemiologic data. J Clin Epidemiol 44(3):221–232. doi: 0895-4356(91)90033-6 [pii]

Uher R, McGuffin P (2007) The moderation by the serotonin transporter gene of environmental adversity in the aetiology of mental illness: review and methodological analysis. Mol Psychiatry 13(2):131–146. doi: 10.1038/sj.mp. 4002067

Uher R, McGuffin P (2010) The moderation by the serotonin transporter gene of environmental adversity in the etiology of depression: 2009 update. Mol Psychiatry 15(1):18–22. doi: mp2009123 [pii], 10.1038/mp. 2009.123

van Os J, Rutten BP, Poulton R (2008) Gene-environment interactions in schizophrenia: review of epidemiological findings and future directions. Schizophr Bull 34(6):1066–1082. doi: 10.1093/schbul/sbn117

van Winkel R, Henquet C, Rosa A, Papiol S, Fañanás L, De Hert M, Peuskens J, van Os J, Myin-Germeys I (2008) Evidence that the COMTVal158Met polymorphism moderates sensitivity to stress in psychosis: an experience-sampling study. Am J Med Genet Part B: Neuropsychiatr Genet 147B(1):10–17. doi: 10.1002/ajmg.b.30559

Wacholder S, Chanock S, Garcia-Closas M, El Ghormli L, Rothman N (2004) Assessing the probability that a positive report is false: an approach for molecular epidemiology studies. J Natl Cancer Inst 96(6):434–442

Witte JS, Gauderman WJ, Thomas DC (1999) Asymptotic bias and efficiency in case-control studies of candidate genes and Gene-environment interactions: basic family designs. Am J Epidemiol 149(8):693–705

Xie P, Kranzler HR, Poling J, Stein MB, Anton RF, Brady K, Weiss RD, Farrer L, Gelernter J (2009) Interactive effect of stressful life events and the serotonin transporter 5-HTTLPR genotype on posttraumatic stress disorder diagnosis in 2 independent populations. Arch Gen Psychiatry 66(11):1201–1209. doi: 66/11/1201 [pii], 10.1001/archgenpsychiatry.2009.153

Yang Q, Khoury MJ (1997) Evolving methods in genetic epidemiology III. Gene-environment interaction in epidemiologic research. Epidemiol Rev 19(1):33–43

Zammit S, Wiles N, Lewis G (2010a) The study of gene-environment interactions in psychiatry: limited gains at a substantial cost? Psychol Med 40:711–716. doi: 10.1017/S0033291709991280

Zammit S, Owen MJ, Lewis G (2010b) Misconceptions about Gene-environment interactions in psychiatry. Evid Based Ment Health 13(3):65–68. doi: 13/3/65 [pii], 10.1136/ebmh.13.3.65

Behavioral Genetics of Affective and Anxiety Disorders

Katharina Domschke and Andreas Reif

Abstract As shown by clinical genetic studies, affective and anxiety disorders are complex genetic disorders with genetic and environmental factors interactively determining their respective pathomechanism. Advances in molecular genetic techniques including linkage studies, association studies, and genome-wide association studies allow for the detailed dissection of the genetic influence on the development of these disorders. Besides the molecular genetic investigation of categorical entities according to standardized diagnostic criteria, intermediate phenotypes comprising neurobiological or neuropsychological traits (e.g., neuronal correlates of emotional processing) that are linked to the disease of interest and that are heritable, have been proposed to be closer to the underlying genotype than the overall disease phenotype. These intermediate phenotypes are dimensional and more precisely defined than the categorical disease phenotype, and therefore have attracted much interest in the genetic investigation of affective and anxiety disorders. Given the complex genetic nature of affective and anxiety disorders with an interaction of multiple risk genes and environmental influences, the interplay of genetic factors with environmental factors is investigated by means of gene-environment interaction (GxE) studies. Pharmacogenetic studies aid in the dissection of the genetically influenced heterogeneity of psychotropic drug response and may contribute to the development of a more individualized treatment of

K. Domschke (✉) · A. Reif
Department of Psychiatry, University of Würzburg,
Füchsleinstrasse 15, D-97080, Würzburg, Germany
e-mail: domschke_k@klinik.uni-wuerzburg.de

A. Reif
e-mail: reif_a@klinik.uni-wuerzburg.de

Curr Topics Behav Neurosci (2012) 12: 463–502

DOI: 10.1007/7854_2011_185

Published Online: 4 February 2012

affective and anxiety disorders. Finally, there is some evidence for genetic factors potentially shared between affective and anxiety disorders pointing to a possible overlapping phenotype between anxiety disorders and depression.

Keywords Intermediate phenotype · GWAS · CNV · Gene-environment interaction (GxE) · Pharmacogenetics

Contents

1 Affective Disorders 464
1.1 Clinical Genetics (Family, Twin, Adoption Studies) 465
1.2 Molecular Genetics (Linkage, Association, GWAS, CNV Analysis) 466
1.3 Pharmacogenetics 478
2 Anxiety Disorders 480
2.1 Clinical Genetics 480
2.2 Molecular Genetics 481
2.3 Genetics of Intermediate Phenotypes of Anxiety Disorders 483
2.4 Gene-Environment Interaction 484
2.5 Pharmacogenetics 484
3 Overlapping Phenotypes 484
4 Outlook 486
References 487

1 Affective Disorders

The group of affective disorders comprises both major depressive disorder (MDD; unipolar depression) with various subtypes as well as bipolar disorder (BPD). The latter displays by changes between (hypo-)manic and depressive phases, with intermittent euthymic phases, while the course of MDD is characterized by depression and euthymia. As patients rarely develop their first manic phase only years after their first depression, they might well be initially mischaracterized as MDD patients (so-called "hidden bipolars"), which is a challenge for genetic studies on affective disorders. To overcome this problem, several indicators for the presence of BPD in depression have been suggested, e.g. subthreshold hypomanic symptoms (Angst et al. 2010; Fiedorowicz et al. 2011). This however has not been incorporated in current genetic studies and thus one should always consider that MDD studies might well include a substantial amount of "hidden bipolar" patients, obfuscating MDD-specific findings. Furthermore, in the following section more recent approaches such as genome-wide association studies (GWAS) and copy-number variant (CNV) analyses are reviewed, as there are already plenty of scholarly review articles on linkage and association studies. These issues will therefore be touched upon more briefly; with respect to intermediate phenotype and gene x environment (GxE) studies in affective disorders, the reader is referred to more specialized reviews for the sake of space, as this extensive topic is beyond the scope of this article.

1.1 Clinical Genetics (Family, Twin, Adoption Studies)

There is ample evidence that BPD is a highly genetic condition featuring an estimated heritability of 0.75 as evidenced by numerous clinical genetic studies. The risk for a bipolar patient to have a bipolar first-degree relative is increased about tenfold, and the risk to have an MDD relative is even higher (10- to 15-fold). The largest study to date, that investigated more than two million Swedish nuclear families (Lichtenstein et al. 2009), demonstrated a heritability of bipolar disorder of 59%. Furthermore, relatives of bipolar patients also had a two to fourfold increased risk to suffer from schizophrenia. Environmental influences were mainly due to non-shared environment. In line with these data, twin studies also argue for a heritability ranging from 59 to 87% (for an overview see Shih et al. 2004). As little as two adoption studies on bipolar disorder have been carried out to date (Mendlewicz and Rainer 1977; Wender et al. 1986), also argued for a genetic cause of the disorder. An excellent overview on clinical genetic studies in bipolar disorder can be obtained from Smoller and Finn (2003).

On the other hand, the heritability of MDD is comparatively lower (estimated to be around 0.37 in the most comprehensive review and meta-analysis available to date; Sullivan et al. 2000) and environmental influences (unique, but not shared environment) are considered to play a more important role as compared to BPD. This is also reflected in twin and adoption studies. However, especially older studies do not discriminate between bipolar and unipolar depression and hence explicit data on unipolar depression is in fact quite limited; only five family and six twin studies, but no adoption study meet the stringent inclusion criteria in a comprehensive analysis (Sullivan et al. 2000). Across the five family studies, the summary odds ratio for MDD in first relatives of MDD patients was 2.84 and significant. Interestingly, the odds ratio increased when only considering controls which have been screened for absence of psychiatric disorders. Two of the three reported twin studies argued for a substantial genetic component of MDD, and finally, the twin studies including more than 21,000 individuals yielded a heritability of 37% and an influence of individual-specific environmental effects of 63%. No large differences in heritability indices were found between community and clinical studies, and the influence of shared environment was negligible. Hence, taken together, there is a clear genetic liability towards MDD although it is much smaller as compared to BPD—which has also been taken into account when reviewing studies where both conditions were not carefully treated separately. Furthermore, it is evident for both MDD as well as BPD that these disorders do not follow a strict Mendelian pattern of inheritance, but rather are complex genetic in nature featuring polygenic and oligogenic models ("common variant, common disease" model), but probably also—for a part of the patients—highly penetrant risk genes ("multiple rare variant, common disease" model, see below).

1.2 Molecular Genetics (Linkage, Association, GWAS, CNV Analysis)

1.2.1 Linkage Studies

More than 40 linkage scans for BPD have been published to date which generated plenty of disparate findings. However, quite a number of loci meeting significance criteria were described by two or more groups: 3p12-14, 4p16, 4q31-35, 5q31-33, 6q16/21-25, 8q21-24, 10q25-26, 11p15.5, 12q23-24, 13q14-32, 18p11, 18q21-22 20q13, 21q21-22, 22q11-12, and Xq24-28. Confirmed positional candidate genes however are yet to emerge from these studies. An initially highly promising linkage peak on chromosome 11p15 is considered meanwhile to be due to type I error (Egeland et al. 1987; Berrettini 2001). Also, other loci which have been initially promising could subsequently not be confirmed. A paradigmatic case in this respect is for e.g. the 12q23-24 locus (Dawson et al. 1995; Ewald et al. 1998, 2002; Green et al. 2005), which is noteworthy due to the co-segregation of Darier's disease with BPD (Maziade et al. 2001). It has been shown that, in some families suffering from both disorders, the BPD locus lies outside of the Darier's disease causing ATP2A2 gene, yet is in linkage disequilibrium with this variant (Jones et al. 2002). Hence, there might indeed be intermediately penetrant variants in this region which are exclusive to only a few families and which are lost in noise when combining many families, or cases, respectively. Similarly, a functional mutation in the gene encoding the brain-specific tryptophan hydroxylase 2 (TPH2) has been described which segregates with BPD in three families (Cichon et al. 2008; Grigoroiu-Serbanescu et al. 2008). Findings like these argue for a "common disease, multiple rare variant" model (McCarthy et al. 2008) and underscore the clinical and genetic heterogeneity of BPD. This however does not argue against the concurrent existence of a "common disease, common variant model". As both models most likely are present in clinical samples, this additional level of complexity further hampers the identification of BPD risk genes.

Different meta-analyses found the strongest evidence for BPD susceptibility loci on 13q and 22q (Badner and Gershon 2002; 1228 patients from 353 families), or 9p22.3-21.1, 10q11.21-22.1 and 14q24.1-32.12 (948 to 2437 patients; Segurado et al. 2003). The latter study used the rank-based genome scan (GSAM) method which, together with sample heterogeneity, might account for the different findings as compared to the first study. Finally, in a combined analysis, 6q21-q25 and 8q24 showed genome-wide significance (5179 patients from 1067 families; McQueen et al. 2005). Again, the underlying genes have not yet been identified. Further scholarly reviews on this topic have been provided by Schulze and McMahon

(2003), Serretti and Mandelli (2008), Barnett and Smoller (2009), and Craddock und Sklar (2009).

Also in MDD, numerous linkage scans were carried out (reviewed e.g. by Lohoff (2010)) yet did not point to clear regions of susceptibility, as expected from the lower heritability rate of MDD as compared to BPD. No meta-analysis has been performed to date, which is surprising given the recent efforts to uncover the genetic basis of MDD and hence there is a clear need for further research. There is no meaningful overlap of linkage peaks between studies, although it is noteworthy that two linkage signals have previously been implicated in BPD: one on chromosome 18q (Camp et al. 2005) and the locus mentioned above on 12q23-24 (McGuffin et al. 2005; Abkevich et al. 2003). Therefore, this region appears to be a promising region for affective disorders, yet most likely carries more than just one risk gene.

1.2.2 Association Studies

With regard to candidate gene studies, many genes were shown to be associated with BPD, but none of them has been established as a specific BPD susceptibility gene. Among the best replicated genes are DAOA/G72 (which was associated in a case–control study, but not in a meta-analyses; Muller et al. 2011; Shi et al. 2008), BDNF (again, meta-analyses provided differing data: Kanazawa et al. 2007; Fan and Sklar 2008), DISC1, NRG1, ARNTL/CLOCK, FAT, and GSK3B (Barnett and Smoller 2009; Serretti and Mandelli 2008; Luykx et al. 2010). Not surprisingly, many genes encoding for components of neurotransmitter pathways have been tested for an association with BPD (such as SLC6A3, HTR2A, TPH2, MAOA, COMT, DRD1, and SLC6A4). Coming from GWAS on schizophrenia, the risk gene ZNF804A was demonstrated not to be specific for this disorder, but rather was also associated with BPD (O'Donovan et al. 2008; Williams et al. 2011; Steinberg et al. 2011). The same was true for the GWAS schizophrenia risk loci around the MHC region and NRGN (Williams et al. 2011) and other genes which initially have been described as schizophrenia risk genes: the above-mentioned DISC1 and DAOA/G72 genes, but also NRG1, DTNB1, and NPAS3 (Huang et al. 2010; Pickard et al. 2009).

Most of the association studies published to date suffer from the drawback of small sample sizes and lack of replication, and hence, the combination of large, well described international samples (as done in the Psychiatric GWAS consortium [PGC]) is paramount especially in the search for risk variants assuming a "common variant, common disease" model. Furthermore, meta-analytic treatment of existing data might shed some light on the contribution of suggested BPD risk genes. Such have been performed on considerably sized samples on only a few genes. The gene encoding for methylenetetrahydrofolate reductase (MTHFR) has been tested for an analysis with mood disorders several times. While the first study on 1222 MDD patients yielded negative results (Gaysina et al. 2008), the latest analysis comprising 9648 cases (MDD, BPD and schizophrenia combined) yielded again a significant yet unspecific association of MTHFR with mood disorders (best OR = 1.26; Peerbooms et al. 2011). This is in line with a positive meta-analysis on BPD by Gilbody et al. (2007), but in discrepancy to three further meta-analyses on MTHFR in BPD (Zintzaras 2006, 1415 cases; Chen et al. 2009, 1260 cases; Cohen-Woods et al. 2010, 2584 cases). Given that the original genotyping data which was presented in the Cohen-Woods study (n = 897 BPD patients) was negative, but not included in the Peerbooms study, which demonstrated a significant effect only when all mood disorders were combined, the role of MTHFR in affective disorders seems to be rather unspecific and small.

A meta-analysis on all mood disorders found borderline evidence for an association of the dopamine receptor 2 (DRD2) *Taq1* polymorphism (which in fact localizes to the neighboring gene ANKK1) with affective disorders, yet two more SNPs in DRD2 proved to be negative in much larger data sets so that there is only weak evidence for this gene being associated with BPD or MDD (Zou et al. 2010). The gene for catechol-O-methyltransferase (COMT) that degrades dopamine, features a well described functional polymorphism resulting in a Val to Met transition and which has been shown to be linked to BPD in a meta-analysis on 2944 cases (Zhang et al. 2009b), although this seemed to be more pronounced in Asian populations. Obviously, also genes coding for components of the serotonin system were subjected to meta-analyses, which often were combined with genuine genotyping efforts. A functional SNP in the promoter region of the serotonin receptor gene HTR1A (rs6295) was demonstrated to be significantly associated with BPD (1148 cases, Kishi et al. 2011). In the MAOA gene, three polymorphisms (sample sizes mostly >1000 cases) were meta-analyzed and the main finding was an association of an intronic CA repeat with BPD in Caucasians (Fan et al. 2010). The SLC6A4 promoter polymorphism (5-HTTLPR), which has mainly been studied in MDD GxE (see below) was also included in several meta-analyses that conclusively demonstrated a small yet significant association of the short variant with BPD (Lasky-Su et al. 2005; Cho et al. 2005, 1712 cases). Two meta-analyses focused on the TPH1 gene (Chen et al. 2008, 2011) and in unison came to the conclusion that TPH1 is not associated with MDD (2340 and 1812 patients, respectively), but with BPD (1951 and 2083 cases). Given that in the brain only the TPH2 isoform of tryptophan hydroxylase is expressed, this finding is rather surprising; however, as the foetal brain depends on maternal 5-HT production which is accomplished by placental TPH1 the observed association might in fact be true but rather due to maternal and not case genotype as has been described for rare TPH1 mutations in ADHD (Halmoy et al. 2010).

The lower heritability of MDD as compared to BPD notwithstanding, plenty of case–control association studies have been published thereon as well. A scholarly overview on candidate gene studies is provided by Lohoff (2010). In order to separate the wheat from the chaff, replication is key and meta-analytic aggregation of data is a possible route to success. Accordingly, a thorough meta-analysis on the data available until June 2007 examined 183 papers on 393 polymorphisms (Lopez-Leon et al. 2008). Twenty-two of these polymorphisms have been tested in at least three studies and were thus subjected to further meta-analysis. Here, significant association was demonstrated for the genes APOE2 (827 cases), GNB3 (375 cases), MTHFR (875 cases), SLC6A4 (3752 cases), and SLC6A3 (as little as 151 cases). Negative results were obtained for ACE, BDNF, COMT, DRD3, GABRA3, HTR1A, HTR1B, HTR2A, HTR2C, MAOA, SLC6A2, and TPH1. Further meta-analyses which have been published before this study were performed on ACE, DRD4, HTR2A, MTHFR, SLC6A4, and TH; positive findings were obtained for DRD4 (Lopez Leon et al. 2005; 917 cases). In the years following the meta-analysis by Lopez Leon, only few other meta-analytic studies have been published including those on MTHFR and TPH1 cited above. Furthermore, Franke and associates recently conducted a meta-analysis on the functional BDNF Val66Met polymorphism. As BDNF has been implicated both in the pathogenesis of depression as well as the mechanism of action of anti-depressant treatment (Duman and Monteggia 2006), it is an obvious candidate gene and accordingly was shown to be associated with depression in this meta-analysis of 2812 cases, although the association is sex-specific and only detectable in males (Verhagen et al. 2010).

The largest body of evidence, and by far the largest sample sizes, exists for the gene encoding the serotonin transporter (SLC6A4). Following the seminal finding by Caspi et al. (2003) that environmental influences interact with SLC6A4 genotype to increase the risk toward depression, emphasis has been put on studies aimed to test such GxE interactions adding a further level of complexity. Following positive meta-analyses confirming a main gene effect of SLC6A4 in depression (Furlong et al. 1998; Lopez-Leon et al. 2008; Clarke et al. 2010), also meta-analyses on GxE interaction studies yielded support for the notion that SLC6A4 has a role in the etiology of depression (Karg et al. 2011). As Karg and Sen elaborate in depth on this topic in this book, the reader is referred to their contribution as well as the review articles by Uher and McGuffin (2008, 2010).

Also, genes encoding components of the cortisol pathway have proven to be interesting candidates for GxE in MDD. The glucocorticoid receptor-regulating co-chaperone FKBP5 has first been associated with recurrence of depression and response to antidepressant treatment in 2004 (Binder et al. 2004). This has later been replicated (Lekman et al. 2008) and FKBP5 was shown to interact with HTR2A and GRIK4 in moderating the response to antidepressant treatment (Horstmann et al. 2010). Most interestingly, FKBP5 has been shown to interact with severity of childhood abuse on later-life PTSD symptoms (Binder et al. 2008) which however might be confined to African Americans (Xie et al. 2010). One of the involved SNPs (rs1360780) was later replicated to interact with childhood maltreatment to affect depression measures (BDI-II) in adult life (Appel et al. 2011). This SNP also displayed a main gene effect on suicidal events in depressive adolescents (Brent et al. 2010), a categorical diagnosis of depression (Lekman et al. 2008) and general depressive symptoms (Lavebratt et al. 2010; Zobel et al. 2010; Velders et al. 2011) and interestingly is associated with decreased cortisol levels (Velders et al. 2011) providing a possible pathophysiological mechanism for the association data. Also, rs1360780 went along with smaller right hippocampal volume in patients with depression (Zobel et al. 2010). Taken together, there is good evidence (Binder 2009) that risk genotypes in FKBP5 (especially rs1360780) interact with early life adversity on later life depression, possibly by long-term adaptive changes of the HPA axis and subsequent morphological changes of the hippocampus increasing vulnerability to disease. In line with these findings, it was also shown that rs110402 in the corticotropin-releasing hormone receptor (CRHR1) interacted with child abuse to increase the risk toward later life depression (Bradley et al. 2008), most interestingly in interaction with the serotonin transporter risk genotype mentioned above (Ressler et al. 2010).

It seems to be a common phenomenon that candidate genes rarely replicate when tested for in GWAS (see below) data sets. A thorough study on MDD candidate genes (Bosker et al. 2011) tested 57 genes in the Genetic Association Information Network (GAIN) MDD sample (n = 1862 cases). From 93 selected candidate SNPs, only 18 were present on the array, and a further 47 were imputed. Of those, only five (including an SNP in NPY) were associated in the GAIN sample, all with $p>0.03$. When candidates were tested on a gene-based level, analyzing 4870 SNPs, the TNF and NET genes yielded suggestive evidence. In general, heavy use of imputation might have introduced a further source of noise in this particular study. Likewise, when candidate genes were targeted in other individual GWAS, rarely more than expected by chance replicated.

Taken together, numerous association studies on MDD and BPD—actually too many to mention in this overview—hitherto only presented few convincing findings. Amongst them, associations of affective disorders with MTHFR, BDNF, and SLC6A4 seem to be robust. A length variant in the latter gene shows solid evidence for GxE effect; FKBP5 is another promising candidate for the moderating effects of early life stress regarding depression in adult life. Lessons that can be learned from the plethora of false-positive findings to date are that large and well-characterized samples have to be accrued, with careful evaluation of life events and the assessment of biological measures such as neuroimaging endophenotypes or therapy response.

1.2.3 Genome-Wide Association Studies

Along with autism, attention-deficit hyperactivity disorder (ADHD) and schizophrenia affective disorder comprise the core disorders of the Psychiatric GWAS Consortium (Sullivan 2010; Psychiatric GWAS Consortium [PGC] 2009), and at the time of writing, more than 12,000 cases of BPD and 14,000 cases of MDD are available within the PGC, with 20,000 more cases each expected to be included over the next 2 to 3 years. While these numbers may at first sound impressive, one has to consider that GWAS are the method of choice to pick common risk alleles conveying only small individual risk. Other complex-genetic traits might provide some clues for appropriate sample sizes: it took almost a quarter million people to detect and respectively confirm 32 risk variants for body mass index, explaining as little as 2–4% of genetic variance, and it was estimated that another 284 variants would carry comparable effect sizes and together would explain 6–11% of the genetic variation. The authors assumed that further 730,000 individuals would have to be genotyped to uncover 95% of these variants (Speliotes et al. 2010). Likewise, a recent study on body height (Lango Allen et al. 2010) examined 180,000 subjects and found 180 loci explaining 16% of phenotypic variants. As body weight and height are for sure somewhat easier to determine than depression, which is a heterogeneous condition from the start, one can easily see the obstacles one has to face when dealing with these kinds of studies. Having said this, and thereby also lowering the bar of expectations somehow, the above-mentioned studies are also encouraging and can be seen as proof-of-principle: GWAS can detect novel pathways and provide meaningful results, and thus larger scale studies should be encouraged in order to identify the molecular determinants of affective disorders, as previous linkage and association studies fell short in conclusively delineating these. Concluding these introductory remarks, the recent debate on the missing—or, rather, hidden—heritability has to be mentioned. The discussion whether the major endogenous psychoses are due to the sum of multiple common alleles with small individual effects (e.g. Purcell et al. 2009) or due to many rare variants, also including copy number variants (CNV), and causing "synthetic associations" (Dickson et al. 2010) and resulting in phenocopies (Gershon et al. 2011) is held lively and far from being resolved. These authors' personal view is that both models might exist—which however complicates matters even more.

BPD

GWA studies on BPD, which has a higher heritability and presumably less GxE effects as compared to depression, have (probably due to these facts) provided stronger findings than MDD GWAS. The following section elaborates on the most interesting findings from BPD GWAS at the time of writing (04/2011) with focus on replicated risk genes and pathways, while issues like population admixture, microarray technology, and statistical comments are not further commented upon.

In one of the first published GWAS on BPD, Baum et al. (2008a) reported genome-wide significance of rs10120253 in intron 1 of diacylglycerol kinase eta (DGKH) in a German and US American population using a pooling approach. The gene product of DGKH metabolizes diacylglycerol (DAG), which is produced upon cleavage of PIP2 into IP3 and DAG by phospholipase C. DAG, in turn, activates protein kinase C which phosphorylates a variety of proteins including Dishevelled, an inhibitor of GSK3β (which itself has been considered an outstanding candidate gene for BPD due to several lines of molecular genetic evidence; Luykx et al. 2010). Furthermore, DGKH knockdown in HeLa cells impaired the MEK/ERK pathway, while overexpression activated the pathway (Yasuda et al. 2009). DGKH might therefore be involved in crucial pathways for psychiatric disorders and especially the mechanism of action of lithium. However, replication of DGKH failed in three studies on BPD and/or lithium response, respectively (Manchia et al. 2009; Tesli et al. 2009; Takata et al. 2011) while four other studies were ambiguous or positive (Baum et al. 2008b; Ollila et al. 2009; Squassina et al. 2009; Zeng et al. 2011). A recent study provided evidence for an association of a DGKH risk haplotype with MDD, BPD, and adult ADHD (Weber et al. 2011). Meta-analyses on Caucasian (Weber et al. 2011) as well as Asian (Takata et al. 2011) samples however demonstrated that DGKH is significantly associated with BPD. Furthermore, increased expression of DGKH in BPD (Moya et al. 2010) was demonstrated in human post-mortem tissue, so that DGKH represents one of the most promising candidate genes for BPD to date. Other candidates from the Baum et al. GWAS include NXN, VGCNL1, DFNB31, and SORCS2, the latter two of which were replicated in a later study (Ollila et al. 2009).

The UK Wellcome Trust Case Control Consortium (WTCCC 2007) aims at the investigation of several complex genetic disorders with high prevalence. As evident from the first glance on the Manhattan plots in this paper, there are no "skyscraping" BPD risk SNPs as compared to very clear signals in cardiovascular or metabolic disorders. The WTCCC BPD GWAS provided genome-wide evidence for a non-gene marker next to PALB2, NDUFAB1, and DCTN5; other signals were observed for KCNC2, GABRB1, GRM7, and SYN3, all of which are in pathways previously implicated in BPD. Shortly after the WTCCC report, Sklar et al. reported on the STEP-UCL study and provided significant findings for MYO5B, TSPAN8, CDH7, and EGFR (Sklar et al. 2008). Some of those genes were attempted to replicate using a targeted approach; both TSPAN8 (Scholz et al. 2010) and CDH7 (Soronen et al. 2010) were confirmed in doing so. When the significant signals from the WTCCC and Baum data sets were tested in Sklar's STEP-BD/UCL sample, negative findings were observed for DGKH and PALB2, however, this analysis provided further support for CACNA1C and DFNB31 arguing for the rationale to combine large data sets. Accordingly, the fourth GWAS study, ED-DUB-STEP2 (Ferreira et al. 2008), investigated another 1000 patients and included meta-analytic treatment of the WTCCC and STEP-UCL data sets (total $n = 4387$ cases). In doing so, the holy grail of genome-wide significance was reached for markers in two genes: CACNA1C (alpha-1 subunit of a voltage dependent calcium channel) and ANK3 (ankyrin 3). Other interesting candidate genes from this study include SYNE1, SPRED1, CMTM8 (which interacts with EGFR), NPAS3 (which has previously been suggested to be associated with schizophrenia and bipolar disorder; Pickard et al. 2009), and ARNT2. In a subsequent meta-analysis including two samples from Nordic countries, the breast cancer risk genes PALB2 and BRCA2 were followed up in these as well as the WTCCC and STEP-UCL/ED-DUB-STEP2 samples (total case $n = 5547$). In doing so, variants in both genes were shown to be associated with BPD (Tesli et al. 2010). Not surprisingly nevertheless, CACNA1C and ANK3 drew most attention in follow-up studies.

For both ANK3 and CACNA1C, it is noteworthy that replication attempts not only provided evidence that these genes are associated with BPD; rather, they were demonstrated to be associated with a broad range of disorders across diagnostic boundaries arguing for a more unspecific role of these genes in psychiatric disorders. For example, ANK3 was not only replicated in BPD (Lee et al. 2010; Schulze et al. 2009; Scott et al. 2009; Smith et al. 2009), but also associated with schizophrenia (Athanasiu et al. 2010). Likewise, CACNA1C was again found to be associated with BPD (Keers et al. 2009), but also with schizophrenia (Green et al. 2009; Moskvina et al. 2009; Nyegaard et al. 2010), MDD (Green et al. 2009), and psychopathological features (e.g., agitation) therein (Casamassima et al. 2010). Neuroimaging studies demonstrated an effect of the CACNA1C risk variant rs1006737 on brain structure (Franke et al. 2010; Kempton et al. 2009) as well as function (Erk et al. 2010; Krug et al. 2010; Wessa et al. 2010).

In the last 2 years, several other GWAS and meta-analyses on BPD have been published. A small study from Japan (Hattori et al. 2009), which applied a two-stage design, provided nominal although not corrected significance for markers within AUTS2 (previously implicated in autism), SNAP25 (which is a schizophrenia and ADHD candidate gene), PLXNA2 (which has been found in schizophrenia and anxiety GWAS) and CSMD1, which was already one of the candidates from the Baum et al. study (Baum et al. 2008a). No other top hits from previous BPD GWAS however were replicated. In Han Chinese patients suffering from BPD type I, Lee et al. (2010) likewise did not provide findings on the genome-wide level, yet interestingly also found suggestive association of BPD with SNPs in another voltage-dependent calcium channel subunit, CACNB2 (other highly significant SNPs were located in KCTD12, SP8, and ST8SIA2) pointing again to calcium signalling having a role in BPD. Targeted investigation of previously identified GWAS candidate genes yielded a $p = 10^{-5}$ for an SNP near ANK3, yet no other gene has been replicated.

Scott and colleagues combined two GWAS studies from the US, Canada, and UK (the NIMH/Pritzker and GSK GWAS) and analyzed them separately as well as in conjunction with the WTCCC study (Scott et al. 2009). In doing so, no genome-wide significant finding was observed; yet three regions with a p around 10^{-7} were reported encompassing the genes MCTP1 (which encodes a high-affinity calcium binding protein which is highly expressed in the brain), ITIH1 and GLN3. Furthermore, CTNNA2 was amongst the top hits which also gained support from other hypothesis-free approaches in psychiatric disorders such as ADHD (Lesch et al. 2008). Neither DGKH, ANK3 nor CACNA1C were confirmed in this GWAS, however the latter yielded convincing support upon a fixed-effects meta-analysis including the Ferreira and Schulze studies. In 2009, Kelsoe and associates reported on two GWAS examining US Americans of European and African ancestry (Smith et al. 2009), respectively, with a combined $n = 1346$ BPD cases. Again, no genome-wide significant findings emerged and interestingly, significant findings were discrepant for each subsample. One of the promising top hits in this study is NTRK2, as this gene which encodes a neurotrophin receptor has been implicated in mood disorders previously. When previous GWAS risk genes were tested for, ANK3 yielded further support, while CACNA1C was negative. It should be noted that this sample overlaps with the one tested by Baum et al.; generally, the sample overlap between different studies will become the rule rather than the exception due to the need for international cooperation and large sample sizes. Furthermore, a small GWAS from Norway ($n = 194$), which however was followed up in a larger Icelandic sample (Djurovic et al. 2010), provided suggestive evidence for several interesting candidate genes (e.g. GUCY1B2, SHANK, and CNTNAP5), none of which however was amongst the top hits in previous studies. The largest study to date has employed a two-stage design using a discovery cohort of 682 BPD patients and carried over the top 48 SNPs to replication samples; SNPs surviving this procedure were subjected to a meta-analysis with previous BPD GWAS data sets (Cichon et al. 2011). The total case number investigated in this study was thus $n = 8441$ BPD patients. In doing so, the neurocan (NCAN) gene was identified as a susceptibility factor for BPD with the best SNP yielding a $p = 2.1 \times 10^{-9}$, i.e. genome-wide significance.

In conclusion, GWAS on BPD and subsequent meta-analysis provided evidence that BPD shares risk variants with schizophrenia, MDD, and ADHD; furthermore, calcium and GABA signalling pathways were repeatedly found to be associated with disease, along with genes modifying neuronal plasticity. At the time of writing, CACNA1C, ANK3, and DGKH can be considered the risk genes with the most compelling body of evidence. Accordingly, those are scrutinized more thoroughly and first functional studies already provided evidence for changes in brain function in risk allele carriers.

MDD

While the prevalence of MDD is five to tenfold higher as compared to BPD, its heritability is lower and presumably heterogeneity is even higher. These issues complicate GWAS on this phenotype, probably explaining the lack of genome-wide findings despite the fact that studies on MDD and schizophrenia feature the largest of all disorders analyzed in the PGC.

The NIH sponsored Genetic Association Information Network (GAIN) studies also featured major depression and these studies were amongst the first published GWAS on MDD (Sullivan et al. 2009). Discovery sample patients came from two Dutch longitudinal studies (NESDA and NTR, combined $n > 1700$). The top 25 SNPs featured four SNPs in the PCLO gene, which encodes for a subunit of the presynaptic vesicle fusion complex, although none of them met the criteria for genome-wide significance. Considerable overlap was noted for the mood disorder candidate genes CACNA1C, ANK3, GRM7, and DGKH. While PCLO did not clearly replicate in the Sullivan et al. study, a later reanalysis questioned this initial notion and argued for an association of a non-synonymous coding SNP with MDD in the very same replication cohorts (Bochdanovits et al. 2009). Furthermore, a later population-based study demonstrated an association of PCLO rs2522833 with depressive disorders (Hek et al. 2010), which also held true when a meta-analysis of all published data was conducted especially when only population-based studies were considered ($p = 1.9 \times 10^{-9}$). Most interestingly, in a hypothesis-free approach, PCLO was demonstrated to be differentially expressed and associated with BPD again questioning the diagnostic specificity of GWAS candidate genes (Choi et al. 2011). However, meta-analysis of an MDD and a BPD GWAS did not support a role of PCLO in BPD (Liu et al. 2011).

In a medium-sized GWAS from Germany (Rietschel et al. 2010), there was a suggestive finding for HOMER1 which replicated in an independent sample. In this study, the authors also conducted a genomic imaging study and demonstrated decreased dorsolateral prefrontal cortex activation in the *n*-back task as well as decreased anterior cingulate cortex activation upon anticipation of a monetary reward in risk allele carriers. Especially the latter might be related to anhedonic behavior, one of the key features of depression. Most interestingly, also CACNA1B was amongst the highest ranked genes, again implicating calcium signalling in affective disorders. A larger ($n = 1636$ cases) UK-based GWAS argued for the BICC1 gene in MDD (almost needless to say, genome-wide significance was missed); however, it was not replicated in samples from Munich and Lausanne. When all three studies were treated by meta-analysis, the schizophrenia/autism candidate gene NLGN1 was amongst the most promising findings with a $p_{combined} = 8.5 \times 10^{-6}$ (Lewis et al. 2010). When the STAR*D study was analyzed for the phenotype MDD ($n = 1{,}221$; Shyn et al. 2011), no genome-wide significant findings also emerged. Promising or previously implicated candidate genes—however, all at a $p > 10^{-5}$—included ANKRD46, CTNND2, and CSMD3. Another recent GWAS focused on recurrent early onset MDD, as defined by an onset before the age of 31 (GenRED, $n = 1020$; Shi et al. 2011). Nested candidate gene analysis yielded the lowest p value in CACNA1C; as is common for all mood disorder GWAS, there were no findings meeting the genome-wide significance threshold, but several highly suggestive findings with the top hit in a brain-expressed transcript of unknown function. Other interesting candidates include GDNF, SP4, STIM1, KCNQ1, VAMP4, and CSMD1. Most noteworthy, the SP4 signal (which almost entirely came from female subjects) became stronger when the GenRED sample was treated meta-analytically with the STAR*D and GAIN studies (total $n_{cases} = 3{,}957$; Shyn et al. 2011). This meta-analysis yielded better, although still only suggestive significance levels and also argued for an association of the GRM7 gene. While SP4 encodes a transcription factor orchestrating gene networks implicated in affective disorders (and most notably, as mentioned above, the SP8 transcription factor has been found in a BPD GWAS), GRM7 which encodes a glutamate receptor was not only of suggestive significance in the WTCCC BPD GWAS, but also amongst the top hits of another MDD GWAS (Muglia et al. 2010). This study was performed in two European samples from Southern Bavaria and Lausanne (total $n > 1{,}500$), yet also did not result in genome-wide significant findings or meaningful overlap of top SNPs between both samples. The authors computed a meta-analysis of both samples as well, along with the implementation of gene-wide tests. This interesting method yielded several genes which also survived a correction procedure, the genes with the lowest p-values being SMG7 and NFKB1. Candidate genes from previous studies however did not replicate in this analysis, apart from the glutamate receptor gene GRM7. A similar approach (i.e., discovery GWAS followed by meta-analysis and gene-based tests) was taken by the largest MDD GWAS to date, the MDD2000 + study (Wray et al. 2012). Here, a total of >2400 cases were examined and a meta-analysis was conducted by including the GAIN sample and the UK-based study reported by Lewis et al. (total $n > 5700$ cases). Suggestive findings ($p < 10^{-5}$) include NOS1AP, ADCY3, and the schizophrenia/autism risk gene CNTNAP2 (the latter in males only). The adenylate cyclase ADCY3 gene was ranked second in the gene-based test, and also the gene encoding galanin (which was previously shown to be associated with antidepressant treatment response and disease severity in MDD and anxiety disorders; Unschuld et al. 2010) was amongst the top ten hits in this analysis. From the pre-selected candidate genes, IL10, OPRM1 (being a candidate stemming from the GAIN MDD GWAS), HTT, HTR1B, GRIN1, and the apparently pleiotropic risk gene CACNA1C were associated with disease. Meta-analysis did not yield significant findings and in particular did not support PCLO as a risk gene for MDD.

Trait depression, as assessed with the NEO-PI personality questionnaire (where "depression" is a subscale of the Neuroticism domain) was assessed in two GWAS in the general population from Sardinia and the US (combined $n = 4811$; Terracciano et al. 2010). The top hit, at a $p = 6 \times 10^{-7}$, was an intronic SNP in the RORA gene. Two other noteworthy high-ranking candidates include the glutamate receptor gene GRM8, which hitherto has mainly been associated with cognitive phenotypes, and CDH13, which has been identified in an ADHD linkage scan meta-analysis (Zhou et al. 2008) as well as GWAS on ADHD (Lesch et al. 2008) and substance use disorders.

Cross-Disorder Analyses

In order to yield larger samples, several meta-analyses have been conducted. The largest meta-analysis combined the WTCCC, STEP-BD, NIMH-BD, and the German BPD sample, as well as the GAIN-MDD GWAS on MDD. The total number of cases exceeded 6600, compared against >9000 controls (McMahon et al. 2010). An inherent problem with this kind of study is the use of different genotyping platforms, diagnostic heterogeneity, as well as ethnic heterogeneity; to minimize these limitations, only subjects of European descent have been analyzed. In doing so, six SNPs which were located in the PBRM1 gene met the criteria for genome-wide significance. The best SNP was also significant in the replication sample, yielding a final $p = 1.7 \times 10^{-9}$. When the ED-DUB-STEP2 GWAS was analyzed together with an MDD GWAS (Liu et al. 2011), CACNA1C SNPs passed the hurdle of genome-wide significance while ANK3 was not supported in the meta-analysis, probably suggesting that this gene is more specific to BPD. SYNE1 was one of the candidates where the significance level actually increased upon meta-analysis, and which interestingly also turned up in the primary PGC BPD GWAS meta-analysis. Not surprisingly, also schizophrenia and BPD were treated meta-analytically (Wang et al. 2010). In this study, meta-analysis provided evidence for the genes ASTN2 and CNTNAP2, both of which have been implicated in ADHD, as well as the GABA receptors GABRR1 and GABRR2. When the three large US American GWAS on psychiatric disorders and treatment efficacy—namely, STEP-BD (BPD), CATIE (schizophrenia), and STAR*D (MDD) were analyzed jointly (yielding a total $n_{cases}>3000$; (Huang et al. 2010), one locus met the criteria for being genome-wide significant (near the ADM gene, and apparently being specific for bipolar II disorder). A total of 24 more SNPs reached the defined Omnibus GWAS Test Threshold; however, more than half of them were imputed. Promising candidates are again CTNND2, SP8, ODZ4, and NPAS3.

An alternative rationale is to search for risk variants influencing phenotypic features of mood disorders. Suitable phenotypes include, for example, therapy response (see below) or suicidal ideation. Accordingly, the STEP-BP, WTCCC, UCL, and STAR*D studies were evaluated with respect to the latter (Perlis et al. 2010). None of the 11 loci which were identified in the discovery cohorts however replicated. Also, candidate genes which were selected according to previous data (such as HTR1A or TPH2) did not yield convincing evidence. Meta-analysis of all samples argued for an involvement of SORBS1 and PRKCE, a gene with some a priori biological evidence. Analysis of STAR*D alone, where 90 out of 1953 patients developed treatment-emergent suicidal ideation (Laje et al. 2009), revealed a highly significant association of an SNP in PAPLN and suggestive association of an IL28RA SNP. Additive effects with previous risk alleles for treatment-associated suicidal ideation in the GRIK2 and GRIA3 genes were observed. However, paucity of psychometric data on the suicidal patients along with the very limited sample size bears the high chance of a type-I error. Similar analyses in GENDEP (total $n = 706$, thereof $n = 244$ with treatment-associated suicidal ideation under treatment with either escitalopram or nortriptyline; Perroud et al. 2010) provided some evidence for the genes GDA, KCNIP4, and ELP3 to be associated with escitalopram-associated suicidal ideation. Nested candidate-driven approaches did not yield significant results. A major concern regarding these studies is whether or not treatment-associated suicidal ideation is genetic at all and whether these studies are homogeneous—the striking differences in the percentage of suicidal ideation casts some doubts on this assumption.

Analysis of the complete PGC data set, comprising 12,000 BPD cases and 52,000 controls, yielded 21 SNPs with a corrected $p<0.05$, the best candidate genes being CACNA1C, ODZ4, and two regions of chromosome 11 and 12. As also suggested from earlier cross-disorder analyses demonstrating a significant overlap of common risk variants for BPD and schizophrenia (including also CACNA1C, as well as another voltage-dependent calcium channel and a member of the diacylglycerol kinase family [DGKI]; Moskvina et al. 2009), the latest cross-disorder analysis of the PGC yielded strong evidence for an association of CACNA1C with endogenous psychoses as evidenced by $p = 8.45 \times 10^{-9}$ when BPD and schizophrenia samples were combined. In these analyses, it became also evident that BPD risk genes were highly predictive for schizophrenia, and vice versa. On the other hand, BPD neither predicted MDD nor did MDD predict BPD or schizophrenia. When all three disorders were pooled together (total $n_{cases}>25,000$), six genes met the criteria for genome-wide significance: ITIH3, the HLA/HIST cluster on chromosome 6p21-p22, CACNA1C, TCF4, NT5C2/CNNM2, and IFI44/ELTD1. In MDD alone, the situation is much more frustrating: when combining more than 11,000 cases from the GAIN, GenRED, GSK, mdd2000, MPIP, RADIANT, STAR*D, and NGFN Germany studies, only two SNPs (in the genes NVL and GPHN, which is a highly interesting candidate) came near the level of genome-wide significance. Lower heritability in conjunction with increased heterogeneity might explain the scarcity of solid findings in MDD as compared to BPD.

1.2.4 Copy Number Variations

Currently, there is increasing interest about the role of deletions/duplications of large chunks of the genome (copy number variations, CNVs) in psychiatric disorders. Especially large (>100 kb) and rare CNV harboring many different genes seem to occur more often in schizophrenia, mental retardation, and autism—although there is no diagnostic specificity and some CNVs might underlie any of these three conditions. In any case, these CNVs are rare and can only account for a small percentage of cases, again arguing for a "common disease, multiple rare variant" model. In bipolar disorder, there are only two studies to date which yielded conflicting findings. While the first study (Zhang et al. 2009a; 1001 patients) demonstrated an excess of large (>100 kb) and rare CNVs in BPD, Grozeva et al. (2010) did not find an increased rate of large and rare CNVs in almost 1700 patients suffering from BPD. Hence, even larger samples are needed to unequivocally evaluate the contribution of CNVs in BPD, although from the present data it seems to be clear that at least large and rare CNVs can only account for a very small fraction of BPD cases, if at all.

1.3 Pharmacogenetics

In the treatment of depression, antidepressive pharmacotherapeutic agents have proven to be highly effective for a large proportion of patients. However, two major problems have to be faced: (1) treatment resistance: 30–40% of all patients fail to respond sufficiently to the initial treatment (Fava and Davidson 1996) and (2) treatment intolerance: There is a considerable rate of "treatment emergent adverse effects" associated with antidepressive pharmacotherapy such as hypotension, weight gain, anticholinergic effects, antidepressant-induced mania, or sleep disturbance, which leads to discontinuation of treatment in about 10% of the cases (MacGillivray et al. 2003).

Among multiple reasons underlying non-response to antidepressive pharmacotherapy or differential development of treatment emergent adverse effects under antidepressants, it has been suggested that psychotropic drug response may be heritable with first-degree relative pairs being significantly concordant for antidepressant treatment response (Pare et al. 1971; O'Reilly et al. 1994; Franchini et al. 1998). Pharmacogenetic studies allowing for the detailed dissection of the genetically influenced heterogeneity of psychotropic drug response have revealed several risk genes on a pharmacokinetic as well as on a pharmacodynamic level to drive antidepressant treatment response. On a pharmacokinetic level, variation in the CYP2D6 gene resulting either in poor metabolizers (PM; 7% of the Caucasian population) or in rapid (RM) or even ultrarapid metabolizers (UM; 3% of the Caucasian population) has been reported to be associated with response to tricyclic antidepressants and SSRIs, particularly paroxetine (cf. Kirchheiner et al. 2004). On a pharmacodynamic level, association of response to antidepressants has been observed with variation in candidate genes of depression, especially those involved in the serotonergic system (e.g. SLC6A4, HTR1A, HTR2A, MAOA, TPH1 (for a review see Kato and Serretti 2010; Serretti et al. 2005). Additionally, there is first evidence for differential genetic effects on treatment response specifically in melancholic depression (Baune et al. 2008), depression with comorbid anxiety (Domschke et al. 2008a, 2010) or in gender subgroups (Domschke et al. 2008c). The identification of genetic risk factors for antidepressant treatment response as known to date tremendously helps in better understanding the mechanism of action of antidepressants. These rapidly growing molecular genetic findings might nourish further biochemical, physiological, or pharmacological studies and eventually lead to a personalized medicine with an individually tailored antidepressive pharmacotherapy according to genotype reducing the patients' suffering and lowering healthcare costs at the same time.

Most of the MDD GWAS were embedded in efficacy studies; thus not surprisingly, treatment response GWAS were reported frequently. It should be noted that treatment response of course encompasses a plethora of diverse factors ranging from adherence to medication to exogenous pharmacokinetic influences, adding further noise to already noisy genetic data. The first report was on the German MARS trial and also included an independent German sample as well as the STAR*D study (Ising et al. 2009). Obviously, not only diagnostic but also treatment heterogeneity has to be taken into account in the interpretation. The best signal, which however was not significant on the genome-wide level, came from an SNP in the 5′ region of CDH17; when the 338 best SNPs from the German samples were tested in the STAR*D sample, 46 were associated at the nominal level of significance. Amongst them, interestingly, was HOMER1 (see above). When the level 1 participants of STAR*D, which received citalopram, were analyzed separately (743 remitters versus 608 non-remitters; Garriock and Hamilton 2009a, b), as little as three SNPs were associated with response on the $p<1 \times 10^{-5}$ level. The most interesting finding from this study probably is ARNTL, which is also a member of the PAS superfamily and related to NPAS3. In 2010, the GWAS data of the multicentre European GENDEP study was published ($n = 811$ cases, treated with either nortriptyline or escitalopram; Uher et al. 2010). Analysis of the complete sample did not provide meaningful signals, while analysis of either compound alone pointed to the IL11, UST, and RGL1 genes. Genotype by drug interaction analyses interestingly implicated a region 11 kb downstream of NOL4, which was also one of the four top regions in the STAR*D study. As both studies were published in parallel, this was not mentioned in either paper, yet can be considered a true independent replication.

As compared to MDD, pharmacogenomics studies in BPD are sparser. As treatment response to lithium, the gold standard drug treatment in BPD, is familial, lithium treatment response studies seem to be most worthwhile; as only ca. 40% of all BPD patients can be considered clear lithium responders, and as treatment might go along with considerable side effects in the case of non-response, data on genetic prediction of lithium response would directly translate into the clinical routine. Several case–control association studies comparing responders to non-responders have suggested a variety of risk genes such as the SLC6A4 (seven studies, mixed findings) and other hypothesis-driven targets mainly of neuro-transmitter pathways, intracellular signal transduction pathways involved in the mechanism of action of lithium, and circadian clock genes. GSK3B and CREB1 are amongst the candidates with the best empirical support and also good face validity and hence should be further tested, as evidence is far from being solid. Samples were often rather small and again, most genes lack replication. Scholarly overviews on published association studies can be obtained from McCarthy et al. (2010) and Smith et al. (2010). Although not specifically designed for this question, the STEP-BD trial tested for lithium response as well by means of a GWA study, which was replicated in a second, independent sample ($n = 458$, and $n = 359$ bipolar I or II patients, respectively). Not surprisingly, there were no genome-wide significant findings, however five SNPs associated in the STEP-BD cohort replicated in the second sample including a polymorphism within the GRIA2 gene (Perlis et al. 2009). To specifically search for lithium response genes, the ConLiGen consortium (Schulze et al. 2010) has gathered more than 1200 lithium treated BPD patients whose treatment response has been evaluated using the Alda scale. Genome-wide genotyping has been accomplished and initial data are expected for the second half of 2011.

2 Anxiety Disorders

2.1 Clinical Genetics

Panic disorder has been found to be highly familial with an up to three to fivefold increased prevalence of the disorder in first-degree relatives of patients with panic disorder (Horwath et al. 1995; Maier et al. 1993; Hettema et al. 2001). In relatives of the subgroup of patients with panic disorder and suffocation anxiety, an even higher familiarity has been discerned (Horwath et al. 1997). Furthermore, familiarity of panic disorder seems to depend on the age of onset in the index patient with an onset before the age of 20 years predicting a 17-fold increased risk of panic disorder in first-degree relatives (Goldstein et al. 1997). Also for generalized anxiety disorder and specific phobias a significant familial aggregation was reported (Hettema et al. 2001; Marks and Herst 1970).

Twin studies have identified up to 2–3 times higher concordance rates for panic disorder in monozygotic as compared to dizygotic twins (Skre et al. 1993), with an even higher concordance rate for the subgroup of patients with carbon dioxide-sensitive panic disorder (Bellodi et al. 1998). According to a comprehensive meta-analysis, the contribution of genetic factors has been calculated to be as high as up to 48%, with the remaining 52% being attributable to individual environmental factors. Generalized anxiety disorder has been estimated to have a heritability of about 32%, while the common heritability of phobias was reported to be about

30%, with highest estimates for agoraphobia (67%), blood-injection-phobia (59%), and social phobia (51%). The heritability of posttraumatic stress disorder was reported to be about 20–30% (Kendler et al. 1999; Hettema et al. 2001; Segman and Shalev 2003). Finally, several studies point towards overlapping genetic risk factors for panic disorder and agoraphobia or other phobias, respectively (Kendler et al. 1995; Mosing et al. 2009; Tsuang et al. 2004). In panic disorder and other anxiety disorders, segregation analyses failed to identify a mode of inheritance according to Mendelian patterns, which points to a complex genetic inheritance with an interaction of multiple "vulnerability" or "risk genes", each with only a minor individual influence ("oligo- or polygenic model"), and environmental influences (Vieland et al. 1996).

2.2 *Molecular Genetics*

2.2.1 Linkage Studies

In panic disorder, linkage studies have yielded a variety of potential risk loci on chromosomes 1p, 4q, 7p, 9q, 11p, 15q, und 20p (Crowe et al. 1987, 2001; Knowles et al. 1998; Gelernter et al. 2001; Hamilton et al. 2003; Thorgeirsson et al. 2003; Fyer et al. 2006; Kaabi et al. 2006). In subgroups of patients with panic disorder with comorbid bipolar disorder or kidney/bladder dysfunction, respectively, risk loci on chromosomes 2, 12, 13, and 18 or 13 and 22, respectively, have been described (MacKinnon et al. 1998; Logue et al. 2009; Weissman et al. 2000; Hamilton et al. 2003). In social or specific phobia, linkage studies have excluded a major influence of HTR2A and SLC6A4 loci, with however, some evidence for potential risk loci on chromosomes 16q and 14p (Stein et al. 1998; Gelernter et al. 2003, 2004).

2.2.2 Association Studies

In panic disorder, a variety of association studies has been published so far. Most studies have investigated variation in classical candidate genes for panic disorder as suggested by animal models (e.g. knock-out mice), challenge experiments (e.g. cholecystokinin [CCK] challenge, caffeine challenge), or pharmacological observations (e.g. clinical efficacy of selective serotonin reuptake inhibitors (SSRIs), monoamine oxidase (MAO) inhibitors). Most significant evidence has been yielded for association of variants in the CCKBR (Kennedy et al. 1999; Hösing et al. 2004), MAOA, particularly in female patients (Deckert et al. 1999; Samochowiec et al. 2004; Maron et al. 2005b), COMT (again restricted to female patients, Hamilton et al. 2002; Domschke et al. 2004; Woo et al. 2002; Woo et al. 2004; Domschke et al. 2007; Zintzaras and Sakelaridis 2007), HTR1A (Rothe et al. 2004; Huang et al. 2004) and ADORA2A (Deckert et al. 1998; Hamilton et al.

2004a; Hohoff et al. 2010). Furthermore, there is some evidence for several other potential risk variants to be involved in the pathogenesis of panic disorder such as polymorphisms in HTR2A (Inada et al. 2003; Rothe et al. 2004; Maron et al. 2005a; Unschuld et al. 2007; Yoon et al. 2008), SLC6A4 (Ohara et al. 1998; Maron et al. 2005a, b; Strug et al. 2010; but: Deckert et al. 1997; Hamilton et al. 1999; Blaya et al. 2007), TPH2 (Maron et al. 2007; Kim et al. 2009; but: Mössner et al. 2006), NET (Lee et al. 2005; but: Sand et al. 2002a), CCK (Wang et al. 1998; Hattori et al. 2001), ACE (Olsson et al. 2004; Erhardt et al. 2008; Bandelow et al. 2010; but: Shimizu et al. 2004), the transcription factor CREM (Domschke et al. 2003; Hamilton et al. 2004b), 'regulator of G-protein signaling' (RGS2, RGS7) (Leygraf et al. 2006, Smoller et al. 2008b; Hohoff et al. 2009b) and several hormone receptors (Sand et al. 2002b; Ho et al. 2004; Keck et al. 2008; Hodges et al. 2009). However, since most of these studies either did not withstand replication in independent samples or still warrant replication, these results have to be considered preliminary. The role of the GABA-ergic system in panic disorder remains to be further elucidated on a molecular genetic level with only little evidence so far for the glutamate decarboxylase (GAD) or GABA receptors and transporters, respectively (Crowe et al. 1997; Sand et al. 2000; Hettema et al. 2006; Nakamura et al. 2006; Kobayashi et al. 2007; Thoeringer et al. 2007, 2009; Unschuld et al. 2009). Recently, besides the classic neurotransmitter systems much attention has been paid to the role of neuropeptides in the mediation of anxiety. Significant association of anxiety or panic disorder in particular have been reported for variants in genes for galanin (Unschuld et al. 2008), the neuropeptide Y (NPY) system (Domschke et al. 2008b) and the neuropeptide S receptor (NPSR) (Domschke et al. 2011). Finally, there is preliminary support for possible interactive effects of several genetics variants in the mediation of the genetic risk for panic disorder, e.g. for HTR1A and COMT (Freitag et al. 2006). For social phobia and generalized anxiety disorder, association has been reported with variation in the dopamine transporter (SLC6A3) gene (Rowe et al. 1998), while DRD2 variants seem to play a role in the pathogenesis of posttraumatic stress disorder (Segman and Shalev 2003). Further associations were observed for COMT in specific phobias (McGrath et al. 2004), for HTR2A in social phobia (Lochner et al. 2007) and MAOA in generalized anxiety disorder (Tadic et al. 2003).

In summary, consistent with findings from clinical genetic and linkage studies, molecular genetic association studies point to a complex genetic etiology of anxiety disorders with an additive or rather interactive effect of multiple risk variants.

2.2.3 Genome-Wide Association Studies

The first genome-wide association study in panic disorder in a Japanese sample yielded evidence for several markers in genes, which have not been implicated in the pathogenesis of anxiety before (PKP1, PLEKHG1, TMEM16B, CALCOCO1, SDK2, and CLU) (Otowa et al. 2009). However, these findings could not be replicated in a follow-up GWAS by the same group (Otowa et al. 2010). Another genome-wide association study in three German samples points to a potential role of the TMEM132D gene in the pathogenesis of panic disorder (Erhardt et al. 2011). Currently, another large GWAS on a homogenous sample from Germany is under analysis (Reif et al., in preparation).

2.3 Genetics of Intermediate Phenotypes of Anxiety Disorders

Dimensional markers such as neuroticism, anxiety sensitivity, state or trait anxiety or behavioral inhibition have been proposed as valid intermediate phenotypes of anxiety disorders with aggregation in families and elevated concordance rates in monozygotic twins pointing to a significant heritability (e.g. Rosenbaum et al. 1991; Maier et al. 1992; Stein et al. 1999). Linkage studies have discerned risk loci on chromosomes 8, 18, 20, and 21 for harm avoidance (e.g. Cloninger et al. 1998), and association studies have reported a potential role of genetic variation of e.g. SLC6A4 for harm avoidance and neuroticism (Lesch et al. 1996) and the corticotropin releasing hormone (CRH) for behavioral inhibition (Smoller et al. 2003, 2005), respectively.

Besides neuropsychological markers, more recently, neurobiological traits have been investigated as intermediate phenotypes of anxiety disorders. Here, significant association was observed, for e.g. increased sympathetic activity and ADORA2A as well as COMT gene variation (Hohoff et al. 2009a; Kang et al. 2010), blushing propensity in social phobia with SLC6A4 variation (Domschke et al. 2009), an increased startle response with COMT and SLC6A4 variants (e.g. Montag et al. 2008; Brocke et al. 2006), as well as CO2-sensitivity to panic attacks with again SLC6A4 variation (Schmidt et al. 2000; Schruers et al. 2011).

Another very promising intermediate phenotype of mental disorders in general and affective and anxiety disorders in particular are neuronal activation correlates of emotional processing as captured by functional imaging techniques such as magnetic resonance imaging (fMRI). In panic disorder, first imaging genetics findings may indicate a distorted corticolimbic interaction depending on variants of the COMT and HTR1A (Domschke et al. 2006, 2008d). In patients with social phobia, polymorphisms in SLC6A4 and TPH were found to be associated with increased amygdala excitability (Furmark et al. 2004, 2008, 2009). Finally, markers spanning RGS2 were reported to be associated with childhood behavioral inhibition and with increased limbic activation during emotion processing (Smoller et al. 2008b). In summary, these first imaging genetics findings in panic disorder, social phobia, and anxiety-related traits may indicate that—depending on variants of COMT, SLC6A4, HTR1A, and RGS2—patients with anxiety disorders are prone to impaired cerebral processing of anxiety-related stimuli in cortical regions known to play a crucial role in the evaluation of emotional stimuli and determination of salient events (for a review see Domschke and Dannlowski 2010).

2.4 Gene-Environment Interaction

In contrast to a multitude of gene-environment studies (GxE) available in depression (see 1.2.2., e.g. SLC6A4 and FKBP5), with respect to anxiety disorders as a categorical nosological entity or anxiety-related traits, to the best of our knowledge only few GxE studies have been performed yet (for a review see Klauke et al. 2010). No associations were found between 5-HTTLPR, childhood emotional abuse, and neuroticism (Antypa and Van der Does 2010). An exemplary GxE study with respect to anxiety-related traits has been published by Stein et al. (2008), who observed a significant interaction between levels of childhood maltreatment and the less active 5-HTTLPR S allele on anxiety sensitivity as measured by the anxiety sensitivity index (ASI). 5-HTTLPR S and L_G haplotypes were furthermore reported to be associated with increased anxiety in interaction with daily stressors (Gunthert et al. 2007). Conversely, assessing 5-HTTLPR genotype and environmental adversity at birth (family adversity) and at 19 years of age (stressful life events), Laucht et al. (2009) found an interactive effect of more active 5-HTTLPR LL genotype and high family adversity on anxiety disorders. Other studies have identified association of the ADORA2A with increased anxiety after caffeine administration in healthy volunteers, demonstrating that a panic disorder risk gene might drive the sensitivity to an environmental stimulus and, therefore, the vulnerability to anxiety (Alsene et al. 2003; Childs et al. 2008).

2.5 Pharmacogenetics

In anxiety disorders, so far only three exemplary studies have investigated the impact of genetic variants on response to a pharmacological treatment regime. Two groups reported significant association of 5-HTTLPR with response to SSRI treatment in panic disorder as well as in generalized anxiety disorder (Perna et al. 2005; Stein et al. 2006). Furthermore, SSRI treatment in panic disorder might in part be driven by variation in HTR1A (Yevtushenko et al. 2010).

3 Overlapping Phenotypes

Depression and anxiety are highly comorbid with up to 60% of patients with depression also displaying anxiety (Leckman et al. 1983) and about 58% of those patients actually meeting DSM criteria for anxiety disorders (see review by Lydiard 1991; e.g. de Graaf et al. 2002; Kessler et al. 1996; Zimmerman et al. 2002). Comorbidity of affective and anxiety disorders has a significant impact on the course and treatment of the respective leading disease with a more chronic course and a significantly detrimental effect on treatment response (e.g. Clayton et al. 1991; Liebowitz 1993; Lydiard 1991). Patterns of occurrence allow for both affective and anxiety disorders preceding the respective other disease.

Besides either simultaneous or sequential true comorbidity of anxiety disorders and major depression there is a continuous debate about a possible overlapping phenotype between anxiety disorders and depression. The clinical phenotype of "anxious depression" (Overall et al. 1966; Overall and Zisook 1980) capturing major depression with subthreshold anxious features has been suggested to constitute a diagnostic entity of its own requiring specific diagnostic and therapeutic attention (see Levine et al. 2001; Lydiard and Brawman-Mintzer 1998; Silverstone and von Studnitz 2003). Dual action drugs acting as reuptake inhibitors on both transporters (serotonin and norepinephrine reuptake inhibitors, SNRI) have been suggested to be superior to SSRI only or tricyclic antidepressants TCA in the treatment of anxious depression (Rudolph et al. 1998; Silverstone and Ravindran 1999). While three large meta-analyses discerned similar response rates to antidepressant treatment in highly anxious and less anxious patients with major depression (Levine et al. 2001; Nelson et al. 2009; Papakostas et al. 2008; see Nelson 2008), there is accumulating evidence for anxious features of depression potentially complicating the course of antidepressant treatment (e.g. Altamura et al. 2004; Domschke et al. 2010; Fava et al. 2008; Joffe et al. 1993; see review by Bagby et al. 2002).

Apart from the individual genetic risk for affective and anxiety disorders, both disease entities also exhibit a common familial risk (as reviewed by Middeldorp et al. 2005). There is evidence from twin studies that depression and general anxiety disorder, panic disorder, and post-traumatic stress disorder share a considerable proportion of their genetic risk (Kendler et al. 1992, 2007; Kendler 1996; Roy et al. 1995). Consistently, molecular genetic studies have yielded evidence for specific genetic loci that may generally influence susceptibility across the anxiety-depression spectrum, e.g. on chromosome 18q (cf. Camp et al. 2005; Hettema 2008). In particular, the combined clinical phenotype of anxious depression has been suggested to constitute a specific subtype with an increased familial risk of depression (Clayton et al. 1990, 1991), which points to a possibly increased heritability of anxious depression with a specific set of genetic risk factors mediating the vulnerability for the development of anxious depression. First imaging and pharmacogenetic studies in anxious depression have implied CNR1, NPY, and SLC6A4 to confer parts of antidepressant treatment response particularly in the clinical phenotype of anxious depression, potentially via a dysfunctional cortico-limbic interaction underlying distorted emotional processing (e.g. Baffa et al. 2010; Domschke et al. 2008a; 2010).

These molecular and imaging genetic findings of overlapping genetic variants as well as common brain networks of emotional processing partly driving both clinical phenotypes of anxiety and affective disorders point to similar neurobiological mechanisms underlying these disorders and therefore possibly a common clinical sub-phenotype shared by anxiety and affective disorders. Particularly, the clinical phenotype of "anxious depression" might thus possibly constitute a diagnostic entity of its own requiring specific diagnostic and therapeutic attention (cf. Lydiard and Brawman-Mintzer 1998; Silverstone and von Studnitz 2003). So, back from bench to bedside, genetic and imaging studies might inspire a re-evaluation and refinement of DSM-IV categorized nosological concepts of depression and anxiety. Alternatively, the current and still emerging body of knowledge in the field of neurobiological research in anxiety and depression might have even more far-reaching consequences in the future by challenging the DSM concept in itself in favor of a more neurobiologically oriented taxonomy of mental disorders. As suggested by Smoller et al. (2008a), genetic and imaging research revealing etiological mechanisms of mental disorders might infer a novel nosological concept based on pathogenesis more than phenomenology. To date, however, despite first essential steps having been made, neurobiological knowledge about the pathomechanism of depression and anxiety has still not progressed far enough to provide a reliable and valid fundament for diagnostic decisions in daily clinical practice. So, in summary the presently known vulnerability genes and patterns of affective and anxiety disorders are slowly beginning to challenge the DSM-defined nosological boundaries and might have the potential to evolve into a valuable tool to more precisely delineate the diagnostic system of mental disorders in the future.

4 Outlook

Future research with respect to the genetic dissection of affective and anxiety disorders will have to comprise technical as well as clinical aspects. On a molecular genetic level, more comprehensive analyses such as tagging SNP approaches, haplotype analyses, as well as the investigation of epistasis of several genes constituting relevant biochemical pathways or cascades are warranted. Here, novel genomic techniques such as duplication/deletion analysis using genotyping arrays and next-generation sequencing of the whole exome or genome for point mutation identification might have a large impact on risk gene identification. Furthermore, it will be of utmost importance to analyze the functional consequences of the associated genetic variants and thereby gain more knowledge about the pathomechanism of the disease of interest. Additionally, there is a need for more detailed gene-environment interaction studies potentially also in a genome-wide fashion (cf. Poulton et al. 2008; Thomas 2010) in order to disentangle the interactive effect of genetic and environmental factors conferring risk or resilience, respectively, to affective and anxiety disorders. In this respect, epigenetic studies investigating e.g. DNA methylation or histone modifications regulating gene activity will tremendously contribute to the elucidation of the interplay between environmental and genetic factors in the pathogenesis of affective and anxiety disorders (cf. for bipolar disorder and schizophrenia: Abdolmaleky et al. 2006, 2008).

Besides the more technical aspects as detailed above, future research in the genetics of affective and anxiety disorders will greatly benefit from clinical considerations. Given that—apart from very few results—most linkage and association findings either did not withstand replication in independent samples or still warrant replication and given that genome-wide association studies in affective as well as in anxiety disorders so far fell short of expectations regarding replicating previous candidate genes or generating novel hypotheses, one possible reason might be the great neuropsychological and neurobiological heterogeneity of the investigated phenotypes of categorical nosological entities as defined by DSM- or ICD-criteria. Thus, besides the recruitment of even larger sample sizes, a more precise definition of the clinical phenotype will be key. In the latter respect, the approach of investigating intermediate phenotypes of affective and anxiety disorders will have to be intensified with the search for novel depression- and/or anxiety-related neurophysiological, biochemical, endocrinological, neuroanatomical, cognitive, or neuropsychological endophenotypes (cf. for major depression: Hasler et al. 2004) and their analysis with respect to their genetic basis.

In summary, to date there is some support for several risk genes contributing to the development of affective and anxiety disorders or their intermediate phenotypes and some light has been shed on gene-environment interactions contributing to the disease risk. However, so far the identified genetic risk factors are of no diagnostic or predictive value, which will only change if the entirety of all genetic risk factors interdependent with environmental factors is identified, which is not foreseeable in the near future. Nevertheless, the increasing elucidation of genetic risk factors tremendously helps in better understanding the pathophysiology of affective and anxiety disorders and might nourish the development of innovative pharmacotherapeutic substances in the treatment of these diseases (e.g. Domschke and Zwanzger 2008), preferably in an individually tailored manner according to genotype.

Acknowledgments This paper was supported by the DFG (Grant RE1632/1-1,/1-3 and/5 to AR, KFO 125 TP4 to AR, DE357/4-1 to AR, SFB TRR 58 C2 and Z2 to AR and KD, respectively; RTG 1252, to AR) and the BMBF (Panik-Net Subproject 6, to AR). We thank A. Alttoa for valuable discussions and her kind help in the preparation of the manuscript.

References

Abdolmaleky HM, Cheng KH, Faraone SV et al (2006) Hypomethylation of MB-COMT promoter is a major risk factor for schizophrenia and bipolar disorder. Hum Mol Genet 15: 3132–3145

Abdolmaleky HM, Zhou JR, Thiagalingam S et al (2008) Epigenetic and pharmacoepigenomic studies of major psychoses and potentials for therapeutics. Pharmacogenomics 9:1809–1823

Abkevich V, Camp NJ, Hensel CH et al (2003) Predisposition locus for major depression at chromosome 12q22–12q23.2. Am J Hum Genet 73:1271–1281

Alsene K, Deckert J, Sand P et al (2003) Association between A2a receptor gene polymorphisms and caffeine-induced anxiety. Neuropsychopharmacology 28:1694–1702

Altamura AC, Montresor C, Salvadori D et al (2004) Does comorbid subthreshold anxiety affect clinical presentation and treatment response in depression? A preliminary 12-month naturalistic study. Int J Neuropsychopharmacol 7:481–487

Angst J, Cui L, Swendsen J et al (2010) Major depressive disorder with subthreshold bipolarity in the National Comorbidity Survey Replication. Am J Psychiatry 167:1194–1201
Antypa N, Van der Does J (2010) Serotonin transporter gene, childhood emotional abuse and cognitive vulnerability to depression. Genes Brain Behav 9:615–620
Appel K, Schwahn C, Mahler J, Schulz A, Spitzer C, Fenske K, Stender J, Barnow S, John U, Teumer A, Biffar R, Nauck M, Völzke H, Freyberger HJ, Grabe HJ (2011) Moderation of adult depression by a polymorphism in the FKBP5 gene and childhood physical abuse in the general population. Neuropsychopharmacology 36(10):1982–1991
Athanasiu L, Mattingsdal M, Kahler AK et al (2010) Gene variants associated with schizophrenia in a Norwegian genome-wide study are replicated in a large European cohort. J Psychiatr Res 44:748–753
Badner JA, Gershon ES (2002) Meta-analysis of whole-genome linkage scans of bipolar disorder and schizophrenia. Mol Psychiatry 7:405–411
Baffa A, Hohoff C, Baune BT et al (2010) Norepinephrine and serotonin transporter genes: impact on treatment response in depression. Neuropsychobiology 62:121–131
Bagby RM, Ryder AG, Cristi C (2002) Psychosocial and clinical predictors of response to pharmacotherapy for depression. Rev Psychiatr Neurosci 27:250–257
Bandelow B, Saleh K, Pauls J et al (2010) Insertion/deletion polymorphism in the gene for angiotensin converting enzyme (ACE) in panic disorder: A gender-specific effect? World J Biol Psychiatry 11:66–70
Barnett JH, Smoller JW (2009) The genetics of bipolar disorder. Neuroscience 164:331–343
Baum AE, Akula N, Cabanero M et al (2008a) A genome-wide association study implicates diacylglycerol kinase eta (DGKH) and several other genes in the etiology of bipolar disorder. Mol Psychiatry 13:197–207
Baum AE, Hamshere M, Green E et al (2008b) Meta-analysis of two genome-wide association studies of bipolar disorder reveals important points of agreement. Mol Psychiatry 13:466–467
Baune BT, Hohoff C, Roehrs T et al (2008) Serotonin receptor 1A–1019C/G variant: Impact on antidepressant pharmacoresponse in melancholic depression? Neurosci Lett 436:111–115
Bellodi L, Perna G, Caldirola D et al (1998) CO2-induced panic attacks: a twin study. Am J Psychiatry 155:1184–1188
Berrettini WH (2001) Molecular linkage studies of bipolar disorders. Bipolar Disord 3:276–283
Binder EB (2009) The role of FKBP5, a co-chaperone of the glucocorticoid receptor in the pathogenesis and therapy of affective and anxiety disorders. Psychoneuroendocrinology 34(Suppl 1):186–195
Binder EB, Bradley RG, Liu W et al (2008) Association of FKBP5 polymorphisms and childhood abuse with risk of posttraumatic stress disorder symptoms in adults. JAMA 299:1291–1305
Binder EB, Salyakina D, Lichtner P et al (2004) Polymorphisms in FKBP5 are associated with increased recurrence of depressive episodes and rapid response to antidepressant treatment. Nat Genet 36:1319–1325
Blaya C, Salum GA, Lima MS et al (2007) Lack of association between the Serotonin Transporter Promoter Polymorphism (5-HTTLPR) and panic disorder: a systematic review and meta-analysis. Behav Brain Funct 3:41
Bochdanovits Z, Verhage M, Smit AB et al (2009) Joint reanalysis of 29 correlated SNPs supports the role of PCLO/Piccolo as a causal risk factor for major depressive disorder. Mol Psychiatry 14:650–652
Bosker FJ, Hartman CA, Nolte IM et al (2011) Poor replication of candidate genes for major depressive disorder using genome-wide association data. Mol Psychiatry 16:516–532
Bradley RG, Binder EB, Epstein MP et al (2008) Influence of child abuse on adult depression: moderation by the corticotropin-releasing hormone receptor gene. Arch Gen Psychiatry 65:190–200
Brent D, Melhem N, Ferrell R et al (2010) Association of FKBP5 polymorphisms with suicidal events in the treatment of resistant depression in adolescents (TORDIA) study. Am J Psychiatry 167:190–197

Brocke B, Armbruster D, Muller J et al (2006) Serotonin transporter gene variation impacts innate fear processing: acoustic startle response and emotional startle. Mol Psychiatry 11: 1106–1112
Camp NJ, Lowry MR, Richards RL et al (2005) Genome-wide linkage analyses of extended Utah pedigrees identifies loci that influence recurrent, early-onset major depression and anxiety disorders. Am J Med Genet B Neuropsychiatr Genet 135:85–93
Casamassima F, Huang J, Fava M et al (2010) Phenotypic effects of a bipolar liability gene among individuals with major depressive disorder. Am J Med Genet B Neuropsychiatr Genet 153:303–309
Caspi A, Sugden K, Moffitt TE et al (2003) Influence of life stress on depression: moderation by a polymorphism in the 5-HTT gene. Science 301:386–389
Chen C, Glatt SJ, Tsuang MT (2008) The tryptophan hydroxylase gene influences risk for bipolar disorder but not major depressive disorder: results of meta-analyses. Bipolar Disord 10:816–821
Chen D, Liu F, Yang C, Liang X, Shang Q, He W, Wang Z (2011) Association between the TPH1A218C polymorphism and risk of mood disorders and alcohol dependence: evidence from the current studies. J Affect Disord (in press)
Chen Z, Liu Y, Zhang D (2009) C677T methylenetetrahydrofolate reductase gene polymorphisms in bipolar disorder: an association study in the Chinese population and a meta-analysis of genetic association studies. Neurosci Lett 449:48–51
Childs E, Hohoff C, Deckert J et al (2008) Association between ADORA2A and DRD2 polymorphisms and caffeine-induced anxiety. Neuropsychopharmacology 33:2791–2800
Cho HJ, Meira-Lima I, Cordeiro Q et al (2005) Population-based and family-based studies on the serotonin transporter gene polymorphisms and bipolar disorder: a systematic review and meta-analysis. Mol Psychiatry 10:771–781
Choi KH, Higgs BW, Wendland JR et al (2011) Gene expression and genetic variation data implicate PCLO in bipolar disorder. Biol Psychiatry 69:353–359
Cichon S, Muhleisen TW, Degenhardt FA et al (2011) Genome-wide association study identifies genetic variation in neurocan as a susceptibility factor for bipolar disorder. Am J Hum Genet 88:372–381
Cichon S, Winge I, Mattheisen M et al (2008) Brain-specific tryptophan hydroxylase 2 (TPH2): a functional Pro206Ser substitution and variation in the 5′-region are associated with bipolar affective disorder. Hum Mol Genet 17:87–97
Clarke H, Flint J, Attwood AS et al (2010) Association of the 5- HTTLPR genotype and unipolar depression: a meta-analysis. Psychol Med 40:1767–1778
Clayton PJ (1990) The comorbidity factor: establishing the primary diagnosis in patients with mixed symptoms of anxiety and depression. J Clin Psychiatry 51:35–39
Clayton PJ, Grove WM, Coryell W et al (1991) Follow-up and family study of anxious depression. Am J Psychiatry 148:1512–1517
Cloninger CR, Van-Erdewegh P, Goate A et al (1998) Anxiety proneness linked to epistatic loci in genome scan of human personality traits. Am J Med Genet 81:313–317
Cohen-Woods S, Craig I, Gaysina D et al (2010) The bipolar association case-control study (BACCS) and meta-analysis: no association with the 5, 10-Methylenetetrahydrofolate reductase gene and bipolar disorder. Am J Med Genet B Neuropsychiatr Genet 153: 1298–1304
Craddock N, Sklar P (2009) Genetics of bipolar disorder: successful start to a long journey. Trends Genet 25:99–105
Crowe RR, Goedken R, Samuelson S (2001) Genome wide survey of panic disorder. Am J Med Genet 105:105–109
Crowe RR, Noyes R Jr, Wilson AF (1987) A linkage study of panic disorder. Arch Gen Psychiatry 44:933–937
Crowe RR, Wang Z, Noyes R Jr (1997) Candidate gene study of eight GABAA receptor subunits in panic disorder. Am J Psychiatry 154:1096–1100
Dawson E, Parfitt E, Roberts Q et al (1995) Linkage studies of bipolar disorder in the region of the Darier's disease gene on chromosome 12q23–24.1. Am J Med Genet 60:94–102

de Graaf R, Bijl RV, Smit F et al (2002) Risk factors for 12-month comorbidity of mood, anxiety, and substance use disorders: findings from the Netherlands mental health survey and incidence study. Am J Psychiatry 159:620–629

Deckert J, Catalano M, Heils A et al (1997) Functional promoter polymorphism of the human serotonin transporter: lack of association with panic disorder. Psychiatr Genet 7:45–47

Deckert J, Catalano M, Syagailo YV (1999) Excess of high activity monoamine oxidase A gene promoter alleles in female patients with panic disorder. Hum Mol Genet 8:621–624

Deckert J, Nothen MM, Franke P (1998) Systematic mutation screening and association study of the A1 and A2a adenosine receptor genes in panic disorder suggest a contribution of the A2a gene to the development of disease. Mol Psychiatry 3:81–85

Dickson SP, Wang K, Krantz I (2010) Rare variants create synthetic genome-wide associations. PLoS Biol 8:e1000294

Djurovic S, Gustafsson O, Mattingsdal M (2010) A genome-wide association study of bipolar disorder in Norwegian individuals, followed by replication in Icelandic sample. J Affect Disord 126:312–316

Domschke K, Braun M, Ohrmann P et al (2006) Association of the functional -1019C/G 5-HT1A polymorphism with prefrontal and amygdala activation measured with 3T fMRI in panic disorder. Int J Neuropsychopharmacol 9:349–355

Domschke K, Dannlowski U (2010) Imaging genetics of anxiety disorders. Neuroimage 53(3): 822–831

Domschke K, Zwanzger P (2008) GABAergic and endocannabinoid dysfunction in anxiety—future therapeutic targets? Curr Pharm Des 14:3508–3517

Domschke K, Dannlowski U, Hohoff C (2010) Neuropeptide Y (NPY) gene: impact on emotional processing and treatment response in anxious depression. Eur Neuropsychopharmacol 20:301–309

Domschke K, Dannlowski U, Ohrmann P (2008a) Cannabinoid receptor 1 (CNR1) gene: impact on antidepressant treatment response and emotion processing in major depression. Eur Neuropsychopharmacol 18:751–759

Domschke K, Deckert J, O'Donovan MC (2007) Meta-analysis of COMT val158met in panic disorder: ethnic heterogeneity and gender specificity. Am J Med Genet B Neuropsychiatr Genet 144:667–673

Domschke K, Freitag CM, Kuhlenbäumer G (2004) Association of the functional V158M catechol-O-methyl-transferase polymorphism with panic disorder in women. Int J Neuropsychopharmacol 7:183–188

Domschke K, Hohoff C, Jacob C (2008b) Chromosome 4q31–34 panic disorder risk locus: Association of neuropeptide Y Y5 receptor variants. Am J Med Genet B Neuropsychiatr Genet 147B:510–516

Domschke K, Hohoff C, Mortensen LS (2008c) Monoamine oxidase A variant influences antidepressant treatment response in female patients with major depression. Prog Neuropsychopharmacol Biol Psychiatry 32:224–228

Domschke K, Kuhlenbaumer G, Schirmacher A (2003) Human nuclear transcription factor gene CREM: genomic organization, mutation screening, and association analysis in panic disorder. Am J Med Genet B Neuropsychiatr Genet 117:70–78

Domschke K, Ohrmann P, Braun M et al (2008d) Influence of the catechol-O-methyltransferase val158met genotype on amygdala and prefrontal cortex emotional processing in panic disorder. Psychiatry Res 163:13–20

Domschke K, Reif A, Weber H, Richter J, Hohoff C, Ohrmann P, Pedersen A, Bauer J, Suslow T, Kugel H, Heindel W, Baumann C, Klauke B, Jacob C, Maier W, Fritze J, Bandelow B, Krakowitzky P, Rothermundt M, Erhardt A, Binder EB, Holsboer F, Gerlach AL, Kircher T, Lang T, Alpers GW, Ströhle A, Fehm L, Gloster AT, Wittchen HU, Arolt V, Pauli P, Hamm A, Deckert J (2011) Neuropeptide S receptor gene—converging evidence for a role in panic disorder. Mol Psychiatry 16(9):938–948

Domschke K, Stevens S, Beck B (2009) Blushing propensity in social anxiety disorder: influence of serotonin transporter gene variation. J Neural Transm 116:663–666

Duman RS, Monteggia LM (2006) A neurotrophic model for stress-related mood disorders. Biol Psychiatry 59:1116–1127

Egeland JA, Gerhard DS, Pauls DL (1987) Bipolar affective disorders linked to DNA markers on chromosome 11. Nature 325:783–787

Erhardt A, Czibere L, Roeske D, Lucae S, Unschuld PG, Ripke S, Specht M, Kohli MA, Kloiber S, Ising M, Heck A, Pfister H, Zimmermann P, Lieb R, Pütz B, Uhr M, Weber P, Deussing JM, Gonik M, Bunck M, Kebler MS, Frank E, Hohoff C, Domschke K, Krakowitzky P, Maier W, Bandelow B, Jacob C, Deckert J, Schreiber S, Strohmaier J, Nöthen M, Cichon S, Rietschel M, Bettecken T, Keck ME, Landgraf R, Müller-Myhsok B, Holsboer F, Binder EB (2011) TMEM132D, a new candidate for anxiety phenotypes: evidence from human and mouse studies. Mol Psychiatry 16(6):647–663

Erhardt A, Lucae S, Kern N et al (2008) Association of polymorphisms in the angiotensin-converting enzyme gene with syndromal panic attacks. Mol Psychiatry 13:242–243

Erk S, Meyer-Lindenberg A, Schnell K et al (2010) Brain function in carriers of a genome-wide supported bipolar disorder variant. Arch Gen Psychiatry 67:803–811

Ewald H, Degn B, Mors O et al (1998) Significant linkage between bipolar affective disorder and chromosome 12q24. Psychiatr Genet 8:131–140

Ewald H, Flint T, Kruse TA et al (2002) A genome-wide scan shows significant linkage between bipolar disorder and chromosome 12q24.3 and suggestive linkage to chromosomes 1p22–21, 4p16, 6q14–22, 10q26 and 16p13.3. Mol Psychiatry 7:734–744

Fan J, Sklar P (2008) Genetics of bipolar disorder: focus on BDNF Val66Met polymorphism. Novartis Found Symp 289:60-72; discussion 72-63, 87-93

Fan M, Liu B, Jiang T et al (2010) Meta-analysis of the association between the monoamine oxidase-A gene and mood disorders. Psychiatr Genet 20:1–7

Fava M, Davidson KG (1996) Definition and epidemiology of treatment-resistant depression. Psychiatr Clin North Am 19:179–200

Fava M, Rush AJ, Alpert JE (2008) Difference in treatment outcome in outpatients with anxious versus nonanxious depression: a STAR*D report. Am J Psychiatry 165:342–351

Ferreira MA, O'Donovan MC, Meng YA et al (2008) Collaborative genome-wide association analysis supports a role for ANK3 and CACNA1C in bipolar disorder. Nat Genet 40:1056–1058

Fiedorowicz JG, Endicott J, Leon AC (2011) Subthreshold hypomanic symptoms in progression from unipolar major depression to bipolar disorder. Am J Psychiatry 168:40–48

Franchini L, Serretti A, Gasperini M et al (1998) Familial concordance of fluvoxamine response as a tool for differentiating mood disorder pedigrees. J Psychiatr Res 32:255–259

Franke B, Vasquez AA, Veltman JA, Brunner HG, Rijpkema M, Fernández G (2010) Genetic variation in CACNA1C, a gene associated with bipolar disorder, influences brainstem rather than gray matter volume in healthy individuals. Biol Psychiatry 68(6):586–588

Freitag CM, Domschke K, Rothe C et al (2006) Interaction of serotonergic and noradrenergic gene variants in panic disorder. Psychiatr Genet 16:59–65

Furlong RA, Ho L, Walsh C et al (1998) Analysis and meta-analysis of two serotonin transporter gene polymorphisms in bipolar and unipolar affective disorders. Am J Med Genet 81:58–63

Furmark T, Appel L, Henningsson S et al (2008) A link between serotonin-related gene polymorphisms, amygdala activity, and placebo-induced relief from social anxiety. J Neurosci 28:13066–13074

Furmark T, Henningsson S, Appel L (2009) Genotype over-diagnosis in amygdala responsiveness: affective processing in social anxiety disorder. J Psychiatry Neurosci 34:30–40

Furmark T, Tillfors M, Garpenstrand H et al (2004) Serotonin transporter polymorphism related to amygdala excitability and symptom severity in patients with social phobia. Neurosci Lett 362:189–192

Fyer AJ, Hamilton SP, Durner M (2006) A third-pass genome scan in panic disorder: evidence for multiple susceptibility loci. Biol Psychiatry 60:388–401

Garriock HA, Hamilton SP (2009a) Genetic studies of drug response and side effects in the STAR*D study, part 1. J Clin Psychiatry 70:1186–1187

Garriock HA, Hamilton SP (2009b) Genetic studies of drug response and side effects in the STAR*D study, part 2. J Clin Psychiatry 70:1323–1325

Gaysina D, Cohen S, Craddock N et al (2008) No association with the 5, 10-methylenetetrahydrofolate reductase gene and major depressive disorder: results of the depression case control (DeCC) study and a meta-analysis. Am J Med Genet B Neuropsychiatr Genet 147B:699–706

Gelernter J, Bonvicini K, Page G et al (2001) Linkage genome scan for loci predisposing to panic disorder or agoraphobia. Am J Med Genet 105:548–557

Gelernter J, Page GP, Bonvicini K et al (2003) A chromosome 14 risk locus for simple phobia: results from a genomewide linkage scan. Mol Psychiatry 8:71–82

Gelernter J, Page GP, Stein MB et al (2004) Genome-wide linkage scan for loci predisposing to social phobia: evidence for a chromosome 16 risk locus. Am J Psychiatry 161:59–66

Gershon ES, Alliey-Rodriguez N, Liu C (2011) After GWAS: searching for genetic risk for schizophrenia and bipolar disorder. Am J Psychiatry 168:253–256

Gilbody S, Lewis S, Lightfoot T (2007) Methylenetetrahydrofolate reductase (MTHFR) genetic polymorphisms and psychiatric disorders: a HuGE review. Am J Epidemiol 165:1–13

Goldstein RB, Wickramaratne PJ, Horwath E (1997) Familial aggregation and phenomenology of 'early'-onset (at or before age 20 years) panic disorder. Arch Gen Psychiatry 54:271–278

Green E, Elvidge G, Jacobsen N et al (2005) Localization of bipolar susceptibility locus by molecular genetic analysis of the chromosome 12q23–q24 region in two pedigrees with bipolar disorder and Darier's disease. Am J Psychiatry 162:35–42

Green EK, Grozeva D, Jones I et al (2009) The bipolar disorder risk allele at CACNA1C also confers risk of recurrent major depression and of schizophrenia. Mol Psychiatry 15: 1016–1022

Grigoroiu-Serbanescu M, Diaconu CC, Herms S et al (2008) Investigation of the tryptophan hydroxylase 2 gene in bipolar I disorder in the Romanian population. Psychiatr Genet 18: 240–247

Grozeva D, Kirov G, Ivanov D et al (2010) Rare copy number variants: a point of rarity in genetic risk for bipolar disorder and schizophrenia. Arch Gen Psychiatry 67:318–327

Gunthert KC, Conner TS, Armeli S et al (2007) Serotonin transporter gene polymorphism (5-HTTLPR) and anxiety reactivity in daily life: a daily process approach to gene-environment interaction. Psychosom Med 69:762–768

Halmoy A, Johansson S, Winge I et al (2010) Attention-deficit/hyperactivity disorder symptoms in offspring of mothers with impaired serotonin production. Arch Gen Psychiatry 67:1033–1043

Hamilton SP, Fyer AJ, Durner M et al (2003) Further genetic evidence for a panic disorder syndrome mapping to chromosome 13q. Proc Natl Acad Sci USA 100:2550–2555

Hamilton SP, Heiman GA, Haghighi F et al (1999) Lack of genetic linkage or association between a functional serotonin transporter polymorphism and panic disorder. Psychiatr Genet 9:1–6

Hamilton SP, Slager SL, de Baisre LA et al (2004a) Evidence for genetic linkage between a polymorphism in the adenosine 2A receptor and panic disorder. Neuropsychopharmacology 29:558–565

Hamilton SP, Slager SL, Heiman GA et al (2002) Evidence for a susceptibility locus for panic disorder near the catechol-O-methyltransferase gene on chromosome 22. Biol Psychiatry 51:591–601

Hamilton SP, Slager SL, Mayo D (2004b) Investigation of polymorphisms in the CREM gene in panic disorder. Am J Med Genet B Neuropsychiatr Genet 126:111–115

Hasler G, Drevets WC, Manji HK et al (2004) Discovering endophenotypes for major depression. Neuropsychopharmacology 29:1765–1781

Hattori E, Ebihara M, Yamada K et al (2001) Identification of a compound short tandem repeat stretch in the 5′-upstream region of the cholecystokinin gene, and its association with panic disorder but not with schizophrenia. Mol Psychiatry 6:465–470

Hattori E, Toyota T, Ishitsuka Y et al (2009) Preliminary genome-wide association study of bipolar disorder in the Japanese population. Am J Med Genet B Neuropsychiatr Genet 150B:1110–1117

Hek K, Mulder CL, Luijendijk HJ et al (2010) The PCLO gene and depressive disorders: replication in a population-based study. Hum Mol Genet 19:731–734

Hettema JM (2008) What is the genetic relationship between anxiety and depression? Am J Med Genet C Semin Med Genet 148C:140–146

Hettema JM, An SS, Neale MC et al (2006) Association between glutamic acid decarboxylase genes and anxiety disorders, major depression, and neuroticism. Mol Psychiatry 11:752–762

Hettema JM, Neale MC, Kendler KS (2001) A review and meta-analysis of the genetic epidemiology of anxiety disorders. Am J Psychiatry 158:1568–1578

Ho HP, Westberg L, Annerbrink K (2004) Association between a functional polymorphism in the progesterone receptor gene and panic disorder in women. Psychoneuroendocrinology 2:1138–1141

Hodges LM, Weissman MM, Haghighi F (2009) Association and linkage analysis of candidate genes GRP, GRPR, CRHR1, and TACR1 in panic disorder. Am J Med Genet B Neuropsychiatr Genet 150:65–73

Hohoff C, Domschke K, Schwarte K et al (2009a) Sympathetic activity relates to adenosine A(2A) receptor gene variation in blood-injury phobia. J Neural Transm 116:659–662

Hohoff C, Mullings EL, Heatherley SV, Freitag CM, Neumann LC, Domschke K, Krakowitzky P, Rothermundt M, Keck ME, Erhardt A, Unschuld PG, Jacob C, Fritze J, Bandelow B, Maier W, Holsboer F, Rogers PJ, Deckert J (2010) Adenosine A(2A) receptor gene: evidence for association of risk variants with panic disorder and anxious personality. J Psychiatr Res 44(14):930–937

Hohoff C, Neumann A, Domschke K et al (2009b) Association analysis of Rgs7 variants with panic disorder. J Neural Transm 116:1523–1528

Horstmann S, Lucae S, Menke A et al (2010) Polymorphisms in GRIK4, HTR2A, and FKBP5 show interactive effects in predicting remission to antidepressant treatment. Neuropsychopharmacology 35:727–740

Horwath E, Adams P, Wickramaratne P et al (1997) Panic disorder with smothering symptoms: evidence for increased risk in first-degree relatives. Depress Anxiety 6:147–153

Horwath E, Wolk SI, Goldstein RB et al (1995) Is the comorbidity between social phobia and panic disorder due to familial cotransmission or other factors? Arch Gen Psychiatry 52: 574–582

Hösing VG, Schirmacher A, Kuhlenbäumer G, Freitag C, Sand P, Schlesiger C, Jacob C, Fritze J, Franke P, Rietschel M, Garritsen H, Nöthen MM, Fimmers R, Stögbauer F, Deckert J (2004) Cholecystokinin- and cholecystokinin-B-receptor gene polymorphisms in panic disorder. J Neural Transm Suppl 68:147–156

Huang J, Perlis RH, Lee PH, Rush AJ, Fava M, Sachs GS, Lieberman J, Hamilton SP, Sullivan P, Sklar P, Purcell S, Smoller JW (2010) Cross-disorder genomewide analysis of schizophrenia, bipolar disorder, and depression. Am J Psychiatry 167(10):1254–1263

Huang YY, Battistuzzi C, Oquendo MA et al (2004) Human 5-HT1A receptor C(-1019)G polymorphism and psychopathology. Int J Neuropsychopharmacol 7:441–451

Inada Y, Yoneda H, Koh J et al (2003) Positive association between panic disorder and polymorphism of the serotonin 2A receptor gene. Psychiatry Res 118:25–31

Ising M, Lucae S, Binder EB et al (2009) A genomewide association study points to multiple loci that predict antidepressant drug treatment outcome in depression. Arch Gen Psychiatry 66:966–975

Joffe RT, Bagby RM, Levitt A (1993) Anxious and nonanxious depression. Am J Psychiatry 150:1257–1258

Jones I, Jacobsen N, Green EK et al (2002) Evidence for familial cosegregation of major affective disorder and genetic markers flanking the gene for Darier's disease. Mol Psychiatry 7:424–427

Kaabi B, Gelernter J, Woods SW et al (2006) Genome scan for loci predisposing to anxiety disorders using a novel multivariate approach: strong evidence for a chromosome 4 risk locus. Am J Hum Genet 78:543–553

Kanazawa T, Glatt SJ, Kia-Keating B et al (2007) Meta-analysis reveals no association of the Val66Met polymorphism of brain-derived neurotrophic factor with either schizophrenia or bipolar disorder. Psychiatr Genet 17:165–170

Kang EH, Song YJ, Kim KJ et al (2010) Sympathetic nervous function and the effect of the catechol-O-methyltransferase Val(158)Met polymorphism in patients with panic disorder. J Affect Disord 123:337–340

Karg K, Burmeister M, Shedden K et al (2011) The serotonin transporter promoter variant (5-HTTLPR), stress, and depression meta-analysis revisited: evidence of genetic moderation. Arch Gen Psychiatry 68:444–454

Keers R, Farmer AE, Aitchison KJ (2009) Extracting a needle from a haystack: reanalysis of whole genome data reveals a readily translatable finding. Psychol Med 39:1231–1235

Kato M, Serretti A (2010) Review and meta-analysis of antidepressant pharmacogenetic findings in major depressive disorder. Mol Psychiatry 15:473–500

Keck ME, Kern N, Erhardt A et al (2008) Combined effects of exonic polymorphisms in CRHR1 and AVPR1B genes in a case/control study for panic disorder. Am J Med Genet B Neuropsychiatr Genet 147B:1196–1204

Kempton MJ, Ruberto G, Vassos E et al (2009) Effects of the CACNA1C risk allele for bipolar disorder on cerebral gray matter volume in healthy individuals. Am J Psychiatry 166: 1413–1414

Kendler KS (1996) Major depression and generalised anxiety disorder. Same genes, (partly) different environments—revisited. Br J Psychiatry 30:68–75

Kendler KS, Gardner CO, Gatz M et al (2007) The sources of co-morbidity between major depression and generalized anxiety disorder in a Swedish national twin sample. Psychol Med 37:453–462

Kendler KS, Karkowski LM, Prescott CA (1999) Fears and phobias: reliability and heritability. Psychol Med 29:539–553

Kendler KS, Neale MC, Kessler RC et al (1992) Major depression and generalized anxiety disorder. Same genes, (partly) different environments? Arch Gen Psychiatry 49:716–722

Kendler KS, Walters EE, Neale MC et al (1995) The structure of the genetic and environmental risk factors for six major psychiatric disorders in women. Arch Gen Psychiatry 52:374–383

Kennedy JL, Bradwejn J, Koszycki D et al (1999) Investigation of cholecystokinin system genes in panic disorder. Mol Psychiatry 4:284–285

Kessler RC, Nelson CB, McGonagle KA (1996) Comorbidity of DSM-III-R major depressive disorder in the general population: results from the US National Comorbidity Survey. Br J Psychiatry Suppl 30:17–30

Kim YK, Lee HJ, Yang JC et al (2009) A tryptophan hydroxylase 2 gene polymorphism is associated with panic disorder. Behav Genet 39:170–175

Kirchheiner J, Nickchen K, Bauer M et al (2004) Pharmacogenetics of antidepressants and antipsychotics: the contribution of allelic variations to the phenotype of drug response. Mol Psychiatry 9:442–473

Kishi T, Okochi T, Tsunoka T et al (2011) Serotonin 1A receptor gene, schizophrenia and bipolar disorder: an association study and meta-analysis. Psychiatry Res 185:20–26

Klauke B, Deckert J, Reif A et al (2010) Life events in panic disorder—An update on "candidate stressors". Depress Anxiety 27:716–730

Knowles JA, Fyer AJ, Vieland VJ et al (1998) Results of a genome-wide genetic screen for panic disorder. Am J Med Genet 81:139–147

Kobayashi Y, Akiyoshi J, Kanehisa M et al (2007) Lack of polymorphism in genes encoding mGluR 7, mGluR 8, GABA(A) receptor alfa-6 subunit and nociceptin/orphanin FQ receptor and panic disorder. Psychiatr Genet 17:9

Krug A, Nieratschker V, Markov V et al (2010) Effect of CACNA1C rs1006737 on neural correlates of verbal fluency in healthy individuals. Neuroimage 49:1831–1836

Laje G, Allen AS, Akula N et al (2009) Genome-wide association study of suicidal ideation emerging during citalopram treatment of depressed outpatients. Pharmacogenet Genomics 19:666–674

Lango Allen H, Estrada K, Lettre G et al (2010) Hundreds of variants clustered in genomic loci and biological pathways affect human height. Nature 467:832–838
Lasky-Su JA, Faraone SV, Glatt SJ et al (2005) Meta-analysis of the association between two polymorphisms in the serotonin transporter gene and affective disorders. Am J Med Genet B Neuropsychiatr Genet 133B:110–115
Laucht M, Treutlein J, Blomeyer D et al (2009) Interaction between the 5-HTTLPR serotonin transporter polymorphism and environmental adversity for mood and anxiety psychopathology: evidence from a high-risk community sample of young adults. Int J Neuropsychopharmacol 12:737–747
Lavebratt C, Aberg E, Sjoholm LK et al (2010) Variations in FKBP5 and BDNF genes are suggestively associated with depression in a Swedish population-based cohort. J Affect Disord 125:249–255
Leckman JF, Weissman MM, Merikangas KR et al (1983) Panic disorder and major depression. Increased risk of depression, alcoholism, panic, and phobic disorders in families of depressed probands with panic disorder. Arch Gen Psychiatry 40:1055–1060
Lee MT, Chen CH, Lee C et al (2010) Genome-wide association study of bipolar I disorder in the Han Chinese population. Mol Psychiatry 16:548–556
Lee YJ, Hohoff C, Domschke K et al (2005) Norepinephrine transporter (NET) promoter and 5′-UTR polymorphisms: association analysis in panic disorder. Neurosci Lett 377:40–43
Lekman M, Laje G, Charney D et al (2008) The FKBP5-gene in depression and treatment response–an association study in the sequenced treatment alternatives to relieve depression (STAR*D) cohort. Biol Psychiatry 63:1103–1110
Lesch K, Bengel D, Heils A et al (1996) Association of anxiety-related traits with a polymorphism in the serotonin transporter gene regulatory region. Science 274:1527–1531
Lesch KP, Timmesfeld N, Renner TJ et al (2008) Molecular genetics of adult ADHD: converging evidence from genome-wide association and extended pedigree linkage studies. J Neural Transm 115:1573–1585
Levine J, Cole DP, Chengappa KN et al (2001) Anxiety disorders and major depression, together or apart. Depress Anxiety 14:94–104
Lewis CM, Ng MY, Butler AW et al (2010) Genome-wide association study of major recurrent depression in the U.K. population. Am J Psychiatry 167:949–957
Leygraf A, Hohoff C, Freitag C et al (2006) Rgs 2 gene polymorphisms as modulators of anxiety in humans? J Neural Transm 113:1921–1925
Lichtenstein P, Yip BH, Bjork C et al (2009) Common genetic determinants of schizophrenia and bipolar disorder in Swedish families: a population-based study. Lancet 373:234–239
Liebowitz MR (1993) Depression with anxiety and a typical depression. J Clin Psychiatry 54:10–14
Liu Y, Blackwood DH, Caesar S et al (2011) Meta-analysis of genome-wide association data of bipolar disorder and major depressive disorder. Mol Psychiatry 16:2–4
Lochner C, Hemmings S, Seedat S et al (2007) Genetics and personality traits in patients with social anxiety disorder: a case-control study in South Africa. Eur Neuropsychopharmacol 17:321–327
Logue MW, Durner M, Heiman GA et al (2009) A linkage search for joint panic disorder/bipolar genes. Am J Med Genet B Neuropsychiatr Genet 150:1139–1146
Lohoff FW (2010) Overview of the genetics of major depressive disorder. Curr Psychiatry Rep 12:539–546
Lopez-Leon S, Janssens AC, Gonzalez-Zuloeta Ladd AM et al (2008) Meta-analyses of genetic studies on major depressive disorder. Mol Psychiatry 13:772–785
Lopez Leon S, Croes EA, Sayed-Tabatabaei FA et al (2005) The dopamine D4 receptor gene 48-base-pair-repeat polymorphism and mood disorders: a meta-analysis. Biol Psychiatry 57:999–1003
Luykx JJ, Boks MP, Terwindt AP et al (2010) The involvement of GSK3beta in bipolar disorder: integrating evidence from multiple types of genetic studies. Eur Neuropsychopharmacol 20:357–368

Lydiard RB (1991) Coexisting depression and anxiety: special diagnostic and treatment issues. J Clin Psychiatry 52:48–54

Lydiard RB, Brawman-Mintzer O (1998) Anxious depression. J Clin Psychiatry 59:10–17

MacGillivray S, Arroll B, Hatcher S et al (2003) Efficacy and tolerability of selective serotonin reuptake inhibitors compared with tricyclic antidepressants in depression treated in primary care: systematic review and meta-analysis. BMJ 326:1014

MacKinnon DF, Xu J, McMahon FJ et al (1998) Bipolar disorder and panic disorder in families: an analysis of chromosome 18 data. Am J Psychiatry 155:829–831

Maier W, Lichtermann D, Minges J et al (1992) Personality traits in subjects at risk for unipolar major depression: a family study perspective. J Affect Disord 24:153–163

Maier W, Lichtermann D, Minges J (1993) A controlled family study in panic disorder. J Psychiatr Res 27:79–87

Manchia M, Squassina A, Congiu D et al (2009) Interacting genes in lithium prophylaxis: preliminary results of an exploratory analysis on the role of DGKH and NR1D1 gene polymorphisms in 199 Sardinian bipolar patients. Neurosci Lett 467:67–71

Marks IM, Herst ER (1970) A survey of 1200 agoraphobics in Britain. Soc Psychiatry 5:16–24

Maron E, Lang A, Tasa G et al (2005a) Associations between serotonin-related gene polymorphisms and panic disorder. Int J Neuropsychopharmacol 8:261–266

Maron E, Nikopensius T, Koks S et al (2005b) Association study of 90 candidate gene polymorphisms in panic disorder. Psychiatr Genet 15:17–24

Maron E, Toru I, Must A et al (2007) Association study of tryptophan hydroxylase 2 gene polymorphisms in panic disorder. Neurosci Lett 411:180–184

Maziade M, Roy MA, Rouillard E et al (2001) A search for specific and common susceptibility loci for schizophrenia and bipolar disorder: a linkage study in 13 target chromosomes. Mol Psychiatry 6:684–693

McCarthy MI, Abecasis GR, Cardon LR et al (2008) Genome-wide association studies for complex traits: consensus, uncertainty and challenges. Nat Rev Genet 9:356–369

McCarthy MJ, Leckband SG, Kelsoe JR (2010) Pharmacogenetics of lithium response in bipolar disorder. Pharmacogenomics 11:1439–1465

McGrath M, Kawachi I, Ascherio A et al (2004) Association between catechol-O-methyltransferase and phobic anxiety. Am J Psychiatry 161:1703–1705

McGuffin P, Knight J, Breen G et al (2005) Whole genome linkage scan of recurrent depressive disorder from the depression network study. Hum Mol Genet 14:3337–3345

McMahon FJ, Akula N, Schulze TG et al (2010) Meta-analysis of genome-wide association data identifies a risk locus for major mood disorders on 3p21.1. Nat Genet 42:128–131

McQueen MB, Devlin B, Faraone SV et al (2005) Combined analysis from eleven linkage studies of bipolar disorder provides strong evidence of susceptibility loci on chromosomes 6q and 8q. Am J Hum Genet 77:582–595

Mendlewicz J, Rainer JD (1977) Adoption study supporting genetic transmission in manic–depressive illness. Nature 268:327–329

Middeldorp CM, Cath DC, Van Dyck R et al (2005) The co-morbidity of anxiety and depression in the perspective of genetic epidemiology. A review of twin and family studies. Psychol Med 35:611–624

Montag C, Buckholtz JW, Hartmann P et al (2008) COMT genetic variation affects fear processing: psychophysiological evidence. Behav Neurosci 122:901–909

Mosing MA, Gordon SD, Medland SE et al (2009) Genetic and environmental influences on the co-morbidity between depression, panic disorder, agoraphobia, and social phobia: a twin study. Depress Anxiety 26:1004–1011

Moskvina V, Craddock N, Holmans P et al (2009) Gene-wide analyses of genome-wide association data sets: evidence for multiple common risk alleles for schizophrenia and bipolar disorder and for overlap in genetic risk. Mol Psychiatry 14:252–260

Mössner R, Freitag CM, Gutknecht L et al (2006) The novel brain-specific tryptophan hydroxylase-2 gene in panic disorder. J Psychopharmacol 20:547–552

Moya PR, Murphy DL, McMahon FJ, Wendland JR (2010) Increased gene expression of diacylglycerol kinase η in bipolar disorder. Int J Neuropsychopharmacol 13(8):1127–1128
Muglia P, Tozzi F, Galwey NW et al (2010) Genome-wide association study of recurrent major depressive disorder in two European case-control cohorts. Mol Psychiatry 15:589–601
Muller DJ, Zai CC, Shinkai T et al (2011) Association between the DAOA/G72 gene and bipolar disorder and meta-analyses in bipolar disorder and schizophrenia. Bipolar Disord 13:198–207
Nakamura K, Yamada K, Iwayama Y et al (2006) Evidence that variation in the peripheral benzodiazepine receptor (PBR) gene influences susceptibility to panic disorder. Am J Med Genet B Neuropsychiatr Genet 141B:222–226
Nelson JC (2008) Anxious depression and response to treatment. Am J Psychiatry 165:297–299
Nelson JC, Delucchi K, Schneider LS (2009) Anxiety does not predict response to antidepressant treatment in late life depression: results of a meta-analysis. Int J Geriatr Psychiatry 24: 539–544
Nyegaard M, Demontis D, Foldager L et al (2010) CACNA1C (rs1006737) is associated with schizophrenia. Mol Psychiatry 15:119–121
O'Donovan MC, Craddock N, Norton N et al (2008) Identification of loci associated with schizophrenia by genome-wide association and follow-up. Nat Genet 40:1053–1055
Ohara K, Nagai M, Suzuki Y et al (1998) Association between anxiety disorders and a functional polymorphism in the serotonin transporter gene. Psychiatry Res 81:277–279
Ollila HM, Soronen P, Silander K et al (2009) Findings from bipolar disorder genome-wide association studies replicate in a Finnish bipolar family-cohort. Mol Psychiatry 14:351–353
Olsson M, Annerbrink K, Westberg L et al (2004) Angiotensin-related genes in patients with panic disorder. Am J Med Genet B Neuropsychiatr Genet 127:81–84
O'Reilly RL, Bogue L, Singh SM (1994) Pharmacogenetic response to antidepressants in a multicase family with affective disorder. Biol Psychiatry 36:467–471
Otowa T, Tanii H, Sugaya N et al (2010) Replication of a genome-wide association study of panic disorder in a Japanese population. J Hum Genet 55:91–96
Otowa T, Yoshida E, Sugaya N et al (2009) Genome-wide association study of panic disorder in the Japanese population. J Hum Genet 54:122–126
Overall JE, Hollister LE, Johnson M et al (1966) Nosology of depression and differential response to drugs. JAMA 195:946–948
Overall JE, Zisook S (1980) Diagnosis and the phenomenology of depressive disorders. J Consult Clin Psychol 48:626–634
Papakostas GI, Stahl SM, Krishen A et al (2008) Efficacy of bupropion and the selective serotonin reuptake inhibitors in the treatment of major depressive disorder with high levels of anxiety (anxious depression): a pooled analysis of 10 studies. J Clin Psychiatry 69:1287–1292
Pare CM, Mack JW (1971) Differentiation of two genetically specific types of depression by the response to antidepressant drugs. J Med Genet 8:306–309
Peerbooms OL, van Os J, Drukker M, Kenis G, Hoogveld L, MTHFR in Psychiatry Group, de Hert M, Delespaul P, van Winkel R, Rutten BP (2011) Meta-analysis of MTHFR gene variants in schizophrenia, bipolar disorder and unipolar depressive disorder: evidence for a common genetic vulnerability? Brain Behav Immun 25(8):1530–1543
Perlis RH, Huang J, Purcell S et al (2010) Genome-wide association study of suicide attempts in mood disorder patients. Am J Psychiatry 167:1499–1507
Perlis RH, Smoller JW, Ferreira MA et al (2009) A genomewide association study of response to lithium for prevention of recurrence in bipolar disorder. Am J Psychiatry 166:718–725
Perna G, Favaron E, DiBella D et al (2005) Antipanic efficacy of paroxetine and polymorphism within the promoter of the serotonin transporter gene. Neuropsychopharmacol 30:2230–2235
Perroud N, Uher R, Ng MY, Guipponi M, Hauser J, Henigsberg N, Maier W, Mors O, Gennarelli M, Rietschel M, Souery D, Dernovsek MZ, Stamp AS, Lathrop M, Farmer A, Breen G, Aitchison KJ, Lewis CM, Craig IW, McGuffin P (2010) Genome-wide association study of increasing suicidal ideation during antidepressant treatment in the GENDEP project. Pharmacogenomics J (in press)

Pickard BS, Christoforou A, Thomson PA et al (2009) Interacting haplotypes at the NPAS3 locus alter risk of schizophrenia and bipolar disorder. Mol Psychiatry 14:874–884

Poulton R, Andrews G, Millichamp J (2008) Gene-environment interaction and the anxiety disorders. Eur Arch Psychiatry Clin Neurosci 258:65–68

Purcell SM, Wray NR, Stone JL et al (2009) Common polygenic variation contributes to risk of schizophrenia and bipolar disorder. Nature 460:748–752

Ressler KJ, Bradley B, Mercer KB et al (2010) Polymorphisms in CRHR1 and the serotonin transporter loci: gene x gene x environment interactions on depressive symptoms. Am J Med Genet B Neuropsychiatr Genet 153B:812–824

Rietschel M, Mattheisen M, Frank J et al (2010) Genome-wide association-, replication-, and neuroimaging study implicates HOMER1 in the etiology of major depression. Biol Psychiatry 68:578–585

Psychiatric GWAS Consortium (2009) A framework for interpreting genome-wide association studies of psychiatric disorders. Mol Psychiatry 14:10–17

Rosenbaum JF, Biederman J, Hirshfield DR et al (1991) Further evidence of an association between behavioral inhibition and anxiety disorders: results from a family study of children from a non-clinical sample. J Psychiatr Res 25:49–65

Rothe C, Gutknecht L, Freitag C et al (2004) Association of a functional 1019C>G 5-HT1A receptor gene polymorphism with panic disorder with agoraphobia. Int J Neuropsychopharmacol 7:189–192

Rowe DC, Stever C, Gard JM et al (1998) The relation of the dopamine transporter gene (DAT1) to symptoms of internalizing disorders in children. Behav Genet 28:215–225

Roy MA, Neale MC, Pedersen NL et al (1995) A twin study of generalized anxiety disorder and major depression. Psychol Med 25:1037–1049

Rudolph RL, Entsuah R, Chitra R (1998) A meta-analysis of the effects of venlafaxine on anxiety associated with depression. J Clin Psychopharmacol 18:136–144

Samochowiec J, Hajduk A, Samochowiec A et al (2004) Association studies of MAO-A, COMT, and 5-HTT genes polymorphisms in patients with anxiety disorders of the phobic spectrum. Psychiatry Res 128:21–26

Sand P, Lesch KP, Catalano M et al (2000) Polymorphic MAO-A and 5-HT-transporter genes: analysis of interactions in panic disorder. World J Biol Psychiatry 1:147–150

Sand PG, Mori T, Godau C et al (2002a) Norepinephrine transporter gene (NET) variants in patients with panic disorder. Neurosci Lett 333:41–44

Sand PG, Schlurmann K, Luckhaus C et al (2002b) Estrogen receptor 1 gene (ESR1) variants in panic disorder. Am J Med Genet 114:426–428

Schmidt NB, Storey J, Greenberg BD et al (2000) Evaluating gene x psychological risk factor effects in the pathogenesis of anxiety: a new model approach. J Abnorm Psychol 109:308–320

Scholz CJ, Jacob CP, Buttenschon HN et al (2010) Functional variants of TSPAN8 are associated with bipolar disorder and schizophrenia. Am J Med Genet B Neuropsychiatr Genet 153B(4): 967–972

Schruers K, Esquivel G, van Duinen M, Wichers M, Kenis G, Colasanti A, Knuts I, Goossens L, Jacobs N, van Rozendaal J, Smeets H, van Os J, Griez E (2011) Genetic moderation of CO_2-induced fear by 5-HTTLPR genotype. J Psychopharmacol 25(1):37–42

Schulze TG, Alda M, Adli M et al (2010) The International Consortium on Lithium Genetics (ConLiGen): an initiative by the NIMH and IGSLI to study the genetic basis of response to lithium treatment. Neuropsychobiology 62:72–78

Schulze TG, Detera-Wadleigh SD, Akula N et al (2009) Two variants in Ankyrin 3 (ANK3) are independent genetic risk factors for bipolar disorder. Mol Psychiatry 14:487–491

Schulze TG, McMahon FJ (2003) Genetic linkage and association studies in bipolar affective disorder: a time for optimism. Am J Med Genet C Semin Med Genet 123C:36–47

Scott LJ, Muglia P, Kong XQ et al (2009) Genome-wide association and meta-analysis of bipolar disorder in individuals of European ancestry. Proc Natl Acad Sci USA 106:7501–7506

Segman RH, Shalev AY (2003) Genetics of posttraumatic stress disorder. CNS Spectr 8:693–698

Segurado R, Detera-Wadleigh SD, Levinson DF et al (2003) Genome scan meta-analysis of schizophrenia and bipolar disorder, part III: Bipolar disorder. Am J Hum Genet 73:49–62
Serretti A, Artioli P, Quartesan R (2005) Pharmacogenetics in the treatment of depression: pharmacodynamic studies. Pharmacogenet Genom 15:61–67
Serretti A, Mandelli L (2008) The genetics of bipolar disorder: genome 'hot regions', genes, new potential candidates and future directions. Mol Psychiatry 13:742–771
Shi J, Badner JA, Gershon ES et al (2008) Allelic association of G72/G30 with schizophrenia and bipolar disorder: a comprehensive meta-analysis. Schizophr Res 98:89–97
Shi J, Potash JB, Knowles JA et al (2011) Genome-wide association study of recurrent early-onset major depressive disorder. Mol Psychiatry 16:193–201
Shih RA, Belmonte PL, Zandi PP (2004) A review of the evidence from family, twin and adoption studies for a genetic contribution to adult psychiatric disorders. Int Rev Psychiatry 16:260–283
Shimizu E, Hashimoto K, Kobayashi K et al (2004) Lack of association between angiotensin I-converting enzyme insertion/deletion gene functional polymorphism and panic disorder in humans. Neurosci Lett 363:81–83
Shyn SI, Shi J, Kraft JB et al (2011) Novel loci for major depression identified by genome-wide association study of sequenced treatment alternatives to relieve depression and meta-analysis of three studies. Mol Psychiatry 16:202–215
Silverstone PH, Ravindran A (1999) Once-daily venlafaxine extended release (XR) compared with fluoxetine in outpatients with depression and anxiety. Venlafaxine XR 360 Study Group. J Clin Psychiatry 60:22–28
Silverstone PH, von Studnitz E (2003) Defining anxious depression: going beyond comorbidity. Can J Psychiatry 48:675–680
Sklar P, Smoller JW, Fan J et al (2008) Whole-genome association study of bipolar disorder. Mol Psychiatry 13:558–569
Skre I, Onstad S, Torgersen S (1993) A twin study of DSM-III-R anxiety disorders. Acta Psychiatr Scand 88:85–92
Smith DJ, Evans R, Craddock N (2010) Predicting response to lithium in bipolar disorder: a critical review of pharmacogenetic studies. J Ment Health 19:142–156
Smith EN, Bloss CS, Badner JA et al (2009) Genome-wide association study of bipolar disorder in European American and African American individuals. Mol Psychiatry 14:755–763
Smoller JW, Finn CT (2003) Family, twin, and adoption studies of bipolar disorder. Am J Med Genet C Semin Med Genet 123C:48–58
Smoller JW, Gardner-Schuster E et al (2008a) Genetics of anxiety: would the genome recognize the DSM. Depress Anxiety 25:368–377
Smoller JW, Paulus MP, Fagerness JA et al (2008b) Influence of RGS2 on anxiety-related temperament, personality, and brain function. Arch Gen Psychiatry 65:298–308
Smoller JW, Rosenbaum JF, Biederman J et al (2003) Association of a genetic marker at the corticotropin releasing hormone locus with behavioral inhibition. Biol Psychiatry 54:1376–1381
Smoller JW, Yamaki LH, Fagerness JA et al (2005) The corticotropin releasing hormone gene and behavioral inhibition in children at risk for panic disorder. Biol Psychiatry 57:1485–1492
Soronen P, Ollila HM, Antila M et al (2010) Replication of GWAS of bipolar disorder: association of SNPs near CDH7 with bipolar disorder and visual processing. Mol Psychiatry 15:4–6
Speliotes EK, Willer CJ, Berndt SI et al (2010) Association analyses of 249, 796 individuals reveal 18 new loci associated with body mass index. Nat Genet 42:937–948
Squassina A, Manchia M, Congiu D et al (2009) The diacylglycerol kinase eta gene and bipolar disorder: a replication study in a Sardinian sample. Mol Psychiatry 14:350–351
Stein MB, Chartier MJ, Kozak MV et al (1998) Genetic linkage to the serotonin transporter protein and 5HT2A receptor genes excluded in generalized social phobia. Psychiatry Res 81:283–291
Stein MB, Jang KL, Livesley WJ (1999) Heritability of anxiety sensitivity. Am J Psychiatry 156:246–251

Stein MB, Schork NJ, Gelernter J (2008) Gene-by-environment (serotonin transporter and childhood maltreatment) interaction for anxiety sensitivity, an intermediate phenotype for anxiety disorders. Neuropsychopharmacology 33:312–319

Stein MB, Seedat S, Gelernter J (2006) Serotonin transporter gene promoter polymorphism predicts SSRI response in generalized social anxiety disorder. Psychopharmacology (Berl) 187:68–72

Steinberg S, Mors O, Borglum AD et al (2011) Expanding the range of ZNF804A variants conferring risk of psychosis. Mol Psychiatry 16:59–66

Strug LJ, Suresh R, Fyer AJ et al (2010) Panic disorder is associated with the serotonin transporter gene (SLC6A4) but not the promoter region (5-HTTLPR). Mol Psychiatry 15:166–176

Sullivan PF (2010) The psychiatric GWAS consortium: big science comes to psychiatry. Neuron 68:182–186

Sullivan PF, de Geus EJ, Willemsen G et al (2009) Genome-wide association for major depressive disorder: a possible role for the presynaptic protein piccolo. Mol Psychiatry 14:359–375

Sullivan PF, Neale MC, Kendler KS (2000) Genetic epidemiology of major depression: review and meta-analysis. Am J Psychiatry 157:1552–1562

Tadic A, Rujescu D, Szegedi A et al (2003) Association of a MAOA gene variant with generalized anxiety disorder, but not with panic disorder or major depression. Am J Med Genet B Neuropsychiatr Genet 117:1–6

Takata A, Kawasaki H, Iwayama Y et al (2011) Nominal association between a polymorphism in DGKH and bipolar disorder detected in a meta-analysis of East Asian case-control samples. Psychiatry Clin Neurosci 65:280–285

Terracciano A, Tanaka T, Sutin AR et al (2010) Genome-wide association scan of trait depression. Biol Psychiatry 68:811–817

Tesli M, Athanasiu L, Mattingsdal M et al (2010) Association analysis of PALB2 and BRCA2 in bipolar disorder and schizophrenia in a scandinavian case-control sample. Am J Med Genet B Neuropsychiatr Genet 153B:1276–1282

Tesli M, Kahler AK, Andreassen BK et al (2009) No association between DGKH and bipolar disorder in a Scandinavian case-control sample. Psychiatr Genet 19:269–272

Thoeringer CK, Binder EB, Salyakina D et al (2007) Association of a Met88Val diazepam binding inhibitor (DBI) gene polymorphism and anxiety disorders with panic attacks. J Psychiatr Res 41:579–584

Thoeringer CK, Ripke S, Unschuld PG et al (2009) The GABA transporter 1 (SLC6A1): a novel candidate gene for anxiety disorders. J Neural Transm 116:649–657

Thomas D (2010) Gene-environment-wide association studies: emerging approaches. Nat Rev Genet 11:259–272

Thorgeirsson TE, Oskarsson H, Desnica N et al (2003) Anxiety with panic disorder linked to chromosome 9q in Iceland. Am J Hum Genet 72:1221–1230

Tsuang M, Domschke K, Jerskey BA et al (2004) Agoraphobic behavior and panic attack: a study of male twins. J Anxiety Disord 18:799–807

Uher R, McGuffin P (2008) The moderation by the serotonin transporter gene of environmental adversity in the aetiology of mental illness: review and methodological analysis. Mol Psychiatry 13:131–146

Uher R, McGuffin P (2010) The moderation by the serotonin transporter gene of environmental adversity in the etiology of depression: 2009 update. Mol Psychiatry 15:18–22

Uher R, Perroud N, Ng MY et al (2010) Genome-wide pharmacogenetics of antidepressant response in the GENDEP project. Am J Psychiatry 16:555–564

Unschuld PG, Ising M, Erhardt A et al (2007) Polymorphisms in the serotonin receptor gene HTR2A are associated with quantitative traits in panic disorder. Am J Med Genet B Neuropsychiatr Genet 144:424–429

Unschuld PG, Ising M, Erhardt A et al (2008) Polymorphisms in the galanin gene are associated with symptom-severity in female patients suffering from panic disorder. J Affect Disord 105:177–184

Unschuld PG, Ising M, Roeske D et al (2010) Gender-specific association of galanin polymorphisms with HPA-axis dysregulation, symptom severity, and antidepressant treatment response. Neuropsychopharmacology 35:1583–1592
Unschuld PG, Ising M, Specht M et al (2009) Polymorphisms in the GAD2 gene-region are associated with susceptibility for unipolar depression and with a risk factor for anxiety disorders. Am J Med Genet B Neuropsychiatr Genet 150:1100–1109
Velders FP, Kuningas M, Kumari M, Dekker MJ, Uitterlinden AG, Kirschbaum C, Hek K, Hofman A, Verhulst FC, Kivimaki M, Van Duijn CM, Walker BR, Tiemeier H (2011) Genetics of cortisol secretion and depressive symptoms: a candidate gene and genome wide association approach. Psychoneuroendocrinology 36(7):1053–1061
Verhagen M, van der Meij A, van Deurzen PA et al (2010) Meta-analysis of the BDNF Val66Met polymorphism in major depressive disorder: effects of gender and ethnicity. Mol Psychiatry 15:260–271
Vieland VJ, Goodman DW, Chapman T et al (1996) New segregation analysis of panic disorder. Am J Med Genet 67:147–153
Wang KS, Liu XF, Aragam N (2010) A genome-wide meta-analysis identifies novel loci associated with schizophrenia and bipolar disorder. Schizophr Res 124:192–199
Wang Z, Valdes J, Noyes R et al (1998) Possible association of a cholecystokinin promotor polymorphism (CCK-36CT) with panic disorder. Am J Med Genet 81:228–234
Weber H, Kittel-Schneider S, Gessner A, Domschke K, Neuner M, Jacob CP, Buttenschon HN, Boreatti-Hümmer A, Volkert J, Herterich S, Baune BT, Gross-Lesch S, Kopf J, Kreiker S, Nguyen TT, Weissflog L, Arolt V, Mors O, Deckert J, Lesch KP, Reif A (2011) Cross-disorder analysis of bipolar risk genes: further evidence of DGKH as a risk gene for bipolar disorder, but also unipolar depression and adult ADHD. Neuropsychopharmacology 36(10):2076–2085
Weissman MM, Fyer AJ, Haghighi F et al (2000) Potential panic disorder syndrome: clinical and genetic linkage evidence. Am J Med Genet 96:24–35
Wender PH, Kety SS, Rosenthal D et al (1986) Psychiatric disorders in the biological and adoptive families of adopted individuals with affective disorders. Arch Gen Psychiatry 43:923–929
Wessa M, Linke J, Witt SH, Nieratschker V, Esslinger C, Kirsch P, Grimm O, Hennerici MG, Gass A, King AV, Rietschel M (2010) The CACNA1C risk variant for bipolar disorder influences limbic activity. Mol Psychiatry 15(12):1126–1127
Williams HJ, Craddock N, Russo G et al (2011) Most genome-wide significant susceptibility loci for schizophrenia and bipolar disorder reported to date cross-traditional diagnostic boundaries. Hum Mol Genet 20:387–391
Woo JM, Yoon KS, Choi YH et al (2004) The association between panic disorder and the L/L genotype of catechol-O-methyltransferase. J Psychiatr Res 38:365–370
Woo JM, Yoon KS, Yu BH (2002) Catechol O-methyltransferase genetic polymorphism in panic disorder. Am J Psychiatry 159:1785–1787
Wray NR, Pergadia ML, Blackwood DH, Penninx BW, Gordon SD, Nyholt DR, Ripke S, Macintyre DJ, McGhee KA, Maclean AW, Smit JH, Hottenga JJ, Willemsen G, Middeldorp CM, de Geus EJ, Lewis CM, McGuffin P, Hickie IB, van den Oord EJ, Liu JZ, Macgregor S, McEvoy BP, Byrne EM, Medland SE, Statham DJ, Henders AK, Heath AC, Montgomery GW, Martin NG, Boomsma DI, Madden PA, Sullivan PF (2012) Genome-wide association study of major depressive disorder: new results, meta-analysis, and lessons learned. Mol Psychiatry 17(1):36–48
WTCCC (2007) Genome-wide association study of 14,000 cases of seven common diseases and 3,000 shared controls. Nature 447:661–678
Xie P, Kranzler HR, Poling J et al (2010) Interaction of FKBP5 with childhood adversity on risk for post-traumatic stress disorder. Neuropsychopharmacology 35:1684–1692
Yasuda S, Kai M, Imai S et al (2009) Diacylglycerol kinase eta augments C-Raf activity and B-Raf/C-Raf heterodimerization. J Biol Chem 284:29559–29570

Yevtushenko OO, Oros MM, Reynolds GP (2010) Early response to selective serotonin reuptake inhibitors in panic disorder is associated with a functional 5-HT1A receptor gene polymorphism. J Affect Disord 123:308–311

Yoon HK, Yang JC, Lee HJ et al (2008) The association between serotonin-related gene polymorphisms and panic disorder. J Anxiety Disord 22:1529–1534

Zeng Z, Wang T, Li T et al (2011) Common SNPs and haplotypes in DGKH are associated with bipolar disorder and schizophrenia in the Chinese Han population. Mol Psychiatry 16:473–475

Zhang D, Cheng L, Qian Y et al (2009a) Singleton deletions throughout the genome increase risk of bipolar disorder. Mol Psychiatry 14:376–380

Zhang Z, Lindpaintner K, Che R et al (2009b) The Val/Met functional polymorphism in COMT confers susceptibility to bipolar disorder: evidence from an association study and a meta-analysis. J Neural Transm 116:1193–1200

Zhou K, Dempfle A, Arcos-Burgos M et al (2008) Meta-analysis of genome-wide linkage scans of attention deficit hyperactivity disorder. Am J Med Genet B Neuropsychiatr Genet 147B:1392–1398

Zimmerman M, Chelminski I, McDermut W (2002) Major depressive disorder and axis I diagnostic comorbidity. J Clin Psychiatry 63:187–193

Zintzaras E (2006) C677T and A1298C methylenetetrahydrofolate reductase gene polymorphisms in schizophrenia, bipolar disorder and depression: a meta-analysis of genetic association studies. Psychiatr Genet 16:105–115

Zintzaras E, Sakelaridis N (2007) Is 472G/A catechol-O-methyl-transferase gene polymorphism related to panic disorder? Psychiatr Genet 17:267–273

Zobel A, Schuhmacher A, Jessen F et al (2010) DNA sequence variants of the FKBP5 gene are associated with unipolar depression. Int J Neuropsychopharmacol 13:649–660

Zou YF, Wang F, Feng XL, Li WF, Tian YH, Tao JH, Pan FM, Huang F (2010) Association of DRD2 gene polymorphisms with mood disorders: a meta-analysis. J Affect Disord (in press)

Behavioural Genetics of the Serotonin Transporter

K. Haddley, V. J. Bubb, G. Breen, U. M. Parades-Esquivel and J. P. Quinn

Abstract The serotonin transporter is a key regulator of the bioavailability of serotonin and therefore any modulation in the expression or action of the transporter would be expected to have consequences on behaviour. The transporter has therefore become a target for pharmaceutical intervention in behavioural and mood disorders. The search for polymorphic variants in the transporter that would associate with neurological disorders has been extensive but has become focused on two domains which are both termed variable number tandem repeat (VNTR) polymorphisms. Both of these VNTRs are in non-coding DNA and therefore proposed to be mechanistically involved in a disorder through their ability to modulate transcriptional or post-transcriptional regulation of the transporter. The most extensively studied is in the promoter and is a bi-allelic insertion/deletion found in the 5′ promoter region of the gene 1.2 kb upstream of the transcriptional start site. This VNTR, termed, 5-HTTLPR was initially identified as two variants containing either, 14 (short/deletion) or 16 (long/insertion) copies of a 22 bp repeat. A second widely studied VNTR found in the non-coding region of the transporter is located within intron 2 and comprises 9, 10 or 12 copies of a 16–17 bp repeat termed, STin2.9, STin2.10 and STin2.12, respectively. These VNTR polymorphisms have been associated with a range of behavioural and psychiatric disorders including depression, OCD, anxiety and schizophrenia,

K. Haddley · V. J. Bubb · J. P. Quinn (✉)
Department of Clinical and Molecular Pharmacology,
Institute of Translation Medicine, University of Liverpool,
Liverpool, Merseyside L69 3BX, UK
e-mail: jquinn@liv.ac.uk

G. Breen · U. M. Parades-Esquivel
Social, Genetic and Developmental Psychiatric Centre (SGDP MRC),
Institute of Psychiatry, King's College London, DeCrespigny Park,
London SE5 8AF, UK

Curr Topics Behav Neurosci (2012) 12: 503–535
DOI: 10.1007/7854_2011_186

Published Online: 20 January 2012

however often the lack of reproducibility in different cohorts has led to debate on the actual association of the polymorphisms with this extensive range of neurological conditions. Here we review these two polymorphic VNTRs in depth and relate that to pharmaceutical response, their ability to regulate differential transporter expression, their core involvement in gene-environment interaction and their genetic association with specific disorders.

Keywords Mood disorders • Genetic polymorphism • Serotonin • Serotonin transporter • Variable number tandem repeat • Gene regulation • Monoaminergic signaling

Contents

1 Introduction 504
2 Investigating Polymorphic Variation 506
3 Serotonin Transporter: Variable Number Tandem Repeats and Clinical Association Studies. The Linked Polymorphic Region (5-HTT LPR) 507
4 The Serotonin Transporter Intron 2 VNTR (STin2 VNTR) 512
5 Pharmacogenetics of the 5-HTT Polymorphisms and Selective Serotonin Reuptake Inhibitor (SSRI) Response 515
6 Gene × Environment (G × E) Interaction Role in the SLC6A4 Polymorphism 515
7 Haplotype Analysis of the SLC6A4 Polymorphisms, Gene × Gene (G × G) and Gene × Gene × Environment (G × G × E) Interactions 516
8 Nuclear Imaging of SLC6A4 Polymorphic Variants in Affective Disorders 517
9 Effects of 5-HTTLPR on Serotonin Signalling Systems (MRI) 518
10 Effects of 5HTTLPR on Serotonin Transporter Binding 520
11 Effect of the Serotonin Transporter Polymorphisms on Functional Gene Expression ... 522
12 Lessons from the 5-HTT Polymorphisms and Their Application to Other Gene Polymorphisms 524
13 Summary 524
References 526

1 Introduction

The monoaminergic signaling system is a key component in maintaining physiological control over many aspects of human health including emotional behaviour and psychological states. One of the major players in this system is the neuromodulator, serotonin (5-HT). Disturbances in the 5-HT system have been implicated in the pathophysiology of many CNS-related disorders such as anxiety (Lesch et al. 2003), depression (Ressler and Nemeroff 2000), Obsessive Compulsive Disorder (OCD) (Bengel et al. 1999), aggression and antisocial behaviour (Lesch and Merschdorf 2000), addiction (Kreek et al. 2005a, b) and suicide (Arango et al. 2002). The level of neuromodulatory 5-HT present in the synaptic cleft and available for action at cell surface serotonin receptors is dependent on reuptake by the serotonin transporter (5-HTT) that belongs to the Na^{+}/Cl^{-} dependant solute-carrier family of transporters and is encoded for by the SLC6A4 gene. In the CNS, 5-HT and 5-HTT are mostly localised on the perisynaptic membrane of neurons of the raphe nuclei whose processes innervate many areas of the brain thought to be involved in cognition and behaviour regulation (McLaughlin et al. 1996). Since this transporter regulates the spatiotemporal fine-tuning of 5-HT signalling and is paramount in the processes controlling many aspects of mood and behaviour, it is naturally a major target for several antidepressants, drugs of abuse and potent neurotoxins and is a major focus in the investigation of several affective disorders.

Genetic variability between individuals comes in many guises including single nucleotide polymorphisms (SNPs), which may be coding or non-coding, and micro- or mini-satellites. Micro-satellites such as simple tandem repeats exist as di, tri or tetra nucleotide repeats, whilst mini-satellites consist of 10–100 nucleotides repeated several times in tandem bordered by unique DNA sequences. Mini-satellites are more stable than micro-satellites and are believed to be present in the genome due to unequal crossover (non-allelic homologous recombination) or unequal sister chromatid exchange. Mini-satellites frequently exist as a multi-allelic polymorphism due to variation in the number of the tandem repeat units (reviewed in Haddley et al. 2008). Here we will focus on a subclass of mini-satellite known as a variable number tandem repeat (VNTR). VNTRs show some degree of degeneration where one repeat may be slightly different from the next but overall the core consensus sequence is maintained (Jeffreys et al. 1985; Naslund et al. 2005). The majority of VNTRs are found in non-coding regions of the genome and many are found at higher density in gene enriched areas compared to non-genic regions. VNTRs have the potential to act as transcriptional regulatory domains as they have a number of sequence specific DNA motifs within the repeat that can act as transcription factor binding sites. They may display evolutionary conservation between humans and non-human primates but are often not found in lower mammals (Lesch et al. 1997; Soeby et al. 2005). The SLC6A4 gene locus presents a number of polymorphisms within non-coding regions and it has been shown that the allelic variation in these VNTRs can function in a tissue-specific and stimulus-inducible manner, in vitro and in vivo, to fine-tune gene expression (Lesch et al. 1996; Heils et al. 1996; MacKenzie and Quinn 1999; Hariri et al. 2002; Hranilovic et al. 2004; Roberts et al. 2007). This fine tuning may be correlated, mechanistically, not only with normal physiological function and variation between individuals, but also with a predisposition to behavioural disorders by altering neurotransmitter signaling in response to challenges and stress. Furthermore, if stimulus-inducible expression varies with a specific disorder-associated polymorphism then that may have similar implications in the response of an individual to therapeutics (Fiskerstrand et al. 1999; Lovejoy et al. 2003; Klenova et al. 2004; Roberts et al. 2007). In addition, VNTRs can participate in regulation of gene expression at other levels. For example, it has been suggested that VNTRs located in untranslated (UTR) regions may play an important role in mRNA stability, which can also affect levels of transcript available in the cellular

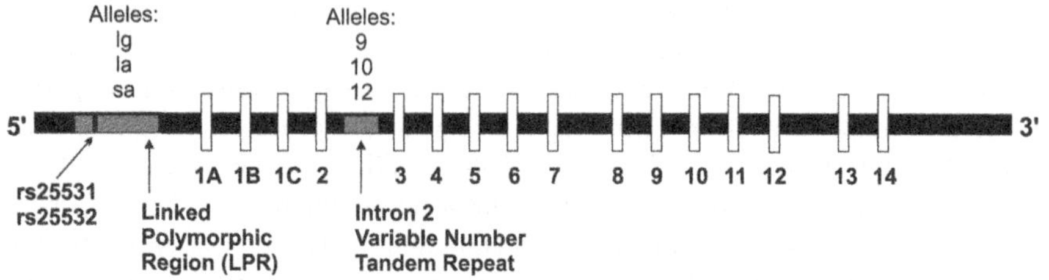

Fig. 1 Schematic representation of the 5-HTTLPR, STin2VNTR and rs25531/2 SNP in the SLC6A4 gene

environment (Nakamura et al. 1998; Zamorano et al. 2006). Further, VNTRs located in exons are known to affect protein folding and have been associated with differential levels of mature RNA availability in the cytoplasm (Nakamura et al. 1998). More recently the role of VNTRs in the epigenetic regulation of response to challenges has been investigated (Olsson et al. 2010; Kinnally et al. 2010; Ali et al. 2010b; Vasiliou et al. 2011; Van Ijzendoorn et al. 2010). We have previously reviewed clinical correlation and function with VNTR genotype in the SLC6A4 gene including the linked polymorphic region (5-HTTLPR) found in the promoter region of the gene and the STin2 VNTR found in intron 2, also sometimes referred to as intron 3, Fig. 1 (Haddley et al. 2008). Here we include more recent findings and create a picture of how these correlations stand to date.

2 Investigating Polymorphic Variation

Beside genome wide association studies, which link a specific polymorphism to a particular disorder, there are a number of applications both in vitro and in vivo, by which the putative molecular function of a polymorphism can be assessed. In vitro techniques that can be applied to investigate gene expression under basal conditions and in response to challenge e.g., drugs (cocaine, citralopram) or transcription factors (CTCF) include: (1) Production of polymorphism-reporter gene constructs to examine the ability of a polymorphism to alter gene expression in clonal or primary cell lines (Klenova et al. 2004; Guindalini et al. 2006; Roberts et al. 2007; Ali et al. 2010a, b; Vasiliou et al. 2011); (2) Quantitative or real-time PCR in lymphoblast cell lines of known genotype to determine relative quantities of gene expression as a function of a specific genotype (Hranilovic et al. 2004); (3) Allelic expression imbalance (AEI) that can determine the relative abundance of transcript derived in an allele-specific manner (Lim et al. 2006; Martin et al. 2007). In vivo techniques that can be applied to examine allele-specific changes include: (1) Chromatin immunoprecipitation (ChIP) in a cell line to examine transcription factor-DNA binding and associated changes in histone modifications over a specific polymorphism (Vasiliou et al. 2011). This can also be performed genome-wide using Geneseq arrays; (2) The generation of "humanised" transgenic mice, in

Fig. 2 **a** Sequence of the 5-HTTLPR. The deletion in the the short variant is represented in *grey*. **b** Sequence of the STin2VNTR. * Denotes repeats absent from both the nine and ten copy variants and ** denotes a repeat missing in the nine copy variant

(a) Linked Polymorphic Region	(b) Intron 2 Variable Number Tandem Repeat
ccctactgcagccctcccagcatccc	ggctgtgacccagggtg
ccctgcaacctcccagcaact	ggctgtgacccggagtg
ccctgtacccctcctaggatcgc	ggctgtcacccggggtg
tcctgcatccccattatccccc	ggctgtgacccgggtg
ccttcaccccctcgcggcatcc	ggctgcgacctggggtg
cccctgcacccccagcatccc	ggctgtgacctgggatg **
ccctgcagcccccccagcatctc	ggctgtgacccgggtg
ccctgcacccccagcatccc	ggctgtgacctggggtg
ccctgcagcccttccagcatcc	ggctgtgacccgggtg
ccctgcacctctcccaggatctc	ggctgtgacctggggtg *
ccctgcaacccccattatccc	ggctgtgacccgggtg *
ccctgcacccctcgcagtatccc	ggctgtgacctgggatg
ccctgcaccccccagcatcccc	
ccatgcacccccggcatccc	
ccctgcacccctccagcattct	
ccttgcaccctaccagtat	

which a VNTR driving marker gene expression for example, β-galactosidase, may be transiently expressed to identify tissues in which the VNTR may be active in response to certain challenges or at particular developmental stages (Mackenzie and Quinn 1999); (3) Formaldehyde assisted identification of regulatory elements (FAIRE) that identifies genome-wide open regions of chromatin allowing assessment of changes in chromatin structure over polymorphic regions in response to challenge (Esquivel et al. 2011); (4) DNA methylation that identifies regions which are not accessible to regulation by transcription factors; (5) Nuclear imaging including single-photon emission computed tomography (SPECT) and positron emission tomography (PET) that allow resolution of events at a molecular level (for example transporter/receptor binding potentials and density) and functional magnetic resonance imaging (fMRI) that allows analysis of a whole system activity and connectivity approach between brain regions (see Sect. 6).

Used in combination these techniques provide a powerful armoury for delineating the functional mechanism of action of polymorphisms with the observed clinical associations for such polymorphisms.

3 Serotonin Transporter: Variable Number Tandem Repeats and Clinical Association Studies. The Linked Polymorphic Region (5-HTT LPR)

The 5-HTT LPR is a biallelic insertion/deletion found in the 5′ promoter region of SLC6A4 approximately 1.4 kb upstream of the transcription start site (Heils et al. 1995, 1996). Initially the LPR was identified as an allelic variant comprising 14 (short) or 16 (long) copies of a 22–23 bp repeat, Fig. 2a. More recently, Nakamura et al. verified the existence of subgroups within these variants to uncover as many as 14 allelic variations of this locus. Most studies focus on the more common biallelic 14 (s) and 16 (l) alleles of the LPR, however, since the discovery of a functional SNP (rs25531) within the long variant of the LPR (an A to G substitution) this is now referred to as the triallelic LPR variant. Inclusion of this SNP (lg) has conferred clinical properties normally associated with the s-allele onto the l-allele (Nakamura et al. 2000; Hu et al. 2005; Kraft et al. 2005; Wendland et al. 2006). However, no significant variation was observed between la and lg variants in a study employing the use of AEI (Martin et al. 2007). The distribution of these genotypes varies, with individuals Hm for the s-allele appearing at lower or similar frequencies to those Hm for the l-allele in the Caucasian population (Heils et al. 1995, 1996). However, allelic frequencies vary among different ethnic groups. The frequency of the alleles for the Caucasian population has been shown to be 27.4–28.3% (l/l), 43.4–51.0% (s/l) and 21.6–28.3% (s/s), whilst that of Asian population has been shown to be 4.2–10.0% (l/l), 30.0–39.2% (s/l) and 55.6–60.0% (s/s) (Smits et al. 2004).

We have previously discussed the clinical and functional associations of this variant with a number of affective disorders and behaviour traits (Table 1) and levels of gene expression in vitro and in vivo (Table 2) (extensively reviewed in Haddley et al. 2008). In general the l-allele of 5-HTTLPR has been associated with increased levels of reporter gene activity, mRNA abundance and transporter binding. We will now extrapolate on these associations using more recent data and meta-analyses.

Early and promising associations were made between 5-HTT polymorphisms and a number of affective disorders with the Hm s-allele believed to be predisposing to affective disorders. However, these initial findings were not always consistent and even the result of meta-analysis found some of these association to be lacking in power. More recent investigations have sought to clarify the specifics of these associations including more detailed parameters such as age, population, environment and co-morbidities, as well as the mechanisms by which these variants may mediate their effect. Recent meta-analysis found a significant association of the Hm s-allele genotype with an increased risk of major depressive disorder (MDD) in Caucasians but not in Asians (Kiyohara and Yoshimasu 2010). In agreement with the latter, when studied in a Thai population no association could be made between 5-HTTLPR genotype and MDD (Tencomnao and Wongpiyabovorn 2010). A recent investigation in a Northern Swedish population could not find any association of the 5-HTTLPR with bipolar depression (Alaerts et al. 2009).

Overall the picture remains unclear on the exact role of the LPR polymorphism risk associations particularly in Asian populations, where the s-allele is present at a higher frequency than the l-allele, but taking into account other factors, especially gene x environment (G $\times$ E) interactions and treatment response, may go some way to clarifying current understanding.

Homozygosity for the s-allele has conferred threefold higher odds of a patient suffering from post-stroke depression (PSD) compared to those with one or more l-allele (Kohen et al. 2008). Similarly, in a Chinese population, homozygosity for the s-allele has been associated with a higher incidence of PSD (Fang et al. 2011). The mechanism(s) by which these alleles may confer susceptibility to mood

Table 1 A sample of previously reported clinical associations for the 5-HTT LPR polymorphism

Author	Disorder	Cohort/ethnicity	Predisposing genotype
5-HTTLPR genotype positively associated with affective disorder			
Cervilla et al. (2006)	Depression	European (Spanish)	s/s
Zalsman et al. (2006)	Depression	Caucasian (European)	s/s
Collier et al. (1996a)	Unipolar depression	European	s/s
Hauser et al. (2003)	Unipolar depression	European	s/s
Collier et al. (1996a)	Bipolar depression	European	s/s
Cho et al. (2005)[a]	Bipolar depression	Mixed	s/s
Lotrich and Pollock (2004)[a]	Bipolar depression	–	s/s
Lasky-Su et al. (2005)[a]	Bipolar depression	Mixed	s/s
Hauser et al. (2003)	Bipolar depression	European	s/s
Lotrich and Pollock (2004)[a]	Major depressive disorder	–	s/s
Lesch et al. (1996)	Anxiety	Caucasian (German)	s/s
Mazzanti et al. (1998)	Anxiety	European (Finnish)	s/s
Katsuragi et al. (1999)	Anxiety	Japanese	s/s
Du et al. (2000)	Anxiety	Caucasian (predominantly)	s/s
Melke et al. (2001)	Anxiety	Caucasian	s/s
Feinn et al. (2005)[a]	Substance abuse (alcohol)	Mixed	s
Caspi et al. (2003)	Suicide	Caucasian (non-Maori)	s/s
Bengel et al. (1999)	OCD	Caucasian	l/l
Malhotra et al. (1998)	Psychosis/ hallucinations	Mixed (N.American)	l/l
Dubertret et al. (2005)	Schizophrenia	Mixed	I
5-HTTLPR genotype not positively associated with affective disorder			
Lasky-Su et al. (2005)[a]	Unipolar depression	Mixed	
Mendlewicz et al. (2004)	Unipolar depression	European	
Mansour et al. (2005)	Bipolar depression	Caucasian (N.America)	
Meira-Lima et al. (2005)	Bipolar depression	Brazilian	
Mendlewicz et al. (2004)	Bipolar depression	European	
Ospina-Duque et al. (2000)	Bipolar depression	Columbian	
Arango et al. (2003)	Major depressive disorder	–	
Frisch et al. (1999)	Major depressive disorder	European (Jewish)	
Middledorp et al. (2007)	Anxiety	European (Dutch)	
Deary et al. (1999)	Anxiety	British	
Mendlewicz et al. (2004)	Suicide	European	
Arango et al. (2003)	Suicide	–	
Geijer et al. (2000)	Suicide	Caucasian (European)	
Serretti et al. (2002)	Major Psychosis	European	
Dickel et al. (2007)	OCD	Mixed	

(continued)

Table 1 (continued)

Author	Disorder	Cohort/ethnicity	Predisposing genotype
Rao et al. (1998)	Schizophrenia	Caucasian (N. American)	
Rao et al. (1998)	Schizophrenia	African-American	
Rao et al. (1998)	Schizophrenia	Caucasian (Swedish)	

Where cohort states European or British, ethnicity was either mixed or not stated
[a] meta-analysis

disorders in unknown, however, Gillihan et al. (2010) suggested that individuals with an s-allele display greater amygdala activity during intentional mood regulation and suggested that hyperactivity of the amygdala during mood recovery may be a potential mechanism by which the s-allele could increase the risk of depression. In addition, the chronicity of depression has been associated with homozygosity for the l-allele in a Korean population who showed a higher rate of chronicity, but not recurrent tendency, compared to those with at least one s-allele (Myung et al. 2010).

The observation that genotype can affect amygdala response would imply a role for the polymorphism in mood and anxiety disorders and higher anxiety scores have been shown to be significantly associated with s-allele or lg-allele homozygosity in a Turkish population concomitant with improved academic performance (Calapoglu et al. 2011). Interestingly, it has also been demonstrated that these anxiety-related associations can have an influence as early on as foetal developmental. Maternal anxiety during pregnancy and negative emotionality in early infancy was significant in infants carrying one or more copies of the s-allele (Pluess et al. 2011). Allelic variation at the LPR has also been shown to influence mother-offspring bonding; mothers with an s- or lg-allele showed higher sensitivity and greater attachment during mother-infant interactions (Mileva-Seitz et al. 2011).

The importance of tighter definition of sub-group classification can be seen in two recent studies investigating genotype interactions and OCD, for example, meta-analysis suggests that the I-allele may be associated with childhood-onset OCD and OCD in Caucasians (Bloch et al. 2008). A further study found little effect of the triallelic LPR in OCD in a mostly male population and the authors suggest that further replication in a larger cohort with an equal number of female subjects might be more predictive of risk associations with OCD (Tibrewal et al. 2010).

The s-allele of the 5-HTTLPR has also shown considerable associations in relation to alcohol dependence (McHugh et al. 2010; Wang et al. 2011), consumption in adolescents (Van der Zwaluw et al. 2010) and relapse in recovering alcoholics (Pinto et al. 2008). Furthermore, the risk association of the LPR with substance abuse has been demonstrated to display age-dependant bias: prior to the age of 15 individuals Hm for the s-allele did not show any difference to those with an l-allele, however, post 15 years of age they showed higher tobacco use and in early adulthood they were more active alcohol, drug and tobacco users (Merenakk et al. 2011).

A recent multi-centre study paired with meta-analysis failed to detect any association with the 5-HTTLPR polymorphism and attention deficit hyperactivity

Table 2 A sample of previously reported functional effects for the 5-HTT LPR polymorphism

Author	Model	Experimental detail	Gene expression bias
Heils et al. (1996)	Placental choriocarcinoma cell line (JAr)	VNTR within the promoter (reporter)	l > s
Lesch et al. (1996)	Lymphoblast cell lines	VNTR with the promoter (reporter)	l > s
Sakai et al. (2002)	Immortalised serotonergic raphe neurons (RN46A)	Isolated VNTR (reporter)	l = s
	Cos-7 cell lines		
	PC12 cell lines		
Mortensen et al. (1999a, b)	Placental choriocarcinoma cell line (JAr)	1.8 kb fragment upstream to the cap site including VNTR and a novel internal 379 bp fragment (reporter)	l = s
Mortensen et al. (1999a, b)	Immortalised serotonergic raphe neurons (RN46A)	1.8 kb fragment upstream to the cap site including VNTR and a novel internal 379 bp fragment (reporter)	l > s
Lesch et al. (1996)	Lymphoblast cell lines	mRNA quantification	l/l > l/s or s/s
Hranilovic et al. (2000)		mRNA quantification	l/l > l/s or s/s
Bradley et al. (2005)		mRNA quantification	l/l > l/s or s/s
Lim et al. (2006)	Human pons tissue post mortem	mRNA quantification	l = s
Lesch et al. (1996)	Lymphoblast cell lines	5-HT binding	l/l > l/s or s/s
Lesch et al. (1996)	Lymphoblast cell lines	5-HT uptake rate	l/l > l/s or s/s
Greenberg et al. (1999)	Blood platelets from healthy controls	5-HT uptake rate	l/l > l/s or s/s
Hanna et al. (1998)	Blood platelets OCD patients	5-HT uptake rate	l/l > l/s or s/s
Nobile et al. (1999)	Blood platelets from depressed patients	5-HT uptake rate	l/l > l/s or s/s

disorder (ADHD) in a Norwegian sample (Landaas et al. 2010). Other studies have also shown a lack of association in schizophrenia, neurocognition or core psychotic symptoms (Konneker et al. 2010), migraine in Europeans and Asians (Schurks et al. 2010a) and bulima nervousa (Lee and Lin 2010). Recent associations have been made between the presence of an s-allele and insomnia (Deuschle et al. 2010) and anorexia nervosa (Lee and Lin 2010).

4 The Serotonin Transporter Intron 2 VNTR (STin2 VNTR)

The STin2VNTR comprises 9, 10 or 12 copies of a 16-17 bp repeat giving rise to six possible genotypes 9/9, 10/10, 12/12, 9/10, 9/12 and 10/12 based on copy number alone without the added complexity of SNPs within each VNTR domain, Fig. 2B (Lesch et al. 1994). We have previously discussed the clinical and functional associations of this variant with a number of affective disorders and behaviour traits (Table 3) and levels of gene expression in vitro and in vivo (Table 4) (extensively reviewed in Haddley et al. 2008).

Clinical associations with the STin2VNTR have been less prevalent than those of the 5-HTTLPR over the last 5 years. In contrast to the 5-HTTLPR polymorphism, a significant effect of the STin2VNTR genotype on susceptibility to MDD was seen with the 10-allele in a Chinese Han population (Pan et al. 2009). In addition, individuals with the 9/12 or 12/12 genotype displayed a fourfold higher incidence of PSD compared with those Hm for the 10-allele (Kohen et al. 2008). Furthermore, chronicity and recurrent tendency of depression were not associated with STin2VNTR (Myung et al. 2010). Meta-analysis in a migraine population of European descent found a protective effect in the absence of the 12-allele (Schurks et al. 2010b).

Individually the serotonin polymorphisms have various effects across affective disorders and populations and more often than not results are conflicting or no associations can be made. However, when taken in combination with other polymorphisms both intra- and intergenic or when considering haplotypes, distinct associations have been made, for example, in a recent migraine study the STin2VNTR genotype had no significant effect, however when taken as a haplotype in conjunction with a serotonin receptor SNP, which also showed no solo association, a positive correlation was seen with the two polymorphisms having a synergistic influence on susceptibility to migraine (Joshi et al. 2010). Many haplotype associations for affective disorders are now being considered investigating not only interactions between the serotonin transporter LPR and VNTR but also other SNPs within the SLC6A4 gene locus and between other genes such as the dopamine receptor 4 (DRD4), serotonin precursor genes, and brain derived neurotrophic factor gene (BDNF). Such haplotype associations are seen with other gene synergies in which individual polymorphisms have no association with a particular condition whereas when addressed together they do demonstrate significant association (Bellivier et al. 1998; Wendland et al. 2007; Miyajima et al. 2008; Hiio et al. 2011).

Table 3 A sample of previously reported clinical associations for the STin2VNTR polymorphism

Author	Disorder	Cohort/ethnicity	Predisposing genotype
StIN2 genotype positively associated with an affective disorder			
Ogilvie et al. (1996)	Major depression	British	9
Battersby et al. (1996)	Uni/bipolar depression	British	9
Evans et al. (1997)	Anxiety	British	9
Ogilvie et al. (1998)	Migraine with aura	British	9/9
Hranilovic et al. (2003)	Suicide	Croatian/southern Slavic	10/10
Jernej et al. (2004)	Suicide	Croatian/southern Slavic	10/10
Gutierrez et al. (1998b)	Major depression with melancholia	Caucasians (Spanish)	10
Collier et al. (1996b)	Bipolar depression	Caucasian (English, European and other)	12
Kirov et al. (1999)	Bipolar depression	Caucasian (British)	12
Ohara et al. (1999)	Anxiety + OCD	Asian	12
Baca-Garcia et al. (2007)	OCD	Caucasians (Spanish)	12
Liu et al. (1999)	Schizophrenia	–	12
Hranilovic et al. (2000)	Schizophrenia	Croatian/southern Slavic	12
Kaiser et al. (2001)	Schizophrenia	Caucasian (German)	12
Ogilvie et al. (1998)	Migraine without aura	British	12/12
StIN2 genotype not associated with an affective disorder			
Furlong et al. (1998)[a]	Unipolar depression	Caucasian (English and European)	
Collier et al. (1996b)[a]	Unipolar depression	Caucasian (English, European and other)	
Furlong et al. (1998)[a]	Bipolar depression	Caucasian (English and European)	
Stober et al. (1996)	Uni/Bipolar depression	Caucasian (German)	
Bellivier et al. (1998)	Bipolar depression	Caucasian (French)	
Kunugi et al. (1996)[a]	Uni/Bipolar depression	Japanese	
Lasky-Su et al. (2005)[a]	Uni/Bipolar depression	Mixed	
Lotrich and Pollock (2004)[a]	Bipolar and major depression	–	
Gutierrez et al. (1998a)	Bipolar depression	Caucasians (Spanish)	
Hoehe et al. (1998)	Bipolar and major depression	Western European	
Rees et al. (1997)	Bipolar and major depression	Caucasian (British)	
Anguelova et al. (2003)[a]	Suicide	–	
(Collier et al. 1996b)	Schizophrenia	Caucasian (English, European and other)	

Where cohort states European or British, ethnicity was either mixed or not stated
[a] Meta-analysis

Table 4 A sample of previously reported functional effects for the STin2VNTR polymorphism

Author	Model	Experimental detail	Gene expression bias
Fiskerstrand et al. (1999)	Embryonic stem cells	VNTR in isolation cloned into a reporter vector showed responses following differentiation	12 > 10
Lovejoy et al. (2003)	Murine embryonic stem cells	Trimeric oligonucleotides composed of individual repeat elements cloned into reporter vector showed varied responses following differentiation.	
Lovejoy et al. (2003)	JAr human clonal cell line	VNTR in isolation	9 > 12 > 10
Klenova et al. (2004)	Monkey Cos 7 clonal cells Human HEK293 clonal cells	Differential expression and regulation by transcription factors CTCFand YB1	12>10>9 9 > 10 > 12
MacKenzie and Quinn (1999)	Transient transgenic mice	Differential expression driven by the human StlN2VNTR 10 or 12 variant supporting LacZ expression	
Hranilovic et al. (2003)	Lymphoblasts	Allelic expression imbalance/semi-quantitative real time PCR	12/12 >10
Roberts et al. (2007)	Stably transfected human JAr clonal cell line. Human HEK293T clonal cells	Differential binding of CTCF and YB-1 transcription factors to the allelic variants in response to lithium chloride	
Kaiser et al. (2002)	Human blood platelets	The maximum rate of 5-HT uptake by platelets was unaffected although the affinity of uptake varied as a function of genotype	12/12 < 9/10

5 Pharmacogenetics of the 5-HTT Polymorphisms and Selective Serotonin Reuptake Inhibitor (SSRI) Response

There is vast variability in the response of patients to antidepressant treatment which is believed to be in part due to heritable characteristics. This would suggest that polymorphisms that may affect the expression of "mood-related genes" may also account for variation in treatment response. A number of studies have investigated such variability in treatment response as a function of genotype particularly in relation to 5-HT specific reuptake inhibitors (SSRIs) e.g., citralopram, escitalopram, paroxetine and fluoxetine. The SSRIs selectively block the 5-HTT maintaining the concentration of 5-HT in the synaptic cleft. There are a myriad of published papers that analyse the effect of the 5-HTT polymorphisms on antidepressant response and tolerance in patients with affective disorders and in particular MDD (extensively reviewed in Keers and Aitchison 2011 and Porcelli et al. 2011). Initially in MDD patients a better response to SSRIs was associated with those individuals with an l-allele in the 5-HTTLPR (Serretti et al. 2007) however, a recent meta-analysis of 5408 participants found no correlation between 5-HTTLPR genotype and drug response (Taylor et al. 2010). In the recent STAR*D analysis an association was found between adverse events to citralopram and the triallelic lg-allele (Hu et al. 2007). Other studies have also pointed to 5-HTT triallelic variant affecting not only tolerance to antidepressants but also interaction with other polymorphisms such as the STin2VNTR (Kraft et al. 2005; Smeraldi et al. 2006). It has also been reported that genotype and response associations may be dependent on (1) the minor allele frequencies between populations because a significant association between genotype (l-allele) and treatment response was seen only in non-Hispanic Caucasians (Mrazek et al. 2009), (2) gender differences (Keers and Aitchison 2010) and (3) environment (stress) (Keers et al. 2011).

Fewer studies have assessed the effect of the STin2VNTR genotype however having a 10/12 genotype has been associated with a poor response and tolerance (Kim et al. 2000; Peters et al. 2004; Mrazek et al. 2009; Min et al. 2009) although these observations were not replicated in other studies (Ito et al. 2002; Ham et al. 2005; Smits et al. 2007). Finally, the recent review by Keers and Atchison, highlights the difficulties in coming to a standard agreement on the effect of polymorphisms due to the heterogeneity between studies but also the probability that an amalgamation of small genotypic effects could be responsible for variation in response and that gene x environment interactions must also be considered in these analyses (Keers and Aitchison 2010).

6 Gene × Environment (G × E) Interaction Role in the SLC6A4 Polymorphism

Personality traits are suggested to be generated by a complex interaction of environmental and genetic factors. Various studies suggest that the effect of the SLC6A4 polymorphisms might only be unmasked following specific environmental cues, for example SLC6A4 Knockout (KO) mice displayed more fearful behaviour compared to wild-type (WT) only in response to stress (Murphy et al. 2001; Wellman et al. 2007). The potential interaction between the 5-HTTLPR and early experiences was demonstrated in non-human primates; rhesus macaques with an s-variant analogous to that of human s-allele, displayed decreased serotonergic function but only in animals reared under stressful conditions (Bennett et al. 2002). Furthermore, peer-reared females with the s-variant were more vulnerable to alcoholism and showed more aggressive behaviour whereas those reared by their mothers were not differentiated by genotype (Champoux et al. 2002; Barr et al. 2003, 2004). Furthermore, in agreement with a seminal study by Caspi et al. demonstrating a greater risk of depression and suicide in those individuals Hm for the s-allele dependant on the number of past stressful life events and childhood maltreatment, it has been found that individuals in a Spanish population Hm for the s-allele acquired a higher risk of depression after considerably fewer stressful life episodes (Caspi et al. 2003; Cervilla et al. 2007). Moreover, in an adult twin study, individuals Hm for the s-allele displayed increased sensitivity of depressogenic effects of stressful-life events than those with one copy of the l-allele (Kendler et al. 2005).

Recently, a potential role for the 5-HTTLPR polymorphisms has been investigated in post-traumatic stress disorder (PTSD) as a function of a gene × environment interaction. Having one or more copies of the s-allele predisposed individuals who had suffered both childhood adversity and adult traumatic events to develop lifetime PTSD (Xie et al. 2009). A similar effect has also been seen in patients with a copy of the la-allele with more than 60% of carriers developing PTSD following three or more stressful life events (Grabe et al. 2009). At first,- these results would seem conflicting, however, other environmental factors may need to be considered, for example the s-allele has been associated with decreased risk of PTSD in low-risk environments (low crime/unemployment rates) but increased risk of PTSD in high-risk environments (Koenen et al. 2009).

These studies emphasise the significant impact of the environment and early experiences on genetic polymorphisms and suggest that with a particular genotype an individual may be more vulnerable to environmental stress and have higher tendency to develop psychiatric disorders, depending on their genotype. The interaction between SLC6A4 polymorphisms and stressful life events is striking and one that has stood up to scrutiny particularly in longitudinal studies (reviewed in Uher and McGuffin 2008).

7 Haplotype Analysis of the SLC6A4 Polymorphisms, Gene × Gene (G × G) and Gene × Gene × Environment (G × G × E) Interactions

Greater associations have been made between genotype and phenotype now that focus has begun to shift towards investigating polymorphisms in gene-gene interactions or as haplotypes rather than as standalone entities. According to the international HapMap project the SLC6A4 gene exists as two haplotype blocks. A recent study in a Chinese Han population found that when assessed alone no association could be made between the STin2VNTR and schizophrenia, however, significant association was found when the locus was studied as a haplotype in conjunction with a number of tagging SNPs, identifying two haplotypes with positive association (Lin et al. 2009). Similar results were also found in a North European population, in that schizophrenia was associated with a haplotype block but not with single loci (Zaboli et al. 2008). Furthermore, haplotype analysis has revealed that having the l-allele and the STin2VNTR 10-allele predisposes Russian females, but not males to suicide (Gaysina et al. 2006). In a Turkish population predisposition towards suicidal behaviour was not observed for either the 5-HTTLPR or STin2VNTR polymorphisms, however when combined it was found that those with an s/10 or l/12 genotype were predisposed to suicidal behaviour (Akar et al. 2010). Moreover, in a study of German alcoholics the sa/10 and lg/12 haplotypes were more prevalent in type 2 alcoholics compared with controls (Reese et al. 2010).

Polymorphic interactions are not limited to those in-cis and evidence is emerging for in-trans interaction between genes as predisposing factors for disease. A recent investigation by Lorenzi et al. (2010) found that in an Italian sample of patients with Alzhiemers linked dementia that s-allele homozygosity was particularly abundant and the 5-HTTLPR synergistically interacted with a polymorphism in the saitohin gene (Q7R) to increase susceptibility. In addition, interactions between the Hm s-allele 5-HTTLPR genotype and polymorphisms within the cannabinoid receptor 1 locus have been associated with increased anxiety scores in Caucasian Hungarians (Lazary et al. 2009). Finally, Armbruster et al. found a significant interaction between the 5-HTTLPR and DRD4 polymorphisms in response to stress in a German population with individuals Hm for the 5-HTTLPR la-allele and one copy of the DRD4 7R allele having a lower cortisol response compared to other genotypes (Armbruster et al. 2009).

In addition to two-way interactions many studies have identified three-way interactions, so called G $\times$ G $\times$ E interactions. Ressler et al. conducted an analysis of SLC6A4 interaction with corticotropin-releasing hormone haplotypes as a function of child abuse in an African-American cohort. They found that the 5-HTTLPR s-allele interacted with CRHR1 haplotypes and child abuse to predict current depressive symptoms (Ressler et al. 2010). Similarly, Conway et al. (2010) found that in individuals Hm for the catechol-o-methyltransferase (COMT) val158 mutation, a Hm 5-HTTLPR l-allele genotype exerted a protective effect in adolescents in response to stress-related depression.

8 Nuclear Imaging of SLC6A4 Polymorphic Variants in Affective Disorders

Over the last decade advances in in vivo imaging techniques have enabled further investigation into the in vitro and genome-wide associations made between polymorphic variation, changes in brain activity and affective disorders, in the hope of clarifying not only whether these associations are accurate but also in the hope of determining biological endophenotypes associated with genotype. Such investigations have included examining differences in brain structure, brain metabolism, transporter and receptor-ligand binding and the signaling interactions between brain regions.

The three most widely used techniques in brain imaging for affective disorders are SPECT, positron emission tomography (PET) and fMRI. MRI is used to give a whole system overview of brain activity, whereas SPECT and PET provide resolution of events at a molecular level allowing the investigation of binding of a radiolabelled ligand to its target receptor or transporter. MRI is rapid but gives low resolution and is generally used as a functional measure of activity. Both SPECT and PET are used to determine binding potential, target density or distribution, and volume ratios (Innis et al. 2007). The most commonly used ligands for assessing binding potentials and distribution for the monoamine transporters are summarised in Table 5. Many studies concentrating on affective disorders and polymorphisms using these techniques analyse multiple brain regions, whereas others focus on specific regions for example the midbrain or striatum which are both rich in serotonin or the amygdala and limbic systems which are central regions involved in the control of mood, anxiety and emotion. For this reason and in combination with the different techniques available there is still conflict over the role of polymorphisms and their associations however a clearer picture is emerging. The results of these investigations, with a focus on the serotonin LPR are discussed below.

9 Effects of 5-HTTLPR on Serotonin Signalling Systems (MRI)

Functional MRI studies have been undertaken to determine the effects of the 5-HTTLPR polymorphism on brain organisation and structure in brain regions associated with affective disorders. Healthy individuals with an s or lg-allele displayed reduced grey matter volumes, when compared to those with a la-allele, suggesting that the polymorphism plays a role in development of brain structure. Interestingly, the same study found that patients with depression suffered from reduced bilateral hippocampal volumes and this was particularly pronounced in those who were Hm for the la-allele, compared to those who were lg/lg, lg/s or s/s suggesting that individuals Hm for the la-allele are more susceptible to morphological changes during depressive episodes (Frodl et al. 2004, 2008a, b). A further study assessed hippocampal volumes in an elderly cohort in a similar manner but on the basis of the age of onset of depression. This study determined that those individuals Hm for the l-allele and with late-onset depression had significantly smaller hippocampal volumes than those with early-onset depression or healthy controls. In addition they found that in individuals Hm for the s-allele early onset of depression was associated with smaller hippocampal volumes suggesting that genotype and age interact to affect the timing of depression (Taylor et al. 2005).

Tracers used in nuclear imaging studies of the 5-HTT

Ligand	A.K.A		Selectivity	Application
[123I](−)-2β-Carbomethoxy-3β-(4-iodophenyl) tropane	[123I]β-CIT	Cocaine analogue	DAT1 5-HTT NET	SPECT
[I-123]-N-omega-fluoropropyl-2beta -carbomethoxy-3beta -(4-iodophenyl)nortropane	[1231] FP-CIT	[123I]-β-CIT analogue	DAT1 5-HTT NET	SPECT
[11C]3-amino-4-[2-[(di(methyl)amino)methyl]phenyl] sulfanylbenzonitrile	[11C]DASB		5-HTT	PET
[llC](+)-6β-(4-Methylthiophenyl)-1,2,3,5,6α, 10β-hexahydropyrrolo[2,1 -a] isoquinoline	[11C]McN 5652		5-HTT	PET
[F-18]fluoro-2-deoxy-D-glucose	FDG		Glucose metabolism	PET

Functional MRI has been used by both Hariri et al. and Heinz et al. in healthy individuals to demonstrate that those with at least one s-allele where found to have stronger amygdala reactivity and stronger amygdala-prefrontal cortex coupling in response to aversive pictures or uncertain contexts, which could be considered stressful (Hariri et al. 2006; Heinz et al. 2005, 2007). Furthermore, in individuals exposed to fearful faces those Hm for the s-allele showed greater activity in the right fusiform gyrus (FG), and greater positive functional connectivity between right amygdala and FG and between right FG and right ventrolateral prefrontal cortex, compared to those with other genotypes (Surguladze et al. 2008). These replications are a good indicator that the 5-HTTLPR polymorphism can contribute to differential hyper-responsiveness of the amygdala, the control centre for emotions, and go some way towards defining a mechanism for the clinical and in vitro observations that the s-allele is associated with a predisposition to anxiety-depression spectrum disorders. It has also been suggested that these findings may also be key in other psychiatric disorders because those Hm for the l-allele displayed what could be considered "hypo-responsiveness" a trait that may be important where a lack of emotion is a key factor e.g., psychopathy (Surguladze et al. 2008; reviewed in Glenn 2011). The effect of 5-HTTLPR genotype on connectiveness between the ventral anterior cingulate cortex and the amygdala has also been examined as this pathway is hyporesponsive in patients with bipolar disorder. When patients with bipolar depression or healthy controls were presented with happy and fearful facial imagery those with bipolar disorder showed decreased activity in this pathway as expected. Furthermore, in healthy individuals and in those with bipolar depression Hm for the s-allele, this pathway was found to be hypoactive compared to those with an l-allele (Shah et al. 2009). Taken together these data strongly suggest an effect of the 5-HTTLPR polymorphism, in particular the s-allele, on amygdala responsiveness and connectivity although the molecular mechanisms behind these differences cannot be delineated using MRI.

10 Effects of 5HTTLPR on Serotonin Transporter Binding

A SPECT study in 22 individuals, 14 of which were alcoholics assessed the effect of the 5-HTTLPR polymorphisms on transporter-ligand binding using [123I]β-CIT. In healthy individuals Hm for the l-allele the binding capacity of the 5-HTT in the midbrain was almost twice that of those with an s-allele. By contrast, the same study noted a modest reduction in binding in alcoholics Hm for the l-allele compared to those with an s-allele (Heinz et al. 2000). These initial results in healthy individuals were not confirmed by further SPECT studies using [I-123]β-CIT that showed no differences in binding on consideration of genotype (Jacobsen et al. 2000; Willeit et al. 2001). In a further study the predominant findings were opposed to those of Heinz et al. in that these subjects Hm for the s-allele displayed significantly higher ligand-binding potential compared to any l-allele containing genotypes (Van Dyck et al. 2004), however, in this assessment all s-allele homozygotes were much younger than non-s-allele homozygotes and it should be noted that decreased ligand-binding in nuclear imagining has been associated with increasing age (Pirker et al. 2000). Supporting a lack of function in vivo seen with SPECT, quantitative autoradiography in the prefrontal cortex of post-mortem brains of suicide victims also failed to show any correlation between 5-HTT binding and genotype (Mann et al. 2000).

Decreased levels of 5-HTT binding have been observed across a number of brain regions in a PET study using [C-11]DASB (Selvaraj et al. 2011) or [11C]McN 5652 (Parsey et al. 2006a) in depressed individuals. The activity of serotonin re-uptake and metabolism in the context of LPR polymorphisms was assessed in healthy individuals using PET with fludeoxyglucose F18 (FDG). The results of this study showed genotype dependant differences in serotonin metabolism across the brain. In particular, individuals Hm for the s-allele ($n = 6$) showed higher activity in the limbic system, prefrontal and temporal cortices, whilst in those Hm for the l-allele ($n = 6$) higher activity was noted in bilateral frontal structures. Overall the authors suggest that a more widespread effect could be the foundation for individuals Hm for the s-allele being more susceptible to anxiety-depression spectrum disorders (Graff-Guerrero et al. 2005).

Interestingly, high-resolution PET FDG imaging showed that monkeys with the s-allele of the LPR variant displayed increased brain activity in orbitofrontal cortical regions, amygdala and the bed nucleus of the stria terminalis in response to stress (relocation to a foreign environment) which provides further support that the s-allele may confer a susceptibility to stress by mediating aberrant activity in limbic structures (Kalin et al. 2008).

Similar to investigations using SPECT imaging, correlation between genotype and binding has also been negative in PET studies utilising the 5HTT-selective radioligand [11C]McN 5652 (Parsey et al. 2006b; Shioe et al. 2003). However, PET studies utilising the more sensitive and preferable 5HTT-selective tracer, [11C]DASB, have demonstrated positive correlations between genotype and binding capacity. These studies analysed the triallelic variants of the 5-HTTLPR (la, lg, s). Replicable studies found that, when compared to non-la-allele homozygotes, those Hm for the la-allele demonstrated higher 5HTT-ligand binding capacity in brain regions associated with high transporter expression in the putamen, caudate and the midbrain (Kalbitzer et al. 2009; Praschak-Rieder et al. 2007; Reimold et al. 2007, respectively). It is worth noting that the study of Reimold et al. found a significant effect of genotype on 5-HTT binding in the midbrain but no effect in the amygdala or thalamus, further stressing the importance of experimental design in assessing genotype effects, as these effects are likely to be tissue-specific or context dependant.

In contrast, in a study using the selective 5-HTT ligand, [11C]DASB, in alcoholic subjects and healthy controls, Martinez et al. (2009) did not find any significant association between genotype and 5-HTT binding across a number of brain regions.

The majority of genotypic effects examined for 5-HTT have been in the midbrain but little evidence exists for other regions. Since the origin of serotonergic projections is the midbrain this may point to an important affect of polymorphism during development which may have an impact on the circuitry projections to other brain regions which may affect signaling in later life and be the basis for the effect of a transporter polymorphism. It follows that a lack of polymorphism-dependant effects on binding potential in limbic areas may not necessarily be seen in adults despite differential metabolic activity being observed by FDG PET.

It is particularly interesting to note that a number of studies have found differential effects on ligand-binding dependant on the season of the year at which the experiments where performed with higher binding in healthy controls noted in the autumn and winter months in both PET (Praschak-Rieder et al. 2008) and SPECT (Ruhe et al. 2009) scanning.

11 Effect of the Serotonin Transporter Polymorphisms on Functional Gene Expression

The regulation of expression of a given gene can vary between individuals because of epigenetics and polymorphic variation in regulatory domains where for example, transcription factors may bind. Many studies have shown that cis-regulatory loci (regulatory polymorphisms) in promoters and other non-coding regions of a gene, in addition to transcription factors (trans-acting modulators) regulate allele-specific expression (Heils et al. 1996; Yan 2002; Bray et al. 2003; Lo et al. 2003; Klenova et al. 2004; Roberts et al. 2007). In the absence of these cis-regulatory domains, paternal and maternal alleles of a gene are equally expressed, unless one allele is imprinted. However, when an individual is heterozygous for a cis-regulatory domain, mRNA expression level may vary from each allele; termed differential allelic gene expression. Different combinations of these polymorphisms within our genome create the 'genetic fingerprint' that contributes to determining phenotypic diversity and leads in part to individuality amongst us. Nearly three decades ago, King and Wilson pointed out that changes in the mechanisms controlling the gene expression, rather than the DNA sequence itself account for the morphological, behavioural and cognitive differences between human beings and other primates (King and Wilson 1975). Therefore, it is suggested that the person we are is not solely determined by our genes but how we control their expression, hence drugs such as cocaine vary our behaviour in part by modulating gene expression.

We and others have previously demonstrated the transcriptional capacity of the VNTRs to support tissue-specific and stimulus inducible gene expression (Table 4) (reviewed in Haddley et al. 2008). We further extended our studies to investigate the possible combinatorial interaction of these polymorphisms with one another in regulating gene expression and recently demonstrated that the 5-HTTLPR and STin2VNTR can act in concert, in the context of a transient or stable reporter gene assay in human 5-HTT-expressing JAr cells and primary rodent neurons, to demonstrate differential reporter gene expression which was not a simplistic additive effect of the individual domains (Ali et al. 2010a).

One characteristic feature of psychiatric disorders, particularly depression, is the long-lasting nature of the disease and the slow response to treatment. Recent psychiatric research suggests that these long-lasting changes in gene expression might be due to modulation of the epigenetic mechanisms regulating neuronal gene expression. A better understanding of how epigenetic changes associate with psychiatric disorders, polymorphims and drug addiction will help in the understanding of the neurobiology of these diseases. A study by Tsankova et al. revealed robust and long lasting histone modifications associated with the promoter of the BDNF gene in the hippocampus of a mouse model of depression. Interestingly, the negative-histone modifications marks were present for 4 weeks following ending of defeat-stress and were not reversed by anti-depressant treatment, indicating the stability of these 'acquired marks', the anti-depressant drugs exerted their effect by establishing positive-histone marks; methylation of H3K4 and acetylation of histone H3 across the promoter (Tsankova et al. 2004). Similarly, research suggests that drugs of abuse and drug-associated cues or stress, exert long lasting changes in gene expression through changes in epigenetic mechanisms (Levine et al. 2005; Kumar et al. 2005; Renthal et al. 2007). Transcription factors play an early role in regulating gene expression and histone modifications and one mechanism by which drugs of abuse and antipsychotic drugs can exert their effect on chromatin modification is via modulating intracellular signalling cascades, for example, cocaine elicits phosphorylation of the cAMP response element (CREB) initiating a transcriptional DNA-binding cascade that results in histone modifications leading to further recruitment of transcriptional activators (Carlezon et al. 1998). Acute cocaine treatment was also found to increase H4 histone acetylation in cfos and fosb promoters within 30 min which disappeared by 3 h (Kumar et al. 2005). Cocaine was also found to regulate the specific histone methyl-transferases and demethylases (Maze et al. 2010).

The SLC6A4 gene is an important modulator of addictive behaviour (Hall et al. 2004), and studies have demonstrated the role of the serotonergic system in cocaine addiction (Cabrera-Vera et al. 2000; Uhl et al. 2002). Little et al. (1998) suggested that SLC6A4 expression could be regulated in a region-specific pattern and that this could be affected by the genotype of the LPR variant. Although performed in post-mortem material, this would be consistent with the action of these domains as tissue-specific or stimulus-induced regulators (Haddley et al. 2008; Ali et al. 2010a). We hypothesised that cocaine could differentially modulate epigenetic markers on the SLC6A4 gene based on the genotype of the polymorphisms. We focused on the transcription factor CTCF, which has been implicated in epigenetic remodelling (Ishihara et al. 2006), and which we had previously characterised as a regulator of both the 5-HTTLPR and STin2VNTR (Klenova et al. 2004; Roberts et al. 2007; Ali et al. 2010a). The identification of CTCF suggested epigenetic changes operational at these loci might also involve MeCP2, known to bind to methylated DNA, which we confirmed occurs at the LPR domain. Specifically, we demonstrated that exposure of JAr cells to acute cocaine treatment resulted in CTCF binding only to the l-allele of the 5-HTTLPR whereas MeCP2 was observed to bind only to the s-allele and to recruit the histone deacetylase complex (HDAC) to this allele. These changes correlated with an increase in the association of positive histone marks and RNA polymerase II binding within the SLC6A4 promoter. Similarly, we have also demonstrated that the mood stabiliser lithium chloride alters the binding of CTCF to the STin2VNTR dependent on genotype (Roberts et al. 2007). These observations suggest a potent mechanism by which genotype can effect gene regulation and indirectly the level of 5-HTT protein present by altering the transcription factor complexes bound by allelic variants under basal conditions and in response to stimuli and may provide further insight into the mechanism(s) behind genotypic influence on susceptibility to affective disorders.

12 Lessons from the 5-HTT Polymorphisms and Their Application to Other Gene Polymorphisms

We have focused in this chapter on two of the many polymorphisms present in the SLC6A4 gene there are many others including SNPs and other mini-satellites present and more are being identified as deep sequencing of the human genome takes place. All genes contain what could be considered functional SNPs and these are not only present in protein coding regions. SLC6A4 is obviously not the only gene involved in the plethora of psychosis and affective disorders and many other genes contain functional and disorder-associated polymorphisms. For example, the widely studied DAT1 (SLC6A3) gene from the same family as the 5-HTT contains VNTRs in non-coding regions that have been associated with a range of disorders including ADHD and drug addiction (Guindalini et al. 2006; Yang et al. 2007). The SLC6A3 polymorphisms have thus been investigated with as much vigour as the SLC6A4 and the conclusions of such studies have been similar with association both negatively and positively correlated with disorders, conflicting imaging studies and differential frequency distribution of alleles in different populations. Nevertheless, molecular models have shown that these polymorphisms support differential gene expression and the lessons that can be learnt from studies with one group of polymorphisms can be applied to other polymorphisms for example the SLC6A4 and SLC6A3 VNTRs share similar transcription factor binding sites within their sequences implying that they may be functional on similar pathways and in response to similar challenges such as stress. The same can be said for many other VNTRs within other neuronal genes such as those in the dopamine R4 receptor (DRD4), monoamine oxidase A (MAOA), neuronal restrictive silencer factor (NRSF) and N-methyl-D-aspartic acid (NMDA) glutamate receptor (GRIN1).

13 Summary

After almost two decades of research the importance of the link between individual genetic polymorphisms as standalone agents and their relation to disease susceptibility remains somewhat controversial and tenuous for many polymorphisms. Initial twin studies pointed towards a genetic component for many psychiatric disorders making the polymorphisms a likely candidate. Early indications linking the serotonin transporter polymorphisms to a range of psychiatric disorders including depression, OCD, anxiety, schizophrenia and treatment response while at first positive have fallen under recent scrutiny in the light of many negative associations. Amongst most researchers, however, there is little doubt that these functional polymorphisms do have a role to play in the aetiology of psychiatric disorders but the key to unlocking these roles is more complex than a simple 'one polymorphism equals one disorder model'. It has been apparent for a long time that the same polymorphism occurs at different frequencies within distinct populations and that an 'at risk' allele in a Caucasian population of European descent will not be an at risk allele for the same disorder in, for example, a Chinese Han population. This points to a more complex network of other factors being involved and recent investigations have begun to expand on this idea looking not only for single polymorphism associations but examining haplotype associations where one polymorphism may only exert an effect in the presence of a distinct polymorphism in the same or another gene, as has been shown on many occasions for the BDNF Val66Met polymorphism (Miyajima et al. 2008; Wells et al. 2010; Hiio et al. 2011). In addition to haplotype analysis it is also apparent that G $\times$ G $\times$ E interactions play as big a role as the polymorphisms themselves. A polymorphism can be a marker or predictor for susceptibility but its effect may only become apparent under certain circumstances; it may only have an adverse (or in some cases positive) effect within a specific context for example following stressful-life events (Caspi et al. 2003).

Advances in sequencing and array technologies are allowing rapid and cost effective analysis of the thousands of human polymorphisms that exist allowing large scale mapping projects worldwide in different populations such as the 1,000 man genome project and the international HAPMAP project. Genome wide association studies are being supported by functional genome wide studies such as ChIP-Geneseq and FAIRE which expand current knowledge on how polymorphisms may regulate gene expression at a molecular level by interaction with transcription factors and by epigenetic remodelling. Perhaps most useful in human functional studies are the advances made in nuclear medicine. Imagining is certainly an effective tool for assessing genotype effects on binding and possible function although like all methods compounding factors must also be considered and not just those inherent in the experimental design e.g., tracer sensitivity or brain region under study. Environmental effects on endogenous transporter expression must also be taken into account for example, underlying pathologies, seasonal changes, age and lifestyle, as these may mask any underlying mechanisms apparent due to polymorphic variation. A large proportion of imaging studies failed to detect any effect of a polymorphism on transporter function, not solely in the serotonin transporter but also in other monoaminergic transporters such as the dopamine transporter. This could be due to the heterogeneity of such studies, as of yet there are still no clear parameters in place that make comparing studies wholly effective. Studies have, however, found differences in brain region volumes, connectivity and activity and it has been suggested by Keers and

Atchinson, that polymorphisms may be involved in indirect effects such as skewing the hypothalamic-pituitary-adrenal (HPA) axis, the neuroendocrine system that mediates mood and which is severely affected in psychiatric disorders, or by affecting neurogenesis, which when suppressed can lead to dysfunction in the HPA axis (Schloesser et al. 2009). Further imaging studies of whole brain function in conjunction with haplotype analysis and epigenetic investigations may help to further elucidate indirect effects of polymorphisms that until recently have been neglected. There is still much to be uncovered and understood about how genetic variation can affect an individual's mental state and capacity, and how small changes in genetic sequence can have major effects on complex systems. Within the next 10 years it is hoped that with the rapid expansion in post-genomic analysis, functional assessments and more concerted effort on implementing a degree of homogeneity between association studies that the true clinical value of polymorphisms as markers of disease and markers and targets of therapeutic response will be revealed.

References

Akar T, Sayin A et al (2010) Investigation of serotonin transporter gene promoter (5-HTTLPR) and intron 2 (variable number of tandem repeats) polymorphisms with suicidal behavior in a Turkish population. DNA Cell Biol 29(8):429–434

Alaerts M, Ceulemans S et al (2009) Detailed analysis of the serotonin transporter gene (SLC6A4) shows no association with bipolar disorder in the Northern Swedish population. Am J Med Genet B Neuropsychiatr Genet 150B(4):585–592

Ali FR, Vasiliou SA et al (2010a) Combinatorial interaction between two human serotonin transporter gene variable number tandem repeats and their regulation by CTCF. J Neurochem 112(1):296–306

Ali FR, Haddley K et al (2010b) Assessing the impact of genetic variation on transcriptional regulation in vitro. Methods Mol Biol 628:195–214

Anguelova M, Benkelfat C et al (2003) A systematic review of association studies investigating genes coding for serotonin receptors and the serotonin transporter: II. Suicidal behavior. Mol Psychiatry 8(7):646–653

Arango V, Underwood MD et al (2002) Serotonin brain circuits involved in major depression and suicide. Prog Brain Res 136:443–453

Arango V, Huang YY et al (2003) Genetics of the serotonergic system in suicidal behavior. J Psychiatr Res 37(5):375–386

Armbruster D, Mueller A et al (2009) Interaction effect of D4 dopamine receptor gene and serotonin transporter promoter polymorphism on the cortisol stress response. Behav Neurosci 123(6):1288–1295

Baca-Garcia E, Vaquero-Lorenzo C et al (2007) Association between obsessive-compulsive disorder and a variable number of tandem repeats polymorphism in intron 2 of the serotonin transporter gene. Prog Neuro-Psychopharmacol Biol Psychiatry 31(2):416–420

Barr CS, Newman TK et al (2003) Serotonin transporter gene variation is associated with alcohol sensitivity in rhesus macaques exposed to early-life stress. Alcohol Clin Exp Res 27(5): 812–817

Barr CS, Newman TK et al (2004) Interaction between serotonin transporter gene variation and rearing condition in alcohol preference and consumption in female primates. Arch Gen Psychiatry 61(11):1146–1152

Battersby S, Ogilvie AD et al (1996) Structure of a variable number tandem repeat of the serotonin transporter gene and association with affective disorder. Psychiatr Genet 6(4): 177–181
Bellivier F, Henry C et al (1998) Serotonin transporter gene polymorphisms in patients with unipolar or bipolar depression. Neurosci Lett 255(3):143–146
Bengel D, Greenberg BD et al (1999) Association of the serotonin transporter promoter regulatory region polymorphism and obsessive-compulsive disorder. Mol Psychiatry 4(5): 463–466
Bennett AJ, Lesch KP et al (2002) Early experience and serotonin transporter gene variation interact to influence primate CNS function. Mol Psychiatry 7(1):118–122
Bloch MH, Landeros-Weisenberger A et al (2008) Association of the serotonin transporter polymorphism and obsessive-compulsive disorder: systematic review. Am J Med Genet B Neuropsychiatr Genet 147B(6):850–858
Bradley SL, Dodelzon K et al (2005) Relationship of serotonin transporter gene polymorphisms and haplotypes to mRNA transcription. Am J Med Genet B Neuropsychiatr Genet 136B(1): 58–61
Bray NJ, Buckland PR et al (2003) Cis-acting variation in the expression of a high proportion of genes in human brain. Hum Genet 113(2):149–153
Cabrera-Vera TM, Garcia F et al (2000) Neurochemical changes in brain serotonin neurons in immature and adult offspring prenatally exposed to cocaine. Brain Res 870(1–2):1–9
Calapoglu M, Sahin-Calapoglu N et al (2011) Serotonin transporter bi- and triallelic genotypes and their relationship with anxiety and academic performance: a preliminary study. Neuropsychobiology 63(2):103–111
Carlezon WA, Thome J Jr et al (1998) Regulation of cocaine reward by CREB. Science 282(5397): 2272–2275
Caspi A, Sugden K et al (2003) Influence of life stress on depression: moderation by a polymorphism in the 5-HTT gene. Science 301(5631):386–389
Cervilla JA, Rivera M et al (2006) The 5-HTTLPR s/s genotype at the serotonin transporter gene (SLC6A4) increases the risk for depression in a large cohort of primary care attendees: the PREDICT-gene study. Am J Med Genet B Neuropsychiatr Genet 141B(8):912–917
Cervilla JA, Molina E et al (2007) The risk for depression conferred by stressful life events is modified by variation at the serotonin transporter 5HTTLPR genotype: evidence from the Spanish PREDICT-Gene cohort. Mol Psychiatry 12(8):748–755
Champoux M, Bennett A et al (2002) Serotonin transporter gene polymorphism, differential early rearing, and behavior in rhesus monkey neonates. Mol Psychiatry 7(10):1058–1063
Cho HJ, Meira-Lima I et al (2005) Population-based and family-based studies on the serotonin transporter gene polymorphisms and bipolar disorder: a systematic review and meta-analysis. Mol Psychiatry 10(8):771–781
Collier DA, Arranz MJ et al (1996a) The serotonin transporter is a potential susceptibility factor for bipolar affective disorder. Neuroreport 7(10):1675–1679
Collier DA, Stober G et al (1996b) A novel functional polymorphism within the promoter of the serotonin transporter gene: possible role in susceptibility to affective disorders. Mol Psychiatry 1(6):453–460
Conway CC, Hamman C et al (2010) Interaction of chronic stress with serotonin transporter and catechol-o-methyltransferase polymorphisms in predicting youth depression. Depression Anxiety 27(8):737–745
Deary IJ, Battersby S et al (1999) Neuroticism and polymorphisms in the serotonin transporter gene. Psychol Med 29(3):735–739
Deuschle M, Schredl M et al (2010) Association between a serotonin transporter length polymorphism and primary insomnia. Sleep 33(3):343–347
Dickel DE, Veenstra-VanderWeele J et al (2007) Association studies of serotonin system candidate genes in early-onset obsessive-compulsive disorder. Biol Psychiatry 61(3):322–329
Du LS, Bakish D et al (2000) Gender differences in association between serotonin transporter gene polymorphism and personality traits. Psychiatr Genet 10(4):159–164

Dubertret C, Hanoun N et al (2005) Family-based association study of the 5-HT transporter gene and schizophrenia. Int J Neuropsychopharmacol 8(1):87–92

Esquivel UP, Haddley K et al (2011) Finding gene expression regulators implicated in major depression using formaldehyde assisted isolation of regulatory elements (FAIRE). Neuropsychopharmacology 21:S28–S29, Abstract

Evans J, Battersby S et al (1997) Association of short alleles of a VNTR of the serotonin transporter gene with anxiety symptoms in patients presenting after deliberate self harm. Neuropharmacology 36(4–5):439–443

Fang J, Yan WH et al (2011) Serotonin transporter gene polymorphism in Chinese patients with poststroke depression a case-control study. Stroke 42(5):1461–1463

Feinn R, Nellissery M et al (2005) Meta-analysis of the association of a functional serotonin transporter promoter polymorphism with alcohol dependence. Am J Med Genet B Neuropsychiatr Genet 133B(1):79–84

Fiskerstrand CE, Lovejoy EA et al (1999) An intronic polymorphic domain often associated with susceptibility to affective disorders has allele dependent differential enhancer activity in embryonic stem cells. FEBS Lett 458(2):171–174

Frisch A, Postilnick D et al (1999) Association of unipolar major depressive disorder with genes of the serotonergic and dopaminergic pathways. Mol Psychiatry 4(4):389–392

Frodl T, Meisenzahl EM et al (2004) Reduced hippocampal volumes associated with the long variant of the serotonin transporter polymorphism in major depression. Arch Gen Psychiatry 61(2):177–183

Frodl T, Koutsouleris N et al (2008a) Reduced gray matter brain volumes are associated with variants of the serotonin transporter gene in major depression. Mol Psychiatry 13(12):1093–1101

Frodl T, Zill P et al (2008b) Reduced hippocampal volumes associated with the long variant of the tri- and diallelic serotonin transporter polymorphism in major depression. Am J Med Genet B Neuropsychiatr Gene 147B(7):1003–1007

Furlong RA, Ho L et al (1998) Analysis and meta-analysis of two serotonin transporter gene polymorphisms in bipolar and unipolar affective disorders. Am J Med Genet 81(1):58–63

Gaysina D, Zainullina A et al (2006) The serotonin transporter gene: polymorphism and haplotype analysis in Russian suicide attempters. Neuropsychobiology 54(1):70–74

Geijer T, Frisch A et al (2000) Search for association between suicide attempt and serotonergic polymorphisms. Psychiatr Genet 10(1):19–26

Gillihan SJ, Rao H et al (2010) Serotonin transporter genotype modulates amygdala activity during mood regulation. Soc Cogn Affect Neurosci 5(1):1–10

Glenn AL (2011) The other allele: exploring the long allele of the serotonin transporter gene as a potential risk factor for psychopathy: a review of the parallels in findings. Neurosci Biobehav Rev 35(3):612–620

Grabe HJ, Spitzer C et al (2009) Serotonin transporter gene (SLC6A4) promoter polymorphisms and the susceptibility to posttraumatic stress disorder in the general population. Am J Psychiatry 166(8):926–933

Graff-Guerrero A, De La Fuente-Sandoval C et al (2005) Frontal and limbic metabolic differences in subjects selected according to genetic variation of the SLC6A4 gene polymorphism. Neuroimage 25(4):1197–1204

Greenberg BD, Tolliver TJ et al (1999) Genetic variation in the serotonin transporter promoter region affects serotonin uptake in human blood platelets. Am J Med Gent 88(1):83–87

Guindalini C, Howard M et al (2006) A dopamine transporter gene functional variant associated with cocaine abuse in a Brazilian sample. Proc Natl Acad Sci (USA) 103(12): 4552–4557

Gutierrez B, Arranz MJ et al (1998a) Serotonin transporter gene and risk for bipolar affective disorder: an association study in Spanish population. Biol Psychiatry 43(11):843–847

Gutierrez B, Pintor L et al (1998b) Variability in the serotonin transporter gene and increased risk for major depression with melancholia. Hum Genet 103(3):319–322

Haddley K, Vasiliou AS et al (2008) Molecular genetics of monoamine transporters: relevance to brain disorders. Neurochem Res 33(4):652–667

Hall FS, Sora I et al (2004) Molecular mechanisms underlying the rewarding effects of cocaine. Ann NY Natl Acad Sci 1025:47–56
Ham BJ, Lee MS et al (2005) No association between the tryptophan hydroxylase gene polymorphism and major depressive disorders and antidepressant response in a Korean population. Psychiatr Genet 15:299–301
Hanna GL, Himle JA et al (1998) Serotonin transporter and seasonal variation in blood serotonin in families with obsessive-compulsive disorder. Neuropsychopharmacology 18(2):102–111
Hariri AR, Mattay VS et al (2002) Serotonin transporter genetic variation and the response of the human amygdala. Science 297(5580):400–403
Hariri AR, Drabant EM et al (2006) Imaging genetics: perspectives from studies of genetically driven variation in serotonin function and corticolimbic affective processing. Biol Psychiatry 59(10):888–897
Hauser J, Leszczynska A et al (2003) Association analysis of the insertion/deletion polymorphism in serotonin transporter gene in patients with affective disorder. Eur Psychiatry 18(3):129–132
Heils A, Teufel A et al (1995) Functional promoter and polyadenylation site mapping of the human serotonin (5-HT) transporter gene. J Neural Transm Gen Sect 102(3):247–254
Heils A, Teufel A et al (1996) Allelic variation of human serotonin transporter gene expression. J Neurochem 66(6):2621–2624
Heinz A, Jones DW et al (2000) A relationship between serotonin transporter genotype and in vivo protein expression and alcohol neurotoxicity. Biol Psychiatry 47(7):643–649
Heinz A, Braus DF et al (2005) Amygdala-prefrontal coupling depends on a genetic variation of the serotonin transporter. Nat Neurosci 8(1):20–21
Heinz A, Smolka MN et al (2007) Serotonin transporter genotype (5-HTTLPR): effects of neutral and undefined conditions on amygdala activation. Biol Psychiatry 61(8):1011–1014
Hiio K, Merenakk L et al (2011) Effects of serotonin transporter promoter and BDNF Val66Met genotype on personality traits in a population representative sample of adolescents. Psychiatr Genet 21(5):261–264
Hoehe MR, Wendel B et al (1998) Serotonin transporter (5-HTT) gene polymorphisms are not associated with susceptibility to mood disorders. Am J Med Genet 81(1):1–3
Hranilovic D, Schwab SG et al (2000) Serotonin transporter gene and schizophrenia: evidence for association/linkage disequilibrium in families with affected siblings. Mol Psychiatry 5(1): 91–95
Hranilovic D, Stefulj J et al (2003) Serotonin transporter gene promoter (5-HTTLPR) and intron 2 (VNTR) polymorphisms in Croatian suicide victims. Biol Psychiatry 54(9):884–889
Hranilovic D, Stefulj J et al (2004) Serotonin transporter promoter and intron 2 polymorphisms: relationship between allelic variants and gene expression. Biol Psychiatry 55(11):1090–1094
Hu X, Oroszi G, Chun J et al (2005) An expanded evaluation of the relationship of four alleles to the level of response to alcohol and the alcoholism risk. Alcohol Clin Exp Res 29:8–16
Hu XZ, Rush AJ, Charney D et al (2007) Association between a functional serotonin transporter promoter polymorphism and citalopram treatment in adult outpatients with major depression. Arch Gen Psychiatry 64(7):783–792
Innis RB, Cunningham VJ et al (2007) Consensus nomenclature for in vivo imaging of reversibly binding radioligands. J Cereb Blood Flow Metab 27(9):1533–1539
Ishihara K, Oshimura M et al (2006) CTCF-dependent chromatin insulator is linked to epigenetic remodeling. Mol Cell 23(5):733–742
Ito K, Yoshida K et al (2002) A variable number of tandem repeats in the serotonin transporter gene does not affect the antidepressant response to fluvoxamine. Psychiatry Res 111:235–239
Jacobsen LK, Staley JK et al (2000) Elevated central serotonin transporter binding availability in acutely abstinent cocaine-dependent patients. Am J Psychiatry 157(7):1134–1140
Jeffreys AJ, Wilson V et al (1985) Hypervariable 'minisatellite' regions in human DNA. Nature 314(6006):67–73
Jernej B, Stefulj J et al (2004) Intronic polymorphism of tryptophan hydroxylase and serotonin transporter: indication for combined effect in predisposition to suicide. J Neural Transm 111(6):733–738

Joshi G, Pradhan S et al (2010) No direct association of serotonin transporter (STin2 VNTR) and receptor (HT 102T>C) gene variants in genetic susceptibility to migraine. Dis Markers 29(5): 223–229

Kaiser R, Tremblay PB et al (2001) Serotonin transporter polymorphisms: no association with response to antipsychotic treatment, but associations with the schizoparanoid and residual subtypes of schizophrenia. Mol Psychiatry 6(2):179–185

Kaiser R, Muller-Oerlinghausen B et al (2002) Correlation between serotonin uptake in human blood platelets with the 44-bp polymorphism and the 17-bp variable number of tandem repeat of the serotonin transporter. Am J Med Genet 114(3):323–328

Kalbitzer J, Frokjaer VG et al (2009) The personality trait openness is related to cerebral 5-HTT levels. Neuroimage 45(2):280–285

Kalin NH, Shelton SE et al (2008) The serotonin transporter genotype is associated with intermediate brain phenotypes that depend on the context of eliciting stressor. Mol Psychiatry 13(11):1021–1027

Katsuragi S, Kunugi H et al (1999) Association between serotonin transporter gene polymorphism and anxiety-related traits. Biol Psychiatry 45(3):368–370

Keers R, Aitchison KJ (2010) Gender differences in antidepressant drug response. Int Rev Psychiatry 22(5):485–500

Keers R, Aitchison KJ (2011) Pharmacogenetics of antidepressant response. Expert Rev Neurother 11(1):101–125

Keers R, Uher R et al (2011) Interaction between serotonin transporter gene variants and life events predicts response to antidepressants in the GENDEP project. Pharmacogenomics J 11(2):138–145

Kendler KS, Kuhn JW et al (2005) The interaction of stressful life events and a serotonin transporter polymorphism in the prediction of episodes of major depression: a replication. Arch Gen Psychiatry 62(5):529–535

Kim DK, Lim SW et al (2000) Serotonin transporter gene polymorphism and antidepressant response. Neuroreport 11:215–219

King MC, Wilson AC (1975) Evolution at two levels in humans and chimpanzees. Science 188(4184):107–116

Kinnally EL, Capitanio JP et al (2010) Epigenetic regulation of serotonin transporter expression and behavior in infant rhesus macaques. Genes, Brain Behav 9(6):575–582

Kirov G, Rees M et al (1999) Bipolar disorder and the serotonin transporter gene: a family-based association study. Psychol Med 29(5):1249–1254

Kiyohara C, Yoshimasu K (2010) Association between major depressive disorder and a functional polymorphism of the 5-hydroxytryptamine (serotonin) transporter gene: a meta-analysis. Psychiatr Genet 20(2):49–58

Klenova E, Scott AC et al (2004) YB-1 and CTCF differentially regulate the 5-HTT polymorphic intron 2 enhancer which predisposes to a variety of neurological disorders. J Neurosci 24(26):5966–5973

Koenen KC, Aiello AE et al (2009) Modification of the association between serotonin transporter genotype and risk of posttraumatic stress disorder in adults by county-level social environment. Am J Epidemiol 169(6):704–711

Kohen R, Cain KC et al (2008) Association of serotonin transporter gene polymorphisms with poststroke depression. Arch Gen Psychiatry 65(11):1296–1302

Konneker TI, Crowley JJ et al (2010) No association of the serotonin transporter polymorphisms 5-HTTLPR and RS25531 with schizophrenia or neurocognition. Am J Med Genet B Neuropsychiatr Genet 153B(5):1115–1117

Kraft JB, Slager SL et al (2005) Sequence analysis of the serotonin transporter and associations with antidepressant response. Biol Psychiatry 58(5):374–381

Kreek MJ, Bart G et al (2005a) Pharmacogenetics and human molecular genetics of opiate and cocaine addictions and their treatments. Pharmacol Rev 57(1):1–26

Kreek MJ, Nielsen DA et al (2005b) Genetic influences on impulsivity, risk taking, stress responsivity and vulnerability to drug abuse and addiction. Nat Neurosci 8(11):1450–1457

Kumar A, Choi KH et al (2005) Chromatin remodeling is a key mechanism underlying cocaine-induced plasticity in striatum. Neuron 48(2):303–314
Kunugi H, Tatsumi M et al (1996) Serotonin transporter gene polymorphism and affective disorder. Lancet 347(9011):1340
Landaas ET, Johansson S et al (2010) An international multicenter association study of the serotonin transporter gene in persistent ADHD. Genes Brain Behav 9(5):449–458
Lasky-Su JA, Faraone SV et al (2005) Meta-analysis of the association between two polymorphisms in the serotonin transporter gene and affective disorders. Am J Med Genet B Neuropsychiatr Genet 133B(1):110–115
Lazary J, Lazary A et al (2009) Promoter variants of the cannabinoid receptor 1 gene (CNR1) in interaction with 5-HTTLPR affect the anxious phenotype. Am J Med Genet 150B(8): 1118–1127
Lee Y, Lin PY (2010) Association between serotonin transporter gene polymorphism and eating disorders: a meta-analytic study. Int J Eat Disord 43(6):498–504
Lesch KP, Balling U et al (1994) Organization of the human serotonin transporter gene. J Neural Transm Gen Sect 95(2):157–162
Lesch KP, Merschdorf U (2000) Impulsivity, aggression, and serotonin: a molecular psychobiological perspective. Behav Sci Law 18(5):581–604
Lesch KP, Bengel D et al (1996) Association of anxiety-related traits with a polymorphism in the serotonin transporter gene regulatory region. Science 274(5292):1527–1531
Lesch KP, Meyer J et al (1997) The 5-HT transporter gene-linked polymorphic region (5-HTTLPR) in evolutionary perspective: alternative biallelic variation in rhesus monkeys. Rapid communication. J Neural Transm 104(11–12):1259–1266
Lesch KP, Zeng Y et al (2003) Anxiety-related traits in mice with modified genes of the serotonergic pathway. Eur J Pharmacol 480(1–3):185–204
Levine AA, Guan Z et al (2005) CREB-binding protein controls response to cocaine by acetylating histones at the fosB promoter in the mouse striatum. Proc Natl Acad Sci USA 102(52):19186–19191
Lim JE, Papp A et al (2006) Allelic expression of serotonin transporter (SERT) mRNA in human pons: lack of correlation with the polymorphism SERTLPR. Mol Psychiatry 11(7):649–662
Lin CW, Tang W et al (2009) Haplotype analysis confirms association of the serotonin transporter (5-HTT) gene with schizophrenia in the Han Chinese population. Neurosci Lett 453(3):210–213
Little KY, McLaughlin DP et al (1998) Cocaine, ethanol, and genotype effects on human midbrain serotonin transporter binding sites and mRNA levels. Am J Psychiatry 155(2): 207–213
Liu W, Gu N et al (1999) Tentative association of the serotonin transporter with schizophrenia and unipolar depression but not with bipolar disorder in Han Chinese. Pharmacogenetics 9(4):491–495
Lo H, Wang SZ et al (2003) Allelic variation in gene expression is common in the human genome. Genome Res 13(8):1855–1862
Lorenzi C, Marcone A et al (2010) Serotonin transporter and saitohin genes in risk of Alzheimer's disease and frontotemporal lobar dementia: preliminary findings. Neurol Sci 31(6):741–749
Lotrich FE, Pollock BG (2004) Meta-analysis of serotonin transporter polymorphisms and affective disorders. Psychiatric Genet 14(3):121–129
Lovejoy EA, Scott AC et al (2003) The serotonin transporter intronic VNTR enhancer correlated with a predisposition to affective disorders has distinct regulatory elements within the domain based on the primary DNA sequence of the repeat unit. Eur J Neurosci 17(2):417–420
MacKenzie A, Quinn J (1999) A serotonin transporter gene intron 2 polymorphic region, correlated with affective disorders, has allele-dependent differential enhancer-like properties in the mouse embryo. Proc Natl Acad Sci USA 96(26):15251–15255
Malhotra AK, Goldman D et al (1998) A functional serotonin transporter (5-HTT) polymorphism is associated with psychosis in neuroleptic-free schizophrenics. Mol Psychiatry 3(4):328–332

Mann JJ, Huang YY et al (2000) A serotonin transporter gene promoter polymorphism (5-HTTLPR) and prefrontal cortical binding in major depression and suicide. Arch Gen Psychiatry 57(8): 729–738
Mansour HE, Talkowski ME et al (2005) Serotonin gene polymorphisms and bipolar I disorder: focus on the serotonin transporter. Ann Med 37(8):590–602
Martin J, Cleak J et al (2007) Mapping regulatory variants for the serotonin transporter gene based on allelic expression imbalance. Mol Psychiatry 12(5):421–422
Martinez D, Slifstein MR et al (2009) Positron emission tomography imaging of the serotonin transporter and 5-HT1A receptor in alcohol dependence. Biol Psychiatry 65(2):175–180
Maze I, Covington JH 3rd et al (2010) Essential role of the histone methyltransferase G9a in cocaine-induced plasticity. Science 327(5962):213–216
Mazzanti CM, Lappalainen J et al (1998) Role of the serotonin transporter promoter polymorphism in anxiety-related traits. Arch Gen Psychiatry 55(10):936–940
McHugh RK, Hofmann SJ et al (2010) The serotonin transporter gene and risk for alcohol dependence: a meta-analytic review. Drug Alcohol Depend 108(1–2):1–6
McLaughlin DP, Little KY et al (1996) Expression of serotonin transporter mRNA in human brainstem raphe nuclei. Neuropsychopharmacology 15(5):523–529
Meira-Lima I, Michelon L et al (2005) Allelic association analysis of the functional insertion/deletion polymorphism in the promoter region of the serotonin transporter gene in bipolar affective disorder. J Mol Med 27(2):219–224
Melke J, Landen M et al (2001) Serotonin transporter gene polymorphisms are associated with anxiety-related personality traits in women. Am J Med Genet 106(1):458–463
Mendlewicz J, Massat I et al (2004) Serotonin transporter 5HTTLPR polymorphism and affective disorders: no evidence of association in a large European multicenter study. Eur J Hum Genet 12(5):377–382
Merenakk L, Maestu J et al (2011) Effects of the serotonin transporter (5-HTTLPR) and α(2A)-adrenoceptor (C-1291G) genotypes on substance use in children and adolescents: a longitudinal study. Psychopharmacology (Berl) 215(1):13–22
Middledorp CM, de Geus EJC et al (2007) Family based association analyses between the serotonin transporter gene polymorphism (5-HTTLPR) and neuroticism, anxiety and depression. Behav Genet 37(2):294–301
Mileva-Seitz V, Kennedy J et al (2011) Serotonin transporter allelic variation in mothers predicts maternal sensitivity, behavior and attitudes toward 6-month-old infants. Genes Brain Behav 10(3):325–333
Min W, Li T et al (2009) Monoamine transporter gene polymorphisms affect susceptibility to depression and predict antidepressant response. Psychopharmacology (Berl) 205:409–417
Miyajima F, Quinn JP et al (2008) Additive effect of BDNF and REST polymorphisms is associated with improved general cognitive ability. Genes Brain Behav 7(7):714–719
Mortensen OV, Kristensen AS et al (1999a) Molecular cloning, expression and characterization of a bovine serotonin transporter. Brain Res Mol Brain Res 71(1):120–126
Mortensen OV, Thomassen M et al (1999b) Functional analysis of a novel human serotonin transporter gene promoter in immortalized raphe cells. Brain Res Mol Brain Res 68(1–2):141–148
Mrazek DA, Rush AJ et al (2009) SLC6A4 variation and citalopram response. Am J Med Genet B Neuropsychiatr Genet 150B:341–351
Murphy DL, Li Q et al (2001) Genetic perspectives on the serotonin transporter. Brain Res Bull 56(5):487–494
Myung W, Lim SW et al (2010) Serotonin transporter gene polymorphisms and chronic illness of depression. J Korean Med Sci 25(12):1824–1827
Nakamura Y, Koyama K et al (1998) VNTR (variable number of tandem repeat) sequences as transcriptional, translational, or functional regulators. J Hum Genet 43(3):149–152
Nakamura M, Ueno S et al (2000) The human serotonin transporter gene linked polymorphism (5-HTTLPR) shows ten novel allelic variants. Mol Psychiatry 5(1):32–38
Naslund K, Saetre P et al (2005) Genome-wide prediction of human VNTRs. Genomics 85(1):24–35

Nobile M, Begni B et al (1999) Effects of serotonin transporter promoter genotype on platelet serotonin transporter functionality in depressed children and adolescents. J Am Acad Child Adolesc Psychiatry 38(11):1396–1402

Ogilvie AD, Battersby S et al (1996) Polymorphism in serotonin transporter gene associated with susceptibility to major depression. Lancet 347(9003):731–733

Ogilvie AD, Russell MB et al (1998) Altered allelic distributions of the serotonin transporter gene in migraine without aura and migraine with aura. Cephalalgia 18(1):23–26

Ohara K, Suzuki Y et al (1999) A variable-number-tandem-repeat of the serotonin transporter gene and anxiety disorders. Prog Neuropsychopharmacol Biol Psychiatry 23(1):55–65

Olsson CA, Foley DL et al (2010) Prospects for epigenetic research within cohort studies of psychological disorder: a pilot investigation of a peripheral cell marker of epigenetic risk for depression. Biol Psychol 83(2):159–165

Ospina-Duque J, Duque C et al (2000) An association study of bipolar mood disorder (type I) with the 5-HTTLPR serotonin transporter polymorphism in a human population isolate from Colombia. Neurosci Lett 292(3):199–202

Pan JY, Ma P et al (2009) Association of serotonin transporter gene polymorphisms and major depressive disorder in Chinese Han population. Neural Regen Res 4(8):635–640

Parsey RV, Hastings RS et al (2006a) Lower serotonin transporter binding potential in the human brain during major depressive episodes. Am J Psychiatry 163(1):52–58

Parsey RV, Hastings RS et al (2006b) Effect of a triallelic functional polymorphism of the serotonin-transporter-linked promoter region on expression of serotonin transporter in the human brain. Am J Psychiatry 163(1):48–51

Peters EJ, Slager SL et al (2004) Investigation of serotoninrelated genes in antidepressant response. Mol Psychiatry 9:879–889

Pinto E, Reggers J et al (2008) The short allele of the serotonin transporter promoter polymorphism influences relapse in alcohol dependence. Alcohol Alcohol 43(4):398–400

Pirker W, Asenbaum S et al (2000) Imaging serotonin and dopamine transporters with 123I-beta-CIT SPECT: binding kinetics and effects of normal aging. J Nucl Med 41(1):36–44

Pluess M, Velders FP et al (2011) Serotonin transporter polymorphism moderates effects of prenatal maternal anxiety on infant negative emotionality. Biol Psychiatry 69(6):520–525

Porcelli S, Drago A et al (2011) Pharmacogenetics of antidepressant response. J Psychiatry Neurosci 36(2):87–113

Praschak-Rieder N, Kennedy J et al (2007) Novel 5-HTTLPR allele associates with higher serotonin transporter binding in putamen: a [(11)C] DASB positron emission tomography study. Biol Psychiatry 62(4):327–331

Praschak-Rieder N, Willeit M et al (2008) Seasonal variation in human brain serotonin transporter binding. Arch Gen Psychiatry 65(9):1072–1078

Rao D, Jonsson EG et al (1998) Schizophrenia and the serotonin transporter gene. Psychiatr Genet 8(4):207–212

Rees M, Norton M et al (1997) Association studies of bipolar disorder at the human serotonin transporter gene (hSERT; 5HTT). Mol Psychiatry 2(5):398–402

Reese J, Kraschewski A et al (2010) Haplotypes of dopamine and serotonin transporter genes are associated with antisocial personality disorder in alcoholics. Psychiatr Genet 20(4):140–152

Reimold M, Smolka MN et al (2007) Midbrain serotonin transporter binding potential measured with [11C]DASB is affected by serotonin transporter genotype. J Neural Transm 114(5):635–639

Renthal W, Maze I et al (2007) Histone deacetylase 5 epigenetically controls behavioral adaptations to chronic emotional stimuli. Neuron 56(3):517–529

Ressler KJ, Nemeroff CB (2000) Role of serotonergic and noradrenergic systems in the pathophysiology of depression and anxiety disorders. Depression Anxiety 12(1):2–19

Ressler KJ, Bradley B et al (2010) Polymorphisms in CRHR1 and the serotonin transporter loci: gene × gene × environment interactions on depressive symptoms. Am J Med Genet B Neuropsychiatr Genet 153B(3):812–824

Roberts J, Scott AC et al (2007) Differential regulation of the serotonin transporter gene by lithium is mediated by transcription factors, CCCTC binding protein and Y-box binding

protein 1, through the polymorphic intron 2 variable number tandem repeat. J Neurosci 27(11): 2793–2801
Ruhe HG, Booij J et al (2009) Serotonin transporter binding with [123I]beta-CIT SPECT in major depressive disorder versus controls: effect of season and gender. Eur J Nucl Med Mol Imaging 36(5):841–849
Sakai K, Nakamura M et al (2002) The silencer activity of the novel human serotonin transporter linked polymorphic regions. Neurosci Lett 327(1):13–16
Schloesser RJ, Manji HK et al (2009) Suppression of adult neurogenesis leads to an increased hypothalamo-pituitary-adrenal axis response. Neuroreport 20(6):553–557
Schurks M, Rist PM et al (2010a) 5-HTTLPR polymorphism in the serotonin transporter gene and migraine: a systematic review and meta-analysis. Cephalalgia 30(11):1296–1305
Schurks M, Rist PM et al (2010b) STin2 VNTR polymorphism in the serotonin transporter gene and migraine: pooled and meta-analyses. J Headache Pain 11(4):317–326
Selvaraj S, Murthy NV et al (2011) Diminished brain 5-HT transporter binding in major depression: a positron emission tomography study with [11C]DASB. Psychopharmacology (Berl) 213(2–3):555–562
Serretti A, Lilli R et al (2002) Serotonin transporter gene (5-HTTLPR) and major psychoses. Mol Psychiatry 7(1):95–99
Serretti A, Kato M et al (2007) Meta-analysis of serotonin transporter gene promoter polymorphism (5-HTTLPR) association with selective serotonin reuptake inhibitor efficacy in depressed patients. Mol Psychiatry 12(3):247–257
Shah MP, Wang F et al (2009) Role of variation in the serotonin transporter protein gene (SLC6A4) in trait disturbances in the ventral anterior cingulate in bipolar disorder. Neuropsychopharmacology 34(5):1301–1310
Shioe K, Ichimiya T et al (2003) No association between genotype of the promoter region of serotonin transporter gene and serotonin transporter binding in human brain measured by PET. Synapse 48(4):184–188
Smeraldi E, Serretti A et al (2006) Serotonin transporter gene linked polymorphic region: possible pharmacogenetic implications of rare variants. Psychiatr Genet 16:153–158
Smits KM, Smits LJ et al (2004) Influence of SERTPR and STin2 in the serotonin transporter gene on the effect of selective serotonin reuptake inhibitors in depression: a systematic review. Mol Psychiatry 9(5):433–441
Smits K, Smits L et al (2007) Serotonin transporter polymorphisms and the occurrence of adverse events during treatment with selective serotonin reuptake inhibitors. Int Clin Psychopharmacol 22:137–143
Soeby K, Larsen SA et al (2005) Serotonin transporter: evolution and impact of polymorphic transcriptional regulation. Am J Med Genet B Neuropsychiatr Genet 136B(1):53–57
Stober G, Heils A et al (1996) Serotonin transporter gene polymorphism and affective disorder. Lancet 347(9011):1340–1341
Surguladze SA, Elkin A et al (2008) Genetic variation in the serotonin transporter modulates neural system-wide response to fearful faces. Genes Brain Behav 7(5):543–551
Taylor WD, Steffens DC et al (2005) Influence of serotonin transporter promoter region polymorphisms on hippocampal volumes in late-life depression. Arch Gen Psychiatry 62(5):537–544
Taylor MJ, Sen S et al (2010) Antidepressant response and the serotonin transporter gene-linked polymorphic region. Biol Psychiatry 68(6):536–43
Tencomnao T, Wongpiyabovorn J (2010) An investigation of the relationship between serotonin transporter gene promoter polymorphism and psoriasis susceptibility in a Thai population. Genet Mol Res 9(4):2275–2282
Tibrewal P, Kumar HB et al (2010) Association of serotonin transporter gene polymorphisms with obsessive-compulsive disorder (OCD) in a south Indian population. Indian J Med Res 132(6):690–695
Tsankova NM, Kumar A et al (2004) Histone modifications at gene promoter regions in rat hippocampus after acute and chronic electroconvulsive seizures. J Neurosci 24(24):5603–5610

Uher R, McGuffin P (2008) The moderation by the serotonin transporter gene of environmental adversity in the aetiology of mental illness: review and methodological analysis. Mol Psychiatry 13(2):131–146

Uhl GR, Hall FS et al (2002) Cocaine, reward, movement and monoamine transporters. Mol Psychiatry 7(1):21–26

Van der Zwaluw CS, Engels RC et al (2010) A serotonin transporter polymorphism (5-HTTLPR) predicts the development of adolescent alcohol use. Drug Alcohol Dependance 112(1–2): 134–139

Van Dyck CH, Malison RT et al (2004) Central serotonin transporter availability measured with [123I]beta-CIT SPECT in relation to serotonin transporter genotype. Am J Psychiatry 161(3):525–531

Van Ijzendoorn MH, Caspers K et al (2010) Methylation matters: interaction between methylation density and serotonin transporter genotype predicts unresolved loss or trauma. Biol Psychiatry 68(5):405–407

Vasiliou SA, Ali FR et al (2011) The SLC6A4 VNTR genotype determines transcription factor binding and epigenetic variation of this gene in response to cocaine in vitro. Addict Biol 17(1):156–170

Wang XJ, Zhong SR et al (2011) Association of polymorphism in the serotonin transporter gene promote with the susceptibility to alcohol dependence in Yunnan Han population. Yi Chuan 33(1):48–53

Wellman CL, Izquierdo A et al (2007) Impaired stress-coping and fear extinction and abnormal corticolimbic morphology in serotonin transporter knock-out mice. J Neurosci 27(3):684–691

Wendland JR, Martin BJ et al (2006) Simultaneous genotyping of four functional loci of human SLC6A4, with a reappraisal of 5-HTTLPR and rs25531. Mol Psychiatry 11:224–6

Wendland JR, Kruse MR et al (2007) A large case–control study of common functional SLC6A4 and BDNF variants in obsessive compulsive disorder. Neuropsychopharmacology 32:2543–2551

Wells TT, Beevers CG et al (2010) Serotonin transporter and BDNF genetic variants interact to predict cognitive reactivity in healthy adults. J Affect Disord 126(1–2):223–229

Willeit M, Stastny J et al (2001) No evidence for in vivo regulation of midbrain serotonin transporter availability by serotonin transporter promoter gene polymorphism. Biol Psychiatry 50(1):8–12

Xie P, Kranzler HR et al (2009) Interactive effect of stressful life events and the serotonin transporter 5-HTTLPR genotype on posttraumatic stress disorder diagnosis in 2 independent populations. Arch Gen Psychiatry 66(11):1201–1209

Yan QS (2002) Reduced serotonin release and serotonin uptake sites in the rat nucleus accumbens and striatum after prenatal cocaine exposure. Brain Res 929(1):59–69

Yang B, Chen RC et al (2007) A meta-analysis of association studies between the 10-repeat allele of a VNTR polymorphism in the 3′-UTR of dopamine transporter gene and attention deficit hyperactivity disorder. Am J Med Genet B Neuropsychiatr Genet 144B(4):541–550

Zaboli G, Joensson EG et al (2008) Haplotype analysis confirms association of the serotonin transporter (5-HTT) gene with schizophrenia but not with major depression. Am J Med Genet B Neuropsychiatr Genet 147B(3):301–307

Zalsman G, Huang YY et al (2006) Association of a triallelic serotonin transporter gene promoter region (5-HTTLPR) polymorphism with stressful life events and severity of depression. Am J Psychiatry 163(9):1588–1593

Zamorano R, Suchindran S et al (2006) 3′-untranslated region of the type 2 bradykinin receptor is a potent regulator of gene expression. Am J Physiol Renal Physiol 290(2):F456–F464

Genetic Factors Modulating the Response to Stimulant Drugs in Humans

Amy B. Hart, Harriet de Wit and Abraham A. Palmer

Abstract Individuals vary in their responses to stimulant drugs, and several lines of evidence suggest that the basis for this variation is at least partially genetic in origin. Association studies have examined the effects of polymorphisms in specific genes on acute and chronic responses to stimulant drugs. Several of these genetic polymorphisms are also associated with other psychiatric dimensions and disorders. This chapter examines the evidence for genetic associations between the genes that have been most carefully examined for their influence on the response to stimulant drugs.

Keywords Stimulants · Inter-individual variation · Drug response · Candidate gene · Genetic association · Genetic polymorphism

Contents

1 Introduction 538
2 Twin Studies of Stimulant Drug Phenotypes 539
3 Association Studies 540
4 Monoamine Transporters 541
5 Neurotransmitter Receptors 558

A. B. Hart · A. A. Palmer (✉)
Department of Human Genetics, University of Chicago,
920 E 58th St. CLSC-507D, Chicago, IL 60637, USA
e-mail: aap@uchicago.edu

H. de Wit · A. A. Palmer
Department of Psychiatry and Behavioral Neuroscience,
University of Chicago, Chicago, IL 60637, USA

Curr Topics Behav Neurosci (2012) 12: 537–577

DOI: 10.1007/7854_2011_187

Published Online: 20 January 2012

6 Biosynthetic Enzymes 561
7 Miscellaneous 564
8 Exploratory Studies 566
9 Genome-Wide Association Studies 566
10 Closing Remarks 566
References 567

1 Introduction

A growing body of evidence suggests that genetic variation underlies inter-individual variability in the response to drugs, including stimulant drugs. Stimulant drugs produce behavioral and subjective effects that include increased vigilance and attention, and feelings of energy and euphoria (Lamb and Henningfield 1994; Martin et al. 1971; Sevak et al. 2009). However, there is substantial individual variation in these responses (Brown et al. 1978; de Wit et al. 1986; Holdstock and de Wit 2001), and some of this variation has been shown to be heritable (Crabbe et al. 1983; Nurnberger et al. 1982). Genetic variation in the response to stimulants may contribute to the potential to develop drug abuse (Fergusson et al. 2003; Haertzen et al. 1983), and may also be relevant for therapeutic uses of stimulants.

In this review we will focus on five prototypic stimulant drugs that have been studied most intensively: amphetamine, methamphetamine, methylphenidate, cocaine, and bupropion. All of these drugs inhibit the reuptake of dopamine, norepinephrine, and serotonin, thereby increasing the levels of these monoamine neurotransmitters in the synaptic cleft. These drugs differ in their potency, synaptic actions, degree of reuptake blockade, pharmacokinetic properties and their specificity and actions on other neurotransmitter systems. Some of the drugs (e.g., amphetamine) also cause the reuptake transporters to work in reverse, leading to non-impulse dependent release of neurotransmitters. Additionally, some stimulants inhibit the monoamine oxidase enzymes (MAO-A and MAO-B), thus preventing the degradation of monoamines (Seiden et al. 1993). Finally, there is evidence that stimulant drugs disrupt the function of the vesicular monoamine transporter type 2 (VMAT2), which transports monoamine neurotransmitters from stores in the cytoplasm to synaptic vesicles (Uhl et al. 2000). Genetic variation in these genes as well as in their up- or down-stream neighbors make them likely candidates to contribute to inter-individual variation in the response to stimulant drugs.

We will focus on two types of genetic studies in this review, twin and association studies. Twin studies estimate the heritability of traits, whereas association studies examine specific polymorphisms in relation to a phenotype. We will not include family-based linkage studies of drug abuse or animal studies on genetic determinants of stimulant effects, both which have been carefully reviewed elsewhere (Phillips et al. 2008; Kreek et al. 2005; Uhl 2006). We will also not discuss the stimulant caffeine, because it acts by distinct neurochemical mechanisms and because we have recently reviewed the genetics of caffeine elsewhere

(Yang et al. 2010). We will discuss two types of association studies: candidate gene studies and genome-wide association studies. Candidate gene studies draw on prior pharmacological knowledge of how stimulants affect the brain to select 'candidate' genes that are likely to be the source of genetic differences. In these studies, polymorphisms within or near the candidate gene are tested to see if they are statistically associated with relevant phenotypes. Phenotypes might include measures of acute response, patterns of drug use or therapeutic response in patients, or clinical diagnoses of drug dependence or abuse. Other studies use a genome-wide association (GWAS) approach to examine many or most of the common polymorphisms in the genome; these studies do not depend on prior hypotheses about which genes might be important.

In all of the studies discussed: twin, candidate gene association, and genome-wide association, we will address two types of genetic polymorphisms: single nucleotide polymorphisms (SNPs) and variable number tandem repeats (VNTRs). SNPs are sites at which a single nucleotide differs among individuals within a population. SNPs can either occur in the coding sequence of a gene and thus alter amino acid sequence (termed non-synonymous) or, more commonly, they may be outside the coding sequence and alter gene regulation. In both cases, a SNP may alter a biological function itself (coding or gene expression), or be linked to another polymorphism that is functionally significant. VNTRs are polymorphisms in which a variable number of short repetitive sequences (tandem repeats) are present at a given locus; as with SNPs they may have direct functional consequences or may be linked to some other functionally significant polymorphism.

We will attempt to synthesize genetic studies of acute, sub-chronic, and chronic administration of stimulant drugs in this review. First, we will discuss overall heritability of stimulant drug-related phenotypes based on twin studies. Next, we will discuss candidate gene and genome-wide association studies that implicate specific genes in modulating responses. Last, we will highlight the key conclusions and identify future directions for study.

2 Twin Studies of Stimulant Drug Phenotypes

Two early twin studies provide strong evidence for the heritability of acute responses to stimulant drugs (Nurnberger et al. 1982; Crabbe et al. 1983). Twin studies estimate heritability by comparing the concordance rate between monozygotic twins, who share a familial environment and all genes, to dizygotic twins, who share the same environment but only half of their genes. Typically, biometric modeling is used to explain variability due to genetic or environmental effects. Heritability estimates can range from 0 (no variation contributed by genetic sources) to 1 (all variation contributed by genetic sources). Nurnberger et al. (1982) administered *d*-amphetamine intravenously (0.3 mg/kg) to 13 pairs of monozygotic twins and 3 pairs of dizygotic twins and measured physiological and subjective effects of the drug. Responses to the drug in monozygotic twins were highly concordant for a subjective measure of excitation, as well as growth hormone and prolactin release, suggesting a large genetic component underlying these traits. Consistent with this, Crabbe et al. (1983) administered 10 mg *d*-amphetamine sulfate to six pairs of monozygotic twins and found less variation in physiological and subjective responses within pairs than between pairs, which suggests these traits are heritable.

More recently, twin studies have examined the heritability of lifetime stimulant use, dependence, and abuse using liability threshold model fitting (Kendler et al. 1999, 2003, 2000). In their most recent study, Kendler et al. (2005) estimated heritability for lifetime use of stimulant drugs excluding cocaine to be 0.42, and heritability for lifetime use of cocaine to be 0.70. They also found substantial familial environmental contributions to the variance (i.e., 0.20) for lifetime use of other stimulant drugs, but not for cocaine use. Specific environmental effects, which are distinguished from shared familial environmental effects, were estimated at 0.38 for other stimulant use and 0.30 for cocaine use. In addition to this study, a study of male twin war veterans yielded similar results (Tsuang et al. 1996). That study estimated the heritability of stimulant abuse (including cocaine abuse) based on *DSM-III-R* criteria (American Psychiatric Association 1987) to be 0.44, but with 0.49 of the variance contributed by specific environmental effects, and no contribution of familial environment in the best-fit model. Although this was a slightly higher estimate of heritability than by Kendler et al. (2005), both studies suggest a fairly large contribution of genetic factors to heritability of stimulant drug abuse.

3 Association Studies

Genetic determinants of response to acute, sub-chronic, and chronic administration of stimulant drugs have been examined using association studies in several different populations. We have conducted a series of association studies of the acute responses to amphetamine (e.g., Lott et al. 2005; Dlugos et al. 2010). In this double-blind, placebo-controlled study, healthy young adults (ranging from $n = 99$ to $n = 152$, depending on the analysis) received oral doses of *d*-amphetamine (10 and 20 mg) and placebo over the course of three different sessions. Measures of subjective, cardiovascular and behavioral responses were obtained at regular intervals. The subjective measures included the Profile of Mood States (POMS; McNair et al. 1971), Drug Effects Questionnaire (DEQ; Johanson and Uhlenhuth 1980) and Addiction Resource Center Inventory (ARCI; Martin et al. 1971). In children diagnosed with ADHD, methylphenidate is typically administered daily or several times a day in a therapeutic context, and the outcome measure is usually therapeutic response as measured by the Clinical Global Impression-Severity scale (CGI-S; Guy 1976) and the ADHD Rating Scale-IV (ARS; DuPaul et al. 1998). In both studies of healthy adults and studies of children diagnosed with ADHD, neuroimaging has also been used to examine variation in brain responses to acute amphetamine and methylphenidate administration. These include techniques such as single photon emission computerized tomography (SPECT), functional magnetic resonance imaging (fMRI) and positron emission tomography (PET) to assess the binding of specific ligands (Warwick 2004; Cabeza and Nyberg 2000).

Sub-chronic (i.e., <8 weeks) or chronic (>8 weeks) stimulant responses have been measured in patients receiving methylphenidate for ADHD or bupropion for smoking cessation. Other studies have examined the phenotypes of dependence, abuse, and consequences of abuse such as drug-induced psychosis. Although it is not possible to accurately determine the doses in studies of active drug abusers, it is safe to assume that they involve higher doses than are found in acute or therapeutic studies. In addition, drug users commonly self-administer drugs via injection, inhalation (smoking) or intranasal routes, where the acute laboratory studies and therapeutic studies most commonly involve oral administration.

Studies of responses to stimulant drugs have focused on a small number of candidate genes that have also been associated with other psychiatric phenotypes. Many of these genes are the direct targets of stimulant drugs or other known psychoactive agents. Table 1 lists these studies and they are also discussed in the following sections. Genes that have been less well studied both in relation to stimulant drugs and other psychiatric phenotypes are only briefly mentioned in the "Exploratory Studies" section and summarized in Table 2.

4 Monoamine Transporters

SLC6A3

The dopamine transporter (*SLC6A3*; *DAT1; DAT)* is a direct target of stimulant drugs and thus a logical candidate gene for studies or stimulant sensitivity. Polymorphisms in this gene have been extensively studied in relation to stimulant drug responses and susceptibility to ADHD (reviewed in Banaschewski et al. 2010). In particular, attention has focused on a 40 bp variable nucleotide tandem repeat (VNTR) located in the 3′-untranslated region (UTR) of the gene, with a range of 3–11 repeats. The two most common alleles have either 9, or more commonly, 10 repeats (Vandenbergh et al. 1992). The 10-repeat allele has been associated with both higher and lower expression of the dopamine transporter in the brain (Fuke et al. 2001; van de Giessen et al. 2009; Van Dyck et al. 2005). The evidence described below suggests that polymorphisms in *SLC6A3* play an important role in responses to stimulant drugs, although the direction of the association is inconsistent across studies.

In analyses examining associations of amphetamine response with the *SLC6A3* 3′-UTR VNTR, we found that individuals homozygous for the 9-repeat allele showed decreased responses to amphetamine at 20 mg, compared to heterozygous (9/10) and homozygous (10/10) individuals (Lott et al. 2005). Using a larger

Table 1 Candidate genes associated with responses to stimulant drugs and associated SNPs

Strong Candidate Genes Gene name	Locus symbol	Administration type	Study	Phenotype	Number	Ethnicity	Group	Dose	Associated polymorphisms	Notes
Dopamine transporter	*SLC6A3; DAT1; DAT*	Chronic	Bellgrove et al. (2005)	Methylphenidate therapeutic response	43	Irish	Children with ADHD	–	3′-UTR VNTR	10-repeat allele associated with better treatment response
		Acute	Cheon et al. (2005)	Methylphenidate therapeutic response	11	Korean	Children with ADHD	Up to 0.7 mg/kg/day	3′-UTR VNTR	10/10 associated with poor treatment response
		Chronic	Gelernter et al. (1994)	Cocaine-induced paranoia	58	Caucasian, African–American	Cocaine users	–	3′-UTR VNTR	9-repeat allele associated with increased risk
		Acute	Hamidovic et al. (2010a)	Amphetamine subjective response	152	Caucasian	Healthy volunteers	Placebo, 10 mg, 20 mg	rs460000	C/C increased Euphoria and Stimulation at 10 and 20 mg
		Acute	Kooij et al. (2008)	Methylphenidate therapeutic response	42	–	Adults with ADHD	Placebo, 0.5 mg/kg/day by week 1, 0.75 mg/kg/day by week 2, up to 1.0 mg/kg/day by week 3	3′-UTR VNTR	10/10 associated with poor response

(continued)

Table 1 (continued)

Strong Candidate Genes Gene name	Locus symbol	Administration type	Study	Phenotype	Number	Ethnicity	Group	Dose	Associated polymorphisms	Notes
		Sub-chronic	Joober et al. (2007)	Methylphenidate therapeutic response	159	–	Children with ADHD	Placebo, 0.5 mg/kg/day	3′-UTR VNTR	9/9 associated with poor treatment response
		Acute	Lott et al. (2005)	Amphetamine subjective response	101	67 Caucasian, 20 African–American, 12 Asian, 2 unknown	Healthy volunteers	Placebo, 10 mg, 20 mg	3′-UTR VNTR	9/9 associated with poor response at 20 mg
		Acute	Purper-Ouakil et al. (2008)	Methylphenidate therapeutic response	141	132 Caucasian; 8 Other	Children with ADHD	10–60 mg/day	3′-UTR VNTR	10/10 associated with poor treatment response
		Acute	Rohde et al. (2003)	Methylphenidate therapeutic response	8	–	Children with ADHD (male)	0.35–0.7 mg/kg	3′-UTR VNTR	10/10 associated with higher cerebral blood flow
		Sub-chronic	Stein et al. (2005)	Methylphenidate therapeutic response	47	–	Children with ADHD	Placebo, 18 mg, 36 mg, 54 mg	3′-UTR VNTR	9/9 associated with poor response
		Chronic	Ujike et al. (2003)	Methamphetamine-induced psychosis	124 cases 160 controls	Japanese	Methamphetamine dependence/psychosis patients	–	3′-UTR VNTR	Possession of allele with 8 = 9 or fewer repeats associated with prolonged psychosis

(continued)

Table 1 (continued)

Strong Candidate Genes Gene name	Locus symbol	Administration type	Study	Phenotype	Number	Ethnicity	Group	Dose	Associated polymorphisms	Notes
Serotonin transporter	*SLC6A4; 5-HTT*	Acute	Lott et al. (2006)	Amphetamine subjective response	101	67 Caucasian, 20 African–American, 12 Asian, 2 unknown	Healthy volunteers	Placebo, 10 mg, 20 mg	5-HTTLPR & Intron 2 VNTR haplotype	L/L and 12/12 genotype combination associated with weak subjective response
		Chronic	Ezaki et al. (2008)	Methamphetamine-induced psychosis	166 cases 197 controls	Japanese	Methamphetamine-induced psychosis patients	–	5-HTTLPR	Frequency of S allele higher in patients with prolonged psychosis
		Sub-chronic	Thakur et al. (2010)	Methylphenidate therapeutic response	157	–	Children with ADHD	Placebo, 0.5 mg/kg/day	5-HTTLPR triallelic	L_G/L_G associated with best treatment response

(continued)

Table 1 (continued)

Strong Candidate Genes Gene name	Locus symbol	Administration type	Study	Phenotype	Number	Ethnicity	Group	Dose	Associated polymorphisms	Notes
Norepinephrine transporter	*SLC6A2; NET*	Acute	Dlugos et al. (2007)	Amphetamine subjective response	99	65 Caucasian, 20 African–American, 12 Asian, 2 unknown	Healthy volunteers	Placebo, 10 mg, 20 mg	rs36017, rs47958 rs36017-rs2270935-rs47958 haplotype	rs36017 C/C, G–C–C haplotype associated with increased positive mood; rs47958 C/C associated with increased elation
		Acute	Dlugos et al. (2009a)	Amphetamine subjective response	159	Caucasian	Healthy volunteers	Placebo, 10 mg, 20 mg	rs36017, rs1861647 rs36017-rs10521329-rs3785155 haplotype	rs36017 C/C, rs1861647 A/A associated with increased elation and vigor; C–C–G haplotype associated with increased vigor

(continued)

Table 1 (continued)

Strong Candidate Genes Gene name	Locus symbol	Administration type	Study	Phenotype	Number	Ethnicity	Group	Dose	Associated polymorphisms	Notes
Dopamine D2 receptor	*DRD2*	Sub-chronic	David et al. (2007)	Bupropion therapeutic response	722	Caucasian	Smokers	Placebo, 150 mg/day first 3 days, then 300 mg/day	Taq1A (rs1800497; T/C)	A2/A2 3x more likely to abstain from smoking after treatment; no effect for A1/A1, A1/A2
		Acute	Hamidovic et al. (2009)	Impulsivity amphetamine response	93	Caucasian	Healthy volunteers	Placebo, 10 mg, 20 mg	rs12364283	rs12364283 A/G, G/G increased reaction time at 10 mg
		Chronic	Ujike et al. (2009)	Methamphetamine-induced psychosis	202 cases 243 controls	Japanese	Methamphetamine dependence patients	–	Taq1A (rs1800497; T/C) *DRD2*-141 Ins/Del	A2/A2, A1/A2 associated with prolonged methamphetamine-induced psychosis Del/Del, Ins/Del associated with rapid onset

(continued)

Table 1 (continued)

Strong Candidate Genes Gene name	Locus symbol	Administration type	Study	Phenotype	Number	Ethnicity	Group	Dose	Associated polymorphisms	Notes
Dopamine D4 receptor	*DRD4*	Acute	Lee et al. (unpublished data)	Amphetamine subjective response	100	66 Caucasian, 20 African–American, 12 Asian, 2 unknown	Healthy volunteers	Placebo, 10 mg, 20 mg	Exon III VNTR	Single 7-repeat allele associated with increased elation, friendliness, heart rate, decreased dysphoria and anxiety at 10 and 20 mg
		Chronic	Chen et al. (2004a)	Methamphetamine abuse	416 cases 435 controls	Taiwanese	Methamphetamine abusers	–	Exon III VNTR	Higher frequency of 6-repeat allele in cases
		Chronic	Hamarman et al. (2004)	Methylphenidate therapeutic response	45	–	Children with ADHD	–	Exon III VNTR	Subjects with 7-repeat allele require higher doses of methylphenidate
		Chronic	Li et al. (2004)	Methamphetamine abuse	416 cases 435 controls	Taiwanese	Methamphetamine abusers	–	Promoter VNTR + exon III VNTR haplotype	Higher frequency of 7-repeat allele in cases
μ-opioid receptor	*OPRM1*	Acute	Dlugos et al. (2010)	Amphetamine subjective response	162	Caucasian	Healthy volunteers	Placebo, 10 mg, 20 mg	rs510769, rs2281617	rs510769 G/G A/G, rs2281617 C/C associated increased euphoria and energy at 10 mg
		Chronic	Ide et al. (2004)	Methamphetaime psychosis	138 cases 213 controls	Japanese	Methamphetamine dependence/ psychosis patients	–	A118G (rs1799971)	G/G associated with psychosis within 3 years of first use
		Chronic	Ide et al. (2006)	Methamphetamine abuse/ psychosis	128 cases 232 controls	Japanese	Methamphetamine dependence/ psychosis patients	–	IVS2 + G691C (rs2075572)	G/G associated with methamphetamine abuse and psychosis

(continued)

Table 1 (continued)

Strong Candidate Genes Gene name	Locus symbol	Administration type	Study	Phenotype	Number	Ethnicity	Group	Dose	Associated polymorphisms	Notes
Adenosine A_{2A} receptor	*ADORA2A*	Acute	Hohoff et al. (2005)	Amphetamine subjective response	99	65 Caucasian, 20 African–American, 12 Asian, 2 unknown	Healthy volunteers	Placebo, 10 mg, 20 mg	1976C/T (rs5751876) 2592C/T ins (rs3032740)	rs5751876 & rs3032740 associated with increased anxiety at 10 & 20 mg
Catechol O-methyl transferase	*COMT*	Acute	Hamidovic et al. (2010b)	Cognitive response to amphetamine	161	Caucasian	Healthy volunteers	Placebo, 10 mg, 20 mg	val158met (rs4680; G/A)	Val/Val, Val/Met performance improves with amphetamine; Met/Met no change
		Chronic	Li et al. (2004)	Methamphetamine abuse	416 cases 435 controls	Taiwanese	Methamphetamine abusers	–	val158met (rs4680; G/A)	Val allele frequency higher in abusers
		Chronic	Lohoff et al. (2008)	Cocaine dependence	330	African–American	Cocaine dependence patients	–	val158met (rs4680; G/A)	Met allele frequency higher in abusers
		Acute	Mattay et al. (2003)	Amphetamine brain response	123	–	Healthy volunteers	Placebo, 0.25 mg/kg of body weight	val158met (rs4680; G/A)	Val/Val performance improves with amphetamine; Met/Met perform worse with amphetamine
		Chronic	Suzuki et al. (2006)	Methamphetamine-induced psychosis	143 cases 200 controls	Japanese	Methamphetamine-induced psychosis patients	–	val158met (rs4680; G/A)	Met allele frequency higher in cases with spontaneous relapse

(continued)

Table 1 (continued)

Strong Candidate Genes Gene name	Locus symbol	Administration type	Study	Phenotype	Number	Ethnicity	Group	Dose	Associated polymorphisms	Notes
Monoamine oxidase A	*MAOA*	Chronic	Nakamura et al. (2009)	Methamphetamine-induced psychosis	118 cases 199 controls	Japanese	Methamphetamine-induced psychosis patients	–	MAOA-u VNTR	4-repeat allele associated with prolonged (vs. transient) psychosis in males
Dopamine beta-hydroxylase	*DBH*	Chronic	Cubells et al. (2000)	Cocaine-induced paranoia	45	European–American	Cocaine dependence/paranoia patients	–	Del-A haplotype	Frequency higher in cases
	DBH	Acute	Kalayasiri et al. (2007)	Cocaine-induced paranoia	31	Euopean–American, African–American	Cocaine dependence patients	0, 8, 16, and 32 mg/70 kg body weight	−1021C→T (rs1611115)	T/T genotype associated with increased risk for paranoia
Tryptophan hydroxylase 2	*TPH2*	Acute	Manor et al. (2008)	Methylphenidate therapeutic response	498	–	Children with ADHD	0.3–1 mg/kg	rs1386488-rs2220330-rs1386495-rs1386494-rs6582720-rs1386492-rs4760814-rs1386497 haplotype	C–G–C–A– A–G–A–C "Yang" haplotype associated with better response to methylphenidate
Fatty acid amide hydrolase	*FAAH*	Acute	Dlugos et al. (2009b)	Amphetamine subjective response	72	Caucasian	Healthy volunteers	Placebo, 10 mg, 20 mg	rs3766246, rs2295633; rs3766246-rs324420-rs2295633 haplotypes	C/C genotype for both SNPs associated with increased arousal and decreased fatigue at 10 mg

(continued)

Table 1 (continued)

Strong Candidate Genes Gene name	Locus symbol	Administration type	Study	Phenotype	Number	Ethnicity	Group	Dose	Associated polymorphisms	Notes
Brain-derived neurotrophic factor	*BDNF*	Acute	Flanagin et al. (2006)	Amphetamine subjective response	100	66 Caucasian, 20 African–American, 12 Asian, 2 unknown	Healthy volunteers	Placebo, 10 mg, 20 mg	val66met (rs6265; G/A)	Val/Val associated with higher response at 10 mg
Casein kinase 1 epsilon	*CSNK1E*	Acute	Veenstra-VanderWeele et al. (2006)	Amphetamine subjective response	101	67 Caucasian, 20 African–American, 12 Asian, 2 unknown	Healthy volunteers	Placebo, 10 mg, 20 mg	rs135745 (C/G)	C/C associated with greater sensitivity at 10 mg
Adenosine A_{2A} receptor	*ADORA2A*	Acute	Hohoff et al. (2005)	Amphetamine subjective response	99	65 Caucasian, 20 African–American, 12 Asian, 2 unknown	Healthy volunteers	Placebo, 10 mg, 20 mg	1976C/T (rs5751876) 2592C/T ins (rs3032740)	rs5751876 & rs3032740 associated with increased anxiety at 10 & 20 mg
Cytochrome P450 2D6	*CPY2D6*	Chronic	Otani et al. (2008)	Methamphetamine dependence	202 cases 337 controls	Japanese	Methamphetamine dependence patients	–	CYP2D6*10, CYP2D6 *14	*10 and *14 alleles at lower frequency in cases

(continued)

Table 1 (continued)

Strong Candidate Genes

Gene name	Locus symbol	Administration type	Study	Phenotype	Number	Ethnicity	Group	Dose	Associated polymorphisms	Notes
Glycoprotein endo-alpha-1, 2-mannosidase	*MANEA*	Chronic	Farrer et al. (2009)	Cocaine-induced paranoia	1,612 individuals for family-based; 1,921 unrelated subjects for case–control replication	Euopean–American, African–American	Cocaine dependence/ paranoia patients	–	rs9387522 and rs9387522 in all populations; rs6937479 from family based	rs9387522 A allele associated with cocaine induce paranoia in all populations

Genes are discussed in the following order: transporters, receptors, biosynthetic enzymes, and miscellaneous

Table 2 Association studies of less commonly studies genes with responses to stimulant drugs

Possible candidate genes Gene name	Locus symbol	Administration type	Study	Phenotype	Number	Ethnicity	Group	Dose	Associated polymorphisms	Notes
Glutathione S-transferase Mu 1	*GSTM1*	Chronic	Nakatome et al. (2009)	Methamphetamine abuse	100 cases 150 controls	Japanese	Methamphetamine users	–	Null allele	Null/null associated with abuse in females
Glutathione S-transferase theta-1	*GSTT1*	Chronic	Nakatome et al. (2009)	Methamphetamine abuse	100 cases 150 controls	Japanese	Methamphetamine users	–	Null allele	Higher liablity to abuse in combination with *GSTM1* null genotype
Glutathione S-transferase P	*GSTP1*	Chronic	Hashimoto et al. (2004)	Methamphetamine abuse/ psychosis	189 cases 199 controls	Japanese	Methamphetamine-induced psychosis patients	–	Ile105Val (rs947894; A/G)	G/G, A/G associated with abuse and psychosis
Prodynorphin	*PDYN*	Chronic	Nomura et al. (2006)	Methamphetamine dependence	143 cases 206 controls	Japanese	Methamphetamine dependence patients	–	Promoter VNTR	3- and 4-repeat alleles associated with increased risk
Prokineticin receptor 2	*PROKR2*	Chronic	Kishi et al. (2010)	Methamphetamine dependence	199 cases 337 controls	Japanese	Methamphetamine dependence patients	–	rs3746684, rs6085086, rs3746684 and rs17721321-rs6085086-rs3746684-rs3746682-rs4815787 haplotype	Individual SNPs and G–G–G–C–G haplotype associated with dependence
Solute carrier family 22 member 3	*SLC22A3*	Chronic	Aoyama et al. (2006)	Methamphetamine dependence	213 cases 443 controls	Japanese	Methamphetamine-induced psychosis patients	–	rs509707-rs4709426 haplotype	Associated with polysubstance abuse
Glycine transporter	*GLYT1*	Chronic	Morita et al. (2008)	Methamphetamine dependence/ psychosis	204 cases 210 controls	Japanese	Methamphetamine users	–	IVS3 + 411C > T (rs2486001); rs2486001-rs2248829 haplotype	T/T and C/T genotypes associated with methamphetamine use disorder
Alpha-synuclein	*SNCA*	Chronic	Kobayashi et al. (2004)	Methamphetamine dependence/ psychosis	170 cases 161 controls	Japanese	Methamphetamine dependence/ psychosis patients	–	rs1372520, rs3756063, rs3756059	Association only in females
Gamma-aminobutyric acid receptor subunit gamma-2	*GABRG2*	Chronic	Nishiyama et al. (2005)	Methamphetamine use disorder	178 cases 288 controls	Japanese	Methamphetamine use disorder patients	–	315 C > T-1128-99C > A haplotypes	Increased susceptibility to methamphetamine use disorder

(continued)

Table 2 (continued)

Possible candidate genes Gene name	Locus symbol	Administration type	Study	Phenotype	Number	Ethnicity	Group	Dose	Associated polymorphisms	Notes
Dystrobrevin-binding protein 1	*DTNBP1*	Chronic	Kishimoto et al. (2008a)	Methamphetamine-induced psychosis	197 cases 243 controls	Japanese	Methamphetamine-induced psychosis patients	–	rs3213207, rs2619538; rs2619539-rs3213207-rs2619538 haplotype	rs3213207 G allele, rs3213207 T allele at higher frequency in cases
Frizzled-3	*FZD3*	Chronic	Kishimoto et al. (2008b)	Methamphetamine-induced psychosis	288 cases 240 controls	Japanese	Methamphetamine-induced psychosis patients	–	rs2241802-rs2323019-rs352203-rs880481 haplotype	G–A–T–G and A–G–C–A haplotypes negative risk factors for psychosis
D-amino acid oxidase activator	*G72*	Chronic	Kotaka et al. (2009)	Methamphetamine-induced psychosis	209 cases 291 controls	Japanese	Methamphetamine-induced psychosis patients	–	rs778293 (A/G); rs3916965-rs2391191 (G-A) and rs947267-rs1421292 (T–T)	rs778293 G/G associated with Methamphetamine-induced psychosis
Glutamate receptor, metabotropic 2	*GRM2*	Chronic	Tsunoka et al. (2010)	Methamphetamine-induced psychosis	196 cases 802 controls	Japanese	Methamphetamine dependence/ psychosis patients	–	rs3821829–rs12487957–rs4687771 haplotype	C–C–A, C–T–T haplotypes associated with Methamphetamine-induced psychosis
NAD(P)H dehydrogenase, quinone 2	*NQO2*	Chronic	Ohgake et al. (2005)	Methamphetamine-induced psychosis	191 cases 207 controls	Japanese	Methamphetamine dependence/ psychosis patients	–	Promoter indel	Del/Del associated with prolonged psychosis
PRKCA-binding protein	*PICK1*	Chronic	Matsuzawa et al. (2007)	Methamphetamine-induced psychosis	208 cases 218 controls	Japanese	Methamphetamine abusers	–	rs713729, rs2076369	rs713729 methamphetamine abuse; rs713729 and rs2076369 spontaneous relapse of psychosis

(continued)

Table 2 (continued)

Possible candidate genes Gene name	Locus symbol	Administration type	Study	Phenotype	Number	Ethnicity	Group	Dose	Associated polymorphisms	Notes
Superoxide dismutase 2	*SOD2*	*Chronic*	Nakamura et al. (2006)	Methamphetamine-induced psychosis	116 cases 189 controls	Japanese	Methamphetamine-induced psychosis patients	–	Ala/Val exon 2 (rs4880)	Ala/Val associated *with* prolonged psychosis; Ala allele at higher frequency
Carboxylesterase 1	*CES1*	Sub-chronic	Nemoda et al. (2009)	Methylphenidate therapeutic response	122	Hungarian	Children with ADHD	10–30 mg (by body weight) twice daily (0.22 to 0.95 mg/kg/day)	Gly143Glu (rs71647871)	Glu/Glu, Glu/Gly associated with better treatment response

The genes here are not as well studied in relation to the phenotypes discussed in this review; they are therefore denoted 'possible candidate genes'. The associations described have primarily been performed for methamphetamine use-related phenotypes sample from the same study, Hamidovic et al. (2010a) tested 4 additional SNPs near the 5′ end of *SLC6A3* that were not in linkage disequilibrium with the 3′-UTR VNTR, and found a significant association between the rs460000 C/C genotype and increased scores on the ARCI Euphoria and Stimulation composite scales after 10 and 20 mg amphetamine administration, when compared to A/A and A/C individuals. Interestingly, this SNP is in perfect linkage disequilibrium with rs463379, a SNP that has been associated with increased risk of ADHD (Friedel et al. 2007), which suggests that variants underlying acute response to amphetamine may also underlie disease risk. Overall, these results suggest that both the 3′-UTR VNTR and SNPs in *SLC6A3* affect acute response to amphetamine.

The 3′-UTR VNTR polymorphism has also been studied in relation to acute responses to methylphenidate in children that have been diagnosed with ADHD. In a pilot study, eight children received acute doses of methylphenidate before SPECT neuroimaging (Rohde et al. 2003). Children homozygous for the 10-repeat allele exhibited greater cerebral blood flow, which may reflect higher transporter activity, in the frontal and basal ganglia brain areas in response to methylphenidate than children who were not homozygous for the 10-repeat allele.

SLC6A3 has also been the analyzed for its role in modulating therapeutic responses to sub-chronic (8 weeks) administration of methylphenidate. Similar to the Rohde et al. (2003) study discussed above, Cheon et al. (2005) used SPECT to measure dopamine transporter density in several brain regions (basal ganglia, right basal ganglia, and occipital cortex) of 11 children diagnosed with ADHD treated with varying doses of methylphenidate, and measured both clinical response and transporter density. Children with at least one copy of the 9-repeat allele ($n = 4$) showed a better therapeutic response to methylphenidate than 10/10 individuals (4 out of 4 responders versus 2 out of 7 responders in the 10/10 group), and significantly lower dopamine transporter density than 10/10 individuals. This suggests that 10/10 individuals may require more methylphenidate due to increased dopamine transporter density within the brain.

The association between the 3′-UTR VNTR and response to sub-chronic methylphenidate has also been examined in a larger cohort of children and adolescents with ADHD (Purper-Ouakil et al. 2008). Individuals received placebo and varying doses of methylphenidate until no further clinical improvement or limiting side effects occurred (mean dose 31.19 mg/day). They were phenotyped with the ARS, the Stroop test (Stroop 1935), the Trail Making Test (Reitan 1958), and the Continuous Performance Test (Rosvold et al. 1956). Subjects that were homozygotes for the 10-repeat allele had significantly lower treatment responses. However, the final methylphenidate doses reached did not differ across genotype group. These results suggest that the maximal response to methylphenidate may be lower in 10/10 homozygotes and cannot be overcome with larger doses. Similarly, another study of adults with ADHD found that the 10/10 genotype was associated with non-statistically significant lower therapeutic response to 3 weeks of treatment with methylphenidate as assessed by the CGI-S and ARS (Kooij et al. 2008).

Although the results discussed above suggest that the 10-repeat allele may be associated with poor treatment response to methylphenidate, other studies have reported that therapeutic responses are poorer in patients with the 9-repeat allele. In a double-blind study, Stein et al. (2005) found that children with the 9/9 genotype showed poor response to methylphenidate treatment as measured by ARS and CGI-S when they received higher (36 and 54 mg) but not lower (placebo or 18 mg) doses of methylphenidate. These results suggest that individuals with at least one copy of the 10-repeat allele responded better when given higher doses of methylphenidate. Joober et al. (2007) also performed a double-blind, placebo-controlled fixed-dose study, in which children diagnosed with ADHD were treated with methylphenidate for two weeks (0.5 mg/kg/day) and the effect of 3′-UTR VNTR was assessed. Similar to the results of Stein et al. (2005), children with the 9/10 and 10/10 genotypes showed significant improvement in ADHD symptoms following treatment as measured by the CGI-Parents scale when compared to 9/9 children; no difference was observed between the 9/10 and 10/10 groups. The 10-repeat allele was also associated with better treatment response to methylphenidate following longer term treatment in children diagnosed with ADHD using family-based association testing (Bellgrove et al. 2005). In addition to these studies, three studies have observed no association between the 3′-UTR VNTR polymorphism and methylphenidate response (da Silva et al. 2010; McGough et al. 2006; Roman et al. 2001). Therefore, the reported effects of the 3′-UTR VNTR on treatment response are contradictory and warrant further investigation.

The dopamine transporter has also been studied in relation to stimulant drug abuse. In these studies, the exact doses of drug are unknown and presumed to be high relative to laboratory or clinical studies. A case–control study examined methamphetamine dependence and drug-induced psychosis in a cohort of individuals with methamphetamine use disorder and psychosis (Ujike et al. 2003). No associations were found between the four polymorphisms tested and methamphetamine dependence or psychosis. However, when the patients with prolonged psychosis were analyzed separately from those with transient psychosis, there was a strong association indicating that individuals with nine or fewer repeat alleles of the 3′-UTR VNTR had prolonged psychosis. The 9-repeat allele was also marginally associated with increased risk of cocaine-induced paranoia in cocaine dependent subjects when the frequency of the 9-repeat allele was compared in individuals with and without psychosis (Gelernter et al. 1994). The reasons for these associations are unclear, given the lack of information about the doses that were ingested. The findings may indicate that drug users with the 9-repeat allele of the 3′-UTR VNTR polymorphism are more sensitive to stimulant-induced psychotic phenotypes, or, alternatively, they may have ingested higher doses of the drug to achieve their desired effect, thus increasing their risk for psychosis.

Lastly, the effects of an additional polymorphism in *SLC6A3* and cocaine-related behaviors have also been examined. Guindalini et al. (2006) examined the *SLC6A3* Intron 8 VNTR (Int8 VNTR), which consists of either five or six repeats and is in moderate to low linkage disequilibrium with the 3′-UTR VNTR (Asherson et al. 2007), and found that the 6-repeat allele was associated with cocaine abuse in a case-control study of cocaine abusers. When the 5- and 6-repeat alleles were cloned into expression vectors and transfected into a dopaminergic cell line, the 6-repeat allele showed increased *SLC6A3* expression following treatment with various stimuli (including cocaine), while the 5-repeat allele showed no change in expression. This study, along with Hamidovic et al. (2010a), suggests that polymorphisms other than the 3′-UTR VNTR may also contribute to responses to stimulant drugs.

SLC6A4

The serotonin transporter (*SLC6A4; 5-HTT*) is another direct target of stimulant drugs, and has been associated with numerous psychiatric phenotypes including obsessive-compulsive disorder (Bloch et al. 2008), autism (Huang and Santangelo 2008), and depression (Brown and Harris 2008; Kato and Serretti 2008; Risch et al. 2009). The 5-hydroxytryptamine transporter gene-linked polymorphic region (5-HTTLPR), which may be the most widely studied polymorphism in all of psychiatric genetics, has at least two common alleles: a short (S) 14-repeat and a long (L) 16-repeat allele (Nakamura et al. 2000; Rausch 2005). The 5-HTTLPR S allele has been associated with reduced gene expression (Hranilovic et al. 2004) and increased risk for psychiatric phenotypes like depression. The long allele has been further refined into two alleles (L_A and L_G) distinguished by an A→G polymorphism within the first repeat; the L_G allele is reported to have equivalent expression to the S allele (Hu et al. 2006). In additional to 5-HTTLPR, a VNTR in Intron 2 of *SLC6A4* has also been described, which consists of either a 10- or 12-repeat allele. The 12-repeat allele has been associated with increased gene expression (Hranilovic et al. 2004).

We have examined both of the *SLC6A4* polymorphisms mentioned above to determine whether they influence acute responses to amphetamine using our sample of healthy volunteers described in the previous section (Lott et al. 2006). When these two polymorphisms were analyzed separately, individuals homozygous for the 10-repeat allele of the Intron 2 VNTR showed a stronger euphoric response. When the polymorphisms were analyzed jointly, no significant association was observed, but trends in the predicted directions were observed—subjects homozygous for the low expressing alleles (S and 10-repeat) had the strongest responses to amphetamine (for the POMS Anxiety, DEQ Feel Drug, and ARCI Euphoria scales). These data identify a non-significant trend towards decreased expression of the serotonin transporter being associated with increased responses to stimulants.

The serotonin transporter has also been investigated for its role in sub-chronic methylphenidate treatment response in children diagnosed with ADHD (Thakur et al. 2010). In a 2 week trial, children received placebo and methylphenidate at varying doses. Subjects homozygous for the higher expressing L_A had the worst response to placebo but the best response to methylphenidate, heterozygotes were intermediate, and individuals homozygous for the L_S and L_G alleles exhibited the best response to placebo but deterioration with methylphenidate treatment as measured by the CGI-Parents subscale. Thus, lower expression of *SLC6A4* is associated with stronger euphoric, but weaker therapeutic responses to stimulant drugs.

The 5-HTTLPR polymorphism has also been associated with adverse responses in methamphetamine abusers. The 5-HTTLPR polymorphism was tested for association with prolonged methamphetamine-induced psychosis in a case-control study of methamphetamine abusers (Ezaki et al. 2008), utilizing the sample and methods described in the previous section (Ujike et al. 2003). *SLC6A4* was chosen as a candidate gene for this phenotype based on previous neuroimaging findings from the same group showing that methamphetamine abusers had lower serotonin transporter density than non-abusers (Sekine et al. 2006). The lower expressing S allele was significantly associated with methamphetamine psychosis, and in particular with prolonged psychosis. Thus, the S allele (and the L_G allele in the one instance where it was differentiated from the L_A allele) appears to be associated with more intense acute and chronic responses to various different stimulants. Furthermore, it appears that the lower expressing alleles of both the dopamine and the serotonin transporters are associated with greater propensity to methamphetamine-induced psychosis.

SLC6A2

The norepinephrine transporter (*SLC6A2; NET*) is a third direct target of stimulant drugs (Sulzer et al. 2005). Variants in *SLC6A2* have been associated with psychiatric phenotypes including depression (Haenisch et al. 2009; Min et al. 2009), antidepressant response (Min et al. 2009), and panic disorder (Lee et al. 2005). Using our sample of healthy adults (Dlugos et al. 2007) we found that rs36017 (C/C genotype) was associated with increased positive mood, as measured by a composite scale composed of "elation" minus "depression", following amphetamine administration (20 mg), while rs47958 (C/C genotype) was associated with increased elation. Haplotypes containing these SNPs (rs36017-rs10521329-rs3785155, G–C–C and C–C–A) were also associated with increased positive mood. A follow-up study with a larger number of individuals replicated the association between rs36017 and the POMS Elation scale, as well as with the POMS Vigor scale, and identified a new association between the A/A genotype at rs1861647 and these scales (Dlugos et al. 2009a). In addition, a different haplotype than those mentioned above, which was also constructed from rs36017, rs10521329, and rs3785155 (C–C–G), was associated with increased vigor following amphetamine administration. In sum, these studies suggest that multiple variants in *SLC6A2* influence responses to acute amphetamine administration.

5 Neurotransmitter Receptors

DRD2

The dopamine D2 receptor (*DRD2*) has been studied in relation to a variety of psychiatric traits, including schizophrenia (Allen et al. 2008), smoking cessation (David and Munafò 2008) and impulsivity (Eisenberg et al. 2007; Rodriguez-

Jimenez et al. 2006). In particular, the historically named *Taq*1A SNP (also known as rs1800497) has been widely studied. Historical nomenclature describes the A1 (T) and the A2 alleles (C). This polymorphism was initially described as being located in *DRD2*, but is now recognized to be located in a neighboring gene, *ANNK1* (Neville et al. 2004). However, this polymorphism appears to influence the expression of the *DRD2* gene and thereby alter *DRD2* function. In a study of *DRD2* polymorphisms and gene expression in postmortem human brain tissues, rs1800497 was not associated with *DRD2* expression, but the G allele of another SNP in *DRD2*, rs12364283, was found to be associated with enhanced expression (Zhang et al. 2007).

Variation in *DRD2* has been investigated in relation to amphetamine-induced impulsive behavior. Amphetamine increases behavioral inhibition, and the degree of this inhibition varies across individuals (de Wit et al. 2000). We examined the role of several SNP polymorphisms in *DRD2* and the effects of amphetamine on behavioral inhibition utilizing the sample of healthy human subjects that is described above. Individuals were phenotyped with the stop task (Logan et al. 1984), a measure of behavioral inhibition, and genotyped at 12 SNPs in *DRD2* (but not the Taq1A polymorphism). One SNP, rs12364283, was significantly associated with better performance on the stop task following 10 mg amphetamine administration in the G/G and A/G groups. This G allele of this SNP, as discussed earlier, is associated with increased *DRD2* expression (Zhang et al. 2007); increased *DRD2* expression may be related to better task performance (Cropley et al. 2006).

Variation in *DRD2* has been investigated for its effect on bupropion treatment response (David et al. 2007). David et al. (2007) found that smokers with the A2/A2 genotype at *Taq1A* who received bupropion (150 mg/day for the first 3 days, then 300 mg/day) were three times more likely to abstain from smoking at the end of the trial compared to A2/A2 subjects receiving placebo. This difference was not observed for the other genotypic groups.

Finally, *DRD2* has also been investigated for association with methamphetamine dependence and methamphetamine-induced psychosis. Šerý et al. (2001) found no association between polymorphisms in *DRD2* and methamphetamine dependence. Ujike et al. (2009) tested 3 SNPs in *DRD2*: 141C insertion/deletion (rs1799732; -/C), Ser311Cys (rs1801028; C/G) and *Taq1A*, and found that the *Taq1A* A2 (C) allele, which was associated with good response to bupropion for smoking cessation, was associated with prolonged methamphetamine psychosis (A2/A2, A1/A2 genotypes). Additionally, they found that the Del/Del and Ins/Del genotypes of 141C insertion allele (rs1799732) were associated with rapid onset of psychotic symptoms (within 3 years of initial abuse). Therefore, the *Taq1A* polymorphism, along with other polymorphisms in *DRD2*, may influence development of stimulant-induced psychosis as well as drug dependence.

DRD4

The dopamine D4 receptor (*DRD4*) has been associated with smoking behavior (Laucht et al. 2008), schizophrenia (Shi et al. 2008), novelty seeking (Munafò et al. 2008) and impulsivity (Munafò et al. 2008), as well as responses to stimulant drugs. The exon III VNTR polymorphism of *DRD4* is the most commonly studied. This polymorphism has 8 alleles, varying from 2 to 8 and 10 repeats (Lichter et al. 1993). The 7-repeat allele has been associated with ADHD in some studies (Brookes et al. 2006; Faraone et al. 2001; Li et al. 2006), but not in all (Johansson et al. 2008). Because of its association with ADHD, the 7-repeat allele has been the focus of most studies involving stimulant drug responses.

We have investigated the effect of the exon III VNTR polymorphism and acute subjective and physiological responses to amphetamine in our sample (Lee et al. 2006), and found that individuals with a single copy of the 7-repeat allele experienced more rewarding effects of the drug, with increased scores on the POMS Friendliness, POMS Elation, and heart rate scales at 20 mg and on the DEQ More scale at 10 mg and decreased scores on the ARCI Dysphoria and POMS Anxiety scales at 20 mg when compared to individuals lacking the 7-repeat allele. Therefore, the 7-repeat allele may modulate subjective and physiological responses to acute amphetamine administration.

Variation in *DRD4* has also been tested for association with therapeutic response following sub-chronic methylphenidate administration in children diagnosed with ADHD (Hamarman et al. 2004). Children with the 7-repeat allele of the exon III polymorphism required higher doses of methylphenidate (about two times more) to achieve the same therapeutic response over a 2-week period as measured by the CGI-S. Cheon et al. (2006) undertook a similar study and found that the 4-repeat allele was associated with the best treatment responses based on the ARS and the Kiddie-Schedule for Affective Disorders and Schizophrenia-Present and Lifetime Version (K-SADS-PL; Kaufman et al. 1997). Only one individual in that study had the 7-repeat allele, and therefore the previous finding could not be examined; conversely, the 4-repeat allele was not analyzed in the Hamarman et al. (2004) study. Therefore, these two sub-chronic studies found associations with different alleles of the exon III polymorphism and different measures of methylphenidate response.

DRD4 has also been investigated for association with methamphetamine dependence. Chen et al. (2004a) reported a higher frequency of the 7-repeat exon III VNTR allele in methamphetamine abusers as compared to controls, although the association did not reach statistical significance. A subsequent study utilizing the same population examined another VNTR polymorphism in the promoter region of *DRD4*, with a long (240 bp) and short allele (120 bp) (Li et al. 2004). Although no association between methamphetamine abuse and the promoter VNTR was found, when the promoter and exon III VNTR polymorphisms were analyzed as a haplotype, there was a significant association of the 7-repeat allele exon III VNTR and short promoter VNTR allele with methamphetamine abuse. Therefore, the 7-repeat allele of the exon III VNTR increases the dose of methylphenidate needed for therapeutic effects and may also increase the risk for development of methamphetamine abuse.

OPRM1

The μ-opioid receptor, *OPRM1,* has been investigated for a role in the subjective response to acute amphetamine. Stimulation of μ-opioid receptors by endogenous beta-endorphins in the ventral tegmental area increases dopamine release (Spanagel et al. 1992); therefore, variation in *OPRM1* may influence dopamine release and therefore response to stimulants. A commonly studied coding polymorphism within *OPRM1*, Asp40Asn (A118G; rs1799971; A/G), was first described by Bergen et al. (1997) and has been associated with numerous phenotypes, including heroin addiction (Drakenberg et al. 2006), schizophrenia (Šerý et al. 2010), and naltrexone treatment response (Oroszi et al. 2009). This SNP has also been shown to alter binding of beta-endorphin to the receptor (Bond et al. 1998). We examined associations between several SNPs (including rs1799971) and the acute responses to amphetamine on the ARCI Euphoria, Energy and Stimulation scales (Dlugos et al. 2010), and found that two polymorphisms (rs510769 G/G, A/G and rs2281617 C/C) were associated with increased euphoric responses to amphetamine at 10 mg; rs510769 is in strong linkage disequilibrium with rs1799971. The results indicate that variation in *OPRM1* may be related to variation in positive, euphoric responses to amphetamine.

Rs1799971 has been associated with the duration of methamphetamine-induced psychosis in a case-control study of individuals with methamphetamine dependence/psychosis; G/G individuals were more likely to become psychotic within 3 years of their first methamphetamine intake (Ide et al. 2004). In a more recent study using the same population, the previous association did not replicate, but the G/G genotype of an additional SNP in *OPRM1* (IVS2 + G691C; rs2075572; G/C) was associated with both methamphetamine dependence and psychosis (Ide et al. 2006). These results suggest the possibility that additional variants underlie chronic methamphetamine response, although rs1799971 has been the main target of the literature in investigating acute and chronic stimulant responses.

ADORA2A

The adenosine A_{2A} receptor (*ADORA2A*) is a major target of caffeine (Daly and Fredholm 1998) and forms a heterodimer with the dopamine D_2 receptor (Fuxe et al. 2005). We investigated the effect of *ADORA2A* polymorphisms on the anxiety response to amphetamine in healthy human subjects (Hohoff et al. 2005) and found that two polymorphisms (rs5751876; T/T and rs3032740; T_{ins}/T_{ins}), were associated with increased anxiety at the 10 and 20 mg doses.

6 Biosynthetic Enzymes

COMT

Various enzymes involved in the synthesis and breakdown of neurotransmitters have also been analyzed for associations with acute response to stimulant drugs.

One major candidate gene is catechol *O*-methyl transferase (*COMT*). This gene codes for an enzyme that preferentially metabolizes dopamine in the prefrontal cortex, rather than in the limbic and striatal brain regions, where the dopamine transporter is more important for clearance (Chen et al. 2004b). *COMT* contains a *val→met* coding polymorphism (*val*^{158}*-met;* rs4680; G(*val*)/A(*met*)) that has been associated with a variety of phenotypes including personality, cognition, risk for psychiatric disorders (Tunbridge et al. 2006), and pain sensitivity (Andersen and Skorpen 2009). The *met* allele has been associated with 3–4 times lower activity compared to the *val* allele (Lachman et al. 1996).

Polymorphisms in *COMT* have been associated with acute amphetamine response. In a landmark study (Mattay et al. 2003), healthy volunteers completed the Wisconsin Card Sorting Test (WCST; Heaton et al. 1993) after placebo or amphetamine, and underwent fMRI while performing the N-back working memory task (Kirchner 1958). Amphetamine administration reduced prefrontal cortical activity at all working memory loads when compared to placebo in individuals homozygous for the *val* allele. This was interpreted as a more efficient physiological response; this reduction was associated with improved reaction time and no decrease in accuracy on the task. Conversely, amphetamine increased prefrontal cortical activity in the most difficult part of the working memory task in *met* homozygotes. This was interpreted as a reduction in efficiency, because increased prefrontal cortical activity was associated with increased reaction time and decreased accuracy. The authors proposed that amphetamine increased dopamine levels above an optimum level in *met/met* individuals and thus negatively impacted cortical function. In contrast, in *val/val* individuals, amphetamine increased the lower pre-drug dopamine levels so that they were closer to optimal and thus enhanced function. The WCST also showed genotypic effects, with *val/val* subjects improving following amphetamine administration and *met/met* individuals performing worse. Taken together, the results of this study suggest that an optimum level of dopamine in the prefrontal cortex is necessary for efficient prefrontal cortex function, and that this efficiency is in part mediated by *val*^{158}*-met* genotype. Another study (Hamidovic et al. 2010b) evaluated the effect of *val*^{158}*-met* on a processing speed task (Digit Symbol Substitution Test; Wechsler 1958) and a reaction time test measuring attention lapses (Deviation from the Mode; de Wit 2009) and reported that when compared to placebo, amphetamine improved DSST performance in *val* homozygotes and heterozygotes, but not in *met* homozygotes. However, *val* homozygotes and heterozygotes exhibited more lapses in attention under placebo conditions, suggesting that the drug improved a preexisting deficit in one genotypic group. Taken together, these findings suggest that individuals with the *met* allele are less sensitive to the cognitive enhancing effects of stimulant drugs.

COMT has been investigated for association with chronic stimulant drug phenotypes. The *val* allele has been associated with polysubstance abuse (Vandenbergh et al. 1997), and in a later case–control study, Li et al. (2004) found that the *val/val* genotype was significantly associated methamphetamine abuse. In addition, Lohoff et al. (2008) found that cocaine-dependent individuals possessed the lower activity *met* allele more often than controls, suggesting that this polymorphism may play some role in dependence. These results differ from the Vandenbergh et al. (1997) and Li et al. (2004) studies described above; Lohoff et al. suggest their finding of the *met* allele association may also be due to linkage disequilibrium or population stratification.

MAOA

Monoamine oxidase A (*MAOA*) degrades monoamine neurotransmitters and has been most commonly associated with aggressive behavior (reviewed in Grigorenko et al. 2010). *MAOA* contains a 30 bp VNTR polymorphism within its promoter that consists of 3, 3.5, 4, or 5 repeats (MAOA-u VNTR); the 3.5 and 4 repeat alleles are transcribed more efficiently than the 3- or 5-repeat alleles (Sabol et al. 1998). Nakamura et al. (2009) tested the MAOA-u VNTR polymorphism for association with methamphetamine-induced psychosis, and found that in males, the 4-repeat allele was associated with having prolonged psychosis rather than transient; this effect was not seen in females.

DBH

Dopamine beta-hydroxylase (*DBH*) catalyzes the conversion of dopamine to norepinephrine (Kaufman and Friedman 1965) and has been associated with ADHD (reviewed in Banaschewski et al. 2010); additionally, a locus near *DBH* was recently associated with smoking cessation (Furberg et al. 2010). Cubells et al. (2000) investigated the effects of *DBH* polymorphisms on cocaine-induced paranoia in patients with cocaine dependence/paranoia, and identified a haplotype found more frequently in cocaine-dependent individuals reporting paranoia, that was also associated with low DBH plasma levels in normal subjects (−4784–4803del-444A→G; del-A haplotype). Individuals from the same group later sequenced *DBH* in eight individuals at the phenotypic extremes for DBH plasma levels, and identified a putative functional polymorphism (−1021C→T; rs1611115; C/T) accounting for 35–52% of phenotype variance in DBH activity (Zabetian et al. 2001); the T allele was associated with low DBH plasma levels, and found to be in positive linkage disequilibrium with the del and A alleles associated with low DBH plasma levels and paranoia in the Cubells et al. (2000) study (Zabetian et al. 2003). A controlled laboratory study was later conducted in which 31 cocaine-using volunteers acutely self-administered intravenous doses of the cocaine (0, 8, 16, and 32 mg/70 kg body weight) over four sessions (Kalayasiri et al. 2007). The previously identified −1021C→T polymorphism was tested for association with cocaine-induced paranoia, and the low DBH plasma level T/T genotype group reported higher paranoia when compared to the C/T and C/C groups.

TPH2

Tryptophan hydroxylase 2 (*TPH2*) catalyzes the rate-limiting step in the synthesis of serotonin, and *TPH2* polymorphisms have been investigated for association with depression, suicidal behavior, bipolar disorder (Zhang et al. 2006) and ADHD (Banaschewski et al. 2010; Brookes et al. 2006). Manor et al. (2008) tested for association of *TPH2* SNPs and acute therapeutic response in children diagnosed with ADHD, and identified an 8-SNP haplotype (rs1386488, rs2220330, rs1386495, rs1386494, rs6582720, rs1386492, rs4760814, rs1386497; C–G–C–A–A–G–A–C) that was significantly associated with better acute therapeutic response following methylphenidate treatment.

FAAH

Fatty acid amide hydrolase (*FAAH*) has also been studied in relation to acute response to stimulants; FAAH degrades several endocannabinoids that bind to cannabinoid receptors (McKinney and Cravatt 2005). Recently, polymorphisms in *FAAH* have been associated with brain response to cannabis (Filbey et al. 2009) and other phenotypes such as obesity (Engeli 2008). Much of the work associating variants in *FAAH* with stimulant response has been done in animal models (e.g., Madsen et al. 2006), and has shown that response to stimulants may also be influenced by the cannabinoid system. We investigated the effect of *FAAH* polymorphisms on subjective responses to acute amphetamine, using the sample of healthy volunteers that has been previously described (Dlugos et al. 2009b). Two SNPs, rs3766246 (C/C genotype) and rs2295633 (C/C genotype), were associated with amphetamine-induced arousal and decreased fatigue at 10 mg but not 20 mg, supporting the hypothesis that the endocannabinoid system may influence the acute response to low doses of amphetamine.

7 Miscellaneous

BDNF

Brain-derived neurotrophic factor (*BDNF*) is a growth factor that is involved in developmental processes (Hyman et al. 1991) and has been implicated in depression (Kato and Serretti 2008), smoking behavior (Furberg et al. 2010), obesity (den Hoed et al. 2010), and body mass index (Shugart et al. 2009). *BDNF* contains a *val→met* substitution (*val^{66}-met;* rs6265; G/A); the *met* allele produces a protein that is improperly secreted from neurons and has been reported to affect memory and hippocampal function (Egan et al. 2003). We examined the effects of *BDNF val^{66}-met* genotype on subjective responses to acute amphetamine (Flanagin et al. 2006) and found that individuals homozygous for the *val* allele possessed increased feelings of arousal and energy when administered 10 mg amphetamine relative to the other genotype groups. Therefore, the *BDNF val^{66}-met* genotype may also contribute to subjective responses to stimulant drugs.

CSNK1E

Another enzyme showing association with acute stimulant drug responses is casein kinase 1 epsilon (*CSNK1E*). *CSNK1E* encodes an enzyme that phosphorylates dopamine and cAMP-regulated phosphoprotein (DAARP-32; PPP1R1B), a second messenger that integrates dopaminergic and glutamatergic signaling

(Greengard 2001). We have previously reported that a quantitative trait locus (QTL) for methamphetamine sensitivity co-maps with an expression QTL for *Csnk1e* in mice, suggesting that the latter might cause the former (Palmer et al. 2005). Furthermore, pharmacological inhibition of *Csnk1e* blocks the locomotor response to methamphetamine in mice (Bryant et al. 2009) and rats (unpublished data). Mice with a null allele for *Csnk1e* exhibit paradoxically higher response to methamphetamine (unpublished data). We have also investigated the effects of SNPs in *CSNK1E* on subjective amphetamine response in humans (Veenstra-VanderWeele et al. 2006). The C allele of rs135745, which is located in the 3′-UTR of the gene, was associated with the response to amphetamine at 10 mg (but not 20 mg) as measured by the DEQ Feel Drug and ARCI Euphoria scales. Interestingly, other SNPs in *CSNK1E* have been associated with both heroin addiction in a case-control study (Levran et al. 2008) and ADHD in a family-based genome-wide association study (Mick et al. 2010), suggesting another possible link between ADHD, sensitivity to stimulants and drug abuse.

CYP2D6

Cytochrome P450 2D6 (*CYP2D6*) is a p450 enzyme that metabolizes methamphetamine (Lin et al. 1997). Otani et al. (2008) investigated the effect of *CYP2D6* polymorphisms in methamphetamine dependent patients in a case-control study, and found the lower activity *CYP2D6*10* and *CYP2D6*14* alleles were under-represented in patients, suggesting that these alleles are protective against development of methamphetamine dependence.

MANEA

An association between rs1133503, located in the 3′-UTR of glycoprotein endo-alpha-1,2-mannosidase (*MANEA*), and cocaine-induced paranoia was the strongest result in a genome-wide scan for variants associated with cocaine dependence and cocaine-induced paranoia (Yu et al. 2008). In a later study from the same group, 11 SNPs in *MANEA* were studied further using a family-based approach in two separate populations (African-–American; AA and European–American; EA). Nine of these SNPs were at least nominally associated with cocaine-induced paranoia in the AA population, whereas six of them at least nominally associated in the EA population (Farrer et al. 2009). Additionally, the rs9400554-rs6937479-rs9387522 haplotype (T–T–A) was associated with cocaine-induced paranoia in the pooled AA/EA family-based sample, and the C–A–C haplotype was associated with decreased risk of cocaine-induced paranoia in the pooled sample. No significant associations were found for the cocaine dependence phenotype. Two separate case–control samples (EA and AA) were used for replication. The individual SNP associations from the family-based studies did not replicate in either population sample. However, two associated haplotypes in the replication study contained SNPs that were found in associated haplotypes in the family-based analysis (rs900554 and rs9387522; C–A; T–A); the C–A haplotype was associated with increased risk of cocaine-induced paranoia and cocaine dependence in EA, and the T–A haplotype was associated with increased risk of cocaine dependence in the AA replication sample.

8 Exploratory Studies

A number of association studies have focused on genes that are not as well studied as those mentioned in the previous section; these more exploratory studies are summarized in Table 2. These studies mostly focus on methamphetamine abuse-related phenotypes. These genes include *CES1* (Nemoda et al. 2009); *DTNBP1* (Kishimoto et al. 2008a); *FZD3* (Kishimoto et al. 2008b); *G72* (Kotaka et al. 2009); *GABRG2* (Nishiyama et al. 2005); *GLYT1* (Morita et al. 2008); *GRM2* (Tsunoka et al. 2010); *GSTM1* (Nakatome et al. 2009); *GSTP1* (Hashimoto et al. 2004); *GSTT1* (Nakatome et al. 2009); *NQO2* (Ohgake et al. 2005); *PDYN* (Nomura et al. 2006); *PICK1* (Matsuzawa et al. 2007); *PROKR2* (Kishi et al. 2010); *SLC22A3* (Aoyama et al. 2006); *SNCA* (Kobayashi et al. 2004) and *SOD2* (Nakamura et al. 2006).

9 Genome-Wide Association Studies

Presently, only three GWAS has been conducted investigating responses to stimulant drugs. One study investigated the therapeutic response to methylphenidate (Mick et al. 2008). Children diagnosed with ADHD ($n = 187$), received varying doses of methylphenidate over a 5-week period, and were phenotyped with the ADHD-RS IV scale and then genotyped at about 300,000 SNPs. No associations reached genome-wide significance, and the strongest associations were not in genes examined in any previous candidate gene studies. Two studies have examined genetic variation underlying stimulant abuse. Uhl et al. (2008) used a case–control nested study to investigate genetic variants underlying methamphetamine dependence, with the goal of identifying genes with an unexpected accumulation of low p-values, and Yu et al. (2008) took a family-based association test (FBAT) approach and scanned evenly spaced markers for association with cocaine dependence and cocaine-induced paranoia. As previously mentioned, the association between *MANEA* SNP (rs1133503) and cocaine-induced paranoia was the most significant result in this study, but after correction for multiple testing this result was not statistically significant.

10 Closing Remarks

The main focus of this review was to investigate sources of genetic variation across a range of different phenotypes and prototypic stimulant drugs. Some of the data are specific to acute or chronic administration and therapeutic or non-therapeutic responses. Some genes, such as *OPRM1*, were associated with both acute subjective amphetamine liking and methamphetamine dependence. Understanding how these polymorphisms influence various phenotypes has important implications for the study of drug abuse and for our understanding of the pathophysiology and treatment of disorders such as ADHD. Interestingly, several genes discussed here are implicated in both ADHD and ADHD treatment response.

Many of these studies test multiple hypotheses and in some cases underpowered to do so, therefore there is a need for replication of these results. For example, our investigation of healthy human subjects has been used to test many different genetic hypotheses, so that if corrections for multiple testing across all of these different genes were applied, several results would not be considered statistically significant. In an effort to examine the reliability of our results we now have collected almost 400 subjects and are in the process of performing replication analyses.

There is still much to be learned about the polymorphisms reviewed here. The *SLC6A3* 3′-UTR VNTR is particularly illustrative of this, since it has been associated with many phenotypes in quite a number of studies, yet the results appear inconsistent or even contradictory. Resolving these apparent inconsistencies across studies is an urgent goal to better understand the function of the 3′-UTR VNTR and its relation to drug response.

There is significant promise in future technologies that will become available shortly. Genome-wide association studies will remain important in providing unbiased answers. As whole-genome re-sequencing becomes more readily available, it will be possible to study rare variants in more sophisticated ways. Copy number variation is another relatively unexplored type of genetic variation that has been suggested to be important in pharmacogenomics (Johansson and Ingelman-Sundberg 2008). Finally, there are substantial opportunities and challenges in integrating the findings from genetic studies in humans and model organisms (Phillips et al. 2008).

Acknowledgements This work was supported by NIH grants DA02812, DA027545 and DA021336.

References

Allen N, Bagade S, McQueen M, Ioannidis J, Kavvoura F, Khoury M, Tanzi R, Bertram L (2008) Systematic meta-analyses and field synopsis of genetic association studies in schizophrenia: the SzGene database. Nat Genet 40:827–834

American Psychiatry Association (1987) Diagnostic and statistical manual of mental disorders, revised 3rd edn. American Psychiatric Association, Washington DC

Andersen S, Skorpen F (2009) Variation in the COMT gene: implications for pain perception and pain treatment. Pharmacogenomics 10:669–684

Aoyama N, Takahashi N, Kitaichi K, Ishihara R, Saito S, Maeno N, Ji X, Takagi K, Sekine Y, Iyo M, Harano M, Komiyama T, Yamada M, Sora I, Ujike H, Iwata N, Inada T, Ozaki N (2006)

Association between gene polymorphisms of SLC22A3 and methamphetamine use disorder. Alcohol Clin Exp Res 30:1644–1649
Asherson P, Brookes K, Franke B, Chen W, Gill M, Ebstein R, Buitelaar J, Banaschewski T, Sonuga-Barke E, Eisenberg J (2007) Confirmation that a specific haplotype of the dopamine transporter gene is associated with combined-type ADHD. Am J Psychiatry 164:674–677
Banaschewski T, Becker K, Scherag S, Franke B, Coghill D (2010) Molecular genetics of attention-deficit/hyperactivity disorder: an overview. Eur child adolescent psychiatry 19:237–257
Bellgrove MA, Hawi Z, Kirley A, Fitzgerald M, Gill M, Robertson IH (2005) Association between dopamine transporter (DAT1) genotype, left-sided inattention, and an enhanced response to methylphenidate in attention-deficit hyperactivity disorder. Neuropsychopharmacology 30:2290–2297
Bergen A, Kokoszka J, Peterson R, Long J, Virkkunen M, Linnoila M, Goldman D (1997) Mu opioid receptor gene variants: lack of association with alcohol dependence. Mol Psychiatry 2:490–494
Bloch M, Landeros-Weisenberger A, Sen S, Dombrowski P, Kelmendi B, Coric V, Pittenger C, Leckman J (2008) Association of the serotonin transporter polymorphism and obsessive-compulsive disorder: systematic review. Am J Med Genet Part B Neuropsychiatr Genet 147:850–858
Bond C, LaForge K, Tian M, Melia D, Zhang S, Borg L, Gong J, Schluger J, Strong J, Leal S (1998) Single-nucleotide polymorphism in the human mu opioid receptor gene alters β-endorphin binding and activity: possible implications for opiate addiction. Proc Natl Acad Sci U S A 95:9608–9613
Brookes K, Xu X, Chen W, Zhou K, Neale B, Lowe N, Aneey R, Franke B, Gill M, Ebstein R (2006) The analysis of 51 genes in DSM-IV combined type attention .deficit hyperactivity disorder: association signals in DRD4, DAT1 and 16 other genes. Mol Psychiatry 11:934–953
Brown G, Harris T (2008) Depression and the serotonin transporter 5-HTTLPR polymorphism: a review and a hypothesis concerning gene-environment interaction. J Affect Disord 111:1–12
Brown WA, Corriveau DP, Ebert MH (1978) Acute psychologic and neuroendocrine effects of dextroamphetamine and methylphenidate. Psychopharmacology (Berl) 58:189–195
Bryant CD, Graham ME, Distler MG, Munoz MB, Li D, Vezina P, Sokoloff G, Palmer AA (2009) A role for casein kinase 1 epsilon in the locomotor stimulant response to methamphetamine. Psychopharmacology (Berl) 203:703–711
Cabeza R, Nyberg L (2000) Neural bases of learning and memory: functional neuroimaging evidence. Curr Opin Neurol 13:415–421
Chen C, Hu X, Lin S, Sham P, Loh E, Li T, Murray R, Ball D (2004a) Association analysis of dopamine D2-like receptor genes and methamphetamine abuse. Psychiatr Genet 14:223–226
Chen J, Lipska B, Halim N, Ma Q, Matsumoto M, Melhem S, Kolachana B, Hyde T, Herman M, Apud J (2004b) Functional analysis of genetic variation in catechol-O-methyltransferase (COMT): effects on mRNA, protein, and enzyme activity in postmortem human brain. Am J Human Genet 75:807–821
Cheon K, Ryu Y, Kim J, Cho D (2005) The homozygosity for 10-repeat allele at dopamine transporter gene and dopamine transporter density in Korean children with attention deficit hyperactivity disorder: relating to treatment response to methylphenidate. Eur Neuropsychopharmacol 15:95–101
Cheon K, Kim B, Cho S (2006) Association of 4-repeat allele of the dopamine D4 receptor gene exon III polymorphism and response to methylphenidate treatment in Korean ADHD children. Neuropsychopharmacology 32:1377–1383
Crabbe J, Jarvik L, Liston E, Jenden D (1983) Behavioral responses to amphetamines in identical twins. Acta Genet Med Gemellol (Roma) 32:139–149
Cropley V, Fujita M, Innis R, Nathan P (2006) Molecular imaging of the dopaminergic system and its association with human cognitive function. Biol Psychiatry 59:898–907

Cubells J, Kranzler H, McCance-Katz E, Anderson G, Malison R, Price L, Gelernter J (2000) A haplotype at the DBH locus, associated with low plasma dopamine β-hydroxylase activity, also associates with cocaine-induced paranoia. Mol Psychiatry 5:56–63

da Silva M, Cordeiro Q, Louza M, Vallada H (2010) Lack of association between a 30UTR VNTR 776 polymorphism of dopamine transporter gene (SLC6A3) and ADHD in a Brazilian sample of 777 adult patients. J Atten Disord 15:305–309

Daly J, Fredholm B (1998) Caffeine—an atypical drug of dependence. Drug Alcohol Depend 51:199–206

David S, Munafò M (2008) Genetic variation in the dopamine pathway and smoking cessation. Pharmacogenomics 9:1307–1321

David S, Strong D, Munafo M, Brown R, Lloyd-Richardson E, Wileyto P, Evins E, Shields P, Lerman C, Niaura R (2007) Bupropion efficacy for smoking cessation is influenced by the DRD2 Taq1A polymorphism: analysis of pooled data from two clinical trials. Nicotine Tob Res 9:1251–1257

de Wit H (2009) Impulsivity as a determinant and consequence of drug use: a review of underlying processes. Addict Bio 14:22–31

de Wit H, Uhlenhuth E, Johanson C (1986) Individual differences in the reinforcing and subjective effects of amphetamine and diazepam. Drug Alcohol Depend 16:341–360

de Wit H, Crean J, Richards JB (2000) Effects of d-amphetamine and ethanol on a measure of behavioral inhibition in humans. Behav Neurosci 114:830–837

den Hoed M, Ekelund U, Brage S, Grontved A, Zhao JH, Sharp SJ, Ong KK, Wareham NJ, Loos RJ (2010) Genetic susceptibility to obesity and related traits in childhood and adolescence; influence of loci identified by genome-wide association studies. Diabetes 59:2980–2988

Dlugos A, Freitag C, Hohoff C, McDonald J, Cook E, Deckert J, de Wit H (2007) Norepinephrine transporter gene variation modulates acute response to D-amphetamine. Biol Psychiatry 61:1296–1305

Dlugos AM, Hamidovic A, Palmer AA, Wit H (2009a) Further evidence of association between amphetamine response and SLC6A2 gene variants. Psychopharmacology (Berl) 206:501–511

Dlugos AM, Hamidovic A, Hodgkinson CA, Goldman D, Palmer AA, Hd Wit (2009b) More aroused, less fatigued: fatty acid amide hydrolase gene polymorphisms influence acute response to amphetamine. Neuropsychopharmacology 35:613–622

Dlugos A, Hamidovic A, Hodgkinson CA, Pei-Hong S, Goldman D, Palmer A, de Wit H (2010) OPRM1 gene variants modulate amphetamine-induced euphoria in humans. Genes Brain Behav 10:199–209

Drakenberg K, Nikoshkov A, Horváth M, Fagergren P, Gharibyan A, Saarelainen K, Rahman S, Nylander I, Bakalkin G, Rajs J (2006) μ Opioid receptor A118G polymorphism in association with striatal opioid neuropeptide gene expression in heroin abusers. Proc Nat Acad Sci 103:7883–7888

DuPaul G, Power T, Anastopoulos A, Reid R (1998) ADHD rating scales-IV: checklists. Norms and Clinical Interpretation, Guilford

Egan MF, Kojima M, Callicott JH, Goldberg TE, Kolachana BS, Bertolino A, Zaitsev E, Gold B, Goldman D, Dean M, Lu B, Weinberger DR (2003) The BDNF val66met polymorphism affects activity-dependent secretion of BDNF and human memory and hippocampal function. Cell 112:257–269

Eisenberg DTA, Mackillop J, Modi M, Beauchemin J, Dang D, Lisman SA, Lum JK, Wilson DS (2007) Examining impulsivity as an endophenotype using a behavioral approach: a DRD2 TaqI A and DRD4 48-bp VNTR association study. BBF 3:2

Engeli S (2008) Dysregulation of the endocannabinoid system in obesity. J Neuroendocrinol 20:110–115

Ezaki N, Nakamura K, Sekine Y, Thanseem I, Anitha A, Iwata Y, Kawai M, Takebayashi K, Suzuki K, Takei N (2008) Short allele of 5-HTTLPR as a risk factor for the development of psychosis in Japanese methamphetamine abusers. Ann N Y Acad Sci 1139:49–56

Faraone S, Doyle A, Mick E, Biederman J (2001) Meta-analysis of the association between the 7-repeat allele of the dopamine D4 receptor gene and attention deficit hyperactivity disorder. Am J Psychiatry 158:1052–1057

Farrer L, Kranzler H, Yu Y, Weiss R, Brady K, Anton R, Cubells J, Gelernter J (2009) Association of variants in MANEA with cocaine-related behaviors. Arch Gen Psychiatry 66:267

Fergusson D, Horwood L, Lynskey M, Madden P (2003) Early reactions to cannabis predict later dependence. Arch Gen Psychiatry 60:1033–1039

Filbey F, Schacht J, Myers U, Chavez R, Hutchison K (2009) Individual and additive effects of the CNR1 and FAAH Genes on brain response to marijuana cues. Neuropsychopharmacology 35:967–975

Flanagin B, Cook E Jr, de Wit H (2006) An association study of the brain-derived neurotrophic factor Val66Met polymorphism and amphetamine response. Am J Med Genet Part B Neuropsychiatr Genet 141:576–583

Friedel S, Saar K, Sauer S, Dempfle A, Walitza S, Renner T, Romanos M, Freitag C, Seitz C, Palmason H (2007) Association and linkage of allelic variants of the dopamine transporter gene in ADHD. Mol Psychiatry 12:923–933

Fuke S, Suo S, Takahashi N, Koike H, Sasagawa N, Ishiura S (2001) The VNTR polymorphism of the human dopamine transporter (DAT1) gene affects gene expression. Pharmacogenomics J 1:152–156

Furberg H, Kim Y, Dackor J, Boerwinkle E, Franceschini N, Ardissino D, Bernardinelli L, Mannucci P, Mauri F, Merlini P (2010) Genome-wide meta-analyses identify multiple loci associated with smoking behavior. Nat Genet 42:441–447

Fuxe K, Ferre S, Canals M, Torvinen M, Terasmaa A, Marcellino D, Goldberg S, Staines W, Jacobsen K, Lluis C (2005) Adenosine A2A and dopamine D2 heteromeric receptor complexes and their function. J Mol Neurosci 26(2):209–220

Gelernter J, Kranzler HR, Satel SL, Rao PA (1994) Genetic association between dopamine transporter protein alleles and cocaine-induced paranoia. Neuropsychopharmacology 11:195–200

Greengard P (2001) The neurobiology of slow synaptic transmission. Science 294:1024–1030

Grigorenko E, DeYoung C, Eastman M, Getchell M, Haeffel G, af Klinteberg B, Koposov R, Oreland L, Pakstis A, Ponomarev O (2010) Aggressive behavior, related conduct problems, and variation in genes affecting dopamine turnover. Aggress Behav 36:158–176

Guindalini C, Howard M, Haddley K, Laranjeira R, Collier D, Ammar N, Craig I, O'Gara C, Bubb VJ, Greenwood T, Kelsoe J, Asherson P, Murray RM, Castelo A, Quinn JP, Vallada H, Breen G (2006) A dopamine transporter gene functional variant associated with cocaine abuse in a Brazilian sample. Proc Natl Acad Sci U S A 103:4552–4557

Guy W (1976) Clinical global impressions. In: ECDEU assessment manual for psychopharmacology, revised. National Institute of Mental Health, Rockville

Haenisch B, Linsel K, Brüss M, Gilsbach R, Propping P, Nöthen M, Rietschel M, Fimmers R, Maier W, Zobel A (2009) Association of major depression with rare functional variants in norepinephrine transporter and serotonin1A receptor genes. Am J Med Genet Part B Neuropsychiatr Genet 150:1013–1016

Haertzen C, Kocher T, Miyasato K (1983) Reinforcements from the first drug experience can predict later drug habits and/or addiction: results with coffee, cigarettes, alcohol, barbiturates, minor and major tranquilizers, stimulants, marijuana, hallucinogens, heroin, opiates and cocaine. Drug Alcohol Depend 11:147–165

Hamarman S, Fossella J, Ulger C, Brimacombe M, Dermody J (2004) Dopamine receptor 4 (DRD4) 7-repeat allele predicts methylphenidate dose response in children with attention deficit hyperactivity disorder: a pharmacogenetic study. J Child Adolescent Psychopharmacology 14:564–574

Hamidovic A, Dlugos A, Skol A, Palmer AA, de Wit H (2009) Evaluation of genetic variability in the dopamine receptor D2 in relation to behavioral inhibition and impulsivity/sensation seeking: an exploratory study with d-amphetamine in healthy participants. Exp Clin Psychopharmacol 17:374–383

Hamidovic A, Dlugos A, Palmer AA, Wit H (2010a) Polymorphisms in dopamine transporter (SLC6A3) are associated with stimulant effects of d-amphetamine: an exploratory pharmacogenetic study using healthy volunteers. Behav Genet 40:255–261

Hamidovic A, Dlugos A, Palmer A, De Wit H (2010b) Catechol-O-methyltransferase val158met genotype modulates sustained attention in both the drug-free state and in response to amphetamine. Psychiatr Genet 20:85–92

Hashimoto K, Shimizu E, Iyo M (2004) Critical role of brain-derived neurotrophic factor in mood disorders. Brain Res Rev 45:104–114

Heaton RK, Chelune GJ, Talley JL, Kay GG, Curtiss G (1993) Wisconsin card sorting test manual (Rev. edn.). Psychological Assessment Resources, Odessa

Hohoff C, McDonald J, Baune B, Cook E, Deckert J, De Wit H (2005) Interindividual variation in anxiety response to amphetamine: possible role for adenosine A2A receptor gene variants. Am J Med Genet Part B Neuropsychiatr Genet 139:42–44

Holdstock L, de Wit H (2001) Individual differences in responses to ethanol and d-amphetamine: a within-subject study. Alcohol Clin Exp Res 25:540–548

Hranilovic D, Stefulj J, Schwab S, Borrmann-Hassenbach M, Albus M, Jernej B, Wildenauer D (2004) Serotonin transporter promoter and intron 2 polymorphisms: relationship between allelic variants and gene expression. Biol Psychiatry 55:1090–1094

Hu X, Lipsky R, Zhu G, Akhtar L, Taubman J, Greenberg B, Xu K, Arnold P, Richter M, Kennedy J (2006) Serotonin transporter promoter gain-of-function genotypes are linked to obsessive-compulsive disorder. Am J Human Genet 78:815–826

Huang C, Santangelo S (2008) Autism and serotonin transporter gene polymorphisms: a systematic review and meta-analysis. Am J Med Genet Part B: Neuropsychiatr Genet 147:903–913

Hyman C, Hofer M, Barde Y, Juhasz M, Yancopoulos G, Squinto S, Lindsay R (1991) BDNF is a neurotrophic factor for dopaminergic neurons of the substantia nigra. Nature 350:230–232

Ide S, Kobayashi H, Tanaka K, Ujike H, Sekine Y, Ozaki N, Inada T, Harano M, Komiyama T, Yamada M, Iyo M, Ikeda K, Sora I (2004) Gene polymorphisms of the mu opioid receptor in methamphetamine abusers. Ann N Y Acad Sci 1025:316–324

Ide S, Kobayashi H, Ujike H, Ozaki N, Sekine Y, Inada T, Harano M, Komiyama T, Yamada M, Iyo M, Iwata N, Tanaka K, Shen H, Iwahashi K, Itokawa M, Minami M, Satoh M, Ikeda K, Sora I (2006) Linkage disequilibrium and association with methamphetamine dependence/psychosis of mu-opioid receptor gene polymorphisms. Pharmacogenomics J 6:179–188

Johanson C, Uhlenhuth E (1980) Drug preference and mood in humans: diazepam. Psychopharmacology (Berl) 71:269–273

Johansson I, Ingelman-Sundberg M (2008) CNVs of human genes and their implication in pharmacogenetics. Cytogenet Genome Res 123:195–204

Johansson S, Halleland H, Halmøy A, Jacobsen K, Landaas E, Dramsdahl M, Fasmer O, Bergsholm P, Lundervold A, Gillberg C (2008) Genetic analyses of dopamine related genes in adult ADHD patients suggest an association with the DRD5-microsatellite repeat, but not with DRD4 or SLC6A3 VNTRs. Am J Med Genet Part B Neuropsychiatr Genet 147:1470–1475

Joober R, Grizenko N, Sengupta S, Amor LB, Schmitz N, Schwartz G, Karama S, Lageix P, Fathalli F, Torkaman-Zehi A, Ter Stepanian M (2007) Dopamine transporter 3′-UTR VNTR genotype and ADHD: a pharmaco-behavioural genetic study with methylphenidate. Neuropsychopharmacology 32:1370–1376

Kalayasiri R, Sughondhabirom A, Gueorguieva R, Coric V, Lynch WJ, Lappalainen J, Gelernter J, Cubells JF, Malison RT (2007) Dopamine beta-hydroxylase gene (DbetaH) -1021C->T influences self-reported paranoia during cocaine self-administration. Biol Psychiatry 61:1310–1313

Kato M, Serretti A (2008) Review and meta-analysis of antidepressant pharmacogenetic findings in major depressive disorder. Mol Psychiatry 15:473–500

Kaufman S, Friedman S (1965) Dopamine-β-hydroxylase. Pharmacol Rev 17:71–100

Kaufman J, Birmaher B, Brent D, Rao UMA, Flynn C, Moreci P, Williamson D, Ryan N (1997) Schedule for affective disorders and schizophrenia for school-age children-present and lifetime version (K-SADS-PL): initial reliability and validity data. J Am Acad Child Adolesc Psychiatry 36:980–988

Kendler K, Karkowski L, Prescott C (1999) Hallucinogen, opiate, sedative and stimulant use and abuse in a population-based sample of female twins. Acta Psychiatr Scand 99:368–376

Kendler KS, Karkowski LM, Neale MC, Prescott CA (2000) Illicit psychoactive substance use, heavy use, abuse, and dependence in a US population-based sample of male twins. Arch Gen Psychiatry 57:261–269

Kendler KS, Jacobson KC, Prescott CA, Neale MC (2003) Specificity of genetic and environmental risk factors for use and abuse/dependence of cannabis, cocaine, hallucinogens, sedatives, stimulants, and opiates in male twins. Am J Psychiatry 160:687–695

Kendler K, Gardner C, Jacobson K, Neale M, Prescott C (2005) Genetic and environmental influences on illicit drug use and tobacco use across birth cohorts. Psychol Med 35:1349–1356

Kirchner WK (1958) Age differences in short-term retention of rapidly changing information. J Exp Psychology 55:352–358

Kishi T, Kitajima T, Tsunoka T, Okumura T, Okochi T, Kawashima K, Inada T, Ujike H, Yamada M, Uchimura N, Sora I, Iyo M, Ozaki N, Iwata N (2010) PROKR2 is associated with methamphetamine dependence in the Japanese population. Prog Neuropsychopharmacol Biol Psychiatry 34:1033–1036

Kishimoto M, Ujike H, Motohashi Y, Tanaka Y, Okahisa Y, Kotaka T, Harano M, Inada T, Yamada M, Komiyama T, Hori T, Sekine Y, Iwata N, Sora I, Iyo M, Ozaki N, Kuroda S (2008a) The dysbindin gene (DTNBP1) is associated with methamphetamine psychosis. Biol Psychiatry 63:191–196

Kishimoto M, Ujike H, Okahisa Y, Kotaka T, Takaki M, Kodama M, Inada T, Yamada M, Uchimura N, Iwata N (2008b) The frizzled 3 gene is associated with methamphetamine psychosis in the Japanese population. Behav Brain Funct 4:37

Kobayashi H, Ide S, Hasegawa J, Ujike H, Sekine Y, Ozaki N, Inada T, Harano M, Komiyama T, Yamada M, Iyo M, Shen H-W, Ikeda K, Sora I (2004) Study of association between alpha-synuclein gene polymorphism and methamphetamine psychosis/dependence. Ann N Y Acad Sci 1025:325–334

Kooij J, Boonstra A, Vermeulen S, Heister A, Burger H, Buitelaar J, Franke B (2008) Response to methylphenidate in adults with ADHD is associated with a polymorphism in SLC6A3 (DAT1). Am J Med Genet Part B: Neuropsychiatr Genet 147:201–208

Kotaka T, Ujike H, Okahisa Y, Takaki M, Nakata K, Kodama M, Inada T, Yamada M, Uchimura N, Iwata N, Sora I, Iyo M, Ozaki N, Kuroda S (2009) G72 gene is associated with susceptibility to methamphetamine psychosis. Prog Neuropsychopharmacol Biol Psychiatry 33:1046–1049

Kreek M, Nielsen D, Butelman E, LaForge K (2005) Genetic influences on impulsivity, risk taking, stress responsivity and vulnerability to drug abuse and addiction. Nat Neurosci 8:1450–1457

Lachman H, Papolos D, Saito T, Yu Y, Szumlanski C, Weinshilboum R (1996) Human catechol-O-methyltransferase pharmacogenetics: description of a functional polymorphism and its potential application to neuropsychiatric disorders. Pharmacogenet Genomics 6:243–250

Lamb RJ, Henningfield JE (1994) Human d-amphetamine drug discrimination: methamphetamine and hydromorphone. J Exp Anal Behav 61:169–180

Laucht M, Becker K, Frank J, Schmidt MH, Esser G, Treutlein J, Skowronek MH, Schumann G (2008) Genetic variation in dopamine pathways differentially associated with smoking progression in adolescence. J Am Acad Child Adolesc Psychiatry 47:673–681

Lee Y, Hohoff C, Domschke K, Sand P, Kuhlenbäumer G, Schirmacher A, Freitag C, Meyer J, Stöber G, Franke P (2005) Norepinephrine transporter (NET) promoter and 5′-UTR polymorphisms: association analysis in panic disorder. Neurosci Lett 377:40–43

Lee A, Kim SJ, Lott DC, Cook EH, de Wit H, McGough JJ (2006) DRD4-7R mediated response to amphetamine in normal subjects. In: Scientific proceedings of the 53rd annual meeting of the American Academy of child and adolescent Psychiatry, San Diego

Levran O, Londono D, O'Hara K, Nielsen D, Peles E, Rotrosen J, Casadonte P, Linzy S, Randesi M, Ott J (2008) Genetic susceptibility to heroin addiction: a candidate gene association study. Genes Brain Behav 7:720–729

Li T, Chen C, Hu X, Ball D, Lin S, Chen W, Sham P, Loh E, Murray R, Collier D (2004) Association analysis of the DRD4 and COMT genes in methamphetamine abuse. Am J Med Genet Part B Neuropsychiatr Genet 129:120–124

Li D, Sham P, Owen M, He L (2006) Meta-analysis shows significant association between dopamine system genes and attention deficit hyperactivity disorder (ADHD). Hum Mol Genet 15:2276–2284

Lichter J, Barr C, Kennedy J, Van Tol H, Kidd K, Livak K (1993) A hypervariable segment in the human dopamine receptor D4 (DRD4) gene. Hum Mol Genet 2:767–773

Lin L, Di Stefano E, Schmitz D, Hsu L, Ellis S, Lennard M, Tucker G, Cho A (1997) Oxidation of methamphetamine and methylenedioxymethamphetamine by CYP2D6. Drug Metab Dispos 25:1059–1064

Logan G, Cowan W, Davis K (1984) On the ability to inhibit simple and choice reaction time responses: a model and a method. J Exp Psychology 10:276–291

Lohoff F, Weller A, Bloch P, Nall A, Ferraro T, Kampman K, Pettinati H, Oslin D, Dackis C, O'Brien C (2008) Association between the catechol-O-methyltransferase Val158Met polymorphism and cocaine dependence. Neuropsychopharmacology 33:3078–3084

Lott D, Kim S, Cook E, de Wit H (2005) Dopamine transporter gene associated with diminished subjective response to amphetamine. Neuropsychopharmacology 30:602–609

Lott D, Kim S, Cook E Jr, de Wit H (2006) Serotonin transporter genotype and acute subjective response to amphetamine. Am J Addict 15:327–335

Madsen MV, Peacock L, Werge T, Andersen MB (2006) Effects of the cannabinoid CB1 receptor agonist CP55, 940 and antagonist SR141716A on d-amphetamine-induced behaviours in Cebus monkeys. J Psychopharmacol (Oxford) 20:622–628

Manor I, Laiba E, Eisenberg J, Meidad S, Lerer E, Israel S, Gritsenko I, Tyano S, Faraone S, Ebstein R (2008) Association between trypotphan hydroxylase 2, performance on a continuance performance test and response to methylphenidate in ADHD participants. Am J Med Genet Part B Neuropsychiatr Genet 147:1501–1508

Martin WRSJW, Sapira JD, Jainski DR (1971) Physiologic, subjective and behavioral effects of amphetamine, methamphetamine, ephedrine, phenmetrazine and methylphenidate in man. Clin Pharmacol Therap 12:245–258

Matsuzawa D, Hashimoto K, Miyatake R, Shirayama Y, Shimizu E, Maeda K, Suzuki Y, Mashimo Y, Sekine Y, Inada T, Ozaki N, Iwata N, Harano M, Komiyama T, Yamada M, Sora I, Ujike H, Hata A, Sawa A, Iyo M (2007) Identification of functional polymorphisms in the promoter region of the human PICK1 gene and their association with methamphetamine psychosis. Am J Psychiatry 164:1105–1114

Mattay V, Goldberg T, Fera F, Hariri A, Tessitore A, Egan M, Kolachana B, Callicott J, Weinberger D (2003) Catechol O-methyltransferase val158-met genotype and individual variation in the brain response to amphetamine. Proc Natl Acad Sci U S A 100:6186–6191

McGough J, McCracken J, Swanson J, Riddle M, Kollins S, Greenhill L, Abikoff H, Davies M, Chuang S, Wigal T, Wigal S, Posner K, Skrobala A, Kastelic E, Ghuman J, Cunningham C, Shigawa S, Moyzis R, Vitiello B (2006) Pharmacogenetics of methylphenidate response in preschoolers with ADHD. J Am Acad Child Adolesc Psychiatry 45:1314–1322

McKinney MK, Cravatt BF (2005) Structure and function of fatty acid amide hydrolase. Annu Rev Biochem 74:411–432

McNair D, Lorr M, Droppleman L (1971) POMS manual for profile of mood states. EDITS, San Diego

Mick E, Neale B, Middleton F, McGough J, Faraone S (2008) Genome-wide association study of response to methylphenidate in 187 children with attention-deficit/hyperactivity disorder. Am J Med Genet Part B Neuropsychiatr Genet 147:1412–1418

Mick E, Todorov A, Smalley S, Hu X, Loo S, Todd R, Biederman J, Byrne D, Dechairo B, Guiney A (2010) Family-based genome-wide association scan of attention-deficit/hyperactivity disorder. J Am Acad Child Adolesc Psychiatry 9:898–905

Min W, Li T, Ma X, Li Z, Yu T, Gao D, Zhang B, Yun Y, Sun X (2009) Monoamine transporter gene polymorphisms affect susceptibility to depression and predict antidepressant response. Psychopharmacology (Berl) 205:409–417

Morita Y, Ujike H, Tanaka Y, Kishimoto M, Okahisa Y, Kotaka T, Harano M, Inada T, Komiyama T, Hori T, Yamada M, Sekine Y, Iwata N, Iyo M, Sora I, Ozaki N, Kuroda S (2008) The glycine transporter 1 gene (GLYT1) is associated with methamphetamine-use disorder. Am J Med Genet B Neuropsychiatr Genet 147B:54–58

Munafò M, Yalcin B, Willis-Owen S, Flint J (2008) Association of the dopamine D4 receptor (DRD4) gene and approach-related personality traits: meta-analysis and new data. Biol Psychiatry 63:197–206

Nakamura M, Ueno S, Sano A, Tanabe H (2000) The human serotonin transporter gene linked polymorphism (5-HTTLPR) shows ten novel allelic variants. Mol Psychiatry 5:32–38

Nakamura K, Chen C-K, Sekine Y, Iwata Y, Anitha A, Loh E-W, Takei N, Suzuki A, Kawai M, Takebayashi K, Suzuki K, Minabe Y, Tsuchiya K, Yamada K, Iyo M, Ozaki N, Inada T, Iwata N, Harano M, Komiyama T, Yamada M, Sora I, Ujike H, Ball DM, Yoshikawa T, Lin S-K, Mori N (2006) Association analysis of SOD2 variants with methamphetamine psychosis in Japanese and Taiwanese populations. Hum Genet 120:243–252

Nakamura K, Sekine Y, Takei N, Iwata Y, Suzuki K, Anitha A, Inada T, Harano M, Komiyama T, Yamada M, Iwata N, Iyo M, Sora I, Ozaki N, Ujike H, Mori N (2009) An association study of monoamine oxidase A (MAOA) gene polymorphism in methamphetamine psychosis. Neurosci Lett 455:120–123

Nakatome M, Miyaji A, Mochizuki K, Kishi Y, Isobe I, Matoba R (2009) Association between the GST genetic polymorphisms and methamphetamine abusers in the Japanese population. Leg Med (Tokyo) 11(1):S468–S470

Nemoda Z, Angyal N, Tarnok Z, Gadoros J, Sasvari-Szekely M (2009) Carboxylesterase 1 gene polymorphism and methylphenidate response in ADHD. Neuropharmacology 57:731–733

Neville M, Johnstone E, Walton R (2004) Identification and characterization of ANKK1: a novel kinase gene closely linked to DRD2 on chromosome band 11q23. 1. Hum Mutat 23:540–545

Nishiyama T, Ikeda M, Iwata N, Suzuki T, Kitajima T, Yamanouchi Y, Sekine Y, Iyo M, Harano M, Komiyama T, Yamada M, Sora I, Ujike H, Inada T, Furukawa T, Ozaki N (2005) Haplotype association between GABAA receptor gamma2 subunit gene (GABRG2) and methamphetamine use disorder. Pharmacogenomics J 5:89–95

Nomura A, Ujike H, Tanaka Y, Otani K, Morita Y, Kishimoto M, Morio A, Harano M, Inada T, Yamada M, Komiyama T, Sekine Y, Iwata N, Sora I, Iyo M, Ozaki N, Kuroda S (2006) Genetic variant of prodynorphin gene is risk factor for methamphetamine dependence. Neurosci Lett 400:158–162

Nurnberger J, Gershon E, Simmons S, Ebert M, Kessler L, Dibble E, Jimerson S, Brown G, Gold P, Jimerson D (1982) Behavioral, biochemical and neuroendocrine responses to amphetamine in normal twins and 'well-state' bipolar patients. Psychoneuroendocrinology 7:163–176

Ohgake S, Hashimoto K, Shimizu E, Koizumi H, Okamura N, Koike K, Matsuzawa D, Sekine Y, Inada T, Ozaki N (2005) Functional polymorphism of the NQO2 gene is associated with methamphetamine psychosis. Addict Bio 10:145–148

Oroszi G, Anton R, O'Malley S, Swift R, Pettinati H, Couper D, Yuan Q, Goldman D (2009) OPRM1 Asn40Asp predicts response to naltrexone treatment: a haplotype-based approach. Alcohol Clin Exp Res 33:383–393

Otani K, Ujike H, Sakai A, Okahisa Y, Kotaka T, Inada T, Harano M, Komiyama T, Hori T, Yamada M (2008) Reduced CYP2D6 activity is a negative risk factor for methamphetamine dependence. Neurosci Lett 434:88–92

Palmer AA, Verbitsky M, Suresh R, Kamens HM, Reed CL, Li N, Burkhart-Kasch S, McKinnon CS, Belknap JK, Gilliam TC, Phillips TJ (2005) Gene expression differences in mice divergently selected for methamphetamine sensitivity. Mamm Genome 16:291–305

Phillips T, Kamens H, Wheeler J (2008) Behavioral genetic contributions to the study of addiction-related amphetamine effects. Neurosci Biobehav Rev 32:707–759

Purper-Ouakil D, Wohl M, Orejarena S, Cortese S, Boni C, Asch M, Mouren M, Gorwood P (2008) Pharmacogenetics of methylphenidate response in attention deficit/hyperactivity disorder: association with the dopamine transporter gene (SLC6A3). Am J Med Genet Part B Neuropsychiatr Genet 147:1425–1430

Rausch J (2005) Initial conditions of psychotropic drug response: studies of serotonin transporter long promoter region (5-HTTLPR), serotonin transporter efficiency, cytokine and kinase gene expression relevant to depression and antidepressant outcome. Prog Neuropsychopharmacol Biol Psychiatry 29:1046–1061

Reitan R (1958) Validity of the trail making test as an indicator of organic brain damage. Percept Mot Skills 8:271–276

Risch N, Herrell R, Lehner T, Liang K, Eaves L, Hoh J, Griem A, Kovacs M, Ott J, Merikangas K (2009) Interaction between the serotonin transporter gene (5-HTTLPR), stressful life events, and risk of depression: a meta-analysis. JAMA 301:2462–2471

Rodriguez-Jimenez R, Ãvila C, Ponce G, Ibanez M, Rubio G, Jimenez-Arriero M, Ampuero I, Ramos J, Hoenicka J, Palomo T (2006) The TaqIA polymorphism linked to the DRD2 gene is related to lower attention and less inhibitory control in alcoholic patients. Eur Psychiatry 21:66–69

Rohde L, Roman T, Szobot C, Cunha R, Hutz M, Biederman J (2003) Dopamine transporter gene, response to methylphenidate and cerebral blood flow in attention-deficit/hyperactivity disorder: a pilot study. Synapse 48:87–89

Roman T, Schmitz M, Polanczyk G, Eizirik M, Rohde L, Hutz M (2001) Attention-deficit hyperactivity disorder: a study of association with both the dopamine transporter gene and the dopamine D4 receptor gene. Am J Med Genet Part B: Neuropsychiatr Genet 105:471–478

Rosvold HE, Mirsky AF, Sarason I, Bransome ED Jr, Beck LH (1956) A continuous performance test of brain damage. J Consult Psychology 20:343–350

Sabol SZ, Hu S, Hamer D (1998) A functional polymorphism in the monoamine oxidase A gene promoter. Hum Genet 103:273–279

Seiden L, Sabol K, Ricaurte G (1993) Amphetamine: effects on catecholamine systems and behavior. Annu Rev Pharmacol Toxicol 33:639–676

Sekine Y, Ouchi Y, Takei N, Yoshikawa E, Nakamura K, Futatsubashi M, Okada H, Minabe Y, Suzuki K, Iwata Y (2006) Brain serotonin transporter density and aggression in abstinent methamphetamine abusers. Arch Gen Psychiatry 63:90–100

Šerý O, Vojtová V, Zvolský P (2001) The association study of DRD2, ACE and AGT gene polymorphisms and metamphetamine dependence. Physiol Res 50:43–50

Šerý O, Přikryl R, Častulík L, Šťastný F (2010) A118G polymorphism of OPRM1 gene is associated with schizophrenia. J Mol Neurosci 41:219–222

Sevak RJ, Stoops WW, Hays LR, Rush CR (2009) Discriminative stimulus and subject-rated effects of methamphetamine, d-amphetamine, methylphenidate, and triazolam in methamphetamine-trained humans. J Pharmacol Exp Ther 328:1007–1018

Shi J, Gershon E, Liu C (2008) Genetic associations with schizophrenia: meta-analyses of 12 candidate genes. Schizophr Res 104:96–107

Shugart Y, Chen L, Day I, Lewis S, Timpson N, Yuan W, Abdollahi M, Ring S, Ebrahim S, Golding J (2009) Two British women studies replicated the association between the Val66Met polymorphism in the brain-derived neurotrophic factor (BDNF) and BMI. Eur J Hum Genet 17:1050–1055

Spanagel R, Herz A, Shippenberg TS (1992) Opposing tonically active endogenous opioid systems modulate the mesolimbic dopaminergic pathway. Proc Natl Acad Sci 15:2045–2050

Stein M, Waldman I, Sarampote C, Seymour K, Robb A, Conlon C, Kim S, Cook E (2005) Dopamine transporter genotype and methylphenidate dose response in children with ADHD. Neuropsychopharmacology 30:1374–1382

Stroop J (1935) Studies of interference in serial verbal reactions. J Exp Psychology 18:643–662

Sulzer D, Sonders M, Poulsen N, Galli A (2005) Mechanisms of neurotransmitter release by amphetamines: a review. Prog Neurobiol 75:406–433

Suzuki A, Nakamura K, Sekine Y, Minabe Y, Takei N, Suzuki K et al (2006) An association study between catechol-O-methyl transferase gene polymorphism and methamphetamine psychotic disorder. Psychiatr Genet 16:133–138

Thakur G, Grizenko N, Sengupta S (2010) The 5-HTTLPR polymorphism of the serotonin transporter gene and short term behavioral response to methylphenidate in children with ADHD. BMC Psychiatry 10:50

Tsuang MT, Lyons MJ, Eisen SA, Goldberg J, True W, Lin N, Meyer JM, Toomey R, Faraone SV, Eaves L (1996) Genetic influences on DSM-III-R drug abuse and dependence: a study of 3,372 twin pairs. Am J Med Genet 67:473–477

Tsunoka T, Kishi T, Kitajima T, Okochi T, Okumura T, Yamanouchi Y, Kinoshita Y, Kawashima K, Naitoh H, Inada T, Ujike H, Yamada M, Uchimura N, Sora I, Iyo M, Ozaki N, Iwata N (2010) Association analysis of GRM2 and HTR2A with methamphetamine-induced psychosis and schizophrenia in the Japanese population. Prog Neuropsychopharmacol Biol Psychiatry 34:639–644

Tunbridge E, Harrison P, Weinberger D (2006) Catechol-o-methyltransferase, cognition, and psychosis: Val158Met and beyond. Biol Psychiatry 60:141–151

Uhl G (2006) Molecular genetics of addiction vulnerability. NeuroRx 3:295–301

Uhl G, Li S, Takahashi N, Itokawa K, Lin Z, Hazama M, Sora I (2000) The VMAT2 gene in mice and humans: amphetamine responses, locomotion, cardiac arrhythmias, aging, and vulnerability to dopaminergic toxins. FASEB J 14:2459–2465

Uhl GR, Drgon T, Liu Q-R, Johnson C, Walther D, Komiyama T, Harano M, Sekine Y, Inada T, Ozaki N, Iyo M, Iwata N, Yamada M, Sora I, Chen C-K, Liu H-C, Ujike H, Lin S-K (2008) Genome-wide association for methamphetamine dependence: convergent results from 2 samples. Arch Gen Psychiatry 65:345–355

Ujike H, Harano M, Inada T, Yamada M, Komiyama T, Sekine Y, Sora I, Iyo M, Katsu T, Nomura A, Nakata K, Ozaki N (2003) Nine- or fewer repeat alleles in VNTR polymorphism of the dopamine transporter gene is a strong risk factor for prolonged methamphetamine psychosis. Pharmacogenomics J 3:242–247

Ujike H, Katsu T, Okahisa Y, Takaki M (2009) Genetic variants of D2 but not D3 or D4 dopamine receptor gene are associated with rapid onset and poor prognosis of methamphetamine psychosis. Prog Neuropsychopharmacol Biol Psychiatry 33:625–629

van de Giessen E, de Win M, Tanck M, van den Brink W, Baas F, Booij J (2009) Striatal dopamine transporter availability associated with polymorphisms in the dopamine transporter gene SLC6A3. J Nucl Med 50:4552

Van Dyck C, Malison R, Jacobsen L, Seibyl J, Staley J, Laruelle M, Baldwin R, Innis R, Gelernter J (2005) Increased dopamine transporter availability associated with the 9-repeat allele of the SLC6A3 gene. J Nucl Med 46:745–751

Vandenbergh D, Persico A, Hawkins A, Griffin C, Li X, Jabs E, Uhl G (1992) Human dopamine transporter gene (DAT1) maps to chromosome 5p15. 3 and displays a VNTR. Genomics 14:1104–1106

Vandenbergh DJ, Rodriguez LA, Miller IT, Uhl GR, Lachman HM (1997) High-activity catechol-O-methyltransferase allele is more prevalent in polysubstance abusers. Am J Med Genet 74:439–442

Veenstra-VanderWeele J, Qaadir A, Palmer A, Cook E, De Wit H (2006) Association between the casein kinase 1 epsilon gene region and subjective response to D-amphetamine. Neuropsychopharmacology 31:1056–1063

Warwick JM (2004) Imaging of brain function using SPECT. Metab Brain Dis 19:113–123
Wechsler D (1958) The measurement and appraisal of adult intelligence. Acad Med 33:706
Yang A, Palmer A, de Wit H (2010) Genetics of caffeine consumption and responses to caffeine. Psychopharmacology (Berl) 211:245–257
Yu Y, Kranzler HR, Panhuysen C, Weiss RD, Poling J, Farrer LA, Gelernter J (2008) Substance dependence low-density whole genome association study in two distinct American populations. Hum Genet 123:495–506
Zabetian C, Anderson G, Buxbaum S, Elston R, Ichinose H, Nagatsu T, Kim K, Kim C, Malison R, Gelernter J (2001) A quantitative-trait analysis of human plasma-dopamine [beta]-hydroxylase activity: evidence for a major functional polymorphism at the DBH locus. Am J Human Genet 68:515–522
Zabetian C, Buxbaum S, Elston R, Köhnke M, Anderson G, Gelernter J, Cubells J (2003) The structure of linkage disequilibrium at the DBH locus strongly influences the magnitude of association between diallelic markers and plasma dopamine [beta]-hydroxylase activity. Am J Human Genet 72:1389–1400
Zhang X, Beaulieu J, Gainetdinov R, Caron M (2006) Functional polymorphisms of the brain serotonin synthesizing enzyme tryptophan hydroxylase-2. Cell Mol Life Sci 63:6–11
Zhang Y, Bertolino A, Fazio L, Blasi G, Rampino A, Romano R, Lee M, Xiao T, Papp A, Wang D (2007) Polymorphisms in human dopamine D2 receptor gene affect gene expression, splicing, and neuronal activity during working memory. Proc Nat Acad Sci 104:20552–20557

The Cognitive Genetics of Neuropsychiatric Disorders

A. Corvin, G. Donohoe, A. Hargreaves, L. Gallagher and M. Gill

Abstract Classification in psychiatry is heavily dependent on clinical symptoms and illness course. This ignores the critical role that cognitive problems play in neuropsychiatric disorders affecting different domains across the lifespan, from ADHD and autism to schizophrenia and Alzheimers disease. At this point, it is unclear whether cognitive mechanisms are specific to disorders, whether multiple processes can contribute to the same disorder, or whether aberrant neural processing can result in many different phenotypic outcomes. Understanding this would allow us to better grasp normal as well as pathological brain function. This could inform diagnostics based on understanding of neurophysiological processes and the consequent development of new therapeutics. Genetics, and the development of genomic research, offers real opportunities to understand the molecular mechanisms relevant to cognition. This chapter defines and describes the main cognitive phenotypes, which are investigated in psychiatric disorders. We review evidence for their heritability and early progress in the field using cytogenetic, linkage and candidate gene-based research methodologies. With high-throughput genomics it is now possible to explore novel common and rare risk variants for psychiatric disorders and

A. Corvin (✉) · G. Donohoe · A. Hargreaves · L. Gallagher · M. Gill
Department of Psychiatry, Trinity College Dublin, Dublin, Ireland
e-mail: acorvin@tcd.ie

G. Donohoe
e-mail: donoghug@tcd.ie

A. Hargreaves
e-mail: hargrea@tcd.ie

L. Gallagher
e-mail: lgallagh@tcd.ie

M. Gill
e-mail: mgill@tcd.ie

Curr Topics Behav Neurosci (2012) 12: 579–613
DOI: 10.1007/7854_2011_188

Published Online: 26 February 2012

their role in cognitive function at a genome-wide level. We review the results of early genomic studies and discuss the novel insights that they are starting to provide. Finally, we review the analysis of whole-genome DNA sequence data and the challenges that this will bring for cognitive genomics research.

Keywords Genetics · Phenotypes · Cognition · Genomics

Contents

1 Introduction 580
2 Selection of Cognitive Deficits in Neuropsychiatric Genetics 581
2.1 Social Cognition and Psychiatric Disorder 582
3 Testing Cognitive Deficits in Neuropsychiatric Genetics 588
4 Cytogenetics and Cognitive Phenotypes 589
4.1 Cytogenetics of Childhood-Onset Neuropsychiatric Disorders 589
4.2 Cytogenetics of Adult-Onset Neuropsychiatric Disorders 591
5 Linkage and Candidate Gene-Based Approached to Cognitive Genetics 593
5.1 Linkage and Candidate Genes Studies for Disorders of Childhood 594
5.2 Linkage and Candidate Genes Studies for Disorders of Adult-Onset 597
6 Application of Genomics Methods 599
6.1 Genome-Wide Association Studies 600
6.2 Studies of Structural Genomic Variation in Neuropsychiatric Disorders 603
7 Future Directions: New Phenotypes and New Approaches 605
References 606

1 Introduction

Classification in psychiatry is heavily dependent on clinical symptoms and illness course. This ignores the critical role that cognitive problems play in neuropsychiatric disorders affecting different domains across the lifespan, from ADHD and autism to schizophrenia and Alzheimer's disease. At this point, it is unclear whether aberrant cognitive mechanisms are specific to disorders, whether multiple disturbances in cognitive processes can contribute to the same disorder, or whether aberrant neural processing can result in many different phenotypic outcomes. Understanding this would allow us to better grasp normal as well as pathological brain function. This could inform diagnostics based on understanding of neurophysiological processes and the consequent development of new therapeutics.

Genetics, and the development of genomic research, offers real opportunities to understand the molecular mechanisms relevant to cognition. This chapter defines and describes the main cognitive phenotypes, which are investigated in psychiatric disorders. We review evidence for their heritability and early progress in the field using cytogenetic, linkage and candidate gene-based research methodologies. With high-throughput genomics it is now possible to explore novel common and rare risk variants for psychiatric disorders and their role in cognitive function at a genome-wide level. We review the results of early genomic studies and discuss the novel insights that they are starting to provide. Finally, we review the analysis of whole-genome DNA sequence data and the challenges that this will bring for cognitive genomics research.

2 Selection of Cognitive Deficits in Neuropsychiatric Genetics

Historically, no aspect of cognition (or any human trait for that matter) has received more attention in terms of its underlying genetics basis than general cognitive ability or intelligence. While this history has not been without controversy (cf. the eugenics movement), the genetic basis of intelligence has been studied widely in both the general population and across psychiatric disorders.

A key question in the selection of specific cognitive phenotypes is what extent performance on a given task is heritable. Differing degrees of relationships within families, associated with more or less sharing of genetic material (e.g. monozygotic twins share 100% of genes, dizygotic twins/siblings 50%, and half-siblings 25%) allow estimation of the proportion of individual differences in performance in a population at a given time that are due to genetic differences (termed heritability (h^2)). The availability of twin and population-based disease register data has confirmed the importance of heritability for many psychiatric disorders. Less information has been available to allow interpretation of the heritability of cognitive deficits within twin samples. Because of this practical sampling constraint, most epidemiological studies have used a family-based design to investigate cognitive deficit in healthy relatives of patients. The finding that this group has higher rates of a particular cognitive deficit is taken as suggestive evidence of heritability.

The total variance in general intelligence that can be attributed to genetic influences range from 30 to 80%. Broad domains of cognitive ability, such as verbal and perceptual abilities show similar measures of genetic influence (Posthuma et al. 2001), although the genetic influence on subdomains such as memory tends to be smaller, due to measurement error and variance, but being highly correlated with general intelligence, genetic effects overlap. The heritability of general intelligence increases with age to 70–80% in adulthood. Variations in brain structures such as the density and the volume of grey and white matter, amygdala, and hippocampus, and overall brain volume are thought to be endophenotypes for intelligence. Therefore, genes involved in intelligence might be more closely associated with variations in brain structure and function than to measures of intelligence.

By comparison, selection of specific cognitive functions for analysis in genetic studies has been heavily influenced by the kinds of deficits observed within specific disorders. This selection has also been strongly influenced by the perceived potential of a cognitive function or process as 'intermediate phenotypes' or 'endophenotypes' for the disorder. The 'endophenotype' concept in psychiatry (described by Gottesman and Gould 2003) relates to the identification of heritable quantifiable characteristics, which may be useful targets for genetic studies as they represent some intermediate stage between genotype and clinical disorder. The authors suggested that the potential utility of an endophenotype can be judged against a set of criteria:

1. Association with illness in the population
2. Co-segregation of the endophenotype with illness in families
3. Evidence that this is genetically mediated (i.e. heritable)
4. Evidence that the endophenotype is state-independent (i.e. present whether illness is active or not)
5. That the endophenotype can be measured accurately and reliably
6. The endophenotype is also shared by nonaffected family members of the proband, this may be advantageous for genetic studies.

Across disorders, deficits in executive function, memory function and attentional control have each been a particular focus for research. Clinical awareness of these impairments has increased as it has been established that such deficits are predictive of psychiatric morbidity. In the case of schizophrenia for example, cognitive deficits are present from an early stage of the disorder and often predate the emergence of clinical symptoms (Erlenmeyer-Kimling et al. 2000). They are relatively stable over time and closely related to functional outcome (Green et al. 2004). This includes deficits in general cognitive ability (Donohoe et al. 2011b) and specific deficits in working and episodic memory (Donohoe et al. 2011a) and attentional control (Bellgrove and Mattingley 2008; Donohoe et al. 2011a). Genetic epidemiological research using family and twin studies indicates that some of these deficits may themselves have a substantial genetic component (Goldberg et al. 1990, 1995; Cannon et al. 2000). While cognitive deficits are somewhat correlated with clinical symptoms (for example, negative symptoms in schizophrenia) the amount of variance shared by these variables appears to be small, and cognitive function often emerges as a separate factor from clinical symptoms in factor analysis (Donohoe and Robertson 2003). Taken together these data indicate that cognitive phenotypes in schizophrenia are heritable, trait-like and generally independent of clinical symptomatology.

Mood disorder researchers, investigating bipolar disorder and recurrent depressive disorder have focused less on general cognitive ability and more on memory related phenotypes (Thomas and Elliott 2009; Frodl et al. 2009). In attention deficit hyperactivity disorder (ADHD), attention has inevitably been a particularly important focus, both in terms of orienting attention, sustaining attention, and inhibiting processing of task-irrelevant stimuli (Bellgrove and Mattingley 2008). In autism, impairments in overall cognitive ability are found in 50% of affected individuals. In higher functioning individuals deficits in executive functioning (specifically attention orienting, response inhibition, and set shifting) are reported. The most widely investigated area of cognition in autism relates to impairments in social cognition. Social cognition is the ability to process social information, thought to be fundamentally impaired in autism.

2.1 Social Cognition and Psychiatric Disorder

A significant development in cognitive genetic studies of psychiatric disorders has been the increased focus on the genetics of social cognition that began with research in autism. Social cognition is the sum of those processes that allow individuals of the same species (conspecifics) to interact with one another (Frith and Frith 2007), specifically it refers to the set of skills that allow us to understand the thoughts and intentions of others and frequently involves the investigation of social information processing, especially its encoding, storage, retrieval, and application to social situations. Essentially it depends upon the exchange of signals, such as speech, facial expression, body posture, and eye gaze (Frith and Frith 2007). Signals such as these can be socially informative in that they tell us what someone may be feeling (Vuilleumier and Pourtois 2007), where they are focusing their attention, and what they are intending to do (Frith and Frith 2006). The ability to process, comprehend and act appropriately to these signals is tantamount to social success and relies on coordination of several cortical regions, e.g. dorsomedial and dorsolateral prefrontal cortices, the paracingulate cortex and the right and left temporoparietal junctions and amygdala (Mitchell 2009). In several psychiatric disorders this ability is impaired leading to social misperceptions, unexpected reactions to and from the person, and social withdrawal (Green et al. 2004), resulting in difficulties with maintaining friendships, employment, and general community functioning (Penn et al. 2008). Alongside these clinical and outcome goals, there is increasing interest in identifying the neural basis underlying social cognitive deficits in psychiatric illness. As such, further research in this domain is viewed as highly valuable (Green et al. 2004).

Currently, social cognition is investigated both in behavioural terms and in terms of neuroanatomy. Some of the tests that fall into the former category include theory of mind (TOM) tests (such as mind in the eyes, Faux pas and hinting task), the hotel task, multiple errands task, Iowa gambling task, the social cognitive skills test (SCST), Mayer-Salovey-Caruso emotional intelligence test (MSCEIT), the Penn emotional recognition task, facial affect recognition, attribution test, and tests of self-control (see Table 1 for task summaries). The latter category however is most often assessed using magnetic resonance imaging (MRI) and involves performing simple social cognitive tests such as facial aspect recognition, and imaging the brain to see what areas are activated during task performance.

To investigate the utility of these measures of social cognition as psychiatric 'endophenotypes' (based on the criteria outlined above) we undertook a review of the main tests found in the literature. Search included test name, heritability, twin, sibling, state, independent, co-segregation, brain region, and endophenotype. Secondary search terms used in the event of the primary terms yielding no results included family, trait, dependent, gene, psychiatric, schizophrenia, and autism. A list of the common measures of social cognition reviewed and evidence of their feasibility as endophenotypes is presented in Table 2. As with neuropsychological measures most of the evidence of heritability for each of the measures of social cognition reviewed is derived from evidence that family members of psychiatrically affected probands also show deficits at a level intermediate between cases and healthy controls. These deficits have been associated with a number of candidate risk genes for psychiatric disorders, including 5HT2A, oxytocin, and vasopressin.

Table 1 Definition and usage of various social cognitive tests

	Test	Definition/Usage
Emotion recognition	MSCEIT	Measures emotional intelligence, namely 4 areas: perceiving emotions, facilitating thought, understanding emotions, managing emotions
	Penn emotional recognition task	A computer-based test that includes 96 colour photographs of facial expression of evoked—or felt—emotions: happy, sad, angry, fearful, disgusted, and nonemotional or neutral
	Facial affect recognition	Facial emotion identification test (FEIT) and the facial emotion discrimination test (FEDT). Both use black and white photographs of facial emotions that are presented on DVD
Theory of mind	TOM	A "Theory of mind" (often abbreviated to TOM) is a specific cognitive ability to understand others as intentional agents, that is, to interpret their minds in terms of theoretical concepts of intentional states such as *beliefs* and *desires*
	Mind in the eyes	Measure of adult mentalising. The ability to deduce emotion from looking at photographs of people's eyes
	Faux pas test	The faux pas test ascertained the participants' ability to identify and understand a social faux pas, and to understand the mental states of the characters (the speaker and the recipient) in a conversation with a social faux pas
	Hinting task	Mental state reasoning, inferring. Measures ability of subject to 'read between the lines' in a social context
Attribution	Attribution test	How one assigns causality in a social context
Social perception	Social cognitive skills test (SCST)	Focuses on the assessment of social reasoning skills
	Iowa gambling task	Decision making abilities. Looks at trust and the ability to analyse intentions
Self regulation	Self control test (implicit association test (IAT))	Measure the strength of automatic association between mental representations of objects (concepts) in memory

Table 2 Suitability of social cognition measures as endophenotypes for schizophrenia

Test	First author	N	Associated with illness	Heritability/ Present in other family members	State independent	Co-segregates with illness
Penn emotional recog task	Greenwood et al. (2007)	183 nuclear families ascertained through probands with schizophreni	Yes	Yes		
	Eack et al. (2010)	70 first-degree relatives of schizophrenia probands, 63 healthy controls	Yes	Yes		
	Calkins et al. (2010)	SZ/Szaff = 610, Relatives = 928, HC = 334	Yes	Yes		Yes
	Gur et al. (2002)	14 patients with schizophrenia and 14 matched comparison subjects	Yes			
	Bediou et al. (2007)	Drug-naive patients with first-episode schizophrenia ($n = 40$) and their unaffected siblings ($n = 30$) compared with controls ($n = 26$)	Yes	Yes	Yes	
Facial affect recognition	Erol et al. (2010)	Patients with schizophrenia ($n = 57$), their unaffected biological siblings ($n = 58$) and healthy controls ($n = 58$)	Yes	Yes		
TOM task	McKinnon et al. (2010)	14 bipolar, 14 control	Yes			
	Sabbagh and Seamans (2008)	46 3-year-olds and their parents		Yes		
	Sprong et al. (2007)	29 studies (combined $n = 1518$)			Yes	
	de Achával et al. (2010)	20 Schizophrenia patients, 20 healthy age- and gender-matched individuals, 20 unaffected first-degree relatives of the schizophrenia patients, and 20 healthy individuals matched for age and gender	Yes	Yes		
Hinting task	Janssen et al. (2003)	43 patients with schizophrenia or schizoaffective disorder, 41 first degree non-psychotic relatives and 43 controls from the general population	Yes	Yes	Yes	
Mind in the eyes	Losh and Piven (2007)	48 parents of individuals with autism (13 of whom were identified as 'aloof'), and 22 control parents	Yes	Yes	.	Yes

(continued)

Table 2 (continued)

Test	First author	N	Associated with illness	Heritability/ Present in other family members	State independent	Co-segregates with illness
	Bora et al. (2005)	Forty-three euthymic bipolar patients and 30 controls			Yes	
Faux pas	de Achával et al. (2010)	20 Schizophrenia patients 20 healthy age- and gender-matched individuals, 20 unaffected first-degree relatives of the schizophrenia patients, and 20 healthy individuals	Yes	Yes		
	Shamay-Tsoory et al. (2009)	19 patients, 20 controls			Yes	
Attribution test (IPSAQ)	Lau and Eley (2008)	Childhood(birth–12 years) (100), School Age (6–12 years) (180), Adolescence (13–17 years) (200), Adulthood(18 years plus) (300), Young Adult(18–29 years) (320)	Yes	Yes	Yes	
	Janssen et al. (2006)	23 patients with psychosis, 36 first-degree relatives of patients with psychosis, 31 subjects with subclinical psychotic experiences and 46 normal controls	Yes			
Social cognitive skills test (SCST)	Scourfield et al. (1999)	Population-based sample of twins aged 5–17		Yes		
	Scourfield et al. (2004)	A population-based sample of twins aged 5–17		Yes	Yes	
	Coleman et al. (2008)	8 children with ASD and 8 control children aged 5–13 years	Yes			
Iowa gambling task	Viswanath et al. (2009)	25 unaffected siblings of OCD probands with familial OCD, and 25 individually matched healthy controls	Yes	Yes		

(continued)

Table 2 (continued)

Test	First author	N	Associated with illness	Heritability/ Present in other family members	State independent	Co-segregates with illness
	Cavedini et al. (2010)	35 pairs of OCD probands and unaffected first-degree relatives and 31 pairs of HC subjects without a known family history of OCD and their relatives	Yes	Yes		
Implicit association test	Egloff and Schmukle (2002)	41 introductory psychology students (33 women and 8 men)	Yes			
	Sasaki et al. (2010)	Individuals scoring High ($n = 26$) and Low ($n = 18$) on Social Anxiety	Yes			
	Glashouwer and de Jong (2010)	Patients ($n = 2{,}329$) and non-clinical controls ($n = 652$)	Yes		Yes	

While many tests show good evidence that poorer performance is both associated with illness and heritable (at least based on assessments in first-degree relatives) both classical twin studies and studies of co-segregation in families are currently lacking

3 Testing Cognitive Deficits in Neuropsychiatric Genetics

Cognitive neuroscience approaches to investigating both cognitive performance in the healthy population and cognitive disability associated with psychiatric disorders have included both behavioural and neuroimaging paradigms. Behavioural measures include traditional neuropsychological paper and pen tasks (e.g. the child and adult versions of the Wechsler intelligence scales, the Wechsler memory scales) and computer-based tasks (e.g. various versions of the continuous performance task (CPT); the Cambridge automated test battery). The main advantage of these neuropsychological tests is that they generally have well-established psychometric properties that can be easily administered to large groups of varying ability, a key requirement for cognitive genetics research. Imaging modalities have included both functional magnetic resonance imaging (fMRI) during tasks of memory, attention and—more recently—affective processing, with the high spatial resolution it offers, and electroencephalogram recordings (EEG), with the high temporal resolution it provides. While, previously, individual groups tended to specialise in one main approach, groups working in this area increasingly aim to integrate findings across modalities. In our own group for example, we have used high density EEG and structural MRI to follow up on specific associations between individual genetics variants and neuropsychological test performance, such as the contribution of abnormal sensory level processing to cognitive performance deficits (Donohoe et al. 2008).

Across disorders, available family and twin studies generally support the concept that the cognitive deficits associated with these disorders are themselves inherited. In the case of schizophrenia, the heritability of a number of specific deficits have been confirmed by twin studies, including general cognitive ability, working memory, and episodic memory (Goldberg et al. 1990, 1995; Cannon et al. 2000; Kremen et al. 2006; Toulopoulou et al. 2007, 2010). The heritability of deficits in general cognitive ability and working memory in particular appear to overlap strongly with the heritability of illness risk in schizophrenia (Toulopoulou et al. 2007) although not to the point of suggesting an identity between the genetic architecture of schizophrenia and cognition. In autism research, deficits in general cognitive ability, executive functioning, and processing of socially relevant information (e.g. face recognition) have consistently been reported (Adolphs et al. 2001). The most consistent familial traits are language and communication skills, insistence on sameness and non-verbal IQ (Szatmari et al. 2007). In ADHD, cognitive studies find widespread abnormalities in children and adults with the disorder, particularly in the executive function domains of response inhibition/delay aversion and sustained attention (Willcutt et al. 2005) and in reaction time variability (Klein et al. 2006), one of the most discriminating between ADHD and controls. Attention, and in particular, sustained attention, deficits are prominent in ADHD, with unaffected siblings performing better than their affected siblings but worse than healthy controls (Slaats-Wilemse et al. 2005). Sustained attention has been examined in preschool twins using the Amsterdam neuropsychological tasks set with correlations higher for MZ compared to DZ twins suggesting a heritability of 0.46–0.72 for this measure (Groot et al. 2004). Yet many proposed endophenotypes in ADHD have not been examined in classical population-based twin studies. It is not yet clear if these impairments are related or perhaps have separate aetiological pathways. Kuntsi et al. (2010) examined this in a large sample of ADHD probands and their siblings using a multivariate familial factor analysis approach. The results suggest that two familial phenotypes, mean reaction time/reaction time variability and omission/commission errors on the go/no-go task reflect 85–98 and 13% of the familial variance of ADHD, respectively. The findings for response time variability reflect recent population twin data (Wood et al. 2010).

4 Cytogenetics and Cognitive Phenotypes

Cytogenetic methods to investigate chromosomes microscopically, initially allowed gross examination of chromosomes, incorporating staining methods which identified cytogenetic bands on chromosomes; these evolved to molecular hybridization methods, known as fluorescent in situ hybridization (FISH). Cytogenetics methods contribute to localising susceptibility genes by identifying chromosomal abnormalities such as deletions or translocations, which segregate with a disorder in families or are found more commonly in cases than in control populations.

A number of these (described further below) have been identified for childhood disorders affecting cognitive ability and mental health, including sex chromosome aneuploidies such as 47XYY, Klinefelters and Turner's syndrome; fragile X, Williams' syndrome, and autism. In adult-onset disorders, two cytogenetic abnormalities have provided consistent evidence that the chromosomal regions involved contribute to disease risk. One is a balanced translocation t(1;11)(q42.1;q14.3) that co-segregates with schizophrenia within a large Scottish pedigree. The other involves one of the commonly known chromosomal abnormalities, small interstitial deletions of chromosome 22q11. The latter cause velo-cardio facial syndrome (VCFS), which increases the risk of psychosis by at least 20-fold, and also presents with cranio-facial dysmorphology and congenital heart disease. In carriers of these cytogenetic abnormalities, both general and specific cognitive effects are associated, including language deficits (Williams syndrome, autism), specific learning deficits (e.g. mathematical ability in Turner's syndrome, Williams syndrome), and memory deficits (Schizophrenia); these are discussed next.

4.1 Cytogenetics of Childhood-Onset Neuropsychiatric Disorders

Sex chromosome aneuploidies (SCAs), where there is variation in the number of sex chromosomes, are the most common chromosomal abnormalities occurring in 1 per 400 births. The commonest syndromes include the 47XYY syndrome,

Klinefelter syndrome (47XXY), and Turner's syndrome (XO). Frequently deficits in speech and language, motor skills, and educational achievement are reported in relation to all SCAs although in general affected individuals usually live independently and the severity of cognitive deficits is not clear (Leggett et al. 2010). Males with 47XYY frequently present normally and the phenotype may go undetected. In general in SCAs, the presence of an additional X chromosome is frequently associated with lowered verbal IQ. Klinefelter syndrome is associated with executive function deficits (Samango-Sprouse 2010). The absence of an X chromosome in girls is also associated with cognitive deficits. Turner syndrome (XO), an SCA affecting girls is characteristically associated with visual-spatial and executive function deficits and social cognitive impairments (Hong et al. 2009). Mathematical learning difficulties are prevalent in Turner syndrome which may relate to deficits in executive function or spatial deficits although this is not clear (Mazzocco 2009). Non-verbal learning deficits are present in 80% of girls with Turner syndrome. The syndrome is also associated with social cognitive deficits such as facial emotion recognition and gaze perception that are also reflected by reported neuroanatomical abnormalities in the amygdala (Burnett et al. 2010).

William Beurens Syndrome (WBS) is a contiguous gene syndrome arising from a deletion of approximately 1.5–1.8 Mb affecting 28 genes on chromosome 7q. It is associated with a characteristic physical phenotype and a cheerful, sociable manner. Individuals with WBS frequently have intellectual disability but present with superficially good language ability. They present with poor visuo-spatial ability (Pober 2010). A microduplication syndrome of the same region has also been described with somewhat contrasting phenotypic characteristics; intellectual disability is also reported; speech and language deficits are more common while visuo-spatial functioning is spared (Merla et al. 2010). Both the duplication and deletion syndromes are reported to be associated with ADHD and autistic type deficits; social deficits and aggression are more characteristic of the duplication syndrome. A clear relationship between specific genes within the affected region and the cognitive phenotype has not been determined, however, several interesting candidate genes exist. Frizzled drosophila homologue of 9 (FZD9) is expressed in the hippocampus and null mouse mutants have defects in memory and learning although WBS-related phenotypic features are not universally reported (Ranheim et al. 2005). The Syntaxin 1A (STX1A) gene is a brain-expressed protein implicated in presynaptic vesicle docking. Expression levels of this gene are correlated with intelligence in WBS (Gao et al. 2010). LIM kinase 1 (LIMK1) is a serine protein kinase predominantly expressed in the nervous system and implicated in synapse formation (Scott and Olson 2007). LIMK1 knockout mice have abnormalities in the dendritic spine morphology and have impaired fear conditioning and spatial learning (Meng et al. 2002). CAP-Gly domain-containing linker protein 2 (CLIP2) is expressed widely in the nervous system, including the hippocampus. Knockout mice have growth deficits, motor coordination deficits and altered synaptic functioning in the hippocampus (Hoogenraad et al. 2002). Clinically, patients with deletions not including CLIP2 have reduced visuospatial impairments and less deficits in gross and fine motor skills (Ferrero et al. 2010).

Fragile X syndrome, is the commonest cause of inherited intellectual disability. Not specifically a cytogenetic abnormality, it is caused by a trinucleotide repeat expansion (CGG) in the FMR-1 protein on the X chromosome. The condition is associated with autistic type difficulties in social interaction, e.g. gaze avoidance, social anxiety, poor pro-social behaviour, and peer relationships. Decreased pre-frontal brain activation is reported and reduced frontal and temporal cortical volumes are reported compared with control subjects or individuals with idio-syncratic autism (Hoeft et al. 2011). Declines in IQ and adaptive function are also reported in longitudinal studies with greater declines in males (Fisch et al. 2010). Declines in central executive and verbal working memory are also reported in later life in males. Working memory deficits correlate with the length of the CGG expansion (Cornish et al. 2009). Attentional difficulties, anxiety, aggression, and mood symptoms are also reported (Boyle and Kaufmann 2010).

Prader-Willi syndrome (PWS) is a rare genetic syndrome associated with nonexpression of a set of genes on the paternal chromosome 15q11–q13. This may be due to deletion of the region on the paternal chromosome (70%) or uniparental disomy (UPD) (25%) involving the maternal chromosome. The region is imprinted and also associated with Angelman's syndrome caused by lack of expression of the *UBE3A* gene originating on the maternal chromosome. Individuals with PWS have a characteristic physical phenotype and usually intellectual disability (approximately 70% of cases), however, they often have strengths in visual perception, reading, and vocabulary. Strengths in verbal IQ are reported in UPD compared with carriers of the deletion (Copet et al. 2010). Deficits in auditory processing are reported as are weaknesses in mathematics, visual, and auditory memory and auditory attention. Larger deletions in the region tend to be associated with greater impairments in cognitive ability (Milner et al. 2005). In Angelman syndrome, deletions are associated with greater cognitive impairment than UPD and cognitive skills are stronger than motor or language skills with relative strengths in the area of receptive language (Gentile et al. 2010).

Down syndrome is caused by Trisomy for all or part of chromosome 21. Associated with learning disability and early onset dementia, the cognitive profile of individuals with Down syndrome can be heterogeneous, but relative strengths in receptive language compared with expressive language are reported.

4.2 Cytogenetics of Adult-Onset Neuropsychiatric Disorders

Cytogenetic studies have also implicated both a relatively common deletion of chromosome 22q11.2 and a complex, rare translocation as increasing risk for psychotic disorders. The gene disrupted-in-schizophrenia-1 (DISC1) was identified at a balanced translocation between chromosome one and eleven, which strongly co-segregates with mental illness in a Scottish family. Although the index case had a diagnosis of conduct disorder, within the family 18 of 29 (70%) translocation carriers had a major mental illness (schizophrenia, bipolar disorder or major depressive disorder), whereas none of 38 non-translocation carriers had such a diagnosis (Blackwood et al. 2001). Outside this family, there is equivocal support for involvement of the gene as a risk factor for major mental disorders (Porteous et al. 2006) and there has been substantial investigation of its role in neurodevelopment. At a cognitive level, within the affected Scottish family there was no difference in mean IQ between 12 relatives with the translocation and eight with a normal karyotype. Despite this, unaffected, as well as affected, translocation carriers had abnormalities on event-related potential (ERP) P300-typical of schizophrenia and bipolar disorder-suggesting that specific cognitive processing deficits may be important.

Using methods (described in Sect. 5), Finnish researchers have identified linkage between a locus containing the DISC1 gene and impaired working memory dysfunction (Gasperoni et al. 2003) and visual working memory (Hennah et al. 2005). Two subsequent studies have reported association between markers at the DISC1 locus and impairments in verbal learning and memory (Cannon et al. 2005); working memory (Callicott et al. 2005); and reduced hippocampal volumes (Cannon et al. 2005; Callicott et al. 2005). In assessing the evidence implicating hippocampal dysfunction in DISC1 it is worth noting that different genetic variants have been investigated, with potentially different phenotypic effects (e.g. conflicting evidence for involvement of a DISC1 SNP and cognitive ageing) but none of these have been confirmed as functionally causal.

The 22q11.2 deletion syndrome (22q11.2DS; also known as velo-cardio-facial syndrome (VCFS)) is caused by the most common large micro deletion in the human genome and has an incidence of one in ~4000 live births. The phenotype is highly variable and can affect multiple organs and tissues, but carriers have a 30-fold increased risk of schizophrenia and an increased rate of other psychiatric phenotypes including ADHD and autistic spectrum disorders. Many, but not all, carriers fall into the lower than average IQ (FSIQ 70–75) range or have mild learning disability. Most studies, particularly in children, report higher scores on verbal than non-verbal tasks. Investigation of these non-verbal deficits across age groups indicate impairments in comparisons of magnitude and time duration, which implicate parietal and frontal circuitry underlying attentional and numerical cognition (reviewed, Karayiorgou et al. 2010). However, the ability to inhibit processing of extraneous information is also critical and impaired performance on inhibition tasks is also reported in affected children. Impairments in other aspects of inhibition including prepulse inhibition and reduced frontal lobe activation during the mismatch negativity paradigm have also been reported. These findings indicate that how information is selected or inhibited in attentional processing is important in the non-verbal deficits evident in this syndrome. Some brain regions are either structurally enlarged or reduced in children with 22q11.2DS compared to healthy controls. There appears to be cortical thinning and reduced cortical gyrus complexity in frontal and parietal cortices. The few reported functional and connectivity studies broadly confirm the findings of the neurocognitive tests with performance on arithmetic and spatial attention tasks correlating with frontal and parietal connectivity and functional measures.

There are a number of ways in which loss of genes at this locus could contribute to the expression of these cognitive phenotypes. First, under-expression of a single gene at this locus could exert a major effect on the phenotype. Second, the effect could be the result of under-expression of a number of proximally located genes. Finally, the microdeletion could unmask the effects of one or more recessive mutations. How these effects can be functionally investigated in an animal (e.g. mouse model) is beyond the scope of this chapter, but is reviewed in Karayiorgou et al (2010). The example of 22q11.2DS identified several key issues in cognitive genetics. Unlike the risk variants reported in the association studies of DISC1, 22q11.2DS is uncommon, so ascertainment of sufficient samples for neurocognitive studies is challenging. Compared to DISC1, we are more informed as to the potential molecular mechanisms involved making this locus more tractable for functional studies in model systems. Unlike the DISC1 example, which implicates one gene, the 22q11.2DS locus potentially involves more than 25 genes, which is more biologically challenging. In the next section, we consider how advances in genotyping technology have expanded the range of risk loci available for investigation in cognitive genetic studies of neuropsychiatric disorders.

5 Linkage and Candidate Gene-Based Approached to Cognitive Genetics

The identification of common polymorphic genetic markers shared in individuals within populations was a key step for molecular genetics. This made possible two new approaches to identify genes or genetic loci, which caused or contributed to phenotypes or traits. The first, genetic linkage analysis, capitalised on genetic maps of markers across the genome to investigate large single pedigrees or multiple families affected with a given disorder. Statistical evidence that specific markers co-segregated with illness, could be used to map loci linked to this disorder. Linkage analysis is most effective when there is a strong correlation between the phenotype being measured and the inferred genotype at the markers being tested for co-segregation. Further fine-mapping of linked loci, could then implicate specific risk genes. This approach was highly successful for Mendelian disorders, including many with cognitive phenotypes, but less successful for disorders with a more complex genetic aetiology and weaker correlation between phenotype and inferred genotype.

The second strategy, which could use either case-control or family-based designs, directly targeted genetic markers at specific genes which were implicated in a disorder through understanding of the disease process (functional candidate gene studies) or targeted likely candidates within a region implicated by linkage or cytogenetic studies (positional candidate genes studies). Association studies are based on populations instead of pedigrees, and compare frequencies of marker alleles in affected individuals versus unaffected individuals from the general population. Challenges in identifying such candidate genes include uncertainty regarding the biological aetiology of the disorder, and limited study power to detect common alleles of small effects. This power issue becomes even more relevant for the genome-wide association studies (GWAS) described in the Genomics section.

5.1 Linkage and Candidate Genes Studies for Disorders of Childhood

5.1.1 ADHD

Genetic studies of the ADHD clinical phenotype have followed the traditional pathway of twin and adoption studies to establish heritability, followed by genetic linkage studies, association studies based on candidate genes and more recently genome wide association studies (GWAS) and analysis of rare structural variants (detailed below).

Results from the handful of published linkage studies using the ADHD diagnosis as a phenotype (Fisher et al. 2002; Arcos-Burgos et al. 2004; Asherson et al. 2008; Rommelse et al. 2008), show some degree of overlap for regions on chromosomes 5p, 9q, 16q, and 17p if nominally significant findings are considered; however, no regions have achieved genome-wide significance using strict criteria. Attempts have been made to include neurocognitive measures in the linkage analysis of ADHD to identify quantitative trait loci linked to these traits. This approach assumes that the neurocognitive impairments in ADHD index a latent trait, or traits, that overlap, at least in part, with the heritable pathophysiology of ADHD. Taking this approach, Rommelse et al. (2008) examined candidate endophenotypes in a genome-wide search for susceptibility loci for ADHD. This study found strong evidence for linkage to 2q21.1 and 13q12.11 for measures of motor timing and digit span measures, respectively, incorporating ADHD symptoms as covariates. Doyle et al. (2008) identified a region on 3q13 showing suggestive evidence for linkage to several neurocognitive traits and inattention symptoms in ADHD. None of these studies have produced convincing evidence for linkage, making the presence of one or a small number of gene variants with a large effect on a given trait measure unlikely.

Early candidate gene studies focused on a range of candidate genes with some a priori evidence for a potential role in ADHD pathophysiology. As with the linkage studies described above, putative ADHD risk variants at candidate genes have been tested for association against clinical and cognitive variables. Kebir et al. (2009) reviewed 29 studies examining 10 genes (DRD4, DAT1, COMT, DBH, MAOA, DRD5, ADRA2A, GRIN2A, BDNF and TPH2) in relation to neuropsychological traits relevant for ADHD. For DAT1, there are conflicting results in relation to omission and commission errors, but more consistent findings that increased reaction time variability (Bellgrove et al. 2005a) and abnormalities in spatial attention (Bellgrove et al. 2005b) are associated with the ADHD associated

10-repeat variant. Against what might have been expected, several studies (Manor et al. 2002; Bellgrove et al. 2005a, b, c), reported better performance on tests of attention in children with the 7-repeat DRD4 variant previously shown to be associated with ADHD. This is a similar finding to studies in psychoses where the GWAS identified *ZNF804A* risk variant (discussed below) may identify a patient subgroup with relatively spared cognitive performance, suggesting that the DRD4 risk variant indexes a pathophysiological pathway to ADHD not mediated by poor performance on cognitive measures. The effect of the 7-repeat variant was confined to children with ADHD and not seen in controls and in a more recent study (Johnson et al. 2008), spectral analysis of reaction time variability supported the hypothesis that the association of greater variability was with the absence of the 7-repeat allele and was also specific to ADHD. Other ADHD candidate genes examined in relation to cognitive function include the X-linked steroid sulfatase gene where case reports of deletions have been found in cases with neurodevelopmental disorders associated with abnormal cognitive function and ADHD (Doherty et al. 2003). Stergiakouli et al. (2011) showed that ADHD associated risk variants were associated with inattentive symptoms and poor performance on verbal IQ and comprehension subtests in ADHD subjects but not controls.

5.1.2 Autism Spectrum Disorders

A number of linkage regions, replicated in two or more studies have been identified in autism 2q21–33, 3q25–27, 3p25, 4q32, 6q14–21, 7q22, 7q31–36, 11p12–13, 17q11–21 (reviewed by (Freitag et al. 2010). The chromosome 2q region had marginally stronger evidence for linkage in individuals with language delays (Buxbaum et al. 2001). A further study stratifying linkage analyses found nonsignificant evidence of linkage in individuals with IQ > 70 and those with delayed language (Liu et al. 2008). Multiple candidate gene studies have been conducted but none have been reliably replicated. Consequently, we have focused here on genetic studies in social cognition where a convergence of evidence appears to support the role of neuropeptides oxytocin and vasopressin.

Nonapeptides and Social Cognition

Oxytocin (OXT) and vasopressin (AVP) are highly conserved neuropeptides with marked diversity in the regulation of their receptors. Modulated significantly by sex steroids, they are likely to have sexually dimorphic effects and are therefore of interest for further investigation in disorders such as autism, which show marked gender bias (M:F ~ 4:1). Considerable evidence from animal literature has implicated nonapeptides oxytocin and vasopressin in social behaviour (Insel 2010) relating particularly to pair bonding, maternal care, social recognition, and response to threat (reviewed by (Donaldson and Young 2008). Administration of oxytocin to humans is associated with reduced anxiety, alteration in parenting behaviour, increases in prosocial behaviour (e.g. trust, generosity, altruism and betrayal aversion), reduction in gaze aversion, improved mentalisation, and differential amygdala activity in fMRI in response to face perception and changes in social memory (Skuse and Gallagher 2011). Studies reporting association with OXT and autism are inconsistent and no evidence for association has emerged from GWAS studies. However, beneficial effects of exogenous oxytocin on core ASD symptoms have been reported offering the potential possibility of new therapeutics (Green and Hollander 2010). Vasopressin has been implicated in aggression, social recognition, and pair bonding. The AVP receptor 1A gene is highly conserved. It contains genetic variation reported to influence species-specific differences in pair bonding in animal studies (Wang and Aragona 2004). Genetic variation in AVPR1A has been investigated in ASD with variable reports of association (Kim et al. 2002; Wassink et al. 2004; Yirmiya et al. 2006). One of the variants in humans has demonstrably reduced expression (Tansey et al. 2011) possibly demonstrating a functional route for genetic association with the gene in autism. The wide-ranging effects of these neuropeptides on human social behaviour are perhaps not specific to autism and may potentially have utility in a wider range of psychiatric disorders with social cognitive deficits.

5.1.3 Intellectual Disability

Notwithstanding the high heritability of intelligence, little progress has been made in identifying loci reliably linked or associated with intelligence in normal population samples. There are exceptions, such as the association, predominantly in older people, between ApoE variants and general cognitive ability, episodic memory, and executive function, and the weak associations reported with COMT and BDNF variants accounting, if true for only a very small proportion of the variance in intelligence. In contrast, several hundred genes are known to be associated with intellectual disability (Chelly et al. 2006).

Genetic forms of ID are divided into syndromic ID, characterised by associated clinical, radiological, metabolic or biological features, and non-syndromic ID in which cognitive impairment represents the only manifestation of the condition. The distinction might be helpful for clinical purposes, but recent phenotype–genotype studies are blurring the distinction. Causes of ID are extremely heterogeneous and include environmental forms (e.g. premature birth, perinatal brain ischaemia or foetal alcohol syndrome), disorders due to chromosomal abnormalities (including sub-microscopic copy number variation discussed below) and conditions due to monogenic causes or dysregulation of imprinted genes. About 50% of cases with moderate or severe ID have a definable cause with a lower percentage in milder cases. Taken together, the emerging genetic findings in ID are suggesting a neurobiology around synaptogenesis, synaptic activity, and plasticity with aberrant length and density of dendritic spines a frequent histological finding. Genetic defects and biochemical abnormalities have been described in several pathways that feed into synaptic function including the RhoGTPase signal transduction pathway; with loss of function at its components PAK3, OPHN1, TM4SF2, and FMRP leading to LD and the Ras/MAPK transcription signaling cascade; with the genes NF1, RKS2, CBP, and PAK3 involved in ID. Many of these genes when mutated in animal models affect learning and memory processes that require gene transcription and translation of proteins.

5.2 Linkage and Candidate Genes Studies for Disorders of Adult-Onset

5.2.1 Schizophrenia

A meta-analysis of more than 30 schizophrenia linkage studies by Ng et al. (2009), suggests the involvement of multiple chromosomal loci in schizophrenia susceptibility. Few investigations focussed on cognitive phenotypes in linkage analysis of such families have been reported. One example, reported by Almasy et al. 2008, investigated 43 families and identified significant linkage to the chromosome 5q region for the cognitive phenotypes of abstraction and mental flexibility. A more recent study of 557 sibling pairs of Han Chinese ethnicity identified association with the 12q24.32 locus and undegraded CPT hit rate (Lien et al. 2010). Rather than focussing on individual neurocognitive phenotypes, Hallmayer et al. (2005) identified families with co-segregation of more pervasive cognitive deficits and identified linkage in these families to chromosome 6p24, a region that had previously been implicated in schizophrenia risk across multiple studies (Straub et al. 1996). Individual markers at this locus have also been associated with deficits in CPT performance in the Han Chinese population detailed above (Lin et al. 2009). The lack of consistency in measured phenotypes makes replication, and final interpretation of these results difficult. Very few studies of this type have been reported for other psychiatric disorders. For example, in bipolar disorder, where many large-scale linkage analyses have also been reported the focus has been on dividing families according to clinical rather than cognitive covariates.

Candidate gene studies of cognitive phenotypes have received much wider attention. This was prompted by studies of functional variants at two candidate genes, the catechol-O-methyltransferase (COMT) and brain derived neurotrophic factor (BDNF) genes. From both animal and human studies it is known that reduced dopamine in prefrontal cortex is associated with impaired performance on cognitive testing. Deficits in working memory can be reversed with dopamine agonists, but both very low and very high levels of dopamine activity are associated with impaired prefrontal cortex function. The COMT gene appears to be key to dopamine catabolism in the prefrontal cortex (PFC) and is a logical candidate for investigation in disorders such as schizophrenia as well as for studies of cognition. Numerous association studies for neuropsychiatric phenotypes have been performed with equivocal results, with several large meta-analyses failing to find association with schizophrenia (Munafo et al. 2005). In the first cognitive study, Egan et al. (2001) reported that the high-activity val allele was associated with poorer performance on the wisconsin card sort test (WCST) and reduced efficiency of physiological response of the dorsolateral prefrontal cortex during a working memory task. This finding has received consistent replication and a meta-analysis of 12 studies supports the original WCST finding (Barnett et al. 2007). The WCST is a complex problem solving task with many cognitive components and several authors have tried to identify simpler tests for specific components of this task involving cognitive stability and flexibility. Because COMT influences the ratio of activation of D1/D2 receptors and D4 receptors are known to have an effect on PFC function, variants in these three genes have also been investigated in cognitive studies of adults with no clear findings emerging.

BDNF is known to have an important role in learning and regulates activity-dependent synaptic plasticity necessary for short- and long-term memory storage (Alonso et al. 2002). Studies of BDNF have focused on a valine (val) to methionine (met) substitution in the 5′ region of the gene, which decreases BDNF activity-dependent secretion. Association between the variant and neuropsychiatric clinical phenotypes has been reported, although replication of these findings has been inconsistent. A meta-analysis of studies across multiple phenotypes reported that the met allele was associated with risk for eating disorders and schizophrenia and a protective effect for substance-related disorders (Gratacos et al. 2007). This may have represented a publication bias. On the basis of BDNF's known role in hippocampal function, it has been suggested that met allele carriers may have impaired performance on memory tasks. Supporting this hypothesis Egan et al. (2003) identified poorer episodic memory performance, a disruption in normal hippocampal fMRI findings during a working memory task and reduced hippocampal levels of a marker for neuronal function in schizophrenia patients and healthy controls. The same year, Hariri et al. (2003) showed that met-carriers had reduced hippocampal engagement during encoding and retrieval of a spatial task and also made more recognition errors on the task. Subsequently, it was reported that val/met heterozygotes have lower hippocampal volumes than carriers of the val/val genotype (Pezawas et al. 2004). Within schizophrenia patients, met-carriers are also reported to have poorer medial temporal lobe-related performance and correspondingly smaller temporal and occipital lobar grey matter volume (Ho et al. 2006). Although negative studies have been reported, most of the available data supports a modest association between the met allele and reduced cognitive performance.

With the emergence of putative candidate genes from the schizophrenia literature these were also systematically investigated for cognitive phenotypes. The list of investigated genes includes Neuregulin-1, Dysbindin, and DAO. Unlike the case of COMT and BDNF, no clear functional variants have been identified at these genes and multiple studies have reported different risk variants, alleles, or haplotypes. This makes direct comparison of studies, which have often also examined different phenotypes, difficult (reviewed in Gill et al. 2010).

5.2.2 Cognitive Ageing and Psychiatric Disorders

Association between an increased risk for Alzheimers disease and the ε4 allele of the apolipoprotein E (ApoE) gene is one of the most robust findings in complex disorder genetics. Carriers of one copy of this allele are 3–4 times more likely to develop late-onset Alzheimers disease (LOAD), but carriers of two copies have a more than 10-fold increase in risk (Farrer et al. 1997). Multiple studies have shown that APOE ε4 is associated with cognitive decline in patients with AD. More recently it has been demonstrated that this allele has an effect on cognitive performance in non-patient groups as well. Carriers of the risk allele perform significantly poorer on tests of episodic memory, global cognitive ability, executive functioning, and perceptual speed although the effect sizes are small. There was no difference between carriers and non-carriers for tests of attention, primary memory, verbal ability, and visuo-spatial skill. For the domains where differences were detected these differences became more significant with increased age (Wisdom et al. 2011). These data lead researchers to explore the effects that other candidate genes may have on cognitive ageing. A study investigating 10 candidate genes (including BDNF, COMT and DISC1) for cognitive function failed to identify association with performance and cognitive ageing in the Lothian birth cohort of over a 1,000 Scottish 70-year old individuals (Houlihan et al. 2009).

6 Application of Genomics Methods

Genome-wide association studies (GWAS), by combining advances in high-throughput genotyping platforms and understanding of common genetic variation in populations, allow most common variation in the genome to be tested in a single, usually case–control experiment. Testing all genes is a powerful hypothesis-free approach and GWAS have proven remarkably successful at identifying common risk variants for complex human disease. However, this comes with a significant multiple testing burden and requires large sample sizes—in the thousands or tens of thousands—to identify what are typically modest gene effects (Corvin et al. 2009).

In psychiatry as with other medical specialties, this requirement has driven collaboration, for example, the formation of the psychiatric GWAS consortium, which is currently performing meta-analysis of GWAS data for schizophrenia, bipolar disorder, autism, recurrent major depression, and ADHD (Psychiatric GWAS Consortium Coordinating Committee 2009). For many of these disorders novel susceptibility loci have been identified (detailed below). GWAS can also inform on the genetic architecture of psychiatric disorders: identified loci appear in many cases to increase risk across traditional diagnostic boundaries; a substantial proportion of schizophrenia and bipolar disorder risk may involve thousands of overlapping gene variants of small effect (International Schizophrenia Consortium 2009a, b) whereas smaller numbers of variants of large effect appear involved in autism susceptibility.

GWAS platforms were designed to assay common genetic variation and SNPs with a population frequency of at least 5%, however, rare or even unique genetic variants are much more frequent in the human genome. Until now our ability to test for involvement of this type of variation in human disease 'the rare variant common disease hypothesis' has been very limited. GWAS platforms and custom-designed microarrays using comparative genomic hybridization (CGH) have allowed investigation of one class of rare genetic variation, namely, copy number variation (CNV). This provides some insight into likely challenges for cognitive research in analysing rare variants: an issue that will become more relevant with the increasing availability of whole-genome sequence data.

6.1 Genome-Wide Association Studies

6.1.1 GWAS for Adult-Onset Psychiatric Disorders

Schizophrenia

Nine schizophrenia loci have been identified: the zinc finger protein 804A (ZNF804A) gene on chromosome 2q32; at the major histocompatibility complex (MHC) region on chromosome 6p21–22; upstream of the neurogranin (NRGN) gene on chromosome 11q24; at the transcription factor 4 (TCF4) gene on chromosome 18q21; downstream from microRNA miRNA137 on chromosome 1p21.3; a 0.5 Mb gene-rich region on chromosome 10q24.32; an intronic SNP in the CUB and sushi multiple domains 1 (CSMD1) gene; and common variants in gene deserts on chromosomes 2q32.3 and 8p21.3. Additionally, there is substantial evidence for overlap in particular between schizophrenia and mood disorders, as the schizophrenia risk variant at ZNF804A has also been implicated in bipolar disorder (Williams et al. 2011) and the CACNA1C variant, identified in bipolar disorder has also been implicated in schizophrenia and recurrent major depression (Green et al. 2010). A logical next step is for cognitive studies to test whether specific neural mechanisms underlie this susceptibility and its clinical expression.

The psychosis risk variant at gene *ZNF804A* has received the most attention to date. Esslinger et al. 2009 investigated the influence of the risk variant (rs1344706) on cortical activity within, and connectivity between, regions during working memory (N-back task) and emotional recognition task performance in a sample of 115 healthy controls. Differences in functional connectivity, but not regional activation, were observed. They reported reduced connectivity in the dorso-lateral prefrontal cortex (DLPFC) between and within hemispheres, but also increased connectivity between the hippocampal formation (HF) and the DLPFC, and between the amygdala and the HF, orbitofrontal cortex and prefrontal cortex. They have subsequently reported evidence for involvement of the variant in aberrant brain activation during social information processing using a theory of mind (TOM) task (Walter et al. 2011) and in state-independent inter-hemispheric processing (Esslinger et al. 2009). Their interpretation that the risk allele has a deleterious effect on cognitive performance has been questioned by several more recent studies. First et al. (2010) found and replicated evidence for better cognitive performance on working memory and episodic memory tasks—which involve the DLPFC and HF—in patient carriers of the risk allele. This effect was not present in controls. In a subsequent study, of a different patient group, the authors found relatively larger hippocampal volumes in risk allele carriers (Donohoe et al. 2011a, b). These data suggest that the *ZNF804A* risk variant may identify a patient subgroup with relatively spared cognitive performance, but possibly more social deficits. Although we note that several smaller equivocal studies have also been reported (Lencz et al. 2010; Balog et al. 2010). Further studies particularly in the domain of social cognition would be useful.

Of the other identified schizophrenia loci, some are large and implicate many genes (e.g. the MHC region) and some map to gene deserts which are not obvious candidates for involvement in cognitive functioning. Of the identified genes, neurogranin (NRGN) is the most compelling target as it plays an important role in calcium–calmodulin signaling, is abundantly expressed in hippocampus, and NRGN knockout mice have severe deficits in hippocampus-dependent tasks. However, a recent study by Donohoe et al. (2011b) failed to identify a strong relationship between the risk allele and neuropsychological performance in either patient or control populations on general cognitive ability, verbal episodic and working memory, spatial episodic or working memory or attentional control.

Bipolar Disorder

In bipolar disorder the best supported loci are the calcium channel, voltage-dependent, L type, alpha 1C subunit (CACNA1C), and the ankyrin 3, node of Ranvier (ANK3) genes (Ferreira et al. 2008). Perceived wisdom is that cognitive deficits are less prominent in bipolar disorder, although a range of abnormalities have been reported including altered identification of emotional stimuli (e.g. facial expression), processing speed, working memory, and impairments in sustained attention (Arts et al. 2010). Cognitive phenotypes in bipolar disorder may vary with mood state and have received less attention in family studies (to estimate heritability) than equivalent studies in schizophrenia. Of the common risk variants identified in GWAS studies, the *CACNA1C* gene has received the most attention as non-synonymous mutations of *CACNA1C* cause Timothy syndrome, a multi-organ disorder, which includes cognitive impairments (Splawski et al. 2005). *Cacna1c* heterozygous female mice also demonstrate mood-related phenotypes including reduced risk-taking behaviour and increased anxiety (Dao et al. 2010).

Association studies have been reported in case and control populations between the risk allele at rs1006737 across different neuropsychological testing paradigms and imaging studies. The first reported study (Krug et al. 2010) found reduced semantic verbal fluency with increased activation of the left inferior frontal gyrus and left precuneus in healthy male subjects who carried the risk variant. The authors acknowledged that the data to suggest reduced verbal fluency in euthymic bipolar disorder is limited and their results require independent replication. To date five imaging studies have been published, with somewhat mixed results. An initial report of reduced grey matter volume in healthy UK carriers of this risk variant (Kempton et al. 2009) did not replicate in a much larger German control sample (Franke et al. 2010). The latter reported association between genetic variation at the gene and reduced brainstem volume, but this was with different SNPs at the gene and requires independent replication. Studies using blood-oxygenation level-dependent (BOLD) functional magnetic resonance imaging (fMRI) have targeted circuits potentially implicated in bipolar disorder. A study of patients and controls, by Bigos et al. (2010) implicated circuits putatively involved in bipolar disorder and schizophrenia. They identified a trend for increased hippocampal activity during emotional processing and also greater prefrontal cortical activity during a working memory paradigm, a pattern previously associated with putative schizophrenia risk variants. A smaller study of healthy controls targeted increased limbic activity as a bipolar disorder phenotype and identified association with increased amygdala activity in response to reward. Erk S. Meyer-Lindenberg et al. (2010) in studying brain activation during a declarative memory task identified reduced bilateral hippocampal activation during episodic memory recall and reduced coupling between left and right hippocampal regions in 110 healthy subjects.

Alzheimers Disease

Until recently, APOE was the only gene known to increase risk of the common form of Alzheimer's disease with late-onset. A number of new susceptibility variants have been identified by GWAS including novel loci for late-onset Alzheimers disease (AD) implicating the genes clusterin (CLU), the phospatidlyinositol-binding clathrin assembly protein gene (PICALM), the complement receptor gene (CR1), the bridging integrator 1 gene (BIN1), the ATP-binding cassette (ABC) transporter gene (ABCA7), MS4A gene, CD2AP, CD33 and EPHA1 (Hollingworth et al. 2011). These genes appear to be involved in different processes, with five being linked to immune function; four to cell membrane endocytosis; and three being involved in lipid processing. As of yet none have been tested for involvement in specific aspects of cognitive functioning or their role in cognitive ageing.

6.1.2 GWAS for Childhood-Onset Psychiatric Disorders

Data from GWAS studies to date have not generally supported the linkage regions previously identified for childhood-onset disorders. A region on chromosome 5p.14 harbouring cadherin genes CDH9 and CDH10 showed evidence for association in one study (Wang et al. 2009). A SNP at 5p.15 was located close to a taste receptor gene (TASR1) and a member of the semaphorin family (SEMA5A), the latter family of genes are implicated in axonal guidance (Weiss et al. 2009). Genome-wide evidence for association at the MACROD2 gene, a gene of uncertain function, was detected in a further study (Anney et al. 2010). Stratification in the latter analysis based on IQ and verbal status did not reveal statistically significant genome-wide evidence of association.

Results of the PGC consortium ADHD meta-analysis have yet to be reported and smaller GWAS studies have yet to provide genome-wide significant evidence of association.

6.2 Studies of Structural Genomic Variation in Neuropsychiatric Disorders

As recently as 2004, it was discovered that submicroscopic deletions or duplications involving the gain or loss of entire DNA segments (e.g. from a thousand to several million bases) are common (Sebat et al. 2004) and may encompass more than 10% of the average human genome. This technology also made it possible to test for the involvement of inherited or de novo CNVs in disease. It rapidly became evident that CNVs play an important role in susceptibility to neurodevelopmental disorders including autism (Sebat et al. 2009), learning disability, schizophrenia (Walsh et al. 2008), and ADHD (Williams et al. 2010). For autism, CNVs have been identified in at least 10% of cases, implicating a large number of novel genomic loci and risk genes (reviewed Betancur 2011). In schizophrenia, the seven most established CNVs collectively account for ~2–4% of susceptibility (reviewed Sebat et al. 2009). An excess of CNVs have also been identified in ADHD including a duplication of chromosome 16p13.11. Data for other neuropsychiatric disorders is more equivocal, with the exception of bipolar disorder where a large study of 1,697 cases and 2,806 controls found no evidence of either an increased total burden or association with individual CNVs (Grozeva et al. 2010).

6.2.1 Cognitive Studies of Structural Genomic Variants and Rare Mutations Implicated in Psychiatric Disorders

The results thus far challenge many preconceptions about the clinical entities being investigated. The same CNVs are being implicated in different disorders. For example, in autism although 70% of affected individuals have learning disability, almost all of the implicated CNVs have also been associated with learning disability. Many of the CNVs identified in schizophrenia have also been implicated in autism: the duplication reported in ADHD has been reported across all three disorders, being most common in ADHD patients with co-morbid learning disability. This diverse phenotypic expression extends beyond psychiatric phenotypes. The 1q21.1 deletion reported in schizophrenia (International Schizophrenia Consortium 2009a, b) is now known to be associated with a broad array of paediatric developmental abnormalities including autism, but also heart defects and cataracts (Mefford et al. 2008). These findings suggest that at least a subset of patients with clinical disorders have underlying rare genomic disorders.

Candidate gene studies of cognitive phenotypes have received much wider attention. This was prompted by studies of functional variants at two candidate genes, the catechol-O-methyltransferase (COMT) and brain derived neurotrophic factor (BDNF) genes. From both animal and human studies it is known that reduced dopamine in prefrontal cortex is associated with impaired performance on cognitive testing. Deficits in working memory can be reversed with dopamine agonists, but both very low and very high levels of dopamine activity are associated with impaired prefrontal cortex function. The COMT gene appears to be key to dopamine catabolism in the prefrontal cortex (PFC) and is a logical candidate for investigation in disorders such as schizophrenia as well as for studies of cognition. Numerous association studies for neuropsychiatric phenotypes have been performed with equivocal results, with several large meta-analyses failing to find association with schizophrenia (Munafo et al. 2005). In the first cognitive study, Egan et al. (2001) reported that the high-activity val allele was associated with poorer performance on the wisconsin card sort test (WCST) and reduced efficiency of physiological response of the dorso-lateral prefrontal cortex during a working memory task. This finding has received consistent replication and a meta-analysis of 12 studies supports the original WCST finding (Barnett et al. 2007). The WCST is a complex problem solving task with many cognitive components and several authors have tried to identify simpler tests for specific components of this task involving cognitive stability and flexibility. Because COMT influences the ratio of activation of D1/D2 receptors and D4 receptors are known to have an effect on PFC function, variants in these three genes have also been investigated in cognitive studies of adults with no clear findings emerging.

Extrapolating from examples of rare genomic disorders that have already been classified we could suspect that some will share core phenotypic features (e.g. Williams syndrome and Prader-Willi/Angelman syndrome) but others (e.g. chromosomal deletions involving 1q21.1 and 22q11.21) may have such a wide range of phenotypic expression as to encompass several clinical syndromes (Lee and Scherer 2010). Not all CNVs are causative: some are likely to have more modest effects on risk and probably interact with other genetic or environmental risk factors.

What does this mean for cognitive studies? We know that some of the implicated loci can have a profound effect on cognitive functioning leading to significant general learning disability. Would a general screen of IQ in clinical populations identify these CNV carriers? Or do some CNVs cause more subtle cognitive deficits? Plausibly, specific CNVs may impact on, and be extremely informative about, discrete aspects of cognitive functioning. Performing such studies is problematic because of the numbers involved.

Lessons can be learned from investigation of chr22q11.21, but many of the validated CNVs have a frequency of less than one in 500 in case samples, having large-scale collaboration and common assessment methods will be essential. Some consensus on batteries of tests is also essential to allow comparison across CNVs, which may be important in identifying where the phenotypic effects may be a consequence of involvement of the same molecular mechanism or pathway. Obvious targets for investigation are loci where a single, or small number of genes are disrupted as these are currently most tractable for other functional studies. CNVs are often complex and both gain or loss of function at a locus may need to be considered. For example, mutation of the gene encoding methyl-CpG-binding protein-2 (MECP2) causes Rett syndrome a neurodevelopmental disorder almost exclusively found in females, however, duplications or triplications of the gene are associated with developmental delay or learning disability in males.

7 Future Directions: New Phenotypes and New Approaches

A major development in cognitive genetics disorders has, as with psychiatric genetics studies, been the move from single gene studies to genome-wide studies of the genetic architecture of cognition. Examples of these studies are already published (e.g. Need et al. 2009; Davis et al. 2010) with several more in progress. These studies are likely to replicate difficulties found in genome-wide association studies (GWAS) of psychiatric disorder, in particular the low power to detect small effects in samples. Current estimates (Visscher and Montgomery 2009) suggest the need for more than 10,000 samples to detect small effects in psychiatric disorders; power to detect variants with an odds ratio of 1.1–1.2 (the effect size associated with already identified common variants) are likely to require even larger samples. If the effect size of 'cognition' genes is similar this requires us to plan experiments on a scale that is far beyond what was traditionally thought of as large. One example of an attempt to achieve the required scale is the COGENT consortium, which has to date amassed data on ~7,000 neuropsychological phenotyped healthy participants for the purposes of a genome-wide association study of general cognitive ability or 'g'. In advance of the results of these experiments, the evidence that cognitive phenotypes are unlikely to be much less complex in its genetic architecture than seen in psychiatric disorders should cause us to expect meaningful but small advances in our understanding of cognitive genetics.

At the same time, the increased ability to investigate rare genetic effects is identifying many interesting candidate genes for further cognitive studies. These may be particularly important as they may each be associated with significant risk. For studies of CNV carriers, the logistics of performing studies of sufficient statistical power is challenging, although not insurmountable as we know from the 22q11.2DS studies. One major question will be whether specific CNVs have distinct cognitive or clinical effects. This will be addressed by ongoing large studies in Europe where carriers of these variants are being targeted for neurocognitive and neuroimaging studies (Meyer-Lindenberg 2010).

This type of analysis will be even more challenging when there are large data resources available with full genome sequencing information. We already know that in the human population, many disease genes can show dozens or even hundreds of independent mutations. Mutations that have more severe functional consequences (i.e. to the production of the gene product) may be lethal or associated with severe phenotypes, such as microcephaly. Research from single gene cognitive disorders suggests that, depending on the functional consequence of the mutation, there may be a range of phenotypic outcomes, which include much more subtle phenotypes (Walsh and Engle 2010). Many of the genes known to be building blocks for important neurodevelopmental processes are likely to harbour mutations with these types of genetic effects. This offers a potential framework for understanding and potentially grouping molecular mechanisms at the level of the gene, or in pathways based on understanding of molecular neurodevelopment. The potential to perform cellular and animal studies as well as the ability to return to families, which carry mutations should offer fascinating insights into the genomic underpinnings of cognitive function. In performing such studies we must not lose sight of the fact that the molecular programmes that govern and modify neurodevelopment are affected by stochastic variation and actively influenced by the world that surrounds us. To maximise what we can learn about cognitive genomics we will need to understand the impact of environment.

References

Adolphs R, Sears L, Piven J (2001) Abnormal processing of social information from faces in autism. J Cogn Neurosci 13:232–240

Almasy L, Gur RC, Haack K, Cole SA, Calkins ME (2008) A genome screen for quantitative trait loci influencing schizophrenia and neurocognitive phenotypes. Am J Psychiatry 165:1185–1192

Alonso M, Vianna MR, Izquierdo I, Medina JH (2002) Signaling mechanisms mediating BDNF modulation of memory formation in vivo in the hippocampus. Cell Mol Neurobiol 22:663–674

Anney R, Klei L, Pinto D, Regan R, Conroy J et al (2010) A genome-wide scan for common alleles affecting risk for autism. Hum Mol Genet 19(20):4072–4082

Arcos-Burgos M, Castellanos FX, Pineda D, Lopera F, Palacio JD et al (2004) Attention-deficit/ hyperactivity disorder in a population isolate: linkage to loci at 4q13.2, 5q33.3, 11q22, and 17p11. Am J Hum Genet 75:998–1014

Arts B, Jabben N, Krabbendam L, van Os J (2010) A 2-year naturalistic study on cognitive functioning in bipolar disorder. Acta Psychiatr Scand 123:190–205

Asherson P, Zhou K, Anney RJ, Franke B, Buitlaar J et al (2008) A high-density SNP linkage scan with 142 combined subtype ADHD sib pairs identifies linkage regions on chromosomes 9 and 16. Mol Psychiatry 13:514–521

Balog K, Kiss I, Keri S (2010) ZNF804A may be associated with executive control of attention. Genes Brain Behav 10:223–227

Barnett JH, Jones PB, Robbins TW, Muller U (2007) Effects of the catechol-O-methyltrasnferase Val158Met polymorphism on executive function: a meta-analysis of the wisconsin card sort test in schizophrenia and healthy controls. Mol Psychiatry 12:502–509

Bediou B, Asri F, Brunelin J, Krolak-Salmon P, D'Amato T, Saoud M, Tazi I (2007) Emotion recognition and genetic vulnerability to schizophrenia. Br J Psychiatry 191:126–130

Bellgrove MA, Hawi Z, Kirley A, Gill M, Robertson IH (2005a) Dissecting the attention deficit hyperactivity disorder (ADHD) phenotype: sustained attention, response variability and spatial attentional asymmetries in relation to dopamine transporter (DAT1) genotype. Neuropsychologia 43:1847–1857

Bellgrove MA, Hawi Z, Kirley A, Fitzgerald M, Gill M, Robertson IH (2005b) Association between dopamine transporter (DAT1) genotype, left-sided inattention, and an enhanced response to methylphenidate in attention-deficit hyperactivity disorder. Neuropsychopharmacology 30:2290–2297

Bellgrove MA, Hawi Z, Lowe N, Kirley A, Roberston IH, Gill M (2005c) DRD4 gene variants and sustained attention in attention deficit hyperactivity disorder (ADHD): effects of associated alleles at the VNTR and -521 SNP. Am J Med Genet B Neuropsychiatr Genet 136B:81–86

Bellgrove MA, Mattingley JB (2008) Molecular genetics of attention. Ann N Y Acad Sci 1129:200–212

Betancur C (2011) Etiological heterogeneity in autism spectrum disorders: more than 100 genetic and genomic disorders and still counting. Brain Res 1380:42–77

Bigos KL, Mattay VS, Callicott JH, Straub RE, Vakkalanka R et al (2010) Genetic variation in CACNA1C affects brain circuitry related to mental illness. Arch Gen Psychiatry 67:939–945

Blackwood DH, Fordyce A, Walker MT, St Clair DM, Porteous DJ, Muir WJ (2001) Schizophrenia and affective disorders-co segregation with a translocation at chromosome 1q42 that directly disrupts brain-expressed genes: clinical and P300 findings in a family. Am J Hum Genet 69:428–433

Bora E, Vahip S, Gonul AS, Akdeniz F, Alkan M, Ogut M, Eryavuz A (2005) Evidence for theory of mind deficits in euthymic patients with bipolar disorder. Acta Psychiatr Scand 112(2): 110–116. http://www.ncbi.nlm.nih.gov/pubmed?term=%22Bora%20E%22%5BAuthor%5D, http://www.ncbi.nlm.nih.gov/pubmed?term=%22Vahip%20S%22%5BAuthor%5D, http://www.ncbi.nlm.nih.gov/pubmed?term=%22Gonul%20AS%22%5BAuthor%5D

Boyle L, Kaufmann WE (2010) The behavioral phenotype of FMR1 mutations. Am J Med Genet C Semin Med Genet 154C(4):469–476

Burnett AC, Reutens DC et al (2010) Social cognition in Turner's syndrome. J Clin Neurosci 17(3):283–286

Buxbaum JD, Silverman JM, Smith CJ, Kilfarski M, Reichert J (2001) Evidence for a susceptibility gene for autism on chromosome 2 and for genetic heterogeneity. Am J Hum Genet 68:1514–1520

Calkins ME, Tepper P, Gur RC, Ragland JD, Klei L, Wiener HW (2010) Project among African-Americans to explore risks for schizophrenia (PAARTNERS): evidence for impairment and heritability of neurocognitive functioning in families of schizophrenia patients. Am J Psychiatry 167(4):459–472

Callicott JH, Straub R, Pezawas L, Egan MF, Mattay VS et al (2005) Variation in DISC1 affects hippocampal structure and function and increases risk of schizophrenia. Proc Natl Acad Sci USA 102:8627–8632

Cannon TD, Huttenen MO, Lonnqvist J, Tuulio-Henriksson A, Pirkola T et al (2000) The inheritance of neuropsychological dysfunction in twins discordant for schizophrenia. Am J Hum Genet 67:369–382

Cavedini P, Zorzi C, Piccinni M, Cavallini MC, Bellodi L (2010) Executive dysfunctions in obsessive-compulsive patients and unaffected relatives: searching for a new intermediate phenotype. Biol Psychiatry 67(12):1178–1184. [Epub: 8 Apr, 2010]

Cannon TD, Hennah W, van Erp TG, Thompson PM, Lonnqvist J et al (2005) Association of DISC1/TRAX haplotypes with schizophrenia, reduced prefrontal gray matter and impaired short- and long-term memory. Arch Gen Psychiatry 62:1205–1213

Chelly J, Khelfaoui M, Francis F, Cherif B, Bienvenu T (2006) Genetics and pathophysiology of mental retardation. Eur J Hum Genet 14:701–713

Coleman N, Hare DJ, Farrell P, Van Manen T (2008) The use of the Social Cognitive Skills Test with children with autistic spectrum disorders. J Intellect Disabil 1:49–57

Copet P, Jauregi J et al (2010) Cognitive profile in a large French cohort of adults with Prader-Willi syndrome: differences between genotypes. J Intellect Disabil Res 54(3):204–215

Cornish KM, Kogan CS et al (2009) Lifespan changes in working memory in fragile X premutation males. Brain Cogn 69(3):551–558

Corvin A, Craddock N, Sullivan PF (2009) Genome-wide association studies: a primer. Psychol Med 40:1063–1077

Dao DT, Mahon PB, Cai X, Kovacsics CE, Blackwell RA, Arad M et al (2010) Mood disorder susceptibility gene CACNA1C modifies mood-related behaviours in mice and interacts with sex to influence behaviour in mice and diagnosis in humans. Biol Psychiatry 68:801–810

Davis OS, Butcher LM, Docherty SJ, Meaburn EL, Curtis CJ et al (2010) A three-stage genome-wide association study of general cognitive ability: hunting the small effects. Behav Genet 40:759–767

de Achával D, Costanzo EY, Villarreal M, Jáuregui IO, Chiodi A, Castro MN, Fahrer RD, Leiguarda RC, Chu EM, Guinjoan SM (2010) Emotion processing and theory of mind in schizophrenia patients and their unaffected first-degree relatives. Neuropsychologia 48(5): 1209–1215

Doherty MJ, Glass IA, Bennett CL, Cotter PD, Watson NF et al (2003) An Xp; Yq translocation causing a novel contiguous gene syndrome in brothers with generalized epilepsy, ichythosis, and attention deficits. Epilepsia 44:1529–1535

Donaldson ZR, Young LJ (2008) Oxytocin, vasopressin, and the neurogenetics of sociality. Science 322(5903):900–904

Donohoe G, Robertson IH (2003) Can specific deficits in executive functioning explain the negative symptoms of schizophrenia? a review. Neurocase 9:97–108

Donohoe G, Morris DW, De Sanctis P, Magno E, Montesi JL et al (2008) Early visual processing deficits in dysbindin-associated schizophrenia. Biol Psychiatry 63:484–489

Donohoe G, Rose E, Frodl T, Morris D, Spoletini I et al (2011a) ZNF804A risk allele is associated with relatively intact gray matter volume in patients with schizophenia. Neuroimage 54:2132–2137

Donohoe G, Walters J, Morris DW, Da Costa A, Rose E, Hargreaves A et al (2011b) A neuropsychological investigation of the genome-wide associated schizophrenia risk variant NRGN rs12807809. Schizophr Res 125:304–306

Doyle AE, Ferreira MA, Sklar PB, Lasky-Su J, Petty C et al (2008) Multivariate genome wide linkage scan of neurocognitive traits and ADHD symptoms: suggestive linkage to 3q13. Am J Med Genet B Neuropsychiatr Genet 147B:1399–1411

Eack SM, Mermon DE, Montrose DM, Miewald J, Gur RE, Gur RC, Sweeney JA, Keshavan MS (2010) Social cognition deficits among individuals at familial high risk for schizophrenia. Schizophr Bull 36(6):1081–1088

Egan MF, Kojima M, Callicott JH, Goldberg TE, Kolachana BS et al (2003) The BDNF vel66met polymorphism affects activity-dependent secretion of BDNF and human memory and hippocampal function. Cell 112:257–269

Egan MF, Goldberg TE, Kolachana BS, Callicott JH et al (2001) Effect of the COMT Val108/158 Met genotype on frontal lobe function and risk for schizophrenia. Proc Natl Acad Sci USA 98:6917–6922

Egloff B, Schmukle SC (2002) Predictive validity of an implicit association test for assessing anxiety. J Pers Soc Psychol 83(6):1441–1455

Erk S. Meyer-Lindenberg A, Schnell K, Opitz von Boberfeld C, Esslinger C et al (2010) Brain function in carriers of a genome-wide supported bipolar disorder variant. Arch Gen Psychiatry 67:803–811

Erlenmeyer-Kimling L, Rock D, Roberts SA, Janal M et al (2000) Attention, memory and motor skills as childhood predictors of schizophrenia-like psychoses: the New York high-risk project. Am J Psychiatry 157:1416–1422

Erol A, Mete L, Sonmez I, Unalin E (2010) Facial emotion recognition in patients with schizophrenia and their siblings. Nord J Psychiatry 64(1):63–67

Esslinger C, Walter H, Kirsch P, Erk S, Schnell K et al (2009) Neural mechanisms of a genome-wide supported psychosis variant. Science 324:605

Farrer LA, Cupples LA, Haines JL, Hyman B, Kukull WA (1997) Effects of age, sex, and ethnicity on the association between apolipoprotein E genotype and Alzheimer disease. A meta-analysis. J Am Med Assoc 278:1349–1356

Ferreira MA, O'Donovan MC, Meng YA, Jones IR, Ruderfer DM et al (2008) Collaborative genome-wide association analysis supports a role for ANK3 and CACNA1C in bipolar disorder. Nat Genet 40:1056–1058
Ferrero GB, Howald C et al (2010) An atypical 7q11.23 deletion in a normal IQ Williams-Beuren syndrome patient. Eur J Hum Genet 18(1):33–38
Fisch GS, Carpenter N et al (2010) The course of cognitive-behavioral development in children with the FMR1 mutation, Williams-Beuren syndrome, and neurofibromatosis type 1: the effect of gender. Am J Med Genet A 152A(6):1498–1509
Fisher SE, Francks C, McCracken JT, McGough JJ, Marlow AJ (2002) A genomewide scan for loci involved in attention-deficit/hyperactivity disorder. Am J Hum Genet 70:1183–1196
Franke B, Vasques AA, Veltman JA, Brunner HG, Rijpkema M, Fernandez G (2010) Genetic variation in CACNA1C, a gene associated with bipolar disorder, influences brainstem rather than gray matter volume in healthy individuals. Biol Psychiatry 68:568–568
Freitag CM, Staal W et al (2010) Genetics of autistic disorders: review and clinical implications. Eur Child Adolesc Psychiatry 19(3):169–178
Frith CD, Frith U (2006) The neural basis of mentalizing. Neuron 50:531–534
Frith CD, Frith U (2007) Social cognition in humans. Curr Biol 17:R724–R732
Frodl T, Bodke AL, Scheuerecker J, Lisiecka D, Schoepf V et al (2009) Functional connectivity bias in the orbitofrontal cortex in drug-free patients with major depression. Biol Psychiatry 67:161–167
Gao MC, Bellugi U et al (2010) Intelligence in Williams syndrome is related to STX1A which encodes a component of the presynaptic SNARE complex. PLoS One 5(4):e10292
Gasperoni TL, Ekelund J, Huttenen M, Palmer CG, Tuulio-Henriksson A et al (2003) Genetic linkage and association between chromosome 1q and working memory function in schizophrenia. Am J Med Genet B Neuropsychaitr Genet 116B:8–16
Gentile JK, Tan WH et al (2010) A neurodevelopmental survey of Angelman syndrome with genotype-phenotype correlations. J Dev Behav Pediatr 31(7):592–601
Gill M, Donohoe G, Corvin A (2010) What have the genomics ever done for the psychoses? Psychol Med 40:529–540
Glashouwer K, de Jong PJ (2010) Disorder-specific automatic self-associations in anxiety and depression: results of the Netherlands study of depression and anxiety. Psychol Med 40: 1101–1111
Goldberg TE, Ragland JD, Torrey EF, Gold JM, Bigelow LB et al (1990) Neuropsychological assessment of monozygotic twins discordant for schizophrenia. Arch Gen Psychiatry 47: 1066–1072
Goldberg TE, Torrey EF, Gold JM, Bigelow LB, Ragland RD et al (1995) Genetic risk of neuropsychological impairment in schizophrenia: a study of monozygotic twins discordant and concordant for schizophrenia. Schizopr Res 17:77–84
Gottesman II, Gould TD (2003) The endophenotype concept in psychiatry: ethymology and strategic intentions. Am J Psychiatry 160:636–645
Gratacos M, Gonzalez JR, Mercader JM, de Cid R et al (2007) Brain-derived neurotrophic factor val 66met and psychiatric disorders: meta-analysis of case-control studies confirm association to substance-related disorders, eating disorders and schizophrenia. Biol Psychiatry 61:911–922
Green JJ, Hollander E (2010) Autism and oxytocin: new developments in translational approaches to therapeutics. Neurotherapeutics 7(3):250–257
Green EK, Grozeva D, Jones I, Jones L, Kirov G et al (2010) The bipolar disorder risk allele at CACNA1C also confers risk of recurrent major depression and of schizophrenia. Mol Psychiatry 15:1016–1022
Green MF, Kern RS, Heaton RK (2004) Longitudinal studies of cognition and functional outcome in schizophrenia: implications for MATRICS. Schizophr Res 72:41–51
Greenwood TA, Braff DL, Light GA, Cadenhead KS, Calkins ME et al (2007) Initial heritability analyses of endophenotypic measures for schizophrenia: the consortium on the genetics of schizophrenia. Arch Gen Psychiatry 64(11):1242–1250

Groot AS, de Sonneville LM, Stins JF, Boomsma DI (2004) Familial influences on sustained attention and inhibition in preschoolers. J Child Psychol Psychiatry 45:306–314

Grozeva D, Kirov G, Ivanov D, Jones IR, Jones L et al (2010) Rare copy number variants: a point of rarity in the genetic risk for bipolar disorder and schizophrenia. Arch Gen Psychiatry 67:318–327

Gur RE, McGrath C, Chan RM, et al (2002) An fMRI study of facial emotion processing in patients with schizophrenia. Am J Psychiatry 159:1992–1999

Hallmayer JF, Kalaydjieva L, Badcock J, Dragovic M, Howell S et al (2005) Genetic evidence for a distinct subtype of schizophrenia characterized by pervasive cognitive deficit. Am J Hum Genet 77:468–476

Hariri AR, Goldberg TE, Mattay VS, Kolachana BS, Callicott JH et al (2003) Brain-derived neurotrophic factor val66met polymorphism affects human memory-related hippocampal activity and predicts memory performance. J Neurosci 23:6690–6694

Hennah W, Tuulio-Henriksson A, Paunio T, Ekelund J, Varilo T et al (2005) A haplotype within the DISC1 gene is associated with visual memory functions in families with high density schizophrenia. Mol Psychiatry 10:1097–1103

Ho BC, Milev P, O'Leary DS, Librant A, Andreasen NC (2006) Cognitive and MRI brain morphometric correlates of brain-derived neurotrophic factor Val66Met gene polymorphism in patients with schizophrenia and controls. Arch Gen Psychiatry 63:731–740

Hoeft F, Walter E et al (2011) Neuroanatomical differences in toddler boys with fragile x syndrome and idiopathic autism. Arch Gen Psychiatry 68(3):295–305

Hollingworth P, Harold D, Sims R, Gerrish A, Lambert JC et al (2011) Common variants at ABCA7, MS4A6A/MS4A4E, EPHA1, CD33 and CD2AP are associated with Alzheimer's disease. Nat Genet 43:429–435

Hong D, Scaletta Kent J et al (2009) Cognitive profile of Turner syndrome. Dev Disabil Res Rev 15(4):270–278

Hoogenraad CC, Koekkoek B et al (2002) Targeted mutation of Cyln2 in the Williams syndrome critical region links CLIP-115 haploinsufficiency to neurodevelopmental abnormalities in mice. Nat Genet 32(1):116–127

Houlihan LM, Harris SE, Luciano M, Gow AJ, Starr JM et al (2009) Replication study of candidate genes for cognitive abilities: the Lothian Birth Cohort 1936. Genes Brain Behav 8:238–247

Insel TR (2010) The challenge of translation in social neuroscience: a review of oxytocin, vasopressin, and affiliative behavior. Neuron 65(6):768–779

International Schizophrenia Consortium (2009a) Rare chromosomal deletions and duplications increase risk of schizophrenia. Nature 455:237–241

International Schizophrenia Consortium (2009b) Common polygenic variation contributes to risk of schizophrenia and bipolar disorder. Nature 460:748–752

Janssen I, Krabbendam L, Jolles J, van Os J (2003) Alterations in theory of mind in patients with schizophrenia and non-psychotic relatives. Acta Psychiatr Scand 108(2):110–117. http://www.ncbi.nlm.nih.gov/pubmed?term=%22Janssen%20I%22%5BAuthor%5D

Janssen I, Vermissen D, Campo JÀ, Myin-Germeys I, Van Os J, Krabbendam L (2006) Attribution style and psychosis: evidence for an externalizing bias in patients but not in individuals at high risk. Psychol Med 36:771–778

Johnson KA, Kelly SP, Robertson IH, Barry E, Mulligan A et al (2008) Absence of the 7-repeat variant at the DRD4 VNTR is associated with drifting sustained attention in children with ADHD but not controls. Am J Med Genet Neuropsychiatr Genet 147B:927–937

Karayiorgou M, Simon TJ, Gogos JA (2010) 22q11.2 microdeletions: linking DNA structural variation to brain dysfunction and schizophrenia. Nat Rev Neurosci 11:402–416

Kebir O, Tabbane K, Sengupta S, Joober R (2009) Candidate genes and neuropsychological phenotypes in children with ADHD: review of association studies. J Psychiatry Neurosci 34:88–101

Kempton MJ, Ruberto G, Vassos E, Tatarelli R, Girardi P et al (2009) Effects of the CACNA1C risk allele for bipolar disorder on cerebral gray matter volume in healthy individuals. Am J Psychiatry 166:1413–1414

Kim SJ, Young LJ et al (2002) Transmission disequilibrium testing of arginine vasopressin receptor 1A (AVPR1A) polymorphisms in autism. Mol Psychiatry 7(5):503–507
Klein C, Wendling K, Huettner P, m Ruder H, Peper M (2006) Intra-subject variability in attention-deficit hyperactivity disorder. Biol Psychiatry 60:1088–1097
Kremen WS, Lyons MJ, Boake C, Xian H, Jacobson KC et al (2006) A discordant twin study of premorbid cognitive ability in schizophrenia. J Clin Exp Neuropsychol 28:208–224
Krug A, Nieratschker V, Markov V, Krach S, Jansen A et al (2010) Effect of CACNA1C rs1006737 on neural correlates of verbal fluency in healthy individuals. Neuroimage 49:1831–1836
Kuntsi J, Wood AC, Rijsdijk F, Johnson KA, Andreou P et al (2010) Separation of cognitive impairments in attention-deficit/hyperactivity disorder into 2 familial factors. Arch Gen Psychiatry 67:1159–1167
Lau JYF, Eley TC (2008) Attributional style as a risk marker of genetic effects for adolescent depressive symptoms. J Abnorm Psychol 117:849–859
Lee C, Scherer SW (2010) The clinical context of copy number variation in the human genome. Expert Rev Mol Med 12:e8
Leggett V, Jacobs P et al (2010) Neurocognitive outcomes of individuals with a sex chromosome trisomy: XXX, XYY, or XXY: a systematic review. Dev Med Child Neurol 52(2):119–129
Lencz T, Szeszko PR, Rosse P, Burdick KE, Bromet EJ et al (2010) A schizophrenia risk gene, ZNF804A, influences neuroanatomical and neurocognitive phenotypes. Neuropsychopharmacology 35:2284–2291
Lien YJ, Liu CM, Faraone SV, Tsuang MT, Hwu HG et al (2010) A genome-wide quantitative trait loci scan of neurocognitive performance in families with schizophrenia. Genes Brain Behav 9:695–702
Lin SH, Liu CM, Liu YL, Shen-Jang Fann C, Hsiao PC et al (2009) Clustering by neurocognition for fine mapping of the schizophrenia susceptibility loci on chromosome 6p. Genes Brain Behav 8:785–794
Liu XQ, Paterson AD et al (2008) Genome-wide linkage analyses of quantitative and categorical autism subphenotypes. Biol Psychiatry 64(7):561–570
Losh M, Piven J (2007) Social-cognition and the broad autism phenotype: identifying genetically meaningful phenotypes. J Child Psychol Psychiatry 48(1):105–112
Manor I, Tyano S, Eisenberg J, Bachner-Melman R, Kotler M, Ebstein RP (2002) The short DRD4 repeats confer risk to attention deficit hyperactivity disorder in a family-based design and impair performance on a continuous performace test (TOVA). Mol Psychiatry 7:790–794
Mazzocco MM (2009) Mathematical learning disability in girls with Turner syndrome: a challenge to defining MLD and its subtypes. Dev Disabil Res Rev 15(1):35–44
McKinnon MC, Cusi AM, Macqueen GM (2010) Impaired theory of mind performance in patients with recurrent bipolar disorder: moderating effect of cognitive load. Psychiatry Res 177(1–2):261–262
Mefford HC, Sharp AJ, Baker C, Itsare A, Jiang Z, Buysse K et al (2008) Recurrent rearrangements of chromosome 1q21.1 and variable pediatric phenotypes. N Engl J Med 359: 1685–1699
Meng Y, Zhang Y et al (2002) Abnormal spine morphology and enhanced LTP in LIMK-1 knockout mice. Neuron 35(1):121–133
Merla G, Brunetti-Pierri N et al (2010) Copy number variants at Williams-Beuren syndrome 7q11.23 region. Hum Genet 128(1):3–26
Meyer-Lindenberg A (2010) From maps to mechanisms through neuroimaging of schizophrenia. Nature 468:194–202
Milner KM, Craig EE et al (2005) Prader-Willi syndrome: intellectual abilities and behavioural features by genetic subtype. J Child Psychol Psychiatry 46(10):1089–1096
Mitchell JP (2009) Inferences about mental states. Philos Trans R Soc Lond B Biol Sci 364(1521):1309–1316
Munafo MR, Bowes L, Clark TG, Flint J (2005) Lack of association of the COMT (Val158/08Met) gene and schizophrenia: a meta-analysis of case-control studies. Mol Psychiatry 10: 765–770

Need AC, Attix DK, McEvoy JM, Cirulli ET et al (2009) A genome-wide study of common SNPs and CNVs in cognitive performance in the CANTAB. Hum Mol Genet 18:4650–4661

Ng MY, Levinson DF, Faraone SV, Suarez BK, deLisi LE et al (2009) Meta-analysis of 32 genome-wide linkage studies in schizophrenia. Mol Psychiatry 14:774–785

Penn DL, Sanna LJ, Roberts DL (2008) Social cognition in schizophrenia: an overview. Schizophr Bull 34:408–411

Pezawas L, Verchinski BA, Mattay VS, Cacllicott JH, Kolachana BS et al (2004) The brani-derived neurotrophic factor val66met polymorphism and variation in human cortical morphology. J Neurosci 24:10099–10102

Pinto D, Pagnamenta AT, Klei L, Anney R, Merico D et al (2010) Functional impact of global rare copy number variation in autism spectrum disorders. Nature 466:368–372

Pober BR (2010) Williams-Beuren syndrome. N Engl J Med 362(3):239–252

Porteous DJ, Thomson P, Brandon NJ, Millar JK (2006) The genetics and biology of DISC1-an emerging role in psychosis and cognition. Biol Psychiatry 60:123–131

Posthuma D, de Geus EJ, Boomsma DI (2001) Perceptual speed and IQ are associated through common genetic factors. Behav Genet 313:593–602

Psychiatric GWAS Consortium Coordinating Committee (2009) Genomewide association studies: history, rationale, and prospects for psychiatric disorders. Am J Psychiatry 166:540–546

Ranheim EA, Kwan HC et al (2005) Frizzled 9 knock-out mice have abnormal B-cell development. Blood 105(6):2487–2494

Rommelse NN, Arias-Vasquez A, Altink ME, Buschgens CJ, Fliers E et al (2008) Neuropsychological endophenotype approach to genome-wide linkage analysis identifies susceptibility loci for ADHD on 2q21.1 and 13q12.11. Am J Hum Genet 83:99–105

Sabbagh MA, Seamans EL (2008) Intergenerational transmission of theory-of-mind. Dev Sci 11:354–360

Sasaki S, Iwanaga M, Kanai Y, Seiwa H (2010) Implicit and explicit associations in the fear structure of social anxiety. Percept Mot Skills 110(1):19–32

Samango-Sprouse C (2010) Expansion of the phenotypic profile of the young child with XXY. Pediatr Endocrinol Rev 8 (Suppl 1):160–168

Scott RW, Olson MF (2007) LIM kinases: function, regulation and association with human disease. J Mol Med 85(6):555–568

Scourfield J, Martin N, Lewis G, McGuffin P (1999) Heritability of social cognitive skills in children and adolescents. Br J Psychiatry 175:559–564

Scourfield J, Martin N, Eley TC, McGuffin P (2004) The genetic relationship between social cognition and conduct problems. Behavior Genet 34(4):377–383

Sebat J, Lakshmi B, Troge J, Alexander J, Young YJ et al (2004) Large-scale copy number polymorphism in the human genome. Science 305:525–528

Sebat J, Levy DL, McCarthy SE (2009) Rare structural variants in schizophrenia: one disorder, multiple mutations; one mutation, multiple disorders. Trends Genet 25:528–535

Shamay-Tsoory SG, Shur S, Barcai-Goodman L, Medlovich S, Harari H, Levkovitz Y (2007) Dissociation of cognitive from affective components of theory of mind in schizophrenia. Psychiatry Res 149(1–3):11–23

Slaats-Wilemse D, Swaab-Barneveld H, de Sonneville L, Buitlaar J (2005) Familial clustering of executive functioning in affected sibling pair families with ADHD. J Am Acad Child Adolesc Psychiatry 44:385–391

Skuse DH, Gallagher L (2011) Genetic influences on social cognition. Pediatr Res 69(5 Pt 2): 85R–91R

Splawski I, Timothy KW, Decher N, Kumar P, Sachse FB et al (2005) Severe arrhythmia disorder caused by cardiac L-type calcium channel mutations. Proc Natl Acad Sci USA 102:8089–8096

Sprong M, Schothorst P, Vos E, Hox J, van Engeland H (2007) Theory of mind in schizophrenia: meta-analysis. Br J Psychiatry 191:5–13. Review

Stergiakouli E. Langley K, Willains H, Walters J, Williams NM et al (2011) Steroid sulfatase is a potential modifier of cognition in attention deficit hyperactivity disorder. Genes Brain Behav 10:334–344

Straub RE, MacLean CJ, Kendler KS (1996) The putative schizophrenia locus on chromosome 6p: a brief overview of the linkage studies. Mol Psychiatry 1:89–92
Szatmari P, White J et al (2007) The use of genetic epidemiology to guide classification in child and adult psychopathology. Int Rev Psychiatry 19(5):483–496
Tansey KE, Hill MJ et al (2011) Functionality of promoter microsatellites of arginine vasopressin receptor 1A (AVPR1A): implications for autism. Mol Autism 2(1):3
Thomas EJ, Elliott R (2009) Brain imaging correlates of cognitive impairment in depression. Front Hum Neurosci 3:30
Toulopoulou T, Goldberg TE, Mesa IR, Picchioni M, Rijsdijk F et al (2010) Arch Gen Psychiatry 67:905–913
Toulopoulou T, Picchioni M, Rijsdijk F, Hua-Hall M, Ettinger U et al (2007) Substantial genetic overlap between neurocognition and schizophrenia: genetic modeling in twin samples. Arch Gen Psychiatry 64:1348–1355
Visscher PM, Montgomery GW (2009) Genome-wide association studies and human disease: from trickle to flood. J Am Med Assoc 302:2028–2029
Viswanath B, Reddy YCJ, Kumar KJ, Kandavel T, Chandrashekar CR (2009) Cognitive endophenotypes in OCD: a study of unaffected siblings of probands with familial OCD. Prog Neuropsychopharmacol Biol Psychiatry 33:610–615
Vuilleumier P, Pourtois G (2007) Distribution and interactive brain mechanisms during emotion face perception: evidence from functional neuroimaging. Neuropsychologia 45:174–194
Walsh CA, Engle EC (2010) Allelic diversity in human developmental neurogenetics: insights into biology and disease. Neuron 68:245–253
Walsh T, McClellan JM, McCarthy SE, Addington AM, Pierce SB et al (2008) Rare structural variants disrupt genes in neurodevelopmental pathways in schizophrenia. Science 320:539–543
Walter H, Schnell K, Erk S, Arnold C, Kirsch P et al (2011) Effects of a genome-wide supported psychosis risk variant on neural activation during a theory-of-mind task. Mol Psychiatry 16: 462–470
Walters J, Corvin A, Owen MJ, Williams H, Dragovic M, Quinn EM (2010) Psychosis susceptibility gene ZNF804A and cognitive performance in schizophrenia. Arch Gen Psychiatry 67:692–700
Wang K, Zhang H et al (2009) Common genetic variants on 5p14.1 associate with autism spectrum disorders. Nature 459(7246):528–533
Wang Z, Aragona BJ (2004) Neurochemical regulation of pair bonding in male prairie voles. Physiol Behav 83(2):319–328
Wassink TH, Piven J et al (2004) Examination of AVPR1a as an autism susceptibility gene. Mol Psychiatry 9(10):968–972
Weiss LA, Arking DE et al (2009) A genome-wide linkage and association scan reveals novel loci for autism. Nature 461(7265):802–808
Willcutt EG, Doyle AE, Nigg JT, Faraone SV, Pennington BF (2005) Validity of the executive function theory of attention-deficit/hyperactivity disorder: a meta-analytic review. Biol Psychiatry 57:1336–1346
Williams HJ, Craddock N, Russo G, Hamshere ML, Moskvina V, Dwyer S et al (2011) Most genome-wide significant susceptibility loci for schizophrenia and bipolar disorder reported to date cross-traditional diagnostic boundaries. Hum Mol Genet 20:387–391
Williams NM, Zaharieva I, Martin A, Langley K, Mantripragada K et al (2010) Rare chromosomal deletions and duplications in attention-deficit hyperactivity disorder: a genome-wide analysis. Lancet 376:1401–1408
Wisdom NM, Callahan JL, Hawkins KA (2011) The effects of apolipoprotein E on non-impaired cognitive functioning: a meta-analysis. Neurobiol Aging 32:63–74
Wood AC, Asherson P, van der Meere JJ, Kuntsi J (2010) Separation of genetic influences on attention deficit hyperactivity disorder symptoms and reaction time performance from those of IQ. Psychol Med 40:1027–1037
Yirmiya N, Rosenberg C et al (2006) Association between the arginine vasopressin 1a receptor (AVPR1a) gene and autism in a family-based study: mediation by socialization skills. Mol Psychiatry 11(5):488–494

Behavioral Genetics of Neurodegenerative Disorders

Daniela Galimberti and Elio Scarpini

Abstract Alzheimer's disease (AD) is the most common cause of dementia in the elderly, and is typically characterized by memory loss. In addition, during the disease progression, most patients develop behavioural and psychiatric symptoms of dementia (BPSD). Frontotemporal Lobar Degeneration (FTLD) is the most frequent neurodegenerative disorder with a presenile onset. It is characterized mainly by behavioural disturbances, whereas memory is conserved. The two major neuropathologic hallmarks of AD are extracellular Amyloid beta (Aβ) plaques and intracellular neurofibrillary tangles (NFTs). Conversely, in FTLD the deposition of tau has been observed in a number of cases, but in several brains there is no deposition of tau but instead a positivity for ubiquitin. In some families these diseases are inherited in an autosomal dominant fashion. Genes responsible for familial AD include the Amyloid Precursor Protein (β*–APP*), Presenilin 1 (*PS1*) and Presenilin 2 (*PS2*). The majority of mutations in these genes are often associated with a very early onset (40–50 years of age). Regarding FTLD, the first mutations described are located in the Microtubule Associated Protein Tau gene (*MAPT*). Tau is a component of microtubules, which represent the internal support structures for the transport of nutrients, vesicles, mitochondria and chromosomes within the cell. Mutations in *MAPT* are associated with an early onset of the disease (40–50 years), and the clinical phenotype is consistent with Frontotemporal Dementia (FTD). Recently, mutations in a second gene, named progranulin (*GRN*), have been identified in some families with FTLD. The pathology associated with these mutations is most frequently characterized by the immunostaining of TAR DNA Binding Protein 43 (TDP-43), which is a transcription factor.

D. Galimberti (✉) · E. Scarpini
Department of Neurological Sciences, "Dino Ferrari" Center, University of Milan,
Fondazione Cà Granda, IRCCS Ospedale Maggiore Policlinico,
Via F. Sforza 35, 20122 Milan, Italy
e-mail: daniela.galimberti@unimi.it

Curr Topics Behav Neurosci (2012) 12: 615–631
DOI: 10.1007/7854_2011_181

Published Online: 20 January 2012

The clinical phenotype associated with *GRN* mutations is highly heterogeneous, including FTD, Progressive Aphasia, Corticobasal Syndrome, and AD. Age at disease onset is variable, ranging from 45 to 85 years of age. The majority of cases of AD and FTLD are however sporadic, and likely several genetic and environmental factors contribute to their development. Concerning AD, it is known that the presence of the ε4 allele of the Apolipoprotein E gene is a susceptibility factor, increasing the risk of about 4 fold. A number of additional genetic factors, including cytokines, chemokines, Nitric Oxide Synthases, contribute to the susceptibility for the disease. Some of them also influence the risk to develop FTLD. Variability in serotonin transporter gene could influence the development of BPSD. In this chapter, current knowledge on molecular mechanisms at the basis of AD and FTLD, as well as the role of genetics, will be presented and discussed.

Keywords Dementia · Frontotemporal lobar degeneration (FTLD) · Behavioural disturbances · Alzheimer's disease (AD) · Behavioural and psychological symptoms of dementia (BPSD)

Contents

1 Alzheimer's Disease and Frontotemporal Lobar Degeneration 616
2 Familial AD 617
3 Sporadic AD 618
3.1 Genes Influencing the Risk to Develop AD 618
3.2 Genes Influencing the Risk to Develop BPSD in AD Patients 622
4 Familial FTLD 623
5 Sporadic FTLD 626
References 626

1 Alzheimer's Disease and Frontotemporal Lobar Degeneration

Alzheimer's disease (AD) is the most common cause of dementia in the elderly, with a prevalence of 5% after 65 years of age. The disease was originally described by Alois Alzheimer and Gaetano Perusini in 1906, and it is clinically characterized by a progressive cognitive impairment, including impaired judgment, decision-making and orientation, often accompanied, in later stages, by behavioural and psychiatric symptoms of dementia (BPSD), including agitation, hallucinations and delusions, as well as language impairment. Loss of noradrenergic and serotonergic neurons contributes to the emergence of BPSD. Conversely, loss of cholinergic neurons is the major contributor to the cognitive impairment of AD (Palmer 1996).

The two major neuropathologic hallmarks of AD are extracellular beta-amyloid (Aβ) plaques and intracellular neurofibrillary tangles (NFTs). The production of

Aβ, which represents a crucial step in AD pathogenesis, is the result of the cleavage of a bigger precursor, named Amyloid precursor protein (APP), which is over-expressed in AD (Griffin 2006). Aβ forms highly insoluble and proteolysis resistant fibrils known as "senile plaques".

Neurofibrillary tangles are composed of the tau protein. In healthy controls, tau is a component of microtubules, which are the internal support structures for the transport of nutrients, vesicles, mitochondria and chromosomes within the cell. Microtubules also stabilize the growing axons, which are necessary for the development and growth of neurites (Griffin 2006). In AD, tau protein is abnormally hyperphosphorylated and forms insoluble fibrils, which originate deposits within the cell.

Frontotemporal lobar degeneration (FTLD) occurs most often in the presenile period, and age at onset is typically 45–65 years, with a mean in the 50 s. Distinctive features in FTLD concern behaviour, including disinhibition, loss of social awareness, overeating and impulsiveness. Despite profound behavioural changes, memory is relatively spared (Hou et al. 2004). Conversely to AD, which is more frequent in women, FTLD has an equal distribution among men and women. The current consensus criteria (Neary et al. 1998) identify three clinical syndromes: Frontotemporal Dementia (FTD), Progressive nonfluent Aphasia (PA) and Semantic Dementia (SD), which reflect the clinical heterogeneity of FTLD. Frontotemporal dementia is characterized by behavioural abnormalities, whereas PA is associated with progressive loss of speech, with hesitant, nonfluent speech output (Scarpini et al. 2006), and SD is associated with loss of knowledge about words and objects (Neary et al. 1998). This variability is determined by the relative involvement of the frontal and temporal lobes, as well as by the involvement of right and left hemispheres (Rosen et al. 2002).

Despite the majority of AD and FTLD are sporadic and likely caused by the interaction between genetic and environmental factors, so far it was observed that clinically typical AD and FTLD can cluster in families and be inherited in an autosomal dominant fashion, suggesting a genetic cause.

2 Familial AD

Autosomal dominant AD forms are characterized by mutations in three genes: Amyloid Precursor Protein (*β–APP*; Goate et al. 1991), Presenilin 1 (*PS1;* Sherrington et al. 1995) and Presenilin 2 (*PS2;* Levy-Lahad et al. 1995).

In 1987, a region of linkage with AD was reported on the long arm of chromosome 21, which encompassed a region harboring the *β–APP* gene, a compelling candidate for AD (Tanzi et al. 1987). The gene is located at chromosome 21q21.22 and encodes for a transmembrane protein that is normally processed into amyloid fragments. In 1991, the first missense mutation in *β–APP* was reported (Goate et al. 1991). Since then, 32 different mutations have been described in the *β–APP* gene in 89 families (http://molgen-www.uia.ac.be). All these mutations cause amino acid changes in putative sites for the cleavage of the protein, thus altering the APP processing, such that more pathological Aβ42 is produced (Hardy and Selkoe 2002). Interestingly, the chromosome 21, in which *β–APP* resides, is triplicated in Down syndrome and most of the cases manifest also AD by the age of 50. Post-mortem analyses of Down's patients who die young show diffuse intraneuronal deposits of Aβ, suggesting that its deposition is an early event in cognitive decline. The recent discovery of an extra copy of the *β–APP* gene in familial AD (Rovelet-Lecrux et al. 2006) provides further support that increased Aβ production can cause the disease.

The other two genes causing familial AD are *PS1* (14q24.3) and *PS2* (1q31-q42). Presenilins represent a central component of γ-secretase, the enzyme responsible for originating Aβ from the C-terminal fragment of the APP protein. Mutations in presenilins also alter APP cleavage, leading to an increased production of Aβ42. So far, 179 mutations in *PS1* have been identified and 14 additional mutations have been found in the homologous gene *PS2* (http://molgen-www.uia.ac.be).

Most variants in *PS1* are missense mutations resulting in single amino-acid substitutions. Some are more complex, for example, small deletions or splice mutations. The most severe mutation in *PS1* is a donor–acceptor splice mutation that causes a two-aminoacid substitution and an in-frame deletion of exon 9. However, the biochemical consequences of these mutations for γ-secretase assembly seem to be limited (Bentahir et al. 2006; Steiner et al. 1999). All these clinical mutations are likely to cause a specific gain of toxic function for *PS1,* determined by an increase of the ratio between Aβ42 and Aβ40 amyloid peptides, thus indicating that presenilins might modify the way in which γ-secretase cuts APP.

Mutations in presenilins occur in the catalytic subunit of the protease responsible for determining the length of Aβ peptides therefore generating toxic Aβ fragments. However, presenilins have also non-proteolytic functions (Baki et al. 2004; Huppert et al. 2005), the disruption of which might also contribute to familial AD pathogenesis.

Despite several carriers develop the disease early (40–50 years of age) with a typical AD phenotype, in some cases patients carrying the same mutation develop signs and symptoms resembling FTD instead of AD (Bruni et al. 2010). In addition, other mutations are associated with myoclonus, seizures, bilateral spasticity, parkinsonian features or ataxia (Larner and Doran 2009).

3 Sporadic AD

3.1 Genes Influencing the Risk to Develop AD

Risk genes are likely to be numerous, displaying intricate patterns of interaction with each other as well as with non-genetic variables, and-unlike classical Mendelian ("simplex") disorders-exhibit no simple mode of inheritance. Mainly due to this reason, the genetics of sporadic AD has been labeled "complex" (Bertram and Tanzi 2005).

The gene mainly related to the sporadic forms of AD is the Apolipoprotein E (*APOE*) (Corder et al. 1993), which is located at chromosome 19q13.32 and was initially identified by linkage analysis (Pericack-Vance et al. 1991). The relationship between APOE and AD has been confirmed in more than 100 studies conducted in different populations. The gene has three different alleles, *APOE*2*, *APOE*3* and *APOE*4*. The *APOE*4* allele is the variant associated with AD. Longitudinal studies in Caucasian populations have shown that carriers for one *APOE*4* allele have a two-fold increase in the risk for AD (Raber et al. 2004). The risk increases in homozygous for the *APOE*4* allele, and this allelic variant is also associated with an earlier onset of the disease.

Several linkage studies have been performed, giving rise to additional candidate susceptibility loci at chromosomes 1, 4, 6, 9, 10, 12 and 19. In particular, promising loci have been found at chromosome 9 and 10 (Grupe et al. 2006; Li et al. 2006). Recently, a wide genome analysis identified variants at *CLU* (which encodes clusterin or ApoJ) on chromosome 8 and *PICALM* in chromosome 11 associated with AD (Harold et al. 2009). Data on *CLU* were replicated in an independent study, which, in addition, demonstrated that *CR1*, encoding the complement component (3b/4b) receptor 1and locate on chromosome 1, is associated with AD (Lambert et al. 2009).

Also, a large number of candidate genes studies have been performed in order to search a robust risk factor for the sporadic form of the disease. Several studies were mainly focused in genes clearly involved in the pathogenesis of AD such as genes encoding for inflammatory molecules or involved in the oxidative stress cascade, both considered major factors in AD pathology. One of the strongest evidence of the role played by genetic variants in inflammatory molecules to increase the risk of AD involves the Interleukin-1 (IL1) complex, which map at chromosome 2q14-21 and includes *IL1-α, IL1-β,* and IL1R antagonist protein (*IL-1Ra*), all of which have a number of polymorphisms found to be associated with AD in several case-controls studies carried out in different populations (Du et al. 2000; Grimaldi et al. 2000; Papassotiropoulous et al. 1999). Several polymorphisms in *IL-6,* which is a potent inflammatory cytokine but has also regulatory functions, have been investigated so far. The *IL6* gene is located at chromosome 7p21 and polymorphisms exist in the −174 promoter region and in the region of a variable number of tandem repeats (VNTR), which is located in the 3′untraslated region. Both of them have been found associated with AD in case–controls studies (Licastro et al. 2003; Nicoll et al. 2000). Investigation of Tumor Necrosis Factor-α (*TNFα*) polymorphisms was initiated because genome screening suggested a putative association of AD with a region on chromosome 6p21.3, which lies within 20 centimorgans of the *TNFα* gene. Furthermore, other polymorphisms located in the promoter region of *TNFα* have been associated with autoimmune and inflammatory diseases (Collins et al. 2000).

As with *TNFα*, investigations of the role of α-2macroglobulin (*A2M*) were initiated as a result of screening studies of the genome. In this case, linkage was found in the region of chromosome 1p, where *A2 M* and its low-density lipoprotein receptor are found. Blacker et al. (1998) tested for association of polymorphisms with AD showing a strong involvement of this gene in its pathogenesis.

Moreover, polymorphisms in chemokines have been investigated with regard of susceptibility of AD. Monocyte Chemoattractant Protein-1 (*MCP-1*) and *RANTES* genes have been widely screened in different neurodegenerative diseases (Huerta et al. 2004). The distribution of the *A-2518G* variant was determined in different AD populations with concordant results (Fenoglio et al. 2004; Combarros et al. 2004a) showing no evidence for association of this variant in AD compared with controls. Moreover, Fenoglio et al. (2004) found a significant increase of MCP-1 serum levels in AD carrying at least one *G* polymorphic allele. Therefore, the *A-2518G* polymorphism does not seem to be a risk factor for the development of AD, but its presence correlates with higher levels of serum MCP-1.

RANTES promoter polymorphism *−403 A/G*, found to be associated with several autoimmune diseases, was examined in AD population, failing to find significant differences between patients and controls (Huerta et al. 2004).

CCR2 and *CCR5* genes, encoding for the receptors of MCP-1 and RANTES respectively, have been screened for association with AD. The most promising variants involve a conservative change of a valine with an isoleucine at codon 64 of *CCR2* (*CCR2-64I*) and a 32-bp deletion in the coding region of *CCR5* (*CCR5Δ32*), which leads to the expression of a non-functional receptor. A decreased frequency and an absence of homozygous for the polymorphism *CCR2-64I* were found in AD, thus suggesting a protective effect of the polymorphic allele on the occurrence of the disease (Galimberti et al. 2004); conversely, no different distribution of the *CCR5Δ32* deletion in patients compared with controls were shown (Galimberti et al. 2004; Combarros et al. 2004b).

A mutation scanning of the Interferon-γ-induced protein 10 (IP-10) gene coding region has been performed in AD patients searching for new variants. The analysis demonstrated the presence of two previously reported polymorphisms in exon 4 (*G/C* and *T/C*), which are in complete linkage disequilibrium, as well as a novel rare one in exon 2 (*C/T*). Subsequently these SNPs have been tested in a wide case–control study but no differences in haplotype distribution were found (Venturelli et al. 2006).

Other genes under investigation are related to oxidative stress, a process closely involved in AD pathogenesis. In this regard, genes coding for the nitric oxide synthase (NOS) complex have been screened. The common polymorphism consisting in a *T/C* transition (*T-786C*) in *NOS3,* previously reported to be associated with vascular pathologies, has been tested in AD, but no significant differences with controls were found. Nevertheless, expression of *NOS3* in PBMC either from patients or controls seems to be influenced by the presence of the *C* polymorphic allele, and is likely to be dose dependent, being mostly evident in homozygous for the polymorphic variant. The influence of the polymorphism on *NOS3* expression rate supports the hypothesis of a beneficial effect exerted in AD by contributing to lower oxidative damage (Venturelli et al. 2005).

An additional variant in *NOS3* gene has been extensively investigated in AD patients, although the results are still controversial. It is a common polymorphism consisting in a single base change (*G894T*), which results in an aminoacid substitution at position 298 of *NOS3* (Glu298Asp). Dahiyat et al. (1999) determined the frequency of the Glu298Asp variant in a two-stage case–control study, showing that homozygous for the wild-type allele were more frequent in late onset AD. However, studies in other populations failed to replicate these results (Crawford et al. 2000; Monastero et al. 2003; Sánchez-Guerra et al. 2001; Tedde et al. 2002).

More recently Guidi et al. (2005) correlated this variant with total plasma homocysteine (tHcy) levels in AD patients and controls, demonstrating that the Glu/Glu genotype is correlated with higher levels of tHcy, which represent a known risk factor for AD (Seshadri et al. 2002), and its frequency was increased in AD patients (Guidi et al. 2005). Thus, the mechanism by which this genotype contributes to increase the risk in developing AD could be mediated by an increase of tHcy.

However, *NOS-1* is the isoform most abundantly expressed in the brain. Recent genetic analyses demonstrated that the double mutant genotype of the synonymous *C276T* polymorphism in exon 29 of the *NOS1* gene represents a risk factor for the development of early onset AD (Galimberti et al. 2005), whereas the dinucleotide polymorphism in the 3′UTR of *NOS1* is not associated with AD (Liou et al. 2002). To date, the promoter region of *NOS1*, located approximately 200 kb upstream of these polymorphism, has not been investigated for susceptibility to AD. Due to this reason and to further explore a possible association of *NOS1* polymorphisms with AD, the distribution of a functional polymorphisms and a VNTR was analyzed in a case-control study (Galimberti et al. 2008). The functional variant considered is located in exon 1c, which is one of the nine alternative first exons (named 1a–1i), resulting in *NOS1* transcripts with different 5′-untraslated regions (Wang et al. 1999). Three SNPs have been identified in exon 1c, but only the *G-84A* variant displays a functional effect, as the *A* allele decreases the transcription levels by 30% in in vitro models (Saur et al. 2004). Regarding exon 1f, a VNTR polymorphism has been recently reported in its putative promoter region, termed *NOS1* Ex1f-VNTR. This VNTR is highly polymorphic and consists of different numbers of dinucleotides (B-Q), which, according to their bimodal distribution, have been dichotomized in short (B-J) and long (K-Q) alleles for association studies. Both Ex1c *G-84A* and Ex1f-VNTR are associated with psychosis and prefrontal functioning in a population of patients with schizophrenia (Reif et al. 2006). Notably, both Ex1c and Ex1f transcripts are found in the hippocampus and the frontal cerebral cortex, i.e. brain regions implicated in the pathogenesis of schizophrenia as well as AD. The presence of the short (*S)* allele of *NOS1* Ex1f-VNTR represents a risk factor for the development of AD. The effect is cumulative, as in *S/S* carriers the risk is doubled. Most interestingly, the effect of this allele is likely to be gender specific, as it was found in females only. In addition, the *S* allele was shown to interact with the APOE*4 allele both in males and females, increasing the risk to develop AD by more than 10 fold (Galimberti et al. 2008). Thus, *NOS1* seems to be a risk factor for AD, but only in female population. This could be explained by a possible interaction with other genes or with additional environmental factors present in females but not males.

3.2 Genes Influencing the Risk to Develop BPSD in AD Patients

Serotonin (5-hydroxytryptamine, 5-HT) is involved in a range of behaviors and psychiatric processes, including mood, aggression, impulsivity and anxiety. Several studies have implicated 5-HT dysfunctions in the pathogenesis of psychiatric diseases, such as depression, obsessive–compulsive disorder and schizophrenia. Nevertheless, 5-HT play a role also in the development of BPSD in patients with AD (Cross 1990). Postmortem studies of brain from patients with AD showed decreased levels of 5-HT, 5-HT receptors and 5-HT transporter (5-HTT) (Reinikaininen et al. 1990). The 5-HTT plays a central role in the fine-tuning of 5-HT neurotransmission by determining the duration and amount of 5-HT present in the synaptic cleft. Its encoding gene, named *SLC6A4*, contains several polymorphisms influencing its expression, including a functional polymorphism, named serotonin-transporter linked-polymorphic region (*5-HTTLPR*), which is located upstream of the gene promoter and displays length variations of 14 (short type) or 16 (long type) repeats of a 20- to 23-base pair (bp) element and a 5-HTT VNTR (*5-HTTVNTR*) in the second intron, that contains 9, 10 or 12 copies of a 17-bp element (Lesch et al. 1994). These polymorphisms can influence transcriptional activity of the *SLC6A4* and ultimately the level of 5-HTT transporters available.

Several studies have found a relationship between *5-HTTLPR* and *VNTR* in various psychiatric conditions, including personality traits (Lotrick et al. 2001). Regarding BPSD in patients with AD, the long variant of *5-HTTLPR* has displayed association with aggressive behaviour and psychosis ind 2 out of 5 studies (Assal et al. 2004; Ha et al. 2005; Rocchi et al. 2003; Sukonick et al. 2001; Sweet et al. 2001). No association with *5-HTTVNTR* has been shown in patients with AD and BPSD (Assal et al. 2004; Ueki et al. 2007). Nevertheless, a significant association was found between the presence of the *5-HTTVNTR* allele 10 and BPSD or aggressiveness (Ueki et al. 2007). Conversely, no influence of *5-HTTLPR* polymorphism on BPSD in AD was shown (Albani et al. 2009).

Recently, in a 3-year follow-up study, Angelucci et al. evaluated the association of 5-HT receptor 5-HT2a *102T/C* SNP with psychotic symptom severity and response to treatment with atypical antipsychotics (risperidone, olanzapine and quetiapine) in 80 patients with mild AD, showing a significant difference in the frequency and severity of delusions, measured through the Neuropsychiatric Inventory (NPI), with patients carrying the *TT* genotype most delusional during the follow-up period. Moreover, patients carrying the *T* allele were resistant to treatment with antipsychotics (Angelucci et al. 2009).

Lastly, dopamine D3 receptor (DRD3) has been investigated in AD patients, as it is present in the limbic system, which is thought to regulate affect and cognition. No effect of genetic variability in *DRD3* on the risk to develop AD was shown. However, a significant association was found between the presence of the *DRD3* glycine allele and paranoid and delusional ideation (Sato et al. 2009).

4 Familial FTLD

Frontotemporal Lobar Degeneration is a heterogeneous disease characterized by a strong genetic component in its aetiology as up to 40% of patients report a family history of the disease in at least one extra family member (Snowden et al. 2002). In 1994 an autosomal dominantly inherited form of FTD with parkinsonism was linked to chromosome 17q21.2 (Wilhelmsen et al. 1994). Subsequently, other familial forms of FTD were found to be linked to the same region, resulting in the denomination "frontotemporal dementia and parkinsonism linked to chromosome 17" (FTDP-17) for this class of diseases.

In 1998, *MAPT* gene on chromosome 17q21, which encodes the microtubule associated protein tau was described as the cause of the disease in these families (Hutton et al. 1998; Poorkaj et al. 1998; Spillantini et al. 1998). So far, 44 different mutations in the *MAPT* gene have been described in 132 families (http://molgen-www.uia.ac.be). *MAPT* mutations are either non-synonimous or deletions, or silent mutations in the coding region, or intronic mutations located close to the splice-donor site of the intron after the alternatively spliced exon 10 (Rademakers et al. 2004). Mutations are mainly clustered in exons 9–13, except for two identified mutations in exon 1 (Rademakers et al. 2002). As regards possible effects on *MAPT* mutations, different mechanisms are involved, depending on the type and location of the mutation. Many of them disturb the normal splicing balance, producing altered ratios of the different isoforms. A number of mutations promote the aggregation of tau protein, whereas others enhance tau phosphorylation (Goedert and Jakes 2005).

However, after the discovery for *MAPT* as causal gene for FTDP-17, there were still numerous families with autosomal dominant FTLD genetically linked to the same region of chr17q21 that contains *MAPT* but in which no pathogenic mutations had been identified, despite extensive analysis of this gene (Lendon et al. 1998; Rosso et al. 2001; van der Zee et al. 2006). The neuropathology phenotype in these families was similar to the microvacuolar-type observed in a large proportion of idiopathic FTD cases with ubiquitin immunoreactive neuronal inclusions. Moreover, clinically, the disease in these families was consistent with diagnostic criteria for FTLD (Neary et al. 1998). Sequence analysis of the whole *MAPT* region failed to find a mutation and tau protein appeared normal in these families. Moreover the minimal region containing the disease gene for this group of families was approximately 6.2 Mb in physical distance. This region defined by markers D17S1787 and D17S806 is particularly gene rich, containing around 180 genes. Collectively, these data strongly argued against *MAPT* and pointed to another gene. Systematic candidate gene sequencing of all remaining genes within the minimal candidate region was performed and after sequencing 80 genes, including those prioritized on known function, the first mutation in progranulin gene (*GRN*) was identified. It consists in a 4-bp insertion of *CTGC* between coding nucleotides 90 and 91, causing a frameshift and premature termination in progranulin (C31LfsX34) (Baker et al. 2006). These results have been contemporarily replicated by Cruts et al. (2006), who analyzed other families with a FTLD-U disease without *MAPT* pathology, finding a mutation five base pairs into the intron following the first non coding exon of the *GRN* gene (IVS0 + 5G-C). This is predicted to prevent splicing out of the intron 0, leading the mRNA to be retained within the nucleus and subjected to nuclear degradation (Cruts et al. 2006). At present there is no obvious mechanistic link between the mutations in *MAPT* and *GRN*, currently assuming that their proximity on chromosome 17 is simply a coincidence. Progranulin is known by several different names including granulin, acrogranin, epithelin precursor, proepithelin and prostate cancer (PC) cell derived growth factor (He and Bateman 2003). The protein is encoded by a single gene on chromosome 17q21, which produces a 593 amino acid, cysteine rich protein with a predicted molecular weight of 68.5 kDa. The full-length protein is subjected to proteolysis by elastase and this process is regulated by a secretory leukocyte protease inhibitor (SLPI) (Zhu et al. 2002). Progranulin and the various granulin peptides are implicated in a range of biological functions including development, wound repair and inflammation by activating signaling cascades that control cell cycle progression and cell motility (He and Bateman 2003). Excess progranulin appears to promote tumor formation and hence can act as a cell survival signal. Despite the increasing literature on the function of progranulin, its role in neuronal function and survival remains unclear. In the human brain, *GRN* is expressed in neurons but significantly is also highly expressed in activated microglia (Baker et al. 2006), with the result that *GRN* expression is increased in many neurodegenerative diseases.

Since the original identification of null-mutations in FTLD in 2006, 68 different mutations spanning most exons have been reported so far (http://www.molgen.ua.ac.be/). Interestingly, the majority of mutations identified create functional null alleles, causing premature termination of the *GRN* coding sequence. This leads to the degradation of the mutant RNA by nonsense mediated decay, creating a null allele (Baker et al. 2006; Cruts et al. 2006). The presence of a null mutation causes a partial loss of functional progranulin protein, which in turn leads eventually to neurodegeneration (haploinsufficency mechanism), although how loss of *GRN* causes neuronal cell death remains unclear. Estimates of the frequency of *GRN* mutations in typical FTD patient populations suggests that they account for about 5–10% of all FTD cases, although numbers vary markedly depending on the nature of the populations considered (Cruts et al. 2006; Gass et al. 2006; Snowden et al. 2006).

Neuropathology analysis revealed that ubiquitin immunoreactive neuronal cytoplasmatic and intranuclear inclusions were present in all cases with FTDP-17, where pathological findings were available (Mackenzie et al. 2006). Furthermore, soon after the identification of mutations in *GRN*, biochemical analyses demonstrated that truncated and hyperphosphorylated isoforms of the TAR-DNA binding protein (TDP-43) are major components of the ubiquitin-positive inclusions in families with *GRN* mutations as well as in idiopathic FTD and a proportion of

Amyotrophic Lateral sclerosis (ALS) cases (Neumann et al. 2006). TDP43 is a ubiquitously expressed and highly conserved nuclear protein that can act as a transcription repressor, an activator of exon skipping or a scaffold for nuclear bodies through interactions with survival motor neuron protein. Under pathological conditions, TDP-43 has been shown to relocate from the neuronal nucleus to the cytoplasm, a consequence of which may be the loss of TDP-43 nuclear functions (Neumann et al. 2006). The mechanism by which loss of progranulin leads to TDP-43 accumulation and whether this is necessary for neurodegeneration in this group of diseases is still to be clarified.

In conclusion, the function of progranulin in the brain is currently unclear and why loss of this protein leads to a neurodegenerative diseases in mid-life remains to be established, and its possible role as regulator of a repair activity in the central nervous system, as it is well known to happen in periphery, remains a challenge for science. The gene encoding for TDP-43, named *TARDBP*, has been extensively studied and a number of mutations found in its C-terminal glycine rich region. Unexpectedly, the clinical phenotype of carries was ALS, and aggregates made of TDP-43 have been described in brain and spinal cord of such patients (see Pesiridis et al. 2009 for review).

A recently published collaborative study (Yu et al. 2010) analyzed *GRN* in a population of 434 patients with FTLD, including FTD, PA, SD, FTD/ALS, FTD/Motor Neuron Disease (MND), Corticobasal Degeneration (CBD), Progressive Supranuclear Palsy (PSP). Fifty-eight variants were identified, including 24 pathogenic variants. The frequency of *GRN* mutations was 6.9% of all FTLD-spectrum cases, 21,4% of cases with a pathological diagnosis of FTLD-U, 16% of FILD-spectrum cases with a family history of a similar neurodegenerative disease, and 56.2 of cases of FTLD-U with a family history. Clinical information were available for 31 *GRN* mutation-positive patients from 28 different families. The most common clinical diagnosis was FTD (n = 24); 3 patients were diagnosed with PA, 3 with AD and 1 with CBD. The majority of *GRN* mutations Introduced a premature termination codon, suggesting that their corresponding mRNA will be degraded through nonsense mediated decay, supporting the hypothesis that most *GRN* mutations create functional null allele (Yu et al. 2010).

An additional gene shown to cause familial FTLD is named *CHMP2B* (charged multivescicular body protein 2B), and is part of the endosomal ESCRTIII-complex (Skibinski et al. 2005). Four different mutations in this gene have been so far described in 4 families (http://www.molgen.ua.ac.be/), making *CHMP2B* a rare genetic cause of FTLD.

Lastly, the first evidence of linkage with chromosome 9q21-22 comes from a study carried out in families with FTD/MND (Hosler et al. 2000). Despite the evidence of linkage to chr9q21-22 in several additional FTD-MND families (Le Ber et al. 2009; Morita et al. 2006; Vance et al. 2006), the gene responsible for the disease in this locus has yet to be identified.

5 Sporadic FTLD

The best well-known risk factor for late onset SAD, Apo E4, has also been considered as a risk factor for sporadic FTLD. A number of studies suggested an association between FTLD and APOE*4 allele (Bernardi et al. 2006; Fabre et al. 2001; Farrer et al. 1995; Gustafson et al. 1997; Helisalmi et al. 1996; Stevens et al. 1997). Other Authors however, did not replicate these data (Geschwind et al. 1998; Riemenschneider et al. 2002; Short et al. 2002). Recent findings demonstrated an association between the *APOE*4* allele and FTLD in males, but not females (Srinivasan et al. 2006), possibly explaining the discrepancies previously reported. An increased frequency of the *APOE*4* allele was described in patients with SD compared to those with FTD and PA (Short et al. 2002).

Concerning the *APOE*2* allele in the development of FTLD, heterogeneous data have been obtained in different populations. Bernardi et al. (2006) showed a protective effect of this allele towards FTLD, whereas other Authors failed to do so (Engelborghs et al. 2003; Riemenschneider et al. 2002; Short et al. 2002; Srinivasan et al. 2006). Despite these results, a recent meta-analysis comprising a total of 364 FTD patients and 2671 controls demonstrated an increased susceptibility to FTD in *APOE*2* carriers (Verpillat et al. 2002).

Besides pathogenic mutations, several polymorphisms have been reported to date, both in *MAPT* and *GRN*. An association between Progressive Supranuclear Palsy (PSP) and a dinucleotide repeat polymorphism in the intron between *MAPT* exons 9 and 10 was described in 1997 (Conrad et al. 1997). The alleles at this locus carry 11–15 repeats. Subsequently, two common *MAPT* haplotypes, named H1 and H2, were identified (Baker et al. 1999). They differ in nucleotide sequence and intron size, but are identical at the amino acid level. Homozygosity of the more common allele H1 predisposes to PSP and Corticobasal Degeneration (CBD), but not to AD or Pick Disease (Baker et al. 1999; Di Maria et al. 2000).

Regarding *GRN*, an association of a SNP located in the promoter and an increased risk to develop FTLD in patients who did not carry causal mutations has recent been demonstrated (Galimberti et al. 2010).

In addition, a known polymorphism in *MCP-1 (A-2518G)* has been shown to exert a protective effects towards the development of FTLD (Galimberti et al. 2009), whereas NOS3 G894T (Glu298Asp) and NOS1 C276T SNPs likely Increases the risk to develop FTLD (Venturelli et al. 2008, 2009).

References

Albani D, Prato F, Tettamanti M et al (2009) The serotonin transporter promoter polymorphic region is not a risk factor for Alzheimer's disease related behavioral disturbances. J Alzheimers Dis 18(1):125–130

Angelucci F, Bernardini S, Gravina P et al (2009). Delusion symptoms and response to antipsychotic treatment are associated with the 5-HT2A receptor polymorphism (102T/C) in Alzheimer's disease: a 3-year follow-up longitudinal study. J Alzheimers Dis 17(1):203–211

Assal F, Alarcón M, Solomon EC et al (2004) Association of the serotonin transporter and receptor gene polymorphisms in neuropsychiatric symptoms in Alzheimer disease. Arch Neurol 61(8):1249–1253
Baker M, Litvan I, Houlden H et al (1999) Association of an extended haplotype in the tau gene with progressive supranuclear palsy. Hum Mol Genet 8(4):711–715
Baker M, Mackenzie IR, Pickering-Brown SM et al (2006) Mutations in progranulin cause tau-negative frontotemporal dementia linked to chromosome 17. Nature 442:916–919
Baki L, Shioi J, Wen P et al (2004) PS1 activates PI3 K thus inhibiting GSK-3 activity and tau overphosphorylation: effects of FAD mutations. EMBO J 23:2586–2596.
Bentahir M, Nyabi O, Verhamme J et al (2006) Presenilin clinical mutations can affect γ-secretase activity by different mechanisms. J Neurochem 96:732–742
Bernardi L, Maletta RG, Tomaino C et al (2006) The effects of APOE and tau gene variability on risk of frontotemporal dementia. Neurobiol Aging 27(5):702–709
Bertram L, Tanzi RE (2005) The genetic epidemiology of neurodegenerative disease. J Clin Invest 115(6):1449–1457
Blacker D, Wilcox MA, Laird NM et al (1998) Alpha-2 macroglobulin is genetically associated with Alzheimer disease. Nat Genet 19(4):357–360
Bruni AC, Bernardi L, Colao R et al (2010) Worldwide distribution of *PSEN1* Met146Leu mutation. Neurology 74:798–806
Collins JS, Perry RT, Watson B Jr et al (2000) Association of a haplotype for tumor necrosis factor in siblings with late-onset Alzheimer disease: the NIMH Alzheimer Disease Genetics Initiative. Am J Med Genet 96(6):823–830
Combarros O, Infante J, Llorca J, Berciano J (2004a) No evidence for association of the monocyte chemoattractant protein-1 (–2518) gene polymorphism and Alzheimer's disease. Neurosci Lett 360(1-2):25–28
Combarros O, Infante J, Llorca J et al (2004b) The chemokine receptor CCR5-Delta32 gene mutation is not protective against Alzheimer's disease. Neurosci Lett 366(3):312–314
Conrad C, Andreadis A, Trojanowski JQ et al (1997) Genetic evidence for the involvement of tau in progressive supranuclear palsy. Ann Neurol 41(2):277–281
Corder EH, Saunders AM, Strittmatter WJ et al (1993) Gene dose of apolipoprotein E type 4 allele and the risk of Alzheimer's disease in late onset families. Science 261(5123):921–923
Crawford F, Freeman M, Abdullah L et al (2000) No association between the NOS3 codon 298 polymorphism and Alzheimer's disease in a sample from the United States. Ann Neurol 47(5):687
Cross AJ (1990) Serotonin in Alzheimer-type dementia and other dementing illnesses. Ann NY Acad Sci 600:405–415
Cruts M, Gijselinck I, van der Zee J et al (2006) Null mutations in progranulin cause ubiquitin-positive frontotemporal dementia linked to chromosome 17q21. Nature 442:920–924
Dahiyat M, Cumming A, Harrington C et al (1999) Association between Alzheimer's disease and the NOS3 gene. Ann Neurol 46(4):664–667
Di Maria E, Tabaton M, Vigo T et al (2000) Corticobasal degeneration shares a common genetic background with progressive supranuclear palsy. Ann Neurol 47(3):374–377
Du Y, Dodel RC, Eastwood BJ et al (2000) Association of an interleukin 1 alpha polymorphism with Alzheimer's disease. Neurology 55(4):480–483
Engelborghs S, Dermaut B, Goeman J et al (2003) Prospective Belgian study of neurodegenerative and vascular dementia: APOE genotype effects. J Neurol Neurosurg Psychiatry 74: 1148–1151
Fabre SF, Forsell C, Viitanen M et al (2001) Clinic-based cases with frontotemporal dementia show increased cerebrospinal fluid tau and high apolipoprotein E epsilon4 frequency, but no tau gene mutations. Exp Neurol 168:413–418
Farrer LA, Abraham CR, Volicer L et al (1995) Allele epsilon 4 of apolipoprotein E shows a dose effect on age at onset of Pick disease. Exp Neurol 136:162–170
Fenoglio C, Galimberti D, Lovati C et al (2004) MCP-1 in Alzheimer's disease patients: A-2518G polymorphism and serum levels. Neurobiol Aging 25(9):1169–1173

Galimberti D, Fenoglio C, Cortini F et al (2010) GRN variability contributes to sporadic frontotemporal lobar degeneration. J Alzheimers Dis 19(1):171–177

Galimberti D, Fenoglio C, Lovati C et al (2004) CCR2-64I polymorphism and CCR5Delta32 deletion in patients with Alzheimer's disease. J Neurol Sci 225(1–2):79–83

Galimberti D, Scarpini E, Venturelli E et al (2008) Association of a NOS1 promoter repeat with Alzheimer's disease. Neurobiol Aging 29(9):1359–1365

Galimberti D, Venturelli E, Gatti A et al (2005) Association of neuronal nitric oxide synthase C276T polymorphism with Alzheimer's disease. J Neurol 252:985–986

Galimberti D, Venturelli E, Villa C et al (2009) MCP-1 A-2518G polymorphism: effect on susceptibility for frontotemporal lobar degeneration and on cerebrospinal fluid MCP-1 levels. J Alzheimers Dis 17(1):125–133

Gass J, Cannon A, Mackenzie IR et al (2006) Mutations in progranulin are a major cause of ubiquitin-positive frontotemporal lobar degeneration. Hum Mol Genet 15(20):2988–3001

Geschwind D, Karrim J, Nelson SF, Miller B (1998) The apolipoprotein E epsilon4 allele is not a significant risk factor for frontotemporal dementia. Ann Neurol 44:134–138

Goate A, Chartier-Harlin MC, Mullan M et al (1991) Segregation of a missense mutation in the amyloid precursor protein gene with familial Alzheimer's disease. Nature 349(6311):704–706

Goedert M, Jakes R (2005) Mutations causing neurodegenerative tauopathies. Biochim Biophys Acta 1739:240–250

Griffin WS (2006) Inflammation and neurodegenerative diseases. Am J Clin Nutr 83(suppl): 470S–474S

Grimaldi LM, Casadei VM, Ferri C et al (2000) Association of early-onset Alzheimer's disease with an interleukin-1alpha gene polymorphism. Ann Neurol 47(3):361–365

Grupe A, Li Y, Rowland C, Nowotny P et al (2006) A scan of chromosome 10 identifies a novel locus showing strong association with late-onset Alzheimer disease. Am J Hum Genet 78(1):78–88

Guidi I, Galimberti D, Venturelli E et al (2005) Influence of the Glu298Asp polymorphism of *NOS3* on age at onset and homocysteine levels in AD patients. Neurobiol Aging 26(6): 789–794

Gustafson L, Abrahamson M, Grubb A et al (1997) Apolipoprotein-E genotyping in Alzheimer's disease and frontotemporal dementia. Dement Geriatr Cogn Disord 8:240–243

Ha TM, Cho DM, Park SW, Joo MJ, Lee BJ, Kong BG, Kim JM, Yoon JS, Kim YH (2005) Evaluating associations between 5-HTTLPR polymorphism and Alzheimer's disease for Korean patients. Dement Geriatr Cogn Disord 20(1):31–34

Hardy J, Selkoe DJ (2002) The amyloid hypothesis of Alzheimer's disease: progress and problems on the road to therapeutics. Science 297(5580):353–356

Harold D, Abraham R, Hollingworth P et al (2009) Genome-wide association study identifies variants at CLU and PICALM associated with Alzheimer's disease. Nat Genet 41(10): 1088–1093

He Z, Bateman A (2003) Progranulin (granulin-epithelin precursor, PC-cell-derived growth factor, acrogranin) mediates tissue repair and tumorigenesis. J Mol Med 81:600–612

Helisalmi S, Linnaranta K, Lehtovirta M et al (1996) Apolipoprotein E polymorphism in patients with different neurodegenerative disorders. Neurosci Lett 205:61–64

Hosler BA, Siddique T, Sapp PC et al (2000) Linkage of familial amyotrophic lateral sclerosis with frontotemporal dementia to chromosome 9q21–q22. JAMA 284:1664–1669

Hou CE, Carlin D, Miller BL (2004) Non-Alzheimer's disease dementias: anatomic, clinical, and molecular correlates. Can J Psychiatry 49(3):164–171

Huerta C, Alvarez V, Mata IF et al (2004) Chemokines (RANTES and MCP-1) and chemokine-receptors (CCR2 and CCR5) gene polymorphisms in Alzheimer's and Parkinson's disease. Neurosci Lett 370(2–3):151–154

Huppert SS, Ilagan MX, De Strooper B, Kopan R (2005) Analysis of Notch function in presomitic mesoderm suggests a γ-secretase-independent role for presenilins in somite differentiation. Dev Cell 8:677–688

Hutton M, Lendon CL, Rizzu P et al (1998) Association of missense and 5′-splice-site mutations in tau with the inherited dementia FTDP-17. Nature 393:702–705
Lambert JC, Heath S, Even G et al (2009) Genome-wide association study identifies variants at CLU and CR1 associated with Alzheimer's disease. Nat Genet 41(10):1094–1099
Larner AJ, Doran M (2009) Genotype-phenotype relationships of presenilin-1 mutations in Alzheimer's disease: an update. J Alzheimers Dis 17(2):259–265
Le Ber I, Camuzat A, Berger E et al (2009) Chromosome 9p-linked families with frontotemporal dementia associated with motor neuron disease. Neurology 72:1669–1676
Lendon CL, Lynch T, Norton J et al (1998) Hereditary dysphasic disinhibition dementia: a frontotemporal dementia linked to 17q21–22. Neurology 50:1546–1555
Lesch KP, Balling U, Gross J et al (1994) Organization of the human serotonin transporter gene. J Neural Transm Gen Sect 95(2):157–162
Levy-Lahad E, Wasco W, Poorkaj P et al (1995) Candidate gene for the chromosome 1 familial Alzheimer's disease locus. Science 269(5226):973–977
Li Y, Grupe A, Rowland C et al (2006) DAPK1 variants are associated with Alzheimer's disease and allele-specific expression. Hum Mol Genet 15(17):2560–2568
Licastro F, Chiappelli M (2003) Brain immune responses cognitive decline and dementia: relationship with phenotype expression and genetic background. Mech Ageing Dev 124:539–548
Liou YJ, Hong CJ, Liu HC et al (2002) No association between the neuronal nitric oxide synthase gene polymorphism and Alzheimer's disease. Am J Med Gen 114:687–688
Lotrich FE, Pollock BG, Ferrell RE (2001) Polymorphism of the serotonin transporter: implications for the use of selective serotonin reuptake inhibitors. Am J Pharmacogenomics 1(3):153–164
Mackenzie IR, Baker M, West G et al (2006) A family with tau-negative frontotemporal dementia and neuronal intranuclear inclusions linked to chromosome 17. Brain 129:853–867
Monastero R, Cefalu AB, Camarda C et al (2003) No association between Glu298Asp endothelial nitric oxide synthase polymorphism and Italian sporadic Alzheimer's disease. Neurosci Lett 341:229–232
Morita M, Al-Chalabi A, Andersen PM et al (2006) A locus on chromosome 9p confers susceptibility to ALS and frontotemporal dementia. Neurology 66(6):839–844
Neary D, Snowden JS, Gustafson L et al (1998) Frontotemporal lobar degeneration: a consensus on clinical diagnostic criteria. Neurology 51:1546–1554
Neumann M, Sampathu DM, Kwong LK et al (2006) Ubiquitinated TDP-43 in frontotemporal lobar degeneration and amyotrophic lateral sclerosis. Science 314:130–133
Nicoll JA, Mrak RE, Graham DI et al (2000) Association of interleukin-1 gene polymorphisms with Alzheimer's disease. Ann Neurol 47(3):365–368
Palmer AM (1996) Neurochemical studies of Alzheimer's disease. Neurodegeneration 5:381–391
Papassotiropoulos A, Bagli M, Jessen F et al (1999) A genetic variation of the inflammatory cytokine interleukin-6 delays the initial onset and reduces the risk for sporadic Alzheimer's disease. Ann Neurol 45(5):666-668
Pericak-Vance MA, Bebout JL, Gaskell PC Jr et al (1991) Linkage studies in familial Alzheimer disease: evidence for chromosome 19 linkage. Am J Hum Genet 48(6):1034–1050
Pesiridis G, Lee VMY, Trojanowski JQ (2009) Mutations in TDP-43 link glycine-rich domain functions to amyotrophic lateral sclerosis. Hum Mol Gen 18(2):R156–162
Poorkaj P, Bird TD, Wijsman E et al (1998) Tau is a candidate gene for chromosome 17 frontotemporal dementia. Ann Neurol 43:815–825
Raber J, Huang Y, Ashford JW (2004) ApoE genotype accounts for the vast majority of AD risk and AD pathology. Neurobiol Aging 25(5):641–650
Rademakers R, Cruts M, Dermaut B et al (2002) Tau negative frontal lobe dementia at 17q21: significant finemapping of the candidate region to a 4.8 cM interval. Mol Psychiatry 7:1064–1074
Rademakers R, Cruts M, van Broeckhoven C (2004) The role of tau (MAPT) in frontotemporal dementia and related tauopathies. Hum Mutat 24(4):277–295

Reif A, Herterich S, Strobel A et al (2006) A neuronal nitric oxide synthase (NOS-I) haplotype associated with schizophrenia modifies prefrontal cortex function. Mol Psychiatry 11(3):286–300

Reinikainen KJ, Soininen H, Riekkinen PJ (1990) Neurotransmitter changes in Alzheimer's disease: implications to diagnostics and therapy. J Neurosci Res 27(4):576–586

Riemenschneider M, Diehl J, Muller U et al (2002) Apolipoprotein E polymorphism in German patients with frontotemporal degeneration. J Neurol Neurosurg Psychiatry 72:639–641

Rocchi A, Micheli D, Ceravolo R et al (2003) Serotoninergic polymorphisms (5-HTTLPR and 5-HT2A): association studies with psychosis in Alzheimer disease. Genet Test 7(4):309–314

Rosen HJ, Hartikainen KM, Jagust W et al (2002) Utility of clinical criteria in differentiating frontotemporal lobar degeneration (FTLD) from AD. Neurology 58:1608–1615

Rosso SM, Kamphorst W, de Graaf B et al (2001) Familial frontotemporal dementia with ubiquitin-positive inclusions is linked to chromosome 17q21–22. Brain 124:1948–1957

Rovelet-Lecrux A, Hannequin D, Raux G et al (2006) APP locus duplication causes autosomal dominant early-onset Alzheimer disease with cerebral amyloid angiopathy. Nat Genet 38:24–26

Sánchez-Guerra M, Combarros O, Alvarez-Arcaya A et al (2001) The Glu298Asp polymorphism in the NOS3 gene is not associated with sporadic Alzheimer's disease. J Neurol Neurosurg Psychiatry 70:566–567

Sato N, Ueki A, Ueno H et al (2009) Dopamine D3 receptor gene polymorphism influences on behavioral and psychological symptoms of dementia (BPSD) in mild dementia of Alzheimer's type. J Alzheimers Dis 17(2):441–448

Saur D, Vanderwinden JM, Seidler B et al (2004) Single-nucleotide promoter polymorphism alters transcription of neuronal nitric oxide synthase exon 1c in infantile hypertrophic pyloric stenosis. Proc Natl Acad Sci 101(6):1662–1667

Scarpini E, Galimberti D, Guidi I et al (2006) Progressive, isolated language disturbance: its significance in a 65-year-old-man. A case report with implications for treatment and review of literature. J Neurol Sci 240(1–2):45–51

Seshadri S, Beiser A, Selhub J et al (2002) Plasma homocysteine as a risk factor for dementia and Alzheimer's disease. New Engl J Med 346:476-483

Sherrington R, Rogaev EI, Liang Y et al (1995) Cloning of a gene bearing missense mutations in early-onset familial Alzheimer's disease. Nature 375(6534):754–760

Short RA, Graff-Radford NR, Adamson J et al (2002) Differences in tau and apolipoprotein E polymorphism frequencies in sporadic frontotemporal lobar degeneration syndromes. Arch Neurol 59:611–615

Skibinski G, Parkinson NJ, Brown JM et al (2005) Mutations in the endosomal ESCRTIII-complex subunit CHMP2B in frontotemporal dementia. Nat Genet 37:806–808

Snowden JS, Neary D, Mann DM (2002) Frontotemporal dementia. Br J Psychiatry 180:140–143

Snowden JS, Pickering-Brown SM, Mackenzie IR et al (2006) Progranulin gene mutations associated with frontotemporal dementia and progressive non-fluent aphasia. Brain 129:3091–3102

Spillantini MG, Murrell JR, Goedert M et al (1998) Mutation in the tau gene in familial multiple system tauopathy with presenile dementia. Proc Natl Acad Sci USA 95:7737–7741

Srinivasan R, Davidson Y, Gibbons L et al (2006) The apolipoprotein E epsilon4 allele selectively increases the risk of frontotemporal lobar degeneration in males. J Neurol Neurosurg Psychiatry 77:154–158

Steiner H, Romig H, Grim MG et al (1999) The biological and pathological function of the presenilin-1 dExon 9 mutation is independent of its defect to undergo proteolytic processing. J Biol Chem 274:7615–7618

Stevens M, van Duijn CM, de Knijff P et al (1997) Apolipoprotein E gene and sporadic frontal lobe dementia. Neurology 48:1526–1529

Sukonick DL, Pollock BG, Sweet RA et al (2001) The 5-HTTPR*S/*L polymorphism and aggressive behavior in Alzheimer disease. Arch Neurol 58(9):1425–1428

Sweet RA, Pollock BG, Sukonick DL et al (2001) The 5-HTTPR polymorphism confers liability to a combined phenotype of psychotic and aggressive behavior in Alzheimer disease. Int Psychogeriatr 13(4):401–409

Tanzi RE, Gusella JF et al (1987) Amyloid beta protein gene: cDNA, mRNA distribution, and genetic linkage near the Alzheimer locus. Science 235(4791):880–884

Tedde A, Nacmias B, Cellini E et al (2002) Lack of association between NOS3 polymorphism and Italian sporadic and familial Alzheimer's disease. J Neurol 249:110–111

Ueki A, Ueno H, Sato N, Shinjo H, Morita Y (2007) Serotonin transporter gene polymorphism and BPSD in mild Alzheimer's disease. J Alzheimers Dis 12(3):245–253

van der Zee J, Rademakers R, Engelborghs S et al (2006) A Belgian ancestral haplotype harbours a highly prevalent mutation for 17q21-linked tau-negative FTLD. Brain 129:841–852

Vance C, Al-Chalabi A, Ruddy D et al (2006) Familial amyotrophic lateral sclerosis with frontotemporal dementia is linked to a locus on chromosome 9p13.2–21.3. Brain 129:868–876

Venturelli E, Galimberti D, Fenoglio C et al (2006) Candidate gene analysis of IP-10 gene in patients with Alzheimer's disease. Neurosci Lett 404(1–2):217–221

Venturelli E, Galimberti D, Lovati C et al (2005) The T-786C NOS3 polymorphism in Alzheimer's disease: association and influence on gene expression. Neurosci Lett 382(3): 300–303

Venturelli E, Villa C, Fenoglio C et al (2009) The NOS3 G894T (Glu298Asp) polymorphism is a risk factor for frontotemporal lobar degeneration. Eur J Neurol 16(1):37–42

Venturelli E, Villa C, Scarpini E et al (2008) Neuronal nitric oxide synthase C276T polymorphism increases the risk for frontotemporal lobar degeneration. Eur J Neurol 15(1):77–81

Verpillat P, Camuzat A, Hannequin D et al (2002) Apolipoprotein E gene in frontotemporal dementia: an association study and meta-analysis. Eur J Hum Genet 10:399–405

Wang Y, Newton DC, Robb GB et al (1999) RNA diversity has profound effects on the translation of neuronal nitric oxide synthase. Proc Natl Acad Sci USA 96(21):121505–12155

Wilhelmsen KC, Lynch T, Pavlou E et al (1994) Localization of disinhibition–dementia–parkinsonism–amyotrophy complex to 17q21–22. Am J Hum Genet 55:1159–1165

Yu CE, Bird TD, Bekris LM et al (2010) The spectrum of mutations in progranulin: a collaborative study screening 545 cases of neurodegeneration. Arch Neurol 67(2):161–170

Zhu J, Nathan C, Jin W et al (2002) Conversion of proepithelin to epithelins: roles of SLPI and elastase in host defence and wound repair. Cell 111:867–878

Index

1q21.1 / 15q11.2 / 15q13.3, 432
22q11, 292, 293
22q11.2, 431
5HT, 267
5-HTT knockout mice, 62

A
Acethylcholine, 268
ACh Transporter, 269
Acoustic startle, 13
Advanced Intercross Lines, 36
Aggression, 76, 368–370
AKT1, 218, 261
Alzheimer's disease (AD), 10, 615, 616
Alzheimer's, 288
AMPA, 67, 264, 435
Amphetamine, 538
Amygdala, 65, 364–370
Amyloid-b precursor protein (APP), 11
Androgen receptor, 364, 370
Angiotensin/Renin, 270
Animal models, 68, 140–142, 157, 171
Antisense oligonucleotides, 6
Anxiety, 11, 12, 64, 140–145, 148, 157–171, 364–367
Apolipoprotein E (ApoE), 11
Approaches, 276
Arabidopsis, 6
Association, 395, 400, 402, 403, 405–409, 411–415
Attention deficit/hyperactivity disorder, 395, 396, 399, 405
Attention-deficit/hyperactivity disorder, 345, 355
Autism, 12, 289, 345, 354
Autism spectrum disorders, 395, 401

B
B. glutamate, 264
Backcross, 32
Background strain, 275
Behavior, 441, 442, 444, 451, 456
Behavioral despair, 63
Behavioural and psychological, 616
Behavioural disturbances, 615, 616
Bioinformatics, 328
Biomarker, 255
BMI, 395, 412–414
Brain-derived neurotrophic factor, 66
Bupropion, 538

C
CA1, 66
Calcium Signaling, 282
CaMKII, 435
Candidate gene studies, 539
Candidate Genes, 296
Catechol-O-Methyltransferase, 217
Chromosome Substitution Strains, 33
Chronic social defeat, 67
CNV, 464, 466, 471, 478
Cocaine, 538
Cognition, 189–193, 195–197, 199–201, 203, 205, 580, 584, 582–585, 588, 590, 592, 595–598, 600, 601, 604, 605
Cognitive constructs, 255
Collaborative Cross, 39

Curr Topics Behav Neurosci (2012) 12: 633–636
DOI: 10.1007/978-3-642-27859-4

C (*cont.*)
COMT, 259, 297, 431
Congenic Strains, 35
Consomic Lines, 33
Copy number variants (CNVs), 432
Copy number variations, 223, 228, 345
Cre/loxP, 8
CRF, 271, 295
Cross-species, 255
c-secretase, 11
CSF2RA, 434
CS-US association, 15

D
Dementia, 616, 617, 623
Depression, 11, 60, 140–148, 151–157, 159, 160, 162–164, 166
Development, 63
Developmental disorders, 433
DISC1, 257, 297, 298, 430
Disrupted-in-schizophrenia-1 (DISC-1), 223
DNA microarrays, 6
Dopamine Receptors, 215
Dopamine Transporter, 216, 373
Dopamine, 263, 364, 371–373, 374, 376
Down syndrome, 293
Drug of Abuse, 230
Drug targets, 256
D-Serine, 221
dsRNAs, 7
Dysbindin, 225

E
Elevated plus maze, 65
Embryonic stem cells, 8
Emotionality, 189, 191–193, 195, 197, 199, 201–203, 205, 207
Endocannabinoid, 364, 373
Endophenotype, 253, 302
ENU, 43
Environment, 66
Ethyl-nitroso-urea (ENU), 9

F
Fatty acide amide hydrolase, 374, 375
Fear, 63
Flat worm, 4
FMR1, 290, 291
Forward genetic screens, 277
Forward genetic, 6, 27
Fragile X-syndrome, 290
Frontal cortex, 67
Frontotemporal lobar degeneration (FTLD), 616, 617
Fruit fly, 4

G
GABA, 106
GABA-A receptors, 66
GABA-glutamate, 278
Gal4-UAS, 8
Gating deficit, 253
Gene 9 environment interaction, 209, 210, 232
Gene expression, 6
Gene regulation, 504, 524
Gene, 66
Gene-environment interaction (GxE), 464, 484, 486, 487
Gene–environment, 256
Genetic component, 430
Genetic polymorphism, 504, 516, 524
Genetic, 276, 395–397, 399, 401, 404–415
Genetics, 297, 580–582, 588, 589, 591, 593, 599, 605
Genome-wide association studies, 539
Genomic, 441, 447, 450, 580, 581, 594, 599, 606
Glucose, 284
Glutamate transporter, 266
Glutamate, 189, 190, 195, 196, 198, 199–206, 219–223, 265
Glutamatergic, 189, 190, 195, 198, 199, 201, 207
Glycine, 220, 266
Glycoprotein, 286
Growth factors, 279
GSK-3, 218
GWAS, 464, 466, 467, 470–476, 479, 483

H
Haplotype association mapping, 41
Hearing loss, 275
Hearing, 292
Heatshock, 8
Heritable, 253
Heterogeneous Stocks, 37
Hippocampus, 16, 65
Homologous recombination in embryonic stem cells, 8
Hormones, 287
Huntington's, 291
Hypothalamic-Pituitary-Adrenal Axis, 364

I

Immune activation, 283
Impulsivity, 364, 368, 371–373, 375, 376, 381
Inbred Strains, 27
Influenza infection, 283
Intercross, 32
Inter-individual variability, 538
Intermediate phenotype, 464, 483, 487

K

Knock out technology, 8

L

Life stress events, 66
Linkage studies / candidate gene / genome wide association studies, 430
Linkage, 395, 399, 401–404, 406, 409, 410, 411, 413

M

Maternal care, 63
Metabotropic Glutamate Receptors, 222
Methamphetamine, 538
Methodological Considerations, 275
Methylphenidate, 538
mGLU, 264
mGluRs, 291
microRNA, 7
Mineralocorticoid receptor, 362, 364, 377, 378
Model, 301, 302
Monoaminergic signaling, 504, 525
Mood disorders, 60, 504
Morpholinos, 6
Mouse Phenome Project, 41
Mouse, 61
Muscarinic, 268
Mutagenesis, 43, 298
Mutant model, 210–214, 223, 229, 230, 234, 238, 239
Mutant mouse models, 254
Mylein, 284

N

Neural circuitry, 255
Neuregulin, 226, 258
Neurexin, 260
Neurodevelopmental disorders, 345
Neurodevelopmental, 285, 294
Neurogranin (NRGN) / TCF4 / HLA, 435
Neuropeptide S, 272
Neuropeptide Y, 271
Neuropeptides, 109, 263, 270, 295
Neurotensin, 270
Neurotransmitters, 263
Nicotinic, 268
NMDA, 219, 222, 265, 278, 435
NMDAR, 264
NMDAR Signalling Complex, 222
nNOS, 284
Nocturnal enuresis, 395, 401, 411, 412
NRG1, 258, 297
NRXN1, 432
Nucleotide polymorphisms, 539

O

Obesity, 395, 396, 401, 412–415
Obsessive compulsive disorders, 11
Open field, 65
Orphan receptors, 263, 273
Outbred Mice, 39
Oxidative stress, 285
Oxytocin, 271

P

PACAP, 272
Parkinson's Disease, 10
Pathophysiology, 277
PDE4, 279
PDE4B, 295
Personalized genomics, 332
Pharmacogenetics, 464, 478, 484
Pharmacological, 295
Phenotype, 210–214, 217, 219, 221–228, 230–239, 298, 299
Phenotypes, 580–583, 585, 589, 592–594, 597, 598, 601, 603–606
Phobias, 11
Pitt-Hopkins-Syndrome, 435
Potassium channel signaling, 283
PPI, 302
Prader-Willi/Angelman syndrome, 433
Prepulse inhibition, 13, 252
Presenilins, 11
Presynaptic proteins, 282, 295
Protein kinase, 280
Psychiatric symptoms, 254

P (*cont.*)
Psychosocial stress, 65
Psychotic illness, 210, 211

Q
QTL Mapping, 28
QTL, 27, 298, 299
Quantitative reverse transcriptase polymerase chain reaction (q-RT-PCR), 6

R
Recombinant Inbred Lines, 29
Reelin, 258
Relaxin, 273
Retroviruses, 7
Reverse genetic, 6
Reward, 364, 371–376, 379, 381, 382
RNA-interference, or RNAi, 7

S
Schizophrenia, 12, 210–223, 234, 237, 238, 239, 345, 354, 429
Screen, 255
Sea slug, 4
Selective breeding, 140, 141, 143, 161
Sensorimotor gating, 252
Sequencing technologies, 322
Serotonin transporter, 60, 504, 505, 507, 512, 520, 522, 525
Serotonin turnover, 67
Serotonin, 66, 81, 84, 267, 364–367, 441, 446, 449, 454, 457
Single
SMARCA2, 260
SNPs, 296
SP4, 301
Spatial learning, 16
Spectrum disorders, 12
Spine, 284
Startle magnitude, 275
Startle, 252
States, 81
Stimulant drugs, 538
Stress, 11, 189–207, 364, 376–382, 441, 446, 449, 452, 454–457
Stress exposure, 60
Stress risk factor, 66
Stressed, 191, 192, 194
Stressful, 190, 199
Stressor, 190, 191, 194, 203
Structural genetic variants, 430
Susceptibility gene, 210, 233, 238, 262
Synaptic proteins, 281

T
Tau, 10
Tauopathies, 10
TBX1 / GNB1L, 431
TCF4, 260
Testosterone, 368–370
Tetracycline transactivator, 8
Tetraploidy, 9
Tilling, 6
Trace amines, 274
Trait, 81
Transgenic technologies, 7
Translational, 255
Transposons, 7
Twin studies, 538
Tyrosine hydroxylase, 10

V
Variable number tandem repeat, 504, 505, 507, 539
Vasopressin, 271
Ventral striatum, 364, 371
VIVO-morpholino, 7
Vulnerability, 257, 296

Z
Zinc finger nucleases, 8
ZNF804A, 434

The manufacturer's authorised representative in the EU is Springer Nature Customer Service Centre GmbH, Europaplatz 3, 69115 Heidelberg, Germany. If you have any concerns regarding our products, please contact ProductSafety@springernature.com

Printed and bound by CPI Group (UK) Ltd, Croydon, CR0 4YY
22/09/2026
02220954-0015